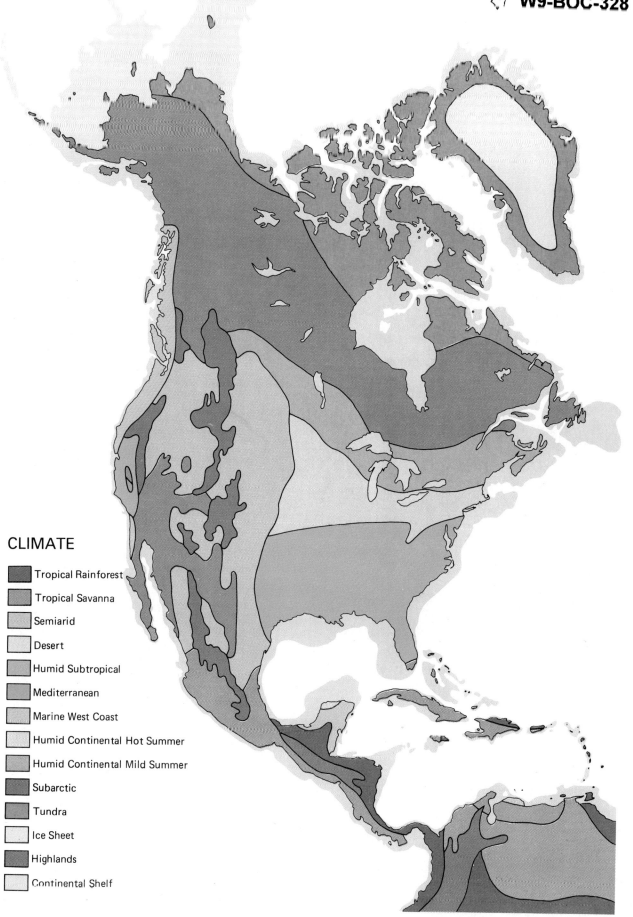

CLIMATE

- Tropical Rainforest
- Tropical Savanna
- Semiarid
- Desert
- Humid Subtropical
- Mediterranean
- Marine West Coast
- Humid Continental Hot Summer
- Humid Continental Mild Summer
- Subarctic
- Tundra
- Ice Sheet
- Highlands
- Continental Shelf

Earth Science and the Environment

Graham R. Thompson, PhD
University of Montana

Jonathan Turk, PhD

Saunders College Publishing
Harcourt Brace Jovanovich College Publishers

Fort Worth   Philadelphia   San Diego   New York   Orlando
San Antonio   Toronto   Montreal   London   Sydney   Tokyo

Text Typeface: Times Roman
Compositor: Monotype Composition, Inc.
Acquisitions Editor: John Vondeling
Developmental Editor: Christine Connelly
Managing Editor: Carol Field
Project Editor: Anne Gibby
Copy Editor: Mary Patton
Manager of Art and Design: Carol Bleistine
Art Director: Christine Schueler
Art Assistant: Caroline McGowan
Text Designer: Rebecca Lemna
Cover Designer: Lawrence R. Didona
Text Artwork: George V. Kelvin/Science Graphics
            Rolin Graphics, Inc.
            J & R Studio
Layout Artwork: Anne O'Donnell
Director of EDP: Tim Frelick
Production Manager: Joanne Cassetti
Marketing Manager: Marjorie Waldron

Cover Credit: Sequoia National Park, California (Galen Rowell/Mountain Light Photography)
Frontispiece: Valley of the Ten Peaks, Canada (Galen Rowell/Mountain Light Photography)

Printed in the United States of America

EARTH SCIENCE AND THE ENVIRONMENT

ISBN 0-03-075446-1

Library of Congress Catalog Card Number: 92-054103

2345   032   987654321

Preface

E arth Science is a study of the world around us. On your way to work or school you may pass through landscapes that were created by geologic processes. Lofty mountain ranges, fertile prairies, and massive glaciers are familiar through travel, books, and film. Nearly half of the world's population lives adjacent to coastlines and many of us have stood along the shore mesmerized by the waves and the beauty of the sculpted rocks along the beach. Sometimes we open a window to see if the weather is warm or cold, rainy or snowy. Finally, who hasn't gazed at the sky and wondered about distant planets, stars, and faraway galaxies?

Many of the changes that occur within the Earth, its oceans, and its atmosphere affect our lives. Every year tragic losses of life and property occur throughout the world when volcanoes erupt or earthquakes topple buildings, twist highways, and destroy cities. Other geological processes form fuel deposits and ore bodies and have obvious impacts on modern society. Ground-water pollution, the effects of CFCs on the ozone layer, the effects of acid rain on forests and lakes, and changing climate all alter our environment.

The primary purpose of this book is to explain fundamental Earth processes to students who have little or no science background. Earth science is not a set of facts to learn or to memorize. It is, rather, the study of an active, changing, and growing *science of the Earth* that will help you to transform casual observations into understanding.

In this book we explain how discoveries are made and how theories develop. Many of the ideas presented have only recently been developed and some hypotheses are controversial. One example of a hotly debated idea is the hypothesis that giant meteorite impacts caused several catastrophic climate changes that were followed by mass extinctions. Possible climatic changes from the greenhouse effect is another example of a modern scientific controversy. We introduce these and other environmental issues not as advocates of any particular side of the debates, but as scientists studying the world around us. We start by carefully separating observations and experimental results from conclusions, and continue by analyzing the reliability of the data. With these objectives in mind, we have created a text that *teaches* Earth science and *relates* it to many important modern issues.

Teaching Options

Earth science is a broad field covering many separate scientific disciplines. During the development of this book we communicated with many geologists, geographers, meteorologists, oceanographers, and astronomers who teach Earth science. The pedagogic strategy of this book was derived from several themes emphasized by these scientists.

We recognize that not all students in introductory Earth science courses have backgrounds in chemistry and physics. Therefore, this text explains the workings of planet Earth accurately, but in a language and style that is readily understood by students with little or no college-level science or mathematics background.

An exhaustive treatment of all the topics introduced in this course would involve several books, not just one. While recognizing the inherent limitations of a survey course, we offer balanced coverage of *all* topics. Few professors will assign the entire book in a quarter or semester course; there is just too much material. In order to accommodate a wide range of course emphases, each of the seven units offers a holistic introduction to the focus of the unit. Both the units and the chapters themselves are written to be as independent as possible so that individual professors can pick the topics that are important to their course requirements. Furthermore, students can pursue any subject by independently reading units or chapters that interest them. This book is designed to be a useful reference for all Earth science topics in addition to being a text for a quarter or semester-long course.

Sequence of Topics

Just as different Earth science courses have many different emphases, a wide variety of logical sequences exist. We have chosen to introduce the Earth's materials and geologic time, and then to start from the Earth's interior and work outward. Thus the book is divided into seven units:

> Unit I: The Earth and Its Materials
>
> Unit II: Internal Processes
>
> Unit III: Surface Processes
>
> Unit IV: The Oceans
>
> Unit V: The Atmosphere
>
> Unit VI: Astronomy
>
> Unit VII: Natural Resources

Some instructors may prefer other sequences. The unit structure of this book allows any alternative topic sequences.

Special Features

Special Topics   In a survey course where many subjects are introduced, it is refreshing to read more in-depth discussions of selected topics. Therefore, interesting special topics are set aside and highlighted in color. These topics are not necessary to the sequential development of each chapter, and they can be ignored without losing the flow of the chapter. However, from our own teaching experience we have learned students are drawn into a subject by specific examples, and at the same time the examples illustrate basic ideas of each chapter. Three types of special topics are included. **Focus On** boxes cover interesting topics in traditional Earth science such as "The Upper Fringe of the Atmosphere." Other boxes, titled, **Earth Science and the Environment** cover currently-active environmental topics such as "Nuclear Waste Storage in North America." Finally, short **Memory Devices** aid the student by explaining relationships between word roots and their modern definitions.

Chapter Review Material   Important words are highlighted in bold type in the text. These **key words** with their corresponding page numbers are then listed at the end of each chapter for review. In addition, a short **summary** of the chapter material is provided at the end of each chapter.

Questions   Two types of end-of-chapter questions are provided. The **review questions** can be answered in a straightforward manner from the material in the text, and allow the student to test himself or herself on how completely he or she has learned the material in the chapter. On the other hand, **discussion questions** challenge students to apply what they have learned to an analysis of situations not directly described in the text. These questions often have no absolute correct answers.

Appendix and Glossary   A **glossary** is provided at the end of the book. In addition, **appendices** cover the ele-ments, mineral classification and identification, metric units, and rock symbols.

Interviews   Often students wonder, what type of life would I lead if I decided to make Earth science my career? With whom would I work and trade ideas? What intellectual rewards would make my life challenging? In order to open a window into these subjects, we have interviewed four prominent scientists whose theories are discussed in the text. These interviews include a brief look at both their professional and nonprofessional lives and are included to encourage students to think about careers in Earth science.

Ancillaries

This text is accompanied by an extensive set of support materials.

Instructors Manual with Test Bank

The Instructors Manual, written by the authors of the text, provides teaching goals, alternate sequences of topics, answers to discussion questions and a short bibliography. An extensive test bank, written by Christine Seashore, is included in the instructors manual as well. The test bank includes multiple choice, true or false, and completion questions for each chapter of the text.

Computerized Test Bank

A computerized version of the test bank is available for both IBM PC and Macintosh. These versions allow instructors the flexibility to add their own questions or modify existing questions. Easy-to-follow commands make customizing tests and quizzes simple and efficient. Instructors without access to computers may receive customized tests within five working days by calling 1-800-447-9457. FAX service is also available.

Study Guide

The Study Guide, written by Lois Gundrum, provides review and study aids to further enhance the students' understanding of the text. The Study Guide includes chapter objectives, chapter outlines, multiple choice questions that test vocabulary, objective questions that test recall of important factual information, and short answer questions that require students to use and apply important chapter concepts. The student is also asked to think critically about environmental issues or problems.

Saunders Earth Science Videodisc

The Saunders Earth Science Videodisc is a 60 minute videodisc with more than 1500 still images and a collection of video clips from *Encyclopedia Britannica* and other sources, in addition to animated figures from the text. A directory will accompany the videodisc and will include instructions, barcode labels and reference numbers for every still image and video clip, organized according to the table of contents for the text.

Overhead Transparencies

A set of one hundred overhead transparencies includes full-color illustrations from the text for use in the classroom or the laboratory.

35mm Slides

The Saunders Earth Science Slide Set which includes five hundred 35mm slides of illustrations and photographs covering a wide range of topics such as physical and historical geology, oceanography, meteorology, and astronomy is available for use in the classroom or the laboratory.

Acknowledgments

We have not worked alone. The manuscript has been extensively reviewed at several stages and the numerous careful criticisms have helped shape the book and ensure accuracy:

James Albanese, *State University of New York at Oneonta*

John Alberghini, *Manchester Community College*

Calvin Alexander, *University of Minnesota*

Edmund Benson, *Wayne County Community College*

Robert Brenner, *University of Iowa*

Walter Burke, *Wheelock College*

Wayne Canis, *University of Northern Alabama*

Stan Celestian, *Glendale Community College*

Edward Cook, *Tunxis Community College*

James D'Amario, *Harford Community College*

Joanne Danielson, *Shasta College*

John Ernissee, *Clarion University*

Richard Faflak, *Valley City State University*

Joseph Gould, *St. Petersburg Junior College*

Bryan Gregor, *Wright State University*

Miriam Hill, *Indiana University—Southeast*

John Howe, *Bowling Green State University*

DelRoy Johnson, *Northwestern College*

Alan Kafka, *Boston College*

William Kohland, *Middle Tennessee State University*

Thomas Leavy, *Clarion University*

Doug Levin, *Bryant College*

Jim LoPresto, *Edinboro University of Pennsylvania*

Joseph Moran, *University of Wisconsin—Green Bay*

Glenn Mason, *Indiana University—Southeast*

Alan Morris, *University of Texas—San Antonio*

Jay Pasachoff, *Williams College*

Frank Revetta, *Potsdam College*

Laura Sanders, *Northeastern Illinois University*

Barun K. Sen Gupta, *Louisiana State University*

James Shea, *University of Wisconsin—Parkside*

Kenneth Sheppard, *East Texas State University*

Doug Sherman, *College of Lake County*

Gerry Simila, *California State University—Northridge*

Edward Spinney, *Northern Essex Community College*

James Stewart, *Vincennes University*

Susan Swope, *Plymouth State College*

J. Robert Thompson, *Glendale Community College*

Brooke Towery, *Pensacola Junior College*

Amos Turk, *City University of New York*

Jeff Wagner, *Fireland College*

Tom Williams, *Western Illinois University*

J. Curtis Wright, *University of Dubuque*

William Zinsmeister, *Purdue University*

Earth science is a visual science. We can readily observe landforms and weather patterns in our daily lives. Although we cannot see many processes, such as movements of tectonic plates, collisions of air masses, or the violent interior of a galactic nucleus, these events can be visualized through an artist's eye. George Kelvin has painted most of the illustrations in this book. It has been a pleasure to work with him.

We are especially grateful to Cindy Lee Van Dover, Stephen Schneider, Paul Hoffman, and Jack Horner for providing us with insightful interviews.

We would never have been able to produce this book without professional support both here in Montana and at the offices of Saunders College Publishing. Thanks to Christine Seashore and Eloise Thompson for finding photographs, contributing personal photographs, and for logistic collaboration. Special thanks to John Vondeling, our Publisher. One of us, Jonathan Turk, has worked with John Vondeling for over twenty years and has developed a long-lasting friendship and a superb professional relationship with him. Christine Connelly, Developmental Editor, Anne Gibby, Project Editor, Christine Schueler, Art Director, and Mary Patton, Copy Editor have all worked hard and efficiently to produce the finished project.

Graham R. Thompson
Missoula, Montana

Jonathan Turk
Darby, Montana

OCTOBER 1992

Contents Overview

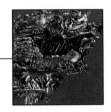
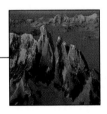
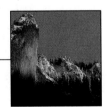

Contents

● U N I T   I

The Earth and Its Materials   25

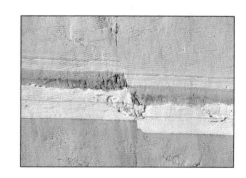

● U N I T I I I

Surface Processes 243

14 The Geology of the Ocean Floor 377

CONVERSATION WITH *Cindy Lee Van Dover* **394**

• U N I T   V

The Atmosphere 399

15 The Earth's Atmosphere 401

16 Weather 429

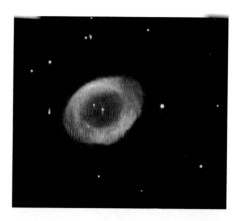

● U N I T V I I

Natural Resources 595

Earth Science and the Earth's Origin

I magine walking along a rocky coast as a storm blows in from the sea. Wind whips the ocean into whitecaps, gulls hurtle overhead, and waves crash onto shore. Before you have time to escape, blowing spray has soaked your clothes. A hard rain begins as you scramble over the last rocks to your car. During this adventure, you have interacted directly with the four major realms of the Earth. The rocks and soil underfoot are the surface of the **solid Earth**. The rain and sea are parts of the **hydrosphere**, the watery part of our planet. The wind is the atmosphere in motion. Finally, you, the gulls, the beach grasses, and all other forms of life in the sea, on land, and in the air are parts of the **biosphere**, the realm of organisms. Earth scientists study all of these realms.

Figure 1–1 shows that the solid Earth is by far the largest of the four realms. The Earth's radius is about 6400 kilometers, 1½ times the distance from New York to Los Angeles. Nearly all of our direct contact with the solid Earth occurs at or very near its surface. The deepest wells ever drilled penetrate only about 10 kilometers, 1/640 of the total distance to the center.

● A storm builds along the Oregon coast.

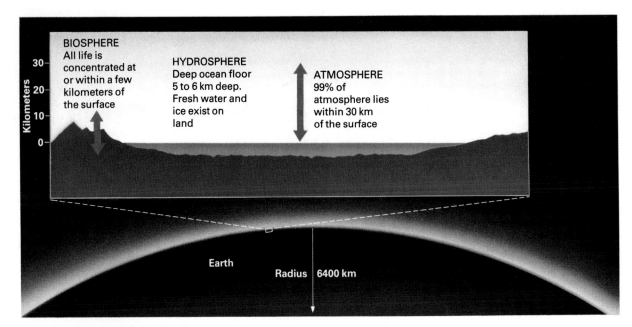

Figure 1–1   A schematic view of Earth showing the solid Earth, the hydrosphere, the atmosphere, and the biosphere.

The hydrosphere includes water in streams, lakes, and oceans; in the atmosphere; and frozen in glaciers. It also includes ground water that soaks soil and rock to a depth of 2 or 3 kilometers.

The atmosphere is a mixture of gases, mostly nitrogen and oxygen. It is held to the Earth by gravity and thins rapidly with altitude. Ninety-nine percent is concentrated in the first 30 kilometers, but a few traces remain even 10,000 kilometers above the Earth's surface.

The biosphere is the thin zone inhabited by life. It includes the uppermost solid Earth, the hydrosphere, and lower parts of the atmosphere. Land plants grow on the Earth's surface, with roots penetrating at most a few meters into soil. Animals live on the surface, fly a kilometer or two above it, or burrow a few meters underground. Sea life also concentrates near the surface, where sunlight is available. Some bacteria live in rock to depths of a few kilometers, and a few wind-blown microorganisms are found at heights of 10 kilometers or more. But even at these extremes, the biosphere is a very thin layer at the Earth's surface.

If you could drive a magical vehicle from the center of the solid Earth to the outer fringe of the atmosphere at 100 kilometers per hour, you would be within the Earth for 64 hours. In another 20 minutes you would pass through nearly all of the atmosphere and would enter the rarefied boundary between Earth and space. You would pass most living organisms in a few seconds, and the entire biosphere in 6 minutes.

An understanding of the Earth is valuable simply because humans are curious creatures; we wonder about the world around us. As you drive on a highway or walk along the seacoast, you experience the world more richly if you understand how the hills and rocks formed and why the sky is stormy or clear. Looking into the sky, you might wonder how far away the stars are or what is our significance in the Solar System or the Universe.

Earth science also has a practical side. We depend on fossil fuels—coal, oil, and natural gas—and mineral resources such as metals, sand, and gravel. We also depend on soil to support crops and other plants. Weather affects us daily, and climate influences agriculture, travel, and land use. Recently, Earth scientists have learned that industrial activities such as burning of fossil fuels and release of pollutants from other sources may now be altering the atmosphere and changing global climate. The oceans provide food and sea lanes for commerce and travel. Clean, fresh water is vital to agriculture, industry, and human consumption.

Earth science is a broad term for several sciences that study the Earth and extraterrestrial bodies: geology, oceanography, climatology, meteorology, and astronomy. Let us briefly consider each.

1.1 Geology and the Solid Earth

Geology is the study of the solid Earth: its rocks and minerals, the physical and chemical changes that occur on its surface and in its interior, and the history of the planet and its life.

Geologists seek to understand both the interior and the surface of the Earth. They study our planet as it exists today and look back at its history.

The Earth's Materials: Rocks and Minerals

Below a thin layer of soil and beneath the ocean water, the outer layers of the Earth are composed entirely of **rock**. Geologists study these rocks, their composition, their formation, and their behavior. Even a casual observer sees that rocks are different from one another: some are soft, others hard, and they come in many colors. Most rocks are composed of tiny, differently colored grains, each of which is a **mineral** (Fig. 1–2).

The Earth's Internal Processes

The Earth is an active planet. Events and processes that occur or originate within the Earth are called **internal processes**. Earthquakes and volcanoes are internal processes that are familiar because they occur rapidly and dramatically. However, the same mechanisms that cause volcanoes to erupt and earthquakes to shake the land also cause slower events such as mountain building and movements of continents. Builders, engineers, and city planners might consult with Earth scientists and ask, "What is the probability that an earthquake will occur in our city? How destructive is it likely to be? Is it safe to build skyscrapers or a nuclear power plant in the area? or What is the likelihood that a volcano will erupt to threaten nearby cities?"

The Earth's Surface Processes

Most of us have seen water running over soil after a heavy rainstorm. You may have noticed that the flowing water dislodges tiny grains of soil and carries them downslope. After a few hours of heavy rain, an exposed hillside may become scarred by gullies. If you stretch your imagination over thousands or millions of years, you can envision

Figure 1–2 Each of the differently colored specks in this rock is a mineral.

flowing water shaping the surface of our planet, enlarging tiny gullies into great valleys and canyons (Fig. 1–3). These and other natural activities that change the Earth's surface are called **surface processes**.

Both theoretical and practical questions arise when we think of surface processes. Why do desert landscapes differ from those in humid climates? How do river valleys change with time? Is a flood likely to destroy a housing development in this valley? How will acid rain affect a lake where people swim and fish? How will soil erosion affect our ability to grow crops and our food supply?

Eruption of Ngauruhoe volcano, New Zealand. *(Don Hyndman)*

Figure 1–3 Over long periods of time running water can carve deep canyons such as the Grand Canyon in the American southwest.

The Earth's History

The Earth is about 4.6 billion years old. In his book *Basin and Range*, about the geology of western North America, John McPhee offers us a metaphor for the magnitude of geologic time. If the history of the Earth were represented by the old English measure of a yard, the distance from the king's nose to the end of his outstretched hand, all of human history could be erased by a single stroke of a file on his middle fingernail. Geologists study the *entire* history of the Earth, from its origins to the present.

Most of us are fascinated by dinosaurs—their size, their abundance, and their sudden disappearance 65 million years ago. If you wished to pursue this fascination, you might become a **paleontologist** and study prehistoric life—either the dinosaurs or organisms that lived before or after.

1.2 The Hydrosphere: The Earth's Water

Hydrology is the study of all the Earth's water, its distribution, and its circulation among oceans, continents, and the atmosphere. Oceans cover 71 percent of the Earth and contain 97.5 percent of its water. Thus, most of the hydrosphere is seawater. The other 2.5 percent of the Earth's water is fresh. Most fresh water is frozen in glaciers. Less than 1 percent of the Earth's fresh water is found in streams, lakes, the atmosphere, and as **ground water** saturating rock and soil of the upper few kilometers of the solid Earth. The distribution of the Earth's water is shown in Figure 1–4.

Reconstruction of a Cretaceous dinosaur from north-central Montana.

(Museum of the Rockies, Montana State University)

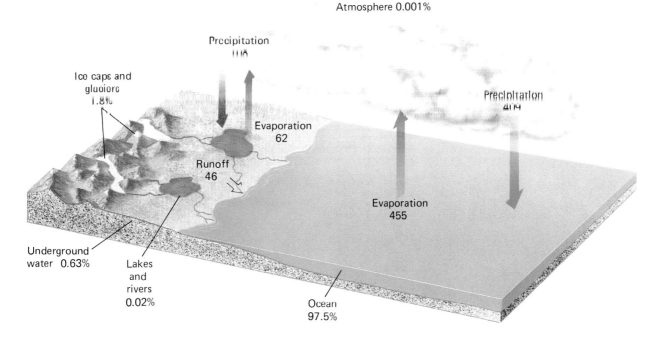

Atmosphere 0.001%

Precipitation
111A

Ice caps and
glaciers
1.8%

Precipitation
414

Evaporation
62

Runoff
46

Evaporation
455

Underground
water 0.63%

Lakes
and
rivers
0.02%

Ocean
97.5%

Figure 1–4 The distribution of water in the hydrosphere. Percentages show the distribution of Earth's water among oceans, glaciers, ground water, lakes and rivers, and the atmosphere. The numbers show thousands of cubic kilometers transferred per year as precipitation, evaporation, and runoff from land.

Oceanography

Oceanography is the study of the world's oceans, including coastlines, topography of the sea floor, the nature of seawater, ocean currents, and marine life.

Ocean currents moderate weather and climate by transporting heat from the equator poleward, cooling the low latitudes and warming regions nearer the poles. The cooling makes the tropics livable, and the warming creates the relatively mild climates of North America and other temperate lands. This moderating effect on temperature is critical to agriculture and human habitation. The seas also provide abundant food and avenues for coastal and intercontinental shipping and travel. For these reasons, more than half the world population lives within 100 kilometers of the seashore.

Coastlines change rapidly as a result of Earth processes. Ocean currents along the shore erode rock and carry sand from place to place. In addition, sea level can rise and fall due to several causes. For example, in the past 18,000 years great continental glaciers have melted, adding enough water to the oceans to raise global sea level by about 100 meters. Enough ice remains in Antarctic and Greenland ice caps to raise sea level by an additional 65 meters if it were to melt completely. Some scientists now think global temperature is rising due to changes in atmospheric composition caused by burning of fossil fuels. An increase of about 5°C would melt most of the

remaining ice. As a result, sea level would rise, flooding many coastal cities and displacing much of the world's population.

Fresh Water

Industrial societies require clean, fresh water for industry and agriculture as well as for drinking. Most modern hydrologists study our fresh water supplies, and many concentrate on environmental and economic aspects of

Surf along the Oregon coast.

Chaffin Creek in western Montana.

water. Fresh water resources are vulnerable to contamination by industrial, agricultural, and other human waste.

Just as cities have grown along coastlines, river banks are also desirable places to live. But streams are dynamic and changeable. They erode their banks, create new channels, and flood their valleys. Although streams and lakes contain most of our visible fresh water, 30 times more water is stored underground, as ground water. More than 50 percent of the people in North America drink ground water. In some regions ground water is extracted and used more rapidly than it is replaced by rain. As a result, reserves are being depleted. In other regions ground water reserves are being polluted.

About 75 percent of the Earth's fresh water is frozen in glaciers. A glacier is a flowing mass of ice. Today glaciers cover about 10 percent of the Earth's land surface. Most modern glacial ice is in the Greenland and Antarctic ice caps. In the past, however, glaciers covered much greater portions of the globe.

1.3 The Atmosphere

The Earth's atmosphere acts as a blanket, retaining heat at night and shielding us from direct solar heating during the day. Additionally, wind transports heat from the equa-

tor toward the poles, cooling equatorial regions and warming temperate and polar zones. In contrast, the Moon, our nearest neighbor, has virtually no atmosphere. As a result, its sunny side is boiling hot and its dark side is frigidly cold.

Weather is the state of the atmosphere at any particular time and place. Climate is the characteristic weather of a region over long periods of time.

You will learn that the atmosphere has changed in the past: evolving life changed it from oxygen-poor to oxygen-rich. If life changed the atmosphere in the past, can it do so again? We will explore effects of air pollution on our atmosphere, climate, and weather.

1.4 Astronomy

Astronomy is the study of the Universe, including its origin and the nature of all celestial bodies and events.

Astronomers in ancient cultures chronicled the movements of the Sun, the Moon, and the stars and used their predictable cycles to create clocks and calendars. The first major problem in astronomy was to interpret motions of heavenly bodies as seen from Earth and to develop a coherent model of the Solar System.

Modern astronomers study the nature and composition of objects in space. They ask how they formed, how hot they are, how far away they are, and how they will evolve in the future.

Although the Earth and the other planets in the Solar System all formed from a single cloud of dust and gas, no two planets are alike today. Mercury, the planet closest to the Sun, has no atmosphere, and its rocky surface is

A thunderhead building over a tropical ocean.

A nearby spiral galaxy (M83), which is similar to our galaxy, the Milky Way. *(Anglo-Australian Telescope Board)*

pockmarked by millions of craters created by meteorites bombarding its surface. Jupiter, which lies farther from the Sun in the cooler regions of the Solar System, is a swirling mass of liquids and gases with no solid surface and a dense atmosphere.

We will study the nuclear reactions that produce the Sun's energy and learn how stars are born, mature, and die. Massive stars die in violent explosions, blasting dust and gas into space to form new stars and perhaps new solar systems. The material left behind contracts to form a black hole that is so dense that even light cannot escape from its gravity.

As our vision extends outward into the far reaches of space toward distant galaxies, we seek answers to questions about the origin of matter, time, and space itself. When did the Universe form? When, if ever, will it end? Is there anything beyond?

1.5 The Biosphere

The biosphere is the thin layer at the Earth's surface containing all life (Fig. 1–5). It will not be discussed in a separate section. However, an important focus of this book is the interactions between humans and the solid Earth, the hydrosphere, and the atmosphere.

Human activity is too insignificant to affect most internal geological processes such as volcanic eruptions

earthquakes, and the movement of continents. However, in the past few decades, human activities have altered the surface of the planet. Today, human-caused changes in the atmosphere, water, and soil are found everywhere: in cities, on farmlands throughout the world, on the fringes

Figure 1–5 The biosphere consists of many different communities of plants and animals, such as this temperate rainforest on the southeast coast of Alaska.

Focus On

Hypothesis, Theory, and Law

On an afternoon field trip you may find several different types of rocks or watch a river flow by. But you can never see the rocks or river as they existed in the past or as they will exist in the future. Yet a geologist could tell you how the rocks formed millions or even a few billion years ago and could predict how the river valley might change in the future.

Scientists not only study events that they have never observed and never will observe, but they also study objects that can never be seen, touched, or felt. In this book we will examine the core of the Earth 6400 kilometers beneath our feet and the surface of quasars 70,000,000,000,000,000,000,000 kilometers above our heads, even though no one has ever visited these places and we are certain that no one ever will.

Much of science is built on inferences about events and objects outside the realm of direct experience. An inference is a conclusion based on thought and reason. How certain are we that a conclusion of this type is correct?

In science, these inferences are called laws, theories, or hypotheses depending on the degree of certainty. A **law** is a formal statement of the way in which events always occur under given conditions. It is considered to be factual and always correct. A law is the most certain of scientific statements. For example, the law of gravity states that all objects are attracted to one another in direct proportion to their masses. We cannot conceive of any realistic contradiction to this principle, and none has been observed. Hence, the principle is called a law.

A **theory** is less certain than a law. It is an interpretation or explanation of some aspect of the world that is supported by experimental or factual evidence, but is not so conclusively proved that it is accepted as a law. For example, the theory of plate tectonics states that the outer layer of the Earth is broken into a number of plates that move horizontally relative to each other. As you will see

in Unit II, this theory is supported by many observations and seems to have no major inconsistencies.

A **hypothesis**, or **model**, is weaker than a theory. It is a tentative explanation of observations that can be tested by comparing it with other observations and experiments. Thus a hypothesis or model is a rough draft of a theory that is tested against the facts. If it explains some of the facts but not all of them, it must be altered, or if it cannot be changed satisfactorily, it must be discarded and a new hypothesis developed.

Scientists develop hypotheses and theories according to a set of guidelines known as the **scientific method**, which involves three basic steps: (1) observation, (2) forming a hypothesis, and (3) testing the hypothesis and developing a theory.

Observation

All modern science is based on observation. Suppose that you observed an ocean wave carrying and depositing sand. If you watched for some time, you would see that the sand accumulates slowly, layer by layer, on the beach. You might then visit Utah or Nevada and see cliffs of layered sandstone hundreds of meters high. Observations of this kind are the starting point of science.

Forming a Hypothesis

Simple observations are only a first step along the path to a theory. A scientist tries to organize his or her observations to recognize patterns, which in turn are formed by fundamental processes. You might note that the sand layers deposited along the coast look just like the layers of sand in the sandstone cliffs. Perhaps you would then infer that the thick layers of sandstone had been deposited by an ancient ocean. You might further conclude that, since the ocean deposits layers of sand very slowly, the thick layers of sandstone must have accumulated over a very long time.

of the Sahara Desert, in the Amazon rainforest, in the central oceans, and at the North Pole. Moreover, some scientists are concerned that these changes are so widespread that they threaten human well-being.

No simple solutions exist for these environmental challenges. In every instance, complex social and technical problems must be addressed. Consider some of the questions that an Earth scientist might be called upon to answer:

At many sites, toxic compounds originating from a variety of human activities have leaked into soil. It is important to know how fast these compounds will contaminate ground water and spread outward. Are drinking water resources threatened? Can the toxic materials be contained or removed?

If a hillside forest is cut and the land terraced and planted to grain, will the terraces hold, or will landslides destroy them?

Thick layers of sandstone formed slowly and attest to the age of the Earth.

If you were then to travel, you would observe that thick layers of sandstone are abundant all over the world. Since thick layers of sand accumulate so slowly, you might infer that a very long time must have been required for all that sandstone to form. From these observations and inferences you might form the hypothesis that the Earth is very old.

Testing the Hypothesis and Forming a Theory

Theories differ widely in form and content, but all obey four fundamental criteria.

1. A theory must be constructed on a series of confirmed observations or experimental results.

2. It must explain all relevant observations or other data.

3. It must not contradict any relevant observations or other established scientific principles.

4. Finally, a theory must be internally consistent. Thus, it must be built from observations and data in a logical manner so that the conclusions do not contradict any of the original premises.

Most theories can be used to predict events that have not yet been observed, and if the theory is a good one, the predictions will be correct. When first proposed in the late 1700s, the hypothesis that the Earth is very old was based only on the observation that sand layers accumulate slowly. In the past 200 years, results of many different measurements and experiments have proved consistent with this hypothesis. Today the idea that the Earth is very old is a firmly grounded theory.

An electric company wants to build a nuclear power plant. Is the proposed site safe, or is it threatened by earthquakes or floods?

Radioactive waste from nuclear power plants and weapons plants must be stored for a long time where it will not escape into the air or into food or water supplies. Storage usually means burial. What types of environments are most stable and therefore safe? What is the probability that an earthquake, volcanic eruption, landslide, or flood will disturb a proposed repository and release radioactive wastes into the environment?

Experts commonly disagree on answers to these questions. In the emotional issue of radioactive waste storage, for example, some Earth scientists feel that the major problems have been solved and that "safe" repositories have been identified. Others feel that important questions about geological hazards at "safe sites" have not yet been answered satisfactorily. When such conflict becomes public, how can you evaluate contradictory statements? No set rules exist, but surely a first step is to understand the basic science behind the human opinions.

1.6 The Origin of the Universe

Before we ask "How did the Universe begin," we must ask an even more fundamental question: "Did it begin at all?" One possibility is that the Universe always existed, and there was no beginning, no start of time. An alternative theory is that the Universe began at a specific time and has been evolving or changing ever since.

In 1929 Edwin Hubble observed that all galaxies are moving away from each other. If this is so, then the galaxies must have started at a common center. By measuring the speeds of galaxies and the distances between them, we can mentally trace their paths in reverse to the time when the entire Universe was compressed into a single, infinitely dense point. According to modern theory, this point exploded. This cataclysmic event, called the **big bang**, marked the beginning of the Universe and the start of time. It was no ordinary explosion. It cannot even be compared with a hydrogen bomb or the catastrophic death of a massive star. On the contrary, this explosion instantaneously created the Universe. Matter, energy, space, and even time came into existence with this single event.

Estimates of the age of the Universe, starting at the big bang, vary from about eight billion to 20 billion years. This wide range occurs because of uncertainty in measuring speeds of galaxies and the distances between them. Most astronomers place the start of time and the origin of the Universe between 15 and 18 billion years ago.

Even though our estimates of the time of the big bang vary by *billions* of years, astronomers have reconstructed a picture of the first few *seconds* after the origin of the

Figure 1–6 A brief pictorial outline of the evolution of the Universe.

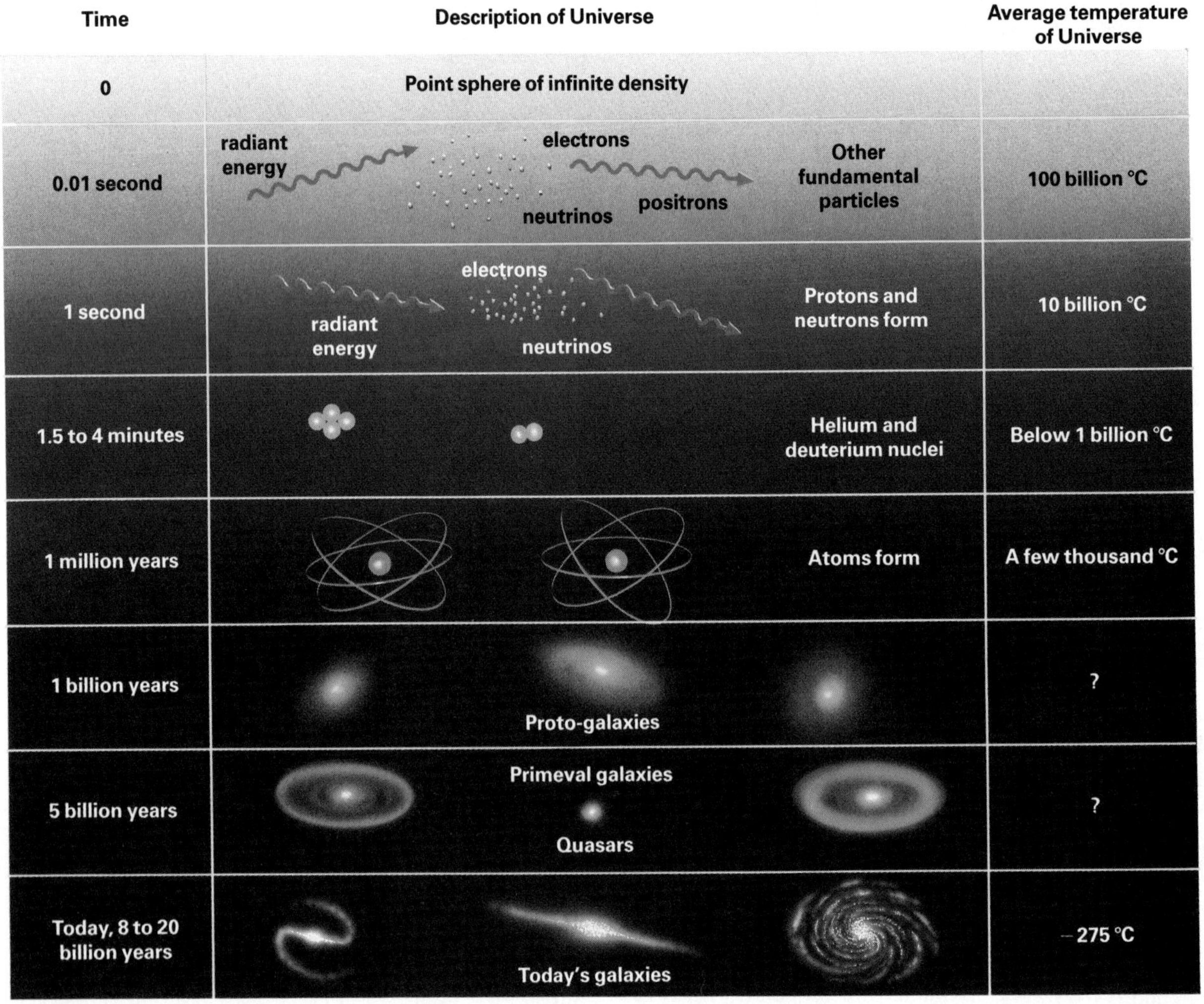

Time	Description of Universe	Average temperature of Universe
0	Point sphere of infinite density	
0.01 second	radiant energy · electrons · neutrinos · positrons · Other fundamental particles	100 billion °C
1 second	radiant energy · electrons · neutrinos · Protons and neutrons form	10 billion °C
1.5 to 4 minutes	Helium and deuterium nuclei	Below 1 billion °C
1 million years	Atoms form	A few thousand °C
1 billion years	Proto-galaxies	?
5 billion years	Primeval galaxies · Quasars	?
Today, 8 to 20 billion years	Today's galaxies	− 275 °C

Focus On

Is the Big Bang Theory Correct?

Several recent astronomical observations cannot be explained by the big bang theory. If the Universe were created in a single explosion, we would expect it to be homogeneous; matter should be distributed uniformly throughout space. However, matter is concentrated into galaxies, and galaxies are clustered into groups. Therefore, the Universe appears to be heterogeneous. Some scientists and several journalists have used this inconsistency to conclude that the big bang theory is incorrect. *Time* magazine ran an article in January 1991 entitled "Bang! A Big Theory May Be Shot." An article in the *New York Times* (January 3, 1991) stated, "How can scientists keep on believing in the Big Bang when they can't understand the details of what went on after?"

The authors of an article in the British journal *Nature** rebutted these charges:

Cosmologists grow used to explaining to their colleagues in other fields why the latest reports of the death of the Big Bang model are premature, to say the least. To the contrary, in the six decades since the formulation of the model, advances in observations and experiments have yielded a considerable body of evidence in support of the big bang and none that convincingly contradicts it.

A healthy degree of skepticism is in order, for we are using a short list of sometimes indirect evidence to arrive at a grand conclusion, that the Universe expanded from a dense hot beginning. Many have given their accounts of the skeptical view. . .

The authors go on to explain that the big bang theory explains most observations well and has predicted others. Even though questions remain, none contradicts the fundamental premises of the theory. It is reasonable to expect that the questions will be answered through modifications of the big bang cosmology. Finally, no feasible alternative theories have been offered.

Debate is part of science. It is important not to accept theories as fact, because sometimes they are wrong. But it is also important not to discard a theory that explains many phenomena just because some questions remain unanswered. As this book was going to press, new data as described in Chapter 21 support the big bang theory.

*P. J. E. Peebles, D. N. Schramm, E. L. Turner, and R. G. Kron, "The Case for the Relativistic Hot Big Bang Theory," *Nature* 352 (August 29, 1991): 769.

Universe. This reconstruction comes from studies of how particles behave when they collide at very high velocities in modern particle accelerators. Other evidence comes from studying particles and radiant energy in space.

Immediately after the big bang, the Universe was extremely hot, about 100 billion degrees Celsius. During the first second, it cooled to about 10 billion degrees, 1000 times the temperature in the center of the modern Sun (Fig. 1–6). At such high temperatures, atoms do not exist. Most of the Universe consisted of a mixture of radiant energy, electrons, and extremely light particles called neutrinos. Protons and neutrons also began to form. After about 1.5 minutes, the temperature fell to 1 billion degrees and a few simple atomic nuclei formed, although the temperature was still too hot for atoms. During the next million years the Universe continued to cool as it expanded. When the temperature dropped to a few thousand degrees, atoms formed and, in a sense, the modern Universe was born. With time, matter collected into galaxies, and within the galaxies stars were born.

At present the galaxies are all flying away from one another. What will happen in the future? Think of a rocket ship taking off from a planet. If the planet's gravitational field is weak enough, the rocket will escape into space and never return. However, if the planet has a large mass and therefore a strong gravitational field, then the rocket will fall back to the surface. In the same manner, if the gravitational force of the universe is sufficient, all the galaxies will eventually slow down, reverse direction, and fall back to the center, forming another point of infinite density. This point may then explode again to form a new universe. In turn the new universe will expand and then collapse, creating a continuous chain of universes. This possibility is called the **oscillating universe cosmology**. The other possibility is that the gravitational force of the Universe is not sufficient to stop the expansion, and the galaxies will continue to fly apart forever. Within each galaxy, stars will eventually consume all their nuclear fuel and stop producing energy. As the stars fade and cool, the galaxies will continue to separate into the cold void. This scenario is called the **forever-expanding cosmology** (Fig. 1–7). Astronomers are attempting to calculate the mass of the Universe in an effort to determine which of the two possibilities is more realistic. However, the measurements are uncertain and the final answer elusive.

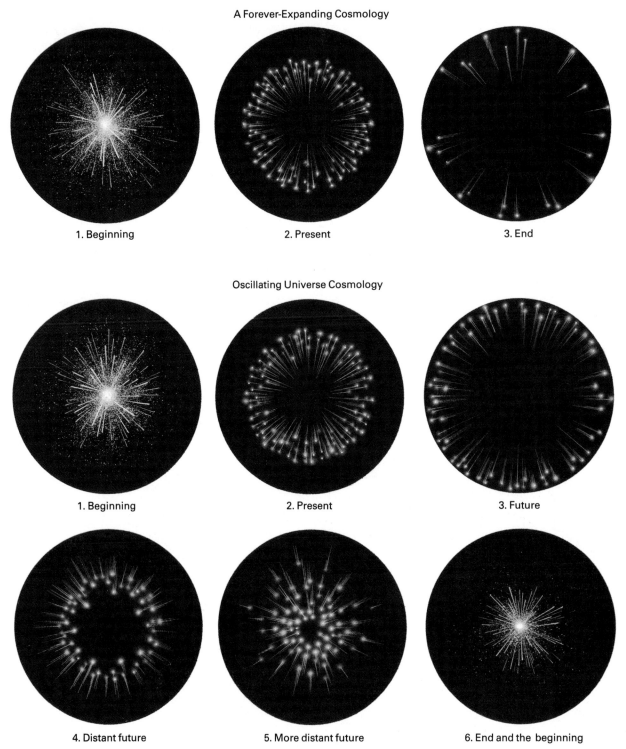

Figure 1–7   A schematic representation of two possible cosmologies, a Forever-Expanding Cosmology and the Oscillating Universe Cosmology.

1.7 Formation of the Solar System and the Earth

Five billion years ago, the matter that eventually became our Solar System formed an immense cloud of dust and gas rotating slowly in space. Its temperature was about –270°C. At this extremely low temperature, particles move so slowly that even a slight force affects them appreciably. Small gravitational attractions among the dust and gas caused the cloud to condense into a sphere (Fig. 1–8). Alternatively, some astronomers suggest that a nearby star may have exploded and that the shock wave triggered the condensation (Fig. 1–9). As the condensation continued, the rotation accelerated and the sphere spread into a disk, as shown in Figure 1–8C.

More than 90 percent of the matter in the cloud gravitated toward the center of the newly formed disk. As atoms were pulled inward, they accelerated under the influence of gravity. Eventually, the center of the disk coalesced to form the **protosun**, the earliest form of the Sun. The protosun was heated by energy from the gravitational collapse of the disk, but it was not a true star because it did not yet generate energy by nuclear fusion.

Formation of the Planets

Heat of the protosun warmed the inner region of the disk. Then, as the gravitational collapse became nearly complete, the disk cooled. Gases condensed to form small aggregates, much as raindrops or snowflakes form when moist air cools in the Earth's atmosphere. The formation of these aggregates was the first step in the evolution of planets.

As the growing cloud rotated around the protosun, aggregates began to stick together as snowflakes sometimes do. Thus, they increased in size, thereby developing stronger gravitational forces and attracting additional particles. This growth continued until a number of small,

Figure 1–8 Formation of the Solar System. (A) The Solar System was originally a diffuse cloud of dust and gas. (B) The dust and gas began to coalesce due to gravity. (C) The shrinking mass began to rotate and formed a disk. (D) The mass broke up into a discrete protosun orbited by large protoplanets. (E) The Sun heated up until fusion temperatures were reached. The heat from the Sun drove most of the hydrogen and helium away from the closest planets, leaving small, solid cores behind. The massive outer planets remain composed mostly of hydrogen and helium.

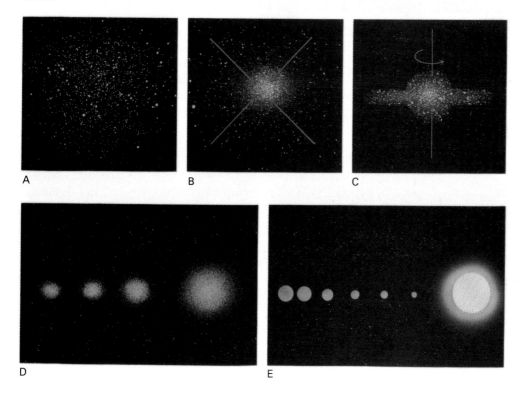

A   B   C

D   E

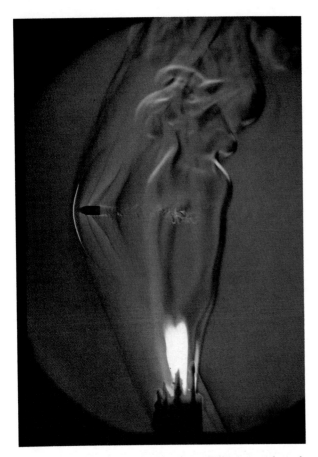

Figure 1–9   A shock wave forms as a bullet passes through hot gases generated by a burning candle. The leading edge of the shock wave is on the left. According to one theory, a shock wave created by an exploding star initiated the collapse of the cloud of dust and gas and thereby was the first step in the formation of the Solar System.   *(Harold E. Edgerton, MIT; Palm Press)*

Formation of the Sun

At the same time that planets were evolving, changes occurred in the protosun. Gravitational attraction pulled the gases inward, creating extremely high pressure and temperature. The protosun became hottest in its center, where atoms accelerated inward most rapidly. The core of the protosun became so hot that hydrogen nuclei began to combine in a process called nuclear fusion. When hydrogen nuclei combine, they form the nucleus of the next heavier element, helium. As fusion began within the Sun, vast amounts of nuclear energy were released in a process comparable to the continuous explosion of millions of hydrogen bombs. The onset of nuclear fusion marked the birth of the modern Sun. The heat and light given off by our Sun are still generated by hydrogen fusion.

The Modern Solar System

In its initial form, the cloud that evolved into the Solar System must have been homogeneous; that is, it was the same throughout. However, as the Sun became hotter, many light gases, such as hydrogen and helium, boiled away from the inner Solar System and collected in the frozen outer regions.

As a result, the four planets closest to the Sun—Mercury, Venus, Earth, and Mars—are now rocky with metallic centers. Most of the gases have been lost. Thus, Mercury has virtually no atmosphere, and the atmospheres surrounding Venus, Earth, and Mars represent a tiny portion of their planetary masses. These four are called the **terrestrial planets** because they are all "Earth-like." In contrast, the four planets beyond this inner circle—Jupiter, Saturn, Uranus, and Neptune—are called the **Jovian planets** and are composed primarily of liquids and gases such as hydrogen, helium, water, ammonia, and methane (Fig. 1–10). Pluto, the outermost planet, is anomalous and will be discussed further in Chapter 21. Figure 1–11 is a schematic representation of the modern Solar System.

The Early Earth

As already explained, the Earth formed and grew by multiple collisions of smaller bodies. The bodies were drawn together by gravity, and this gravitational collapse generated heat. Additional heat was released by radioactivity. Some elements in the Earth are radioactive. When a radioactive atom breaks apart, energy is released and converts to heat. Only a tiny amount of heat is created by decay of a single radioactive atom, and radioactive atoms were dispersed throughout the early Earth. Therefore, in a given volume of rock, heat was generated very slowly.

rocky spheres formed. Their gravitational forces caused them to collide and coalesce to form mini-planets, called **planetesimals**, that ranged in size from a few kilometers to about 100 kilometers in diameter. The entire process, from the disk to the planetesimals, occurred relatively quickly in geological terms and probably required only about 10,000 to 100,000 years. As the planetesimals grew, they were attracted to one another by gravity and collided. Many small spheres coalesced to a few large ones, including the Earth. The Earth formed about 4.6 billion years ago. Since then, additional planetesimals and small chunks of rock called **meteoroids** continued to slam into its surface, adding to its mass. Meteoroids continue to bombard the Earth to this day. However, bombardment is now so infrequent that the mass of the Earth has remained essentially constant for three to four billion years.

Focus On

Scientific Evidence to Support the Current Theory of the Formation of the Solar System

Observation	Interpretation of the Observation in Terms of the Current Theory for the Evolution of our Solar System
All the planets (except Pluto) orbit in the same plane.	All the planets formed from a common planar disk.
All the planets revolve around the Sun in the same direction, which is also the direction in which the Sun rotates on its axis.	If the planets and the Sun all formed from a single rotating disk they would all retain the direction of motion of the original disk.
The planets closest to the Sun are small and composed primarily of heavy elements. The more distant planets are larger and composed primarily of light gases.	According to our theory, all planets originally had the same composition because they formed from a common cloud. However, the Sun drove the light gases away from the closest planets, leaving small, dense, rocky spheres. This is exactly what we observe.
Flattened disks of dust and gas have been observed around several young stars (see illustration).	We cannot look backward in time to see how our Solar System evolved, but the theory of its evolution would be supported if we could see similar events occurring elsewhere today. The thin disk around the star Beta Pictoris is believed to be similar to the disk that formed the planets of our own Solar System.
Some nearby stars have been observed to wobble as if they were perturbed by the gravitational field of a nearby object, such as a planet.	In principle, a very sensitive telescope could detect a planet in orbit around a nearby star. In practice, the image of the planet is drowned out by the light of the star. However, by observing tiny aberrations in the motion of a star, we can deduce that it is affected by the gravitational field of an orbiting planet. This tells us that planetary evolution is not unique to the Solar System.

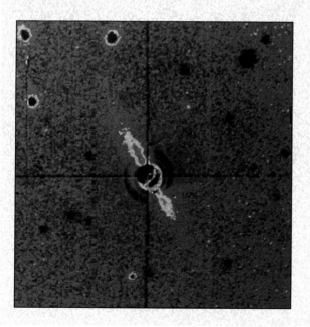

This photograph may be an image of another solar system in the process of formation. The thin disk around the central star, Beta Pictoris, is composed of bits of dust, which are believed to be similar to the material that condensed to form the planets of our own Solar System. *(University of Arizona and Jet Propulsion Laboratories)*

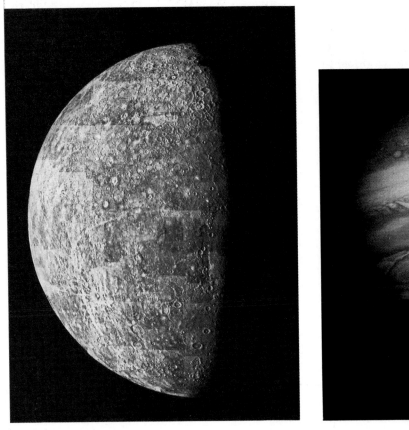

A

B

Figure 1–10   (A) Mercury is a small planet close to the Sun. Because of this proximity, most of its lighter elements have long since been boiled off into space, and today its surface is solid and rocky. (B) Jupiter, on the other hand, is composed mainly of gases and liquids, with a small solid core. This photograph is a close-up of its turbulent atmosphere.   *(NASA)*

For example, ordinary granite contains small but measurable amounts of radioactive materials. If no energy were lost, it would take 500 million years to brew a cup of coffee with the heat released from 1 cubic centimeter of granite. However, the Earth is large and geologic time is long. The heat from gravitational collapse and radioactivity was retained by a thick insulating layer of surface rocks. At the same time, the surface was heated by intense bombardment as the Earth swept up chunks of rock, comets, and other debris floating about in the early Solar System.

A few hundred million years after it formed, the Earth became so hot that it began to melt and separate into layers. Heavy molten iron and nickel seeped toward the center and eventually collected to form a dense, hot core. Light elements floated toward the surface to form the relatively light rocks of the Earth's crust. The remaining rock concentrated between the crust and the core to form the mantle. Shortly after it melted, the Earth began to cool and most of it solidified. Today the Earth continues to cool, but it retains a layered structure.

1.8   The Structure of the Modern Earth

During its formation, the Earth separated into three distinct layers: the crust, the mantle, and the core (Fig. 1–12).

The **crust** is a thin, rigid surface veneer. Its thickness ranges from 7 kilometers under some portions of the oceans to a maximum of about 70 kilometers under the highest mountain ranges. By comparison, the radius of the entire Earth is 6370 kilometers. If you built a model of the Earth 1 meter in radius, the crust would be 1.1 to 11 millimeters thick. If the Earth were the size of an egg, the crust would be thinner than an eggshell. It is made of low-density rocks that floated to the surface when Earth was molten.

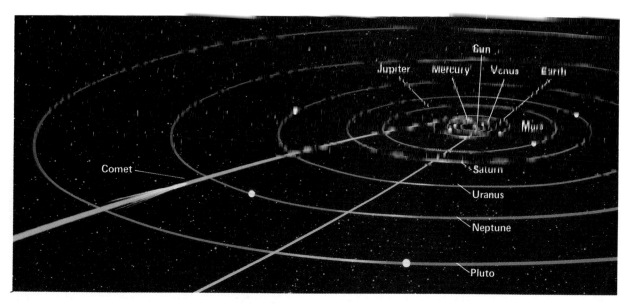

Figure 1–11 A schematic view of the Solar System.

A thick, mostly solid layer called the **mantle** lies beneath the crust and surrounds the core. The mantle extends from the base of the crust to a depth of 2900 kilometers. It contains more than 80 percent of the volume of the Earth.

The uppermost portion of the mantle is relatively cool. Because the rocks are cool, they are strong and brittle. They crack when stressed, much like the rocks of the crust. The cool, strong, brittle outer portion of the Earth, including both the crust and this upper layer of mantle, is called the **lithosphere** (Greek for "rock layer"). The lithosphere extends from the Earth's surface to an average depth of 100 kilometers.

Beneath the lithosphere, but also within the mantle, lies a layer called the **asthenosphere** (Greek for "weak layer"). It extends from a depth of about 100 kilometers to about 350 kilometers below the Earth's surface. The Earth's temperature increases with depth, and the asthenosphere is hot enough that a percent or two of the rock is melted. The remaining rock, although solid, is so hot that it flows slowly without cracking—it behaves plastically. We are familiar with materials that deform plastically, and the asthenosphere has often been compared with road tar or putty. It is solid, yet when stressed it gives and flows as a fluid does.

As already mentioned, the asthenosphere extends to about 350 kilometers below the surface, whereas the base of the mantle lies at a depth of 2900 kilometers. Thus, most of the mantle lies below the asthenosphere. Below the asthenosphere pressure is so great that even though the rock is hot, it is solid and considerably more rigid than the rock in the asthenosphere. However, rigidity is

Figure 1–12 A schematic view of the interior of the Earth. The inset is a view of the outer layers.

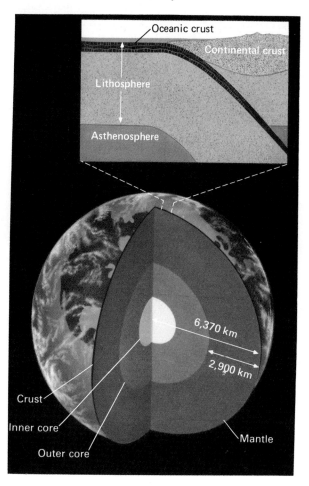

used here in a relative sense. Careful measurements show that the Earth is not perfectly spherical but slightly distorted. The rotation of the Earth causes it to bulge at the equator. Therefore, a person standing on the equator is about 50 kilometers farther from the center of the Earth than a person standing at the South Pole. If you fill a balloon with water and spin it on a tabletop, it will distort in an exaggerated model of the Earth's bulge.

The Earth's **core** is composed primarily of iron and nickel. It is divided into a solid inner core and a liquid outer core. Insulated by 2900 kilometers of overlying rock, it has retained much of the original heat acquired during the formation of the Earth. Today the temperature in the core is about 6000°C, about the same as that at the Sun's surface.

1.9 Consequences of a Hot Earth: Earthquakes, Volcanoes, and Plate Tectonics

Our planet is *not* a static, homogeneous sphere, like a bowling ball. Instead, Earth is active and dynamic. The thin, cool, brittle lithosphere floats on the hot, plastic asthenosphere much as blocks of wood float in a tub of honey. The asthenosphere is solid, not liquid like honey, but because of its plastic nature, it can flow. We will refine our understanding of this important layer throughout the book.

Within the past 30 years the theory of **plate tectonics** has revolutionized Earth science. According to this theory, the lithosphere is segmented into seven major plates and several smaller ones. These lithospheric plates are packed tightly together like the segments of a turtle shell (Fig. 1–13). In this theory, **the plates move by floating on and gliding over the plastic asthenosphere**. The boundaries between plates are zones where one lithospheric plate meets another moving in a different direction. As you might expect, the boundaries are geologically active. When plates rub, jerk, or jump past one another, the Earth shakes. This vibration is called an **earthquake**. In some regions, most commonly near plate boundaries, parts of the asthenosphere and lithosphere melt. The resulting liquid rock is called **magma**. Large quantities of magma rise through the lithosphere and pour out onto the Earth's surface in **volcanic eruptions**.

Three different types of movement occur at plate boundaries (Fig. 1–14):

1. A **divergent boundary** is a zone where two or sometimes three plates separate, or move apart from each other.

2. If plates on a sphere separate in some places, they must collide in others. A zone where plates collide head-on is a **convergent boundary**.

3. At a **transform boundary**, plates slide horizontally past one another.

Shortly after World War II, oceanographers mapped a submarine mountain range running north-south through

Figure 1–13   Major plates of the world.

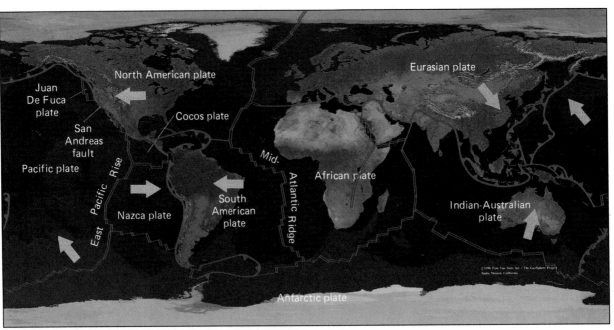

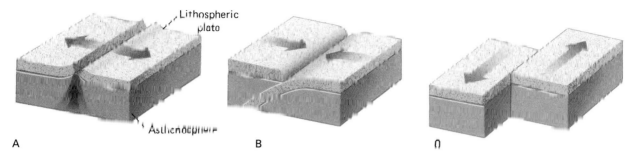

Figure 1–14   Three types of plate motion: (A) divergent, (B) convergent, (C) transform.

the whole length of the Atlantic Ocean, called the **Mid-Atlantic ridge**. A long, narrow, deep cleft, called a **rift valley**, runs along the middle of the ridge. The plates on opposite sides of the rift valley are moving away from each other. Therefore, the Mid-Atlantic ridge is a divergent plate boundary. As the two plates spread, magma from the asthenosphere oozes upward to fill the gap and form new oceanic crust. Extensions of the Mid-Atlantic ridge continue into all other ocean basins. This global submarine mountain chain is called the **mid-oceanic ridge**. It marks divergent boundaries between separating plates in all ocean basins.

At this point you may ask: If the sea floor is spreading at the mid-oceanic ridge, is the Earth's crust growing and is our planet slowly expanding like a marshmallow roasting in a fire? The answer is most certainly no! Again, evidence comes from the sea floor. In several regions, notably on the west coast of South America and in the western Pacific Ocean, the sea floor sinks abruptly, form-ing deep **trenches** (Fig. 1–15). These long, narrow oceanic trenches occur at convergent boundaries where two tectonic plates collide. At a collision zone, one lithospheric plate dives under the other and sinks into the mantle. This downward movement of a lithospheric plate is called **subduction**, and the nearby region is called a **subduction zone**. Volcanoes and earthquakes are common in subduction zones. The average depth of the sea floor is about 5 kilometers below sea level. Trenches are simply places where the sea floor is pulled downward to depths of 10 kilometers or more by a sinking plate.

In other regions, lithospheric plates slide horizontally past each other. This type of motion is occurring today along the San Andreas fault in California. The plate west of the fault is moving northwestward relative to most of North America. This horizontal movement causes most of the earthquakes for which California is famous.

(Text continues on p. 22)

Figure 1–15   The Earth is geologically active. New lithosphere forms at the mid-oceanic ridges and spreads outward. At the same time old lithosphere dives into the asthenosphere at subduction zones.

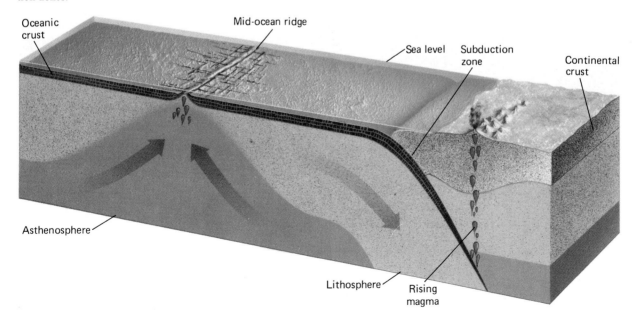

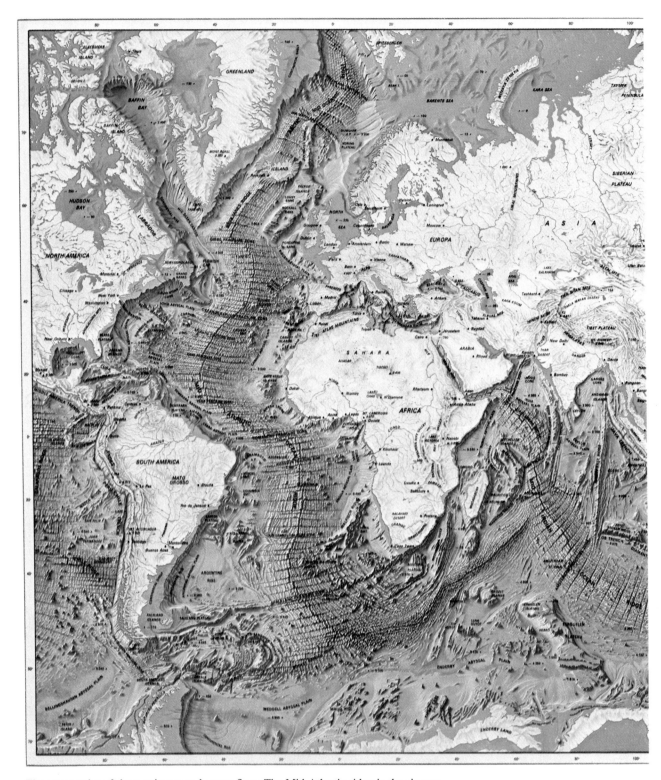

The topography of the continents and ocean floor. The Mid-Atlantic ridge is the sinuous
mountain chain snaking its way down the middle of the Atlantic Ocean. Extensions of this ridge
continue into all other ocean basins. This global submarine mountain chain is called the mid-
oceanic ridge. *(Marie Tharp)*

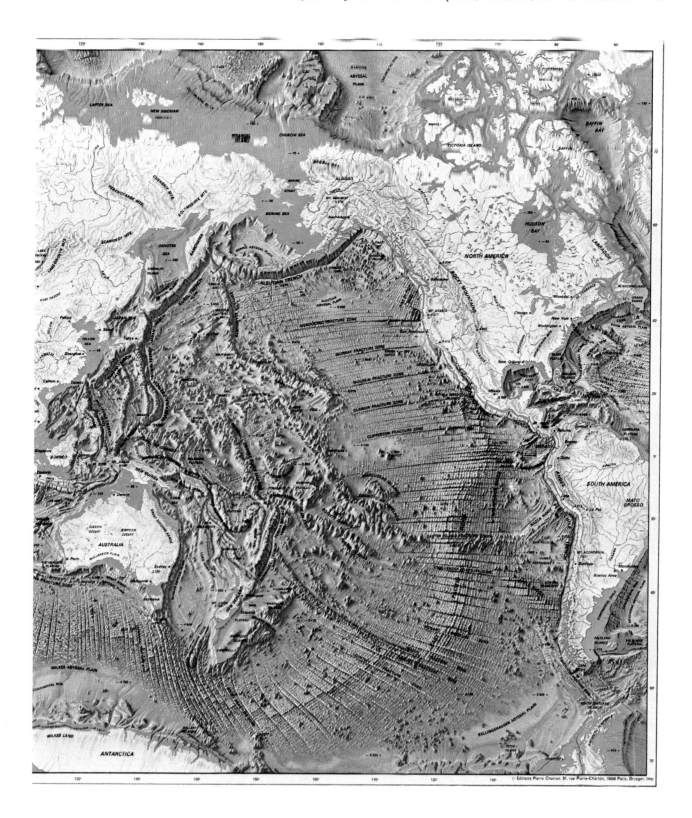

SUMMARY

•

Earth scientists study four realms of the Earth. The rocks and soil underfoot are part of the **solid Earth**. The **hydrosphere** includes all of the Earth's water. The **atmosphere** is the layer of gas surrounding the solid Earth, and the **biosphere** is the zone near the surface where all life is found.

Geology is the study of the solid Earth, its rocks and minerals, the **internal processes** that move continents and cause earthquakes and volcanoes, and the **surface processes** that sculpt mountains and valleys. Geologists study both modern Earth processes and the Earth's history from its beginning 4.6 billion years ago to the present.

Oceans contain 97.5 percent of the Earth's water. Of the remaining 2.5 percent, most is frozen in the ice caps of Antarctica and Greenland, and only 0.65 percent is fresh water in lakes, streams, and **ground water**. **Oceanographers** study the seas, and **hydrologists** study our fresh water.

The Earth's **atmosphere** evolved along with life, and its present composition is a result of organic processes. **Weather** is the state of the atmosphere at any particular time and place. **Climate** is the characteristic weather of a region over a long period of time.

Astronomers study the other planets and moons of the Solar System, distant stars, galaxies, and other extraterrestrial bodies.

Organisms of the **biosphere**, including humans, affect Earth's surface processes and the compositions of the hydrosphere and atmosphere.

According to the **big bang theory**, all matter in the Universe was originally compressed into a single point, about 15 to 20 billion years ago, and this point exploded to form the Universe.

The Solar System formed from a mass of dust and gas that rotated slowly in space. Within the center, the gravitational attraction was so great that the gases were pulled inward with enough velocity to initiate nuclear fusion, the process that still lights the Sun. In the outer disk, planets formed from coalescing dust and gas. Most light gases escaped from the inner planets but were retained by the outer giants.

The primordial Earth was heated and melted by energy released from radioactive decay and by bombardment from outer space. It has since cooled so that most of it is solid, although the inner layers remain hot. The modern Earth is made up of a dense **core** of iron and nickel, a **mantle** of lower density, and a **crust** of yet lower density. The **lithosphere** contains the uppermost portion of the mantle and the crust. The **asthenosphere** is a hot, plastic layer that lies just below the lithosphere.

The **theory of plate tectonics** states that the lithosphere is broken into several plates that move about relative to one another. Separation of plates (**divergence**) occurs along mid-oceanic ridges, and head-on collisions (**convergence**) lead to **subduction** of lithospheric plates. At **transform** boundaries, plates move horizontally past one another.

KEY TERMS

•

REVIEW QUESTIONS

●

1. Describe the relative sizes and locations of the four realms of Earth.

2. What proportion of the Earth's water is in the seas? What proportion is in glaciers?

3. Where is the water that is not part of the oceans or the glaciers?

4. What gases are most abundant in the atmosphere?

5. List as many of the Earth's surface processes as you can think of, and briefly describe each.

6. What are "internal processes," and what are some of the effects of Earth's internal processes?

7. How old is the Earth?

8. What is ground water? Where in the hydrosphere is it located?

9. How did Earth's atmosphere evolve?

10. What are the differences between weather and climate?

11. In what ways do organisms, including humans, change the Earth? What kinds of Earth processes are unaffected by humans and other organisms?

12. Very briefly outline the formation of the Universe.

13. How old was the Universe when our Solar System started to evolve? How long after the start of the evolution of the Solar System did the planets take form?

14. Briefly outline the evolution of the planets.

15. How did the Sun form? How is its composition different from that of the Earth? Explain the reasons for this difference.

16. Compare and contrast the properties of the terrestrial planets with those of the Jovian planets.

17. The entire Earth was molten soon after its formation. Explain why it cooled.

18. Briefly outline the layered structure of the modern Earth.

19. What type of plate motion occurs at the mid-oceanic ridges? What type of activity leads to subduction of plates?

DISCUSSION QUESTIONS

●

1. What would the Earth be like if it (a) had no atmosphere? (b) Had no water?

2. How might Earth be different without life?

3. Only 0.65 percent of Earth's water is fresh and liquid; the rest is salty seawater or is frozen in glaciers. What are the environmental implications of such a small proportion of fresh water?

4. Discuss specific ways in which studies of astronomy might enlighten us about our own planet.

5. Explain how the theory of the evolution of the Solar System explains the following observations: (a) All the planets in the Solar System are orbiting in the same direction. (b) All the planets in the Solar System except Pluto are orbiting in the same plane. (c) The chemical composition of Mercury is similar to that of the Earth. (d) The Sun is composed mainly of hydrogen and helium, but also contains all the elements found on Earth. (e) Venus has a solid surface, whereas Jupiter is mainly a mixture of gases and liquids with a small, solid core.

6. The radioactive elements that are responsible for the heating of the Earth decompose very slowly, over a period of billions of years. How would the Earth be different if these elements decomposed much more rapidly—say, over a period of a few million years? Defend your answer.

7. Explain how the size of a terrestrial planet can affect its surface environment.

The Earth and Its Materials

UNIT

I

• Sunset over Prince Rupert Bay in the Caribbean Sea
(Galen Rowell/Mountain Light)

Minerals

T he Earth's continents are composed mostly of granite. If you look closely at a piece of granite like the one in Figure 2–1, you can see many small, differently colored grains. Some grains may be pink, some black, and others white. Each grain is a separate mineral. Some rocks are made of only one mineral, but granite and most other rocks contain three or four abundant minerals plus small amounts of a few others.

2.1 What Is a Mineral?

Minerals are the substances that make up rocks. This statement is correct, but it does not tell us much about minerals. A more informative definition is that **a mineral is a naturally occurring, inorganic solid with a definite chemical composition and a crystalline structure**. Thus, a mineral has five characteristics: (1) it is natural in origin, (2) it is inorganic, (3) it is solid, (4) it has a distinct chemical composition, and (5) it has a crystalline structure.

The most important properties of a mineral are its chemical composition and its crystalline structure. They distinguish any mineral from all others. Because these two qualities are so important, they are discussed separately in

● Tourmaline crystals.   *(Dane A. Penland, Smithsonian Institution)*

Figure 2–1 Each of the differently colored grains in this granite is a different mineral. The pink grains are feldspar, the white ones are quartz, and the black ones are amphibole.

the following two sections. First, however, the natural, inorganic, and solid aspects of minerals are considered briefly.

The qualification that minerals occur naturally means that mineral-like substances made in laboratories or factories are not true minerals. In one sense this distinction is artificial. A synthetic diamond can be identical to a natural one, yet natural gems are valued more highly than synthetic ones. For this reason, jewelers should always tell their customers which gems are natural and which are synthetic.

Organic substances are those produced by living organisms, or are similar to ones produced by organisms. They differ chemically from inorganic substances. Although coal is a naturally occurring rock, it is not a mineral because it is derived from organisms. Similarly, oil is not a mineral because it is an organic liquid and has neither a crystalline structure nor a definite chemical composition.

2.2 The Chemical Composition of Minerals

Elements, Atoms, and Ions

In the third century B.C., the Greek philosopher Aristotle defined an element, saying "Everything is either an element or composed of elements." Although Aristotle's

TABLE 2–1

The Eight Most Abundant Chemical Elements in the Earth's Crust		
Element	**Chemical Symbol**	**Common Ion(s)**
Oxygen	O	O^{2-}
Silicon	Si	Si^{4+}
Aluminum	Al	Al^{3+}
Iron	Fe	Fe^{2+} and Fe^{3+}
Calcium	Ca	Ca^{2+}
Magnesium	Mg	Mg^{2+}
Potassium	K	K^{1+}
Sodium	Na	Na^{1+}

definition is still correct, a more complete modern definition is that **an element is a fundamental form of matter that cannot be broken into simpler substances by ordinary chemical processes.** Elements are the building blocks of which all other substances are composed. They are the fundamental materials of chemistry. A total of 88 elements occur naturally in the Earth's crust.

Of those 88, only eight elements—oxygen, silicon, aluminum, iron, calcium, magnesium, potassium, and sodium—make up more than 98 percent of the Earth's crust.

All of the elements are listed in Appendix 1. Each element is assigned a one- or two-letter symbol. The symbols for the eight most abundant elements are given in Table 2–1.

An **atom** is the basic unit of an element. It consists of a small, dense, positively charged center called a **nucleus** surrounded by a cloud of negatively charged **electrons** (Fig. 2–2). An electron is a fundamental particle; as far

Figure 2–2 An atom consists of a small, dense, positive nucleus surrounded by a much larger cloud of negative electrons.

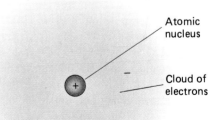

Atomic nucleus

Cloud of electrons

as we know it is not made up of smaller components. The nucleus, however, is made up of two different kinds of particles: (1) positively charged **protons** and (2) **neutrons**, which have no charge. In a neutral atom the number of protons equals the number of electrons. Therefore, the positive and negative charges balance each other so that neutral atoms have no overall electrical charge.

However, in many elements, including all eight of the most abundant ones, the neutral atoms easily lose or gain electrons. When an atom loses one or more electrons, its positive charges then outnumber its negative ones. The atom therefore becomes positively charged. If an atom gains one or more extra electrons, it becomes negatively charged. Atoms with a positive or negative charge are called **ions.**

A positively charged ion is a **cation.** All of the abundant elements except oxygen release electrons to become cations, as shown in Table 2–1. For example, each potassium atom (K) loses one electron to form a cation with a charge of + 1. Silicon atoms lose four electrons each, forming cations with + 4 charges.

In contrast, oxygen *gains* two extra electrons to acquire a − 2 charge. Atoms with negative charges are called **anions.**

Chemical Bonds

Atoms and ions rarely exist as separate, isolated entities. Instead, they unite with other atoms or ions to form **compounds.** Most minerals are compounds. The atoms or ions in a compound are held together by electrical forces called **chemical bonds.**

Four types of chemical bonds hold atoms together to form minerals: ionic bonds, covalent bonds, metallic bonds, and van der Waals forces. Many physical properties of a mineral, including color, hardness, density, and the ability to conduct electricity, depend on the bond type.

The opposite electrical charges of cations and anions attract each other to form **ionic bonds.** When cations and anions bond together to form a mineral, they always combine in proportions so that the negative charges exactly equal the positive ones. Thus, minerals are always electrically neutral. As an example, consider the mineral halite, which is table salt. It is composed of equal numbers of sodium cations and chlorine anions. Sodium is a soft, silvery metal that is extremely chemically reactive. If you held pure sodium in your hand, it would react with moisture in your palm and burn your skin. If you sprinkled powdered sodium into water it would explode. Chlorine is a green poisonous gas used for chemical warfare during World War I. When sodium reacts with chlorine, each sodium atom loses one electron to form a cation, Na^+. The electron is captured by a chlorine atom to form an anion, Cl^- (Fig. 2–3). When the two react, they form halite. The total charge of halite is $+1 - 1 = 0$. A thumb-sized crystal of halite contains about 10^{20} (1 followed by 20 zeros) sodium and chlorine ions, but the proportion is always 1:1.

A **covalent bond** forms when nearby atoms share their electrons. Diamond consists of a three-dimensional

Figure 2–3 When sodium and chlorine atoms combine, sodium loses one electron, becoming the cation Na^{1+}. Chlorine acquires the electron to become the anion Cl^{1-}.

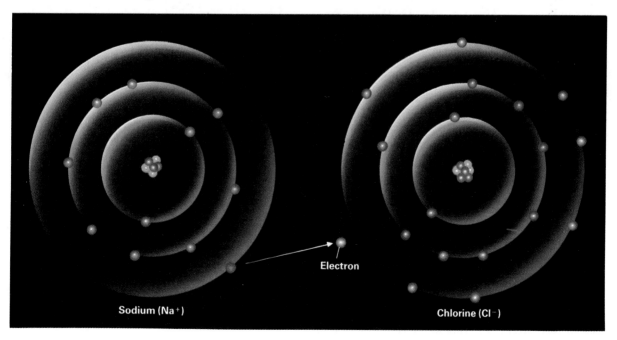

Electron

Sodium (Na⁺)

Chlorine (Cl⁻)

Diamond is the hardest of all minerals. *(Ward's Natural Science Establishment, Inc.)*

network of carbon atoms. Each carbon atom bonds to four neighbors by sharing electrons to form four covalent bonds. The strength of the bonds makes diamond the hardest of all minerals. In most minerals, the bonds between atoms are partly covalent and partly ionic.

In a **metallic bond**, some of the electrons are loose; that is, they are not associated with a particular atom. Thus, the metal atoms sit in a "sea" of electrons that are free to move from one atom to another. That arrangement allows the atoms to pack together as closely as possible, resulting in the characteristic high density of metals. Since the electrons are free to move, metals are excellent conductors of electricity and heat.

Atoms are also attracted to each other by weaker electrical forces called **van der Waals forces**. These weak bonds result from uneven distributions of electrons around individual atoms, so one portion of an atom may have a greater density of negative charge while another portion has a partial positive charge.

The Chemical Compositions of Minerals

Recall that a mineral has a definite chemical composition. This means that **a mineral is made up of elements bonded together in definite proportions**. Therefore, its composition can be expressed with a chemical formula.

A few minerals consist of only a single element. For example, gold (Au) and silver (Ag) are single-element minerals. Most minerals, however, are made up of two to six elements. For example, the formula of quartz is SiO_2, meaning that it consists of one atom of silicon (Si) for every two of oxygen (O). Quartz from anywhere in the

Universe has that exact composition. If it had a different composition, it would be some other mineral. The compositions of some minerals, such as quartz, do not vary even by a fraction of a percent. The compositions of other minerals vary slightly, but the variations occur only within narrowly restricted ranges, as explained in Section 2.6.

Since 88 elements occur naturally in the Earth's crust and these elements can combine in many different ways, we might expect to see an overwhelming variety of minerals on a half-day field trip. In fact, more than 2500 different minerals are known! However, because only eight elements are common, only the nine **rock-forming minerals** are abundant. This small number of common minerals makes the life of a field geologist, who must identify minerals every day, less complicated.

2.3 The Crystalline Nature of Minerals

A crystal is any substance whose atoms are arranged in a regular, orderly, periodically repeated pattern. All minerals are crystals. Halite has the composition NaCl: one sodium ion (Na^+) for every chlorine ion (Cl^-). Figure 2–4 includes a photo of halite crystals and two sketches showing halite's arrangement of sodium and chlorine ions. Figure 2–4A is an "exploded" view that allows you to see into the structure. Figure 2–4B is more realistic, showing the ions in contact. They lie in orderly rows and columns of alternating sodium and chlorine from left to right, top to bottom, and front to back. The rows and columns all intersect at right angles. This orderly arrangement is the **crystalline structure** of halite. All minerals have their atoms in orderly arrangements, although the pattern is not always as obvious as in halite. Any solid with such an orderly, repetitive arrangement of atoms is a crystal.

Think of a familiar object with an orderly, repetitive pattern, such as a brick wall. The rectangular bricks repeat themselves over and over throughout the wall. Therefore, the whole wall also has the shape of a rectangle or some modification of a rectangle. In every crystal, a small group of atoms, like a single brick in a wall, repeats itself over and over.

The shape of a large, well-formed crystal such as the halite in Figure 2–4 is determined by the shape of this small group of atoms and the way in which the groups stack. For example, only certain crystal shapes can develop from a cubic group, as in halite. It is obvious from Figure 2–5A that the stacking of small cubes can produce the large cubic crystal of halite. Figure 2–5B shows that a different kind of stacking of the same cubes can also produce an eight-sided crystal. Halite sometimes crystallizes with this shape. All minerals consist of small groups

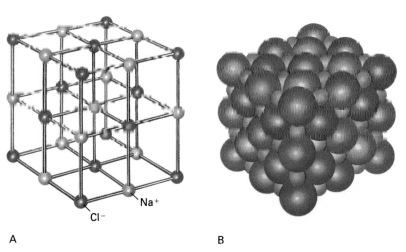

A   B   C

Figure 2–4 (A, B) The orderly arrangement of sodium and chlorine ions in halite. (C) Halite crystals. The crystal model in (A) is exploded so that you can see into it; the ions are actually closely packed as in (B). Note that the arrangement of ions in (A) and (B) is that of a cube, and the crystals in (C) are also cubes. *(Figure C: American Museum of Natural History)*

of atoms stacked up as in halite. Figure 2–5C shows a crystal formed from noncubic groups.

Crystal faces are flat surfaces that form if a crystal grows without obstructions. The halite in Figure 2–4C has well-developed crystal faces. In nature, crystal growth is often hindered by other minerals. For this reason, minerals rarely show perfect crystal faces.

2.4 Physical Properties of Minerals

How does an Earth scientist identify a mineral that he or she finds in the field? Chemical composition and crystal structure distinguish each mineral from all others. For example, halite always consists of sodium and chlorine in a one-to-one ratio, with the atoms arranged in a cubic

Figure 2–5 A and B show that different kinds of stacking of identical cubes form different crystal shapes. (A) A cubic crystal. (B) An octahedron. Both crystal shapes develop from stacking of identical cubes. (C) Stacking of noncubic shapes results in crystals with other shapes.

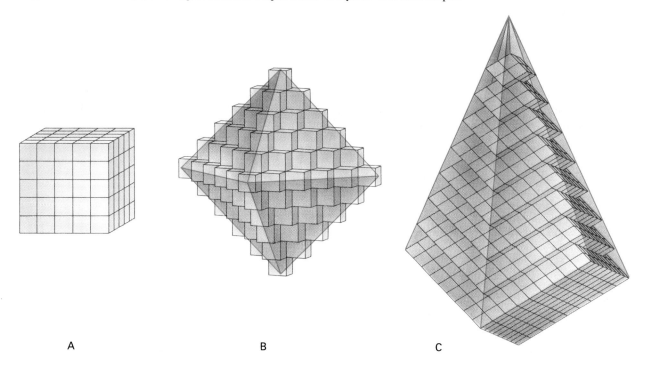

A   B   C

A

B

C

Figure 2–6   (A) *Equant* garnet crystals have about the same dimensions in all directions. (B) Asbestos is *fibrous*. (C) Kyanite forms *bladed* crystals.

fashion. But if you pick up a crystal of halite, you cannot see the ions. You could identify a sample of halite by measuring its chemical composition and crystal structure in laboratory procedures, but they are expensive and time-consuming. Instead, Earth scientists commonly identify minerals by visual recognition, and confirm the identification with simple tests of physical properties.

Most minerals have distinctive appearances. Once you become familiar with common minerals, you will recognize them just as you recognize any familiar object or person. For example, an apple just looks like an apple. In the same way, to an Earth scientist quartz looks like quartz. Just as apples come in many colors and shapes, the color and shape of quartz may vary from sample to

Figure 2–7   (A) *Prismatic* quartz grows as elongated crystals. (B) *Massive* quartz shows no characteristic shape.   *(Geoffrey Sutton)*

A

B

Figure 2–8 A photomicrograph of a thin slice of granite. When crystals grow simultaneously, they commonly develop an interlocking texture and show no characteristic habit. To make this photo, a thin slice of granite was cut with a diamond saw, glued to a microscope slide, and ground to a thickness of 0.02 mm. Most minerals are transparent when such thin slices are viewed through a microscope.

sample, but it still looks like quartz. Some minerals, however, look enough alike that their physical properties must be examined to make a correct identification.

Crystal Habit

Crystal habit is the characteristic shape of a mineral and the manner in which its crystals grow together. If a crystal grows freely, it develops a characteristic shape controlled by the arrangement of its atoms, as in the cubes of halite shown in Figure 2–4C. Three types of crystal habits found in common minerals are described and shown in Figure 2–6.

Some minerals can occur in more than one habit. For example, Figure 2–7A shows quartz with a prismatic

Figure 2–9 Cleavage in mica. This large crystal is the variety of mica called muscovite. *(Geoffrey Sutton)*

habit, and Figure 2–7B shows massive quartz. As mentioned previously, growth of a crystal is often obstructed by other crystals. When that kind of interference occurs, the crystal cannot develop its characteristic habit. Figure 2–8 is a photomicrograph (a photo taken through a microscope) of a thin slice of granite. Notice that the crystals fit like pieces of a jigsaw puzzle. This interlocking texture developed because some crystals grew around others as the granite solidified. Because this type of interference is common, perfectly formed crystals are rare.

Cleavage

Cleavage is the tendency of some minerals to break along flat surfaces. The surfaces are planes of weak bonds in the crystal. Micas show excellent cleavage. You can peel sheet after sheet from a mica crystal as if you were peeling layers from an onion (Fig. 2–9).

Some minerals, such as mica and graphite, have one cleavage plane. Others have two, three, or even four different cleavage planes, as shown in Figure 2–10. Some

Figure 2–10 Some minerals have more than one cleavage plane. (A) Feldspar has two cleavages intersecting at right angles. (B) Calcite has three cleavage planes. (C) Fluorite has four cleavage planes. *(Geoffrey Sutton)*

A

B

C

minerals, like the micas, have excellent cleavage. Others have poor cleavage. Many minerals have no cleavage at all because they have no planes of weak bonds to favor breakage. The number of cleavage planes, the quality of cleavage, and the angles between cleavage planes all help in mineral identification.

It is important to distinguish between a flat surface created by cleavage and a crystal face. They can appear identical because both are flat, smooth surfaces. The difference is that cleavage is repeated by parallel breaks when the crystal is broken, whereas crystal faces are not duplicated by breakage. So, if you are in doubt, break the sample with a hammer unless, of course, you want to save it.

Fracture

Fracture is the way in which a mineral breaks other than along planes of cleavage. Many minerals form characteristic shapes where they break. **Conchoidal** fracture is breakage into smooth, curved surfaces, as shown in Figure 2–11. It is characteristic of quartz and glass. Some minerals break into **splintery** or **fibrous** fragments. Most fracture into **irregular** shapes.

Hardness

Hardness is the resistance of a mineral to scratching. It is controlled by the strength of bonds in the mineral. Thus, it is a fundamental property of a mineral. It is easily measured and commonly used by geologists to identify minerals. Hardness is gauged by attempting to scratch a mineral with a knife or other object of known hardness. If the blade scratches the mineral, the mineral is softer than the knife. If the knife cannot scratch the mineral, the mineral is harder.

To measure hardness more accurately, geologists use a scale based on ten fairly common minerals, numbered

Figure 2–11 Quartz shows smooth, concave, conchoidal fracture. *(Geoffrey Sutton)*

T A B L E 2 – 2

Mohs Hardness Scale	
Minerals of Mohs Scale	**Common Objects**
1. Talc	
2. Gypsum	Fingernail
3. Calcite	Copper penny
4. Fluorite	
5. Apatite	Knife blade
	Window glass
6. Orthoclase	Steel file
7. Quartz	
8. Topaz	
9. Corundum	
10. Diamond	

from 1 through 10. Each mineral is harder than those with lower numbers on the scale, so 10 (diamond) is the hardest and 1 (talc) is the softest. The scale is known as the **Mohs hardness scale**, after F. Mohs, the Austrian mineralogist who developed it in the early nineteenth century.

Table 2–2 shows that a mineral scratched by quartz but not by orthoclase has a hardness between 6 and 7. Because the minerals of the Mohs scale are not always handy, it is useful to know the hardnesses of common materials. A fingernail has a hardness of slightly more than 2, a copper penny about 3, a pocketknife blade slightly more than 5, window glass about 5.5, and a steel file about 6.5. If you practice with a knife and the minerals of the Mohs scale, you can develop a "feel" for hardnesses of minerals 5 and under by how easily the blade scratches them.

Specific Gravity

Specific gravity is the weight of a substance relative to that of an equal volume of water. If a mineral weighs 2.5 times as much as an equal volume of water, its specific gravity is 2.5. You can estimate a mineral's specific gravity simply by hefting a sample in your hand. If you practice with known minerals, you can develop a feel for specific gravity. Most common minerals have specific gravities of about 2.7. Metals have much greater specific gravities; for example, gold has the highest of all minerals, 19. Silver is 10.5 and copper 8.9.

Color

Color is the most obvious property of a mineral but is commonly unreliable for identification. If all minerals were pure and had perfect crystal structures, then color would be reliable. However, both small amounts of chem-

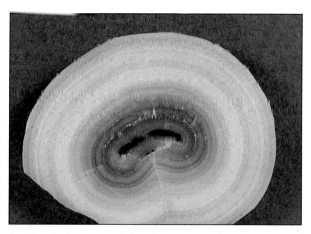

Smithsonite is an ore of zinc that can occur in several different colors. *(Ward's Natural Science Establishment, Inc.)*

ical impurities and imperfections in crystal structure can alter color dramatically. For example, the mineral corundum is aluminum oxide, Al_2O_3. It is normally a cloudy, translucent, brownish to bluish mineral. Addition of a small amount of chromium produces a beautiful, clear, red gem known as ruby. A small quantity of iron or titanium turns corundum into the striking blue gem called sapphire.

Streak

Streak is the color of a fine powder of a mineral. Streak is more reliable than the color of the mineral itself for identification. Streak is measured by rubbing a mineral across a piece of unglazed porcelain known as a streak plate. If the mineral is softer than the porcelain, which has a hardness of about 7, the mineral leaves a streak of powder on the plate.

Luster

Luster is the manner in which a mineral reflects light. A mineral with a metallic look, irrespective of color, has a **metallic luster**. The luster of nonmetallic minerals is usually described by self-explanatory words such as **glassy**, **pearly**, **earthy**, and **resinous**.

Other Properties

Properties such as **reaction to acid**, **magnetism**, **radioactivity**, **fluorescence**, and **phosphorescence** can be characteristic of a mineral and should be noted whenever they are recognized. Fluorescent materials emit visible light when they are exposed to ultraviolet light. Phosphorescent materials continue to emit light after the external stimulus ceases.

2.5 Mineral Classification

Geologists classify minerals according to their anions (negatively charged ions).

A simple anion is a single negatively charged ion such as O^{2-}. Alternatively, two or more atoms can bond firmly together and acquire a negative charge to form a complex anion. Two common examples are silicate, $(SiO_4)^{4-}$, and carbonate, $(CO_3)^{2-}$, complex anions.

Each mineral group is named after the anion in the minerals of the group. For example, the oxides all contain O^{2-}, the silicates contain $(SiO_4)^{4-}$, and the carbonates contain $(CO_3)^{2-}$. Common and useful mineral groups and important minerals in each group are listed in Table 2–3.

2.6 The Rock-Forming Minerals

The nine rock-forming minerals are the most abundant minerals in rocks. Because they are so common, they are the minerals you are most likely to find and identify. Notice that seven are silicates.

Silicates

All minerals containing silicon and oxygen are called **silicates**. The silicate group makes up more than 95 percent of the Earth's crust. Silicate minerals are so abundant for two reasons. First, they are made up principally of the two most plentiful elements in the crust, silicon and oxygen. Second, silicon and oxygen bond together readily. The seven most abundant silicate minerals are feldspar, quartz, pyroxene, amphibole, mica, the clay minerals, and olivine. Except for quartz, each of these minerals

Native gold has a metallic luster. *(Ward's Natural Science Establishment, Inc.)*

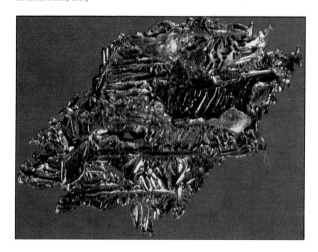

TABLE 2-3

Important Mineral Groups			
Group	**Member**	**Formula**	**Economic Use**
Oxides	Hematite	Fe_2O_3	Ore of iron
	Magnetite	Fe_3O_4	Ore of iron
	Corundum	Al_2O_3	Gemstone, abrasive
	Ice	H_2O	Solid form of water
	Chromite	$FeCr_2O_4$	Ore of chromium
Sulfides	Galena	PbS	Ore of lead
	Sphalerite	ZnS	Ore of zinc
	Pyrite	FeS_2	Fool's gold
	Chalcopyrite	$CuFeS_2$	Ore of copper
	Bornite	Cu_5FeS_4	Ore of copper
	Cinnabar	HgS	Ore of mercury
Sulfates	Gypsum	$CaSO_4 \cdot 2H_2O$	Plaster
	Anhydrite	$CaSO_4$	Plaster
	Barite	$BaSO_4$	Drilling mud
Native elements	Gold	Au	Electronics, jewelry
	Copper	Cu	Electronics
	Diamond	C	Gemstone, abrasive
	Sulfur	S	Sulfa drugs, chemicals
	Graphite	C	Pencil lead, dry lubricant
	Silver	Ag	Jewelry, photography
	Platinum	Pt	Catalyst
Halides	Halite	$NaCl$	Common salt
	Fluorite	CaF_2	Used in steel making
	Sylvite	KCl	Fertilizer
Carbonates	Calcite	$CaCO_3$	Portland cement
	Dolomite	$CaMg(CO_3)_2$	Portland cement
	Aragonite	$CaCO_3$	Portland cement
Hydroxides	Limonite	$FeO(OH) \cdot nH_2O$	Ore of iron, pigments
	Bauxite	$Al(OH)_3 \cdot nH_2O$	Ore of aluminum
Phosphates	Apatite	$Ca_5(F,Cl,OH)(PO_4)_3$	Fertilizer
	Turquoise	$CuAl_6(PO_4)_4(OH)_8 \cdot 4H_2O$	Gemstone
Silicates	(See Fig. 2–13 for silicate minerals.)		

is actually a group whose members have very similar chemical compositions and crystal structures.

To understand the silicate minerals, remember three principles:

1. Every silicon atom surrounds itself with four oxygens. The bonds between the silicon and its four oxygens are very strong.

2. The silicon atom and its four oxygens form a pyramid called the **silica tetrahedron**, with silicon in the center and oxygens at the four corners (Fig. 2–12). **The silica tetrahedron is the fundamental building block of all silicate minerals.** As explained in the accompanying memory device, the silica tetrahedron has a negative charge, forming the $(SiO_4)^{4-}$ complex anion.

3. Silica tetrahedra link together by sharing oxygens. Thus, two tetrahedra share a single oxygen, bonding the two tetrahedra together.

Silicate minerals fall into five classes based on five different ways in which tetrahedra share oxygens. Each class contains at least one of the rock-forming minerals (Fig. 2–13).

5. In the **framework** silicates, each tetrahedron shares all four of its oxygens with adjacent tetrahedra. Because tetrahedra share oxygens in all directions, minerals using the framework structure tend to grow blocky crystals that have the same dimensions in all directions.

Each silica tetrahedron is negatively charged. However, all minerals are electrically neutral. Therefore, cations must enter the structures of most silicate minerals to balance the negative charges. The lone exception is quartz, SiO_2. In quartz the positive charges on the silicons exactly balance the negative ones on the oxygens.

Rock-Forming Silicate Minerals

Feldspar (Fig. 2–14A) makes up more than 50 percent of the Earth's crust and is the most abundant mineral. It is a major component of nearly all common rocks. Feldspar is a group of minerals with similar crystal structures and compositions. Individual minerals within the group are named according to whether they contain potassium, sodium, or calcium. **Orthoclase** is the most common type of potassium feldspar. Feldspar containing calcium and sodium is called **plagioclase**. Plagioclase and orthoclase often look alike and can be difficult to tell apart.

Quartz (Fig. 2–14B) is pure SiO_2. It is the only silicate mineral that contains no cations other than silicon. It is widespread and abundant in continental rocks, but rare in oceanic crust and the mantle.

Pyroxene (Fig. 2–14C), like feldspar, is a group of similar minerals. It is a major component of oceanic crust and the mantle and is abundant in some rocks of the continents. **Amphibole** (Fig. 2–14D) also is a group of

1. In **independant tetrahedra** silicates, adjacent tetrahedra do not share oxygens.

2. In the **single-chain** silicates, each tetrahedron links to two others by sharing oxygens. This forms a continuous chain of tetrahedra.

3. The **double-chain** silicates consist of two single chains cross-linked by the sharing of more oxygens. Minerals using both the single- and double-chain structures grow crystals that are elongate parallel to the chains.

4. In the **sheet** silicates, each tetrahedron links to three others in the same plane, forming a continuous sheet of tetrahedra. Mica is a sheet silicate. All of the atoms in each sheet are strongly bonded, but each sheet is only weakly bonded to those above and below. Therefore, it is easy to peel sheet after sheet from a mica crystal.

Figure 2–12 The silica tetrahedron consists of one silicon atom surrounded by four oxygens. It is the fundamental building block of all silicate minerals. (A) A schematic representation. (B) A proportionally accurate model.

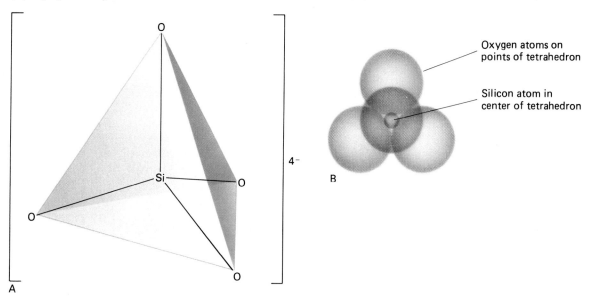

Class	Arrangement of SiO$_4$ tetrahedron	Unit composition	Mineral Examples
A Independent tetrahedra		$(SiO_4)^{4-}$	Olivine: The composition varies between Mg_2SiO_4 and Fe_2SiO_4
B Single chains		$(SiO_3)^{2-}$	Pyroxene: The most common pyroxene is augite, $Ca(Mg, Fe, Al)(Al, Si)_2O_6$
C Double chains		$(Si_4O_{11})^{6-}$	Amphibole: The most common amphibole is hornblende, $NaCa_2(Mg, Fe, Al)_5(Si, Al)_8O_{22}(OH)_2$.
D Sheet silicates		$(Si_2O_5)^{2-}$	Mica, clay minerals, chlorite eg: muscovite $KAl_2(Si_3Al)O_{10}(OH)_2$
E Framework silicates		SiO_2	Quartz: SiO_2 Feldspar: As an example, potassium feldspar is $KAlSi_3O_8$

Figure 2–14 The seven rock-forming silicate minerals.
(A) Feldspar, represented here by orthoclase feldspar.
(B) Quartz. (C) Pyroxene. (D) Amphibole. (E) Black biotite
is one common type of mica. White muscovite (Fig. 2–9) is
the other. (F) Clay. (G) Olivine. *(Geoffrey Sutton)*

Figure 2–13 The five silicate structures are based on sharing of oxygens among silica tetrahedra. (A) Independent tetrahedra share no oxygens. (B) In single chains, each tetrahedron shares two oxygens with adjacent tetrahedra, forming a continuous chain. (C) A double chain is a pair of single chains linked together by sharing additional oxygens. (D) In the sheet silicates, each tetrahedron shares three oxygens with adjacent tetrahedra, forming a sheet of linked tetrahedra. (E) All four oxygens of each tetrahedron are shared with adjacent tetrahedra in the framework silicates. This forms a three-dimensional network.

Earth Science and the Environment

Asbestos and Cancer

Asbestos is an industrial name for a group of minerals that crystallize as long, thin fibers. The two most common types are fibrous habits of the minerals **chrysotile** and **amphibole**. The fibers of chrysotile form tangled, curly bundles, whereas amphibole asbestos occurs as straight, sharply pointed needles.

Asbestos fibers are commercially valuable because they are flameproof, chemically inert, and extremely strong. For example, a chrysotile fiber is eight times stronger than a steel wire of equivalent diameter. Asbestos fibers have been woven into brake linings, protective clothing, insulation, shingles, tile, pipe, and gaskets, but now are allowed only in brake pads, shingles, and pipe.

In the early 1900s, asbestos miners and others who worked with asbestos realized that prolonged exposure to the fibers caused **asbestosis**, an often lethal lung disease. Later, in the 1950s and 1960s, it became clear that asbestos also causes lung cancer and other forms of cancer. One reason that so much time passed before the cancer-causing properties of asbestos were recognized is that cancer commonly does not develop until decades after the first exposure to asbestos.

Experiments have shown that lung diseases are caused by the fibrous nature of asbestos, not by its chemical composition. For example, forms of amphibole of identical composition can occur with both a fibrous and a nonfibrous habit. In a laboratory study, a group of rodents was exposed to fibrous amphibole, and another group to identical amounts of nonfibrous amphibole. The group exposed to the fibrous type developed cancers, but the other group did not.

Another experiment with rodents showed that amphibole asbestos is a more effective cause of lung cancer than is chrysotile. Apparently the curly chrysotile fibers are more easily expelled from the lungs, whereas the sharp amphibole needles remain in the lung. Additionally,

the incidence of cancer among chrysotile workers is proportionally lower than among those working with amphibole asbestos. Although it is not clear how asbestos causes cancer, it is clear that the fibrous habit is important and that the sharp needles of the less common amphibole asbestos are more dangerous than chrysotile.

In response to growing awareness of its health effects, the Environmental Protection Agency (EPA) banned the use of asbestos in construction in 1978. However, the ban did not address the issue of what should be done with the asbestos already installed. In 1986 the EPA passed a ruling called the Asbestos Hazard Emergency Response Act, requiring that all schools be inspected for asbestos. Public response has resulted in hasty programs to remove asbestos from schools and other buildings. EPA estimates that removal of asbestos from schools and public and commercial buildings will cost between $50 and $150 billion. But what is the real level of hazard?

Most asbestos is the less dangerous chrysotile. More important, most asbestos in buildings is already woven tightly into cloth, and often the surface has been further stabilized by painting. Therefore, the fibers are not free to blow around. The levels of airborne asbestos in most buildings are no higher than that in outdoor air. Many scientists argue that asbestos insulation poses no health danger if left alone, but when the material is removed it is disturbed and asbestos dust escapes. Not only are workers endangered, but airborne asbestos persists in the building for months after completion of the project.

Thus, when assessing the health effects of asbestos, we must understand how it is transported and incorporated into living tissue. Asbestos is unquestionably unhealthful or even deadly in a mine where rock is drilled and blasted and dust hangs heavy in the air. However, in a school or commercial building it may be harmless until, in the interest of public safety, workers release fibers as they disturb the insulation during removal.

---

minerals with similar properties. It is common in many rocks of the continents. Pyroxene and amphibole can resemble each other so closely that they are difficult to tell apart.

Mica (Fig. 2–14E) has a platy habit and perfect cleavage. Both result from the sheet linkages of silica tetrahedra. Mica is common in continental rocks. The **clay minerals** (Fig. 2–14F) are similar to mica in structure, composition, and platy habit. Individual clay crystals are so small that they can barely be seen with a good optical microscope. Most clay forms when other minerals weather at the Earth's surface. Thus, clay is abundant at

and near the Earth's surface and is an important component of soil.

Olivine (Fig. 2–14G) occurs in small quantities in both continental and oceanic rocks. However, olivine and pyroxene make up most of the mantle.

Rock-Forming Nonsilicate Minerals

Two minerals that are not silicates—**calcite**, $CaCO_3$, and **dolomite**, $CaMg(CO_3)_2$—are abundant enough to qualify as rock-forming minerals (Figs. 2–15A and B). Both are carbonates, and both are common in near-surface rocks

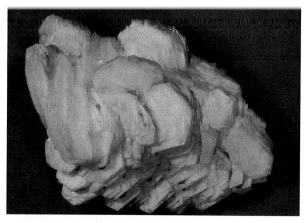

A   B

Figure 2–15   Calcite (A) and dolomite (B) are the only two rock-forming minerals that are not silicates.   *(Ward's Natural Science Establishment, Inc.)*

of the continents. Calcite and dolomite make up the rocks called "carbonate rocks" or sometimes simply "limestones."

Most carbonate rocks start out as shell fragments and other hard parts of marine organisms. When you see limestone cliffs in the Canadian Rockies, in New York State, or anywhere else, you know that the region once lay beneath the sea.

2.7 Other Important Minerals

The rock-forming minerals are important because they make up most of the Earth's crust. A small number of other minerals are important for economic reasons or because they are commonly found in small quantities. They fall into three categories: ore minerals, gems, and accessory minerals. The most important minerals in each of these three groups are listed and described in Appendix 2.

Ore minerals are minerals from which metals or other elements can be profitably recovered. Thus, they are minerals that contain commercially valuable elements or compounds. Native gold and native silver are ore minerals comprised of pure metals. Most other metals exist in nature as compounds. The industrially important metals copper, lead, and zinc are obtained from chalcopyrite, galena, and sphalerite, respectively. Halite is mined for table salt, and gypsum is mined for the manufacture of plaster and sheetrock.

A **gem** is a mineral that is commercially valuable for its beauty rather than for industrial use. Depending on its value, a gem can be either precious or semi-precious.

Emerald is a precious gem.   *(American Museum of Natural History)*

Galena is an ore of lead. Here it occurs as nearly perfect cubic crystals.   *(Ward's Natural Science Establishment, Inc.)*

Precious gems include diamond, emerald, ruby, and sapphire. Several varieties of quartz, including amethyst, agate, jasper, and tiger's eye, are semi-precious gems. Garnet, olivine, topaz, turquoise, and many other minerals sometimes occur as aesthetically pleasing semi-precious gems.

Accessory minerals are minerals that are seen often, but usually only in small amounts. Although common, they are not abundant enough to classify as rock-forming minerals. Chlorite, garnet, limonite, magnetite, and pyrite are among the most common accessory minerals.

SUMMARY

•

Minerals are the substances that make up rocks. A mineral is a naturally occurring inorganic solid with a definite chemical composition and a crystalline structure. Each mineral consists of specific chemical elements bonded together in certain proportions, so that its chemical composition can be given as a chemical formula. The **crystalline structure** of a mineral is the orderly, periodically repeated arrangement of its atoms. The shape of a crystal is determined by the arrangement of its atoms. Every mineral is distinguished from others by its chemical composition and crystal structure.

Most common minerals are easily recognized and identified visually, and identification is aided by observing a few physical properties, including **crystal habit**, **cleavage**, **fracture**, **hardness**, **specific gravity**, **color**, **streak**, and **luster**.

Although more than 2500 minerals are known in the Earth's crust, only the nine **rock-forming minerals** are abundant in most rocks. They are **feldspar**, **quartz**, **pyroxene**, **amphibole**, **mica**, the **clay minerals**, **olivine**, **calcite**, and **dolomite**. The first seven on this list are **silicates**; their structures and compositions are based on the **silica tetrahedron**, in which a silicon atom is surrounded by four oxygens. Silica tetrahedra link together by sharing oxygens to form the basic structures of the silicate minerals. The silicates are the most abundant minerals because silicon and oxygen are the two most abundant elements in the Earth's crust and bond together readily to form the silica tetrahedron.

Ore minerals and **gems** are important for economic reasons. **Accessory minerals** are commonly found, but in small amounts.

KEY TERMS

•

Mineral 27	Cation 29	Crystal 30	Color 34
Element 28	Anion 29	Crystal face 31	Streak 35
Atom 28	Chemical bond 29	Habit 33	Luster 35
Nucleus 28	Ionic bond 29	Cleavage 33	Silicate 35
Electron 28	Covalent bond 29	Fracture 34	Silica tetrahedron 36
Proton 29	Metallic bond 30	Hardness 34	Ore mineral 41
Neutron 29	van der Waals force 30	Mohs hardness scale 34	Gem 41
Ion 29	Rock-forming mineral 30	Specific gravity 34	Accessory mineral 42

ROCK-FORMING MINERALS

•

Feldspar	Amphibole	Clay minerals	Calcite
Quartz	Mica	Olivine	Dolomite
Pyroxene			

REVIEW QUESTIONS

•

1. What properties distinguish minerals from other substances?

2. Explain why oil and coal are not minerals.

3. What does the chemical formula for quartz, SiO_2, tell you about its chemical composition? What does $KAlSi_3O_8$ tell you about orthoclase feldspar?

4. List the eight most abundant chemical elements in the Earth's crust. Are any unfamiliar to you? List familiar elements that are not among the eight. Why are they familiar?

5. What is an atom? An ion? A cation? An anion? What roles do they play in minerals?

6. What is a chemical bond? What role do chemical bonds play in minerals?

7. Quartz is SiO_2. Why does no mineral exist with the composition SiO_3?

8. Every mineral has a "crystalline structure." What does this mean?

9. What factors control the shape of a well-formed crystal?

10. What is a crystal face?

11. What conditions allow minerals to grow well-formed crystals? What conditions prevent their growth?

12. List and explain the most common and useful physical properties of minerals.

13. If you were given a crystal of diamond and another of quartz, how would you tell which is diamond?

14. Why do some minerals have cleavage and others do not? Why do some minerals have more than one plane of cleavage?

15. Why is color often an unreliable property for mineral identification?

16. List the rock-forming minerals. Why are they called "rock-forming?" Which are silicates? Why are so many of them silicates?

17. Make a table with two columns. In the left column list the basic silicate structures. In the right column list one or more rock-forming minerals with that structure.

DISCUSSION QUESTIONS

•

1. Diamond and graphite are two minerals with identical chemical compositions, pure carbon (C). Diamond is the hardest of all minerals, and graphite is one of the softest. If their compositions are identical, why do they have such profound differences in physical properties?

2. Table 2–1 shows that silicon and oxygen together make up nearly 75 percent by weight of the Earth's crust; but silicate minerals make up more than 95 percent of the crust. Explain the difference.

3. Would you expect minerals found on the Moon, Mars, or Venus to be different from those of the Earth's crust? Explain your answer.

Rocks

3

T he Earth is almost entirely rock. A thin layer of soil conceals bedrock in most places on land, but this covering is only a few meters thick on a planet with a radius of more than 6000 kilometers. Beneath the soil, continents are hard, solid rock. If you were to dive to the sea floor and dig through a layer of mud, again you would find solid rock. If you could tunnel 2900 kilometers to the boundary between the mantle and core, you would be excavating hard, solid rock all the way. The outer core is molten metal, but the inner core is solid metal.

Even casual observation reveals that rocks are not all alike. The great peaks and cliffs of the Sierra Nevada in California are hard, strong granite. The rock is made of black and white minerals firmly welded together, giving it a salt-and-pepper appearance. The red cliffs of the Utah desert are soft sandstone. If you scrape the rock with a knife, tiny sand grains pile up at your feet. If you were to climb to the top of Mount Everest you would find rock called limestone, made up of clamshells and the remains of other small marine animals.

● Olo Canyon, a tributary of the Colorado River in the Grand Canyon.

In this chapter we will study rocks: how they form and what they are made of. In later chapters we will use our understanding of rocks to interpret the geological history of the Earth. The limestone containing marine fossils on top of Mount Everest must have formed in the sea. Some force must have pushed it up after it formed. What forces form mountains? Where did the vast amounts of sand in the Utah sandstone come from? Why are the sand grains so easily released from the rock? Why does the sandstone form vertical cliffs and delicate arches? How did the granite of the Sierra Nevada form? All of these questions ask about the nature of rocks, but the answers involve the processes that formed the rocks and the geological history of each region.

3.1 Types of Rocks and the Rock Cycle

Geologists separate rocks into three classes based on how they form: igneous rocks, sedimentary rocks, and metamorphic rocks.

Under certain conditions, rocks of the upper mantle and lower crust melt, forming a hot liquid called **magma**. **Igneous rock forms when magma cools and solidifies.** Igneous rock is the most abundant kind of rock. It makes up much more than half of the Earth's crust. Granite and basalt are the two most common igneous rocks.

Rocks of all kinds decompose, or **weather**, at the Earth's surface. Weathering breaks large rocks into smaller fragments such as gravel, sand, and clay. Some rock dissolves in rainwater as it soaks into the ground. Streams, wind, glaciers, and gravity carry this weathered

A volcanic eruption on Hawaii.   *(United States Geological Survey)*

material, called **sediment**, downhill and deposit it at lower elevations. The sand on a beach and mud on a mud flat form by these processes. **With time, sediment is cemented together to form sedimentary rock.** When the beach sand is cemented, it becomes sandstone; the mud becomes shale. Sedimentary rock makes up less than 5 percent of the Earth's crust. However, because sediment accumulates on the surface of the Earth, sedimentary rocks form a thin veneer covering about 80 percent of the continents. Therefore, it is easy to get the impression that sedimentary rocks are more abundant than they really are. The most common sedimentary rocks are shale, sandstone, and limestone.

Some Earth processes force rocks downward from the surface to deeper regions of the crust. For example, tectonic activity can depress portions of the crust, and rocks can be buried by thick piles of sediment accumulating in the depression. When a rock is buried, both temperature and pressure increase. The higher temperature and pressure cause changes in both the minerals and the texture of the rock. These changes are called **metamorphism**, and the rock is termed a **metamorphic rock**. **Metamorphic rocks form when igneous, sedimentary, or other metamorphic rocks change because of high temperature and/or pressure or are deformed during mountain building.**

Nearby magma or deep burial can heat a rock and cause metamorphism. In addition, high pressure resulting from burial can cause changes in mineralogy and texture. Metamorphism also occurs when rocks are sheared, crushed, or bent where mountains rise as two lithospheric plates grind together. Schist, gneiss, and marble are common metamorphic rocks.

Igneous, sedimentary, and metamorphic rocks seem to be permanent features of the Earth over a human life span and even over the range of human history. Archeologists have used biblical descriptions of rocky peaks to locate ancient ruins. But historical records go back only a few thousand years, whereas geologic time extends back a few billion years. Over this much greater length of time, rocks change. In geologic time it is common for Earth processes to convert a sedimentary rock to a metamorphic rock, or an igneous rock to a sedimentary rock. Thus, no particular rock is permanent over geologic time; instead, all rocks change slowly from one of the three rock types to another. This continuous transformation is called the **rock cycle** (Fig. 3–1).

Although the term rock cycle implies an orderly progression from one type of rock to another, such a regular sequence does not necessarily occur. Shortcuts are common, as shown by the arrows cutting across the circle of Figure 3–1. For example, a sedimentary or metamorphic rock may be uplifted and weather to form sediment. An igneous rock may be metamorphosed. The rock

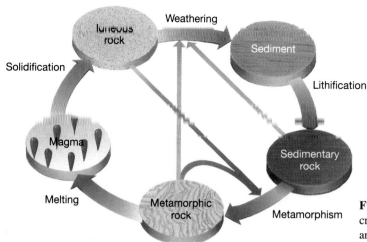

Figure 3–1 The rock cycle shows that rocks of the crust change continuously over geologic time. The arrows show paths that rocks can follow as they change.

cycle simply expresses the important concept that rocks are not permanent, but change continuously over geologic time.

IGNEOUS ROCKS

3.2 Magma: The Source of Igneous Rocks

If you drilled a well deep into the crust, you would find that Earth temperature rises about 30°C for every kilometer of depth. Below the crust, in the upper mantle, temperature continues to rise, but not as rapidly. In the upper mantle, between depths of 100 and 350 kilometers, the temperature is so high that in certain places large amounts of rock melt to form magma. Recall from Chapter 1 that the Earth layer between 100 and 350 kilometers in depth is called the asthenosphere and is weak, soft, plastic rock. It is weak because it is hot and partly melted.

The temperature of magma varies from about 600° to 1400°C, depending on its chemical composition and the depth at which it forms. As a comparison, an iron bar turns red-hot at about 600°C and melts at slightly over 1500°C. Blacksmiths easily heat iron to redness on the glowing embers of a coal forge.

When rock melts to form magma, it expands by about 10 percent. Therefore, magma is of lower density than the solid rock around it. Because of its lower density, magma starts to rise as soon as it forms, just as a hot air balloon rises in the atmosphere. As the magma rises, it enters the cooler, lower-pressure environment near the Earth's surface. When temperature and pressure drop sufficiently, the liquid solidifies to form solid igneous rock.

Because the Earth melted shortly after it formed, the crust began as a mass of igneous rock. Later geological activity has modified the original igneous crust to form sedimentary, metamorphic, and younger igneous rocks. However, about 95 percent of the Earth's crust is still igneous rock or metamorphosed igneous rock. Even though today much of this igneous foundation is buried by a relatively thin layer of sedimentary rock, igneous rocks are easy to find because they make up some of the world's most spectacular mountains (Fig. 3–2).

Figure 3–2 Granite peaks in the Bugaboo Mountains, British Columbia, Canada.

Focus On

Bowen's Reaction Series

N. L. Bowen was a geologist who worked during the first part of this century. He studied the order in which minerals crystallize from a cooling magma. Since magma forms deep within the crust where it cannot be studied, Bowen made artificial magma by heating powdered rock samples in a container called a bomb. A **bomb** is a strong hollow steel cylinder that can be sealed with a threaded cap.

The bomb is heated until the powder melts and is then cooled to the temperature and pressure chosen for the experiment. It is left at that temperature and pressure long enough for minerals to crystallize from the melt. Commonly a few months are required for the minerals to form.

Bowen allowed a sample of artificial magma to cool slowly in his bomb until it had partly solidified. At that point the sample consisted of a mixture of crystals plus the melt that had not yet crystallized. He then plunged the bomb into cold water or oil, causing the sample to cool so rapidly that the crystals were preserved and the uncrystallized melt solidified as glass. He removed the sample from the bomb and identified the minerals with a microscope.

Bowen repeated the experiment many times, allowing his artificial magmas to form crystals at different temperatures before quick-cooling them. By identifying the minerals that had formed at each temperature, he was able to determine the order in which minerals crystallize from a cooling magma.

For example, Bowen found that as basalt magma cools, crystals of olivine and calcium-rich plagioclase form first at high temperatures. After the first crystals form, the crystallization of additional minerals upon further cooling can follow one of two paths, depending on what happens to the first crystals.

Crystals Separating from the Magma

In a natural magma chamber the olivine and the calcium-rich plagioclase may settle to the bottom of the chamber because they are denser than the liquid. As they collect at the bottom of the magma chamber, they become isolated from the magma and cannot react with it.

In this case, because olivine and plagioclase are poor in silica, the remaining melt becomes enriched in silica. Therefore, as the magma cools further, minerals richer in silica form. The final bit of melt to solidify has the composition of granite: potassium feldspar, sodium plagioclase, and quartz. In this way, a cooling basalt magma produces a large amount of basalt, but it also can produce a small amount of granite.

Crystals Mixing with the Magma

Alternatively, in a natural magma chamber, currents in the magma may prevent the crystals from settling to the bottom, thereby keeping them in contact with the liquid. Bowen found that if the crystals remained mixed with the magma, they continued to react with the liquid as it cooled.

Although it may seem contrary to intuition, Bowen discovered that crystals of olivine that had formed at high temperature dissolved back into the melt as it cooled. At the same time, crystals of pyroxene formed. Bowen reasoned that, as the magma cooled, the olivine crystals reacted with the melt to form the pyroxene crystals. At the same time, the early-formed calcium-rich plagioclase crystals reacted with the melt to form plagioclase with less calcium and aluminum and more sodium and silica.

When he started with an artificial basalt magma, Bowen found that all of the magma had solidified by the time most of the olivine had reacted to form pyroxene and the calcium-rich plagioclase had reacted to form plagioclase of intermediate composition. He had produced an artificial basalt consisting mainly of pyroxene and intermediate plagioclase with a small amount of olivine, despite the fact that initially only olivine and calcium-rich plagioclase crystallized.

Bowen also experimented with artificial magmas of intermediate and granitic compositions. He discovered that, as those melts cooled, pyroxene reacted with the melt to form hornblende. Hornblende then reacted to form biotite. Plagioclase continued to react with the melt to become progressively enriched in sodium and silica and depleted in calcium and aluminum with falling temperature. At the lowest temperatures, potassium feldspar, muscovite, and quartz formed as the last of the magma crystallized.

3.3 Classifications of Igneous Rocks

Igneous rocks are divided into two groups on the basis of how they form. **Intrusive igneous rocks** form when magma solidifies *within* the Earth, before it can rise all the way to the surface. Intrusive rocks are sometimes called **plutonic rocks** after Pluto, the ancient Greek god of the underworld.

Extrusive igneous rocks form when magma erupts and solidifies on the Earth's surface. Because extrusive rocks are so commonly associated with volcanoes, they are also called **volcanic rocks**. Vulcan was the Greek god of fire.

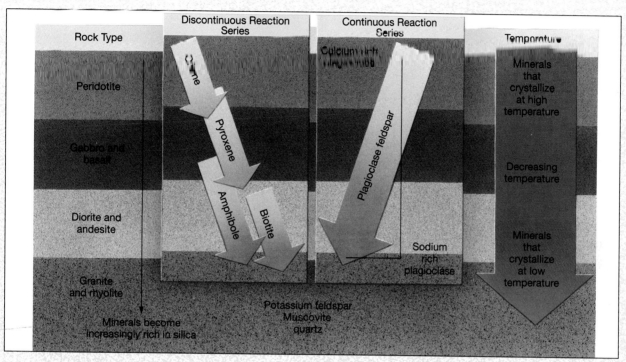

Bowen's reaction series shows the order in which minerals crystallize from a cooling magma and then react with the magma as it cools to form minerals lower on the diagram.

Bowen summarized his work in a Y-shaped figure, now known as **Bowen's reaction series** (see illustration). It is called a reaction series to emphasize that minerals formed at high temperatures react with cooling magma to form other minerals.

The right side of Bowen's reaction series describes the reactions that occur between plagioclase and magma. As the temperature decreases, plagioclase reacts continuously with the melt to become progressively enriched in sodium and silica and depleted in calcium and aluminum. Thus, this arm of the Y is known as a **continuous reaction series**.

The left side of Bowen's reaction series shows that olivine, the first mineral to form, reacts with the melt to form pyroxene as the temperature decreases. Then pyroxene dissolves back into the magma as hornblende (amphibole) forms. The reactions continue until biotite, the last mineral in the series, forms as hornblende dissolves. This arm of the Y is called a **discontinuous reaction series** because each of the minerals has a different crystal structure.

The minerals of both arms of the Y become progressively higher in silica content with decreasing temperature.

The rock names in the illustration are placed on the same level as the minerals that make up each of those rocks. The placement of each rock type also shows that basalt forms at relatively high temperatures, andesite at intermediate temperatures, and granite at relatively low temperatures.

Textures of Igneous Rocks

The **texture** of a rock refers to the the size, shape, and arrangement of its mineral grains, or crystals. In igneous rocks, grain size is the most important factor in rock texture. Some igneous rocks consist of mineral grains that are too small to be seen with the naked eye; others are made up of thumb-size or even larger crystals. The common names for igneous rock textures are summarized in Table 3–1.

One of the most striking differences between volcanic and plutonic rocks is the contrast in their textures. Volcanic rocks are usually fine-grained, whereas plutonic rocks are medium- or coarse-grained. The difference exists because crystals grow slowly as magma solidifies.

T A B L E   3 – 1

Igneous Rock Textures Based on Grain Size	
Grain Size	**Name of Texture**
No mineral grains (obsidian)	Glassy
Too fine to see with naked eye	Very fine-grained
Up to 1 millimeter	Fine-grained
1–5 millimeters	Medium-grained
More than 5 millimeters	Coarse-grained
Relatively large grains in a finer-grained matrix	Porphyry

Volcanic magma cools rapidly on the Earth's surface and solidifies before crystals have time to grow to a large size. In contrast, plutonic magma cools slowly within the crust, and crystals have a long time to grow to larger sizes.

Extrusive (Volcanic) Rocks

If volcanic magma solidifies within a few hours of erupting, volcanic glass, called **obsidian**, may form (Fig. 3–3). Glass has no crystalline structure; the atoms or ions have no orderly arrangement because the magma solidified before they could align themselves to form crystals. If a magma

Figure 3–3 Obsidian is natural volcanic glass. It contains no crystals. *(Geoffrey Sutton)*

Figure 3–4 Basalt is a dark, fine-grained volcanic rock. *(Geoffrey Sutton)*

solidifies somewhat more slowly, over a period of days to a few years, crystals begin to form, but they do not have time to grow to large sizes. The result is a very fine-grained rock, one in which the crystals are too fine to be seen with the naked eye. Basalt is the most abundant example of a very fine-grained volcanic rock (Fig. 3–4).

Intrusive (Plutonic) Rocks

Plutonic igneous rocks form when magma solidifies deep within the crust. Overlying rock insulates the magma like a thick blanket. This keeps the magma hot so that it solidifies slowly, over hundreds of thousands or even millions of years. As a result, crystals have a long time to grow, and they form large grains. Therefore, most plutonic rocks are medium- to coarse-grained. Granite, the most abundant rock in continental crust, is a medium- or coarse-grained, plutonic igneous rock. When you look at granite, you see individual grains of different colors. Each grain is a separate mineral.

If magma rises slowly through the crust, some crystals may grow while most of the magma remains molten. If this mixture of magma and crystals suddenly erupts onto the surface, the magma cools quickly, forming a very

Figure 3–5 Porphyry is igneous rock containing large crystals embedded in a fine-grained matrix. This is a rhyolite porphyry with large, pink feldspar phenocrysts. *(Geoffrey Sutton)*

A   B

Figure 3–6 Although granite (A) and rhyolite (B) contain the same minerals, they have very different textures because granite cools slowly and rhyolite cools rapidly. *(Geoffrey Sutton)*

fine-grained rock with the large, early-formed crystals embedded within it. A **porphyry** is an igneous rock containing large crystals in a fine matrix, and the large crystals are called **phenocrysts** (Fig. 3–5).

Classification Based on Minerals and Texture

Geologists use both minerals and texture to classify and name igneous rocks. For example, **granite** consists mainly of feldspar and quartz, and is medium- or coarse-grained. Any igneous rock with these minerals and texture is granite. But **rhyolite** contains the same minerals. The

difference between granite and rhyolite is not one of composition, but of texture. Granite is medium- to coarse-grained, and rhyolite is very fine-grained (Fig. 3–6). The same magma that erupts onto the Earth's surface to form rhyolite also forms granite if it solidifies slowly within the crust.

Thus, igneous rocks are classified in pairs. The members of each pair contain the same minerals but have different textures. The texture depends mainly on whether the rock is volcanic or plutonic in origin. Figure 3–7 shows the mineralogy and textures of common igneous rocks.

Figure 3–7 The minerals and textures of the most common igneous rocks. A mineral's abundance in a rock is proportional to the thickness of its colored band beneath the rock name. Use the numbers on the left side of the figure to estimate the relative abundance of each mineral in the rock listed at the top of the figure.

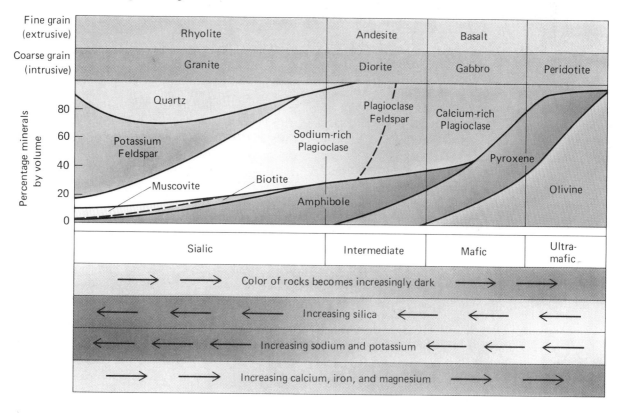

51

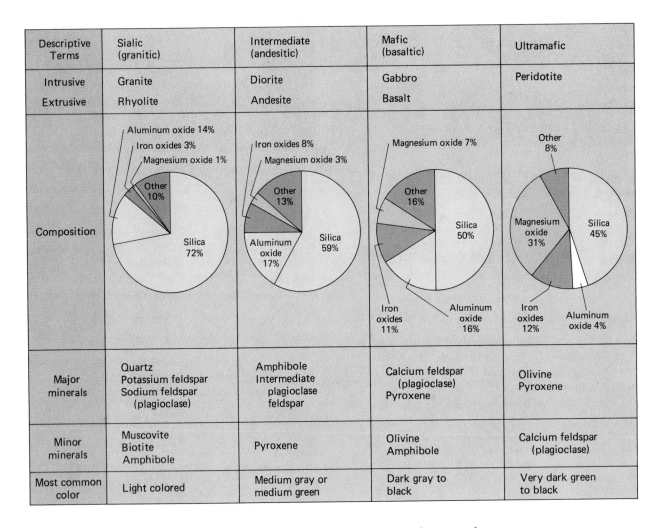

Descriptive Terms	Sialic (granitic)	Intermediate (andesitic)	Mafic (basaltic)	Ultramafic
Intrusive	Granite	Diorite	Gabbro	Peridotite
Extrusive	Rhyolite	Andesite	Basalt	
Composition	Aluminum oxide 14% Iron oxides 3% Magnesium oxide 1% Other 10% Silica 72%	Iron oxides 8% Magnesium oxide 3% Other 13% Aluminum oxide 17% Silica 59%	Magnesium oxide 7% Other 16% Silica 50% Iron oxides 11% Aluminum oxide 16%	Other 8% Magnesium oxide 31% Silica 45% Iron oxides 12% Aluminum oxide 4%
Major minerals	Quartz Potassium feldspar Sodium feldspar (plagioclase)	Amphibole Intermediate plagioclase feldspar	Calcium feldspar (plagioclase) Pyroxene	Olivine Pyroxene
Minor minerals	Muscovite Biotite Amphibole	Pyroxene	Olivine Amphibole	Calcium feldspar (plagioclase)
Most common color	Light colored	Medium gray or medium green	Dark gray to black	Very dark green to black

Figure 3–8   Chemical compositions, minerals, and typical colors of the common igneous rocks.

The same rocks and their chemical compositions are shown in Figure 3–8. Granite and rhyolite contain large amounts of *si*licon and *al*uminum, and so are called **sialic** rocks. Basalt and gabbro are called **mafic** rocks because of their high magnesium and iron contents. The word *mafic* is derived from *ma*gnesium and *fe*rrum, the Latin word for iron. Rocks with especially high magnesium and iron concentrations are called **ultramafic**. Rocks with intermediate compositions are called **intermediate** rocks.

3.4 The Most Common Igneous Rocks

Granite and Rhyolite

Granite is the most common rock in continental crust. It is found nearly everywhere beneath the relatively thin veneer of sedimentary rocks that covers most of the continents. Geologists often call this granite **basement rock** because it makes up the foundations of continents. Granite

is hard and resistant to weathering; it forms steep, shear cliffs in many of the world's great mountain ranges. Such cliffs are sought out by mountaineers for the steepness and strength of the rock (Fig. 3–9).

Rhyolite is made of the same minerals as granite, but has a fine-grained texture because it is volcanic. When granitic magma rises toward the Earth's surface, commonly a portion erupts from a volcano to form rhyolite while the remainder solidifies beneath the volcano, forming granite. Most obsidian forms from magma with a granitic (rhyolitic) composition.

Basalt and Gabbro

Basalt is a dark, very fine-grained volcanic rock. It is about half plagioclase and half pyroxene. Most oceanic crust is basalt. As we will learn in Chapter 5, the basalt of oceanic crust erupts beneath the sea along a great submarine mountain chain known as the mid-oceanic ridge. Basalt magma also erupts in great volumes on continents, forming large **basalt plateaus**. **Gabbro** is mineralogically identical to basalt, but has larger mineral grains because it is a plutonic rock. Gabbro is uncommon at the Earth's surface, although it is abundant in deeper parts of oceanic crust where magma crystallizes at depth.

Andesite and Diorite

Andesite is a volcanic rock intermediate in composition between basalt and granite. It is commonly gray or green and consists of plagioclase and dark minerals (usually biotite, hornblende, or pyroxene). It is named for the Andes Mountains, the volcanic chain on the western edge of South America, which is made up mostly of andesite. Because it is volcanic, andesite typically has a very fine-grained texture. The medium- to coarse-grained plutonic equivalent of andesite is **diorite**. Diorite commonly underlies large areas of volcanic andesites, such as the Andes. It formed from the same magmas that produced the andesite, but which solidified in the crust beneath the volcanoes.

Peridotite

Peridotite is an ultramafic igneous rock that is rare in the Earth's crust. However, most of the upper mantle is peridotite. It is coarse-grained and composed of olivine, and usually also contains pyroxene, amphibole, or mica, but no feldspar. Figures 3–7 and 3–8 show that peridotite has the lowest silica content of all the important igneous rocks.

Recognizing and Naming Igneous Rocks

Once you learn to identify the nine rock-forming minerals, it is easy to name a plutonic rock using Figure 3–7 because the minerals are large enough to be seen.

It is harder to name volcanic rocks because the minerals are usually too small to identify. In these cases, a field geologist often uses color to make a tentative identification. Figure 3–8 shows that rhyolite is usually light in color: white, tan, red, and pink are common. Andesites

Figure 3–9 Rock climbers prize granite cliffs because the rock is strong. Baffin Island, Northwest Territories. *(Steve Sheriff)*

Windblown sand forming a sand dune. *(White Sands National Monument)*

are gray or green, and basalts are black. In many volcanic rocks, the minerals cannot be identified even with a microscope because of their tiny crystal sizes. In this case, definitive naming is based on chemical analysis carried out in the laboratory.

SEDIMENTARY ROCKS

3.5 Sediment

All rocks at the Earth's surface disintegrate slowly by weathering. The processes of weathering are described in detail in Chapter 10. As a result of these processes, rocks decompose to sand, clay, other solid particles, and ions dissolved in water. These weathering products are eroded

and carried off by running water, wind, glaciers, and gravity. Eventually they accumulate in layers of **sediment**. With time, the loose, unconsolidated sediment becomes compacted and cemented to form sedimentary rock.

Sediment refers to all solid particles transported and deposited by water, wind, glaciers, and gravity. It includes weathered rocks and minerals; organic remains such as clamshells, bits of coral, and pieces of plants; and minerals precipitated from solution. (A precipitated mineral is one that has separated as a solid from solution.)

Most kinds of sediment are familiar. Beach sand, pebbles and cobbles in a riverbed, dust in the air, mud in a puddle, and boulders embedded in a glacier are all sediment. Accumulations of shell fragments near a reef in the ocean and salt deposited on the shores of Great Salt Lake in Utah are also sediment.

Clastic sediment consists of fragments of weathered rock or of shells and other organic remains. Sand, clay, boulders, and shell fragments are examples. Clastic sediment and clastic sedimentary rocks are named according to the sizes of the clastic particles (Table 3–2).

Rainwater partly dissolves some rocks as it seeps into the ground. When a rock dissolves, ions are carried off in solution. In certain environments these dissolved ions may precipitate directly to form **chemical sediment**. Large sedimentary deposits of halite (table salt) form in this way.

Erosion and Transport of Sediment

After rocks weather, streams, wind, glaciers, and gravity erode the sediment and carry it downhill. Streams and rivers carry the greatest amount of sediment. Since nearly all streams eventually empty into the oceans, most sediment is transported to coastlines.

Sediment transport is a gravity-driven process (Fig. 3–10). Since streams flow downhill, all sediment and

TABLE 3–2

Sizes and Names of Sedimentary Particles and Clastic Rocks			
Diameter (mm)	**Sediment**		**Clastic Sedimentary Rock**
256 — 64 — 2 —	Boulders Cobbles Pebbles	Gravel (rubble)	Conglomerate (rounded particles) or breccia (angular particles)
1/16 —	Sand		Sandstone
1/256 —	Silt	Mud	Siltstone Claystone ⎫ Mudstone or shale ⎭
	Clay		

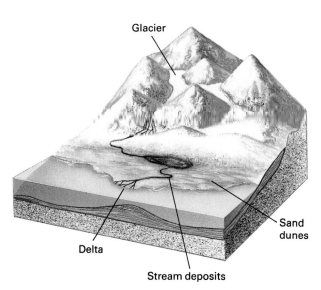

Glacier

Sand dunes

Delta

Stream deposits

Figure 3–10 Streams and glaciers flow downhill and carry sediment to lower elevations. Eventually most sediment reaches the ocean.

Figure 3–11 Cobbles and boulders become rounded as they are carried by a stream.

dissolved ions carried by water move toward lower elevations. Glaciers and their sediment flow downhill. Wind-blown sand and silt may be carried uphill, but eventually this sediment, too, is carried back downhill.

As clastic particles tumble downstream, their sharp edges are worn off and they become **rounded** (Fig. 3–11). Particles ranging in size from coarse silt to boulders are

rounded during transport. Finer particles do not round as effectively because they are so small and light that water, and even wind, cushions them as they bounce along.

Streams and wind separate sediment according to size, a process called **sorting**. Figure 3–12 is a profile of a stream flowing from the mountains to the plains. Near its source, the stream is steep and the water flows rapidly

Figure 3–12 A stream is steepest at its headwaters in the mountains, and its steepness decreases downstream. Large particles are carried and deposited in the steeper headwaters, and smaller ones are deposited on the nearly level plain.

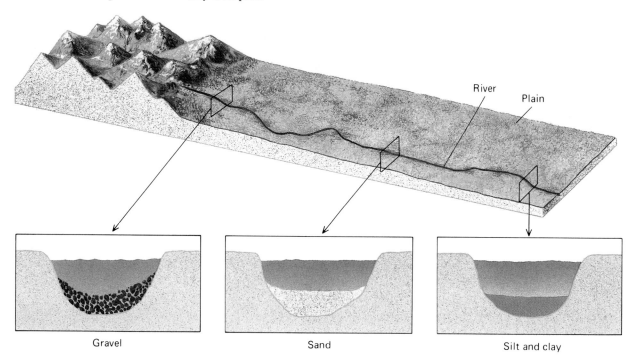

River

Plain

Gravel

Sand

Silt and clay

Figure 3–13 Glaciers deposit poorly sorted sediment containing particles of all sizes, from clay to boulders.

sea or in lakes. Wind may deposit sand and silt on land, forming dunes. Glaciers deposit large volumes of sediment wherever they melt. Calcium dissolved in streams may be carried to a reef in the ocean, where it is taken up by clams, corals, and other organisms to form shells and other hard parts composed of calcite. When the organisms die, the hard parts contribute to the growing pile of sediment around the reef.

Lithification of Sediment

Lithification refers collectively to all the processes that convert loose sediment to hard rock. If you fill a measuring cup with sand, you can still add a substantial amount of water. The water fills empty spaces, or **pore space**, among the sand grains (Fig. 3–14A). Most sediment contains pore space. When sediment is deposited in water,

Figure 3–14 (A) Pore space is the open space between sediment grains. (B) Compaction squashes the grains together, reducing the pore space and lithifying the sediment by interlocking the grains. (C) Cement fills the remaining pore space, lithifying the sediment by gluing the grains together.

and with much energy. Therefore, large boulders are carried and deposited in the upper portion of the stream. A steep mountain stream levels out where it flows into the valley below. As its steepness decreases, its energy diminishes. Here, only smaller particles are carried and deposited. Thus, the largest particles are usually found near the headwaters of a stream, and the sediment becomes progressively smaller downstream.

Wind transports only small particles: sand, silt, and clay. Therefore, sand dunes and other wind-deposited sediment are well sorted. A glacier, in contrast, is solid ice and therefore carries particles of all sizes, from boulders to clay, together. As a consequence, glaciers deposit poorly sorted sediment (Fig. 3-13). Thus, sorting is one criterion for determining how sediment was transported and deposited.

Deposition of Sediment

Deposition of clastic sediment occurs when transport stops, usually because the wind or water slows down and loses energy or, in the case of glaciers, when the ice melts. Deposition of dissolved ions occurs when they precipitate directly from solution or are extracted from solution by an organism to form a shell or skeleton.

Sediment deposition occurs in many environments. Streams deposit sediment in streambeds, on flood plains adjacent to the streambeds, and on deltas where they enter a lake or the ocean. Currents redistribute sediment in the

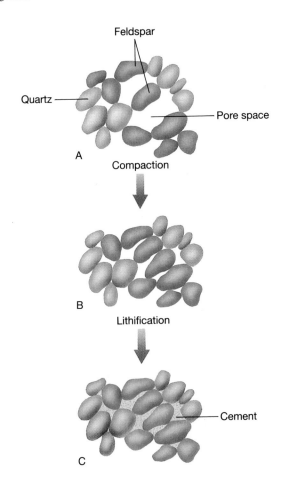

Figure 3–15 Red sandstone in Utah is colored by red iron oxide cement.

the pores fill with water. Commonly, sand and similar sediment have about 20 to 40 percent pore space.

As more sediment accumulates, the weight of the overlying layers compresses the buried sediment. Some of the water is forced out, and the pore space shrinks (Fig. 3–14B). This process is called **compaction**. If the grains have platy shapes, as in clay and silt, compaction alone may lithify the sediment as the platy grains interlock like pieces of a puzzle.

As sediment is buried and compacted, water circulates through the remaining pore space. This water commonly contains dissolved ions. The dissolved materials precipitate in the pore spaces, gluing the clastic grains firmly together to form a hard rock (Fig. 3–14C). This process is **cementation**. Calcite, quartz, and iron oxides are the most common cements in sedimentary rocks. The type of cement affects the nature of the rock. The red sandstone in Figure 3–15 gets its color from red iron oxide cement. Quartz cement forms the toughest sedimentary rocks.

The time required for lithification of loose sediment varies greatly, depending on the availability of cement and water to carry the cement. In some heavily irrigated areas of southern California, calcite precipitated from irrigation water has cemented soils within a few decades. In the Rocky Mountains, some glacial deposits less than 20,000 years old are cemented by calcite. In contrast, sand and gravel deposited in southwestern Montana between 30 and 40 million years ago can still be dug with a hand shovel.

3.6 Types of Sedimentary Rock

Sedimentary rocks are broadly divided into three categories based on the type of sediment of which they are made.

1. **Clastic sedimentary rocks** are composed of fragments of preexisting rocks and other particles that have been physically transported and deposited. This category includes rocks made up of broken shells and other organic fragments, called **bioclastic rocks**. The "bio" portion of "bioclastic" refers to their biological origin.

2. **Organic sedimentary rocks** consist of the lithified remains of plants and animals. Bioclastic rocks fall into this category as well as the preceding one.

3. **Chemical sedimentary rocks** form by direct precipitation of minerals from solution.

Limestone and dolomite are sedimentary rocks that can form by any of the foregoing three processes; they are discussed separately.

Clastic Sedimentary Rocks

Clastic rocks are the most abundant of the three types, accounting for more than 80 percent of all sedimentary rocks. Table 3–2 shows that clastic rocks are classified according to size of their particles, or **clasts**. The clasts may be rock fragments, mineral grains, or bioclastic fragments.

Figure 3–16   Conglomerate is lithified gravel.

Conglomerate (Fig. 3–16) is lithified gravel. Conglomerates usually consist of rounded rock fragments because their clasts are larger than most individual mineral grains.

Sandstone consists of lithified sand grains (Fig. 3–17). Of the nine rock-forming minerals, quartz is the most resistant to weathering. Feldspar and the other common minerals succumb to chemical attack and physical abrasion during weathering and transport. In contrast, about all that happens to quartz grains during weathering and transport is that they become rounded. Consequently, most sandstones consist of rounded quartz grains.

Shale is a fine-grained clastic sedimentary rock (Fig. 3–18A). It consists of clay minerals and small amounts

Figure 3–17   Sandstone is lithified sand. (A) A sandstone cliff in Zion National Park, Utah. (B) A close-up of the same sandstone. Notice the well-rounded sand grains.

A

B

A

B

Figure 3–18 Shale is made up mostly of platy clays. Therefore, it shows very thin layering called fissility. (A) An outcrop of shale. (B) A close-up of the same shale.

of quartz and feldspar. Shale has thin bedding called **fissility**, along which the rock splits easily (Fig. 3–18B). Clay minerals, like micas, have platy shapes. The plates stack like dishes or sheets of paper. The fissility of shales results from parallel alignment of the clay plates.

Shale makes up about 70 percent of all sedimentary rocks (Fig. 3–19). Its abundance reflects the vast quantities of clay produced by weathering, as described in Chapter 10.

Shale is usually gray to black due to the presence of decayed remains of plants and animals commonly deposited with clay. This organic material in shale is the source of most oil and natural gas.

Organic Sedimentary Rocks

Organic sedimentary rocks such as chert and coal form by lithification of organic sediment.

Chert is pure quartz (Fig. 3–20). Microscopic examination of most chert shows that it is made up of the remains of tiny marine organisms that make their skeletons of silica.

Coal is lithified plant remains. When plants die, they usually decompose by reaction with oxygen. However, in warm swamps and other environments where plants grow rapidly, dead plants can accumulate so quickly that the available oxygen is used up before decay is complete.

Figure 3–19 The relative abundances of sedimentary rocks.

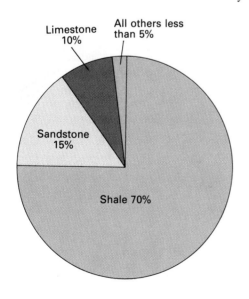

Limestone 10%

All others less than 5%

Sandstone 15%

Shale 70%

Figure 3–20 Red nodules of chert in light-colored limestone.

Earth Science and the Environment

Radon

Radon is a radioactive gas. Invisible, odorless, and tasteless, it occurs naturally in bedrock and soil. It seeps from the ground into homes and other buildings, where it concentrates and causes an estimated 5000 to 20,000 cancer deaths per year among Americans. The risk of dying of radon-caused lung cancer in the United States is about 0.4 percent over a lifetime, much greater than the risk of dying from cancer caused by asbestos, pesticides, or other air pollutants and nearly as high as the risk of dying in an auto accident or from a fall or fire at home.

However, Americans are not all exposed to equal amounts of radon. Some homes contain very low concentrations of the gas; others have high concentrations. The variations in concentration are due to two factors: geology and home ventilation.

Radon is one of a series of radioactive elements formed by the radioactive decay of uranium. Thus, radon forms wherever uranium occurs. Uranium occurs naturally in tiny amounts in all types of rock, but concentrates in some types and occurs in only miniscule amounts in others. It concentrates in granite and shale. For this reason, radon concentrations are highest in poorly ventilated homes built on granite or shale bedrock or on soil derived from these rocks. In other cases, building materials contain rocks with high uranium contents. The highest home radon concentrations ever measured have occurred in houses built on the Reading Prong, a uranium-rich granite pluton that extends from Reading, Pennsylvania, through northern New Jersey and into New York. The air in one home in this area contained 700 times as much radon as the EPA "action level"—the concentration at which the Environmental Protection Agency recommends that

corrective measures be taken to reduce the amount of radon in indoor air.

Thus, a homeowner may ask two questions: "What is the radon concentration in my house?" and "If it is high, what can be done about it?" Because radon is radioactive, it can be measured with a simple detector available at most hardware stores and from local government agencies for about $25.

If the detector indicates excessive radon in a home, measures can be taken to remove it. Radon gas forms by slow radioactive decay of uranium in soil or bedrock beneath a house. After radon seeps from the ground or foundation into the basement, it circulates throughout the house. Thus, with time it accumulates in indoor air. An unventilated house seals out the weather and therefore accumulates radon. Three types of solutions have proved effective. The first is to simply extend a ventilation duct from the basement directly outside the house. In this way, air from the basement does not circulate throughout the entire house, and the basement air is changed frequently so that radon does not accumulate at high levels. The second solution is to ventilate the entire house so that indoor air is continually refreshed. Essentially an open window will suffice. However, an open window allows hot air to escape and thereby increases fuel consumption. A third solution is to pump outside air into the house to keep indoor air at a slightly higher pressure than the outside air. This positive pressure prevents gas from seeping from soil or bedrock into the basement.

It is impossible to avoid exposure to radon completely because it is everywhere, in outdoor air as well as in homes and other buildings. But it is relatively easy and inexpensive to minimize exposure and thus avoid a major cause of lung cancer.

The partly decayed plant remains form **peat**. As peat is buried and compacted by overlying sediment, it converts to coal, a hard, black, combustible rock that commonly contains abundant plant fossils (Fig. 3-21).

Chemical Sedimentary Rocks

When rocks weather, some elements such as calcium, sodium, potassium, and magnesium dissolve. Those dissolved ions are transported by ground water and streams to the oceans or to saline lakes such as Great Salt Lake of Utah. **Evaporites** are sedimentary rocks formed when evaporation of water concentrates dissolved ions to the point where they precipitate from solution. The most

common minerals found in evaporite deposits are **gypsum** ($CaSO_4 \cdot 2H_2O$)* and **halite** (NaCl). Evaporites constitute only a small proportion of all sedimentary rocks but can be important sources of salt and other materials (Fig. 3–22).

Carbonate Rocks: Limestone and Dolomite

Carbonate rocks are made up primarily of the minerals calcite and dolomite. They are called "carbonates" because both minerals contain the carbonate ion, CO_3^{2-}.

---

*The $2H_2O$ in the chemical formula of gypsum means that water is incorporated into the mineral structure.

A   B

Figure 3–21 Coal is lithified plant remains and commonly contains abundant plant fossils. (A) This sample of coal contains numerous needle-shaped plant fossils. (B) This sample of organic-rich shale was taken between two coal beds, and contains well-preserved fossil ferns.

Calcite-rich carbonate rocks are called **limestone**, whereas carbonate rocks rich in the mineral dolomite are also called **dolomite**. Some geologists use the term "**dolostone**" for the rock to distinguish it from the mineral.

Seawater contains much dissolved calcium carbonate. Clams, oysters, corals, some types of algae, and other marine organisms make their shells and other hard body parts of calcium carbonate. Often waves or ocean currents break up and transport fragments of these hard body parts to form clastic sediment. Limestone formed by lithification of such sediment is called **bioclastic limestone**. The name indicates that both biologic and clastic processes were involved in its two-step origin. Most limestones are

Figure 3–22 The Bonneville Salt Flats, Utah.

A

B

Figure 3–23 Most limestone is lithified shell fragments and other remains of marine organisms. (A) A limestone mountain in Alberta, Canada. (B) A graph of shell fragments in limestone.

bioclastic. The bits and pieces of shells and other hard body parts appear as fossils in bioclastic limestone (Fig. 3–23). Bioclastic rocks fit equally well into both the clastic and organic classes of sedimentary rocks. **Coquina** is bioclastic limestone consisting wholly of coarse shell fragments cemented together. **Chalk** is a very fine-grained, soft, earthy, white to gray bioclastic limestone made of the shells and skeletons of microorganisms that spend their lives floating near the surface of the oceans. When they die, their remains sink to the bottom and accumulate to form chalk. The famous white cliffs of Dover in England are made of chalk.

The rock dolomite is widespread. It makes up more than half of all carbonate rocks more than a billion years old and a smaller proportion of younger carbonate rocks. However, there is no place in the world today where dolomite is forming in large amounts. Since it is so abundant in ancient rocks, we would expect dolomite to make up a large proportion of carbonate sediment forming today; yet it does not. This dilemma is known among geologists as the dolomite problem, and it has been the cause of a tremendous amount of research.

The general consensus among geologists is that most dolomite does not originate as a primary sediment or rock. Instead, it forms when magnesium-rich solutions derived from seawater percolate through beds of existing lime-

stone. Magnesium ions replace half of the calcium in the calcite, converting the limestone beds to dolomite.

3.7 Sedimentary Structures

Nearly all sedimentary rocks contain **sedimentary structures**, features that developed during or shortly after deposition of the sediment. These structures often contain important clues that help in identifying how the sediment was transported and deposited.

The most obvious and widespread sedimentary structure is **bedding**, or **stratification**: layering that develops as sediment is deposited (Fig. 3–24). It forms because sediment is almost always deposited in a layer-by-layer process. Bedding may result from differences in texture, mineral composition, color, or cementation between the layers. Most sedimentary beds were originally horizontal because most sediment is deposited on nearly level surfaces.

Flowing water or wind piles sand in small parallel ridges called **ripple marks**. (Fig. 3–25A). Ripple marks are often preserved in sandy sedimentary rocks (Fig. 3–25B). A **sand dune** is a large-scale version of a ripple.

Cross-bedding is an arrangement of small beds lying at an angle to the main sedimentary layering (Fig. 3–

A

B

Figure 3–24 (A) Thick, horizontal sedimentary beds form steep walls along the San Juan River in Utah. (B) Thin sedimentary beds of clay and fine sand that accumulated in a glacial lake in Montana.

Figure 3–25 (A) Modern ripples formed on a mud flat along the Oregon coast. (B) Ripples preserved in billion-year-old mudrocks near Grand Canyon.

A

B

A   B

Figure 3–26 (A) Cross-bedding preserved in lithified ancient sand dunes in Zion National Park, Utah. (B) The development of cross-bedding in sand as a ripple or dune migrates.

26A). Figure 3–26B shows that cross-beds form as sand grains tumble down the steep, downstream face of a sand dune or ripple. Cross-bedding forms in both windblown and water-transported sediments.

Graded bedding is a type of bedding in which the largest grains collect at the bottom of a layer and the grain size decreases toward the top (Fig. 3–27). Graded beds commonly form when sediment containing a mixture of different particle sizes settles to the bottom of a body of water. The larger grains settle rapidly and concentrate at the base of the bed. Finer particles settle more slowly and accumulate in the upper parts of the bed.

Figure 3–27 Three graded beds. The lens cap is in the middle bed. Within each bed, the size of the particles become finer in the upward direction. Above the middle bed is the coarse-grained bottom of another bed, and below is the fine, black clay of the top of a third graded bed. *(Bern Aarons Educational Images)*

Mud cracks are polygonal cracks that form when mud shrinks as it dries (Fig. 3–28). They indicate alternating wetting and drying—for example, on an intertidal mud flat, where the sediment is flooded by water at high tide and exposed at low tide.

Occasionally, very delicate sedimentary structures are preserved in rocks. Geologists have found the imprint of a single raindrop that fell on mud about one billion years ago and the imprint of a cubic salt crystal that formed as a puddle evaporated. Mud cracks, raindrop imprints, and salt crystal imprints all show that the mud must have been deposited in shallow water and was intermittently exposed to air.

Fossils are any remains or traces of a plant or animal preserved in rock—any evidence of past life. They, too, are sedimentary structures; they are discussed in detail in Chapter 4.

METAMORPHIC ROCKS

3.8 Metamorphism

A potter forms a delicate vase from moist clay. The new piece is placed in a kiln and is slowly heated to 1000°C. As the temperature rises, the clay minerals decompose. The atoms from the disintegrating clays unite to form new minerals. The new minerals makes the vase strong and hard. The breakdown of the clay minerals, growth of new minerals, and hardening of the vase all occur without melting. The reactions in a potter's kiln are called **solid-state reactions** because they occur in solid materials.

Metamorphism (from the Greek words for "changing form") is the process by which rocks and minerals change

Figure 3–28 Mud cracks form when mud shrinks as it dries. They are often preserved in ancient rocks.

because of changes in temperature, pressure, or other environmental conditions. Like the potter's vase in the kiln, metamorphism occurs in solid rocks. Small amounts of water and other fluids help metamorphism, but the rock remains solid as it changes. Both the texture and the minerals can change as a rock is metamorphosed.

Textural Changes

A rock is an aggregate of individual mineral grains. As a rock is metamorphosed, the grains grow and their shapes change. For example, fossils give fossiliferous limestone its texture (Fig. 3–29A). Both the fossils and the cement between them are made of small calcite crystals. If the

Figure 3–29 Metamorphism converts fossiliferous limestone (A) to marble (B), which has a very different texture, although both are made of the mineral calcite.

A

B

limestone is buried and heated, the calcite grains grow larger. In the process, the fossiliferous texture is destroyed.

The resulting metamorphic rock, called **marble** (Fig. 3–29B), is still made of calcite, but its texture is now one of large interlocking grains. Although its minerals are the same as those of limestone, marble is a coarse-grained rock that can be polished to create a smooth, lustrous surface and therefore is prized for sculpture.

Mineralogical Changes

In the example of limestone, metamorphism changed its texture but not its minerals. In many other cases, metamorphism forms new and different minerals as the original ones decompose. For example, a typical shale contains clay, quartz, and feldspar (Fig. 3–30A). When heated, the clay minerals decompose, as they do when the potter's vase is fired in the kiln. The atoms of the clay recombine to form new minerals. Figure 3–30B shows a rock called hornfels that is formed when both new textures and new minerals develop during metamorphism of shale.

Thus, two types of metamorphic reactions occur. In one, the original mineral grains simply grow larger. In the other, the original minerals decompose and new minerals grow in their place. Both kinds of reactions occur in the solid state.

As a general rule, when a parent rock (the original rock) contains only one mineral, metamorphism forms a rock composed of the same mineral but with a coarser texture. The metamorphism of limestone to marble is one example. In contrast, metamorphism of a parent rock containing several minerals usually forms new and different minerals. Shales commonly contain clay minerals as well as quartz and feldspar. During metamorphism, shales always grow entirely new minerals as well as new textures.

The Causes of Metamorphism

The outer layers of the Earth move constantly. If you could watch the Earth over hundreds of millions of years, you would see continents move completely around the globe, crashing together and then splitting apart. Huge mountain ranges would rise in the collision zones, only to erode to flatness as time passed. You would see new ocean basins form and old ones disappear. Some of these processes force rocks downward into deeper regions of the crust, burying them under 5, 10, or even 20 kilometers of sediment.

When rocks are buried, they become hotter and the pressure on them increases. These new conditions cause chemical and physical changes in the same way that the heat of a potter's kiln alters moist clay. Let us look briefly at each of the factors that cause metamorphism, to see how they work and what kinds of changes they cause.

Temperature

Recall that the Earth's crust gets hotter by an average of 30°C for each kilometer of depth. Heat causes metamorphism. Think of a layer of clay deposited in a sedimentary basin. If the basin sinks due to tectonic processes, it continues to fill with more sediment. As more sediment accumulates, the clay layer is buried. If several kilometers of sediment pile up on top of the layer of clay, the temperature rises enough to decompose the clay minerals. The atoms from the clays then recombine to grow new minerals.

Figure 3–30 When shale (A) is metamorphosed to hornfels (B), both a new texture and new minerals form. The white spots in (B) are metamorphic minerals. *(Geoffrey Sutton)*

A

B

Mount Sir Sanford in British Columbia, Canada, is made entirely of marble.

Pressure

Minerals are also sensitive to pressure. If atoms in a crystal are squeezed together very tightly, the bonds between the atoms can break. The atoms then reorganize to form a new mineral that is stable under the higher pressure. Most minerals are more sensitive to temperature changes than to pressure variations. Nevertheless, pressure does play an important role in metamorphism.

Migrating Fluids

Recall that sediment commonly contains water in the pore space between the grains. Water is also present in most rocks of the Earth's crust and in magma. This water usually contains dissolved ions and flows slowly through rock. The water and ions can react with rock, decomposing original minerals and replacing them with new ones. In this way, migrating fluids can also cause metamorphism.

Deformation

Tectonic plates move and smash together, creating tremendous forces. Rocks bend and break in response. **Deformation** is the change in shape of rocks in response to tectonic forces. It occurs in tectonically active regions—areas where rocks move.

When metamorphism occurs without deformation, mineral grains grow with random orientations (Fig. 3–31A). In contrast, Figure 3–31B shows originally flat-lying shale beds being squeezed (deformed) into **folds**

during metamorphism. The clays decompose and are replaced with platy minerals such as mica. The mica grows with its flat surfaces perpendicular to the force squeezing the rocks. This parallel arrangement of minerals forms layering called **slaty cleavage** (Figs. 3–31C and D). Notice that the slaty cleavage cuts across the original sedimentary bedding. Rocks with slaty cleavage break neatly along the newly formed planes. Any kind of metamorphic layering, such as slaty cleavage, is called **foliation**.

Metamorphic Grade

The **metamorphic grade** of a rock is the intensity of metamorphism that formed the rock. Temperature is the most important factor in metamorphism, and therefore grade mostly reflects the temperature of metamorphism. Because different minerals form as a rock becomes hotter, the minerals indicate metamorphic grade. Since temperature increases with depth in the Earth, a general relationship exists between depth and metamorphic grade. Low-grade metamorphism occurs at shallow depths, less than 10 kilometers beneath the surface, where temperature is no higher than 300° to 400°C. High-grade conditions are found deep within continental crust and in the upper mantle, 40 to 55 kilometers below the Earth's surface. The temperature here is 600° to 800°C, close to the melting point of rock. High-grade conditions can develop at shallower depths, however, in areas adjacent to rising magma or hot intrusive rocks. For example, today metamorphic rocks are forming beneath Yellowstone Park,

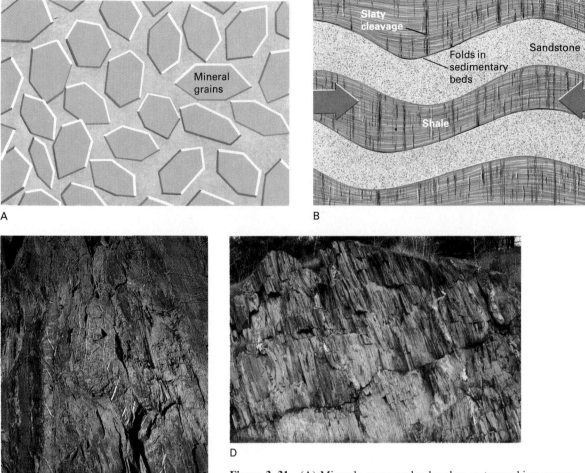

A

B

C

D

Figure 3–31 (A) Minerals grow randomly when metamorphism occurs without deformation. (B) If metamorphism occurs while tectonic forces deform shale into folds, mica flakes grow perpendicular to the force (arrows) that is squeezing shale into folds. The rock then breaks easily parallel to the mica flakes. This breakage is called slaty cleavage. (C) Slaty cleavage cuts across a gray, folded sandstone bed surrounded by brown slate. White calcite veins fill cleavage cracks in the sandstone. Notice the parallel slaty cleavage in the slate. *(Jim Sears)* (D) The development of slaty cleavage has obliterated all traces of original sedimentary bedding in this outcrop.

where hot magma lies close to the Earth's surface. Figure 3–32 summarizes the relationships among depth, pressure, temperature, and metamorphic grade.

3.9 Types of Metamorphism and Metamorphic Rocks

Metamorphism is divided into three general categories on the basis of the cause of metamorphism.

Contact Metamorphism

Contact metamorphism results from the intrusion of hot magma into cooler rocks. The country rock (the rock intruded by the magma) may be of any type—sedimentary, metamorphic, or igneous. The intrusion heats the adjacent rock, causing old minerals to decompose and new ones to form. The highest-grade metamorphic rocks form at the contact, where the temperature is greatest. Lower-grade rocks develop farther out, forming a halo of metamorphism around the intrusion (Fig. 3–33). Contact metamorphic halos can range in width from less than a meter to hundreds of meters, depending on the size and temperature of the intrusion and the effects of water or other fluids.

Because contact metamorphism commonly occurs without deformation, the rocks are **nonfoliated**; that is, they have no metamorphic layering.

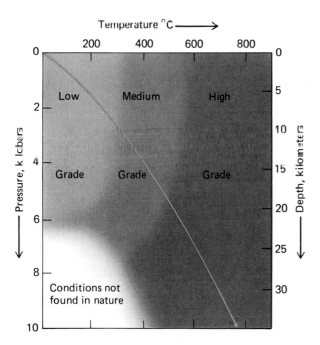

Figure 3–32 Temperatures, pressures, and depths below the Earth's surface at which different grades of metamorphism occur. The blue arrow traces the path of increasing temperature and pressure with depth in a normal part of the crust.

Hornfels (Fig. 3–30B) is a hard, dark, fine-grained rock usually formed by contact metamorphism of shale. **Tactite**, also called **skarn** (Fig. 3–34), forms by contact metamorphism of limestone. Tactite often contains large crystals of calcite, garnet, and pyroxene.

Regional Metamorphism

Regional metamorphism affects broad regions of the Earth's crust, in contrast to contact metamorphism, which affects only the rock immediately surrounding an igneous intrusion. Regional metamorphism is usually accompanied by deformation, so the rocks are foliated. It is the most common and widespread type of metamorphism in areas of mountain building and granite intrusion.

Recall that shale is the most abundant type of sedimentary rock. It consists mostly of platy clay minerals arranged parallel to bedding planes. The clay grains are too small to be seen with the naked eye. Shale changes in a regular sequence as temperature rises and rocks are deformed during regional metamorphism (Fig. 3–35).

As regional metamorphism begins, clay minerals decompose and new minerals grow perpendicular to the squeezing direction, as already described. Thus, slaty cleavage develops. Rock formed in this manner is called **slate** (Fig. 3–35).

With increasing metamorphic grade and continuing deformation, the crystals grow larger and foliation becomes very well developed. Rock of this type is called

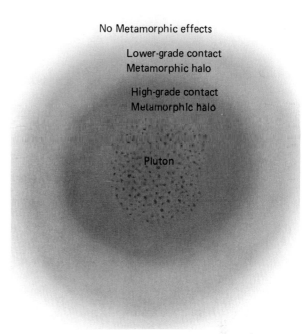

Figure 3–33 A contact metamorphic halo surrounding a pluton.

schist (Fig. 3–35). Schist first forms between low and intermediate metamorphic grades.

At high metamorphic grades, light- and dark-colored minerals often separate into bands a centimeter or more thick, to form a rock called **gneiss** (pronounced "nice") (Fig. 3–35).

Hydrothermal Metamorphism

Hydrothermal metamorphism, also called **hydrothermal alteration**, is the changes in rock caused by migrating hot water and by ions dissolved in the hot water.

Figure 3–34 Tactite containing garnet (brown) and calcite (white).

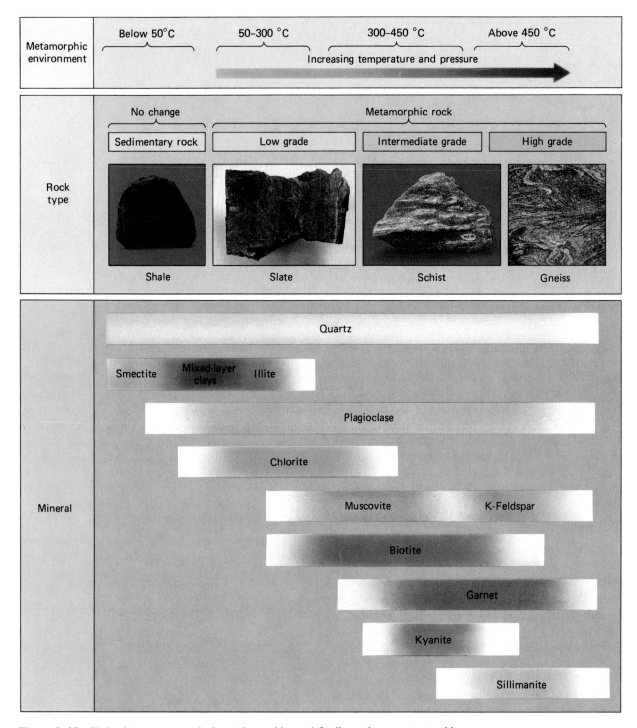

Figure 3–35 Shale changes progressively to slate, schist, and finally gneiss as metamorphic grade increases. The lower part of the figure shows when old minerals decompose and new ones grow as metamorphic grade increases. *(Geoffrey Sutton, Don Hyndman and Hubbard Scientific Co.)*

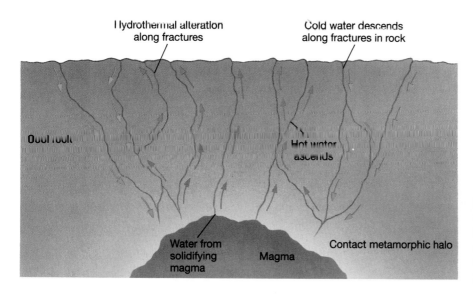

Hydrothermal alteration
along fractures

Cold water descends
along fractures in rock

Cool rock

Hot water
ascends

Water from
solidifying
magma

Magma

Contact metamorphic halo

Figure 3–36 Ground water descending through fractured rock is heated by magma and ascends through other fractures, causing hydrothermal metamorphism in nearby rock.

Most hydrothermal alteration is caused by circulating **ground water**—water contained in soil and bedrock. Cold ground water sinks through fractures in bedrock to depths of a few kilometers, where it is heated by the hotter rocks. Shallow magma or a hot, shallow pluton enhances the heating of shallow ground water. Water is a chemically active fluid that attacks and dissolves many minerals. If the water is hot, it decomposes minerals even more rapidly. In some hydrothermal environments, hot water reacts with sulfur or chloride minerals in the rock to form strong acids such as sulfuric or hydrochloric acid, making the solution even more corrosive. Upon heating, the water expands and rises back toward the surface through other fractures, altering the rocks adjacent to the fractures as it goes (Fig. 3–36).

Quartz veins often form during hydrothermal alteration as silica dissolved from the original minerals precipitates in fractures. Metals such as gold, silver, copper, lead, and zinc sometimes concentrate with quartz in **hydrothermal veins** to form ore deposits.

SUMMARY

•

Geologists divide rocks into three groups depending upon how they formed. **Igneous rocks** solidify from **magma**. **Sedimentary rocks** form by **lithification** of **clay**, **sand**, **gravel**, and other loose **sediment** that collects at the Earth's surface. **Metamorphic rocks** form when any rock is changed by temperature, pressure, or deformation. The **rock cycle** summarizes processes by which rocks continuously recycle in the outer layers of the Earth, forming new rocks from old ones.

Intrusive, or **plutonic**, igneous rocks are medium- to coarse-grained rocks that solidify within the Earth's crust. **Extrusive**, or **volcanic**, igneous rocks are fine-grained rocks that solidify from magma erupted onto the Earth's surface. **Granite** and **basalt** are the two most common igneous rocks.

Sediment forms by weathering of rocks and minerals. It includes all solid particles such as rock and mineral fragments, organic remains, and precipitated minerals. It is **transported** by streams, glaciers, wind, and gravity; **deposited** in layers; and eventually **lithified** to form sedimentary rock. **Shale**, **sandstone**, and **limestone** are the most common kinds of sedimentary rock.

When a rock is heated, when pressure increases, or when the rock is deformed, both its minerals and its textures change in a process called **metamorphism**. **Contact metamorphism** affects rocks heated by a nearby igneous intrusion. **Regional metamorphism** affects large regions of the crust during mountain building. **Hydrothermal metamorphism** is caused by hot solutions soaking through rocks and is often associated with emplacement of ore deposits. **Slate**, **schist**, **gneiss**, and **marble** are common metamorphic rocks.

KEY TERMS

•

Magma   *46*

Igneous rock   *46*

Sedimentary rock   *46*

Metamorphic rock   *46*

Rock cycle   *46*

Intrusive igneous rock   *48*

Plutonic rock   *48*

Extrusive igneous
   rock   *48*

Volcanic rock   *48*

Porphyry   *51*

Phenocryst   *51*

Sialic   *52*

Mafic   *52*

Sediment   *54*

Clastic sediment   *54*

Rounding   *55*

Sorting   *55*

Deposition   *56*

Lithification   *56*

Pore space   *56*

Compaction   *57*

Cementation   *57*

Sedimentary structure   *62*

Slaty cleavage   *67*

Foliation   *67*

Metamorphic grade   *67*

Contact metamorphism   *68*

Regional
   metamorphism   *69*

Hydrothermal
   alteration   *69*

COMMON IGNEOUS ROCKS

•

Obsidian

Granite
Rhyolite

Diorite
Andesite

Gabbro
Basalt

Peridotite

COMMON SEDIMENTARY ROCKS

•

Sandstone

Shale

Limestone

Dolomite

COMMON METAMORPHIC ROCKS

•

Marble

Slate

Schist

Gneiss

REVIEW QUESTIONS

•

1. What are the three main kinds of rock in the Earth's crust?

2. How do the three main types of rock differ from each other?

3. What is magma, and where does it originate?

4. Describe how igneous rocks are classified and named.

5. What are the two most common kinds of igneous rock?

6. Describe and explain the differences between a plutonic and a volcanic rock.

7. What is sediment? How does it form?

8. How do sedimentary grains become rounded?

9. How do sedimentary grains become sorted?

10. Where in your own area would you look for rounded and sorted sedimentary grains?

11. Describe how sedimentary grains become lithified.

12. What are the differences among shale, sandstone, and limestone?

13. Explain why almost all sedimentary rocks are layered, or bedded.

14. How does cross-bedding form?

15. What do mud cracks and raindrop imprints in shale tell you about the water depth in which the mud accumulated?

16. What is metamorphism? What factors cause metamorphism?

17. What kinds of changes occur in a rock as it is metamorphosed?

18. What is metamorphic foliation? How does it differ from sedimentary bedding?

19. Explain the concept of metamorphic grade.

20. How do contact metamorphism and regional metamorphism differ, and how are they similar?

21. Explain what the rock cycle tells us about Earth processes.

DISCUSSION QUESTIONS

•

1. Magma usually begins to rise toward the Earth's surface as soon as it forms. It rarely accumulates as large pools in the upper mantle or lower crust, where it originates. Why?

2. In the San Juan Mountains of Colorado, parts of the range are made up of granite and other parts are volcanic rocks. Explain why these two types of igneous rock are likely to be found together.

3. Suppose you were given a fist-size sample of igneous rock. How would you tell whether it is volcanic or plutonic in origin?

4. How would you tell whether another rock sample is igneous, sedimentary, or metamorphic in origin?

5. Why is shale the most common sedimentary rock?

6. What information can you infer about the history of a sandstone that contains 75 percent quartz grains and 25 percent feldspar grains?

7. One sedimentary rock is composed of rounded grains, and the grains in another are angular. What can you tell about the history of the two rocks from these observations?

8. How would you distinguish between contact metamorphic rocks and regional metamorphic rocks in the field?

9. What happens to bedding when sedimentary rocks undergo regional metamorphism?

10. How can granite form during metamorphism?

Geologic Time, Fossils, Evolution, and Extinction

4

To a large degree, geology is a study of events that occurred in the past. Geologists observe rocks and landforms and ask questions such as: *What* did they look like in the past? *How* did they evolve to their modern forms? *When* did they change? Many of these questions have to do with the age of the landscape and of rocks around us. How old is the Earth? How old are the hills in the distance? How much time is required for a delicate sandstone arch to form in the desert?

4.1 Uniformitarianism

James Hutton was a gentleman farmer who lived in Scotland in the late 1700s. Although trained as a physician, he never practiced medicine and instead turned his thoughts to geology. Hutton observed that sandstone is composed of grains of sand cemented together. He also noted that exposed rocks gradually weather to sand and that streams erode the sand and carry it into the lowlands, where they deposit it in layers. He deduced that sandstone

• Delicate Arch, Arches National Park, Utah.

is composed of sand that originated by weathering and erosion of now-extinct cliffs and mountains.

Hutton tried to deduce how much time was required to form a thick layer of sandstone. He studied rock outcrops and observed that pieces were slowly breaking off. He stared into creek beds and watched sand grains bouncing downstream. Finally he traveled to beaches and river deltas where sand was accumulating layer by layer. Hutton concluded that geological change was occurring before his eyes and that sandstone was forming by the sequence of events that he had observed. He reasoned, based on his observations, that the formation of a thick layer of sandstone must take a long, long time. He wrote:

> On us who saw these phenomena for the first time, the impression will not easily be forgotten. . . . We felt ourselves necessarily carried back to the time . . . when the sandstone before us was only beginning to be deposited, in the shape of sand and mud, from the waters of an ancient ocean. . . . The mind seemed to grow giddy by looking so far into the abyss of time.

Hutton's conclusions were based on a principle now known as **uniformitarianism**. This principle states that **geological processes and scientific laws operating today also operated in the past**. Thus, ancient rocks must have formed by the same processes that we can observe somewhere on Earth today. Sometimes this idea is summarized in the statement "The present is the key to the past." Today we can observe somewhere in the world each individual step that leads to the formation of sandstone. We cannot watch the entire development of a specific layer of sandstone because it takes too long, but we can infer that ancient sandstone formed by the same processes that we see in action today.

The principle of uniformitarianism does *not* exclude catastrophes that cause rapid change. Earthquakes and volcanoes cause rapid geologic changes, and from time to time giant meteorites strike Earth with energy equivalent to that of millions of nuclear bombs. Catastrophic events occur, but the basic premise of all science is that any movement, any change, can be explained by the fundamental laws of chemistry and physics.

If we measure rates of geologic processes that we can see in action today, such as those resulting in the formation of a layer of sandstone, then we must accept the idea that most rocks are much, much older than the span of a human life. We must also conclude that our planet itself has existed for a very long time. Hutton was so overwhelmed by the magnitude of geologic time that he wrote, "We find no vestige of a beginning, no prospect of an end."

4.2 Geologic Time

In Chapter 1 you learned that the Earth is approximately 4.6 *billion* years old. To help comprehend the great length of geologic time, we used John McPhee's metaphor symbolizing the past 4.6 billion years with the old English measure of the yard, the distance from the king's nose to the end of his outstretched hand. All of human history can be erased by a single stroke of a nail file on his middle fingernail, and recorded human history is only about 5000 years old. In contrast, geologists commonly refer to the ages of rocks and geologic events in terms of millions or billions of years before present. How do geologists *measure* the ages of rocks and events that occurred before calendars and clocks were invented?

Geologists measure geologic time in two different ways. **Absolute time** is measured in years. The Earth is 4.6 billion years old. Dinosaurs became extinct 65 million years ago. The Teton Range in Wyoming began rising 6 million years ago. The Pleistocene Ice Age began 1.6 million years ago. Absolute time tells us both the *order* in which events occurred and the *amount* of time separating them.

In contrast, **relative time** simply lists the *order* in which events occurred, without regard to the amount of time separating them. Measurement of relative time is based on a simple principal: in order for an event to affect a rock, the rock must exist first. Thus, the rock must be older than the event. This principal seems too obvious to bother stating, yet it is the basis of much geologic work. As an example, consider relative time as expressed by the folded sedimentary rock shown in Figure 4–1. Sediment normally accumulates in horizontal layers. If you observe a fold in the layers, you can deduce that the folding occurred *after* the sediment was deposited. The order in which rocks and geologic features form can almost always be interpreted by careful observations and logic.

4.3 Absolute Time

How does a geologist measure the absolute age of an event that occurred before calendars, and even before humans evolved to keep calendars?

Think for a moment of how a calendar measures time. The Earth rotates about its axis at a constant rate, once a day. Thus, each time the Sun rises, you know that a day has passed and you check it off on your calendar. If you mark off each day as the Sun rises, you record the passage of time. To know how many days have passed since you started the calendar, you just count the check marks. The measurement of absolute time depends on two factors: *a process that occurs at a constant rate* (e.g., the Earth rotates once every 24 hours) and some way to keep *a cumulative record of that process* (e.g., marking a day on the calendar each time the Sun rises). Measure-

Figure 4–1 A fold in limestone in Costa Rica. The limestone must have been deposited in horizontal layers before the rock folded.

ment of time with a calendar, a clock, an hourglass, or any other device depends on these same two factors.

Geologists have found a natural process that occurs at a constant rate and accumulates its own record. It is the radioactive decay of common elements present in small amounts in most rocks. Thus, rocks have built-in clocks. To measure the ages of rocks and geological events, we must understand radioactivity and learn how to read the clocks.

Radioactivity

Uranium is a well-known radioactive element. With time, some atoms of uranium spontaneously break down, or *decay*, to become atoms of lead. As a uranium atom decays, both subatomic particles and energy fly away at high speeds. This emission of particles and energy is called **radioactivity**. The high-speed particles and energy can cause cancer and mutations; consequently radioactivity is dangerous.

Recall from Chapter 2 that an atom consists of a small, dense nucleus surrounded by a diffuse cloud of electrons. Radioactive decay occurs when an atom's nucleus breaks apart. Therefore, to understand radioactivity, we must first look at the structure of the nucleus.

All atoms of any given element have the same number of protons in the nucleus. However, the number of neutrons may vary. **Isotopes** are atoms of the same element with different numbers of neutrons. For example,

Radiometric age dating can be used to measure the absolute age of granite. Lotus Flower Tower, Cirque of the Unclimbables, Northwest Territories, Canada.

all isotopes of potassium have 19 protons, but one isotope has 21 neutrons and another has 20 neutrons.

Each isotope is given the name of the element followed by the total number of protons *plus* neutrons in its nucleus. For example, potassium-40 is the isotope of potassium with 19 protons and 21 neutrons. Potassium-39 has 19 protons but only 20 neutrons. Different isotopes of the same element are chemically identical, but each has a different nuclear structure.

Many isotopes are not radioactive and are called *stable*; they do not change with time. If you studied a sample of potassium-39 for a second, a year, or even ten billion years, all the atoms would remain unchanged. It is a stable isotope. Other isotopes are radioactive, or *unstable*. Given time, they spontaneously break apart. Potassium-40 decomposes naturally to form two other isotopes, argon-40 and calcium-40 (Fig. 4–2). An unstable isotope such as potassium-40 is known as a **parent isotope**. An isotope created by radioactive decay of a parent, such as argon-40 or calcium-40, is called a **daughter isotope**.

Many common elements, such as potassium, consist of a mixture of radioactive and nonradioactive isotopes. With time, the radioactive isotopes decay, but the nonradioactive ones do not. A few elements, such as uranium, consist only of radioactive isotopes. Given enough time, all of the Earth's uranium will decay to other elements, such as lead.

Figure 4–2   Potassium-40 is a radioactive parent isotope that decays to two daughter isotopes, argon-40 and calcium-40. Eleven percent of the potassium-40 decays to argon-40 as a small, negatively charged subatomic particle is added. The other 89 percent converts to calcium-40 as a small, negatively charged particle is released.

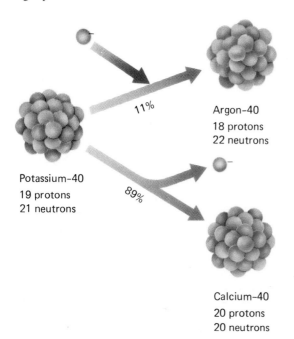

Radioactivity and Half-Life

If you observed a single atom of potassium-40, when would it decompose? This question cannot be answered because any particular potassium-40 atom may or may not decompose at any time. But if you observed a large number of potassium-40 atoms, they would decompose at a rate called the **half-life** of the isotope. The half-life is the time it takes for half of the atoms to decompose. The half-life of potassium-40 is 1.3 billion years. Therefore, if 1 gram of potassium-40 were placed in a container, ½ gram would remain after 1.3 billion years, ¼ gram after 2.6 billion years, and so on. Each radioactive isotope has its own half-life; some half-lives are only fractions of a second and others are measured in billions of years.

The half-life is an *average* value for the life span of all potassium-40 atoms. A single gram of potassium-40 contains 1.5×10^{22} (15 followed by 21 zeros) atoms. The chance that a specific atom will decompose in any given second is small. But an immense number of atoms exist in a gram, so many decompose every second. Consider this analogy: if you buy a lottery ticket, your chance of winning is very small. But someone wins nearly every week because so many tickets are sold.

The Basis of Radiometric Dating

Two aspects of radioactivity are essential to the calendars in rocks. First, the rate at which any parent decays to a daughter (its half-life) is constant. The decay rate is easily measured in the laboratory and is unaffected by geologic processes. So **radioactive decay occurs at a known, constant rate**. Second, as a parent isotope in a rock decays, its daughter accumulates. **The longer the rock exists, the more daughter isotope accumulates.** The accumulation of daughter is analogous to marking off days on a calendar as they pass: it measures the cumulative effect of radioactive decay. Because radioactive isotopes are widely distributed throughout the Earth, most rocks have built-in calendars that allow us to measure their ages. **Radiometric age dating** is the process of determining the ages of rocks, minerals, and other geologic materials by measuring radioactive elements and their daughter isotopes.

Figure 4–3 shows the relationships between age and relative amounts of parent and daughter isotopes. At the end of one half-life, 50 percent of the parent atoms have decayed to daughter. At the end of two half-lives, the mixture is 25 percent parent and 75 percent daughter. To determine the age of a rock, a geologist measures the proportions of parent and daughter isotopes in a sample and compares the ratio to a similar graph. Consider a hypothetical parent-daughter pair having a half-life of one million years. If we determine that a rock contains a mixture of 25 percent parent isotope and 75 percent

Focus On

Carbon-14 Dating

Carbon-14 dating differs from dating with the other parent-daughter pairs shown in Table 4–1 in two ways. First, carbon-14, the parent isotope, has a very short half-life—only 5730 years, in contrast to the millions or billions of years of the other parents. Second, accumulation of the daughter, nitrogen-14, cannot be measured. Nitrogen-14 is the common isotope of nitrogen; it is impossible to distinguish nitrogen-14 produced by decay of carbon-14 from other nitrogen-14 in an object being dated.

The abundant stable isotope of carbon is carbon-12. However, carbon-14 forms continuously in the atmosphere as cosmic radiation converts nitrogen-14 to carbon-14. The carbon-14 then decays back to nitrogen-14. Because carbon-14 is continuously created and decays at

a constant rate, the ratio of carbon-14 to carbon-12 in the atmosphere is constant.

While an organism is alive, it absorbs carbon from the atmosphere. Therefore, the organism contains the same ratio of carbon-14 to carbon-12 as does the atmosphere. However, when the organism dies, it stops absorbing carbon. Therefore, the proportion of carbon-14 in the remains begins to diminish as the carbon-14 decays. The relative amount of carbon-14 steadily decreases. Thus, carbon-14 age determinations are made by measuring the ratio of carbon-14 to carbon-12 in organic remains.

By the time an organism has been dead for 50,000 years, so little carbon-14 remains that accurate measurement is difficult. After about 70,000 years, the method is no longer useful.

---

daughter, Figure 4–3 shows that the age is two half-lives, or two million years.

The half-life of a parent isotope determines the range of age over which the isotope is useful for dating. If the half-life is short, the isotope gives accurate ages for young materials. For example, carbon-14 has a half-life of 5730

years. Carbon-14 dating gives accurate ages for materials younger than 70,000 years. It is useless for older materials because by 70,000 years virtually all the carbon-14 has decayed, and no additional change can be measured. Isotopes with very long half-lives give good ages for old rocks, but in young ones not enough daughter has accu-

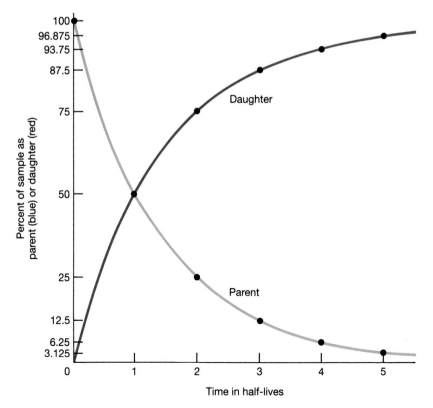

Figure 4–3 As a radioactive parent isotope decays to a daughter, the proportion of parent decreases (blue), and the amount of daughter increases (red). The half-life is the amount of time required for half of the parent to convert to daughter. At time zero, when the radiometric calendar starts, a sample is 100 percent parent. Now look at both curves at the points for one half-life. At the end of one half-life, 50 percent of the parent has converted to daughter. At the end of two half-lives, 25 percent of the sample is parent and 75 percent is daughter. Thus, by measuring the proportions of parent and daughter in a rock, its age in half-life can be obtained. Since the half-lives of all radioactive isotopes are well known, it is then simple to convert age in half-life to age in years.

The Most Commonly Used Isotopes in Radiometric Age Dating				
Isotopes		**Half-Life of Parent (years)**	**Effective Dating Range (years)**	**Minerals and Other Materials That Can Be Dated**
Parent	*Daughter*			
Carbon-14	Nitrogen-14	5730	100–70,000	Anything that was once alive: wood, other plant matter, bone, flesh, or shells; also, carbon dissolved as carbon dioxide in ground water, deep layers of the ocean, or glacier ice.
Potassium-40	Argon-40 Calcium-40	1.3 billion	50,000–4.6 billion	Muscovite Biotite Hornblende Whole volcanic rock
Uranium-238 Uranium-235 Thorium-232	Lead-206 Lead-207 Lead-208	4.5 billion 710 million 14 billion	10 million–4.6 billion	Zircon Uraninite and pitchblende
Rubidium-87	Strontium-87	47 billion	10 million–4.6 billion	Muscovite Biotite Potassium feldspar Whole metamorphic or igneous rock

mulated to be measured. For example, rubidium-87 has a half-life of 47 billion years. In a geologically short period of time—ten million years or less—so little of its daughter has accumulated that it is impossible to measure accurately. Therefore, rubidium-87 is not useful for rocks younger than about ten million years.

Several radioactive isotopes are common in rocks. Some are more useful for age dating than others. The six most commonly used isotopes, their half-lives, the ranges of ages over which they can be used, and minerals and other materials that commonly are dated using them are summarized in Table 4–1.

Radiometric dating works only if neither parent nor daughter isotopes are lost from or added to rocks and minerals after they form. Consider the potassium-argon pair. Potassium-40 is an unstable isotope that decays to argon-40. Argon is an inert gas that is trapped in minerals as it forms. But if a mineral is heated above a certain temperature, the argon gas escapes. Therefore, since all the daughter isotope is lost, the calendar starts all over again at zero. For this reason, a potassium-argon age date tells us the time at which a rock or mineral cooled below the temperature at which argon escapes. That temperature varies from a few hundred degrees Celsius upward, depending on the mineral. Thus, a potassium-argon date from an igneous rock does not tell us when the magma formed or when it solidified, but it does tell us when the

solid rock cooled below a few hundred degrees (Fig. 4–4.)

4.4 Relative Time

Radiometric dating has become common for measuring geologic time only in the second half of this century. Previously, geologists measured **relative time**, placing rocks and events in the order in which they formed or occurred without reference to the absolute time involved. Modern geologists often use relative time rather than going to the expense of radiometric dating. Determination of the order in which rocks formed and events occurred is achieved by a combination of common sense and a few simple principles.

The **principle of original horizontality** of sedimentary rocks is based on the observation that sediment usually accumulates in horizontal beds, and therefore most sedimentary rocks started out with horizontal layering (Fig. 4–5A). If we see sedimentary rocks lying at an angle, as in Figure 4–5B, we can infer that they were originally deposited in horizontal layers and later were tilted. The folded sedimentary rocks shown in Figure 4–1 were also initially deposited as flat-lying sediments and folded at a later date.

The **principle of superposition** states that sedimentary rocks become younger from bottom to top (as long

A

Granite pluton intrudes country rock and solidifies 1 billion years ago. Radiometric calendar starts.

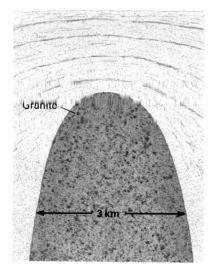

B

Basalt dike intrudes 0.6 billion years ago. Heat from basalt resets calendar to time zero in nearby granite.

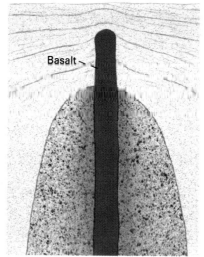

C Uplift and erosion expose granite at Earth's surface

Geologist dates sample from here. Obtains date of 1 billion years.

Geologist dates sample from here. Obtains date of 0.6 billion years.

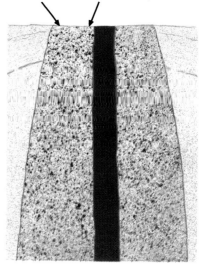

Figure 4–4 A radiometric age date measures the time elapsed since a rock or mineral last cooled down. For example, in (A) this granite magma solidified and cooled one billion years ago. It then began to accumulate argon-40 as potassium-40 decayed, and the radiometric calendar started. (B) Later, 0.6 billion years ago, a large basalt dike intruded the granite, heating adjacent rock back up again. This heat allowed the argon to escape, resetting the calendar to time zero. (C) A geologist samples the granite near the dike and measures its potassium-argon age today at 0.6 billion years. This is the age of the reheating, not of the original formation of the granite.

Figure 4–5 (A) The principle of original horizontality tells us that most sedimentary rocks are deposited with horizontal bedding. When we see tilted rocks (B), we infer that they were tilted after they were deposited.

A

B

Older

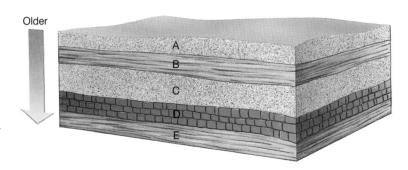

Figure 4–6 The principle of superposition tells us that sedimentary layer E, on the bottom of the sequence, is the oldest, and the top layer, A, is the youngest.

as they have not been turned upside down by tectonic forces). This is because younger layers of sediment always accumulate on top of older layers. In Figure 4–6, the sedimentary layers become progressively younger in the order E, D, C, B, and A.

The 2-kilometer-high walls of Grand Canyon are made of sedimentary rocks lying on older igneous and metamorphic rocks. Their ages range from about 200 million years to nearly two billion years. Thus, they span more than half the age of the Earth. The principle of superposition tells us that the deepest rocks are the oldest and the upper layers are progressively younger. However, the principle of superposition does *not* assure us that the rocks formed continuously from two billion to 200 million years ago. Thus, the rock record may not be complete. Suppose that no sediment was deposited for a period of

Rocks exposed in Grand Canyon represent more than 2.5 billion years of geologic time.

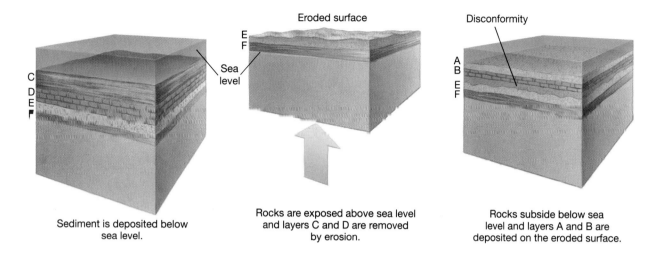

Sediment is deposited below sea level.

Rocks are exposed above sea level and layers C and D are removed by erosion.

Rocks subside below sea level and layers A and B are deposited on the eroded surface.

Figure 4–7 Development of a disconformity.

time, or some sedimentary layers were eroded. In either case a gap would exist in the rock record. We can rightly infer that any rock layer is younger than the layer below it, but without more information we do not know how much younger.

Layers of sedimentary rocks are **conformable** if they were deposited without interruption. An **unconformity** represents an interruption of deposition of sediment, usually of long duration. During the interval when no sediment was deposited, some rock layers may have been eroded.

Several types of unconformities exist. The rock layers above and below a **disconformity** are parallel, as in a conformable sequence (Fig. 4–7). Thus, a disconformity can be difficult to recognize. Figure 4–8 shows a disconformity between sandstone and conglomerate.

In an **angular unconformity**, older sedimentary beds were tilted and eroded before younger layers were deposited (Fig. 4–9). Angular unconformities are easily recognized because the older beds and the younger beds meet at an angle (Fig. 4–10).

A **nonconformity** is an unconformity in which sedimentary rocks lie on older igneous or metamorphic rocks. The nonconformity shown in Figure 4–11 represents a time gap of about one billion years.

The **principle of cross-cutting relationships** is based on the obvious fact that a rock must first exist before anything can happen to it. Figure 4–12 shows sedimentary rocks cut by a basalt dike. Clearly, the sedimentary rocks must have existed first for the dike to have intruded into them. Therefore, the sedimentary rocks are older than the basalt. Figure 4–13 shows sedimentary rocks intruded by three granite dikes. Dike B cuts dike

C, and dike A cuts dike B, so dike C is older than B, and dike A is the youngest. The sedimentary rocks must be older than all of the dikes.

(Text continues on p. 86)

Figure 4–8 A disconformity between horizontally layered sandstone and an overlying layer of conglomerate in Wyoming. Some sandstone layers were eroded away before the conglomerate was deposited.

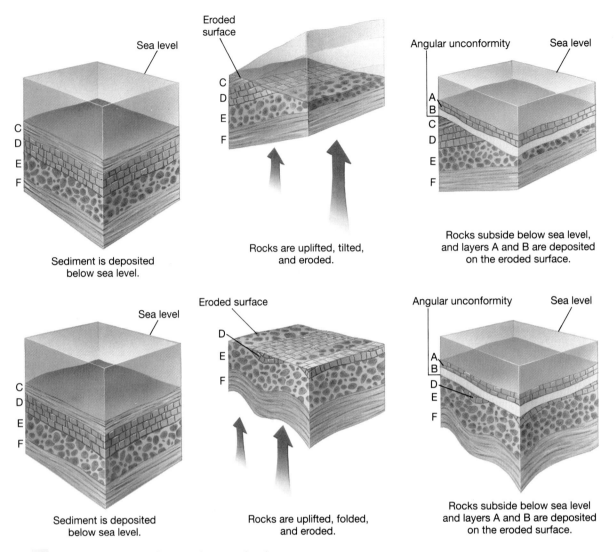

Figure 4–9 Development of an angular unconformity.

Figure 4–10 An angular unconformity near Capitol Reef National Park, Utah.

Figure 4–11 A nonconformity in the cliffs of Grand Canyon, Arizona. You can see sedimentary bedding in the red sandstone above the nonconformity, but none in the igneous and metamorphic rocks below.

Figure 4–12 A basalt dike cutting across sedimentary rocks in the walls of Grand Canyon. The sedimentary rocks must be older than the dike.

Figure 4–13 Three granite dikes cutting sedimentary rocks. The dikes become younger in the order C, B, A. The sedimentary rocks must be older than all three dikes.

4.5 Fossils and Faunal Succession

The study of past life is called **paleontology**. To understand the history of life and evolution, paleontologists study **fossils**, the remains and other traces of prehistoric life. A great variety of different types and ages of fossils exist. The oldest are tiny traces of ancient bacteria-like organisms that lived about 3.5 billion years ago. A more recent fossil is an entire baby mammoth that froze into a Siberian glacier about 44,000 years ago.

A ten-million-year-old spider preserved in amber.   *(American Museum of Natural History)*

The Formation of Fossils

After they die, all plants and animals are susceptible to decomposition and consumption by animals and bacteria. Even hard parts such as shells, bones, or teeth dissolve or weather away to grains of sand. These processes are so efficient that nearly all plants and animals disappear completely after death. However, under special conditions both soft and hard remains are preserved as fossils.

Preservation is a process in which an organism is kept intact with little change. As mentioned previously, a whole mammoth, complete with skin and hair, was found frozen in a glacier. Preserved mammoth meat was once served to geologists at a convention, and biologists have studied genetic information from frozen cells. Insects have been preserved in amber, sabre-toothed tigers in tar, and many mammals by a drying process called **mummification**. Hard parts such as bones, teeth, and shells resist decomposition; in the proper chemical and physical environments they can last for millions of years. Geologists have found unaltered clamshells in sedimentary rocks more than 100 million years old.

Preservation is rare, and fossils of this type are uncommon. It is more common for parts of an organism to be replaced by minerals in a process called **mineralization**. In mineralized fossils, even the most delicate cell structures may remain visible under a microscope, although the original material is completely replaced. Quartz, opal, calcite, and pyrite are common minerals that replace organisms in mineralization. Petrified wood forms by this process when dissolved silica seeps into the tissue of a fallen tree and precipitates as quartz.

Imagine that a shell settles to the bottom of the sea and is buried by sediment. Over time the sediment hardens to rock. Then the shell dissolves, leaving a shell-shaped cavity in the rock. The cavity is called an **external mold** (Fig. 4–14). The mold may be preserved as a cavity or it may fill with mud to form a **cast**. Yet another type of fossil forms when the inside of a shell fills with sediment or precipitated minerals that harden to form an **internal mold**. Some internal molds are especially interesting be-

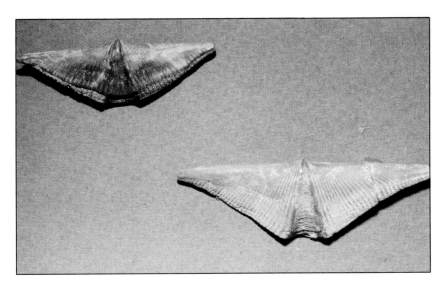

Brachiopods are clam-like marine animals that first appeared about 570 million years ago. They were abundant for about 300 million years, but only a few varieties still live.

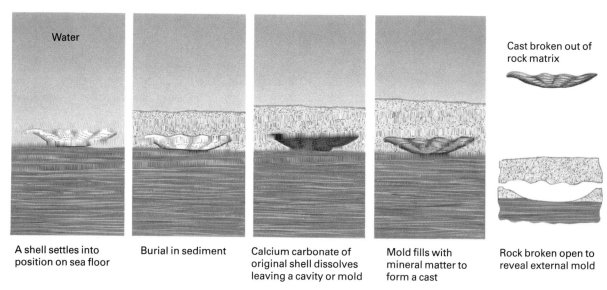

Water

A shell settles into position on sea floor

Burial in sediment

Calcium carbonate of original shell dissolves leaving a cavity or mold

Mold fills with mineral matter to form a cast

Cast broken out of rock matrix

Rock broken open to reveal external mold

Figure 4–14 Development of molds and casts.

cause they retain imprints of places where ligaments were attached and other evidence of soft tissue structures.

Fossils also form by **carbonization**. If soft tissue such as a leaf, a jellyfish, or a worm is rapidly buried and compressed by mud, volatiles are squeezed out, leaving a thin carbon imprint of the organism in the mud.

Trace fossils are not remains of an organism, but instead are tracks, burrows, or other marks made by the organism. Animal tracks have been found in sedimentary rocks. They tell us how the animals moved and whether they stood on two legs or four, and yield other clues as to their habits and behavior.

Interpreting Geologic History from Fossils

If you find a single fossil, you learn that an animal or plant existed. If you can measure the absolute age of the rock from which the fossil came, you can learn when the organism lived. However, by studying all of the different fossils found together in a sedimentary rock layer, you can learn much more.

About 550 million years ago, in Cambrian time, a mud slide occurred on the floor of an ancient sea in what is now western Canada. The mud carried plants and animals down into deeper water and buried them as it settled to the sea floor. Gradually the mud lithified to shale, and much later the shale was uplifted to form mountains. Today this rock layer, called the Burgess Shale, is located in a rugged, mountainous region of British Columbia. Geologists have found fossils of more

Dinosaur tracks preserved in shale. *(American Museum of Natural History)*

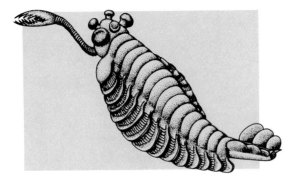

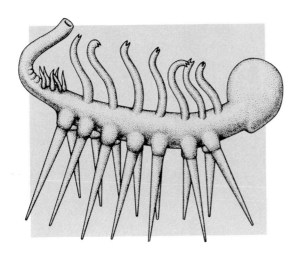

Artist's renditions of Cambrian *Opabinia* and *Hallucigenia.*

than 150 animal species in the Burgess Shale. The list includes some organisms with living descendants, such as jellyfish, sea cucumbers, and sponges. It also includes animals with no known descendants or relatives, such as *Opabinia*, a five-eyed swimming creature with a single claw at the end of a frontal snout, and *Hallucigenia*, an animal that stood in the Paleozoic muck on seven pairs of spines.

Far fewer fossils and fewer kinds of organisms are found in rocks older than the Burgess Shale. Thus, it is clear that a dramatic increase in both diversity and number of animals occurred during Cambrian time as a result of some major evolutionary change. In this way, the fossil record helps us interpret the history of evolution. Fossils also help us interpret the geologic history of rocks and regions of the Earth's crust. For example, the remains of marine animals in the Burgess Shale and in rocks of central Utah, Connecticut, and near the top of Mount Everest tell us that these places once lay submerged beneath the sea.

Fossils are also useful in measuring relative time. Because of evolution, different animals and plants lived at different times in the Earth's history. Trilobites, for example, lived from about 570 million to 245 million years ago. Dinosaurs dominated the Earth from 245 million to 65 million years ago. In sedimentary rocks, different fossils follow each other from bottom to top in the same order in which they appeared and then disappeared through evolution. Rocks containing dinosaur bones must be younger than rocks containing trilobite remains. This method of interpreting relative time is an application of the the **principle of faunal succession**. This principle

Trilobites were abundant lobster-like marine animals that swam and crawled about on the sea floor. They first appeared about 560 million years ago and became extinct about 250 million years ago. The largest were 70 cm in length   *(Ward's Natural Science Establishment, Inc.)*

states, simply, that fossil organisms succeed one another in a definite and recognizable order and that the relative ages of rocks can be determined from their fossil content.

4.6 The Geologic Time Scale

Ideally, geologists would like to develop a complete history of the Earth using information from rocks that formed continuously from the Earth's beginning to the present. Unfortunately, no single place contains rocks that formed continuously throughout the Earth's history. In any single region, erosion has destroyed some rocks, and great lengths of time have passed when no rocks formed. As a result, the rock record in any one place is full of gaps. To assemble a continuous history of the Earth, geologists combine bits and pieces of evidence from all over the Earth. This is done by a process called **correlation**, the matching of rocks of similar ages from different localities. If identical groups of fossils are found in sedimentary rocks from different localities, the rocks must be of the same age. Figure 4–15 shows how geologists use fossils and faunal succession to correlate sedimentary rocks.

Using correlation, geologists have constructed a **geologic column**, a diagram showing rocks from all over the world that together represent most of the Earth's history.

The geologic column is commonly combined with the **geologic time scale** in a single table (Table 4–2). For convenience, geologic time is subdivided. The largest time units are **eons**, which are divided into **eras**. Eras are subdivided in turn into **periods** (the most commonly used time unit), which are further subdivided into **epochs**.

The geologic column and time scale were constructed on the basis of relative age long before radiometric dating made it possible to measure absolute time. With the advent of radiometric dating, absolute ages were added to the geologic column and time scale, as shown in Table 4–2.

Notice that all of the Earth's history is divided into four eons. The most recent is called the **Phanerozoic eon**. The word *Phanerozoic* derives from ancient Greek roots meaning "evident life." The Phanerozoic eon encompasses the most recent 570 million years of geologic time. Although this is only about 12 percent of Earth history, the Phanerozoic eon takes up most of the time scale and contains all of the named subdivisions. Earlier eons are not subdivided at all, even though together they are about nine times as long as the Phanerozoic. Commonly, these earlier eons together are called **Precambrian** time because they preceded the Cambrian period, when fossil remains first became abundant. Why is Phanerozoic time

Figure 4–15 The use of fossils and the principle of faunal succession to correlate sedimentary rocks from different localities. Sedimentary rocks containing identical fossils are interpreted to be of the same age and therefore are correlated. Layer D is correlated among all four localities because it contains identical fossils. Layer B is missing from locality 2 because A sits directly on C; this is a disconformity demonstrated by faunal succession. Either layer B was never deposited here or it was eroded away.

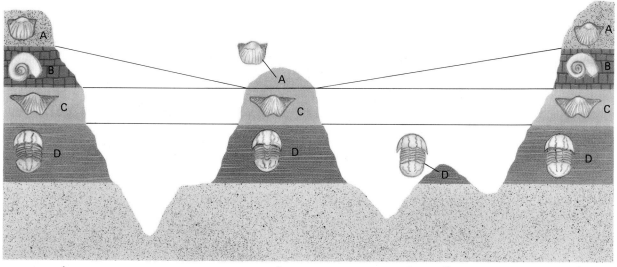

T A B L E   4 – 2

The Geologic Column and Time Scale*

Time Units of the Geologic Time Scale				Distinctive Plants and Animals
Eon	Era	Period	Epoch	
Phanerozoic Eon (*Phaneros* = "evident"; *Zoon* = "life")	Cenozoic Era	Quaternary	Recent or Holocene	Humans ["Age of Mammals"]
			Pleistocene	
		Tertiary — Neogene	Pliocene — 2	Mammals develop and become dominant
			Miocene — 5	
			— 24	
		Tertiary — Paleogene	Oligocene — 37	
			Eocene — 58	Extinction of dinosaurs and many other species
			Paleocene	
			— 66	
	Mesozoic Era	Cretaceous — 144		First flowering plants, greatest development of dinosaurs ["Age of Reptiles"]
		Jurassic — 208		First birds and mammals, abundant dinosaurs
		Triassic — 245		First dinosaurs
	Paleozoic Era	Permian — 286		Extinction of trilobites and many other marine animals ["Age of Amphibians"]
		Carboniferous — Pennsylvanian — 320		Great coal forests; abundant insects, first reptiles
		Carboniferous — Mississippian — 360		Large primitive trees
		Devonian — 408		First amphibians ["Age of Fishes"]
		Silurian — 438		First land plant fossils
		Ordovician — 505		First fish ["Age of Marine Invertebrates"]
		Cambrian — 570		First organisms with shells, trilobites dominant
Proterozoic				First multicelled organisms
		Sometimes collectively called Precambrian		
2500				
Archean				First one-celled organisms
3800				Approximate age of oldest rocks
Hadean				Origin of the Earth
4600 ±				

*Absolute ages of boundaries are based on radiometric dating. The origins of some of the names of the time divisions are given.

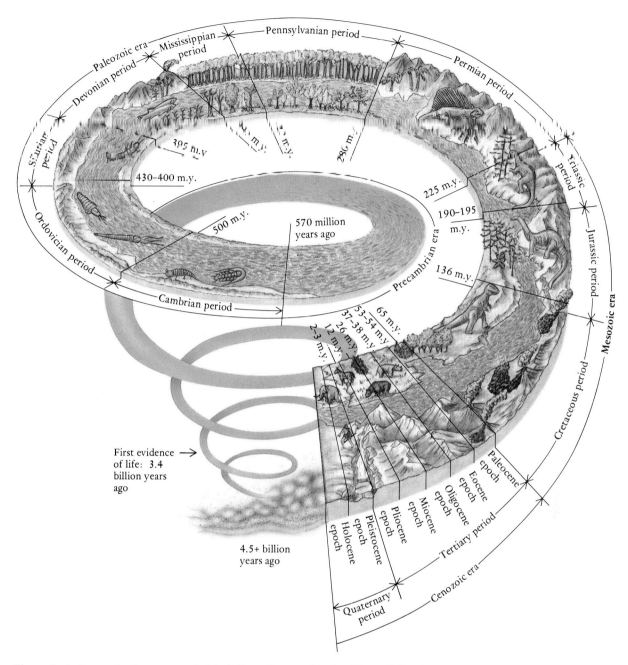

The geologic time scale shown as a spiral to indicate the great length of Precambrian time.

subdivided so finely? Or, conversely, what prevents sub-division of the earlier eons?

The subdivisions of geologic time are based largely upon fossils found in rocks that formed during each inter-val. As we saw in the previous section, the diversity and numbers of plants and animals bloomed suddenly at the beginning of Phanerozoic time, 570 million years ago. In addition, the sizes of plants and animals increased

abruptly at the same time. Phanerozoic time is so finely subdivided because fossils are abundant in rocks of that eon. Precambrian time is less subdivided because life was scarce and, hence, fossils are uncommon. In addition to the abundance of Phanerozoic fossils, there is another reason that the history of Phanerozoic time is better known than that of earlier times. It is a simple matter of probability that older rocks have a greater chance of being metamorphosed, deformed, or eroded, obliterating evidence of their early histories. Phanerozoic rocks are younger and thus more commonly well preserved.

4.7 Earliest Eons of Geologic Time: Precambrian Time

Only a few Earth rocks are known that formed during the **Hadean eon** (Greek for "beneath the Earth"; 4.6 to 3.8 billion years ago). It is the earliest time in the Earth's history. That the Earth even existed during Hadean time is inferred from radiometric age dates from a few Earth rocks and from extraterrestrial rocks. No fossils of Hadean age are known. It may be that traces of Hadean life were destroyed by erosion or metamorphism or that life had not evolved in Hadean time. In any case, Hadean time cannot be subdivided using fossils.

The **Archean eon** (Greek for "ancient"; 3.8 to 2.5 billion years ago) was when nearly all of the oldest rocks on Earth formed. Most rocks of this age are igneous or metamorphic, although a few Archean sedimentary rocks are preserved. Some contain microscopic fossils of bacteria, and a few contain fossil algae. Life apparently began sometime during the Archean eon, although fossils are neither numerous nor well preserved enough to permit fine tuning of Archean time.

The evolution of living organisms from nonliving molecules is improbable. No one has been able to reproduce this transition in the laboratory, and no one has found evidence that it has occurred anywhere else in the Solar System. Yet life evolved relatively early in the history of the Earth. Our planet formed about 4.6 billion years ago. For 100 to 900 million years the surface was too hot for life to exist. The earliest fossils date to about 3.5 billion years ago. Perhaps older fossils exist and have not been found, or existed and were destroyed by Earth processes. In any case, life formed within roughly 300 million to one billion years after the Earth's surface cooled. During the next three billion years, organisms remained predominantly single-celled and microscopic.

Rocks of the **Proterozoic eon** (Greek for "earlier life"; 2.6 to 0.57 billion years ago) contain abundant fossil algae and other simple plants. Animal fossils have been found in the youngest Proterozoic rocks in Australia, British Columbia, and several other parts of the world. The animals lacked hard parts such as shells and skeletons. Consequently, the fossils consist of imprints of their soft bodies in shale. Preservation of such delicate remains requires a gentle environment in which the soft bodies are not consumed by scavengers, lost to decay, or destroyed by sedimentary processes. Such an environment must have been uncommon. Therefore, similar animals may have lived in other places and even at earlier times, but were not preserved.

Then, within a very short time, at the end of the Proterozoic eon and the start of the Phanerozoic eon, large, multicellular plants and animals evolved abruptly, proliferated, and dominated the Earth.

This sudden explosion of life is a puzzle. Why did it take much longer for multicellular organisms to evolve from single-celled organisms than it did for life to evolve from nonliving molecules? Why have multicellular

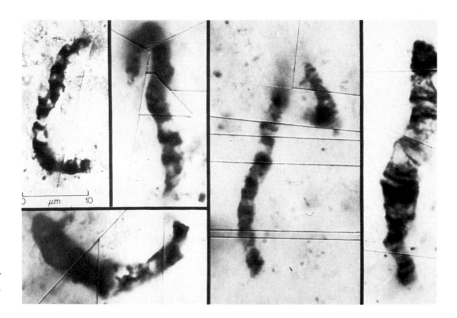

Tiny, 3.5-billion-year-old, bacteria-like fossils from western Australia. *(J. W. Schopf and B. M. Packer)*

organisms existed for only the most recent 15 percent of the time that life has existed on the Earth? One theory is that the Earth's early atmosphere contained only a small amount of oxygen. Although single-celled organisms could thrive in this environment, multicellular ones could not evolve until the oxygen concentration increased to suitable levels. This theory will be discussed further in Chapter 13.

4.8 The Phanerozoic Eon: The Last 570 Million Years

Sedimentary rocks of the **Phanerozoic eon** contain abundant fossils. Four changes occurred at the beginning of Phanerozoic time that greatly improved the fossil record:

1. Animals evolved shells and skeletons.
2. The total number of individual organisms preserved as fossils increased greatly.
3. The total number of species preserved as fossils increased greatly.
4. The sizes of individual organisms increased from microscopic to macroscopic.

Shells and skeletons are much more easily preserved than plant remains and soft body tissues. Thus, in rocks of earliest Phanerozoic time and younger, the most abundant fossils are the hard, tough shells and skeletons.

Apparently something happened about 570 million years ago that resulted in the sudden appearance of animals with shells and other hard parts. No one knows why hard parts developed, but it is easy to speculate that a shell allowed an animal to survive in places where soft organisms could not—for example, in the surf zone of a beach or in a spot surrounded by carnivorous predators. In this scenario, animals with shells had such a survival advantage that they rapidly dominated the world. Perhaps their rise was also related to an increase in atmospheric oxygen.

The subdivision of the Phanerozoic eon into the Paleozoic, Mesozoic, and Cenozoic eras is based on the most abundant plants and animals of each era. The three eras are in turn subdivided into periods, the time unit most commonly used by geologists. The names of periods vary in origin. Some are based on characteristics of rocks that formed during those particular periods. For example, the Cretaceous period is named from the Latin word for "chalk" (*creta*), after chalk beds of this age in Europe. The Mississippian and Pennsylvanian periods together are sometimes called the Carboniferous period for the extensive coal beds that existed worldwide at that time.

Other periods bear names taken from geographic localities where rocks of that age were first studied. For example, the Jurassic period is named for the Jura Mountains of France and Switzerland. The Cambrian period is named for *Cambria*, the Roman term for Wales, where rocks of this age were first described.

The Paleozoic Era

Sedimentary rocks that formed during the **Paleozoic era** (Greek for "old life"; 570 to 245 million years ago) contain fossils of animals and plants that evolved early, including invertebrates, fish, amphibians, reptiles, ferns, and cone-bearing trees. In early Paleozoic time, life was almost completely confined to the oceans. The beds of warm, shallow seas were populated with snail-like gastropods, worms, brachiopods that looked like clams and oysters, colonies of corals, and horseshoe crab–like trilobites. Algae and other simple plants shared the sea floor with these animals. Fish and sharks swam the oceans, increasing in diversity and numbers until they dominated the seas in late Paleozoic time. Amphibians and reptiles evolved in late Paleozoic time. Land plants first appeared about 440 million years ago, and scale trees, ferns, ginkgoes, and conifers evolved rapidly to cover land and form great coal swamps by late Paleozoic time (Fig. 4–16).

The Paleozoic era began with a sudden increase in types of organisms and it ended with a sudden decrease in types of organisms. Thus, the end of Paleozoic time is defined by a **mass extinction**, in which half of all families of organisms suddenly disappeared.

The Mesozoic Era

Sedimentary rocks of the **Mesozoic era** (Greek for "middle life"; 245 to 66 million years ago) contain the remains of life forms that followed those of the Paleozoic. Oysters, clams, corals, fish, and sharks populated Mesozoic seas. Ferns and conifers grew in great forests in Triassic and Jurassic times, and flowering plants, including predecessors of modern hardwoods, appeared in Cretaceous time. A wide variety of insects also emerged during the Cretaceous period. But dinosaurs ruled the Mesozoic landscape. Evolving from small, two-legged Triassic reptiles, they developed rapidly into many species of all sizes and shapes (Fig. 4–17). Although conventional wisdom portrays them as green, scaly, cold-blooded, reptile-like animals, modern evidence suggests that many were warm-blooded and either hairy or covered with feathers like modern birds, and that they cared for their young after

(Text continues on p. 97)

A

Figure 4–16   (A) In Paleozoic time, the sea teemed with a wide variety of life.   *(Smithsonian Institution)*
(B) The land was inhabited by reptiles and amphibians and covered by ferns, gingkoes, and conifers.   *(Ward's Natural Science Establishment)*

B

Restoration of a scene in North America during late Jurassic time. *(Smithsonian Institution)*

Figure 4–17 Evolution of the dinosaurs. *(From Colbert, E. H. Evolution of the Vertebrates. New York: John Wiley & Sons, 1969.)*

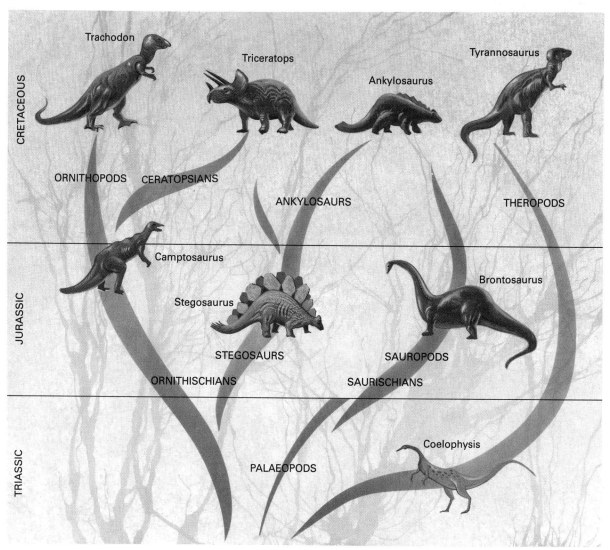

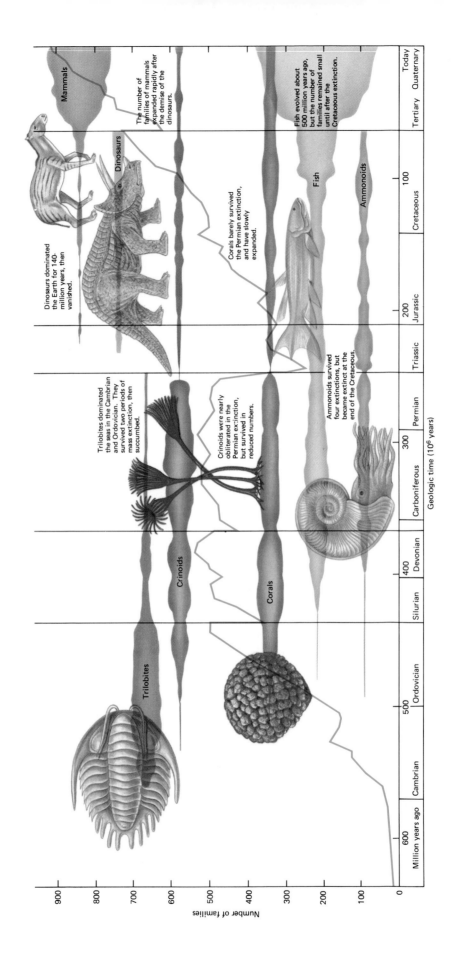

The number of mammals families expanded rapidly after the demise of the dinosaurs.

Fish evolved about 500 million years ago, but the number of families remained small until after the Cretaceous extinction.

Dinosaurs dominated the Earth for 140-million years, then vanished.

Corals barely survived the Permian extinction, and have slowly expanded.

Mammals

Dinosaurs

Fish

Ammonoids

Trilobites dominated the seas in the Cambrian and Ordovician. They survived two periods of mass extinction, then succumbed.

Crinoids were nearly obliterated in the Permian extinction, but survived in reduced numbers.

Ammonoids survived four extinctions, but became extinct at the end of the Cretaceous.

Trilobites

Crinoids

Corals

Today | Quaternary | Tertiary | 100 | Cretaceous | 200 | Jurassic | Triassic | Permian | 300 | Carboniferous | 400 | Devonian | Silurian | Ordovician | 500 | Cambrian | 600 | Million years ago

Geologic time (10^6 years)

Number of families
900 | 800 | 700 | 600 | 500 | 400 | 300 | 200 | 100 | 0

Earth Science and the Environment

Extinction in Modern Times

The fossil record shows us that, throughout the history of life, the average rate of evolution of new species has been slightly greater than the rate of extinction. Therefore, the number of species has gradually increased. No one knows exactly how many different species exist today, but approximately 1.5 million have been identified and many more are thought to exist. Estimates vary from five million to more than 50 million. Insects are by far the most numerous class and outnumber all other types of organisms combined.

Within the past 50 to 100 years, a tiny blink in geologic time, extinction rates have increased dramatically. At present, 300 to 400 species of mammals and 400 to 500 species of birds are listed as endangered. These numbers are dwarfed by the numbers of endangered plants and smaller animals such as insects, mites, and worms. If present rates continue, one million species may become extinct by the year 2000. This is a loss of 100 species per day for the remainder of this century. The actual number will never be known with certainty because many tropical species are becoming extinct even before they can be classified! Some biologists think that the rate of extinction is greater today than at any other time since the mass extinction 65 million years ago that included the demise of dinosaurs.

It is easier to obtain information about the present than about the past. We know that the climate has been relatively stable for the past century and that no global catastrophic events have occurred. In other words, the rapid increase in the rate of extinction appears to have no known natural cause. We are virtually certain that human activities and technological advances are responsible. For example, vast tracts of tropical rainforest are being eliminated. Many species that lived in these regions have already become extinct, and if current trends continue, more will succumb.

What are the effects of these extinctions, beyond the loss of the species? Some ecologists claim that, except for an aesthetic loss, the disappearance of a million species will not measurably affect Earth. Others disagree. Many endangered organisms may be economically important. If they survived, some of the plants could be cross-bred with modern food crops to produce healthier, more productive varieties. Others might produce valuable natural medicines.

Beyond economics, even more disturbing questions arise. For example, is it possible that the deforestation of vast tracts of rainforest may initiate a chain of events that will alter conditions on the Earth's surface? The composition of the Earth's atmosphere is determined partly by respiration of plants and animals. It is conceivable that a loss of large tracts of rainforest might lead to changes in atmospheric composition. In turn, atmospheric composition controls global climate. Thus, loss of large areas of rainforest might ultimately trigger climatic changes. A few decades ago such questions were relegated to science fiction. Today the questions are discussed seriously by ecologists. Our current understanding of biological and physical systems is insufficient to predict the effects of a human-caused mass extinction with any degree of certainty.

---

they hatched. Early mammals and birds also appeared in Mesozoic time, although they were overshadowed by the dinosaurs.

Sixty-five million years ago, all species of dinosaurs suddenly became extinct. This event is remarkable because it included not only hundreds of species of dinosaurs but many other plants and animals as well. At least one fourth of all animal species on Earth, both marine and terrestrial, became extinct at the same time. This

◀————————————————————

Figure 4–18 The green line shows variations in the number of families of organisms during the past 620 million years. Mass extinctions appear as sudden drops in the line. Modern extinctions are not shown.

extinction *defines* the end of Mesozoic time and the beginning of a new era. Figure 4–18 shows that at least four mass extinctions have taken place since the beginning of Paleozoic time. Why do mass extinctions occur?

In 1977, a father-and-son team, Walter and Luis Alvarez, found a sooty, thin, 65-million-year-old clay layer containing 50 to 100 times more of the element iridium than is normal in such rock. Iridium is rare in the Earth's crust but abundant in meteorites. This same 65-million-year-old, iridium-rich, sooty clay has now been found at several locations around the Earth.

Alvarez and Alvarez suggested that 65 million years ago, a giant meteorite 10 to 25 kilometers in diameter hit the Earth. Astronomers calculate that an object of that size would have struck Earth with the energy of 10 to 100 million hydrogen bombs. The impact vaporized both the meteorite and the Earth's crust at the point of impact, igniting massive fires. Soot from the fires and iridium-rich meteorite dust rose into the upper atmosphere, circling the globe. The thick, dark cloud blocked out the sun. Surface waters froze. Many plant and animal species froze and starved to death. Then, the sooty, iridium-rich dust settled to Earth to form the distinctive clay layer.

Other scientists have found evidence of huge meteorite impacts that occurred exactly 65 million years ago both in western India and in the Caribbean Sea. Geologists continue to find evidence of giant meteorite impacts that coincide with other mass extinctions, including the one that marks the end of Paleozoic time.

Many scientists disagree with the meteorite theory of mass extinctions, and several other hypotheses have been offered to explain the sudden and simultaneous disappearances of large numbers of plants and animals. The extinctions occurred on a global scale and, hence, a global explanation must be sought. Most other theories invoke sudden changes in global climate or in sea level.

Global climate is regulated by several factors, including atmospheric composition and the orientation of continents and ocean basins. According to one hypothesis, increased volcanic activity 65 million years ago injected large quantities of ash into the atmosphere. In turn, this ash blocked out significant amounts of sunlight and led to rapid global cooling, which caused the late Mesozoic mass extinction.

Alternatively, some evidence indicates that the Arctic Ocean was landlocked late in the Mesozoic era. At the end of the Mesozoic era the land barrier was broken by movement of tectonic plates. This renewed circulation of ocean waters may have altered global climate and led to mass extinction.

Yet another theory suggests that tectonic activity led to a global lowering of sea level. According to this idea,

Figure 4–19   One interpretation of human evolution.

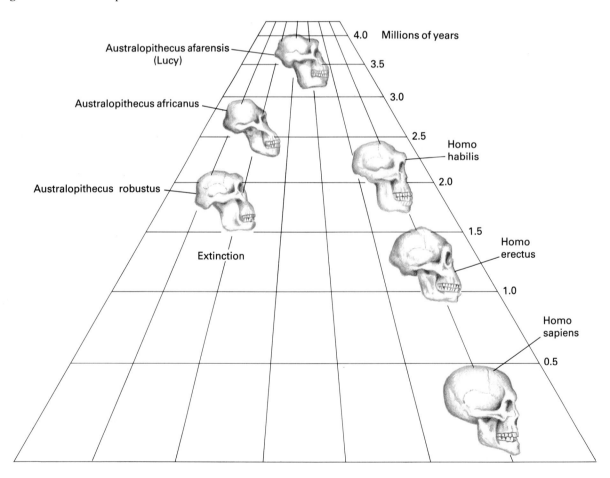

as sea level dropped, coastal regions became dehydrated, food chains were disrupted, and animals died. In addition, the lowering of sea level may have caused a change in climate.

The Cenozoic Era

With the dinosaurs gone at the beginning of the Cenozoic era (Greek for "recent life"; 66 million years ago to the present), mammals evolved rapidly to dominate land. Grasses evolved in Miocene time, creating lush prairies and savannahs that supported large herds of newly evolving grazing animals, such as horses. Carnivores evolved to prey on the expanding herds. Only a few reptiles survived the extinction at the end of Mesozoic time. Birds, possibly the only surviving evolutionary descendants of dinosaurs, spread rapidly. In the seas, reef-forming corals proliferated throughout warm, shallow water regions, and tiny marine microorganisms thrived.

About five million years ago, near the end of Miocene time, creatures resembling modern humans more than apes appeared in India. Later, between three and one million years ago, several separate human-like lineages developed and are preserved as fossils in East Africa (Figure 4–19). Primitive men and women had evolved between a million and 500,000 years ago and had spread across Europe and Africa by 400,000 years ago. They used fire, made tools, and lived in caves. Neanderthal people roamed Europe, Africa, and the Middle East by 100,000 years ago. They made fine tools and had brains larger than those of modern humans. By 30,000 years ago, early modern humans, called Cro-Magnon people, had replaced the Neanderthal people in Europe. They crafted well-made weapons and tools, developed religions, and created art. They are our immediate ancestors.

SUMMARY

•

Geologists measure time in two different ways. **Absolute time** is measured in years. **Relative time** lists the order in which events occurred, without regard to the amount of time separating them. **Radiometric age dating** measures absolute time, based on the decay of a radioactive **parent isotope** to a **daughter isotope** at a constant, known rate expressed as the **half-life** of the parent. Absolute time can be measured because the daughter accumulates in rocks, and thus the amount of daughter is proportional to the age of the rock. Measurements of relative time are based on the principles of **original horizontality, superposition, cross-cutting relationships**, and **faunal succession**.

Sedimentary rocks are **conformable** if they were deposited without interruption; an **unconformity** represents an interruption of deposition of sediment. Types of unconformities include **disconformities, angular unconformities**, and **nonconformities**.

Paleontology is the study of past life. **Fossils** are the remains and traces of prehistoric life. They form by **preservation, mineralization**, formation of **molds** and **casts**, and **carbonization**. A **trace fossil** is any track, burrow, or other mark made by an organism. Much can be learned about the geologic history of rocks by studying their fossils.

Correlation is the matching of rocks of similar ages from different localities. Worldwide correlation of rocks of all ages has allowed geologists to construct the **geologic column**, a composite record of rocks formed throughout the Earth's history. The **geologic time scale** is divided into **eons, eras, periods**, and **epochs**. The **Phanerozoic eon** is the most recent and covers the last 570 million years. It is easily subdivided on the basis of abundant fossils found in rocks that formed during that time. The three earlier eons, the **Hadean, Archean**, and **Proterozoic** eons, contain fewer subdivisions because of their scarcity of fossils. They are collectively called **Precambrian** time.

KEY TERMS

•

Uniformitarianism   *76*
Absolute time   *76*
Relative time   *76*
Radioactivity   *77*

Isotope   *77*
Parent isotope   *78*
Daughter isotope   *78*
Half-life   *78*

Radiometric age
  dating   *78*
Principle of original
  horizontality   *80*

Principle of
  superposition   *80*
Unconformity   *83*
Disconformity   *83*

REVIEW QUESTIONS

•

1. What are two ways in which geologic time is expressed? How do they differ, and how are they similar?

2. What is radioactivity?

3. What is a stable isotope? An unstable isotope?

4. Describe the similarities between measurement of time with a calendar and measurement of time by radiometric age dating.

5. What is the half-life of a radioactive isotope? How is the half-life used in radiometric dating?

6. Some radioactive isotopes are useful for measuring geologically short times; others can be used only to measure longer times. Why?

7. Make a simple sketch of a cross-section showing how the principle of original horizontality can be used to determine relative time in folded sedimentary rocks.

8. Make another sketch showing the use of the principle of superposition in unfolded sedimentary rocks.

9. What important assumption is made when using the principle of superposition?

10. Sketch cross-sections showing a disconformity, an angular unconformity, and a nonconformity.

11. Describe several different ways in which a fossil can form.

12. Explain how the principle of faunal succession can be used to correlate sedimentary rocks from two different continents.

13. Construct a simple table listing the four eons of geologic time. Make the space allotted to each eon proportional to its length. Which eon is longest? Which is shortest?

14. Explain why the Phanerozoic eon contains so many subdivisions, whereas Precambrian time, which is about nine times longer, has so few.

15. Describe the Earth during Hadean time.

16. Describe the Earth during Archean time.

17. Describe the Earth during Proterozoic time.

18. Describe the Earth during Phanerozoic time.

19. In what ways do the amounts of detail in your answers to the four previous questions vary from Hadean to Phanerozoic time? Why does the amount of detail change?

20. Describe the Earth's life during Paleozoic, Mesozoic, and Cenozoic times.

DISCUSSION QUESTIONS

•

1. Devise one or two metaphors or analogies for geologic time in addition to that in the text. Locate some of the important time boundaries in your analogies.

2. What geologic events are represented by a radiometric age date from a biotite flake found in (a) granite, (b) schist, (c) sandstone?

3. What information could be obtained from a fossil formed by preservation that could not be obtained from one formed by mineralization?

4. What kinds of information regarding the physical characteristics and habits of a particular species of dinosaur might be obtained from its tracks and other trace fossils?

5. What sequence of geologic events is recorded by an angular unconformity? A disconformity? A nonconformity? What can you infer about the relative timing of each set of events?

6. If you were attempting to correlate a sandstone bed in New York with another layer of sandstone in Ohio, what tools and criteria would you use to determine whether the two sandstone beds were equivalent?

7. Why are the periods of the Phanerozoic eon of such different lengths?

8. No traces of life are known in Hadean rocks, and abundant fossils are found in rocks of the Proterozoic eon. Therefore, life must have originated sometime during the Archean eon. Discuss what the Archean environment must have been like in order for life to have evolved in that time.

John Horner

John R. Horner was born in 1946 in Shelby, Montana. His father operated a large gravel quarry and his mother was a homemaker. He developed an early interest in fossils and geology during youthful forays on the High Plains around Shelby, where the remains of marine animals, dinosaurs, and Ice Age mammals are plentiful. His study of geology at the University of Montana was interrupted by service in Vietnam with the U.S. Marine Corps from 1966 to 1968. He returned to the University of Montana in 1968. He was Research Assistant in the Department of Geological Sciences at Princeton University from 1975 to 1982 and also was Museum Scientist at the American Museum of Natural History in New York City from 1978 to 1982. While at Princeton and the American Museum, he returned occasionally to the area near his home in Montana during summer field seasons to explore and recover the fossils of large animals that he had noticed as a child. Many of his discoveries are now prized specimens in the Princeton and American Museum of Natural History collections. During this time, he began to concentrate on dinosaurs, and he realized that the rocks near Shelby contain abundant and well-preserved dinosaur remains. In 1982 he was appointed

> All our evidence suggests that dinosaurs were warm-blooded just like we are.

Curator of Paleontology at the Museum of the Rockies in Bozeman, Montana, where he has continued his research on the nature and living habits of dinosaurs. He was awarded an Honorary Doctorate of Science from the University of Montana in 1986.

He is author of many publications, most dealing with dinosaurs and their social habits, and of two popular books about dinosaurs. He has received several large National Science Foundation grants in support of his research. In 1986 he was awarded the MacArthur Foundation Fellowship, commonly known as the "Genius Award." It is a large cash grant given to unusually creative and productive individuals. He has been featured in many national news and science magazines, including *Readers' Digest*, *Natural History*, *Omni*, *Life*, *National Geographic*, *People*, and *U.S. News and World Report*.

●

What initiated your interest in geology and paleontology?

I really don't know how my interest in geology and paleontology started. My father owned a gravel quarry, and he says that when I was very young, I was always sorting the rocks out of the big gravel pile into what I thought were groups of different kinds of rocks.

My father would take me to places that he had ridden horseback when he was a rancher, and he would show me areas where he had found what he thought were dinosaur bones. I was eight years old when he first took me to one of these places, and I collected a couple of bone fragments. Then we went back to the same area when I was in high school, and I collected two partial dinosaur skeletons.

Through high school, I was really interested in science. I spent most of my time working on science projects. In my senior year I did a project on fossils. I was trying to figure out why the dinosaur remains found in the Judith River Formation in Montana were different from the ones found in the Judith River Formation in Canada. On the Canadian side, there are articulated dinosaurs and lots of duck-billed bones. But even at that time I knew that most of my dinosaur bones from Montana were flatheaded kinds and were strewn all over. I never could find an articulated one. I saw this difference as a problem, but I was unable to resolve it then. In fact, I didn't resolve it until I published a paper just two years ago on it. Now we know that the dinosaur bone beds in Alberta are of a slightly different age from those in Montana, so they are stratigraphically not the same. They also had different environments of deposition.

•

You are now one of the most highly visible and best known paleontologists/geologists in the country. Yet in high school and as a university student you got low grades. How do you reconcile your success as a scientist with your grades as a student?

My academic record at the university was even worse than my high school grades. After my first year and a half at the University of Montana, my cumulative grade point average was a 0.06. If I went to college now, I would still have the same problem. I have a learning disability called dyslexia. It's a problem that didn't stop me from wanting to learn, but it did stop me from

being able to. It made it almost impossible for me to absorb information that was assigned to me. I just couldn't assimilate the material fast enough. It got to the point where I didn't care what the grade was. I would learn as much as I could, and if at the end of the course that was D work for the professor, then I got a D. If I felt that the course seemed really interesting and if it was something that I thought I could get more out of, I'd just take it again. If I thought I knew enough for what I wanted to do, then I wouldn't take it again regardless of my grade. I didn't know what the problem was at the time, but I did know that I really wanted to learn.

I spent a year and a half at the university, and then in 1966 I got drafted into the Marine Corps. I served with the Special Forces in Vietnam. I got out of the Marine Corps in February 1968 and went back to the University of Montana. My grades were a lot better when I came back, and I started taking zoology and geology courses. But my grades were still lousy by university standards, and that's when they started throwing me out of school every quarter. Each time I could always demonstrate that I was bringing my grades up so they had to let me back in. I took every geology course that was offered and all the zoology courses that looked appropriate, as well as botany, physical geography, and a few anthropology courses. I never finished my undergraduate degree, but I did eventually get a doctorate from the University of Montana.

•

What did you do after college? What was your first job?

After I took all the courses I was interested in, I went back to Shelby where my brother and I bought the gravel company from my father,

who was retiring. I had been there for about a year and a half when I began sending letters to all the museums in the English-speaking world to apply for jobs from janitor up to curator. I didn't really care what it was, I would have taken anything. I got three responses: from the Los Angeles County Museum for the position of Chief Preparator, from the Royal Ontario Museum in Toronto for Assistant Curator, and from Princeton for a Preparator and Research Assistant. I took the job at Princeton and worked there for seven years.

In 1979, I was talking to the Director of the Museum of the Rockies and he told me that he had heard about all the dinosaur eggs we were finding in Montana around Choteau and Shelby. He asked if we would donate some to the Museum of the Rockies so I donated a clutch of eggs. In 1981, I saw the Director again, and I told him I was from Montana and that I really wanted to work there. He offered me the Curator of Paleontology position at the Museum of the Rockies and I came home.

•

Tell us about your work at the Museum of the Rockies.

The paleontology crew at the museum includes one curator (me), four full-time preparers, a thin-section histology technician, an illustrator, a collection manager, a computer illustrator who does all of the mathematical simulation, six graduate students, and a staff of about 15 part-time people.

My research is primarily on dinosaur behavior, ecology, and evolution. My graduate students are all geologically oriented to do field studies in stratigraphy and sedimentology. One of the students is doing comparative studies of different kinds of bone beds we have found. A bone bed is one geologic horizon on which lots of specimens occur together. For example, one at Choteau appears to be a volcanic ash kill, and our evidence shows that over 10,000 animals are buried there. All of the bone beds we work on cover at least one square mile.

•

Much of your work has changed the traditional views of dinosaurs, how they behaved, and even what they were. Tell us what the traditional view is and how your work has affected this view.

Originally, scientists decided, on the basis of certain cranial features, that the dinosaurs should be classified as reptiles. Once they had been placed in this group, there were certain characteristics, certain little labels, that were automatically assigned to dinosaurs simply because they were called reptiles. For example, modern reptiles are cold-blooded, therefore dinosaurs were considered cold-blooded. Modern reptiles are slow moving, so dinosaurs were probably slow. Modern reptiles drag their tails, so dinosaurs must have dragged their tails.

You have to realize that in the early days of dinosaur discoveries, people didn't really study dinosaurs; they collected their remains for museums. It wasn't until the last 15 years or so that people started to consider how dinosaurs actually lived. In those 15 years we've come to find out that the original classification is probably wrong. Dinosaurs don't belong to the reptile group. They are much more like birds, and it is likely that modern day birds evolved from dinosaurs.

Most reptiles dig a hole in the ground, lay their eggs in it, cover it up, and then leave. So it was assumed dinosaurs did the same thing. But we now know that dinosaurs, like modern birds, put a lot of time

and care into building a nest, laying the eggs, and guarding the eggs. Our evidence shows that dinosaurs even guarded their young after they hatched by herding or flocking in large groups. These large groups of dinosaurs were very similar to what we observe in modern herding animals or flocking birds, which are not just aggregations of animals, but actually structured groups with certain individuals in charge.

The animals that we find in large groups, such as the duck-billed dinosaurs and the horned dinosaurs, all have some type of cranial display features such as horns on their heads. We know that the horns of modern mammals are a primary adaptation for determining hierarchies within a society. Modern horned animals, such as elk, live in big groups, and generally the males

use their horns for male-to-male combat to determine the individual hierarchy. We see a similar thing with the horned dinosaurs.

•

What would large herds of dinosaurs eat? How could so many of them live in a square mile? What was their environment like?

Well, one of the interesting things is that the dinosaurs that lived in large herds, hadrosaurs and ceratopsians, did not evolve until Late Cretaceous time. That coincides with the evolution of the angiosperms, deciduous plants that can be stripped one season and still grow back the next. So, I don't think the large groups or herds existed before angiosperms appeared. But once these

plants existed as a food source, then all that's required of these big groups is for them to migrate with the seasons. About 75 to 80 million years ago, the western part of North America was a linear continent extending north to south. The Rocky Mountains were young and were actively building up. So there was a mountain barrier on the west and an ocean to the east with a coastal plan in between. The dinosaurs would migrate north to south on that coastal plan with the seasons. For their size and stride lengths, it appears that they could have easily walked 1500 to 2000 miles each year following seasonal shifts in temperature and food supply.

The only constraint on the migrating dinosaurs was the nesting period. The dinosaurs had to wait at least a month somewhere for the incubation of their eggs. Most of the incubation, or growth of the fetuses, may have occurred in the mothers' bodies prior to egg laying. After the eggs were laid, it was probably a relatively short time (possibly only two or three weeks) before the eggs hatched. Another three or four weeks at the most would have been needed for the nesting period. During this short time the young would grow large enough to walk with the adults. We're guessing from preliminary information that duck-billed dinosaurs hatched out of their eggs at about 18 to 20 inches long and grew to about 45 to 50 inches long, possibly more, by the time they left the nest. But at that same growth rate, they would have grown to 9 to 12 feet the first year. The maximum size of an adult is about 35 feet. This suggests that the growth rate was a little faster than an ostrich, but a lot slower than most birds.

●

What do you think about the traditional view of dinosaurs being green and scaly as opposed to more recent suggestions that they were furry, hairy, or covered with feathers?

I think the babies had some kind of downy cover. All our evidence suggests that dinosaurs were warm-blooded just like we are. In fact, they were probably warmer blooded that we are. Their fast growth rate suggests a very high metabolism, suggesting a relatively warm internal body temperature. So, the babies had to have some kind of insulatory mechanism to keep the heat in. But our bone histology shows that the rapid growth ended at about 20 feet in length. Then we see rest lines in the bones suggesting that the metabolism slowed way down. At this point, the dinosaurs were still creating heat, but much less than before.

So this means that their rate of food consumption decreased when they became large. This would address those who argue that dinosaurs could not be warm-blooded because there wouldn't be enough food for many animals of that size. If there are hundreds of dinosaurs or possibly more nesting in one area, there would have to be a large food supply for the young, and the adults would be eating too. Well, I don't think the adults were eating at all. It's like when birds are feeding their young—they don't eat. They just haul in food for the babies. So all that was needed was a supply large enough for the young, and it really makes little difference how far the adults had to walk to get food. For example, penguins often go a hundred miles to get food for their babies.

●

What good is a study of geology to someone who doesn't plan to become a geologist?

I think our environment is going to be the next century's biggest topic. And you cannot have a good handle on the environment unless you understand both biology and geology. So as the environment becomes more and more of an issue, there are going to be more and more jobs in the field of environmental geology. I think we are going to see the pendulum swing towards fields that figure out how to save our world. We're going to have to understand animals as animals, how rivers work, and similar concepts. A strong understanding of geology and biology is what most people are going to need to address environmental concerns.

This interview was conducted by Graham Thompson, University of Montana, and appears in *Modern Physical Geology*, Saunders College Publishing, 1991.

Internal Processes

● Cerro Torre at Dawn from the summit of Fitz Roy in Patagonia   *(Galen Rowell/Mountain Light)*

Plate Tectonics

5

S cience usually creeps forward by innumerable little discoveries, each won by months or years of hard work in the field or laboratory. Occasionally, however, scientists gather all the little advances into a new idea, or a new way of looking at old ideas, and a major scientific revolution occurs. Modern Earth scientists are lucky to be in the midst of such a revolution. Its effects are as exciting and important to Earth science as Einstein's theory of relativity was to physics in the early part of this century.

The theory that has evolved from this revolution is called the **plate tectonics theory**. Briefly, it states that the Earth's outer layer is a 100-kilometer-thick shell of brittle, rigid rock called the lithosphere, which is broken into independent segments called **plates** (Fig. 5–1). The plates float on a layer of hot, soft, plastic rock called the asthenosphere. They move horizontally across the Earth's surface by gliding over the asthenosphere like sheets of ice floating back and forth on a pond.

● A satellite photo of the north end of the Red Sea and the Sinai Peninsula, northeastern Africa. The Red Sea is a spreading center that splits into the Gulf of Suez on the left, and the Gulf of 'Aqaba on the right. *(NASA)*

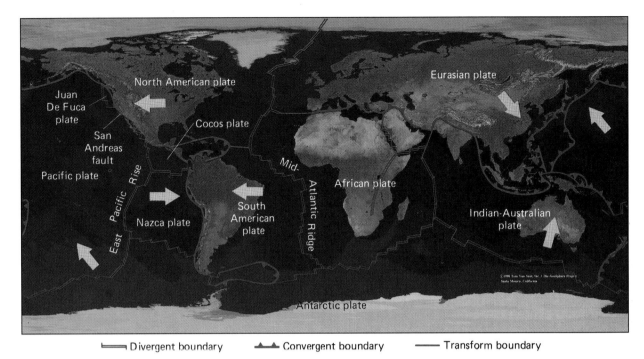

═══ Divergent boundary   ▲─▲ Convergent boundary   ─── Transform boundary

Figure 5–1 Lithospheric plate boundaries are shown in red. The major plates are the African, Eurasian, Indian-Australian, Antarctic, Pacific, North American, and South American. A few of the smaller plates are also shown. Gray arrows indicate directions of plate movement.

A **fault** is any fracture in rock along which movement has occurred. A **plate boundary** is a fault separating one plate from another. Since the plates move relative to one another as they float over the asthenosphere, they collide at some of the plate boundaries. At other boundaries, two plates move apart from one another; and at yet other boundaries, two plates move horizontally past one another. California's famous San Andreas fault is an example of this type. Because 100-kilometer-thick plates bump and grind together at a plate boundary, you can easily imagine that a boundary is a geologically active place. Earthquakes, volcanoes, mountain building, and similar geological activities and features concentrate near plate boundaries.

5.1 Alfred Wegener and the Origin of an Idea

In the early twentieth century, a young German scientist named Alfred Wegener noticed that the African and South American coastlines on opposite sides of the Atlantic Ocean seemed to fit as if they were adjacent pieces of a jigsaw puzzle (Fig. 5–2). He then launched an idea whose pursuit became one of the most fascinating chronicles in the history of science.

Born in 1880, Wegener began his career studying meteorology. While a university student, he took part in an exploratory expedition to Greenland in 1906. He developed a fascination for that ice-covered island in the North Atlantic that continued throughout his life and was eventually responsible for his premature death.

Figure 5–2 The African and South American coastlines appear to fit together like adjacent pieces of a jigsaw puzzle.

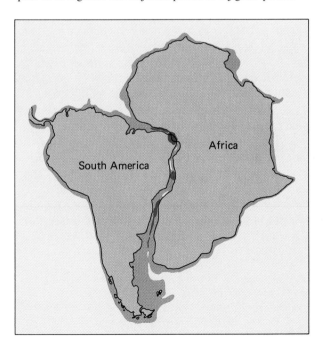

Fossil sand dunes are preserved in these desert sandstones of Zion National Park in Utah.

rocks of Ordovician through Triassic age (505 through 208 million years ago) from Africa and South America contained identical fossils. Rocks younger than Triassic age, however, showed development of different species on the two continents. Wegener reasoned that the Atlantic Ocean basin had begun to open and tear Pangaea apart at the end of the Triassic period. Rocks that had formed before this separation contained identical fossils because the plants and animals had evolved and spread throughout Pangaea. Rocks formed after Pangaea split up contained different fossils because evolution had followed different paths on the separated continents. Wegener was unable to determine the absolute time when the Atlantic Ocean began opening because radiometric dating had not yet been developed. However, we now know that the end of the Triassic period, when Pangaea started to split up, was slightly more than 200 million years ago.

Evidence from Paleoclimatology

Paleoclimatology is the study of ancient climates. As both a meteorologist and a geologist, Wegener knew that certain types of sedimentary rocks form in certain climatic zones of the Earth. Glaciers and glacial sediment, for example, concentrate in high latitudes. Deserts and the rocks that form in deserts cluster around latitudes 30° north and south. Coral reefs and coal swamps thrive in near-equatorial tropical climates. Thus, sedimentary rocks reflect the latitudes at which they formed (Fig. 5–5).

Reefs produce fossil-rich limestone beds in tropical climates. *(Larry Davis)*

Figure 5–5 Climatic zones and many rock-forming environments are closely related to latitude.

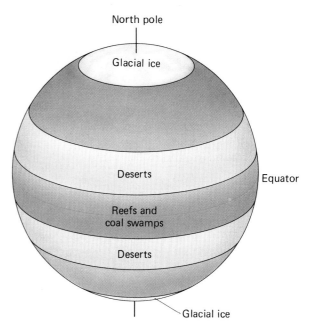

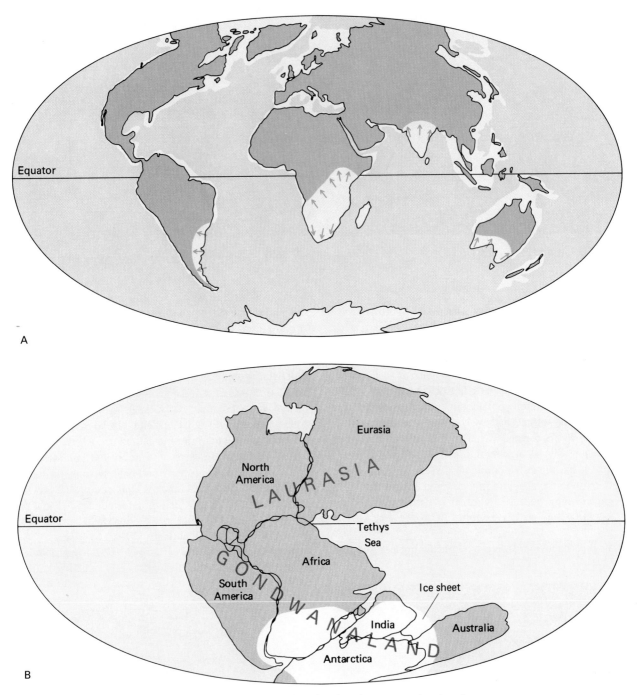

Figure 5–6 (A) Three-hundred-million-year-old glacial deposits plotted on a map showing the modern distribution of continents. Arrows show directions of ice movement. (B) The same glacial deposits plotted on a map of Pangaea.

Wegener plotted pre-Triassic sedimentary rocks that indicated climate and latitude on maps showing the modern distribution of continents. Figure 5–6A shows his map of 300-million-year-old glacial deposits. The light blue area shows how large the ice mass would have been if the continents had been in their present positions. Notice that the glacier would have crossed the equator, and glacial deposits would have formed in tropical and subtropical zones. Figure 5–6B shows the same glacial deposits plotted on Wegener's Pangaea map. Here they are neatly clustered about the South Pole. Other types of climate-sensitive sedimentary rocks are plotted on a Pangaea map in Figure 5–7. As in the case of glacial deposits, the rock distribution is more sensible on the Pangaea map than on a map showing the modern distribution of continents.

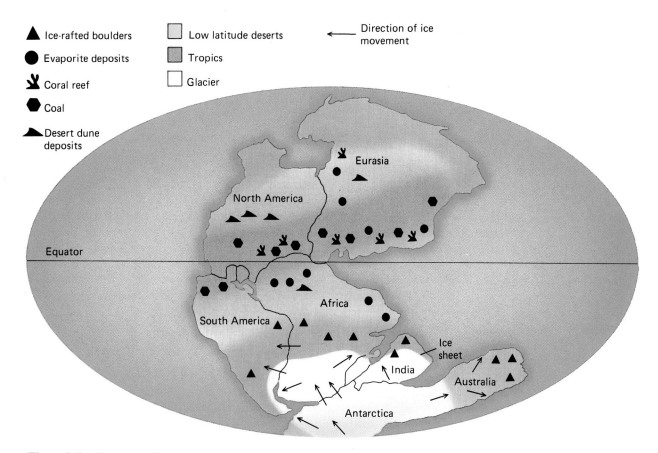

Figure 5–7   Climate-sensitive sedimentary rocks plotted on a map of Pangaea.

Geologic Evidence

Wegener noticed several instances in which an uncommon rock type or a distinctive sequence of rocks on one side of the Atlantic Ocean was identical to rocks on the other side. When he plotted the rocks on a Pangaea map, those on the east side of the Atlantic were continuous with their counterparts on the west side. For example, the Cape Fold Belt of South Africa consists of a sequence of deformed rocks similar to rocks found in the Buenos Aires Province of Argentina. Plotted on a Pangaea map, the two sequences of rocks appear as a single, continuous belt of folded rocks. Figure 5–8 shows this relationship along with two other localities of geologic continuity between South America and Africa.

Wegener's Mechanism for Continental Drift: The Tragic Flaw

Wegener's theory of continental drift was so revolutionary that skeptical scientists demanded an explanation of *how* continents could move. They wanted an explanation of the *mechanism* of continental drift. Wegener had concentrated on developing evidence *that* continents had

Figure 5–8   Locations of distinctive rock types in South American and Africa, plotted on a portion of a Pangaea reconstruction.

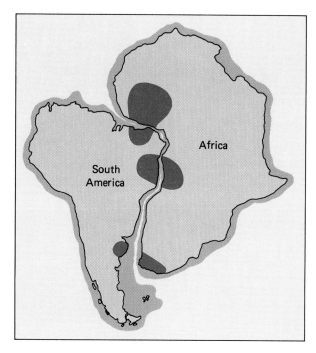

drifted, not on *how* they moved. Finally, perhaps out of exasperation and largely as an afterthought to what he considered the important part of his theory, Wegener suggested two alternative possibilities: first, that continents plow their way through oceanic crust, shoving it aside as a ship plows through water; and second, that continental crust slides over oceanic crust. These suggestions turned out to be an ill-considered and fatal step for his theory. Physicists immediately proved that both of Wegener's mechanisms were impossible. Oceanic crust is too strong for continents to plow through it. The attempt would be like trying to push a match-stick boat through heavy tar. The boat, analogous to the continents, would break apart. Furthermore, frictional resistance is too great for continents to slide over oceanic crust.

The calculations of the physicists were quickly adopted by most scientists as proof that Wegener's theory of continental drift was wrong. Notice, however, that those calculations only proved that the *mechanism* proposed by Wegener was incorrect. They did not disprove, or even consider, the huge mass of evidence that he had collected to indicate that the continents were once joined together. This obvious distinction appears to have been missed by many. During the 30-year period from about 1930 to 1960, a few geologists continued to debate the continental drift theory, but it was largely ignored by most of the scientific community.

Wegener's fascination for Greenland led him to undertake a third expedition to the ice cap in 1930. One of the shortest routes of the newly available air travel from northern Europe to North America involved flying over Greenland. To make the flights safer, it was necessary to have weather information from Greenland. Part of Wegener's mission was to establish a weather station near the center of the ice cap. In the late summer and autumn of 1930, Wegener and his companions freighted supplies and equipment by dogsled to establish the station, which was to be manned by a single meteorologist through the arctic winter. Upon looking at the meager-seeming pile of supplies and considering the long winter ahead, the man who was to stay announced that the supplies were insufficient and that he would not stay unless he had more food and gear. The days were becoming short, and winter was coming rapidly. To save the expedition, Wegener and his assistant, an Eskimo named Rasmus Willimsen, set off on Wegener's fiftieth birthday to obtain additional supplies. Wegener was never seen again. Willimsen's body was found, frozen, the following summer by a search party.

An Assessment of Wegener's Theory

As you will see in the following sections, much of the theory of continental drift is similar to plate tectonics theory. Modern evidence indicates that the continents *were* together, just as Wegener's Pangaea map showed. Furthermore, Wegener's interpretation of when Pangaea started to split up—at the end of Triassic time, about 200 million years ago—is also validated by recent data. He accumulated a vast amount of accurate data that firmly

The granite of Yosemite Valley contains iron-bearing minerals that record the orientation of the Earth's magnetic field at the time the rock cooled. *(Don Hyndman)*

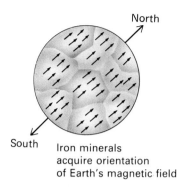

Figure 5–9 Iron-bearing minerals in a cooling igneous rock acquire a permanent magnetic orientation parallel to the Earth's magnetic field.

supported his theory that the continents were once joined together to form Pangaea and later had rifted apart. The results of his work are a contribution to science of which any geologist would be proud.

Why, then, was the theory of continental drift nearly universally rejected? Most scientists of the day were apparently stuck in their intuitive notion that rocks are too hard and solid to permit the movement of continents. When physicists proved Wegener's *mechanism* impossible, the scientific community was only too willing to reject his *data* as well. Therefore, his careful work and well-thought-out conclusions were ignored.

As we shall see in the following sections, the plate tectonics theory resurrected Wegener's theory of continental drift and offers a reasonable mechanism for the movement of continents. The resurrection occurred unexpectedly in the 1960s as a result of new information about the Earth's magnetic field and the way in which rocks preserve a record of the Earth's magnetic history. Today, most geologists recognize the excellence of Wegener's contributions.

5.2 Rock and Earth Magnetism

Early navigators learned that no matter where they sailed, a needle-shaped magnet aligned itself in a north-south orientation. Thus, they learned that the Earth exhibits magnetic behavior and has a magnetic north pole and a magnetic south pole.

All iron-bearing minerals are permanent magnets. Although their magnetism is much weaker than that of a magnet used to stick cartoons on your refrigerator door, it is strong enough to be measured with laboratory instruments.

When magma solidifies, iron-bearing minerals crystallize and become permanent magnets. When an iron-

bearing mineral cools within the Earth's magnetic field, its magnetic field aligns parallel to the Earth's field (Fig. 5–9). That is, the mineral's magnetic field points toward the North Pole just as a compass needle does. Thus, minerals in an igneous rock record the orientation of the Earth's magnetic field at the time the rock cooled.

Many sedimentary rocks also preserve a record of the orientation of the Earth's magnetic field at the time the sediment was deposited. As sedimentary grains settle through water in lakes or oceans, iron-bearing grains tend to settle with their magnetic axes parallel to the Earth's field. Even silt settling through air orients parallel to the external magnetic field.

Reversals of the Earth's Magnetic Field

The **polarity** of a magnetic field is the orientation of its positive, or north, end and of its negative, or south, end. Because many rocks record the orientation of the Earth's magnetic field at the time the rocks formed, we can construct a history of the Earth's field by studying magnetic orientations in rocks from many different ages and places. When geologists constructed a history of the Earth's magnetic field in this way, they arrived at an astonishing conclusion: **the Earth's magnetic field has reversed polarity throughout geologic history**. When a **magnetic reversal** occurs, the north magnetic pole becomes the south magnetic pole, and vice versa. The Earth's polarity has reversed about 130 times during the past 65 million years, an average of once every half-million years. The

These shales of southern Utah record the orientation of the Earth's magnetic field at the time the mud was deposited.

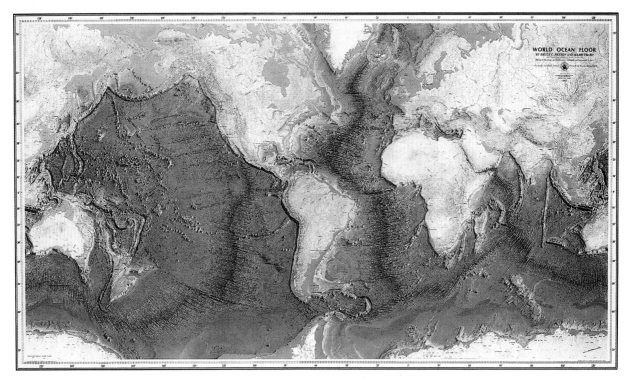

Figure 5–10   The Mid-Atlantic ridge is a submarine mountain chain in the middle of the Atlantic Ocean. It is a segment of the mid-oceanic ridge, which circles the globe like the seam on a baseball.   *(Marie Tharp)*

orientation of the Earth's field at present is referred to as **normal**, and that during a time of opposite polarity is called **reversed**.

5.3 Sea-Floor Spreading

Shortly after World War II, scientists began to explore the ocean floor. Although these sea-floor studies ultimately led to the theory of plate tectonics, they were initially conducted for military and economic reasons. Defense strategists wanted maps of the sea floor for submarine warfare, and the same information was needed to lay undersea telephone cables. As they mapped the sea floor, oceanographers discovered a long submarine mountain range in the middle of the Atlantic Ocean. This range is called the **Mid-Atlantic ridge** (Fig. 5–10). Further studies showed that the Mid-Atlantic ridge is part of a continuous submarine mountain chain called the **mid-oceanic ridge**, which girdles the entire globe and is by far the Earth's largest and longest mountain chain.

One remarkable feature of the Mid-Atlantic ridge is that it lies right in the middle of the Atlantic Ocean basin, halfway between Europe and Africa to the east and North and South America to the west. As you learned in Chapter 3, most oceanic crust is basalt. Basalt is an iron-rich igneous rock and therefore becomes magnetic as it cools.

Thus, sea-floor basalt records the orientation of the Earth's magnetic field at the time the basalt cooled.

Figure 5–11 shows magnetic orientations of sea-floor rocks along a portion of the Mid-Atlantic ridge known as the Reykjanes ridge. The black stripes represent rocks with **normal polarity**; that is, their magnetic orientation parallels that of the Earth today. The white stripes represent rocks with **reversed polarity**, in which the magnetic orientation is exactly opposite that of the Earth today. Notice that the stripes form a pattern of alternating normal and reversed polarity. Note also that this pattern is symmetrical about the axis of the ridge. The central stripe is black, indicating that the rocks of the ridge axis have a magnetic orientation parallel to that of the Earth today.

What caused this pattern of alternating stripes of normal and reversed polarity in the rocks of the sea floor, and why are they symmetrically distributed about the mid-oceanic ridge? In the mid-1960s, geologists suggested that a sequence of events creates the symmetrical pattern of magnetic stripes on the sea floor:

1. New sea floor continuously forms as basaltic magma rises beneath the ridge axis. It solidifies to form oceanic crust and spreads outward from the mid-oceanic ridge. This movement is analogous to two broad conveyor belts moving away from one another.

2. As the new sea floor cools, it acquires the orientation of the Earth's magnetic field.

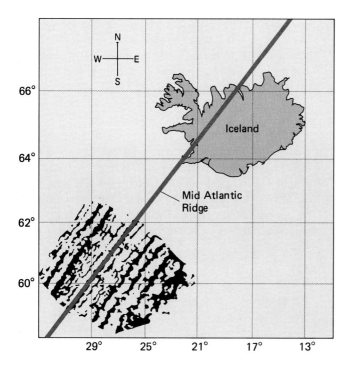

Figure 5–11   The Reykjanes ridge, near Iceland, is part of the Mid-Atlantic ridge. Magnetic orientations of sea-floor rocks at the Reykjanes ridge are shown in the lower left portion of the map. The black stripes represent sea-floor rocks with normal magnetic polarity, and the white stripes represent rocks with reversed polarity. Notice that the stripes form a symmetrical pattern of alternating normal and reverse polarity on each side of the ridge. *(After Heirtzler, et al., 1966, Deep-Sea Research, Vol. 13.)*

3. The Earth's magnetic field reverses orientation periodically—on an average of every half-million years.

4. Thus, the alternating magnetic stripes on the sea floor simply record the succession of reversals in the Earth's magnetic field that take place as the sea floor spreads away from the ridge (Fig. 5–12).

    To return to our analogy of the conveyor belt, imagine that a can of white spray paint is mounted above a black conveyor belt. Alternating white and black stripes across the width of the belt are produced if paint is sprayed intermittently as the conveyor belt moves beneath it.

When the paint sprayer is off, a black stripe forms, analogous to a time of normal polarity. When the paint sprayer is on, a white stripe forms, symbolizing a time of reversed polarity.

    At about the same time that oceanographers discovered the magnetic stripes on the sea floor, they also began to sample the mud covering the deep-sea floor. Mud is thinnest at the Mid-Atlantic ridge and becomes progressively thicker at greater distances from the ridge. If mud falls to the sea floor at about the same rate everywhere, the fact that it is thinnest at the ridge confirms the idea that the sea floor is youngest there and becomes progressively older away from the ridge.

    Oceanographers soon recognized similar magnetic stripes and sediment trends along other portions of the mid-oceanic ridge. As a result, they proposed the theory of **sea-floor spreading** as a model for the origin of all oceanic crust. However, the sea-floor spreading theory

Figure 5–12   As new oceanic crust cools at the mid-oceanic ridge, it acquires the magnetic orientation of the Earth's field. Alternating stripes of normal (blue) and reversed (green) polarity record reversals in the Earth's magnetic field that occur as the crust spreads outward from the ridge.

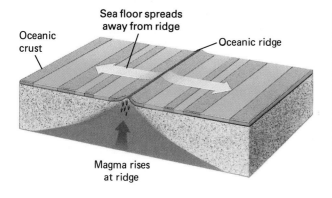

● M E M O R Y   D E V I C E

Tectonics is the study of large-scale deformation and structures in the outer portion of the Earth. Plate tectonics is the study of the motion of the lithospheric plates and the consequences of those motions. The word tectonic is derived from the Greek root *tekton-ikos*, pertaining to construction. Tectonics, then, has to do with the construction of the outer part of the Earth.

offered no further enlightenment regarding Wegener's evidence for continental drift, nor did it explain other geological phenomena such as earthquakes, volcanic activity, and mountain building. In a short time, however, geologists combined the sea-floor spreading theory with Wegener's evidence and other data to develop the plate tectonics theory. The plate tectonics theory offers us a single model to explain how and why continents move, sea floor spreads, mountains rise, earthquakes shake our planet, and volcanoes erupt.

The Earth is a layered planet, and the theory of plate tectonics depends on an understanding of the Earth's layers.

5.4 The Earth's Layers

The Crust

Figure 5–13 shows a cross-sectional view of the Earth, and Table 5–1 summarizes properties of its layers. The thinnest layer is the outer shell, called the **crust**. The crust is not even or homogeneous. Both its thickness and its composition vary. Oceanic crust ranges from 7 to 10 kilometers in thickness and is composed mostly of basalt.

Continental crust is much thicker than oceanic crust. Not only do the continents rise above the ocean floor, they also extend below it. Thus, continents have roots that protrude into the mantle. In mid-continent regions, the crust is about 20 to 40 kilometers thick, whereas under major mountain ranges, it increases to as much as 70 kilometers. Most continental crust is granitic.

The granitic rock of continental crust is rigid and brittle. It is important to understand that rigidity is relative. Imagine that you could cut out the state of Kansas from the crust to a depth of 35 kilometers and place it on top of a neighboring state. It could neither support its own weight nor hold its shape. The edges would fracture and crumble, and flow out over the Great Plains like honey.

The Mantle

The **mantle** is almost 2900 kilometers thick and makes up about 80 percent of the Earth's total volume. Its chemical composition is nearly constant throughout. However, the Earth's temperature and pressure increase with depth, and these changes cause the physical properties and minerals of the mantle to vary with depth. The upper part of the mantle consists of two layers.

TABLE 5–1

The Layers of the Earth				
	Layer	**Composition**	**Depth**	**Properties**
Crust	Oceanic crust Continental crust	Basalt Granite	7–10 km 20–70 km	Cool, rigid, and brittle Cool, rigid, and brittle
Lithosphere	Lithosphere includes the crust and the uppermost portion of the mantle	Varies; the crust and the mantle have different compositions	About 100 km	Cool, rigid, and brittle
Mantle	Uppermost portion of the mantle included as part of the lithosphere Asthenosphere Remainder of upper mantle Lower mantle	Entire mantle is untramafic rock. Its mineralogy varies with depth	Extends from 100 to 350 km Extends from 350 to 670 km Extends from 670 to 2900 km	Hot and plastic, 1% or 2% melted Hot, under great pressure, rigid, and brittle High pressure forms minerals different from those of the upper mantle
Core	Outer core Inner core	Iron and nickel Iron and nickel	Extends from 2900 to 5150 km Extends from 5150 km to the center of the Earth	Liquid Solid

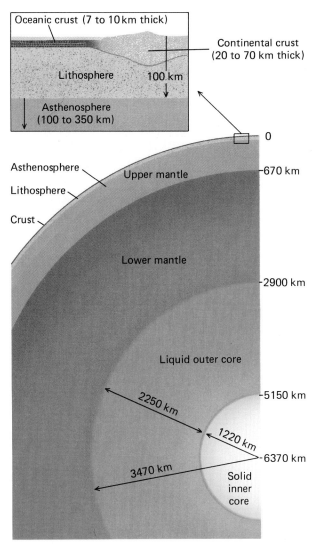

Figure 5–13   The Earth is a layered planet. The insert is drawn on an expanded scale to show near-surface layering.

The Lithosphere

The outer 100 kilometers of the Earth, including both the uppermost mantle and the crust, make up a layer called the **lithosphere** (Greek for "rock layer"). The lithosphere is close enough to the surface that both temperature and pressure are relatively low. Therefore, the rock in this layer is rigid and brittle. Although the compositions of the crust and upper mantle are different, this 100-kilometer-thick zone behaves as a single layer. Most earthquakes originate in the brittle rock of the lithosphere.

The Asthenosphere

At a depth of about 100 kilometers, the rigid, brittle rock of the lithosphere suddenly gives way to hot, soft, plastic rock of the **asthenosphere**. In this layer, rock is so hot and plastic that it flows readily even though it is solid.

To visualize a solid that can flow, think of Silly Putty or road tar on a hot day. The asthenosphere extends from the base of the lithosphere to a depth of about 350 kilometers. At the base of the asthenosphere, the mantle again becomes more rigid and less plastic, and it remains in this state all the way down to the core.

Isostasy

The lithosphere is of lower density than the asthenosphere and it floats on the asthenosphere much as an iceberg or a block of wood floats on water. This concept of a floating lithosphere is called **isostasy**.

To illustrate isostasy, imagine three icebergs of different sizes floating in the ocean. When ice floats in seawater, approximately 10 percent of the berg rises above water level while the remaining 90 percent is under water. Of the three bergs, the largest will have the most material under water, but it will also have the highest peak (Fig. 5–14). The lithosphere behaves in a similar manner. Continents have "roots" that extend into the man-

Figure 5–14   The principle of isostasy. (A) The largest of the three icebergs has the most material underwater and also the most above. (B) In an analogous manner, a stylized diagram of the crust shows that the roots under high mountains extend deeper into the mantle than the roots under lower areas of the continents.

A

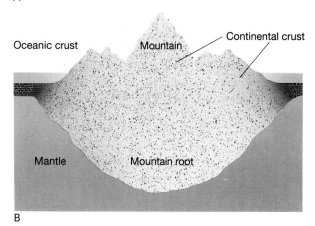

B

tle. The roots beneath major mountain ranges are deeper than the roots beneath continental plains, just as the bottom of a large iceberg is deeper than the base of a smaller one.

If our theory of isostasy is a good one, it should explain why the ocean floor is lower than the continents. Even if both the ocean floor and the continents were made of the same material, the thinner oceanic lithosphere would be expected to float lower in the asthenosphere just as a small iceberg sits lower in the water than a large one does. But, in addition, oceanic crust is made up of basalt, which is denser than granite. The denser oceanic lithosphere settles down even farther in the asthenosphere giving rise to the large difference in elevation between ocean basins and continents.

If you have ever loaded equipment onto a small boat, you may have noticed that it settles lower in the water as cargo is added, whereas when the boat is unloaded, it rises. The lithosphere behaves in a similar manner. But how is "cargo" added or subtracted from the Earth's lithosphere? One example is the growth and melting of large glaciers. When glaciers grow, the added weight of ice forces the continents to sink. The central portion of Greenland, which lies under a 3000-meter-thick glacier, has been depressed so far that it actually lies below sea level. On the other hand, when glaciers melt, the continents rebound or rise up. Geologists have discovered Ice Age beaches along many coastlines in the northern hemisphere that now lie high above sea level because the land rose as the glaciers melted. This vertical movement in response to changing burdens is called **isostatic adjustment** (Fig. 5–15). The largest man-made structures alter the distribution of weight on the lithosphere sufficiently to trigger a detectable isostatic adjustment. When dams are built, the newly formed lakes are heavy enough to cause the lithosphere to sink a small but measurable amount. This sinking often causes a series of small earthquakes.

The Core

Recall from Chapter 1 that during the early history of the Solar System, Earth was a homogeneous mixture of dust and gas. As the planet evolved, it became hotter until it began to melt. Once the Earth melted, most of the heavy elements gravitated toward the core and most of the lighter ones floated upward toward the surface. As a result, the **core** consists mostly of the heavy metals iron and nickel. It is a sphere with a radius of about 3470 kilometers, larger than the planet Mars. The outer core is molten because of the high temperatures near the Earth's center. The inner core is even hotter, but it is solid because the higher pressures in the inner core overwhelm the temperature effect.

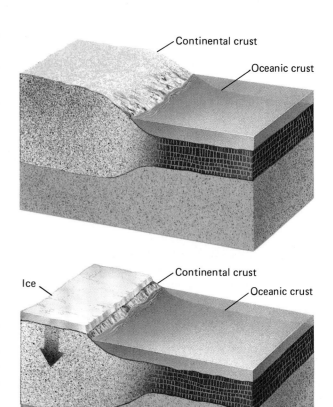

Figure 5–15 Isostatic adjustment of a continent when the crust is weighted down by a continental glacier.

Near its center the core's temperature is about 6000°C, which is as hot as the surface of the Sun. The pressure exceeds 1 million atmospheres.

5.5 Plates and Plate Tectonics

The plate tectonics theory is a model of the Earth in which the brittle lithosphere "floats" on the hot, plastic asthenosphere. The lithosphere is broken into seven large plates and several small ones, resembling segments of a turtle's shell. The plates move across the Earth's surface, each in a different direction from its neighbors. They glide slowly over the weak, plastic asthenosphere at rates ranging from less than 1 to about 18 centimeters per year. **Plate tectonics** is the study of the movement of lithospheric plates and the consequences of those motions.

Because the plates move in different directions, they bump and grind together at their boundaries. Just imagine two 100-kilometer-thick slabs of rock smashing together along a boundary a few thousand kilometers in length! No convenient analogy exists to describe such a collision.

The Peruvian Andes rise along the western edge of South America, where two tectonic plates collide.

If a billion large bulldozers were to drive headlong into a giant city made of all the buildings on Earth, the force would be tiny compared with that resulting from a collision of two tectonic plates. Because of the great forces generated at plate boundaries, mountain building, volcanic eruptions, and earthquakes occur where two plates meet. These events are called **tectonic** activity, from the ancient Greek word for "construction." Tectonism constructs and modifies the Earth's crust. In contrast to plate boundaries, interior portions of lithospheric plates are tectonically quiet because they are far from the zones where two plates interact.

Simple logic tells us that one plate can move relative to an adjacent plate in three different ways (Fig. 5–16).

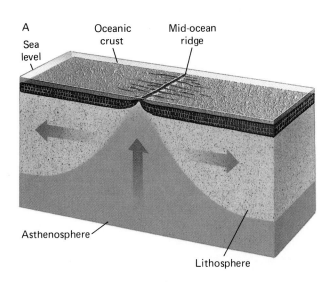

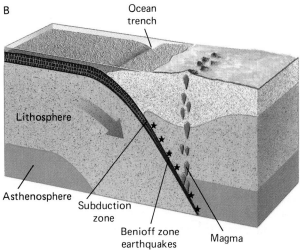

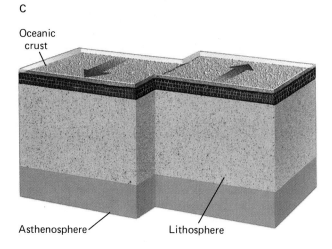

Figure 5–16 Three types of plate boundaries exist. (A) Two plates separate at a divergent plate boundary. New lithosphere forms as hot asthenosphere rises to fill the gap where the two plates spread apart. Note that the lithosphere is relatively thin at this type of plate boundary. (B) Two plates collide at a convergent plate boundary. If one or both plates are capped by oceanic crust, the dense oceanic plate sinks into the mantle in a subduction zone. Here an oceanic plate sinks beneath a less dense continental plate. Magma rises from the subduction zone, and an oceanic trench forms where the subducting plate begins to sink into the mantle. The stars mark earthquakes. (C) At a transform plate boundary, rocks on opposite sides of the fracture slide horizontally past each other.

123

At a **divergent plate boundary**, two plates move apart, or separate. At a **convergent plate boundary**, two plates move toward each other and collide. At a **transform plate boundary**, two plates slide horizontally past each other. Each of these three types of boundaries creates different tectonic features. Table 5–2 summarizes characteristics of each type of plate boundary and lists modern examples of each.

5.6 Divergent Plate Boundaries

A divergent boundary, also called a **spreading center** and a **rift**, occurs where two plates move apart horizontally (Fig. 5–17). As the two plates separate, hot, plastic asthenosphere rock flows upward to cool and form new lithosphere in the gap left by the diverging plates. The rising asthenosphere partly melts, forming basalt magma that oozes to the surface. Rifts occur in both oceanic and continental crust.

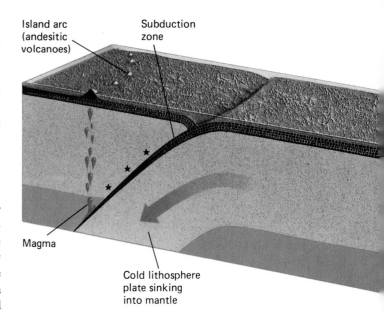

Island arc (andesitic volcanoes)

Subduction zone

Magma

Cold lithosphere plate sinking into mantle

TABLE 5–2

Plate Boundaries				
Type of Boundary	**Types of Plates Involved**	**Topography**	**Geologic Events**	**Modern Examples**
Divergent	Ocean-ocean	Mid-oceanic ridge	Sea-floor spreading, shallow earthquakes, rising magma, volcanoes	Mid-Atlantic ridge
	Continent-continent	Rift valley	Continents torn apart, earthquakes, rising magma, volcanoes	East African rift
Convergent	Ocean-ocean	Island arcs and ocean trenches	Subduction, deep earthquakes, rising magma, volcanoes, deformation of rocks	Western Aleutians
	Ocean-continent	Mountains and ocean trenches	Subduction, deep earthquakes, rising magma, volcanoes, deformation of rocks	Andes
	Continent-continent	Mountains	Deep earthquakes, deformation of rocks	Himalayas
Transform	Ocean-ocean	Major offset of mid-oceanic ridge axis	Earthquakes	Offset of East Pacific rise in South Pacific
	Continent-continent	Small deformed mountain ranges, deformations along fault	Earthquakes, deformation of rocks	San Andreas fault

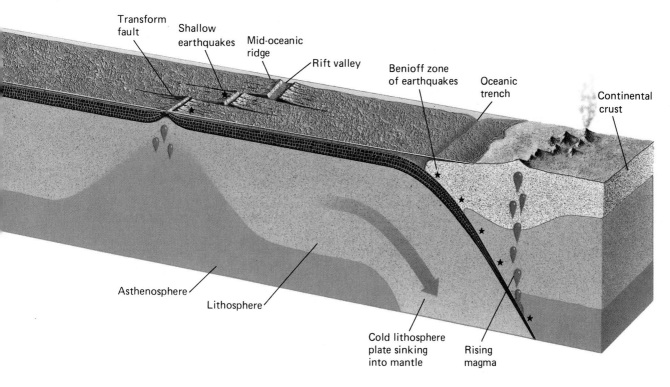

Figure 5–17   A cross-section of the outer few hundred kilometers of the Earth. In the center of the drawing, new lithosphere forms at a spreading center. At the sides of the drawing, old lithosphere sinks into the mantle at subduction zones. The lithospheric plates move away from the spreading center by gliding over the weak, plastic asthenosphere.

Divergent Boundaries in Oceanic Crust: The Mid-Oceanic Ridge

As shown in Figure 5–17, the mid-oceanic ridge is a divergent boundary in oceanic crust. The basalt magma that oozes out onto the sea floor at the ridge forms oceanic crust on top of the new lithosphere. Note that it is not merely oceanic crust that spreads outward from the mid-oceanic ridge, as the sea-floor spreading theory proposed. Rather, the entire lithosphere spreads, carrying the sea floor on top of it in piggyback fashion.

At the spreading center, the new lithosphere is hot and therefore of low density. It then cools and becomes denser as it spreads outward from the ridge axis. The new, hot, low-density rock at the spreading center "floats" at a higher elevation than the older and cooler rock on both sides. Thus, the mid-oceanic ridge is elevated above surrounding sea floor because it is made of the newest, hottest, and lowest-density lithosphere.

Although the mid-oceanic ridge is the Earth's longest mountain chain, we do not normally see it because it lies below sea level. It winds through all of the Earth's ocean basins, much like the seam on a baseball. Occasionally it rises above sea level to form islands such as Iceland.

Divergent Boundaries in Continental Crust: Continental Rifting

A continent can also be pulled apart at a divergent boundary; the process is called **continental rifting**. Continental rifting is occurring today along a north-south fault zone in eastern Africa called the East African rift. If the rifting continues, eastern Africa will eventually separate from the main portion of the continent. Basalt magma will rise to fill the growing gap, forming a new ocean basin between the separating portions of Africa. Continental rifting may also be occurring in North America along the Snake River plain, which extends westward from Yellowstone National Park to southeastern Oregon. Elongate depressions called **rift valleys** develop along continental rifts because continental crust becomes stretched, fractured, and thereby thinned as it is pulled apart. Both the

Tectonic forces are pulling the African continent apart along the East African rift, Kenya. *(Amos Turk)*

East African rift and the Snake River plain form great valleys in continental crust.

5.7 Convergent Plate Boundaries

A convergent boundary develops where two plates are moving horizontally toward each other and therefore are colliding (Fig. 5–16B). Collisions can occur (1) between a plate carrying oceanic crust and another carrying a continent, (2) between two plates carrying oceanic crust, and (3) between two continental plates.

Convergence of Oceanic Crust with Continental Crust

Recall that oceanic crust is denser than continental crust. The difference in density determines what happens in a collision. Think of a boat colliding with a floating log. The log is denser than the boat, so it sinks beneath the boat. When a continental plate collides with a denser oceanic plate, the oceanic plate sinks beneath the continental plate and dives into the mantle. This process is called **subduction**.

A **subduction zone** is a long, narrow belt where a lithospheric plate dives into the mantle. On a worldwide scale, the rate at which old lithosphere sinks into the mantle at subduction zones is equal to the rate at which new lithosphere forms at spreading centers. In this way, a perfect global balance is maintained between the creation of new lithosphere and the destruction of old lithosphere.

Only lithosphere covered with oceanic crust can sink into the mantle at a subduction zone. Continental crust is of lower density than oceanic crust and cannot sink. Attempting to stuff a continent down a subduction zone would be like trying to flush a marshmallow down a toilet: it just would not go because it is too light. The oldest sea-floor rocks on Earth are only about 200 million years old because oceanic crust is continuously recycled back into the mantle at subduction zones. Far older rocks are found on continents because continental crust is not consumed by subduction.

Today, oceanic plates are subducting beneath continental crust along the western edge of South America and the coasts of Oregon, Washington, and British Columbia (Fig. 5–1). Earthquakes, active volcanoes, and rising mountains all characterize these regions. Each of these tectonic activities is discussed in a following chapter. We will describe them briefly here.

Subduction and Earthquakes

As a lithospheric plate sinks toward the mantle, it slips and jerks beneath the opposite plate causing numerous earthquakes. These quakes trace the path of the subducting plate as it sinks into the mantle (Fig. 5–16B). This zone of earthquakes is called a **Benioff zone**, after one of the geologists who first recognized it. The deepest earthquakes known occur in Benioff zones at a depth of about 700 kilometers. Below 700 kilometers, subducting plates must become so hot that they flow in a plastic manner rather than fracturing. Earthquakes are common in western South America and along the coasts of Oregon, Washington, and British Columbia.

Subduction and Volcanoes

Because oceanic crust is covered by the sea, the upper part of a subducting plate consists of water-soaked sea-floor mud and basalt. As a subducting plate sinks, this water escapes into the asthenosphere. As you learned earlier in this chapter, the asthenosphere is so hot that it is soft and plastic—in fact, it is very close to melting. **Addition of water to very hot rock can cause the rock to melt.** Thus, water from the subducting plate causes melting in the asthenosphere (Fig. 5–16B). In this way, *huge* quantities of magma form in subduction zones. The magma rises through the overlying lithosphere. Some solidifies within the crust to form coarse-grained igneous rocks such as granite, while some erupts onto the Earth's surface to form volcanic rocks. Igneous activity is common in the Cascade Range of Oregon, Washington, and British Columbia and in western South America. The relationships between subduction and igneous activity will be discussed further in Chapter 7.

Subduction and Mountain Building

Many of of the world's great mountain chains, including the Andes and parts of the mountains of western North America, formed at subduction zones. Several factors contribute to growth of mountain chains at subduction zones. The great volume of magma rising through the Earth's crust thickens the crust, causing mountains to rise.

Volcanic eruptions, resulting from subduction that occurs off the coasts of Washington and Oregon, have formed Mount Adams and other volcanoes of the Cascade Range.

Volcanic eruptions pour huge amounts of lava onto the surface, constructing chains of volcanoes. Additionally, the Earth's crust crumples and buckles into mountain ranges where two lithospheric plates crash together. The growth of mountains at subduction zones will be described in greater detail in Chapter 8.

Oceanic Trenches

An **oceanic trench** is a long, narrow trough in the sea floor, formed where a subducting plate turns downward to sink into the mantle (Fig. 5–14B). A trench can form wherever subduction occurs—where oceanic crust sinks beneath either continental or oceanic crust. The sinking plate simply drags the sea floor downward. Trenches are the deepest parts of the ocean basins. The deepest point on Earth is in the Mariana trench, in the southwestern Pacific Ocean north of New Guinea, where the sea floor is as much as 10.9 kilometers below sea level.

Convergence of Two Plates Carrying Oceanic Crust

Subduction also occurs where two oceanic plates collide. Recall that new oceanic lithosphere is hot, and therefore of low density, when it first forms at the mid-oceanic ridge. It cools and becomes denser as it ages and spreads away from the ridge. When two oceanic plates collide, the older, cooler, and denser plate subducts into the mantle. Great quantities of magma form and rise toward the Earth's surface in a manner similar to that in a subduction zone adjacent to a continent.

Collision and subduction between two plates carrying oceanic crust is occurring today just south of the western Aleutian Islands between Alaska and Siberia.

Island Arcs

As one oceanic plate dives beneath another, magma forms and rises to erupt onto the sea floor near the subduction zone. Eventually, the submarine volcanoes grow above sea level to form a chain of volcanic islands, called an **island arc**, along the subduction zone. The western Aleutian Islands and many of the island chains in the southwestern Pacific Ocean are island arcs. Both the trenches and island arcs of the southwestern Pacific show up clearly in Figure 5–10.

Convergence of Two Plates Carrying Continental Crust

If two colliding plates are both covered with continental crust, subduction cannot occur because continental crust is too light to sink into the mantle. In this case, the two continents collide and crumple against each other, forming huge mountain chains in the collision zone. The

The great peaks of the Himalayas rose as a result of a collision between India and Asia.

Himalayas, the Alps, and the Appalachians all formed as results of continental collisions. These processes are discussed further in Chapters 8 and 9.

5.8 Transform Plate Boundaries

A transform plate boundary forms where two plates slide horizontally past one another (Fig. 5–14C). California's San Andreas fault is a transform boundary between two major lithospheric plates, the North American plate and the Pacific plate. Although earthquakes are common at this type of boundary, igneous activity is not.

5.9 Anatomy of a Plate

Plate tectonics theory provides a mechanism for the movement of continents that was lacking in Wegener's theory of continental drift. Recall that Wegener proposed that continents plowed through or slid over ocean crust, and that these proposed mechanisms were discredited. Now geologists understand that continents are just large masses of granite sitting on top of some lithospheric plates. When lithospheric plates move, they carry the continents along with them. This is the mechanism that was not understood in Wegener's day. Continents do not plow through or slide over oceanic crust. They simply ride as passengers on the thick lithospheric plates.

The nature of a tectonic plate can be summarized as follows:

1. A plate is a segment of the lithosphere; thus, it includes the uppermost mantle and all of the overlying crust.

2. A portion of a plate with continental crust composing its uppermost layer is thicker than one bearing oceanic crust (Fig. 5–18). The average thickness of the part of a plate carrying oceanic crust is about 75 kilometers, whereas that of a plate bearing continental crust is about 125 kilometers. Lithosphere may be as little as 20 kilometers thick at an oceanic spreading center.

3. A plate is hard, rigid or nearly rigid rock.

4. A plate "floats" on the underlying hot, plastic asthenosphere and glides horizontally over it.

5. A plate behaves like a large slab of ice floating on a pond. It may flex slightly, as thin ice does when a skater goes by, allowing minor vertical movement such as isostatic adjustment. In general, however, each plate moves as a single, large, intact sheet of rock.

6. A plate margin is tectonically active. Earthquakes and faulting are common at all plate boundaries. Igneous activity—volcanic eruptions and intrusion of magma—is frequent at subduction zones and spreading centers. Intense deformation of the Earth's crust occurs at transform boundaries and subduction zones. In contrast, the interior of a lithospheric plate is generally tectonically stable.

Plate Velocity

How rapidly does a lithospheric plate move? Calculations based on several methods show that plates move away from spreading centers at rates that vary from 1 to 18 centimeters per year. In recent years, plate motion has been measured directly by surveying techniques that bounce laser beams off the Moon and satellites and by other methods that use radio waves originating outside our galaxy (Fig. 5–19).

5.10 The Search for a Mechanism

Geologists have accumulated ample evidence *that* lithospheric plates move and can even measure *how fast* they move. But *why* do plates move? At present, one of the most active and exciting areas of research in geology is the search for the cause of plate motions.

Mantle Convection

Although the mantle is solid rock (except for small, partially melted zones in the asthenosphere), it is so hot that over geologic time it flows slowly as a stiff fluid. Mantle rock flows in elliptical or circular patterns. Hot rock from deep in the mantle rises to the base of the lithosphere. At the same time, cooler upper mantle rock sinks. This flow of solid rock is called **mantle convection**.

Mantle convection is thought to be closely related to movement of lithospheric plates. However, it is not clear whether convection of the mantle causes the plates to move or, conversely, movement of the plates is the cause of mantle convection.

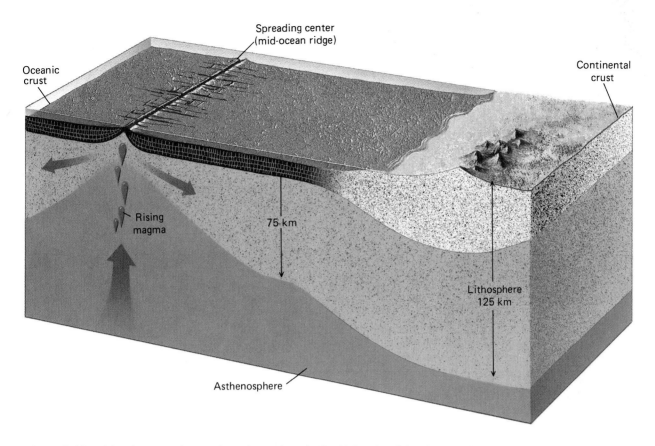

Figure 5–18 Lithosphere carrying continental crust is typically thicker than lithosphere bearing oceanic crust.

Figure 5–19 Velocities of plate movements in centimeters per year. Numbers along the mid-oceanic ridge indicate the rates at which the two plates on opposite sides of the ridge are separating, based on magnetic reversal patterns on the sea floor. Arrows indicate directions of plate motions. The yellow lines connect stations that measure present-day rates of plate motions with satellite laser ranging methods. The numbers followed by "L" are the present-day rates at which the two points connected by the lines are separating, measured by laser. The numbers followed by "M" are the rates of separation measured by magnetic reversal patterns. *(Modified from NASA report, Geodynamics Branch, 1986)*

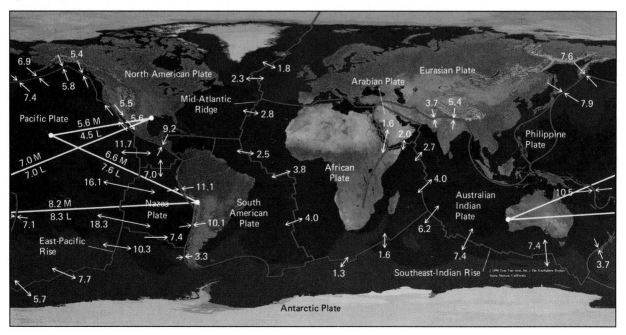

Mantle Convection as the Cause of Plate Movement

Convection occurs when a fluid is heated from below, as in a pot of soup on a stove. The soup at the bottom of the pot expands as it is heated and becomes less dense than the soup at the top. Because it is less dense, it rises. When the hot soup reaches the top of the pot, it flows along the surface as cooler soup sinks to take its place (Fig. 5–20). If the heat source persists, this cool, sinking soup is then warmed. It rises, and the convection continues.

Mantle convection might occur in a manner similar to that in the soup pot. In this model, the base of the mantle is heated from below, perhaps by the hot core. In turn, the heating causes mantle convection. Imagine a block of wood floating on a tub of honey. If you heated the honey from below so that it started to convect, the horizontal flow of honey along the surface would drag the block of wood along with it. Some geologists suggest that lithospheric plates are dragged along in a similar manner by a convecting mantle (Fig. 5–21).

Plate Movement as a Cause of Mantle Convection

Some geologists have suggested that movement of lithospheric plates might be the cause, rather than the effect, of mantle convection. Return to our analogy of the block of wood and the tub of honey. If you dragged the block of wood across the honey, friction between the block and the honey would make the honey flow. Similarly, if some

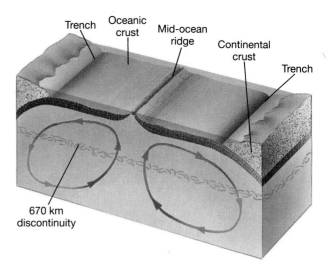

Figure 5–21 In one explanation for the movement of lithospheric plates, the plates are dragged along by mantle convection.

force caused the plates to move, their motion might cause the mantle to flow.

But what force would move the plates? A plate may simply glide downhill, away from a spreading center. Because newly formed lithosphere is hot and of low density, it is thin and sits at relatively high elevation at the spreading center. That is why the mid-oceanic ridge is topographically high. As the lithospheric plate spreads away from the ridge, it cools and thickens. Therefore, both the surface and the base of the lithosphere slope downward from the spreading center.

The average slope of the *surface* of the mid-oceanic ridge is about 0.6 percent. However, because the lithosphere thickens as it spreads away from the ridge (Fig. 5–22), the slope of the *base* of the lithosphere beneath the ridge is much steeper. The average slope of the base of the lithosphere beneath the ridge is about 8 percent, steeper than almost any paved road in North America. Calculations show that if the slope is as slight as 0.03 percent, a plate could glide away from a spreading center at a rate of a few to several centimeters per year.

Mantle Plumes, Meteorite Impacts, and Plate Movement

A recent suggestion for the cause of lithospheric plate motion is that a **mantle plume** initiates plate movement. A mantle plume is a vertical column of plastic rock rising through the mantle like hot smoke from an industrial smokestack (Fig. 5–23). Plumes originate deep within the mantle, perhaps even at the core-mantle boundary. When a plume reaches the base of the lithosphere, it spreads outward, dragging the lithosphere apart and initiating a spreading center.

The suggestion that mantle plumes initiate plate movement, raises another question: what causes mantle

Figure 5–20 Soup convects because it is heated from the bottom of the pot.

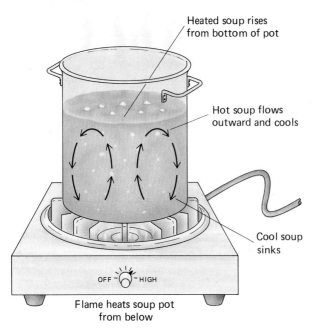

Heated soup rises
from bottom of pot

Hot soup flows
outward and cools

Cool soup
sinks

Flame heats soup pot
from below

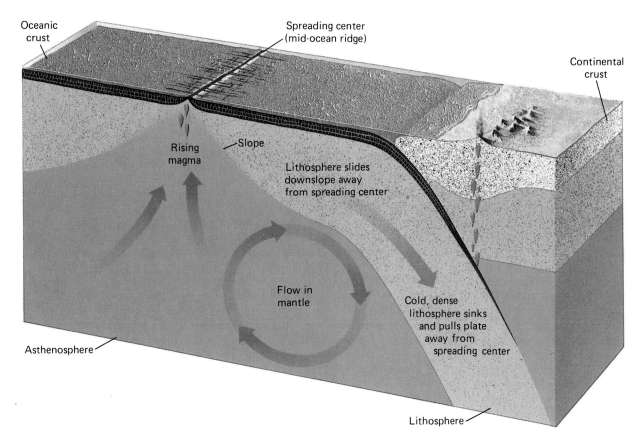

Figure 5–22 Two possible causes, other than mantle convection, for plate movement. (1) A plate glides down an inclined surface on the asthenosphere. (2) A cold, dense plate sinks at a subduction zone, pulling the rest of the plate along with it. In this drawing, both mechanisms are shown operating simultaneously.

Figure 5–23 A mantle plume rises through the mantle as a vertical column and spreads radially at the base of the lithosphere.

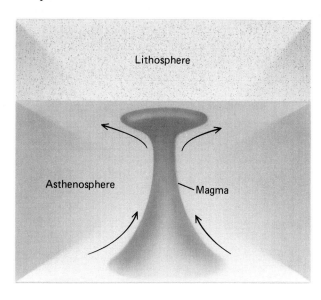

plumes? Geologists have recently suggested that a mantle plume (as well as several other major geologic features) forms when a large meteorite strikes the Earth with enough energy to blast a huge crater in the crust. Mantle rock directly below the crater begins to flow upward in isostatic adjustment to fill the crater, and the upwelling of a mantle plume begins. Once started, such upwelling may continue for millions of years. This model is more fully explored in the box "Effects of Meteorite Impacts on the Earth."

Synthesis of a Single Model for the Cause of Plate Movement

At present, a synthesis of these models is the most appealing explanation for plate movement. In this synthesis, lithospheric spreading is initiated by a mantle plume. The mantle plume may have begun rising because of an

(Text continues on p. 134)

Earth Science and the Environment

Effects of Meteorite Impacts on the Earth

The effects of large meteorites striking the Earth have received serious consideration by geologists only in the past few years. Recall from Chapter 4 that Luis W. and Walter Alvarez suggested that the sudden extinction of one-fourth of the Earth's animal species 65 million years ago was caused by the impact of a large meteorite. In their model, the impact threw up such a huge cloud of dust that solar radiation was blocked, causing a short but severe climate change. Four sudden major extinctions and numerous smaller ones have occurred in the past 570 million years, and some or all of them may have been due to meteorite impacts.

Despite the fact that the Alvarez theory of extinctions has gained wide acceptance, most geologists feel that meteorite impacts have little geologic significance, save for the occasional mass extinction. Recently, however, other geologists have suggested that meteorite impacts play an important role in plate tectonics. Furthermore, these geologists claim to have identified the location of the meteorite impact that caused the 65-million-year-old mass extinction, as well as the sites of several other major meteorite strikes.

Astronomers estimate that meteorites with diameters greater than 10 kilometers should strike the Earth on the average of once every 40 million years. More than 100 craters formed by impacts of smaller objects are well known on the Earth's surface. But where are the geologic features formed by impacts of the large meteorites? At a rate of one every 40 million years, about 100 major impacts should have occurred since the planet formed. Their geologic effects should be even more striking than those of the smaller meteorites, yet none is easily recognizable. Perhaps the features created by impacts of large meteorites are obvious and have been included on geologic maps for more than a hundred years, but we have simply failed to recognize them for what they are.

When a large meteorite strikes the Earth, it does so with enough energy to blast a huge hole in the lithosphere. The mantle below then flows upward to fill the hole, initiating a mantle plume. As the mantle plume rises, pressure decreases and partial melting forms basalt magma. Thus, large amounts of basalt magma form directly beneath the site of impact. The magma then rises to the Earth's surface, where it erupts in vast lava flows to form a basalt plateau. In this model, the great lava plateaus of the world mark the sites of large meteorite impacts.

One such lava plateau, the Deccan Plateau of western India, is 1000 kilometers in diameter and contains a 2-kilometer-thick sequence of basalt flows. The lavas are 65 million years old, precisely the same age as the extinction of the dinosaurs. Thus, the Deccan Plateau may be the site of the meteorite impact responsible for that extinction.

Recently geologists have located evidence of another large meteorite impact that took place 65 million years ago, in the Yucatan Peninsula of Mexico. At first glance, this additional information may seem to be the start of a new geological controversy: which of the two 65-million-year-old impacts was really responsible for the extinction of the dinosaurs and so many other life forms? However, it is reasonable that a large asteroid coming into the influ-

Basalt lava flows of the Columbia Plateau. (Don Hyndman)

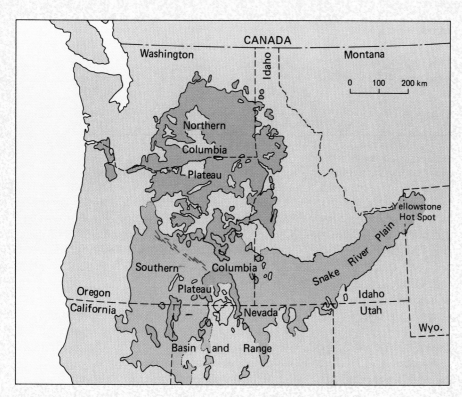

The Columbia Plateau and the Snake River plain, which is the track of the Yellowstone hot spot.

ence of Earth's gravitational forces might break into two or more segments. Two or more simultaneous impacts, all contributing to a single extinction event, are not surprising.

More than ten great lava plateaus similar to the Deccan Plateau have formed during the past 250 million years. It may be that all of them mark sites of impacts of large meteorites. The Columbia Plateau in the American northwest is another example of this type of feature (see figure). Recognition of the meteorite origin of lava plateaus could account for the dozens of "missing" major meteorite impact sites.

The hypothesis does not stop here, however. Some geologists further suggest that the impact of a large meteorite cracks the lithosphere. The crack grows longer as magma rises to form the basalt plateau. The spreading crack is the beginning of a new lithospheric spreading center. If the impact occurs in an ocean basin, the crack becomes a mid-oceanic ridge. If it occurs on continental crust, it initiates continental rifting. The rifting of the Yellowstone hot spot track along the Snake River plain of southern Idaho may be an example of continental rifting caused by meteorite impact (see figure). Rifting in this re-

gion began about 17 million years ago, the same time that basaltic volcanism began in the Columbia Plateau. Many of the world's spreading centers can be traced back, both geographically and in time, to lava plateaus. The rifting of Pangaea may have been initiated by a meteorite impact, traces of which are preserved as the deeply eroded remains of a 200-million-year-old lava plateau in South Africa.

Will North America rift apart along the trace of the Columbia Plateau–Snake River plain within the next 100 million years to become two separate continents riding on two separate tectonic plates? Will a new ocean basin develop in the growing gap? Will Africa similarly separate along the East African rift? The evolution of such continental rifts into full-blown tectonic plate boundaries is a possibility. However, in other cases, lithospheric rifts seem to have started up and then, after a few tens of millions of years, died out without further development. Such failed rifts are thought to have formed the St. Lawrence lowlands along the St. Lawrence River in southeastern Canada as well as the path of the Rio Grande River from southeastern Colorado through Albuquerque, New Mexico, and down to El Paso, Texas.

unusually hot region near the core-mantle boundary or, alternatively, in response to a meteorite impact.

Once lithospheric spreading begins, the lithosphere-asthenosphere boundary steepens and the plates continue to slide downhill away from the new spreading center (Fig. 5–22). At this point, subduction must also begin; if the plate is growing at one end, it (or some other lithospheric plate) must be consumed back into the mantle elsewhere. As the old, cold lithosphere sinks into the mantle

at a subduction zone, it pulls on the rest of the plate like a weight on the edge of a tablecloth.

In this synthesis, plate motion is *initiated* by a mantle plume, which may be the result of deep heating or a meteorite impact. However, once the plate begins to move over the asthenosphere, the motion becomes self-perpetuating and may continue for tens or even hundreds of millions of years.

SUMMARY

•

The **plate tectonics theory** is the concept that the **lithosphere**, the outer, 100-kilometer-thick layer of the Earth, is segmented into seven major **plates**, which move relative to one another by gliding over the **asthenosphere**. Most of the Earth's major geologic activity occurs at huge fractures called **plate boundaries**. Alfred Wegener's theory of **continental drift** preceded the plate tectonics theory by four decades and was similar in many ways. Although it was based on accurate data, Wegener's theory was rejected because of a faulty explanation of how continents move.

The Earth's magnetic field reverses its orientation about every half-million years. **Rock magnetism** records the orientation of the Earth's field at the time rocks form. A pattern of alternating **normal** and **reverse magnetic polarity** in sea-floor rocks is arranged symmetrically about the mid-oceanic ridge. Comparison of this pattern with prehistoric reversals of the Earth's magnetic field led to the theory of **sea-floor spreading**. This theory states that new oceanic crust continuously forms at mid-oceanic ridges and spreads outward. The sea-floor spreading theory was expanded into the plate tectonics theory, which states that the entire lithosphere, not merely oceanic crust, forms and spreads outward from the mid-oceanic ridge.

The **core** is mostly iron and nickel and consists of a liquid outer layer and a solid inner sphere. The **mantle**

extends from the base of the crust to a depth of 2900 kilometers, where the core begins. The Earth's crust is its outermost layer and varies from 7 to 70 kilometers in thickness. The **lithosphere** is the cool, brittle outer 100 kilometers of the Earth, and it includes all of the crust and the uppermost mantle. The lithosphere "floats" on the hot, plastic **asthenosphere**, which extends from 100 to 350 kilometers in depth.

The concept that the lithosphere floats on the asthenosphere is called **isostasy**. When weight is added or subtracted from portions of the crust, it rises or falls. This vertical movement in response to changing burdens is called **isostatic adjustment**.

Tectonic plates move at rates that vary from 1 to 18 centimeters per year. Three types of plate boundaries exist: (1) new lithosphere forms and spreads outward at a **divergent boundary**, or **spreading center**; (2) two lithospheric plates collide at a **convergent boundary**, which develops into a **subduction zone** if at least one plate carries oceanic crust; and (3) two plates slide horizontally past each other at a **transform plate boundary**. Interior parts of lithospheric plates are tectonically stable. The cause or causes of plate motion are not well understood at present. A **mantle plume** may initiate plate movement. A plate may then continue to move because it slides downhill from a spreading center as its cold leading edge sinks into the mantle.

KEY TERMS

•

REVIEW QUESTIONS

•

1. Summarize the important aspects of the plate tectonics theory.

2. Describe the lithosphere and the asthenosphere.

3. How many major tectonic plates exist? List them.

4. Describe the three types of tectonic plate boundaries.

5. Explain why tectonic plate boundaries are geologically active and the interior regions of plates are geologically stable.

6. Describe the similarities and differences among the theories of continental drift, sea-floor spreading, and plate tectonics.

7. Why did Wegener consider that fossils of *Mesosaurus* found on both sides of the Atlantic Ocean constituted good evidence that the two fossil localities were once joined as a single continental mass?

8. Explain how Wegener's fossil evidence indicated when Pangaea began to rift apart.

9. How does a sedimentary rock record the orientation of the Earth's magnetic field at the time the sediment was deposited?

10. Describe the Mid-Atlantic ridge and the mid-oceanic ridge.

11. What is the magnetic orientation of a rock with normal polarity? What is reverse polarity?

12. What do rocks with normal and reverse polarity tell us about the history of the Earth's magnetic field?

13. Explain how a magnetic pattern in sea-floor rocks led to the theory of sea-floor spreading.

14. Draw a cross-sectional view of the Earth. List all the major layers and the thickness of each.

15. Describe the physical properties of each of the Earth's layers.

16. What properties of the asthenosphere allow the lithospheric plates to glide over it?

17. How is it possible for the solid rock of the mantle to flow and convect?

18. How might a mantle plume fracture the lithosphere?

19. Explain how convection of the mantle could cause movement of lithospheric plates. Explain how movement of lithospheric plates could cause mantle convection.

DISCUSSION QUESTIONS

•

1. Describe the similarities between Wegener's continental drift theory and modern plate tectonics theory. What are the major differences between the two theories?

2. Discuss the various mechanisms that have been suggested for the movement of tectonic plates. Attempt to decide which mechanism, if any, is preferable.

3. Although most earthquakes occur at plate margins, occasionally very large earthquakes occur within lithospheric plates. How might this happen?

4. Discuss the geologic effects of the impact of a large meteorite on the Earth's surface. Can you think of any major effects not mentioned in the box "Effects of Meteorite Impacts on the Earth"?

5. The interior of the Moon is much cooler than that of the Earth, and the lunar crust is thicker than that of Earth. Compare the probable geologic effects of the impact of a large meteorite on the Moon with the effects of a similar impact on the Earth.

Earthquakes and the Earth's Structure

6

E arthquakes are among the most dramatic of all geological events. They frequently shake many parts of the world, causing death, misery, and billions of dollars in damage. Yet geologists cannot reliably predict when or where a quake will occur.

People in the San Francisco area were startled when the ground began to shake on October 17, 1989. One resident driving home from work noticed that the road suddenly started to roll like sea waves. Mesmerized, she continued driving until the pavement fractured and rose in front of her. She slammed on the brakes to avoid crashing into it. Others were not so lucky. In Oakland the upper tier of a double-deck freeway collapsed, crushing motorists on the lower tier. Another resident reading quietly in his room heard the dishes on the table begin to rattle. Pictures on the wall swung back and forth, and the floor began to shake. The motion was so foreign to his sense of normal earthly stability that he did not recognize what was happening. He wondered: "Who is shaking my house?" Within a few moments, however, he realized: "The Earth is shaking; it's an earthquake!" Structural damage varied greatly throughout the Bay Area. Most

● Fractures in Loma Prieta, east of Santa Cruz, near the epicenter of the 1989 earthquake in the San Francisco Bay area.

buildings survived, but some were destroyed. Sixty-five people died.

About a million earthquakes occur worldwide every year. Most are too mild to be felt and are detected only with sensitive instruments. Many shake houses and rattle windows but cause little damage. A few destroy buildings or even entire cities. An average of about 10,000 earth-

quake fatalities occur every year. Major historical earthquakes are listed in Table 6–1.

6.1 What Is an Earthquake?

An **earthquake** is a sudden motion or trembling of the Earth. The motion is caused by the release of slowly

T A B L E 6 – 1

Major Historical Earthquakes					
Year	**Date**	**Region**	**Deaths**	**Richter Magnitude***	**Comments**
1556		China, Shensi	830,000		More deaths than in any other natural disaster in history
1663	Feb. 5	Canada, St. Lawrence River	?		Chimneys broken as far away as Massachusetts
1811	Dec. 16	Missouri, New Madrid	Several		Three shocks; largest historical earthquake in contiguous 48 states
1857	Jan. 9	California, Fort Tejon	?		San Andreas fault rupture
1868	Aug. 16	Ecuador and Colombia	Ecuador, 40,000 Colombia, 30,000		
1886	Aug. 31	South Carolina, Charleston-Summerville	About 60		Last major quake in eastern United States
1896	June 15	Japan, Riku-Ugo	22,000		Giant tsunami
1906	April 18	California, San Francisco	650	8.25†	San Francisco fire
1923	Sept. 1	Japan, Kwanto	143,000	8.2†	Great Tokyo fire
1960	May 21–30	Southern Chile	5700	8.5–8.7	Largest sequence of earthquakes ever recorded
1964	March 28	Alaska	131	8.6	Damaging tsunami
1970	May 31	Peru	66,000	7.8	Great rock slide destroyed the town of Yungay
1971	Feb. 9	California, San Fernando	65	6.5	$550 million damage
1976	July 27	China, Tangshan	250,000	7.6	Great economic damage; not predicted
1985	Sept. 19	Mexico, Michoacan	9500	7.9	More than $3 billion damage; 30,000 injured; small tsunami
1988	Dec. 7	Armenia	24,000	6.9	Death toll large due to poor construction
1989	Oct. 17	California, Loma Prieta	65	6.9	May be harbinger of additional Bay Area quakes

*Richter magnitudes were not recorded prior to 1935.
†Estimated magnitudes based on reported damage.

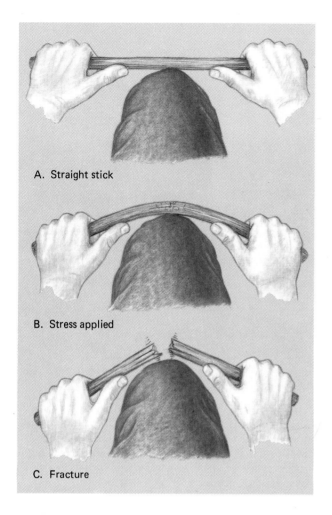

A. Straight stick

B. Stress applied

C. Fracture

Figure 6–1 When you bend a stick over your knee, elastic energy is stored as the stick bends, then is released as it snaps.

accumulated energy in rocks. What is the source of this energy, how does it accumulate in rocks, and why does it suddenly cause the Earth to shake?

Recall from Chapter 5 that the lithosphere is a layer of brittle rock, about 100 kilometers thick, that "floats" on the hot, plastic rock of the asthenosphere. The entire lithosphere is broken into seven large pieces and several smaller ones, called **tectonic plates**. As the plates glide over the hot, plastic asthenosphere, they slip past one another along immense fractures called plate boundaries. California's famous San Andreas fault is this kind of fracture. Although tectonic plates move continuously at rates between 1 and 18 centimeters per year, slippage along the fractures is not smooth and continuous.

Commonly, as two lithospheric plates move past one another, rocks along the fracture are held tightly by friction and cannot move. Thus, although the two plates continue to move past each other, no slipping occurs at the fault. **The interior of a plate can move while its edges do not because rock can stretch like an elastic band.** We do not commonly think of rock as elastic, but

if you drop a rock onto a cement floor, it bounces. Although the magnitude of the bounce is small, the rock behaves like other elastic objects, such as tennis balls.

Imagine that you are trying to break a stick by gradually bending it over your knee (Fig. 6–1). At first the two ends of the stick move, but the section bent over your knee is stationary. If you continue to pull, the stick suddenly snaps in two. The energy that you put into the stick by bending it was stored as **elastic energy**. When the stick breaks, that energy converts to motion as the two halves of the stick spring back to become straight again.

In a similar manner, when two tectonic plates move past one another, rock near the plate boundary stretches and stores elastic energy. When the energy overcomes the friction that is keeping the rocks from moving, the rocks jump along the fault (Fig. 6–2). The rocks may move from a few centimeters to a few meters, depending on the amount of stored energy. The sudden movement releases the elastic energy, and the rock springs back to its original shape. This rapid motion sets up vibrations that travel through the Earth like vibrations in a bell struck

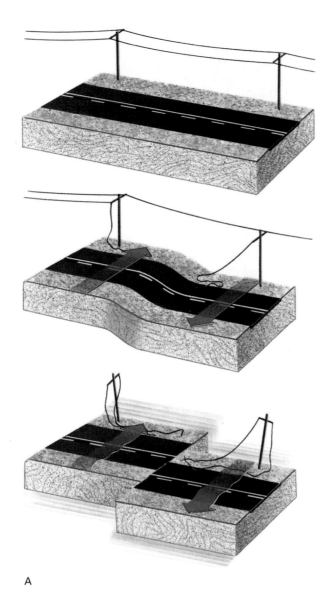

B

Figure 6–2 (A) When rock is stretched by a tectonic force, it stores elastic energy. Eventually the rock fractures and snaps back to its original shape but in a new position, creating an earthquake. (B) Fractures in a roadway in Santa Cruz following 1989 quake.

Figure 6–3 The Earth's major earthquake zones coincide with tectonic plate boundaries. Each yellow dot represents an earthquake that occurred between 1961 and 1967. Note the concentration of earthquakes along the San Andreas fault.

A

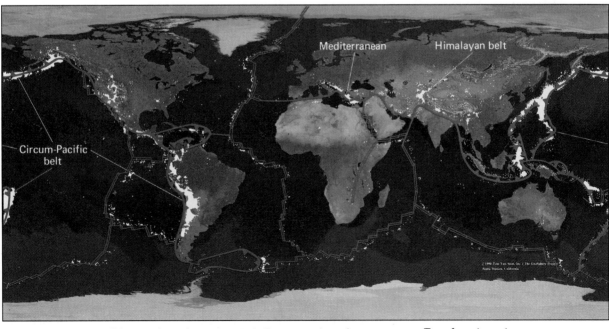

⊏⟹ Divergent boundary ▲▲ Convergent boundary ⎯⎯ Transform boundary

Earth Science and the Environment

Earthquake Danger in the Central and Eastern United States

No major earthquakes have occurred in the central or eastern United States in the past 100 years,* and no lithospheric plate boundaries are known in these regions. Therefore, one might conclude that earthquake danger is insignificant. Indeed, building codes in most major central and eastern cities do not require earthquake-resistant construction.

Today this complacency is being questioned. According to one seismologist:

> We were told by theory that plate interiors should be quiet areas. Then people started building nuclear power plants and started to look at the details of data and discovered, in fact, that there is a lot of seismicity in interior plates and that we know very little about it.[†]

The largest historical earthquake sequence in the contiguous 48 states occurred not in California, but in Missouri. In 1811 and 1812, three shocks (with estimated Richter magnitudes between 7.3 and 7.8) altered the course of the Mississippi River and made church bells ring 1500 kilo-

meters away in Washington, D.C. Another large quake (with an estimated Richter magnitude of 7.0) occurred in Charleston, South Carolina, in 1886.

Geologists have re-measured distances between old survey pins near New York City and found that the pins have moved significantly during the past 50 to 100 years. This motion indicates that the crust in this region is being deformed. If a major quake were to occur near New York City today, the consequences could be disastrous.

Causes of earthquakes in plate interiors are not as well understood as those at plate boundaries, and interior quakes are nearly impossible to predict. Some occur where thick piles of sediment were deposited to form great river deltas. The weight of sediment cannot be supported by the underlying lithosphere, and the lithosphere fractures. Other interior quakes may be caused by plate movement. As the plates glide horizontally over the asthenosphere, they may undergo some vertical motion above irregularities, or "bumps," in that plastic zone. Alternatively, under special circumstances, stress from a plate boundary may be transmitted hundreds of kilometers into the interior of a plate to produce an earthquake.

*A major earthquake is defined as one with a magnitude of 7.0 or higher on the Richter scale.

[†]Leonardo Seeber of Columbia University, as quoted in *The New York Times*, March 1, 1988.

---

by a hammer. The vibrations are felt as an earthquake. Figure 6–3 shows that most earthquakes happen along tectonic plate boundaries, where two plates slip past one another. Long, quiet intervals separate the sudden movements.

6.2 Earthquakes and Tectonic Plate Boundaries

Recall from Chapter 5 that a fault is a fracture in rock along which movement has occurred. In most cases, an earthquake starts on an old fault that has moved many times in the past and will move again in the future, simply because it is easier for rocks to move along an old fracture than for tectonic forces to create a new fault. Therefore, earthquakes occur over and over again in the same places, along the same faults. Although many faults occur within tectonic plates, the largest and most active faults are the boundaries between tectonic plates. Therefore, as Figure 6–3 shows, most earthquakes occur along plate boundaries.

Earthquakes and Transform Plate Boundaries: The San Andreas Fault

The San Andreas fault runs parallel to the California coast (Fig. 6–3). It is the boundary between two plates that are sliding horizontally past one another. The Pacific plate is moving northwest relative to the North American plate at a rate of about 3.5 centimeters per year. Along some portions of the fault, rocks on opposite sides slip past one another at a continuous, snail-like pace. This type of movement is called **fault creep**. In Hollister, California, old houses were inadvertently built directly over the fault. Slowly, millimeter by millimeter, fault creep has torn the houses in two. The movement occurs without violent and destructive earthquakes because the rocks move only a little bit at a time, continuously and slowly.

Along other portions of the San Andreas fault, friction binds rocks on opposite sides of the fault together, keeping those parts of the fault motionless for decades. Nevertheless, the plates continue to move past one another. Elastic energy accumulates as the movement deforms rock near the fault. Because the plates move past

The San Andreas fault slices the Earth's surface in San Luis Obispo County, California. *(R. E. Wallace, USGS)*

one another at 3.5 centimeters per year, 3.5 meters of elastic deformation accumulate over a period of 100 years. When the elastic energy exceeds the friction binding the fault, the rock suddenly slips along the fault and snaps back to its original shape. During the San Francisco earthquake of 1906, rocks along the fault suddenly moved 4.5 to 6 meters.

Subduction Zone Earthquakes

In a subduction zone, a cold, brittle lithospheric plate slowly sinks into the mantle as it dives beneath another plate. In most places the subducting plate does not sink smoothly by fault creep, but with intermittent slips and jerks, giving rise to numerous earthquakes. The earthquakes concentrate in the Benioff zone along the upper part of the sinking plate, where it scrapes past the opposing plate. Thus, earthquake distribution in a subduction zone enables geologists to locate the top of the subducting plate. Subduction zone earthquakes of this type occur frequently along the west coast of South America, in Japan, Alaska, the Aleutian Islands, and in other places where subduction is active today. Many of the world's most destructive earthquakes occur in subduction zones.

A long time is required for a cold subducting plate to become hot and plastic as it sinks into the mantle.

The 1964 Alaska earthquake destroyed much of Anchorage. *(Ward's Natural Science Establishment, Inc.)*

Therefore, a sinking plate remains brittle and generates earthquakes even when it has sunk to a depth of a few hundred kilometers. For this reason, the deepest earthquakes occur in subduction zones. Although very few occur below about 350 kilometers, the deepest earthquake ever recorded occurred at a depth of 700 kilometers. The fact that no deeper earthquakes occur indicates that at greater depths rocks become too hot and plastic to fracture. Instead, they flow as a very viscous fluid.

Recall from Chapter 5 that subduction is occurring today along the coasts of Oregon, Washington, and southern British Columbia in the Pacific Northwest. Here, an oceanic plate is diving beneath the continent at a rate of 3 to 4 centimeters per year. Magma generated in the subduction zone rises to erupt from Mount St. Helens, Mount Rainier, and other Cascade Range volcanoes. One would also expect many earthquakes at such an active plate boundary. Small earthquakes occur occasionally in the region, but no large ones have occurred in the past 150 to 200 years.

Why are earthquakes relatively uncommon in this tectonically active region? The answer to this question is important because the Pacific Northwest is densely populated and heavily industrialized. Two possible answers exist. Subduction may be occurring slowly and continuously by fault creep. Alternatively, rocks along the fault may be locked together by friction, accumulating a huge amount of elastic energy that will be released in a giant, destructive quake sometime in the future.

Recently geologists have discovered evidence of great prehistoric earthquakes in the Pacific Northwest. A major coastal earthquake commonly creates violent sea waves that deposit a layer of mud and sand along the coast. Six such layers have formed in this region within the past 7000 years. Other evidence indicates that the coastline suddenly dropped by 0.5 to 2 meters as each of the sediment layers formed. This information suggests that continuous movement of plates along that subduction zone is not accommodated by creep or frequent small earthquakes. Instead, friction has locked the plates together for an average of 1150 years between large, devastating earthquakes. Thus, many geologists anticipate another major, destructive earthquake in the Pacific Northwest sometime in the future.

On April 25th, 1992, a magnitude 6.9 earthquake rocked northern California and southern Oregon. Centered near Eureka, California, it caused damage locally and was felt as far away as Reno, Nevada. Two aftershocks of magnitudes 6.5 and 6.0 followed on the next day. These earthquakes occurred in the area where the Pacific Northwest subduction zone joins the San Andreas fault system. Although they caused some damage, these quakes were not the destructive "major" events predicted for the region.

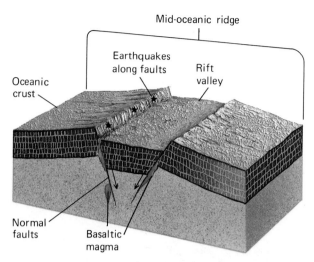

Figure 6–4 Blocks of oceanic crust drop downward, forming a rift valley in the center of the mid-oceanic ridge, as two lithospheric plates spread apart. Earthquakes are common along the faults.

Earthquakes at Spreading Centers

Many earthquakes occur along the mid-oceanic ridge as a result of fractures and faults that form as the two plates separate. Blocks of oceanic crust drop downward along the faults, forming the rift valley in the center of the ridge (Fig. 6–4). Only shallow earthquakes occur along the mid-oceanic ridge because here the asthenosphere rises to within 20 to 30 kilometers of the Earth's surface and is too hot and plastic to fracture.

6.3 Earthquake Waves

If you have ever bought a watermelon, you know the challenge of trying to pick a ripe, juicy one without being able to look inside. One trick is to tap the melon gently with your knuckle. If you can hear a sharp, clean sound, it is probably ripe; a dull thud indicates that it may be overripe and mushy. This illustrates two points that can also be applied to the Earth: (1) the energy of your tap is transmitted through the melon, and (2) the quality of the sound is affected by the nature of the interior of the melon.

Waves transmit energy from one place to another. Thus, a drumbeat travels through air to your ear as waves, and the Sun's heat travels to Earth as waves. Similarly, a tap travels through a watermelon in waves. Waves that travel through rock are called **seismic waves**. Seismic waves are initiated naturally by earthquakes. They can also be produced artificially by explosions. **Seismology** is the study of earthquakes and of the nature of the Earth's interior based on evidence provided by seismic waves.

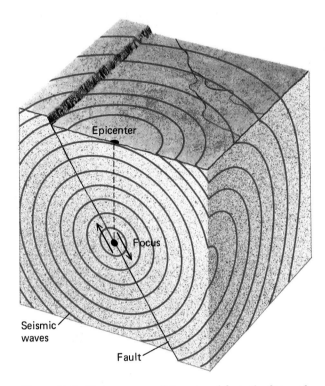

Figure 6–5 Body waves radiate outward from the focus of an earthquake.

An earthquake produces several different types of seismic waves. **Body waves** travel through the Earth's interior. They start at the earthquake's **focus**, where rocks move along a fault, and radiate outward in concentric spheres (Fig. 6–5). The point on the Earth's surface directly above the focus is the **epicenter**. During an earthquake, **surface waves** radiate away from the epicenter along the surface of the Earth like the waves that form when you throw a rock into a calm lake.

Body Waves

Two main types of body waves travel through the Earth's interior. A **primary wave**, or **P wave**, forms by alternate compression and expansion of the rock (Fig. 6–6A). Consider a long spring such as the popular Slinky toy. If you stretch a Slinky and strike one end, a compressional wave travels along its length. Sound also travels as a compressional wave. A ringing bell produces a sound wave in air, which is a type of P wave. Liquids and solids also transmit P waves. Next time you take a bath, immerse your head until your ears are underwater and listen as you tap the sides of the tub with your knuckles. In a similar manner the music from a neighbor's radio in the apartment next

Figure 6–6 Two different types of body waves travel through the Earth. Their respective characteristics are shown by a spring and a rope. (A) A compressional, or P, wave travels along the spring. The particles in the Slinky move parallel to the direction in which the wave itself travels. (B) An S wave travels along a rope, but the particles in the rope move perpendicular to the direction in which the wave travels.

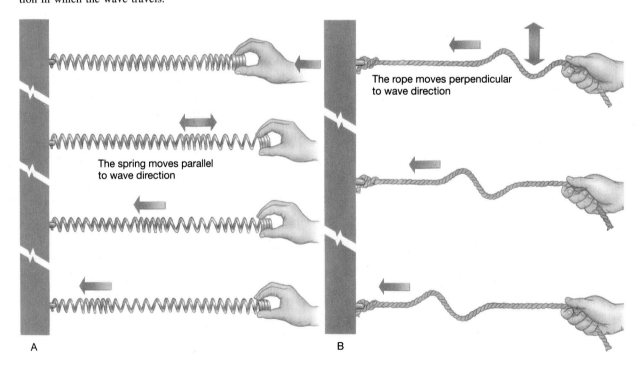

door travels easily through the walls when you are trying to study geology.

P waves travel at speeds between 4 and 7 kilometers per second in the Earth's crust and at about 8 kilometers per second in the upper mantle. As a comparison, the speed of sound in air is only 0.34 kilometer per second, and the fastest jet fighters fly at about 0.85 kilometer per second. Therefore, even the slowest P waves in the Earth travel more than ten times faster than the speed of sound in air, and more than five times faster than a jet fighter. P waves are called primary waves because they are so fast that they are the first, or *primary*, waves to reach an observer.

A second type of body wave, called a **secondary wave** or **S wave**, can be illustrated by tying a rope to a wall, holding the end, and giving it a sharp up-and-down jerk (Fig. 6–6B). An S-shaped wave moves from your hand to the wall.

S waves are slower than P waves, traveling at speeds between 3 and 4 kilometers per second in the crust. As a result, they arrive after P waves and are the *secondary* waves to reach an observer on Earth.

Unlike P waves, S waves move *only through solids.* Because molecules in liquids and gases are only weakly bound together, they slip past each other and thus cannot transmit S waves. As we shall see in later sections, this difference is important for geologists studying the interior of the Earth. For example, we know that the Earth's outer core is liquid because P waves travel through it but S waves do not.

Surface Waves

Surface waves, called **L waves** (for *l*ong waves), travel more slowly than either type of body wave. Two types of surface waves occur simultaneously (Fig. 6–7). One is an up-and-down motion. Recall from the introduction that a woman driving home during the 1989 San Francisco earthquake noticed that the road was rolling like an ocean wave. This motion was one form of L wave. The second type of L wave is a side-to-side vibration. Think of a

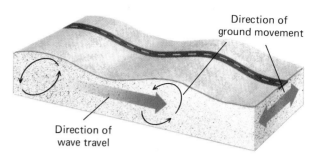

Figure 6–7 Surface waves move up and down, like ocean waves, and also from side to side.

snake writhing from side to side as it crawls along the ground. Surface waves cause the most property damage because they generate more ground movement than body waves.

Measurement of Seismic Waves

Earthquakes are detected and measured with a device called a **seismograph**. To understand how a seismograph works, imagine writing a letter while riding in an airplane. If the plane hits turbulence, your handwriting becomes wiggly. Because the paper is on a tray that in turn is connected to the frame of the aircraft, the paper moves with the plane when the plane bounces. But your hand is connected to your body by a series of movable joints that flex when the plane lurches. Inertia keeps your hand stationary as the plane moves back and forth beneath it,

Surface waves cause the greatest ground motion, and therefore the greatest destruction, during an earthquake.

● MEMORY DEVICE

To help in recalling the different types of earthquake waves, think of a P wave as a *p*ressure, or compressional, wave. As S wave transmits *s*hearing forces through rock and can be thought of as a *s*hear wave. Surface waves are L waves (for *l*ong waves). They are the *l*ast type of wave to arrive because they travel more slowly than other waves.

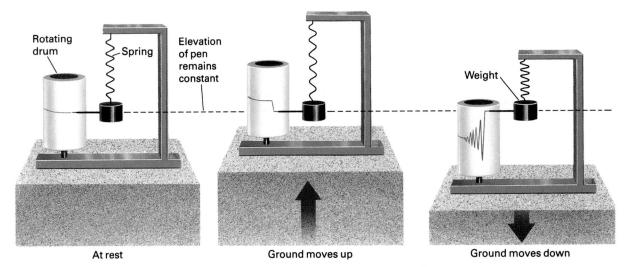

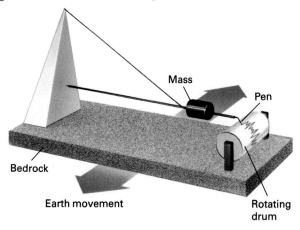

Figure 6–8 A seismograph records ground motion during an earthquake. When the ground is stationary, the pen draws a straight line across the rotating drum. When the ground rises abruptly during an earthquake, it carries the drum up with it. But the spring stretches so that the weight and the pointer hardly move. Therefore, the pointer points lower on the scale. Conversely, when the ground sinks, the pen rises. During an earthquake, the pen traces a jagged line on the rotating drum, thus recording the up-and-down movement of the ground.

so your hand does not move as erratically as the plane. The paper jiggles back and forth beneath a relatively motionless hand, and your handwriting becomes erratic.

One type of seismograph works on the same principle. Imagine a weight suspended from a spring. A pointer attached to the weight is aimed at the zero mark on a scale (Fig. 6–8). The scale is attached firmly to bedrock by solid metal braces, but the weight and pointer hang from the flexible spring. During an earthquake, the scale jiggles up and down, but inertia keeps the suspended

Figure 6–9 Another type of seismograph measures horizontal ground motion. The weight and pen are suspended by a thin wire. During an earthquake, the drum and paper move from side to side, and the stationary pen scribes a record of ground movement on the rotating drum.

weight stationary. As a result, the scale moves up and down beneath the pointer. If the scale is replaced by a rotating drum, and a pen is mounted on the pointer, then the pen records earthquake motion on the rotating drum. This record of Earth vibration is called a **seismogram**.

To measure horizontal motion, the weight and pen are suspended on a pendulum (Fig. 6–9). If the Earth shifts toward the left, the rotating drum moves left, but inertia keeps the weight stationary. As a result, a line is drawn toward the right. During an earthquake, the Earth jiggles back and forth, producing a wiggly line on the paper.

An earthquake measuring station generally has at least three seismographs. Two horizontal seismographs measure east-west and north-south movements. A third instrument records vertical vibrations. In this manner, waves of different types and with different directions of motion are recorded.

6.4 Locating the Source of an Earthquake

If you have ever watched an electrical storm, you may have used a simple technique for estimating the distance between you and the place where the lightning is striking. You watch for the flash of the lightning bolt and then count the seconds until you hear the thunder. When an electrical discharge occurs, the thunder and lightning are produced simultaneously. But light travels so rapidly that it reaches you virtually instantaneously, whereas sound travels much more slowly, at 340 meters per second.

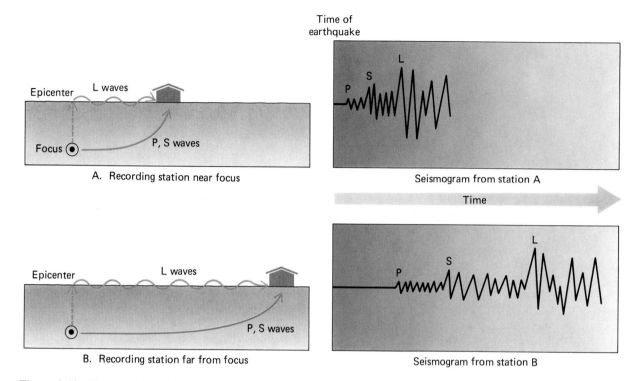

Figure 6–10 The time intervals between arrivals of P, S, and L waves at a recording station increase with distance from the focus of an earthquake.

Therefore, if the time interval between the flash and the thunder is 1 second, then the lightning struck about 340 meters away.

The same principle is used to determine the distance from a recording station to an earthquake. Recall that P waves travel faster than S waves and that L waves are slower yet. If a seismograph is located close to an earthquake epicenter, the different waves will be separated by only short times for the same reason that the thunder and lightning come close together when a storm is close. On the other hand, if a seismograph is located farther away from the earthquake, the S waves will arrive at correspondingly later times after the P waves arrive, and the L waves will be even farther behind, as shown in Figure 6–10.

Geologists use **time-travel curves** to quantify this general relationship between distance from an earthquake epicenter and arrival times of the different types of waves. To *make* a time-travel curve you must know both when and where an earthquake started. A number of seismic stations at different locations on the Earth then record the time of arrival of earthquake waves, and a graph such as Figure 6–11 is drawn. This graph can then be *used* to measure the distance between a recording station and an earthquake whose epicenter is unknown. Time-travel

Figure 6–11 A time-travel curve. With this graph you can calculate the distance from a seismic station to the epicenter of an earthquake. In the example shown, a 3-minute delay between the first arrivals of P waves and S waves corresponds to an earthquake with an epicenter 1900 kilometers from the seismic station.

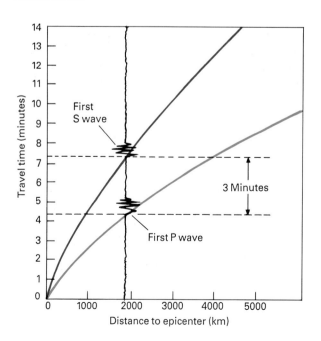

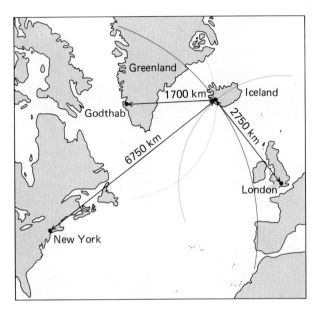

Figure 6–12 Locating an earthquake's epicenter. The distance from each of three seismic stations to the epicenter is determined from time-travel curves. The three distance circles intersect at the epicenter.

curves were first drawn from data obtained from natural earthquakes. However, a problem with using a natural earthquake is that its location and time are not always known precisely. In the 1950s and 1960s, geologists studied seismic waves from underground atomic bomb tests to refine their data. During these tests, both the location and timing of the events were known.

Figure 6–11 shows us that if the first P wave arrives 3 minutes before the first S wave, the recording station is about 1900 kilometers from the epicenter. But this distance by itself gives no information about direction; it does not indicate whether the earthquake originated to the north, south, east, or west. To solve this problem and to pinpoint the location of an earthquake, geologists compare data from three separate recording stations. Imagine that a seismic station in New York records an earthquake with an epicenter 6750 kilometers away. From this information, geologists know that the epicenter must lie on a circle surrounding New York at a distance of 6750 kilometers. Part of this circle is shown by the red arc in Figure 6–12. But the same epicenter is also reported to be 2750 kilometers from a seismic station in London

Destruction caused by the Manjil, Iran, earthquake of June 21 and 22, 1990. The earthquake had a magnitude of 7.7 on the Richter scale and killed more than 50,000 people. *(M. Shandiz/Sygma)*

and 1700 kilometers from one in Godthab, Greenland. If three such circles are drawn, one for each recording station, the arcs intersect at only one point, and that is the epicenter of the quake.

6.5 Earthquakes and Humans

Measurements of Earthquake Strength

Earthquakes vary from gentle tremors that cannot be detected without a seismograph, to destructive giants that create large-scale movements of the Earth's surface and catastrophic losses of life and property. Before seismographs were in common use, earthquake strength was evaluated on scales based on human experience and damage to buildings. One such scale, devised in 1902, is called the **Mercalli scale** (Table 6–2). This scale of earthquake **intensity** measures the effects of an earthquake at a particular place. Although useful, the Mercalli scale is qualitative and subjective. For example, some people exaggerate their experiences and others underestimate them. Therefore, it may be difficult to decide whether a given quake was a level V ("felt by nearly everyone") or a level VI ("many frightened and run outdoors"). In addition, human experience and structural damage depend on a variety of factors including distance from the focus, the rock and soil where the quake occurs, and the quality of construction in the area.

A quantitative scale based on seismograph measurements was first refined by Charles Richter in 1935, and today the **Richter scale** is used almost universally. It measures the **magnitude** of an earthquake from the amplitude (height)* of the largest wave recorded on a seismograph. Because seismic stations from all around the globe must be able to calculate the same magnitude for an earthquake, seismologists have designed a specific type of seismograph as the standard recording device. Adjustments are made for the distance from each recording station to the earthquake.

The Richter scale is logarithmic: an increase of one unit on the scale represents a ten-fold increase in the amplitude of a seismic wave. Thus, the amplitude of the largest seismic wave produced by a magnitude 7 quake is ten times greater than that from a magnitude 6 quake and 100 times greater than that from a magnitude 5 quake. An increase of one unit—for example, from 7 to 8—on the Richter scale corresponds approximately to a 30-fold increase in energy released during the quake. Thus, a magnitude 8 quake releases 30 times as much energy as

TABLE 6–2

Modified Mercalli Intensity Scale of 1931 (Abridged)

I. Not felt except by a very few under especially favorable circumstances.

II. Felt only by a few persons at rest, especially on upper floors of buildings.

III. Felt quite noticeably indoors, especially on upper floors, but many people do not recognize it as an earthquake. Vibration like passing truck.

IV. During the day felt indoors by many, outdoors by few. At night some awakened. Dishes, windows, doors disturbed; walls make cracking sound. Sensation like heavy truck striking building.

V. Felt by nearly everyone; many awakened. Some dishes, windows, and so on, broken; a few instances of cracked plaster; unstable objects overturned. Disturbance of trees, poles, and other tall objects sometimes noticed. Pendulum clocks may stop.

VI. Felt by all; many frightened and run outdoors. Some furniture moved; a few instances of damaged chimneys. Damage slight.

VII. Everybody runs outdoors. Damage *negligible* in buildings of good construction, *slight* to moderate in well-built ordinary structures, *considerable* in poorly built or badly designed structures.

VIII. Damage *slight* in specially designed structures, *considerable* in ordinary substantial buildings, *great* in poorly built structures. Fall of chimneys, factory stacks, columns, monuments, walls.

IX. Damage *considerable* in specially designed structures; well-designed frame structures thrown out of plumb; damage *great* in substantial buildings, with partial collapse. Ground cracked conspicuously. Underground pipes broken.

X. Some well-built wooden structures destroyed; most masonry and frame structures destroyed; ground badly cracked. Considerable landslides from river banks and steep slopes.

XI. Few if any (masonry) structures remain standing; bridges destroyed. Broad fissures in ground; underground pipelines completely out of service. Earth slumps and land slips in soft ground.

XII. Damage total. Waves seen on ground surface. Lines of sight and level distorted. Objects thrown upward into the air.

a magnitude 7 quake and 900 times as much as a magnitude 6 quake. An earthquake with a magnitude of 6.5 has an energy of about 10^{21} (10 followed by 21 zeros) ergs.[†]

---

*Amplitude is the height of a wave, as measured by the distance from the flat zero point to the top of the crest.

[†]An erg is the standard unit of energy in scientific usage. One erg is a small amount of energy. Approximately 3×10^{12} ergs are needed to light a 100-watt light bulb for 1 hour. However, 10^{21} ergs is a very large number and represents a considerable amount of energy.

The atomic bomb dropped on the Japanese city of Hiroshima at the end of World War II released about this much energy. The upper limit of the magnitude of an earthquake is determined by the strength of rocks. A strong rock can store more elastic energy before it fractures than a weak rock. The largest earthquakes ever observed had magnitudes of 8.5 to 8.7, about 900 times greater than the energy released by the Hiroshima bomb.

Earthquake Damage

Ground Motion

During an earthquake, waves travel both along the surface of the Earth and through subterranean rock. The ground undulates; buildings and bridges may topple and roadways fracture. Most of the people killed during an earthquake are crushed by falling debris.

As previously mentioned, earthquake damage depends not only on the magnitude of the quake and its proximity to population centers but also on rock and soil types and the quality of the construction in the affected area. The quake in Armenia in 1988 and the San Francisco quake in 1989 both measured 6.9 on the Richter scale. More than 28,000 people died in the Armenian quake, whereas the death toll in the Bay Area was only 65. The tremendous mortality in Armenia occurred because buildings were not engineered to withstand earthquakes. Proper engineering and construction in an earthquake-prone area demand both a foundation on stable rock or soil and a structure that can withstand movement.

Figure 6–13 Horizontal displacement along a fault offset this fence during the 1906 San Francisco earthquake. *(G. K. Gilbert/USGS)*

Figure 6–14 A fire caused by the 1989 San Francisco earthquake. *(Michael Williamson/Sygma)*

Permanent Alteration of Landforms

As previously explained, the Earth's surface moves both horizontally and vertically during an earthquake. As much as 6 meters of horizontal displacement occurred during the 1906 San Francisco quake (Fig. 6–13). The New Madrid, Missouri, earthquake of 1811 changed the course of the Mississippi River. Permanent displacement obvi-

Figure 6–15 The landslide that buried the town of Yungay. *(USGS)*

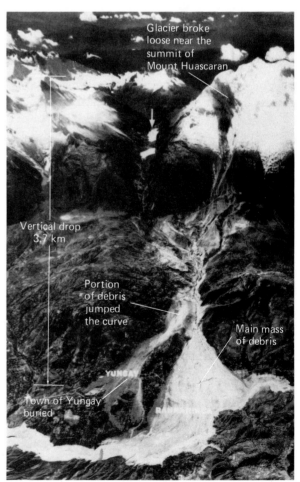

ously destroys any structures built across a fault. Although relatively few buildings are likely to be situated right on a fault, many roadways and pipelines cross faults and are destroyed by such displacements.

Fire

When rock moves on a fault, buried gas mains and electrical wires crossing the fault may be torn apart, leading to fire (Fig. 6–14). In addition, water mains may rupture, so fire fighters may not have water. Much of the damage from the 1906 San Francisco earthquake resulted from fires.

Landslides

When the Earth trembles, hillsides may shake loose, and large quantities of rock and soil can slide downslope (Fig. 6–15). An earthquake in Peru in 1970 triggered a landslide that buried the town of Yungay, killing 17,000 people.

Tsunamis

Imagine an earthquake epicenter beneath the sea. When part of the sea floor drops, as shown in Figure 6–16, the water drops with it. Almost immediately, water from the surrounding area rushes in to fill the depression, forming a wave. Sea waves produced by an earthquake are often called "tidal waves," but because they have nothing to do with tides, geologists prefer to call them by their Japanese name, **tsunami**.

In the open sea, a tsunami is so flat and spread out that it is barely detectable. Typically, the crest may be only 1 to 3 meters high, and successive crests may be

Figure 6–16 A tsunami develops when part of the sea floor drops during an earthquake. Water rushes in to fill the low spot, but the inertia of the rushing water forces too much water into the area, creating a bulge in the water surface. The long, shallow waves can build up into destructive giants when they reach shore.

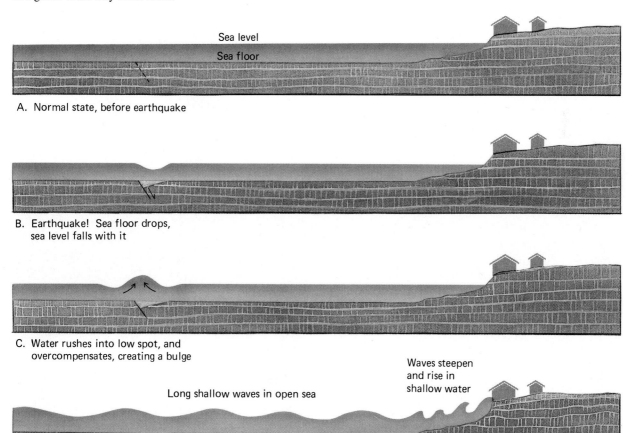

A. Normal state, before earthquake

B. Earthquake! Sea floor drops, sea level falls with it

C. Water rushes into low spot, and overcompensates, creating a bulge

D. Tsunami generated

An earthquake in Chile in May 1960 created a tsunami that flooded Hilo and other parts of the Hawaiian coast, halfway across the Pacific Ocean. *(Ward's Natural Science Establishment, Inc.)*

more than 100 to 150 *kilometers* apart. However, tsunamis travel rapidly and are very destructive when they reach land. A tsunami may attain speeds of 750 kilometers per hour (450 miles per hour). When the wave approaches shore, its base drags against the ocean floor and slows down. This compresses the wave, and the distance between successive crests decreases as the wave height increases rapidly. The rising wall of water then flows inland. A tsunami can flood the land for as long as 5 to 10 minutes before it withdraws.

One of the worst tsunamis in history struck the eastern coast of Japan in 1896. The wave was probably formed by a submarine earthquake in the western Pacific Ocean. As it approached shore, the water rose about 30 to 35 meters (about 100 feet) above high-tide level, flooding villages and killing 26,000 people. The same wave was 3 meters high when it reached Hawaii, where it destroyed buildings near the coast; minor effects were felt on the west coast of North America. The wave then bounced off the coast and sped westward, back across the Pacific Ocean, until it reached New Zealand and Australia. By then it had lost enough energy that it was no longer destructive.

6.6 Earthquake Prediction

About 10 percent of the U.S. population and industrial resources are located in California. About 85 percent of the people and industry in California are concentrated along the west coast from San Francisco to San Diego. This area straddles the great San Andreas fault zone.

The fault zone is the boundary between the North American lithospheric plate to the east and the Pacific Ocean plate to the west. At present, the Pacific plate is moving northwest relative to the North American plate at a rate of about 3.5 centimeters per year. This motion has produced many earthquakes in recent times. A few have been severe. One occurred near Los Angeles in 1857, and another destroyed San Francisco in 1906. The most recent large quake occurred in 1989 and was centered south of San Francisco. Hundreds of thousands of smaller quakes have occurred along the fault zone. Geologists of the United States Geological Survey recorded 10,000 earthquakes in 1984 alone, although most could be detected only with seismographs.

The fact that the San Andreas fault zone is part of a major boundary between lithospheric plates tells us that more great earthquakes are inevitable in the future. Can the next large one be predicted with enough reliability to evacuate threatened areas and reduce loss of life?

Long-Term Prediction

Motion along a fault is not uniform. Three types of motion can occur:

1. On segments of the fault where creep occurs, the plates slip past one another smoothly and without major earthquakes.
2. In other segments of the fault, the plates pass one another in a series of small "hops," causing numerous small, nondamaging earthquakes.

This map shows potential earthquake damage in the United States. The predictions are based on records of frequency and magnitude of historical earthquakes. *(Ward's Natural Science Establishment, Inc.)*

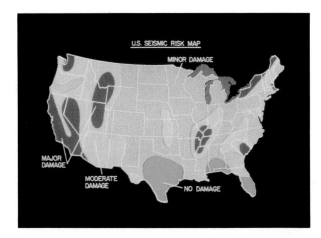

3. In still other segments of the fault, plates become locked together for tens to hundreds of years and then produce catastrophic earthquakes when they break free.

The basic concept of long-range earthquake prediction is straightforward: all segments of a fault such as the San Andreas eventually must move by the same amount because the fault is the boundary between two tectonic plates that are moving past each other. To understand how this concept applies to earthquake prediction, consider the following example. In 1857 a major earthquake occurred along the Mojave segment of the San Andreas fault, near Los Angeles. This segment is shown in pink in Figure 6–17. Most of the stress on this segment must have been relieved when the earthquake occurred and the rocks rebounded to a relaxed state. The Mojave segment has not moved since 1857, but the same fault to the north and south has moved 4.6 meters since 1857 by fault creep and frequent small quakes. Thus, the accumulated displacement of rock in the Mojave segment must be about 4.6 meters. This stretching is stored as elastic energy, waiting for a sudden fracture to be released. Therefore, a major earthquake is expected there sometime soon. Such an immobile region of a fault bounded by moving segments is called a **seismic gap**.

Studies of landforms also suggest that a major earthquake is due in the Mojave segment. Horizontal movement of a fault creates offsets in streams, ridges, and other landforms. By studying offset landforms, geologists have found evidence for 12 major earthquakes in the Mojave district in the past 1400 years. The time between successive quakes has ranged from 50 to 300 years, with an average interval of 117 years. As of 1992, 135 years have passed since the last major earthquake.

The advantage of this type of analysis is that it makes it relatively easy to identify zones of high earthquake hazard. Since geologists are nearly certain that a major earthquake will occur in the Mojave district "in the near future," stringent building codes have been established. All commercial and industrial facilities must be built to withstand earthquakes, and structures such as nuclear power plants are prohibited in the most vulnerable areas. The disadvantage of such an analysis is that the phrase "in the near future" is too vague. As shown in Figure 6–17, the probability is greater than 40 percent that a major earthquake will occur in the Mojave district by the year 2018.

In April 1992, a magnitude 6.0 earthquake occurred in the Mojave Desert near Palm Springs, California. Then, two more earthquakes measuring 7.4 and 6.5 on the Richter scale were recorded in July in the same general area. All of these quakes occurred along the San Andreas fault

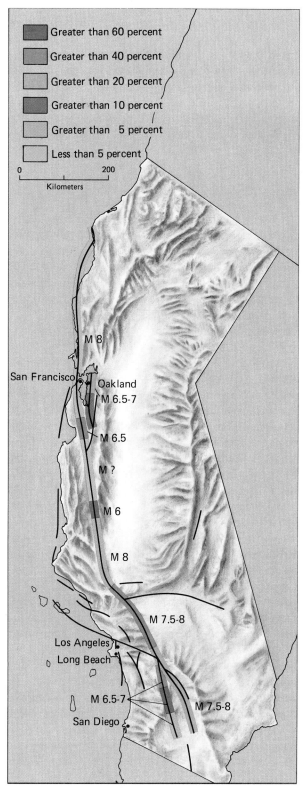

Figure 6–17 The probability of an earthquake during the next 30 years along segments of the San Andreas fault. *(R. L. Wesson and R. E. Wallace, "Predicting the Next Great California Earthquake," Scientific American, Feb. 1985.)*

zone, but not on the San Andreas fault itself. Tom Henyey, Director of the Southern California Earthquake Center, expressed concern that this activity may be a precursor for a "major" earthquake in southern California in the near future. The practical definition of a "major" quake depends both on magnitude and location. The 7.4 tremor in the desert caused relatively little damage and only one death, compared to the 6.9 magnitude San Francisco earthquake of 1989 that caused $4 billion damage and left 65 dead.

Short-Term Prediction

Long-term predictions give enough information to identify hazard areas and to establish building codes, but they do not provide a warning for evacuation of an area just prior to an earthquake. Short-term prediction depends on a reliable early warning system—a signal or group of signals that immdiately precedes an earthquake.

Foreshocks are small earthquakes that precede a large quake by an interval ranging from a few seconds to a few weeks. The cause of foreshocks can be explained by a simple analogy. If you try to break a stick by bending it slowly, you may hear a few small cracking sounds just before the final snap. If foreshocks consistently preceded major earthquakes, they would be a reliable tool for short-term prediction. However, only about half of the major earthquakes in recent years were preceded by a significant number of foreshocks. At other times, swarms of small shocks that could have been foreshocks were recorded, but a large quake did not follow.

Another approach to earthquake prediction is to measure changes in the shape of the land surface. California seismologists monitor rising bulges and other unusual Earth movements with tiltmeters and laser surveying instruments. The concept behind these studies is that distortions of the crust precede major earthquakes. Some earthquakes have been successfully predicted with this method, but in other instances predicted quakes did not occur, or quakes occurred that had not been predicted.

Chinese scientists reported that just prior to the 1975 quake in the city of Haicheng, snakes crawled out of their holes, chickens refused to enter their coops, cows broke their halters and ran off, and even well-trained police dogs became restless and refused to obey commands. Some researchers in the United States have attempted to quantify the relationship between animal behavior and earthquakes. In one study near San Francisco, scientists asked animal trainers and zoo keepers to report unusual behavior. The scientists received a flurry of calls—*after* a magnitude 5.7 earthquake—from people who reported that they had neglected to call earlier, but now that they thought about it, they remembered that some animals had behaved strangely. However, there was no significant increase in calls *before* the quake, so no predictions were made.

Social and Economic Factors in Earthquake Prediction

In the early 1970s, Chinese geophysicists used seismic gap theory and historical data to issue a long-range warning for a major earthquake near the city of Haicheng. Seismic stations were established in and around the city. In January 1975, swarms of foreshocks were recorded. In addition, tiltmeters measured unusual bulges and other changes in the shape of the land. When the earthquake swarms became especially intense on February 1, authorities ordered an evacuation of portions of the city. The evacuation was completed on the morning of February 4, and in the early evening of the same day, a large earthquake destroyed houses, apartments, and factories, but very few deaths occurred.

Immediately after that success, geologists hoped that a new era of quake prediction had begun. But a year later, Chinese scientists failed to predict an earthquake in the adjacent city of Tangshan. This major quake was *not* preceded by a swarm of foreshocks, no warning was given, and at least 250,000 pople died.* Shortly after that failure, a quake was predicted in a third city and the city was evacuated, but the earthquake did not occur.

Not only is short-term earthquake prediction a formidable scientific problem, but it also involves political, social, and economic issues. Imagine that geologists issued a warning to the city of San Francisco that a major earthquake was imminent. The scientists might be right or wrong. What are officials to do: ignore the warning, or shut down businesses, close trade and commerce, and evacuate millions of people? Such draconian measures would cost billions of dollars, and what if nothing happened? On the other hand, if a prediction were given and nothing were done, and the earthquake did occur, many people would die needlessly. Either way, the consequences of the wrong choice are severe.

6.7 Studying the Earth's Interior

Let us return to our image of tapping a watermelon to decide if it is ripe. Just as sound waves tell us about the interior of a watermelon, geologists use seismic waves to study the Earth's interior. In fact, it was by studying the behavior of seismic waves passing through the Earth that geologists discovered the Earth's layers. To understand how seismic waves tell us about the Earth's interior, we must consider several properties of waves.

*Accurate reports of the death toll are unavailable. Published estimates range from 250,000 to 650,000.

Figure 6–18 If you place a pencil in water, the pencil appears bent. It actually remains straight, but our eyes are fooled because light rays bend, or refract, as they cross the boundary between air and water.

1. The velocity of a seismic wave depends on the kind of rock it travels through. In addition, temperature, density, rigidity, and other properties of rock affect wave velocity.

2. A wave **refracts** (bends) and sometimes **reflects** (bounces back) as it passes from one transmitting medium into another. If you place a pencil in a glass half filled with water, the pencil appears bent. Of course, the pencil does not really bend; it is the light rays that bend. Light rays slow down when they pass from air to water, and as the velocity changes they refract (Fig. 6–18). A mirror reflects light from your face when you look into it. Seismic waves both refract and reflect as they pass from one medium into another.

3. P waves are compressional waves and travel through all media—gases, liquids, and solids—whereas S waves can travel only through solids.

During the late 1800s and the early 1900s, geologists discovered that earthquake waves both refract and reflect sharply at certain depths within the Earth. These changes show that the Earth is composed of three major layers: the core, the mantle, and the crust.

Discovery of the Core

If the entire Earth were composed of one type of rock, then P and S waves would travel everywhere and would be detected anywhere on the planet. However, Figure 6–19 shows that direct S waves reach the Earth's surface only up to 105° from the epicenter of an earthquake. Thus, geologists reasoned, the inner part of the Earth must be unable to transmit S waves. Since S waves travel through solids but not through liquids, the Earth's outer core must be liquid.

Neither S nor P waves arrive in a "shadow zone" between 105° and 140° from an epicenter. Beyond 140°,

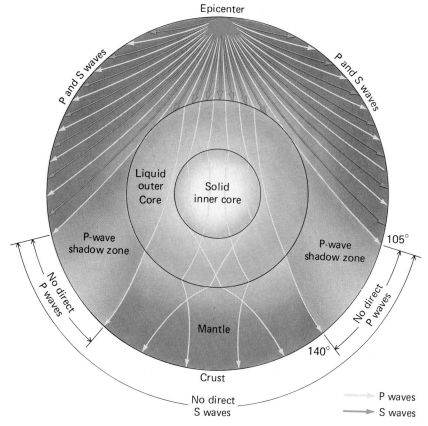

Figure 6–19 Seismic waves curve gently as they pass through the Earth. They also bend sharply where they cross major layer boundaries in the Earth's interior. The blue S waves do not travel through the liquid outer core, and therefore direct S waves are observed only within 105° of the epicenter. The yellow P waves bend sharply at the core-mantle boundary to create a shadow zone of no direct P waves from 105° to 140°.

Earth Science and the Environment

Two Hundred Years of Earthquake Activity in the San Francisco Bay Area

San Francisco lies on a segment of the San Andreas fault. To the east, across San Francisco Bay, the Hayward fault runs past San Jose and through Oakland. Both faults are active. A devastating earthquake struck San Francisco on April 18, 1906. The Richter scale did not exist in 1906, but geologists estimate that the quake was of magnitude 8.3. The maximum movement along the fault was horizontal and exceeded 6 meters. Although the ground motion toppled many buildings, the major damage was caused by fire (Fig. 1). Underground gas lines were ripped apart by the moving Earth, and flames spread through the city, totally destroying the downtown area. Between 500 and 600 people were killed, and 250,000 were left homeless.

On October 17, 1989, a magnitude 6.9 earthquake left 65 people dead in the city and surrounding areas. Its epicenter was in Loma Prieta, east of Santa Cruz and about 90 kilometers south of San Francisco (Fig. 2). Ground motion destroyed much of the Santa Cruz business district, segments of both the Bay Bridge and Interstate Highway 880 collapsed (Fig. 3), and damage was heavy in the Marina district, which had been built on an old landfill. The total damage was estimated at more than $4 billion.

Historical data suggest that the 1989 quake may be followed by a more devastating one. In the 1800s, two significant *pairs* of earthquakes occurred. A quake on the eastern Hayward fault in 1836 was followed by one on the San Andreas fault, on the San Francisco Peninsula, two years later. Then a second quake on the peninsula in 1865 was followed three years later by a quake on the Hayward fault. Thus, a quake on the San Andreas fault was followed by one on the Hayward fault, and vice versa. If this pattern repeats itself, we might expect an earthquake to occur on the Hayward fault near Oakland within the next few years.

Figure 1 A portion of downtown San Francisco following the 1906 earthquake and fire.

direct P waves arrive, but direct S waves do not (Fig. 6–19). The shadow zone, too, results from the change from solid rock to molten liquid at the core-mantle boundary. Earthquake waves curve gently as they pass through the Earth. But P waves refract sharply as they pass from the mantle into the core. As a result, no P waves arrive in the shadow zone.

P waves refract sharply again when they pass from the outer core to the inner core, indicating another radical change in physical properties of the Earth's interior. In this case, the change in direction results from an abrupt transition from the molten outer core to the solid inner core. Thus, seismic data tell us that the core is composed of an inner solid sphere surrounded by an outer liquid shell. The entire core is composed mostly of iron and nickel.

Discovery of the Crust-Mantle Boundary

In 1909 a seismologist, Andrija Mohorovičíc, discovered that seismic wave velocities increase sharply at a depth varying from 7 to 70 kilometers beneath the Earth's sur-

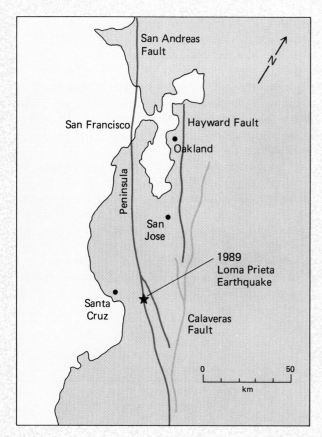

Figure 2 Active faults in the San Francisco Bay area.

Figure 3 Collapse of Interstate Highway 880 during the 1989 earthquake. *(Paul Scott/Sygma)*

Two factors reduced its impact. First, the epicenter was located in a sparsely populated area, not in one of the densely populated Bay Area cities. Second, although a magnitude 6.9 quake is substantial, it is *not* a great earthquake. It released only 1/60 as much energy as the magnitude 8.3 quake in 1906.

If a magnitude 7 or 8 earthquake occurred in San Francisco or Oakland, the damage would eclipse that of the 1989 quake. At least two scenarios are possible. Many homes in the San Francisco Bay area are wood-frame structures that resist collapse. Therefore, if a quake were to strike at night or during the weekend, experts predict a low mortality, perhaps 1000 out of a total population of two million in the Bay Area. However, if the quake struck late on a summer afternoon, when bridges and subways were jammed with commuters and the streets were packed with late shoppers, the death toll would be higher.

As discussed in the text, this type of prediction is uncertain. Yet it is important to ask how much damage and loss of life we should expect from another major earthquake in the Bay Area. The Loma Prieta quake caused death and destruction, but it was not a major catastrophe.

face. This sudden velocity change indicates an abrupt transition in rock type at that depth. The upper layer is the Earth's crust, and the lower layer is the mantle. The boundary between the two is called the **Mohorovičíc discontinuity** or the **Moho,** in honor of its discoverer.

The Lithosphere-Asthenosphere Boundary

Recall that the lithosphere is the 100-kilometer-thick brittle outer shell of the Earth. It includes both the crust and

the uppermost part of the mantle and floats on the hot, plastic asthenosphere. Geologists discovered the boundary between the two layers when they realized that both the velocity and strength of seismic waves decreases abruptly at a depth of 100 kilometers. The sudden changes occur because the cool, brittle rock of the lithosphere carries seismic waves more effectively than the hot, soft rock of the asthenosphere.

Thus, the Earth's layers, and the nature of the material in each, are known from the behavior of seismic waves as they travel through our planet.

SUMMARY

•

An **earthquake** is a sudden motion or trembling of the Earth caused by the release of **elastic energy** stored in rocks. About a million earthquakes occur each year, but only a small fraction are strong enough to be felt. Most earthquakes occur along the boundaries between moving lithospheric plates. Earthquake energy is released at the **focus** of the earthquake, where rocks move abruptly. The **epicenter** of a quake is the point on the Earth's surface directly over the focus. Earthquake energy travels through rock as **seismic waves**, which occur as **body waves** and **surface waves**. Two main types of body waves transmit earthquake energy. A **P wave** travels through solids, liquids, and gases, but an **S wave** is transmitted only through solids, and travels more slowly than a P wave. Surface waves cause most earthquake damage because they travel along the Earth's surface.

An earthquake is detected and measured by a **seismograph**. The epicenter and focus of an earthquake can be identified using a **time-travel curve** to determine the distances from the earthquake source to three or more seismograph stations.

The **Mercalli scale** measures earthquake **intensity** on the basis of damage and human response to a quake. The **Richter scale** measures earthquake **magnitude** on the basis of the amplitude of ground movement caused by a quake. Earthquake damage is caused by ground motion, alteration of the Earth's surface, fire, landslides, and **tsunamis**.

Long-term prediction of earthquakes is based on studies of the earthquake history of an area and on identification of **seismic gaps** along a fault zone. Short-term prediction is based on occurrences of **foreshocks**, bulges and other changes in the land surface, and animal behavior.

The Earth's internal structure and properties are known by studies of earthquake wave velocities and **refraction** and **reflection** of seismic waves as they pass through the Earth.

KEY TERMS

•

Earthquake *138*
Elastic energy *139*
Fault creep *141*
Seismic wave *141*
Seismology *143*
Body wave *144*
Focus *144*

Epicenter *144*
Surface wave *144*
Primary (P) wave *144*
Secondary (S) wave *145*
Surface (L) wave *145*
Seismograph *145*
Seismogram *146*

Time-travel curve *147*
Mercalli scale *149*
Intensity *149*
Richter scale *149*
Magnitude *149*
Tsunami *151*
Seismic gap *153*

Foreshock *154*
Refraction *155*
Reflection *155*
Mohorovičíc discontinuity
   (Moho) *157*

REVIEW QUESTIONS

•

1. Explain how energy is stored in rocks and then released during an earthquake.

2. Why do most earthquakes occur at the boundaries between tectonic plates?

3. Explain why fault creep along a segment of a fault provides evidence that an earthquake may soon occur in a nearby segment where creep is *not* occurring.

4. Describe the differences among P waves, S waves, and L waves.

5. Why is most earthquake damage caused by surface waves?

6. Explain how a seismograph works. Sketch what a seismogram would look like before and during an earthquake.

7. Describe how the epicenter of an earthquake is located.

8. Describe the similarities and differences between the Mercalli and Richter scales. What does each actually measure, and what information does each provide?

9. List five different mechanisms for earthquake damage. Discuss each briefly.

10. Discuss the scientific reasoning behind long-term and short-term earthquake prediction.

11. What is the Moho? How was it discovered?

12. Describe how the behavior of seismic waves led to the discovery of the Earth's core.

13. Describe how the behavior of seismic waves led to the discovery of the boundary between the lithosphere and the asthenosphere.

DISCUSSION QUESTIONS

1. If rock is caught between two lithospheric plates moving in different directions, it may bend in a plastic manner or, alternatively, it may fracture. Which type of movement is more likely to cause an earthquake? Explain.

2. Using the graph in Figure 6–11, determine how far away from an earthquake you would be if the first P wave arrived 5 minutes before the first S wave.

3. Using a map of the United States, locate an earthquake that is 1000 kilometers from Seattle, 1300 kilometers from San Francisco, and 700 kilometers from Denver.

4. It has been suggested that engineers should inject large quantities of liquids into locked portions of the San Andreas fault. Proponents of the plan believe that these liquids will reduce friction by lubricating the fault, and consequently the fault will creep slowly and a major earthquake will be averted. If you were the mayor of San Francisco, would you encourage or discourage the injection of fluids into the fault? Defend your stance.

5. Significant earthquakes have occurred in Parkfield, California, in 1857, 1881, 1901, 1922, 1934, and 1966. Draw a graph with the dates on the vertical (Y) axis and the numbers of the events (simply 1, 2, 3, and so on) spaced evenly on the X axis. Use your graph to predict when the next earthquake might occur in Parkfield.

6. Imagine that geologists predict a major earthquake in a densely populated region. The prediction may be right or it may be wrong. City planners may heed it and evacuate the city, or ignore it. The possibilities lead to four combinations of predictions and responses, which can be set out in a grid as follows:

Does the predicted earthquake really occur?

		Yes	No
Is the city evacuated?	**Yes**		
	No		

For example, the space in the upper left corner of the grid represents the situation in which the predicted earthquake occurs and the city is evacuated. For each space in the square, outline that consequences of that sequence of events.

Plate Tectonics, Volcanoes, and Plutons

In the spring of 1980, about 1 cubic kilometer of rock, lava, and ash exploded from Mount St. Helens in western Washington. The blast flattened and burned surrounding forests, and the sky grew so dark with pulverized rock that motorists 150 miles from the mountain had to use their headlights at noon. But this volcanic explosion was minor compared with some historical eruptions. For example, in 1815 Mount Tambora in the southwestern Pacific Ocean exploded, throwing out about 100 cubic kilometers of rock and ash and killing 12,000 people; Mount Pelée in the Caribbean exploded in 1902, killing 29,000. Yet even the most violent historical eruptions have been small compared with some prehistoric ones. A volcanic explosion in Yellowstone Park, Wyoming, 1.9 million years ago ejected approximately 2500 cubic kilometers of rock, lava, and ash. In contrast to these violent explosions, Hawaiian volcanoes erupt gently enough that tourists approach closely to photograph lava pouring from volcanic vents and to witness spectacular fire fountains erupting hundreds of meters into the sky.

Volcanoes are common in some regions and unheard of in others. Eighteen recently active volcanoes form

● Pu'u O'o vent during an eruption in June 1986. *(U.S. Geological Survey, J. D. Griggs)*

high peaks in the Cascade Range in northern California, Oregon, and Washington, but no volcanoes exist in the central or eastern United States. Why are some parts of the Earth volcanically active, whereas other parts are not? And why do some volcanoes explode violently but others erupt gently? Answers to both questions are found in this chapter.

7.1 Formation of Magma

As explained in Chapter 3, igneous rock forms in a two-step process. First, part of the upper mantle or crust melts to form magma. The magma then solidifies to form new igneous rock. Magma does not form just anywhere. It forms in three particular geologic environments: subduction zones, spreading centers, and mantle plumes. Let us consider each of these environments to see why and how rock melts to make magma.

Magma Production in a Subduction Zone

At a subduction zone, two 100-kilometer-thick plates of lithosphere collide. One of the plates then dives beneath the other, sinking hundreds of kilometers into the mantle. During subduction, three processes melt portions of the asthenosphere to form great amounts of magma (Fig. 7–1).

Heating

Everyone knows that a solid melts when it become hot enough. Butter melts in a frying pan, and snow melts under the spring Sun. Friction heats rock near a subduc-

Figure 7–1 Magma formation at a subduction zone. A subducting plate covered with oceanic crust dives into the mantle. Friction heats rocks in the subduction zone, water rises from the oceanic crust, and circulation of the asthenosphere decreases pressure on hot rock in the asthenosphere. All three factors contribute to melting of the asthenosphere and production of great amounts of magma.

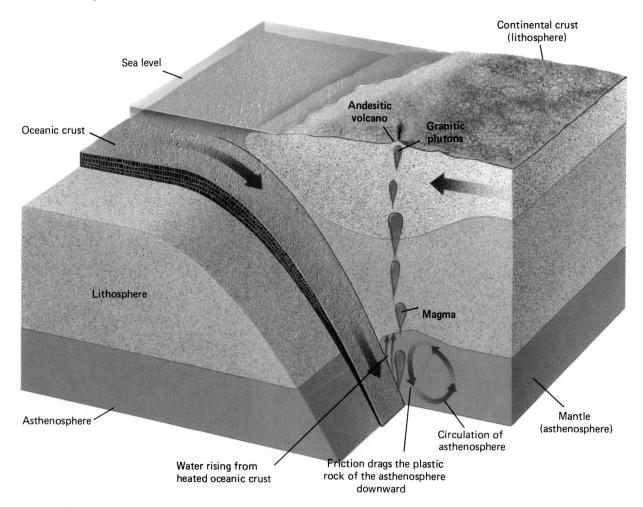

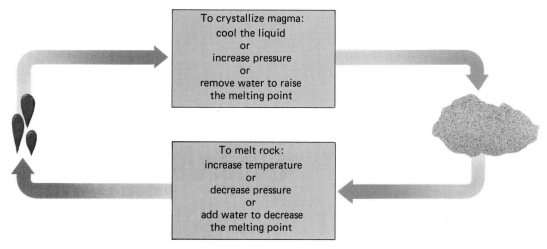

Figure 7–2   Increasing temperature, addition of water, and decreasing pressure all melt rock to form magma.

tion zone as one plate moves downward, scraping past the other plate. Because the asthenosphere is already hot, this additional heat contributes to its melting (Fig. 7–2). Oddly, however, increasing temperature is the least important of the three causes of melting in a subduction zone.

Addition of Water to the Asthenosphere

In general, a wet rock melts at a lower temperature than an identical dry rock. Thus as the water content rises, the melting point decreases. Therefore, addition of water can melt a rock that is already so hot that it is close to its melting point. Recall that the asthenosphere is very hot—so hot that it is plastic. Thus, if water is added to the asthenosphere, it melts.

A subducting plate is covered by oceanic crust, and the mud and fractured basalt of oceanic crust are soaked with seawater. As soaked oceanic crust dives into the mantle, it becomes hotter because Earth temperature increases with depth. Eventually the water of the subducting plate boils to generate steam. The steam rises, adding water to the hot asthenosphere directly above the subducting plate. This addition of hot water melts portions of the asthenosphere forming huge quantities of magma (Fig. 7–1). This process is probably the most important of the three causes of melting in a subduction zone (Fig. 7–2).

Pressure-Relief Melting

In Chapter 3 you learned that a rock expands by about 10 percent when it melts to form magma. If the rock is near the Earth's surface it can expand easily, for there is little pressure preventing it from doing so. However, in the asthenosphere, pressure is so great from the weight of the overlying 100 kilometers of lithosphere that expansion,

and therefore melting, are more difficult. However, if the pressure should somehow decrease, the hot rock in the asthenosphere can expand and melt. In other words, a drop in pressure can melt a hot rock. Melting caused by decreasing pressure is called **pressure-relief melting** (Fig. 7–2).

As a subducting plate descends, it drags plastic asthenosphere rock down with it, as shown by the arrows in Figure 7–1. Rock from deeper in the asthenosphere then flows upward to replace the sinking rock. As this

Molten lava flowing from a natural tunnel on Mauna Loa, Hawaii, March 1984.   *(Scott Lopez, Hawaii Volcanoes National Park)*

rock rises, pressure decreases and thus pressure-relief melting occurs.

To summarize, parts of the asthenosphere above a subducting plate melt by three processes: (1) addition of water from subducting oceanic crust, (2) pressure-relief melting, and (3) frictional heating. These three factors combine to form huge amounts of magma in subduction zones at depths of about 100 kilometers, approximately where the subducting plate passes from the lithosphere into the asthenosphere. Therefore, volcanic and plutonic rocks are both common features of a subduction zone. The volcanoes of the Pacific Northwest, the granite cliffs of Yosemite, and the Andes Mountains are all examples of igneous rocks formed at subduction zones.

Magma Production in a Spreading Center

When two lithospheric plates separate at a spreading center, soft, hot asthenosphere oozes upward to fill the gap (Fig. 7–3). As the hot asthenosphere rises, pressure-relief melting forms vast quantities of magma at the rift zone. Most of the world's rift boundaries are in the ocean basins, where they form the mid-oceanic ridge system. Most of the Earth's oceanic crust is created by pressure-relief melting below the mid-oceanic ridge. The basalt crust *forms* at the mid-oceanic ridge, then *spreads* outward. Thus, all of the Earth's oceanic crust is created at the mid-oceanic ridge. Some rifts occur in continents, and here, too, great amounts of basaltic magma erupt onto the Earth's surface.

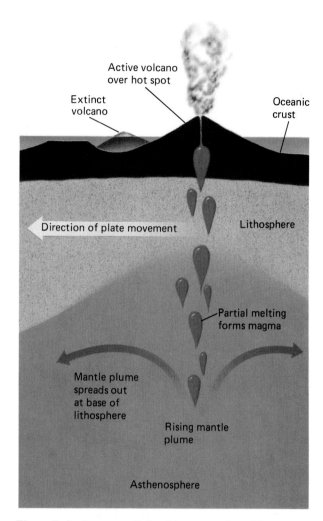

Figure 7–4 Pressure-relief melting occurs in a rising mantle plume, and magma rises to form a volcanic hot spot.

Figure 7–3 At a spreading center, pressure-relief melting occurs where hot asthenosphere rises to fill the gap left by separating lithospheric plates.

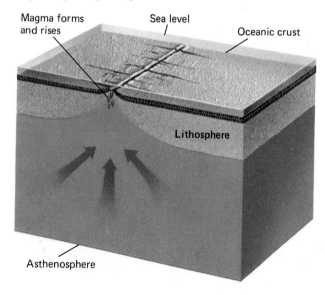

Magma Production at a Hot Spot

The third environment in which magma commonly forms, known as a **hot spot**, is a volcanic region at the Earth's surface directly above a rising plume of hot, plastic mantle rock. As a mantle plume rises, pressure-relief melting forms magma that flows upward to erupt at the Earth's surface (Fig 7–4).

A mantle plume originates deep in the mantle, and its magma then rises through the lithosphere to erupt and form a volcano. If a mantle plume rises beneath the sea, volcanic eruptions build submarine volcanoes and volcanic islands. Because a lithospheric plate continues to move over the relatively stationary asthenosphere, a mantle plume may generate magma continuously as a plate migrates over the plume. This process forms a chain of volcanic islands that becomes progressively younger toward one end, such as the Hawaiian Islands.

7.2 The Behavior of Magma: Why Some Magma Erupts from a Volcano and Other Magma Solidifies Below the Surface

After it forms, magma rises toward the Earth's surface. But not all magma behaves in the same way. Basaltic magma most commonly rises all the way to the Earth's surface to erupt from a volcano. In contrast, granitic magma typically solidifies within the Earth's crust.

The contrasting behavior of granitic and basaltic magmas is a result of their different compositions. Granitic magma contains about 70 percent silica, whereas the silica content of basaltic magma is only about 50 percent. In addition, granitic magma generally contains 10 to 15 percent water, whereas basaltic magma contains only 1 to 2 percent water. These differences are summarized in the following table.

Typical Granitic Magma	Typical Basaltic Magma
70% silica 10–15% water	50% silica 1–2% water

Effects of Silica on Magma Behavior

In the silicate minerals, silica tetrahedra link together to form the chains, sheets, and framework structures described in Chapter 2. In magma, silica tetrahedra link together in a similar manner. They form long chains if silica is abundant in the magma, but shorter chains if less silica is present. Therefore, granitic magma contains longer chains of silica tetrahedra than does basaltic magma. When granitic magma flows, the long chains become tangled, making the magma stiff, or viscous. **Viscosity** is resistance to flow. Basaltic magma, with its shorter silica chains, is less viscous and flows easily. Because of its fluidity it rises rapidly to erupt at the Earth's surface. Granitic magma, in contrast, rises slowly due to its stiffness. Therefore, it commonly solidifies within the crust before reaching the surface.

Effects of Pressure and Water on Magma Behavior

Once magma forms, it starts to rise because it is less dense than surrounding rock. As it rises, two changes occur simultaneously: (1) it cools as it enters shallower and cooler levels of the Earth, and (2) pressure drops because the weight of overlying rock decreases. Cooling and decreasing pressure have opposite effects on rising magma: cooling tends to solidify it, but dropping pressure tends to keep it liquid.

As you just learned, decreasing pressure can melt a hot rock because rock must expand in order to melt. For the same reason decreasing pressure also lowers the temperature at which magma solidifies. Thus, a magma that will solidify at 1000°C at a depth of 20 kilometers may not solidify at a depth of 5 kilometers until it cools to 800°C. Therefore, as magma rises, decreasing pressure tends to keep it liquid.

If rising magma contains no water, the dropping pressure can override the cooling effect, and the magma can remain liquid and rise to erupt at the Earth's surface (Fig 7–2). Basaltic magma contains little water, and therefore decreasing pressure keeps it liquid as it rises. This

The numbers are radiometric ages of volcanic rocks on the Hawaiian Islands. The islands become progressively younger from one end of the island chain to the other.

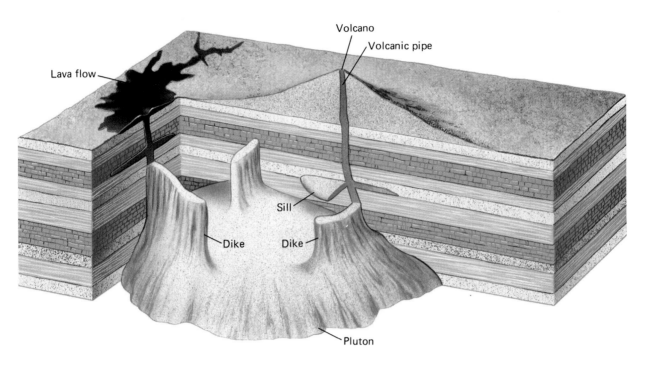

Figure 7–5   Forms of igneous rocks.

factor contributes to the fact that most basaltic magma erupts at the Earth's surface.

Granitic magma does not behave in this manner because it contains 10 to 15 percent water, and water lowers the temperature at which magma solidifies. Thus, if dry granitic magma solidifies at 700°C, the same magma with 10 percent water may remain liquid to 600°C.

The Bugaboo Mountains in British Columbia are a large granite pluton that has been uplifted and exposed.

Water is a volatile substance; at high temperatures it tends to escape from magma as steam. Deep in the crust, where granitic magma forms, high pressure prevents the water from escaping. But when the magma rises, pressure decreases and water begins to escape as steam. Because the water escapes, the solidification temperature of the magma *rises*. Thus, loss of water from a rising granitic magma causes it to solidify within the crust. For this reason, most granitic magmas solidify and stop rising at depths from 5 to 20 kilometers beneath the Earth's surface.

Because basaltic magmas contain only 1 to 2 percent water to begin with, any loss is relatively unimportant. As basaltic magma rises, it remains liquid all the way to the surface. Thus, basalt volcanoes are common.

7.3 Plutons

For reasons already described, granitic magma usually solidifies within the Earth's crust to form a mass of intrusive rock called a **pluton** (Fig. 7–5). Thus a pluton is emplaced in **country rock** (any previously existing rock). Occasionally basaltic magma also soldifies as an intrusive rock. Basalt plutons tend to form thin, sheet-like masses because of the lower viscosity of basaltic magma.

A **batholith** is the largest of all plutons (Fig. 7–6). It is a mass of intrusive rock exposed over more than 100 square kilometers of the Earth's surface. A large batholith may be 20 kilometers thick, but an average one is about 10 kilometers thick. Most batholiths are granite. A **stock**

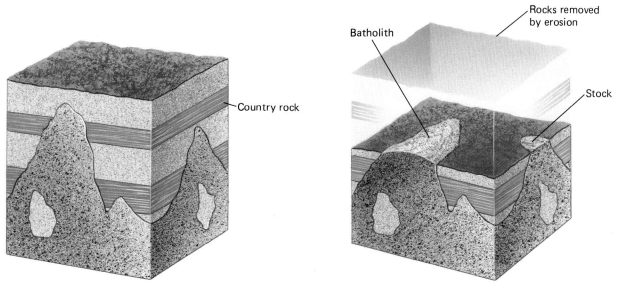

Figure 7–6 A batholith is a pluton with more than 100 square kilometers of exposure at the Earth's surface. It may be part of a larger subterranean pluton. A stock is similar to the batholith but has a smaller surface area.

is similar to a batholith but is exposed over a smaller area of the Earth's surface.

To form a batholith, a huge mass of granitic magma must rise through solid continental crust. How can such a large bubble of magma rise through solid rock? If you place oil and water in a jar, screw the lid on tightly, and shake it, oil droplets disperse throughout the water. Set the jar down and the droplets slowly coalesce to form larger bubbles, which rise to the surface. The water is easily displaced as the bubbles rise. Although the rock of continental crust is solid, at the high temperatures in the lower crust it also behaves in a plastic manner. As magma rises, it shoulders aside the hot, plastic crustal rock, which then slowly flows back to fill in behind the rising bubble.

Many mountain ranges contain large granite batholiths that are the remains of plutons once emplaced deep within the crust, and later uplifted and eroded. Figure 7–7 shows the major batholiths in western North America and the mountain ranges associated with them. Most of the great granite batholiths of the world formed in the immense regions of magma production at subduction zones.

Whereas a batholith pushes country rock aside as it rises, magma may also flow into a fracture or between layers in country rock. A **dike** is a tabular, or sheet-like, intrusive igneous rock that forms when magma oozes into a fracture (Fig. 7–8). Dikes cut across sedimentary layers or other features in country rock and range from less than a centimeter to more than a kilometer thick (Fig. 7–9A). Dikes are commonly more resistant to weathering than surrounding rock. Where this is the case, country rock

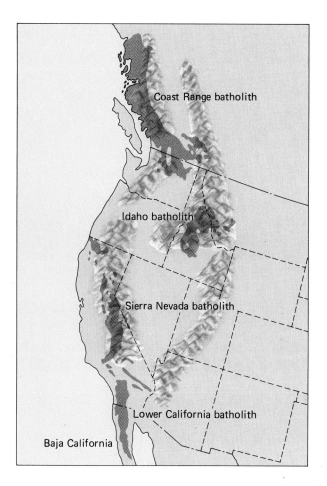

Figure 7–7 Major batholiths in western North America and the mountain ranges associated with them.

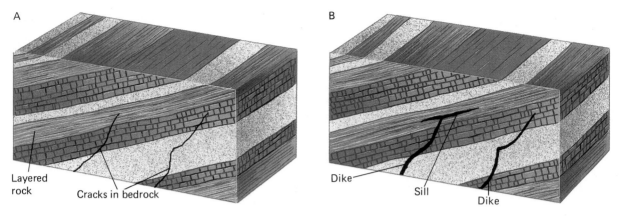

A

B

Layered rock

Cracks in bedrock

Dike

Sill

Dike

Figure 7–8 A dike forms when magma solidifies in a fracture that cuts across the grain of country rock. A sill forms when magma solidifies in a fracture parallel to the grain, or layering, of country rock.

Figure 7–9 (A) A basalt dike cuts across red sandstone and shale in Grand Canyon. (B) This large resistant dike was left standing when softer country rock eroded away. Shiprock, a volcanic neck, is in the background.

A

B

may be eroded away, leaving the dike standing alone on the surface (Fig. 7–9B).

If magma oozes between layers of country rock, a **sill** forms. It is a tabular intrusive rock that lies parallel to the grain, or layering, of country rock rather than cutting across the layers (Fig. 7–8). Sills also range in thickness from less than a centimeter to more than a kilometer and may extend for tens of kilometers in length and width.

7.4 Volcanoes and Other Volcanic Landforms

Volcanic eruptions create a wide variety of landforms, including several different kinds of volcanoes and lava plateaus. Additionally, many islands, including all of the Hawaiian Islands, Iceland, and most of the islands of the southwestern Pacific Ocean, were constructed entirely by volcanic eruptions. Occasionally a violent eruption destroys a volcanic peak, as happened in the 1980 eruption of Mount St. Helens, when most of the mountaintop was blown away.

Structures and Textures of Lava Flows

Hot magma shrinks as it cools and solidifies. The shrinkage pulls the rock apart, forming cracks that grow as the rock continues to cool.

In Hawaii geologists have watched fresh lava cool and solidify. When a solid crust only ½ centimeter thick forms on the surface of the glowing liquid, five- or six-sided cracks begin to develop. As the lava cools and solidifies from the surface down, the cracks grow downward to hotter zones where the last bit of magma is solidifying. Such cracks are common in lava flows and

A

B

Figure 7–10 (A) Columnar joints in basalt in Grand Canyon. (B) A top view of columnar joints in basalt, showing the polygonal fracture pattern. *(Don Hyndman)*

shallow sills, and are called **columnar joints**. They are regularly spaced and intersect to form five- or six-sided polygonal columns (Fig. 7–10).

Lava can develop different textures depending on its viscosity and rate of cooling. Lava with low viscosity may continue to flow as it cools and stiffens, forming basalt with smooth, glassy-surfaced, wrinkled, or "ropy" ridges. This type of lava is called **pahoehoe** (Fig. 7–11). If the viscosity of the lava is higher, it may partially solidify as it flows, to form a slow-moving mixture of

Figure 7–11 A car buried in pahoehoe lava, Hawaii. *(Kenneth Neuhauser)*

Figure 7–12 Aa lava near Shoshone, Idaho, showing numerous gas bubbles frozen into the solid rock.

solid rock and liquid lava. As a result, **aa** lava has a jagged, rubbly, broken surface (Fig. 7–12). Aa lava commonly contains numerous holes and resembles Swiss cheese. The holes are gas bubbles that formed in the magma and became frozen in the rock as it solidified. Pahoehoe and aa are Hawaiian names for the lava types.

When basaltic magma erupts under water, rapid cooling causes it to contract and form spheroidal **pillow lava** (Fig. 7–13). Pillow lavas are abundant in the upper layers of oceanic crust, where they form when basaltic magma oozes onto the sea floor from cracks in the mid-oceanic ridge.

When a volcano erupts explosively, it may eject both liquid magma and solid rock fragments. A rock formed from explosively erupted rock particles or magma is called a **pyroclastic rock**. The smallest particles are glassy pieces of **volcanic ash**, ranging up to 2 millimeters in diameter. Mid-sized particles called **cinders** range from about 2 to 64 millimeters. Still larger fragments called **volcanic bombs** form when blobs of molten lava spin through the air as they solidify and therefore take the form of spindles or spheroids (Fig. 7–14).

Although the words ash and cinder are used to describe these volcanic particles, they are *not* the same as the ashes and cinders produced by a conventional fire. In December 1989, Mount Redoubt volcano, southwest of Anchorage, Alaska, erupted. The pilot of a Boeing 747, with 231 passengers aboard, ignored warnings about the rising ash cloud and flew into it. The abrasive particles were sucked into the jet engines and quickly ground the sharp blades of the compressor into small stubs. Exposed to the intense heat in the combustion chamber, the volcanic ash particles melted. Moments later, the liquid solidified as glass on the cooler turbine blades. As a result, all four engines stalled and the jet fell 4000 meters in 8 minutes. Finally the pilot was able to restart the engines and fly the disabled craft back to Anchorage, where he landed safely.

Flood Basalts

The gentlest, least catastrophic type of volcanic eruption occurs when magma is so fluid that it simply oozes from cracks and flows over the land like water. When a large flow of this type solidifies, it forms a broad plain or

Figure 7–13 A large lava pillow, with a film canister for scale.

Figure 7–14 A volcanic bomb lying in cinders. The streaky surface formed as the blob of magma whirled through the air. *(Jack Herbert)*

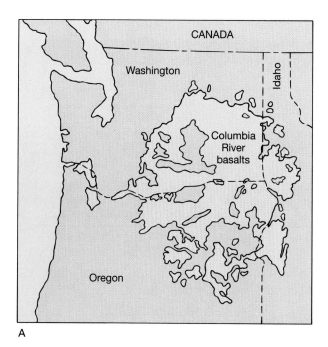

A

B

Figure 7–15 (A) The Columbia River basalt plateau covers much of Washington, Oregon, and Idaho. (B) Columbia River basalt in eastern Washington. Each layer is a separate lava flow.
(Larry Davis)

plateau. Basaltic magma commonly pours out as a **flood basalt**, which is so named because it covers the landscape like a flood. The plateaus created by such lava flows are called **lava plateaus**. The Columbia River plateau in eastern Washington, northern Oregon, and western Idaho is an extensive lava plateau (Fig. 7–15). This sequence of basalt flows contains 350,000 cubic kilometers of rock, is 3000 meters thick in places, and covers 200,000 square kilometers. It formed over a period of three million years, from 17 to 14 million years ago, as layer upon layer of

basaltic magma oozed from fissures in eastern Washington and Oregon. Each flow formed a layer between 15 and 100 meters thick. Similar large basalt plateaus are found in western India, northern Australia, Iceland, Brazil, Argentina, and Antarctica.

Volcanoes

If lava is too viscous to spread out as a lava plateau, it builds up into a hill or mountain called a **volcano**. Vol-

TABLE 7–1

Characteristics of Different Types of Volcanoes					
Type of Volcano	Form of Volcano	Size	Type of Magma	Style of Activity	Examples
Basalt plateau	Flat to gentle slope	100,000 to 1,000,000 km² in area; more than 1800 m thick	Basalt	Gentle eruption from long fissures	Columbia River Plateau
Shield volcano	Slightly sloped 6° to 12°	Up to 9000 m high	Basalt	Gentle, some fire fountains	Hawaii
Cinder cone	Moderate slope	100 to 400 m high	Basalt or andesite	Ejections of pyroclastic material	Paricutín, Mexico
Composite volcano	Alternate layers of flows and pyroclastics	100 to 3500 m high	Variety of types of magmas and ash	Often violent	Vesuvius, Mount St. Helens, Aconcagua
Caldera	Cataclysmic explosion leaving a circular depression called a caldera	Less than 40 km in diameter	Granite	Very violent	Yellowstone, San Juan Mountains

canoes differ widely in shape, structure, and size (Table 7–1). Lava and rock fragments erupt from an opening called a **vent**. In many volcanoes the vent is located in a **crater**, a bowl-like depression at the summit of the volcano.

Shield Volcanoes

If basaltic magma is too viscous to form a lava plateau but still quite fluid, it will heap up slightly to form a gently sloping volcanic mountain called a **shield volcano**

Figure 7–16 Mount Skjoldbreidier in Iceland bears the typical profile and low-angle slopes of a shield volcano. *(Science Graphics, Inc./Wards Natural Science Establishment, Inc.)*

(Fig. 7–16). The sides of a shield volcano generally slope away from the vent at angles between 6° and 12° from horizontal. As a reference, a ski slope for beginning to intermediate skiers is about 10°, and an expert slope may be as steep as 35°. Unless erosion has formed deep gullies, you will find no challenging ski runs on a shield volcano; it is simply not steep enough.

When a shield volcano erupts, the fluid lava usually flows gently over the lip of the crater or from **fissures**, linear cracks in the sides of the volcano. Although the Hawaiian shield volcanoes erupt regularly, the eruptions are normally gentle, rarely life-threatening. Lava flows occasionally overrun homes and villages but the flows advance slowly enough to give people time to evacuate.

Cinder Cones

A **cinder cone** is a small volcano, as high as 400 meters, made up of pyroclastic fragments blasted out of a central vent at high velocity. A cinder cone forms when large amounts of gas accumulate within rising magma. When the gas pressure builds up sufficiently, the entire mass erupts explosively, hurling cinders, ash, and molten magma into the air. The particles then fall back around the vent to accumulate as a small mountain of volcanic debris.

As the name implies, a cinder cone is symmetrical. It also can be quite steep (about 30°), especially near the vent where ash and cinders pile up (Fig. 7–17). A cinder cone is usually active for only a short period of time because once the gas escapes, the driving force behind

the eruption is removed. Since a cinder cone is made up of unconsolidated material, it erodes easily and quickly. Therefore, it is a geologically transient feature of the landscape.

About 350 kilometers west of Mexico City a broad plain is dotted with numerous extinct cinder cones. Prior to 1943, a small hole existed in the ground in one of the level portions of the plain. The hole had been there for as long as anyone could remember, and people grew corn just a few meters away. In February of 1943, as two farmers were preparing their field for planting, smoke and sulfurous gases rose from the hole. As night fell, hot, glowing rocks flew skyward, creating a spectacular series of arcing flares like a giant fireworks display. By morning, a 40-meter-high cinder cone had grown in the middle of the cornfield. For the next five days, pyroclastic material erupted 1000 meters into the sky and the cone grew to 100 meters in height. Within a few months, a fissure opened at the base of the cone extruding lava that buried the town of San Juan Parangaricutiro. Two years later the cone had grown to a height of 400 meters. After nine years the eruptions ended, and today the volcano, called El Paricutin, is dormant.

Composite Cones

Some of the most beautiful and spectacular volcanoes in the world are **composite cones**, sometimes called **strato-volcanoes**. They form by a series of alternating lava flows and pyroclastic eruptions. As explained previously, pyroclastic eruptions tend to form steep but unconsoli-

Figure 7–17 A cinder cone near Mexico City with a lava flow extending from one flank.
(Science Graphics, Inc./Ward's Natural Science Establishment, Inc.)

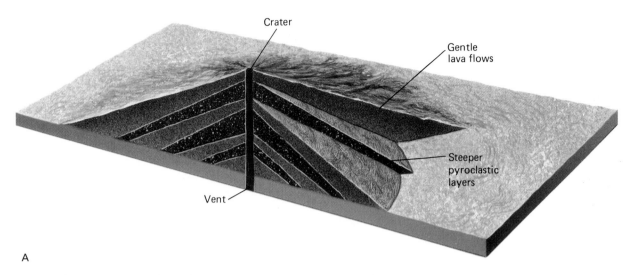

A

B

Figure 7–18 (A) A schematic cross section of a composite cone showing alternating layers of lava and pyroclastic material. (B) Steam and ash pouring from Mount Ngauruhoe, a composite cone in New Zealand. *(Don Hyndman)*

dated slopes. When these loose cinders are covered by lava flows, the hard lava rock protects them from erosion (Fig. 7–18).

A composite cone grows as a result of many eruptions occurring over a long time. Many of the highest mountains of the Andes are composite cones, as are many of the most spectacular mountains of the west coast of North America. Two examples are Mount St. Helens and Mount Rainier (Fig. 7–19). Mount St. Helens erupted in 1980. Mount Rainier has been dormant in recent times but could become active at any moment; repeated eruption is a trademark of a composite volcano.

Volcanic Necks and Pipes

After an eruption, the vent of a volcano may fill with magma that later cools and solidifies. Commonly this **volcanic neck** is harder than surrounding rock. Given enough time, the slopes of the volcano may erode, leaving only the tower-like neck exposed (Fig. 7–20).

In some locations, cylindrical dikes called **pipes** extend from the asthenosphere to the Earth's surface. They are conduits that once carried magma on its way to erupt at a volcano, but they are now filled with the last bit of magma that solidified within the conduit. They are both

Figure 7–19 Mount Rainier is a composite cone. *(Don Hyndman)*

fascinating and economically important. Most known pipes formed between 70 and 140 million years ago. For some unknown reason, most intruded into continental crust older than 2.5 billion years. Pipes are interesting because the rock found in them, called **kimberlite**, originated in the asthenosphere. It is among the few direct samples of the upper mantle available to geologists.

The best evidence indicates that pressure in the asthenosphere was so great that the kimberlite magma shot upward through the Earth's crust at very high, perhaps even supersonic, speed. Under such intense pressure, the small amount of carbon in some pipes crystallized as diamond, making those pipes commercially important. The most famous diamond-rich kimberlite pipes are in South Africa. The diamonds crystallized more than 200 kilometers beneath the surface and were carried upward with the rising magma to depths of a few kilometers or less, where they can be accessed by modern mining technology.

(Text continues on p. 178)

Figure 7–20 Shiprock, New Mexico, is a volcanic neck. The great rock was once the core of a volcano. The softer flanks of the cone have now eroded away. A dike several kilometers long extends to the left. *(Dougal McCarty)*

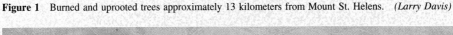

Focus On

The May 18, 1980, Eruption of Mount St. Helens

The 1980 eruption of Mount St. Helens is perhaps the best known volcanic eruption of all time. It was recorded by photographers on the ground and in the air, and it dominated media reports for weeks after it occurred. About two months before the eruption, the volcano, which had last erupted in 1857, began to experience small-to-moderate-size earthquakes. Puffs of steam and volcanic ash were emitted from a newly formed crater on the summit. The activity was great enough to convince geologists to install seismographs and tiltmeters around and on the volcano. The seismographs were emplaced to detect earthquakes caused by moving bodies of magma that might be associated with an impending eruption. Tiltmeters detect small local tilting of the Earth's surface that

might be caused by magma moving upward in the volcano and swelling the flanks.

In early April the emissions of steam and ash ceased, although earthquakes continued. The tiltmeters detected swelling on the north side of the volcano, which by early May had grown to an easily visible, ominous bulge rising as fast as 1.5 meters per day. Eventually it measured nearly 1 by 2 kilometers and had swelled outward by more than 100 meters. Warned by the geologists, officials barred the public from the area around the volcano and evacuated most of the local inhabitants.

Two strong earthquakes rocked Mount St. Helens at 8:27 and 8:31 on Sunday morning, May 18. The bulge had grown so steep that the second earthquake caused it to break away from the mountain, forming an immense landslide of rock, soil, and glacial ice. The landslide roared down the mountain and, in doing so, relieved the

Figure 1   Burned and uprooted trees approximately 13 kilometers from Mount St. Helens.   *(Larry Davis)*

Figure 2 Eruption of Mount St. Helens. *(USGS, R. P. Hoblitt)*

pressure on the gas-charged magma that had been causing the bulge. The magma then exploded out through the side of the mountain where the bulge had been. This horizontal blast flattened trees in a 400-square-kilometer area on the north side of the mountain (Fig. 1). Large trees as far as 25 kilometers away were knocked over. The landslide, combined with large volumes of new volcanic ash from the eruption, poured down the Toutle River and into the Columbia River, filling their channels.

A vertical plume of ash, rocks, and gas was blown upward to a height of 18 kilometers (Fig. 2). Static electricity generated by the billowing plume caused lightning strikes in the surrounding forest. Mountaineers on Mount Adams, 50 kilometers to the east, saw sparks jump from

their ice axes, were pelted with falling debris, and felt the intense heat from the blast. The eruption blew out the entire north side of the volcano, and the top 410 meters of the mountain were removed.

An airborne ash cloud spread directly eastward on the prevailing winds at an average speed of about 60 kilometers per hour. It darkened the sky and turned the air gritty and unpleasant for several days as far away as western Montana. Some ash fall was noted in Minnesota and Oklahoma. The fine, abrasive ash made breathing difficult, destroyed crops, and ruined automobile engines. Damage resulting from the eruption was estimated to be in the hundreds of millions of dollars. A total of 63 people were known or presumed to have been killed.

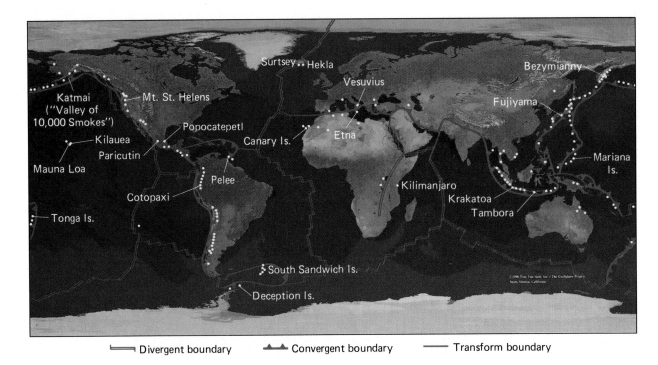

Divergent boundary   Convergent boundary   Transform boundary

Figure 7–21 This map shows the close relationship between subduction zones (convergent boundaries) and major volcanoes, indicated by the dots.

7.5 Igneous Activity and Plate Tectonics

Igneous Activity at Subduction Zones

We learned in Section 7.1 that magma forms abundantly in subduction zones. It then rises to form both volcanoes and plutons directly over the subducting tectonic plate. Figure 7–21 shows that most of the world's volcanoes are located near subduction zones. The "ring of fire" is a zone of concentrated volcanic activity encircling the Pacific Ocean basin. About 75 percent of the Earth's active volcanoes lie in the ring of fire.

All 18 volcanoes of the Cascade Range, from Mount Baker near the Canadian border to Mount Lassen in northern California, have been active in the past two million years, fired by subduction of oceanic plates beneath the continent. Mount St. Helens erupted in 1980 and Mount Lassen in 1915. About 7000 years ago, a Cascade volcano, posthumously named Mount Mazama, exploded, blasting 10 cubic kilometers of rock and ash into the air. Some of the mountain was blown away by the explosion, and the remainder collapsed into the hole left by the ejected rock and ash, leaving what is now Crater Lake, Oregon (Fig. 7–22). The ash is found in soils over much of western and central North America.

Some volcanoes that have not erupted recently show signs of activity today. Mount Baker has recently experienced earthquake swarms, and gas regularly escapes from its crater. Mount Rainier has steam caves under its summit

glaciers. It is reasonable to predict that the Cascades will see additional eruptions similar to those of Mounts St. Helens, Lassen, and Mazama, although it is difficult to predict where, when, and how large they will be. It is important to remember that, despite the loss of life and tremendous damage, the eruption of Mount St. Helens was very small compared with other known eruptions from similar volcanoes.

Explosive eruptions are also common elsewhere in the ring of fire. On June 9, 1991, Mount Pinatubo, 60 miles from Manila in the Philippines, began ejecting a gray-green cloud of ash, rock, and smoke 15 kilometers into the sky. By the end of June, 338 people had been killed and several towns evacuated because of continued explosions. As many as 200 earthquakes a day were felt. Both the U.S. Clark Air Base, 15 kilometers from the volcano, and Subic Bay Naval Station sustained heavy damage. Clark was evacuated because of the threat to personnel and aircraft. As this book goes to press, the eruptions are expected to continue for months or years.

Volcanic eruptions are the most obvious form of igneous activity at a subduction zone. However, huge bodies of magma lie beneath most of these volcanic regions. This magma feeds the volcanoes, but much of it never reaches the Earth's surface. Instead, it slowly cools and solidifies to form large plutons. Eventually mountain building elevates many of these regions, and erosion strips away overlying rock, exposing the plutons as batholiths. The granite batholiths of the Sierra Nevada of California, the Coast Range of western Canada, and the Appalachian Mountains of eastern North America, and most of the rest of the Earth's great batholiths, formed beneath great chains of subduction-zone volcanoes. The volcanic rocks have now eroded away, exposing the granite at the Earth's surface.

Igneous Activity at Rift Zones

The island of Iceland in the North Atlantic Ocean is about the size of Virginia and supports a population of a quarter of a million people. It lies directly over the Mid-Atlantic ridge and thus is on a spreading boundary between two tectonic plates. The island formed by repeated volcanic eruptions at the ridge. Iceland has experienced numerous eruptions since its settlement by Vikings before A.D. 1000. Some have covered villages with ash and cinders, although human injury and death have been rare. Hot springs and hot rock resulting from recent volcanism are used to generate electrical power for the island, and the spectacular volcanic scenery is a major tourist attraction.

Although eruptions on Iceland are readily visible, most of the volcanic activity at the mid-oceanic ridge occurs unseen as submarine lava oozes from cracks in the ridge axis onto the sea floor. The rocks of the sea floor will be described further in Chapter 14.

Rifting can also occur in continental crust. When it does, pressure-relief melting in the mantle forms basaltic magma below the continent. This magma rises into lower granitic continental crust and melts it to form granitic magma. Thus, continental rifting commonly forms both basalt volcanoes and granite plutons. Continental rifting is occurring today along a north-south trend through eastern

An American soldier guards Clark Air Base in the Philippines as Mount Pinatubo erupts steam and ash on June 29, 1991. The Air Base had been evacuated a week before this picture was taken, and was destroyed by later eruptions. *(Wide World Photos, Inc.)*

Africa, in a zone called the East African rift. It may also be occurring in Idaho's Snake River plain and in central Colorado and New Mexico along the Rio Grande River, where the Rio Grande rift is marked by numerous young volcanoes.

Figure 7–22 Crater Lake, Oregon. *(Crater Lake National Park Administration)*

Igneous Activity at Mantle Plumes and Hot Spots

The island of Hawaii is composed of several overlapping volcanoes above a mantle plume. The youngest, Kilauea, frequently erupts basaltic lava, commonly for periods of weeks or months. As in Iceland, lava flows occasionally destroy homes and agricultural land, but they rarely cause injury or death because eruptions are relatively gentle and the flows advance slowly enough that people can evacuate threatened areas. Because the eruptions are comparatively safe, at least from a distance, and because they continue for long periods of time, tourists flock to the island to see real volcanic eruptions.

Mantle plumes also rise beneath continents. As in the case of continental rifting, basaltic magma forms in a rising mantle plume beneath the continent. The magma then ascends into the lower continental crust, melting it and forming great amounts of granitic magma, which in turn rise to create a continental hot spot, such as Yellowstone National Park.

7.6 Violent Magma: Ash-Flow Tuffs and Calderas

As we saw in Section 7.2, granitic magma usually solidifies within the Earth's crust. Under certain conditions, however, granitic magma can rise all the way to the Earth's surface, where it characteristically erupts with great violence. This violent behavior contrasts sharply with that of basaltic magma, which typically erupts with relative calm. Why does some granitic magma rise all the way to the Earth's surface, and why does it erupt with such violence?

The granitic magmas that do rise to the surface and explode probably start out with less water than "normal" granitic magma, perhaps only a few percent, like basaltic magma. With such a low water content, granitic magma would rise to the Earth's surface for the same reasons that basaltic magma does.

"Dry" granitic magma rises more slowly than basaltic magma, however, due to its higher viscosity. As it rises, the pressure decreases, allowing the small amount of water to separate and form gas bubbles in the liquid magma (Fig. 7–23A). The water takes the form of gas rather than liquid because the temperature is high enough to turn the water to steam.

As an analogy, think of a bottle of beer or soda pop. When the cap is on and the contents are under pressure, carbon dioxide gas is dissolved in the liquid. Because the gas is dissolved, no bubbles are visible. If you remove the cap, the pressure is reduced. As a result, the gas escapes from solution and bubbles rise to the surface. If conditions are favorable, the frothy mixture of gas and liquid then erupts through the opening. In a similar manner, gas bubbles form in rising magma. The gas rises to the uppermost layer of magma to create a highly pressurized, frothy, expanding mixture of gas and liquid magma. The temperature of this mixture may be as high as 900°C.

Figure 7–23   (A) When granitic magma rises to within a few kilometers of the surface, it causes the overlying rock to dome upward and fracture. Gas separates from the magma and rises to the upper part of the magma body. (B) The gas-rich magma explodes through fractures, rising in a vertical column of magma, rock fragments, and gas. (C) When the gas is used up, the column collapses and spreads outward as a high-speed ash flow. (D) Because so much material has erupted from the top of the magma chamber, the roof collapses to form a caldera.

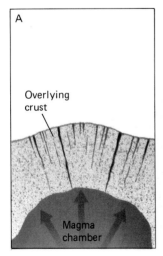  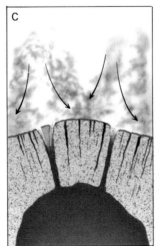 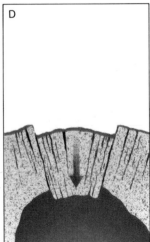

As the entire magma body rises to within a few kilometers of the Earth's surface, it creates a blister, or dome, by uplifting and fracturing overlying rocks. The highly pressurized mixture of gas and magma then explodes through the fractures, blowing great streams of liquid magma, gas, and rocks several kilometers into the sky (Fig. 7–23B). In other cases, the gas-charged magma may simply ooze through the fractures and flow rapidly over the land, like root beer foam overflowing the edge of a mug. During an eruption of this type, some of the frothy magma may solidify to form **pumice**, a rock so full of gas bubbles that it floats on water.

The amount of material ejected in a violent granitic eruption depends on the size of the magma body and the amount of gas available. In a large eruption, the column of rising pyroclastic material may reach a height of 12 kilometers above the Earth's surface and be several kilometers in diameter. A cloud of fine ash may be blown much higher, into the upper atmosphere. The column may be held up for hours or even days by the force of additional material streaming out of the magma chamber. Several recent eruptions of Mount Pinatubo blasted ash columns high into the atmosphere and held them up for hours. In contrast, the small 1980 Mount St. Helens eruption blasted out nearly all of its pyroclastic material nearly instantaneously.

Ash Flows

When the gas in the upper part of the magma is used up, the eruption ceases abruptly. Since no more material is streaming upward to support it, the frothy column of ash, rock, and gas falls back to the Earth's surface (Fig. 7–23C). Although the collapsing column contains some solid particles, it behaves as a fluid falling from a great height.

The falling column spreads outward over the Earth's surface from its point of impact. Because the fluid is denser than air, it flows down stream valleys and other topographically low features. Such a flow is called an **ash flow**. Small ash flows travel at speeds up to 200 kilometers per hour. Large flows have traveled distances exceeding 100 kilometers. One large flow leaped over a 700-meter-high ridge as it crossed from one valley into another. Ash flows racing across the land at night glow brightly because of their high temperature. For this reason, an ash flow is also called a **nuée ardente**, from the French term for "glowing cloud."

When an ash flow comes to a stop, most of the gas escapes into the atmosphere, leaving behind a chaotic mixture of volcanic ash and rock fragments. The rock formed by such a process is an **ash-flow tuff** (Fig. 7–24). **Tuff** includes all pyroclastic rocks—that is, rocks composed of volcanic ash or other material formed in a

Figure 7–24   Ash-flow tuff forms when an ash flow comes to a stop. The fragments in the tuff are pieces of rock that were carried along with the volcanic ash and gas. *(Geoffrey Sutton)*

volcanic explosion. Some ash flows are hot enough to partially melt after they stop moving. A tough, hard rock called **welded tuff** forms when this mixture solidifies. Welded tuff often shows spectacular structures and textures formed by plastic flow of the partly melted ash (Fig. 7–25). Other ash-flow tuffs are unwelded.

A single large eruption may eject a few hundred to a few thousand *cubic kilometers* of ash. To visualize this

Figure 7–25   This welded tuff formed when an ash flow became hot enough to melt and flow as a plastic mass. The streaky texture was created when rock fragments similar to those in Figure 7–24 melted and smeared out as the rock flowed.   *(Geoffrey Sutton)*

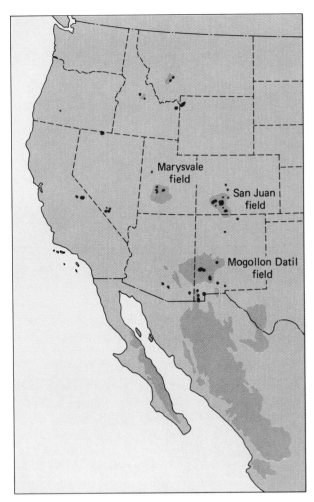

Figure 7–26 Calderas (red dots) and ash-flow tuffs (orange areas) in western North America.

roughly circular when viewed from above. Consequently, roof collapse usually forms a circular depression called a **caldera**. A large caldera may be 40 kilometers in diameter and have walls as much as a kilometer high. Some calderas fill up with ash-flow tuff as the ash column collapses; others show remarkable preservation of the circular depression and steep walls. We usually think of volcanic landforms in terms of gracefully symmetrical peaks. The topographic depression of a caldera is an interesting exception to this notion. Figure 7–26 shows the extent of ash-flow tuffs and related rocks in western North America.

Magma that explodes to form ash-flow tuffs and calderas commonly erupts more than once, with substantial time intervals between eruptions. Following an initial eruption, the remaining upper part of a magma body has been depleted in gas and has lost its explosive potential. However, lower portions of the magma continue to release gas, which rises and builds pressure anew to begin another cycle of eruption. Time intervals between successive eruptions vary from a few thousand to about half a million years.

Yellowstone National Park

Yellowstone National Park in Wyoming and Montana is the oldest national park in the United States. Its geology consists of three large overlapping calderas and the ash-flow tuffs that erupted from them (Fig. 7–27). The oldest eruption occurred 1.9 million years ago and ejected 2500 cubic kilometers of pyroclastic material. The next major eruption occurred 1.3 million years ago. The most recent, 0.6 million years ago, produced the Yellowstone caldera in the very center of the park by ejecting 1000 cubic kilometers of ash and other debris.

Consider the periodicity of the three Yellowstone eruptions. They occurred at 1.9, 1.3, and 0.6 million years ago. Intervals of 0.6 to 0.7 million years separated the eruptions. It has been 0.6 million years since the last eruption. The numerous geysers and hot springs indicate that hot magma still lies beneath the park, and seismographs frequently detect movement of liquid magma near the surface. The periodicity of Yellowstone eruptions, the presence of shallow magma, and the well-known tendency of magma of this type to erupt multiple times all suggest that a fourth eruption may be due. Geologists would not be surprised if an eruption occurred at any time. However, arguments based on periodicity are only approximate. Even if the periodicity were exactly 0.6 million (600,000) years, a 1 percent error is 6000 years. Thus, it is conceivable that the next eruption will not occur for several thousand years. It is also possible that Yellowstone has seen its last eruption.

quantity of material, think of a cube of rock with a volume of 1000 cubic kilometers. The perimeter of its base would be 40 kilometers, or slightly less than the length of a marathon race. A world-class distance runner could circle it in a little over two hours. Its height would be 10,000 meters, 2000 meters higher than Mount Everest. The largest known ash-flow tuff from a single eruption is located in the San Juan Mountains of southwestern Colorado. It has a volume greater than 3000 cubic kilometers—three times the size of the imaginary cube. Another of comparable size has been mapped in southern Nevada.

Calderas

Think of what must happen back in the magma chamber when such an immense volume of material suddenly explodes skyward: nothing remains to hold up the overlying rock. Therefore, the roof of the magma chamber simply collapses (Fig. 7–23D). Most large magma bodies are

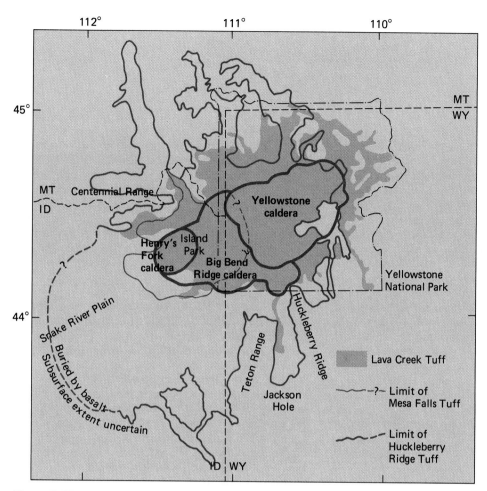

Figure 7-27 Calderas and ash-flow tuffs of Yellowstone Park. The Big Bend Ridge caldera formed during the eruption of the Huckleberry Ridge Tuff 1.9 million years ago. The Henry's Fork Caldera formed during the eruption of the Mesa Falls Tuff 1.3 million years ago. The Yellowstone Caldera formed during the eruption of the 0.6-million-year-old Lava Creek Tuff. Portions of older calderas are obliterated by younger calderas. Dashed boundaries of the Huckelberry Ridge and Mesa Falls Tuff are covered by younger rocks. *(Figure modified from Hildreth and others,* Journal of Geophysical Research *89 [1984])*

7.7 Volcanoes and Human Settlements

Volcanic eruptions have caused death and destruction throughout history. The most destructive eruptions have been of the type described in the previous section, in which gas-charged magma explodes from a shallow granitic magma chamber to form an ash flow. When Mount St. Helens erupted in 1980, it exploded with the force of 500 atomic bombs of the size used on Hiroshima in World War II. Nearly 1 cubic kilometer of rock and ash was ejected, forests were leveled, and, despite geologists' warnings that an eruption was imminent, 63 people were killed. Yet, compared with other volcanic eruptions in recent geologic history, Mount St. Helens was a small

eruption. Table 7-2 summarizes the major known volcanic disasters since A.D. 1000.

Mount Pelée

On May 2, 1902, the coastal town of St. Pierre on the Caribbean Island of Martinique was completely destroyed by an ash flow that erupted from the nearby volcano of Mount Pelée. All but two of the 28,000 residents of the town died nearly instantaneously when an 800°C cloud of gas and volcanic ash roared down the side of the volcano and through the town at speeds up to 100 kilometers per hour. Only one of 18 ships in the harbor escaped, and it lost many crew members. The magma responsible

Earth Science and the Environment

Volcanoes and Climate

When Mount St. Helens erupted, the ash darkened the sky over a wide area. In Yakima, Washington, which is 140 kilometers from the mountain, people had to turn their car headlights on at noon. In Missoula, Montana, 620 kilometers from Mount St. Helens, people observed an eerie darkness, or dry fog. Clouds of volcanic ash and gas reflect light and heat from the Sun out into space, which leads to a cooling and darkening of the Earth's surface. The immediate effects are easy to document. However, it is more difficult to assess the long-term climatological effects of major volcanic eruptions.

The largest volcanic eruption in recent history occurred in 1815 when Mount Tambora in the southwestern Pacific Ocean exploded, ejecting approximately 100 times as much magma and ash into the atmosphere as did Mount St. Helens. The following year, 1816, was one of the coldest years in recent history and has been recorded as the "year without a summer" and "eighteen hundred and froze to death." Crop failures (compounded by the devastation caused by the Napoleonic Wars) led to widespread famine in Europe, a period that has been called "the last great subsistence crisis in the Western world." The question remains: was this cold period caused by the Mount Tambora eruption, or was it merely a coincidence that the two events occurred together?

Effects of volcanic emissions of SO_2 on climate are discussed in Chapter 17. Some scientists think that the recent eruptions of Mount Pinatubo in the Philippines may lead to global cooling because of the SO_2 content of its gases.

The figure shows a plot of global temperatures before and after eight major volcanic events that have occurred in recent times. This graph clearly shows a statistical correlation between global cooling and volcanic eruptions. Meteorological models also demonstrate that high-altitude dust and gas act as an umbrella to prevent sunlight from reaching the Earth.

As explained in the text, historical eruptions have been small compared with some in the more distant past. What would happen if a huge eruption like those that created the Yellowstone calderas were to occur? In a worst-case scenario, dust and sulfur aerosols would block out so much sunlight that daytime would be just a little brighter than a full-moon night for nearly a year after the eruption. However, other calculations indicate that a "bright, sunny day" a year later would be about as bright as a normal overcast day. In either case, the effects on climate and agriculture would be significant, perhaps catastrophic. But on the positive side, only a few years would pass before normal climatological conditions returned.

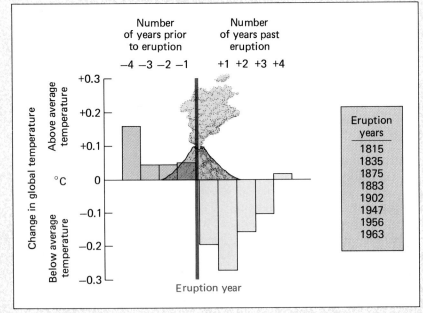

A composite plot of temperature changes in the Northern Hemisphere in the four years immediately before and after eight large eruptions in recent times. *(Michael Rampino,* Annual Review of Earth and Planetary Science *16 (1988):73–99)*

TABLE 7-2

Some Notable Volcanic Disasters Since the Year A.D. 1000 Involving 5000 or More Fatalities							
			Primary Cause of Death and Number of Deaths				
Volcano	*Country*	*Year*	*Pyroclastic Flow*	*Debris Flow*	*Lava Flow*	*Posteruption Starvation*	*Tsunami*
Kelut	Indonesia	1586		10,000			
Vesuvius	Italy	1631			18,000		
Etna	Italy	1669			10,000		
Lakagigar	Iceland	1783				9,340	
Unzen	Japan	1792					15,190
Tambora	Indonesia	1815	12,000			80,000	
Krakatoa	Indonesia	1883					36,420
Pelée	Martinique	1902	29,000				
Santa Maria	Guatemala	1902	6,000				
Kelut	Indonesia	1919		5,110			
Nevada del Ruiz	Colombia	1985		>22,000			

for the eruption was viscous, silica-rich, and charged with gas. Ironically, one of the survivors was a convicted murderer imprisoned in a dungeon when the eruption occurred.

Krakatoa

Before August 1893, Krakatoa was an 800-meter-high island lying between Java and Sumatra in the southwestern Pacific Ocean. It consisted of three volcanic cones within a caldera. The caldera, which was about 6 kilometers in diameter, was the remains of an ancient volcano that had previously erupted and collapsed. The cones had formed by volcanic activity that followed the collapse. In the days before August 27, several small volcanic explosions occurred on the island. On August 27, the entire island exploded with an amount of energy equivalent to 100 million tons of TNT. Although Krakatoa itself was unpopulated, the volcanic explosion formed an immense sea wave called a **tsunami,*** between 35 and 40 meters high, that radiated outward, inundating coastal villages on nearby islands and killing about 35,000 people. Most of the island disappeared in the explosion, leaving in its place a basin 100 meters deep.

The sound of the explosion of Krakatoa was heard in Australia, more than 4000 kilometers away. The dust that was hurled into the upper atmosphere circled the

* A tsunami is sometimes called a "tidal wave," an unfortunate misnomer for this phenomenon because it has nothing to do with tides.

The smoldering ruins of St. Pierre, May 14, 1902, following the May 8 eruption of Mount Pelée. *(Institute of Geological Sciences, London)*

globe for years afterward, turning sunsets red as far away as London and blocking so much sunlight that the Earth's surface temperature decreased by a few degrees for the next few years.

It is thought that much of the explosive energy developed as seawater flowed into cracks formed by the rising magma body. The seawater heated rapidly as it approached the shallow magma beneath the island. The explosion was caused in part by the pressure of the resultant steam.

Mount Vesuvius

In A.D. 79, the Roman cities of Pompeii and Herculaneum and several neighboring villages near what is now Naples, Italy, were destroyed when Mount Vesuvius erupted. Prior to that eruption the volcano had been inactive for more than 2000 years—so long that vineyards had been cultivated on the sides of the mountain, all the way to the summit. During the eruption, a pyroclastic flow streamed down the flanks of the volcanic cone, burying the cities and towns under 5 to 8 meters of hot ash. Pompeii was found and excavated 17 centuries later. The excavations revealed molds of inhabitants trapped by the ash flow as they attempted to flee or find shelter. Some of the molds even appear to have preserved facial expressions of terror. Mount Vesuvius returned to relative quiescence, only to become active again in 1631. It was active frequently from 1631 to 1944, and in this century it has experienced eruptions in 1906, 1929, and 1944.

SUMMARY

•

Magma forms in three geologic environments: subduction zones, spreading centers, and mantle plumes. Heat, addition of water, and **pressure-relief melting** cause rocks to melt and form magma. Granitic magma typically solidifies within the Earth's crust, whereas basaltic magma usually erupts from a **volcano** at the surface. This contrast in behavior of the two types of magma is due to differences in silica and water content.

Any intrusive mass of igneous rock is a **pluton**. A **batholith** is a pluton with more than 100 square kilometers of exposure at the Earth's surface. A **dike** and a **sill** are both sheet-like plutons. Dikes cut across layering in country rock, and sills run parallel to layering.

Magma may flow onto the Earth's surface as **lava flows** or erupt explosively as **pyroclastic** material. Fluid lava forms **lava plateaus** and **shield volcanoes**. A pyroclastic eruption may form a **cinder cone**. Alternating eruptions of fluid lava and pyroclastic material from the same vent form a **composite cone**. About 70 percent of the Earth's volcanic activity occurs along a circle of subduction zones in the Pacific Ocean called the **ring of fire**. When granitic magma rises to the Earth's surface, it usually erupts explosively, forming **ash-flow tuffs** and **calderas.** Eruptions of this type have caused widespread death and destruction throughout history.

KEY TERMS

•

Pressure-relief melting   *163*
Hot spot   *164*
Pluton   *166*
Country rock   *166*
Batholith   *166*
Stock   *166*
Dike   *167*
Sill   *168*

Columnar joint   *169*
Pahoehoe   *169*
Aa   *170*
Pillow lava   *170*
Pyroclastic rock   *170*
Volcanic ash   *170*
Cinder   *170*
Volcanic bomb   *170*
Flood basalt   *171*

Lava plateau   *171*
Volcano   *171*
Vent   *172*
Crater   *172*
Shield volcano   *172*
Fissure   *173*
Cinder cone   *173*
Composite cone   *173*
Stratovolcano   *173*

Volcanic neck   *174*
Pipe   *174*
Kimberlite   *175*
Pumice   *181*
Ash flow   *181*
Nuée ardente   *181*
Tuff   *181*
Welded tuff   *181*
Caldera   *182*
Tsunami   *182*

R E V I E W Q U E S T I O N S

•

1. List and describe the major tectonic environments in which magma forms.

2. List and explain the most important factors in the melting of rock to form magma.

3. What happens to most basaltic magma after it forms?

4. What happens to most granitic magma after it forms?

5. Explain why basaltic magma and granitic magma behave differently as they rise toward the Earth's surface.

6. How large is a batholith?

7. Explain the difference between a dike and a sill.

8. How do columnar joints form in a basalt flow?

9. How do a shield volcano, a cinder cone, and a composite cone differ from one another? How are they similar?

10. Which type of volcanic mountain has the shortest life span? Why is this structure a transient feature of the landscape?

11. How does a composite cone form?

12. What is a volcanic neck? How is it formed?

13. Why do diamonds occur naturally in kimberlite pipes? Can you think of any other natural environment in which they might be found?

14. What is the ring of fire? Why do most of the Earth's volcanoes occur in this zone?

15. Explain why and how granitic magma forms ash-flow tuffs and calderas.

16. What is pumice, and how does it form?

17. How does welded tuff form?

18. How does a caldera form?

19. How much pyroclastic material can erupt from a large caldera?

20. Explain why additional eruptions in Yellowstone Park seem likely. Describe what such an eruption might be like.

21. What is a tsunami?

D I S C U S S I O N Q U E S T I O N S

•

1. If you were handed a sample of rock, but given no information about the geologic environment from which it came, would it be possible to determine whether it originated in a batholith, a dike, or a sill? Defend your answer.

2. How could you distinguish between a sill exposed by erosion and a lava flow?

3. Many mountains are composed of granite that is exposed at the Earth's surface. Does this observation prove that granite forms at the Earth's surface?

4. Are basalt plateaus made of extrusive or intrusive rocks? How could you distinguish between a basalt plateau and an uplifted batholith?

5. How does the huge mass of magma that eventually solidifies to form a batholith make its way upward through the Earth's crust?

6. Much of the surface of the Moon is scoured by meteor craters. However, some regions, called seas or maria, are flat expanses of rock. Outline a plausible geologic explanation for these maria. Be sure to include a chronology of events in your sequence.

7. Sometimes gases dissolve in a liquid, and at other times they form bubbles within it. Discuss the difference between these two conditions and its relevance to the geology of volcanoes.

8. Imagine that you detect a volcanic eruption on a distant planet, but have no other data. What conclusions could you draw from this single bit of information? What types of information would you search for to expand your knowledge of the geology of the planet?

9. Explain why some volcanoes have steep, precipitous faces, but many do not.

10. Parts of the San Juan Mountains of Colorado are composed of granite plutons, and other parts are volcanic rock. Explain why these two types of rock are likely to occur in proximity.

11. Compare and contrast the danger of living 5 kilometers from Yellowstone National Park with the danger of living an equal distance from Mount St. Helens. Would your answer differ for people who live 50 kilometers, or those who live 500 kilometers, from the two regions?

12. Study a geologic map of the place where you grew up. Evaluate the possible threat of a catastrophic volcanic eruption in that area.

Mountains and Geologic Structures

8

8.1 Mountains and Mountain Ranges

8.2 Plate Tectonics and Mountain Building

8.3 Folds, Faults, and Joints

8.4 Island Arcs: Mountain Chains Formed by
Collision of Two Oceanic Plates

8.5 The Building of Two Mountain Chains:
The Andes and the Himalayas

A mong all the Earth's topographic features, mountains and mountain ranges stand out for sheer beauty and their projection of the power of nature. Green forests and alpine meadows blanket the lower slopes of the great ranges. High on the mountainsides the last stunted trees yield to rubble deposited by alpine glaciers only a few thousand years ago. Higher yet, rock walls may rise vertically for 1000 meters or more. The highest peaks sparkle with glacial ice and snow, remnants of the last alpine ice age. Because of this beauty and the feeling of closeness to nature, mountain ranges have become popular vacation areas. The European Alps, the American Rockies, the Sierra Nevada of California, the Tetons of Wyoming, and most other ranges in populous parts of the world draw millions of visitors each year. Even the less accessible Himalayas and South American Andes are standard on the itineraries of more adventurous tourists.

8.1 Mountains and Mountain Ranges

A **mountain** is any part of the Earth's surface sufficiently elevated above its surroundings to have a distinct summit and to be considered worthy of a name. Normally, in order to qualify as a mountain, a peak must rise at least

● Machhapuchhre, a high peak in the Himalayas.

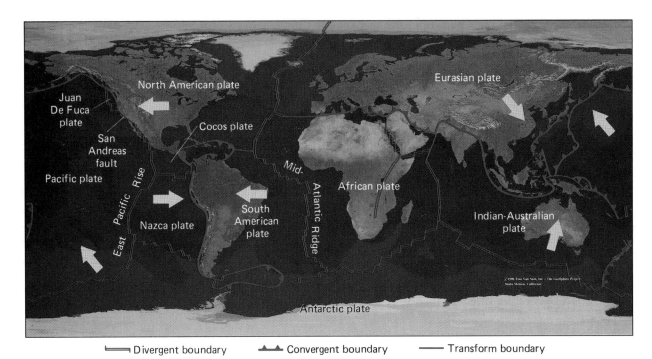

Divergent boundary   Convergent boundary   Transform boundary

Figure 8–1 Lithospheric plate boundaries and major mountain ranges. Arrows indicate directions of plate movements.

300 meters above the surrounding terrain. A **mountain range** is a series of mountains or mountain ridges that are closely grouped and similar in age and mode of formation. Commonly, several mountain ranges cluster together in an elongate zone called a **mountain chain**. Mountains nearly always occur in ranges and chains because the forces that create them operate over large, linear regions of the Earth's crust rather than in small, isolated localities.

8.2   Plate Tectonics and Mountain Building

Most of the Earth's tectonic activity occurs at lithospheric plate boundaries, where two moving plates interact with each other. Tectonic activity builds mountains. Therefore, nearly all of the Earth's mountains and mountain ranges are found at plate boundaries (Fig. 8–1). Mountain building is commonly accompanied by other kinds of tectonic activity that also occur at plate boundaries, including folding and faulting of rocks, earthquakes, volcanic eruptions, intrusion of plutons, and metamorphism.

Recall from Chapter 5 that three different types of boundaries separate tectonic plates: divergent boundaries, transform boundaries, and convergent boundaries.

1. At a divergent boundary, or rift, two plates spread apart as new lithosphere forms between them. Such a boundary can form in either oceanic or continental crust. The world's longest mountain chain, the mid-oceanic ridge, formed in this tectonic environment. The highest mountains on the African continent,

Mounts Kilamanjaro and Kenya, are volcanoes along the East African rift.

2. At a transform boundary, two plates slide horizontally past each other. The San Andreas fault is a transform boundary in continental crust in western California. Movement along this fault has uplifted the San Gabriel Mountains. However, few major mountain ranges develop along transform boundaries.

3. At a convergent boundary, two plates meet in a head-on collision. Most of the Earth's great mountain ranges, with the exception of the mid-oceanic ridge, grew at convergent plate boundaries.

Highly deformed rocks from the core of an ancient mountain range. *(J. M. Harrison, Geological Survey of Canada, KGS 381)*

Before we continue our discussion of mountain building, let us consider what happens to rocks caught between two moving tectonic plates.

8.3   Folds, Faults, and Joints

If you push or pull on any solid object with enough force, it either bends or breaks. If you step on a supple, green stick, it bends: if you step on a dry twig, it snaps. Similarly, tectonic forces bend and fracture rocks. Whether a rock bends or breaks depends on the rock type, temperature, and pressure.

Enormous tectonic forces develop at a convergent plate boundary where two 100-kilometer-thick slabs of lithosphere collide. Rocks trapped between colliding plates bend and fracture. In some cases the collision deforms rocks tens or even hundreds of kilometers away from the actual boundary. In addition, huge masses of magma form and rise through the lithosphere. They rise by shouldering aside hot, plastic country rock, bending and deforming it as they force their way upward through the Earth's crust. Because great mountain chains grow at convergent plate boundaries, rocks in mountainous regions are commonly broken and bent by these processes. Less intense forces also deform rocks at divergent and transform plate boundaries.

A **geologic structure** is any feature produced by deformation of a rock. Folds, faults, and joints are the most common types of structures. A **fold** is a bend in rock (Fig. 8–2). Folds show up best in layered rock, but

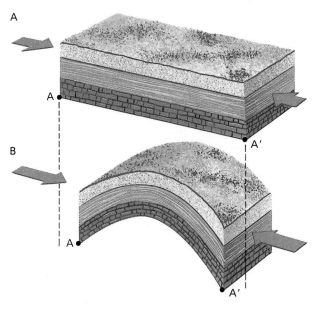

Figure 8–2 (A) Horizontally layered sedimentary rocks. (B) A fold in the same rocks. The forces that folded the rocks are shown by the arrows. Points A and A′ are the same points in the rock in both figures. Notice that they are closer after folding. (C) A fold in shale, Sun River, Montana. *(M. Mudge and the U.S. Geological Survey)*

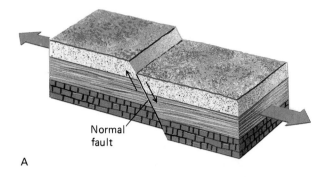

Figure 8–3 (A) A diagram of a fault in sedimentary rocks. The offset of sedimentary layering shows the direction and amount of rock movement along the fault. The forces that caused the fault are shown by the large arrows; the small arrows show direction of rock movement. (B) A small fault in sedimentary rocks near Kingman, Arizona. *(Ward's Natural Science Establishment, Inc.)*

also occur in unlayered rock such as granite. A **fault** is a fracture along which rock on one side has moved relative to rock on the other side (Fig. 8–3). A **joint** is a fracture without movement of rock (Fig. 8–4). Joints and faults often occur as sets of many parallel fractures.

Folds

Figure 8–5 shows that a fold arching upward is called an **anticline**, and one arching downward is a **syncline**.* The sides of a fold are called the **limbs**. Notice that a single limb is shared by an anticline-syncline pair. The backbone of a fold, where the two limbs meet, is called the **axis**.

If you were to walk along the top of a horizontal anticline, you would be walking along a level fold axis. In other folds the axis tilts at an angle, as shown in Figure 8–5B, forming a **plunging fold**. If you walked along the axis of a plunging fold you would be traveling uphill or downhill.

Folds come in different shapes. Many are neatly symmetrical, like those shown in Figure 8–5. A **monocline** is a special kind of fold with only one limb. Figure 8–6 shows a monocline that developed where sedimentary rocks sag over an underlying fault. A circular or

---

*Properly, an upward-arched fold is called an anticline only if the oldest rocks are in the center and the youngest are on the outside of the fold. Similarly, a downward-arched fold is a syncline only if the youngest rocks are at the center and the oldest are on the outside. The age relationships become reversed if the rocks are turned completely upside down and folded. If the age relationships are unknown, as sometimes occurs, an upward-arched fold is called an **antiform** and a downward-arched fold a **synform.**

Figure 8–4 Vertical joints in sandstone, Indian Creek, Utah.

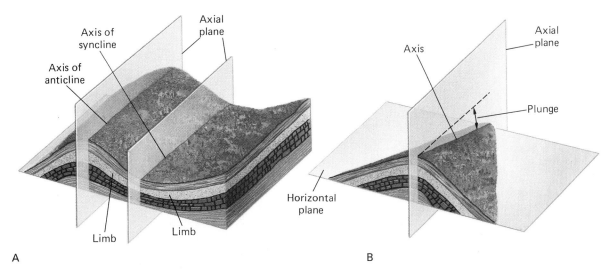

Figure 8–5 (A) An anticline, a syncline, and the parts of a fold. (B) A plunging anticline. (C) A syncline in southern Nevada.

Figure 8–6 (A) A diagram of a monocline formed where near-surface sedimentary rocks sag over a fault. (B) A monocline in southeastern Utah.

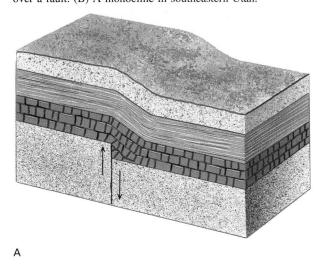

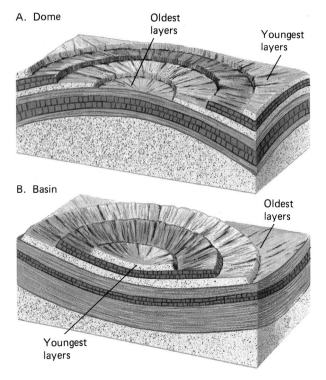

A. Dome

Oldest layers

Youngest layers

B. Basin

Oldest layers

Youngest layers

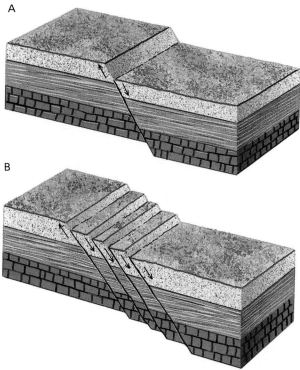

A

B

Figure 8–7 (A) Sedimentary layering dips away from a dome in all directions, and the outcrop pattern is circular or elliptical. (B) Layering dips toward the center of a basin.

Figure 8–8 (A) Movement along a single fracture characterizes faults with relatively small slip. (B) Movement along numerous fractures in a fault zone is typical of faults with large slip.

elliptical anticline resembling an inverted bowl is called a **dome**. Sedimentary layering dips away from the center of a dome in all directions (Fig. 8–7). A similarly shaped syncline is called a **basin**. Domes and basins can be small structures a few kilometers in diameter or less, but commonly are much larger. Large domes and basins result from regional warping of the entire continental crust. The Black Hills of South Dakota are a large structural dome. The Michigan basin covers much of the state of Michigan.

Look again at Figure 8–2. Notice that the distance between the same two points, A and A′, is shorter in the folded rock than before folding. Folding usually results from tectonic compression and always shortens distances in rock.

Faults

If tectonic force fractures the Earth's crust, rocks on opposite sides of the fracture may move past each other to create a fault. If the movement is rapid, it releases stored elastic energy and generates an earthquake, as described in Chapter 6. **Slip** is the distance that rocks on opposite sides of a fault have moved. Some faults are a single fracture in rock. On large faults with hundreds of meters or many kilometers of slip, movement may occur

along numerous closely spaced fractures collectively called a **fault zone** (Fig. 8–8).

Many faults and fault zones move repeatedly for two reasons: (1) tectonic stress commonly continues to be active in the same place over long periods of time, and (2) once a fault forms, it is easier for movement to occur again along the same fracture than for a new fracture to develop nearby.

Ore deposits often concentrate in faults. In the early days of mining, miners dug shafts and tunnels along faults to get at the ore. Most faults are not vertical but dip into the Earth at an angle. Therefore, most faults have an upper and a lower side. Miners referred to the side of a fault that hung over their heads as the **hanging wall**, and the side they walked on as the **footwall**. These names are still commonly used (Fig 8-9).

A fault in which the hanging wall has moved down relative to the footwall is called a **normal fault**. Notice that the horizontal distance between points on opposite sides of the fault, such as A and A′ in Figure 8–9, is greater after normal faulting occurs. Hence, normal faults form where tectonic forces are stretching the Earth's crust, pulling it apart.

Figure 8–10 shows a wedge-shaped block of rock called a **graben** dropped downward between a pair of

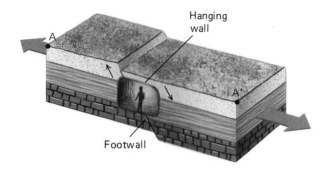

Figure 8–9 A normal fault forms where tectonic forces stretch the Earth's crust as shown by the large arrows. The overhanging side of the fault is called the hanging wall, and the side beneath the fault is called the footwall.

normal faults. The word **graben** comes from the German word for grave (think of a large block of rock settling downward into a grave). If tectonic forces stretch the crust over a large area, many normal faults may develop, allowing numerous grabens to settle downward along the faults. The blocks of rock between the down-dropped grabens then appear to have moved *upward* relative to the grabens; they are called **horsts**.

Normal faults, grabens, and horsts are common where the crust is being pulled apart at spreading centers, such as the mid-oceanic ridge and the East African rift zone. They are also common where tectonic force stretches a single plate, as in the Basin and Range of western North America.

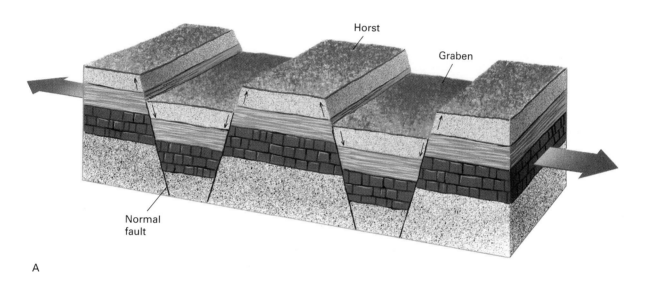

A

B

Figure 8–10 (A) Horsts and grabens commonly form along normal faults where tectonic forces stretch the Earth's crust. (B) In this photo of the Inyo Mountains of California, the valley is a graben and the mountain ranges in the foreground and background are horsts.

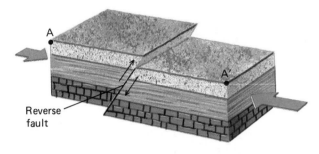

Reverse
fault

A

B

Figure 8–11 (A) A reverse fault accommodates crustal shortening and reflects squeezing of the crust, shown by large arrows. (B) A small reverse fault in Zion National Park, Utah.

In regions where tectonic force squeezes the crust, geologic structures must accommodate crustal shortening. A fold is one structure that shortens crustal rocks. A **reverse fault** is another (Fig. 8–11). In a reverse fault the hanging wall has moved *up* relative to the footwall. The distance between points A and A′ is clearly shortened by the faulting.

A **thrust fault** is a special type of reverse fault that is nearly horizontal (Fig. 8–12). In some thrust faults, the rocks of the hanging wall have moved several tens of kilometers over the footwall rocks. For example, all of the rocks of Glacier National Park in northwestern Montana slid eastward along the Lewis thrust fault to their present location. Their original position, before thrusting,

was 50 to 100 kilometers to the west. The Lewis thrust is one of many thrust faults that formed from about 200 to 45 million years ago as the mountains of western North America were being built. Most of those thrusts moved large slabs of rock, some even larger than that of Glacier Park, from west to east in a zone reaching from Alaska to Mexico.

A **strike-slip fault** is one in which the fracture is vertical, or nearly so, and rocks on opposite sides of the fracture have moved horizontally past each other (Fig. 8–13). A transform plate boundary is a strike-slip fault. The famous San Andreas fault is actually a zone of many parallel strike-slip faults in western California. As explained previously, it is a boundary between two lithospheric plates: the Pacific plate is moving northwestward relative to the North American plate. The true San Andreas fault is just one of the faults in the fault zone.

Joints

A joint is a fracture similar to a fault except that rocks on either side of the fracture have not moved. Columnar joints in basalt (Chapter 7) are examples of joints. Columnar jointing occurs when hot basalt cools and shrinks. The stress set up by the shrinking of the rock fractures it into five- or six-sided columns. In other cases, numerous parallel joints are caused by tectonic forces sufficient to fracture the rock but not to move it and cause faulting.

Joints are important in engineering, mining, and quarrying because they are planes of weakness in otherwise strong rock. Dams constructed in jointed rock often leak, not because the dams themselves have holes, but because water seeps into the joints and flows around the dam through the fractures. You can commonly see seepage caused by such leaks in canyon walls downstream from a dam.

Geologic Structures and Plate Boundaries

Tectonic forces stretch the crust and pull it apart at a divergent boundary, but squeeze and shorten it at a convergent boundary. At a transform boundary, rock is sheared as one plate slips horizontally past another. Because the forces are different, different kinds of geologic structures commonly develop at each type of plate boundary.

Normal faults and grabens are typical of divergent boundaries at mid-oceanic ridges and continental rifts, where the crust is stretched. A valley runs down the middle of the mid-oceanic ridge. It is a graben bounded by normal faults on both sides. Normal faults and grabens are the most common structures in the East African rift zone, as well.

At most convergent plate boundaries, where two plates collide, compressional forces squeeze rocks, form-

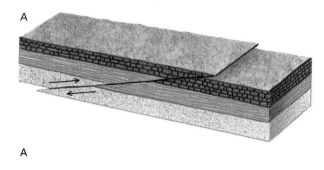

A

B

Figure 8–12   (A) A thrust fault is a low-angle reverse fault. (B) A small thrust fault near Flagstaff, Arizona.   *(Ward's Natural Science Establishment, Inc.)*

ing folds and reverse and thrust faults. For example, broad regions of folded and thrust-faulted rocks formed in the mountains of western North America from about 200 to 45 million years ago, while collision and subduction were occurring along the west coast. Folds and thrust faults are common in the Appalachian Mountains, the Alps, and the Himalayas, all of which developed as the result of collisions between two continents (Fig. 8–14).

At other convergent plate boundaries, however, subduction is accompanied by crustal stretching and normal faulting. The Andes Mountains of western South America are one of the Earth's highest mountain ranges. They formed as a result of subduction of the Pacific plate beneath the western edge of the South American plate. The two plates are converging, yet large grabens—structures that reflect crustal stretching—occur just west of the

Figure 8–13   (A) A strike-slip fault is nearly vertical, but movement along the fault is horizontal. The large arrows show direction of movement. (B) Recent movement along the San Andreas strike-slip fault in southern California has produced distinctive topographic features.   *(R. E. Wallace and the U.S. Geological Survey)*

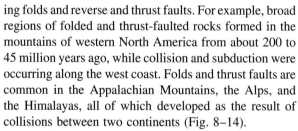

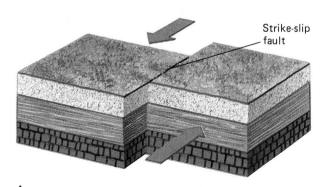

A

B

Figure 8–14   A thrust fault has carried older, light-colored dolomite over younger, dark shale at Long Rock Point on Lake Champlain, near Burlington, Vermont.   *(Rolfe Stanley)*

mountains. Thus, although convergent plate boundaries normally squeeze rocks, forming folds and reverse and thrust faults, this is not always the case.

A transform plate boundary is an immense strike-slip fault that cuts through the entire lithosphere. Frictional drag along the fault may cause folding, faulting, and uplift of nearby rocks. Forces of this type have formed the San Gabriel Mountains along the San Andreas fault zone.

8.4   Island Arcs: Mountain Chains Formed by Collision of Two Oceanic Plates

Where two oceanic plates converge, one subducts beneath the other and dives into the mantle forming an oceanic trench. Magma forms in the subduction zone and erupts onto the sea floor. Eventually submarine volcanoes grow into an **island arc**, a chain of volcanic islands next to the trench. However, the geology of an island arc is not simply that of a string of volcanoes in an ocean basin.

A layer of mud a half kilometer or more thick covers the sea floor. This sediment is saturated with water and consequently is soft and not very dense. Because of its low density, most of it cannot descend into the mantle with the subducting slab. Instead, it is scraped from the subducting slab and jammed against the growing island arc. The process is like a bulldozer blade scraping soil from bedrock. As the bulldozer advances, more and more soil piles up on the blade.

Occasionally, slices of basalt from the oceanic crust, and even pieces of the upper mantle, are scraped off and

mixed in with the sea floor sediment. The ocean-floor sediment and slices of rock are highly deformed, sheared, faulted, and metamorphosed in the bulldozer process. They are collectively termed a **subduction complex**, and become a part of the island arc, as shown in Figure 8–15.

A subduction complex grows by addition of the newest slices at the *bottom* of the complex. Consequently, as more and more slices are added, the complex is pushed upward, forming a depression called a **forearc basin** between the complex and the island arc. The forearc basin fills with sediment eroded from the volcanic islands to become a part of the island arc (Fig. 8–15).

Thus, an island arc is a volcanic mountain chain plus associated sedimentary and metamorphic rocks. Island arcs are abundant in the Pacific Ocean, where convergence of oceanic plates is common. The western Aleutian Islands and most of the island chains of the southwestern Pacific basin are island arcs.

8.5   The Building of Two Mountain Chains: The Andes and the Himalayas

Imagine that you have the opportunity to spend a season mountaineering and trekking through the Andes in western South America and, shortly thereafter, another season traveling through the Himalayas between Asia and India. Initially you might be struck by similarities between the two mountain chains. The immensity and great height of the peaks are the most striking features of both; the high summits are covered with snow and ice. The Andes have

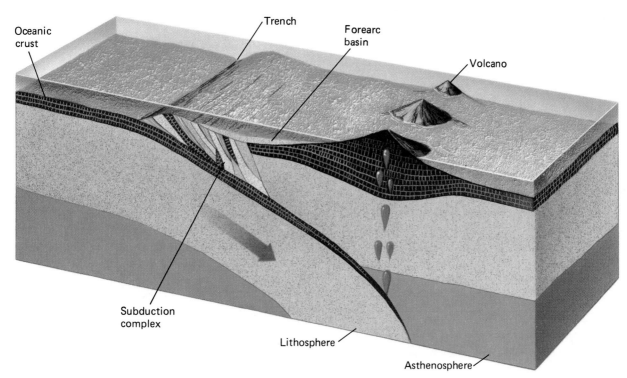

Figure 8–15   An island arc forms during a collision between two plates each carrying oceanic crust. A subduction complex contains slices of oceanic crust and upper mantle scraped from upper layers of a subducting plate.

49 peaks with elevations above 6000 meters (nearly 20,000 feet). The highest is Aconcagua, at 6962 meters. The Himalayas have 14 peaks above 8000 meters (26,000 feet), including Mount Everest, the highest peak on Earth, at 8848 meters.

Another similarity between the two chains is that both rise abruptly from adjacent low-lying regions. The Andes rise almost immediately from the Pacific coast of South America and thus start nearly at sea level. The Himalayas rise from the low plains of India to the south. Furthermore, both mountain chains are deeply eroded by glaciers that were once much more extensive than they are today. Both are still capped by immense alpine glaciers.

Thus, the two mountain chains might seem nearly identical except for local variations in terrain and culture. However, as a student of geology, you can distinguish among igneous, sedimentary, and metamorphic rocks. After a month in the Andes, you would have seen all three, with igneous rocks, particularly volcanics, being the most abundant. In fact, if you hiked the entire length of the Andes and were then asked to summarize the geology in a single phrase, you might reply, "It's a pile of volcanic rocks!"

After a month or two of trekking through the Himalayan chain, again you would have seen igneous, sedimentary, and metamorphic rocks. Here too, one type is most abundant. Folded and thrust-faulted sedimentary rocks

are prominent on many mountain faces and dominate Himalayan geology.

The fact that the Andes is predominantly a chain of volcanic rocks, whereas the Himalayas contain mostly sedimentary rocks, suggests that the two mountain chains must have been built by different geologic processes. Both chains formed at convergent plate boundaries. However, the Andes rose in a collision zone between a tectonic plate carrying oceanic crust and another carrying continental crust, whereas the Himalayas developed from a collision between two plates carrying continental crust.

The Andes: Subduction at a Continental Margin

Recall from Chapter 5 that during Triassic time all of the Earth's continents were assembled as the supercontinent called Pangaea. As Pangaea began to break apart about 200 million years ago, it first split into two large pieces, a southern portion called Gondwanaland and a northern piece called Laurasia. During the next 60 million years those two fragments also broke apart as the Atlantic Ocean began to open. At that time, the lithospheric plate that carried South America started moving westward. To accommodate the westward motion, oceanic lithosphere at the west coast of South America fractured and began to subduct beneath the continent (Fig. 8–16A). A trench

A. Cretaceous 140 million years ago

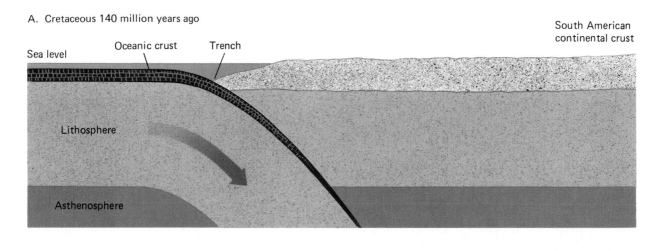

B. Cretaceous 130 million years ago

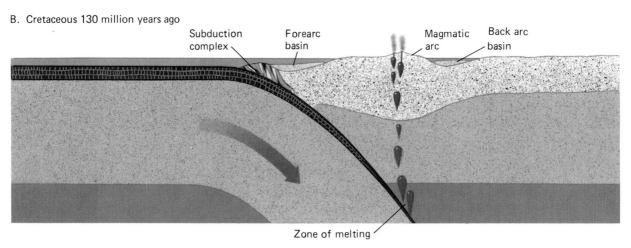

C. Late Cretaceous 90 million years ago

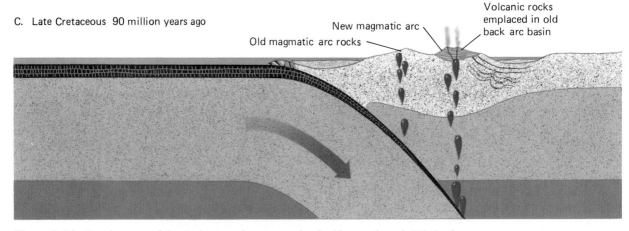

Figure 8–16 Development of the Andes, seen in cross section looking northward. (A) As the South American lithospheric plate began to move westward in early Cretaceous time, about 140 million years ago, subduction began and a trench formed at the west coast of the continent. (B) By 130 million years ago, igneous activity began and a subduction complex, forearc basin, and back arc basin formed (C) In late Cretaceous time, the trench and region of igneous activity had both migrated eastward. As a result, volcanic and plutonic rocks were emplaced in the sedimentary rocks of the early Cretaceous back arc basin. A new back arc basin formed further to the east.

Modern subduction beneath the west coast of South America fires volcanic eruptions throughout the Andes. Here, Chile's Villarrica volcano erupts into a moonlit sky in November, 1980.
(Noel Cramer)

formed off the coast where the down-going slab bent to dive into the mantle.

By about 130 million years ago, the diving plate had reached the lithosphere-asthenosphere boundary (Fig. 8–16B). Melting then began in the asthenosphere above the descending plate, forming vast amounts of basaltic magma. This magma rose to the base of the South American continent where it, in turn, melted the granitic crust. The magmas then rose into the continent. Some solidified within the crust to form batholiths and smaller plutons. However, vast quantities of magma continued upward to erupt at the surface, forming numerous volcanoes. This intrusive and volcanic activity occurred along the entire length of western South America, but in a band only a few tens of kilometers wide, directly over the zone of melting.

As oceanic crust dived beneath the continent, slices of sea-floor mud and rock were scraped from the subducting plate and jammed onto the edge of the continent. This process formed a subduction complex similar to that of an island arc. As in the case of an island arc, growth of the subduction complex created a forearc basin between the complex and the volcanic mountain chain (Fig. 8–16B).

As the Andes rose, the huge quantities of magma greatly increased the weight and thickness of the crust. Continental crust below some parts of the Andes is more than 60 kilometers thick. The entire mountain chain sank isostatically under the added weight and dragged down adjacent thinner crust on both sides of the mountains. In this way the forearc basin became deeper, and a **back arc basin** formed on the east side of the mountains. Both basins filled with sediment, and today thick sequences of sedimentary rocks are found both east and west of the Andes.

As the crust beneath the Andes thickened, it was heated by rising magma and grew weak and soft. Eventually it became so thick and soft that rocks began to spread outward under their own weight, as a mound of cool honey would. The spreading formed a great belt of thrust faults and folds along the east side of the Andes (Fig. 8–16C).

The Andes, then, are a relatively narrow mountain chain consisting predominantly of volcanic and plutonic rocks produced by subduction at a continental margin. The chain also contains extensive sedimentary rocks deposited in basins on both sides of the mountains. Some of those rocks were folded, faulted, and metamorphosed by tectonic forces and high temperature caused by rising magma. The Andes are a good general example of subduction at a continental margin. This type of plate margin is called an **Andean margin**.

The Himalayan Mountain Chain: A Collision Between Continents

The **Himalayan mountain chain** separates the Earth's two most populous nations, China and India (Fig. 8–17).

Figure 8–17   The Himalayas separate the Indian subcontinent from southern Asia.   *(Andy Zimet)*

The world's highest mountains, including Mount Everest and K2, are located in the Himalayas. Today, if you stand on the southern edge of Tibet and look southward, you can see the high peaks of the Himalayas. Beyond this great mountain chain are the rainforests and hot, dry plains of India. If you had been able to stand in the same place 100 million years ago and look southward, you would have seen only ocean. At that time India lay south of the equator, separated from Tibet by thousands of kilometers of open ocean. The Himalayas had not yet begun to rise.

Formation of an Andean-Type Margin

When Pangaea initially split into Laurasia and Gondwanaland, the two were separated by open ocean. About 120 million years ago a large, triangle-shaped piece of lithosphere split off from Gondwanaland. It began drifting northward toward Asia at high speed—geologically speaking—perhaps as fast as 20 centimeters per year (Fig. 8–18A). The northern part of this lithospheric plate carried oceanic crust, but the Indian subcontinent lay on the southern corner (Fig. 8–18B).

Figure 8–19A shows that India and southern Asia were connected by oceanic crust before India began moving northward. As the Indian plate started to move, oceanic crust began to subduct at Asia's southern margin (Fig. 8–19B). As a result, magma formed, volcanoes erupted and granite plutons were emplaced in southern Tibet. At this point southern Tibet was an Andean-type continental margin, and it continued to be so from about 120 to 50 million years ago, while India drew closer to Asia.

Continent-Continent Collision

By about 50 million years ago, all of the oceanic lithosphere between India and Asia had been consumed by subduction (Fig. 8–19C). Then the two continents collided. Since both are continental crust, neither could subduct deeply into the mantle. The collision did not stop the northward movement of India, but it did slow it down. India had been speeding northward at 20 centimeters per year and suddenly slowed to about 5 centimeters per year.

Continued northward movement of India after the collison began was accommodated in two ways. The leading edge of India began to slide under Tibet in a process called **underthrusting**. As a result, the thickness of continental crust in the region doubled (Fig. 8–19C). When underthrusting began, igneous activity in Tibet stopped because the subducting plate was no longer diving into the mantle. As India slid beneath Tibet, the edge of Tibet scraped the soft Indian sedimentary rocks from harder, older basement rock and pushed them into folds and thrust faults. These deformed sedimentary rocks make up the greatest part of the Himalayas (Fig. 8–19D).

The second way in which India continued moving northward after colliding with Tibet was by crushing Tibet and wedging China out of the way along huge strike-slip faults. India has pushed southern Tibet 1500 to 2000 kilometers northward since the beginning of the collision. The forces have created major mountain ranges and basins north of the Himalayas.

The Himalayas Today

The underthrusting of India beneath Tibet and the squashing of Tibet have produced thick continental crust under the Himalayas and the Tibetan Plateau to the north. Consequently, the region isostatically floats at high elevation. Even the valleys lie at elevations of 3000 to 4000 meters,

Wildly folded sedimentary rocks on the Nuptse-Lhotse Wall, from an elevation of 7600 meters on Mount Everest. *(Galen Rowell)*

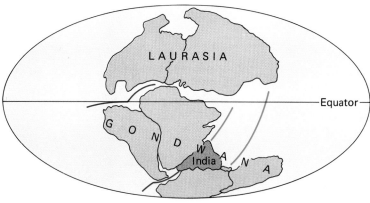

A   200 Million years ago

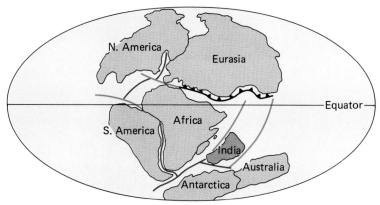

B   120 Million years ago

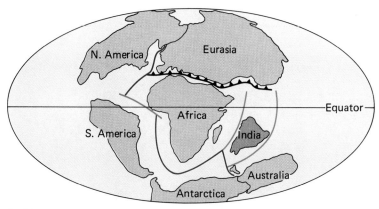

C   80 Million years ago

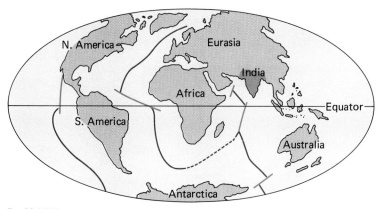

D   40 Million years ago

Figure 8–18 (A) Gondwanaland and Laurasia formed shortly after 200 million years ago as a result of the early breakup of Pangaea. Notice that India was initially part of Gondwanaland. (B) About 120 million years ago, India broke off from Gondwanaland and began drifting northward. (C) By 80 million years ago India was isolated from other continents and was approaching the equator. (D) By 40 million years ago it had moved 4000 to 5000 kilometers northward and collided with Asia.

A

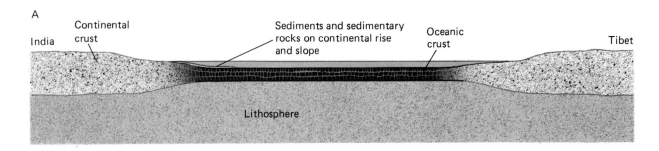

India

Continental crust

Sediments and sedimentary rocks on continental rise and slope

Oceanic crust

Tibet

Lithosphere

B

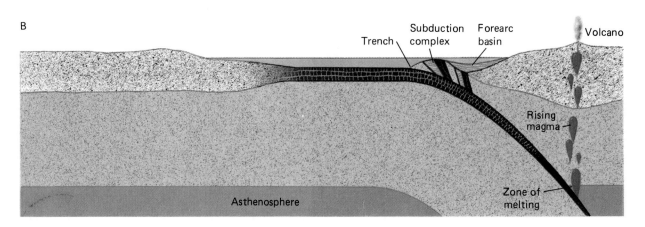

Trench

Subduction complex

Forearc basin

Volcano

Rising magma

Zone of melting

Asthenosphere

C

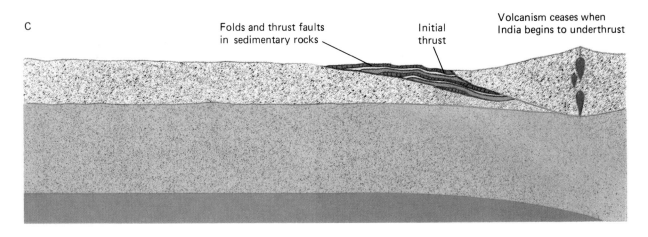

Folds and thrust faults in sedimentary rocks

Initial thrust

Volcanism ceases when India begins to underthrust

D

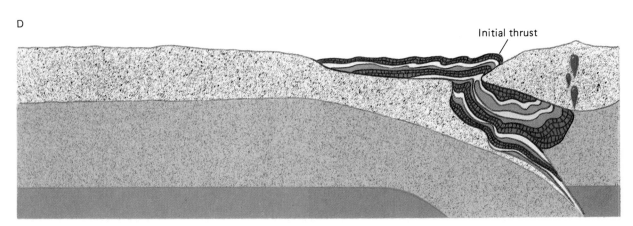

Initial thrust

The Himalayas are an actively rising mountain range today.

and the Tibetan Plateau has an average elevation of 4000 to 5000 meters. One reason the Himalayas contain all of the Earth's highest peaks is simply that the *bases* of the peaks are at such a high elevations. From its base to the summit, Mount Everest is actually smaller than Alaska's Denali (Mount McKinley), North America's highest peak. Mount Everest rises about 3300 meters from base to summit whereas Denali rises about 4200 meters. The difference in elevation of the respective summits lies in the fact that the base of Mount Everest is at about 5500 meters, but Denali's base is at 2000 meters.

Frequent moderate earthquakes and occasional large and destructive ones testify to the continuation of the collision between India and Asia. Several quakes of Richter magnitude 8 have occurred in historical time as India continues to penetrate Asia at a rate of about 5 centimeters per year. Careful measurements of the heights of selected Himalayan peaks have been repeated several times in the last century. Comparisons of older surveys with newer ones show that the Himalayas continue to rise rapidly today—perhaps as fast as 1 centimeter per year, faster than they are eroding away.

Notice that the development of the Himalayan chain occurred in two steps. First, an Andean-type continental margin formed in Tibet as oceanic crust subducted beneath southern Asia. At that time, the geology of southern Asia must have been similar to the present geology of the Andes. Only later, after all the oceanic crust between the two continents was consumed by subduction, did the effects of continent-continent collision begin. The two-step nature of the process must be common to all continent-continent collisions because ocean basins separating continents must first be subducted before the continents can collide.

When two continents collide, they weld into a single mass of continental crust. The junction is called a **suture zone** or **continental suture**. Continental sutures are often recognized by a sudden difference between rock types, ages, and structures on opposite sides of the suture zone. Such a geologic discontinuity occurs because the rocks

Figure 8–19 The cross-sectional views show the Indian and Asian plates before and during the collision between India and Asia. (A) Shortly before 120 million years ago, India, southern Asia, and the intervening ocean basin were parts of the same lithospheric plate. (In this figure, the amount of oceanic crust between Indian and Asian continental crust is abbreviated to fit the diagram on the page.) (B) When India began moving northward, the plate broke and subduction began at the southern margin of Asia. By 80 million years ago an oceanic trench, subduction complex, and forearc basin had formed. Volcanoes erupted, and granite plutons formed in the region now called Tibet. (C) By 40 million years ago India had collided with Tibet. The leading edge of India was underthrust beneath southern Tibet. (D) Continued underthrusting and collision between the two continents has crushed Tibet and created the high Himalayas by folding and thrust-faulting the sedimentary rocks. India continues to underthrust and crush Tibet today.

on one side of the suture originated on one continent, whereas the rocks on the other side formed on the other continent, perhaps at different times and by different processes. Another criterion for identifying a suture is that rocks of the suture zone are commonly deformed and sheared by the collision. In addition, bits and scraps of oceanic crust and mantle occur in suture zones, preserving traces of the oceanic lithosphere that was mostly consumed by subduction.

The Himalayan chain is the best, but not the only, example of mountain building by continent-continent collision. The Appalachian Mountains formed when eastern North America collided with Europe, Africa, and South America between 400 and 250 million years ago. The European Alps formed during repeated collisions between northern Africa and southern Europe beginning about 30 million years ago. The Urals of northwestern Asia formed by a similar process about 250 million years ago.

SUMMARY

•

Mountain chains and ranges form by tectonic activity at boundaries between lithospheric plates. Volcanic eruptions, intrusion of granite, metamorphism, earthquakes, and folding and faulting of rocks commonly accompany growth of a mountain chain.

Folds, faults, and joints are geologic structures that develop as mountains are built. Folds, reverse faults, and thrust faults usually form where tectonic forces squeeze rocks together. Normal faults usually form where rocks are pulled apart. Strike-slip faults form where blocks of crust slip horizontally past each other along vertical fractures.

Mountains form at all three types of plate boundaries. The mid-oceanic ridge and East African rift ranges formed at divergent boundaries. The San Gabriel Mountains of California rose along the San Andreas transform plate boundary. But most of the great continental ranges,

including the Andes, the Appalachians, the Alps, and the Himalayas, grew at convergent plate boundaries. If two converging plates both carry oceanic crust, a volcanic island arc forms. If one plate carries oceanic crust and the other carries continental crust, an Andean margin develops, with numerous plutons and volcanoes. It also contains rocks of a subduction complex and sedimentary rocks deposited in forearc and back arc basins.

When two plates carrying continental crust collide, an Andean margin first develops as oceanic crust separating the two continents subducts. Later, when all oceanic crust is consumed and the continents themselves collide, one continent is underthrust beneath the other. The geology of mountain ranges formed by continental collisions, such as the Himalayas, is dominated by vast regions of folded and thrust-faulted sedimentary and metamorphic rocks and by older plutonic and volcanic rocks.

KEY TERMS

•

Mountain range *190*	Limb *192*	Hanging wall *194*	Island arc *198*
Mountain chain *190*	Axis *192*	Footwall *194*	Subduction complex *198*
Geologic structure *191*	Plunging fold *192*	Normal fault *194*	Forearc basin *198*
Fold *191*	Monocline *192*	Graben *194*	Back arc basin *201*
Fault *192*	Dome *194*	Horst *194*	Andean margin *201*
Joint *192*	Basin *194*	Reverse fault *196*	Underthrusting *202*
Anticline *192*	Slip *194*	Thrust fault *196*	Suture zone *205*
Syncline *192*	Fault zone *194*	Strike-slip fault *196*	Continental suture *205*

REVIEW QUESTIONS

•

1. Explain the differences among a mountain, a mountain range, and a mountain chain.

2. Explain why the interiors of lithospheric plates are normally tectonically inactive, whereas plate boundaries are tectonically active. List the types of tectonic activity that normally occur at plate boundaries.

3. Describe types of tectonic activity that are specific to each of the three different types of plate boundaries. What kinds of tectonic activity are common to all three types of plate boundaries?

4. Describe the tectonic environments of the three different types of convergent plate boundaries.

5. What is a geologic structure? What are the three main types of structures?

6. What is the difference between a fault and a joint?

7. What is the difference between an anticline and a syncline?

8. At what type of tectonic plate boundary would you expect to find normal faults?

9. How are faults related to earthquakes?

10. Why does movement often occur repeatedly on a single fault or along a fault zone?

11. Why are thrust faults, reverse faults, and folds commonly found together?

12. Draw a cross-sectional sketch of an anticline-syncline pair and label the parts of the folds. Use your sketch to explain how folds accommodate crustal shortening and tectonic squeezing of rocks.

13. Draw a cross-sectional sketch of a normal fault. Label the hanging wall and the footwall. Use your sketch to explain how a normal fault accommodates crustal extension. Sketch a reverse fault and show how it accommodates crustal shortening.

14. In what sort of a tectonic environment would you expect to find a strike-slip fault?

15. Give examples of mountain chains formed at each of the three types of convergent plate boundaries.

16. What mountain chain has formed at a divergent plate boundary? Explain the main differences between this chain and those developed at convergent boundaries.

17. What mountain ranges have formed at transform plate boundaries?

18. Where are island arcs forming today? In what tectonic environment do they form?

19. Describe the main geologic features of an island arc.

20. Describe the similarities and differences between the Andes and the Himalayan chain. Why do the differences exist?

21. Sketch a cross section of an Andean-type plate boundary to a depth of several hundred kilometers. Show the positions of the subducting plate, trench, subduction complex, forearc and back arc basins, volcanoes and plutons, and earthquakes.

22. Draw a series of cross-sectional sketches showing the evolution of a Himalayan-type plate boundary. Why does this boundary start out as an Andean margin?

DISCUSSION QUESTIONS

•

1. Why do most major continental mountain chains form at convergent plate boundaries? What topographic and geologic features characterize divergent and transform plate boundaries in continental crust? Where do these types of boundaries exist in continental crust today?

2. Discuss the relationships among types of lithospheric plate boundaries, the kinds of tectonic forces that occur at each type of plate boundary, and the main types of geologic structures you might expect to find in each environment.

3. Discuss how and why forearc and back arc sedimentary basins form at convergent plate boundaries. Would you expect any difference in the sediments deposited in each type of basin?

4. Discuss the rock types, tectonic forces, and geologic structures that you might expect to find in rocks of a subduction complex.

5. Compare and explain the similarities and differences between the Andes Mountains and the Himalayan chain. How would the Himalayas 60 million years ago have compared with the modern Andes?

6. Where would you be more likely to find large quantities of igneous rocks in the Himalayan chain—in the northern parts of the chain, near Tibet, or southward, near India? Discuss why.

7. Explain why sedimentary beds in a plunging fold form a "V" pattern on the Earth's surface.

8. If you were studying photographs of another planet, what features would you look for to determine whether or not the planet is or has been tectonically active?

The Geological Evolution of North America

W e hope that you have become curious about the geology and geologic history of specific places—perhaps about the history of our own continent, the region where you live, or the area surrounding your college or university. Almost certainly your professor has spoken of geologic features on or near your campus. You may have taken field trips to see local rocks and landforms. This chapter describes the geologic evolution of North America. In reading it, you will come to understand the geology and geologic history of your own part of our continent.

The oldest known rocks in North America are the 3.96-billion-year-old Arcasta gneisses near Great Slave Lake, Northwest Territories, Canada. Thus, the geologic history of North America must have begun at least that long ago. In the past 3.96 billion years, tectonic activity has continuously changed our continent. You have read about tectonic processes in preceding chapters of this book. In this chapter you will see how they have built and shaped our continent.

● High summer in the Bugaboos of British Columbia.

As you read this chapter, bear in mind that we are dealing with theories. These theories are based on data—facts that can be observed in the rocks of North America. But the theories are not facts; they are *interpretations* of the data. Ten years from now the theories may be different, because new data will become available as geologists carry out more research, and interpretations of old data may change. It is simply the nature of geology that our understanding of the Earth continues to change as we learn more.

9.1 The Structure of North America

Figure 9–1 shows the topography of North America. If you began driving westward from the Atlantic coast, say from Norfolk, Virginia, you would cross a low coastal plain for the first few hundred kilometers. Then the road would rise a thousand meters or so to cross the rolling Appalachian Mountains. Beyond the Appalachians, you would descend onto a broad plain, heading toward the Mississippi River in the middle of the continent. For two or three days you would drive through this flat landscape, passing rich fields of corn and wheat. Then in the distance you would see the high, snowcapped western mountains. For another few days you would cross mountains, valleys, and deserts until you reached the Pacific Ocean.

As you crossed our continent, you might wonder why the middle of America is flat and the mountains are located near its edges. From what you have learned of tectonics, you know that most mountain ranges form during collisions between tectonic plates. North America is like a car that has been involved in a multiple collision. If a car runs into a telephone pole and is struck from behind by another car, both the front and rear ends of the car become crumpled, but the middle part, where the passengers sit, is protected. Similarly, North America has been involved in several collisions with other tectonic plates. The collisions formed mountain chains along its edges, but the central part of the continent has been relatively unaffected by tectonic activity for almost 2 billion years. During that time, older mountains that once existed in the mid-continent region have eroded to a flat, low-lying plain.

North America is made up of three major geologic regions (Fig. 9–2). The **craton** is the flat interior of the continent (Fig. 9–3). Most parts of the craton have not seen deformation, metamorphism, or hot magma for almost 2 billion years. The craton contains two major subdivisions: the **shield**, where old igneous and metamorphic rocks are exposed at the surface, and the **platform**, where the same kinds of old rocks are covered by a veneer of younger sedimentary rocks.

The second geologic region is the **mountain chains**, which occur at or near the edges of the continents and surround the craton. (Fig. 9–4). Although some of the mountains of North America are hundreds of millions of years old, they are young relative to the craton. The Appalachian mountain chain formed along the east coast from about 490 to 250 million years ago. While the Appalachians were rising, related events constructed the

Figure 9–1 A satellite view of North America. *(© Tom Van Sant, Inc./The Geosphere Project, Santa Monica, CA)*

© 1990 Tom Van Sant,
Inc./The Geosphere Project,
Santa Monica, CA

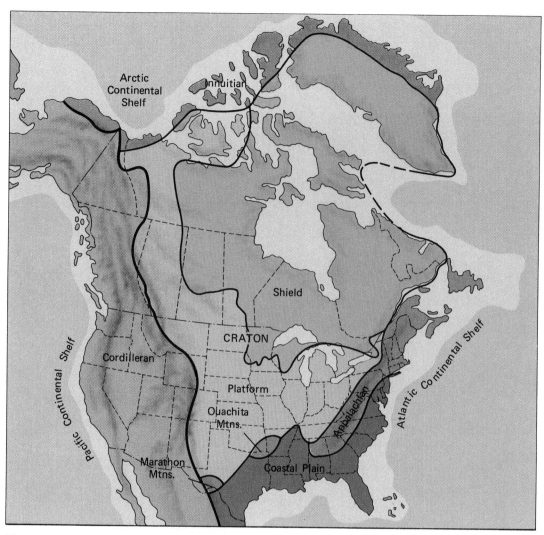

Figure 9–2 The three major geologic regions of North America. (1) The craton makes up the central portion of the continent, and includes the platform and the shield. (2) The Appalachian, Innuitian, and Cordilleran mountain chains occur near the continental margins. (3) The coastal plain and continental shelves separate the older mountains and the craton from oceanic crust.

Innuitian mountain chain along the northern margin of the continent and across northern Greenland. The mountains of western North America are collectively called the Cordilleran mountain chain. They began to form about 180 million years ago, and the region remains tectonically active today.

The third geologic region is the **continental shelves**, which lie between the mountain chains and oceanic crust. In these zones along the seacoasts, young sedimentary rocks have accumulated over older continental crust. On the shelves, continental crust is so thin that it commonly lies below sea level. The **coastal plain** is a region where

similar sedimentary rocks have formed at higher elevations and farther inland during times of higher sea level (Fig. 9–5).

9.2 Supercontinent Cycles and the Construction of the North American Craton

If you could strip away the relatively young sedimentary rocks from the North American craton and examine the old, underlying igneous and metamorphic rocks, you

Figure 9–3 Aerial photograph of a portion of the North American craton, Quebec. *(Canada National Air Photo Library)*

Figure 9–4 Mountain chains occur near the margins of North America.

would discover that the craton consists of several distinct **geologic provinces** (Fig. 9–6). A geologic province is a large region in which rocks share a similar geologic history. The provinces and the boundaries separating them are identified in the following ways:

1. Geologic relationships among rocks within each province are continuous, whereas relationships across province boundaries are discontinuous. For example, within a single province, a sequence of metamorphic rocks may change gradually from low through medium to high grades of metamorphism over a distance of several kilometers. This is a normal progression. In contrast, across a province boundary, rocks of low metamorphic grade may lie directly against high-grade rocks, or high-grade rocks may lie next to unmetamorphosed sedimentary rock. These are not normal relationships.

2. The rocks of each province give different radiometric ages from those of adjacent provinces.

3. Rocks at the boundaries between provinces are commonly sheared and metamorphosed like those in the suture zone between India and Asia.

What caused these differences and discontinuities at the boundaries of the provinces? Recently, Paul Hoffman of the Geologic Survey of Canada has suggested that at least three times during the Earth's history all continents were joined together in a single **supercontinent** like Pangaea and then split apart. Hoffman has estimated that it takes about 300 to 500 million years for a supercontinent to assemble, split apart, and then reassemble.

In this model, rifting breaks up a supercontinent and causes the fragments to spread apart. But because the Earth is a sphere, the continental fragments migrate halfway around the globe and then collide on the far side to reassemble as a new supercontinent. Thus, the breakup of one supercontinent leads to the assembly of a new one.

9.3 Pangaea I: Two Billion Years Ago

The Assembly of the First Supercontinent and the Building of the North American Craton

Hoffman suggests that each of the provinces of the North American craton originated as an individual island arc or a microcontinent. Recall from Chapter 8 that an island arc is a curved chain of volcanic islands rising from the

Figure 9–5 Low-lying swamps of the coastal plain, North Carolina coast.

Figure 9–6 The provinces of the North American craton.

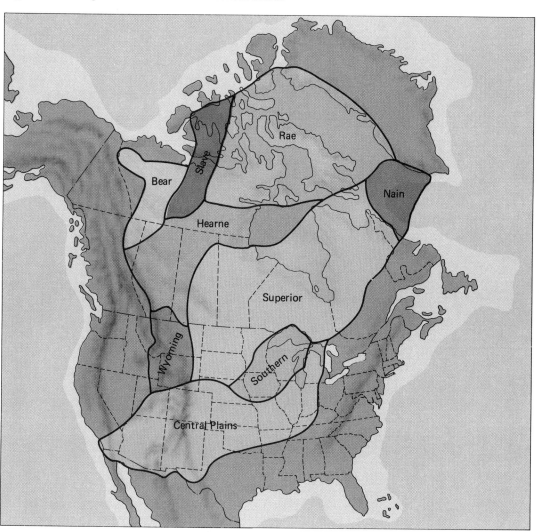

deep-sea floor near a mid-oceanic subduction zone. A **microcontinent** is a small mass of continental crust surrounded by oceanic crust. Japan is a modern microcontinent. It is similar to a continent, but smaller.

Prior to about two billion years ago, these island arcs and microcontinents were scattered about the globe and separated by ocean basins. Then movements of tectonic plates swept all of the island arcs, microcontinents, and other masses of continental crust on the Earth's surface together into a single great landmass. This accretion took about 200 million years to complete. Thus, Hoffman suggests, by 1.8 billion years ago, all of the Earth's continental crust was joined together in a supercontinent that we call **Pangaea I** (after Alfred Wegener's Pangaea, a word

meaning "all lands"). The North American craton was a small part of Pangaea I.

The geologic differences among the provinces of the North American craton exist because each province was isolated from the others when it formed. Each developed as an independent microcontinent or island arc separated from the others by oceanic crust, much as Japan is now separated from the Aleutians. The rocks of each province give different radiometric dates because each microcontinent or island arc formed at a different time from the others. The sheared and faulted boundaries between provinces are suture zones where the microcontinents welded together as the supercontinent assembled between 2 and 1.8 billion years ago.

Figure 9–7 About one billion years ago, western North America was joined to the Siberian portion of modern Asia. Similarities in rock types (green and brown zones) and structural trends (red lines) show continuity of geological features between the North American and Siberian cratons.

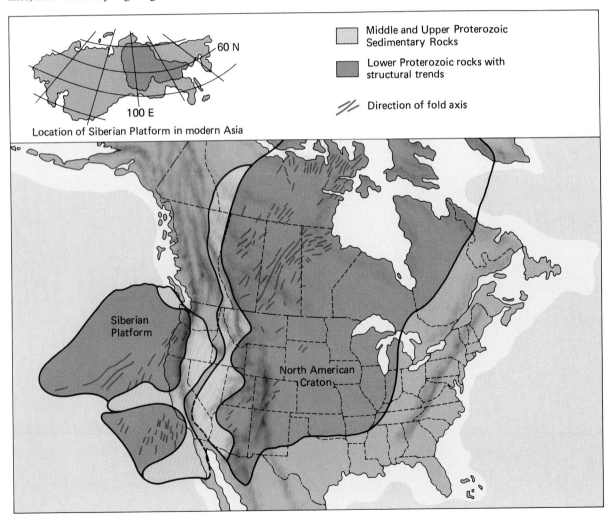

Breakup of the First Supercontinent

In Hoffman's model, the mantle beneath the new super-continent soon began to warm up because the vast layer of continental crust acted as a giant blanket and kept heat from escaping. The warming mantle then started to rise as a mantle plume beneath the supercontinent. Another possibility is that a major meteorite impact initiated a mantle plume. In any case, the plume spread out at the base of the supercontinent, tearing it into several fragments, each riding on its own lithospheric plate. In this way Pangaea I broke apart about 1.3 billion years ago. However, the provinces of the North American craton did not separate from each other, but instead remained welded together. Thus, the North American craton was created during assembly of Pangaea I and has remained essentially intact to this day.

9.4 Pangaea II: One Billion Years Ago

Assembly of the Second Supercontinent

After Pangaea I split up, the separate fragments of continental crust migrated across the Earth and then reassembled, forming a second supercontinent called **Pangaea II**, about 1.0 billion years ago. Thus, it appears that about one billion years ago, western North America occupied an interior portion of a late Precambrian supercontinent that included part of modern Asia (Fig. 9–7). Portions of what is now Europe were attached to eastern North America at the same time.

Breakup of the Second Supercontinent

When Pangaea II began to break apart, the Siberian portion rifted away, leaving a shoreline at the western margin of North America. The western edge of North America, however, did not appear as it does today. Parts of Alaska and western Canada, and much of Washington, Oregon, western Idaho, and western California, had not yet become part of the continent, as shown in Figure 9–8. Instead, the area was ocean.

As Siberia tore away from western North America, about 700 to 600 million years ago, the portion of Europe that had been attached to our east coast also rifted away. This rift opened a new ocean basin, called the Proto-Atlantic Ocean (an early version of the modern Atlantic Ocean), along the east coast. Thus, by the end of Precambrian time, the North American craton became isolated from other continents and was surrounded by oceans (Fig. 9–9A).

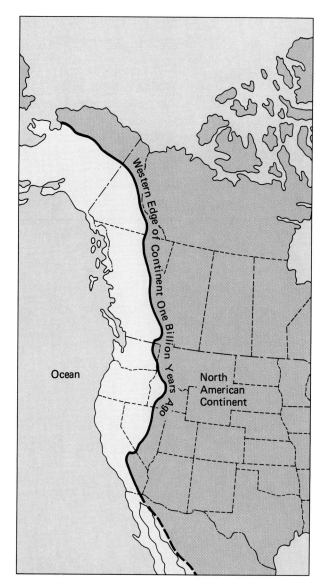

Figure 9–8 Western North America after rifting shortly after one billion years ago. The light blue area shows the modern outline of the continent for comparison.

9.5 Pangaea III: One-Half Billion Years Ago

The Assembly of Pangaea III (Wegener's Pangaea) and the Building of the Appalachian Mountains

When Pangaea II first broke up, North America migrated westward, separating from Europe as the Proto-Atlantic

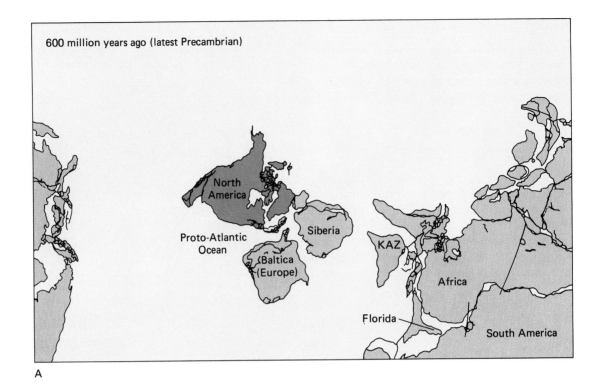

A

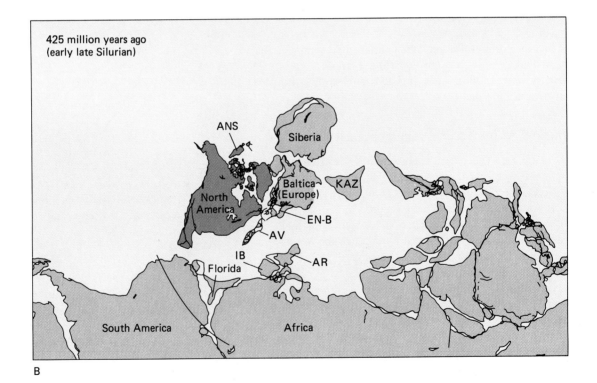

B

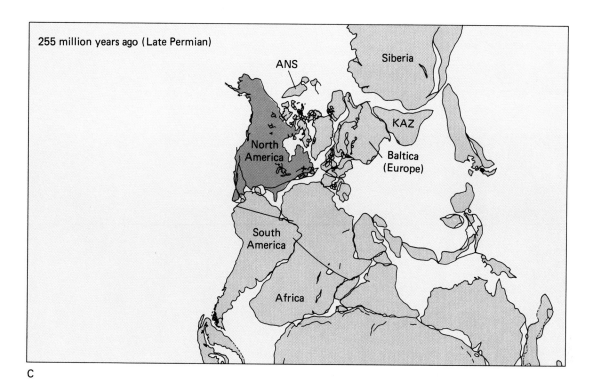

255 million years ago (Late Permian)

C

Figure 9–9 Positions of continents at three different times between 600 million years ago and 255 million years ago. NA = North America, SIB = Siberia, BA = Baltica (the name given to what is now most of Europe), AF = Africa, SA = South America, FL = Florida, ANS = Alaska North Slope, KAZ = Kazakhstania (now part of central Asia), AV = Avalonia (now part of eastern North America), En-B = England-Brabantia, IB = Iberia (now most of Spain), AR = Armorica (now part of Europe). Continents near the centers of the maps are shown fairly accurately, but those near the map margins are distorted. *(From Bally and others,* The Geology of North America—An Overview *[GSA, 1989])*

opened (Fig. 9–10A). By Middle Ordovician time, North America had *reversed* direction and was moving back toward Europe and Africa (Fig. 9–10B). We do not know why our continent changed its direction, but that change put eastern North America on a collision course with Africa and Europe. As a result, subduction began near the east coast of North America, and the Proto-Atlantic Ocean began to close up. Volcanoes erupted, granite plutons rose into the crust, mountains formed, and broad regions of metamorphic rocks developed along the east coast (Figs. 9–10B and C).

Beginning about 400 million years ago, in Early Devonian time, some parts of the Proto-Atlantic Ocean had completely closed up, and eastern North America began colliding with Europe, Africa, and South America (Figs. 9–9B and C, and Fig. 9–10D). The collision and suturing of continents began the assembly of the third supercontinent, **Pangaea III**. This continent-continent collision shoved sedimentary rocks westward, forming tremendous thrust faults and folds (Fig. 9–11). The collision continued until about 250 million years ago.

Thus, subduction led to an initial stage of mountain building at an Andean margin, followed by a second stage caused by the continental collision. This two-step process formed the **Appalachian mountain chain** along the eastern and southeastern margins of North America (Fig. 9–12). Note that the two steps that formed the Appalachians were similar to the more recent collision between India and Asia that formed the Himalayas. At their maximum stage of development, the Appalachians must have been similar to the Himalayas: a huge, high mountain chain characterized by frequent earthquakes, rapidly rising mountains, and folded and thrust faulted rocks. Today, however, erosion has worn the Appalachians down to maximum elevations of less than 2000 meters.

Younger sedimentary rocks of the coastal plain cover the igneous and metamorphic rocks of the Appalachians from northeastern Alabama to Texas. However, the belt of plutons and metamorphic rocks continues beneath those sedimentary rocks for another 1300 kilometers, wrapping around the southeastern margin of North America. The igneous and metamorphic rocks poke up through the sedi-

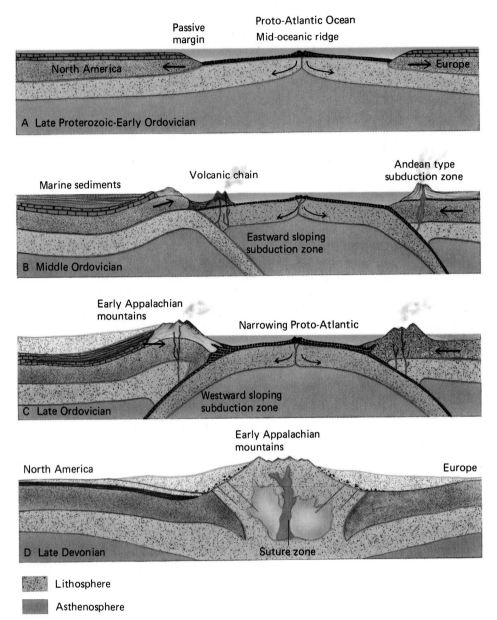

Lithosphere

Asthenosphere

Figure 9–10 A hypothetical sequence of events forming the northern Appalachian mountain chain. (A) At the beginning of Paleozoic time, about 570 million years ago, North America was separated from Europe by a broad ocean basin. (B) By Middle Ordovician time, an eastward-sloping subduction zone had formed near the east coast of North America, creating a chain of volcanic islands. (C) In Late Ordovician time, the Proto-Atlantic Ocean was closing because North America had reversed direction, and was moving toward Europe. The volcanic islands were shoved onto the eastern edge of the continent. The eastward-sloping subduction zone died out and was replaced by westward-sloping subduction. Plutons, volcanoes, and metamorphic rocks formed along the Andean-type continental margin as a result of the westward-sloping subduction. (D) In Late Devonian time, northeastern North America collided with Europe, creating a Himalayan-like continental collision zone. The Proto-Atlantic Ocean in this region had closed completely.

Figure 9–11 Eastern North America began to collide with Africa and Europe 400 million years ago, forming great thrust faults and folds such as these in Nova Scotia. *(Geological Survey of Canada)*

mentary rocks in the Ouachita and Marathon mountains in Arkansas, Oklahoma, and east Texas (Fig. 9–13). The continuity of the Appalachian rocks is concealed by the younger coastal-plain sedimentary rocks, but is known because of deep drilling in oil-rich Texas.

In Late Devonian time, about 360 million years ago, another major mountain building episode occurred across the northern margin of North America. This event formed the **Innuitian mountain chain** (named for the Inuit, the name Eskimos call themselves), which extends from northern Greenland to the northern Yukon Territory. It may have been caused by a continental collision related

to the closing of the Proto-Atlantic Ocean farther to the south.

While western Europe, Africa, and South America collided with eastern North America, forming the Appalachians, the remaining masses of continental crust were added to the southern part of the new supercontinent. Thus, Pangaea III assembled by about 300 million years ago, 70 million years before the appearance of dinosaurs. North America formed the northwestern portion of Pangaea III (Fig. 9–9C). Pangaea III was the supercontinent discovered and described so accurately by Alfred Wegener 65 years ago.

Paleozoic Platform Sedimentary Rocks

When Pangaea II began to break up in late Precambrian time, a new and very long mid-oceanic ridge system

Figure 9–12 Hanging Rock State Park, North Carolina. *Tom Till*

Figure 9–13 The Ouachita Mountains, Arkansas. *(Tom Till)*

Figure 9–14   Flat-lying platform sedimentary rocks in Iowa.   *(Dennis R. Kolata, Illinois Geological Survey)*

must have formed among the spreading fragments of the supercontinent and displaced large volumes of seawater. Thus, sea level rose, flooding low-lying portions of continents, just as the milk in your cereal bowl overflows when

you add too many Cheerios. The new ridge system raised sea level for much of the following 200 million years, flooding the low central portion of North America with shallow seas.

The craton was not continuously flooded during that time, however. Instead, seas advanced and withdrew several times as sea level rose and fell. Each time sea level rose, beaches migrated inland until the entire craton was submerged. As the beaches advanced inland, they spread a blanket of sand across the craton. Shale and limestone deposited offshore from the beaches eventually covered the sand layer. In this way, layers of **platform sedimentary rocks** accumulated on the craton from Cambrian through Pennsylvanian time. They are only a few hundred meters thick and cover the old igneous and metamorphic rocks of the craton. It is common to see these flat-lying beds of shale, limestone, and sandstone as you drive through the central part of our continent (Fig. 9–14).

As the Appalachian Mountains rose, shallow swamps formed just west of the high mountains in a region that is now Pennsylvania, West Virginia, and surrounding areas. The climate must have been wet and warm because the swamps filled with dense, tropical vegetation. The buried remains of this swamp environment became the rich coal deposits of eastern North America.

Figure 9–15   As Pangaea broke up beginning about 200 million years ago, rifting followed the lines of the earlier collision zone that had formed the Appalachians. (A–C) Positions of continents at three times from earliest Cretaceous through early Oligocene.   *(From Bally and others, The Geology of North America—An Overview [GSA, 1989])*

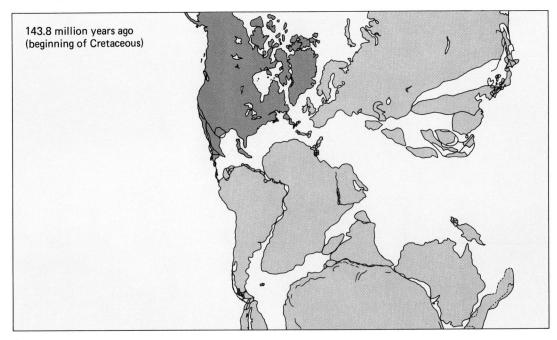

143.8 million years ago (beginning of Cretaceous)

A

9.6 Breakup of Pangaea III

Opening of the Modern Atlantic Ocean

Pangaea III remained intact for about 100 million years, from about 300 to 200 million years ago. Then, about 200 million years ago, it began to rift apart. As North and South America separated from Eurasia and Africa, the modern Atlantic Ocean began to open. When tectonic activity on the eastern margin of North America ceased, the lofty Appalachians stopped rising and began to erode away.

Pangaea III broke up approximately along the sutures where continents had welded together 100 million years previously (Fig. 9–15). The new eastern margin of North America was in about the same place it had been before

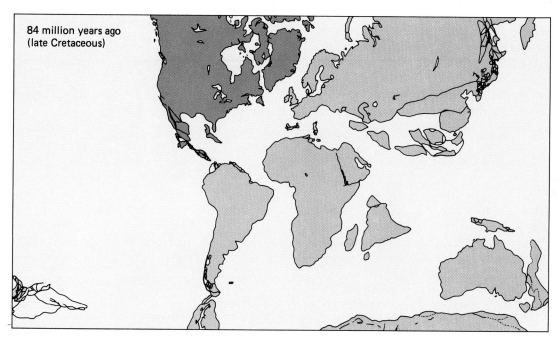

84 million years ago
(late Cretaceous)

B

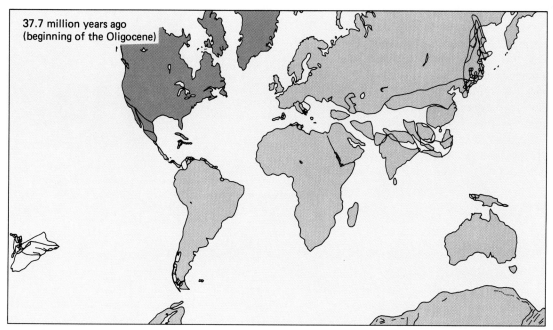

37.7 million years ago
(beginning of the Oligocene)

C

Pangaea III assembled. Suture zones may be lines of weakness within supercontinents like the perforations in tear-out magazine advertisements.

Sedimentary Rocks of the Coastal Plain and Continental Shelf

A huge new spreading center, the Mid-Atlantic ridge, formed as the Atlantic Ocean began to open. The new submarine mountain chain displaced seawater, raising sea level by a hundred meters or more. Low-lying portions of continents were flooded by shallow seas, and continental shelf sediments were deposited well inland along the eastern and southeastern margins of North America. In this way, the sedimentary rocks of the coastal plain accumulated on top of older continental basement rocks.

9.7 Building of the Western Mountains

The **Cordilleran mountain chain** is the long, broad mountain chain of western North America. It reaches from the east face of the Rockies, where the mountains meet the Great Plains, to the west coast. The chain extends from Alaska into Mexico and as much as 1500 kilometers inland from the coast (Fig. 9–16). The name *Cordillera* is the Spanish word for "chain of mountains." The Cordillera includes the Rocky Mountains, the Coast Ranges, and all other mountains and intermountain regions in the western part of our continent.

As the Atlantic Ocean began to open, the lithospheric plate carrying North America began moving westward. Subduction began along the west coast of North America to accommodate this new westward movement, and an Andean-type continental margin developed.

From about 180 to 80 million years ago, the great granite batholiths of western North America formed as magma rose from the asthenosphere above the subducting oceanic plate (Fig. 9–17). Extensive volcanism must have accompanied emplacement of the plutons, but erosion has now removed most of the volcanic rocks and exposed the batholiths. Thus, the formation of the Cordillera started with subduction.

Accreted Terranes

As mentioned earlier, Figure 9–8 shows that one billion years ago the western margin of the continent did not look at all as it does today. Its general outline did not change much from one billion years to about 180 million

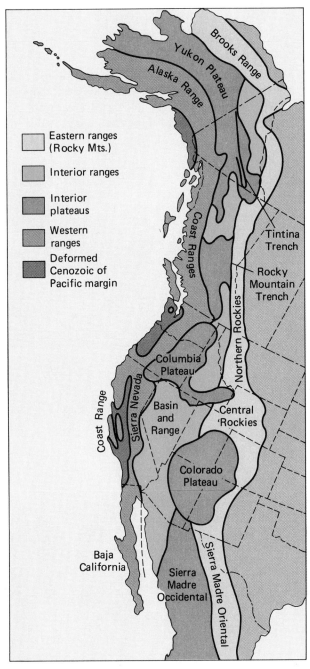

Figure 9–16 The Cordilleran mountain chain includes several subdivisions forming the vast mountainous region of western North America.

years ago. Much of modern western Alaska, the Yukon, British Columbia, Washington, Oregon, western Idaho, and western California were missing.

During the assembly of Pangaea III, several subduction zones must have developed in the Pacific Ocean

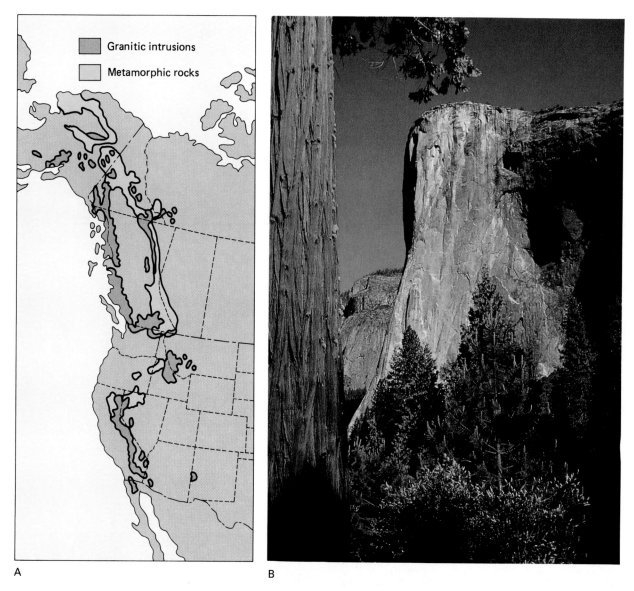

Figure 9–17   (A) Granite batholiths and related metamorphic rocks of western North America formed as a result of subduction along the western margin of the continent, beginning about 200 million years ago. (B) The granite of Yosemite Valley is part of the Sierra Nevada batholith, which formed in this way.   *(Don Hyndman)*

basin, forming island arcs and microcontinents. At that time, parts of the Pacific Ocean may have looked much like the southwestern Pacific does today—full of island arcs and microcontinents. As North America moved westward and the Pacific Ocean plates subducted beneath the western edge of the continent, one by one these island arcs and microcontinents were carried to the subduction zone. Because they were too light to follow the descending plate into the mantle, they were jammed onto the western margin of the continent. This process of accretion

of island arcs onto a continental margin is sometimes called **docking** because of its similarity to a ship making fast alongside a dock.

In this way, western North America grew considerably from about 180 to 80 million years ago. About 35 to 40 individual island arcs and microcontinents that are now parts of western North America are shown in Figure 9–18. They are called **accreted terranes**, referring to the fact that they originated elsewhere and were added onto the continent.

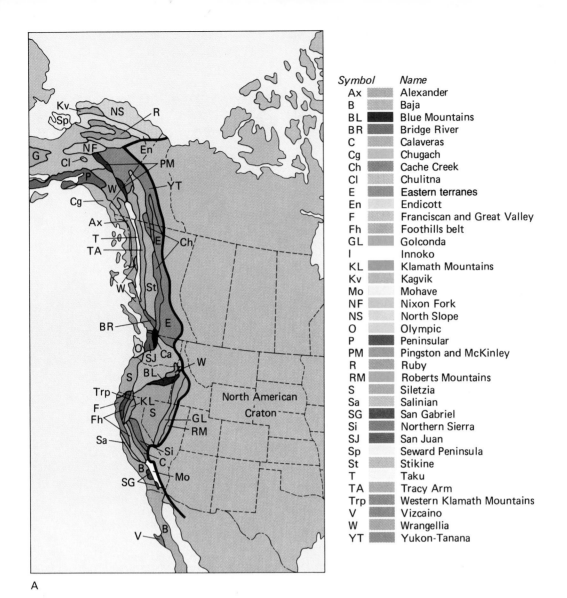

A

Symbol	Name
Ax	Alexander
B	Baja
BL	Blue Mountains
BR	Bridge River
C	Calaveras
Cg	Chugach
Ch	Cache Creek
Cl	Chulitna
E	Eastern terranes
En	Endicott
F	Franciscan and Great Valley
Fh	Foothills belt
GL	Golconda
I	Innoko
KL	Klamath Mountains
Kv	Kagvik
Mo	Mohave
NF	Nixon Fork
NS	North Slope
O	Olympic
P	Peninsular
PM	Pingston and McKinley
R	Ruby
RM	Roberts Mountains
S	Siletzia
Sa	Salinian
SG	San Gabriel
Si	Northern Sierra
SJ	San Juan
Sp	Seward Peninsula
St	Stikine
T	Taku
TA	Tracy Arm
Trp	Western Klamath Mountains
V	Vizcaino
W	Wrangellia
YT	Yukon-Tanana

Figure 9–18 The accreted terranes of western North America were added to the western margin of the continent during the Mesozoic and Cenozoic eras. Some terranes, such as Wrangellia, apparently broke up before they docked and are now found in widely separated places. The names of the terranes are listed in the key. (B) These pillow basalts from western Oregon are a slice of sea floor added to North America during docking of an accreted terrane.

B

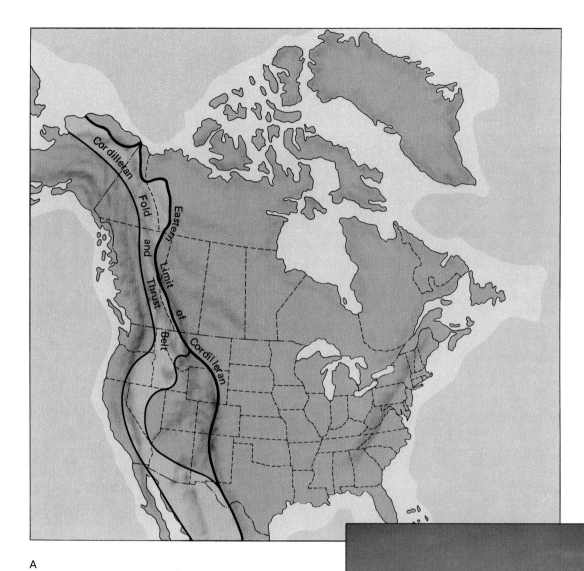

A

Figure 9–19   (A) The Cordilleran fold and thrust belt is a zone of thrust faults and folded rocks that extends for the entire length of the Cordilleran chain. (B) Folded limestone forms this peak near Banff, Alberta.

Folding and Thrust Faulting in the Cordillera

As the terranes crashed into the continent, they crushed rocks near the collision zone, forming a great region of folding and thrust faulting called the **Cordilleran fold and thrust belt**. It is only a few hundred kilometers wide but extends north-south for the entire length of the Cordillera (Fig. 9–19). The last of the terranes docked about 80 million years ago, but folding and thrust faulting continued until about 45 million years ago.

B

Dinosaur Migrations and the Rocky Mountain Seaway

The great weight of the Cordillera pressed down lower parts of the continent just east of the rising mountains. At the same time, global sea level was high because the Mid-Atlantic ridge was displacing great volumes of seawater. As a result of both factors, a large, shallow inland sea flooded much of North America east of the growing mountain chain during Jurassic and Cretaceous time, from about 175 to 100 million years ago (Fig. 9–20). Erosion of the mountains continually filled the seaway with mud and sand, creating a vast, swampy plain fed by streams that meandered from the rising highlands.

Fossils of more than 70 species of dinosaurs, including some of the largest land animals known, have been found in rock formed from this sediment. Bones, eggs, nests, and footprints of the animals contain evidence

Figure 9–20 In this reconstruction, a mother duckbill dinosaur nurtures her babies in their nest. Duckbills lived adjacent to shallow seas and swamps in eastern Montana from 175 to 100 million years ago. *(Museum of the Rockies)*

of their habits and life styles. Jack Horner, a paleontologist at the Museum of the Rockies in Bozeman, Montana, has reconstructed a scene of this vast flood plain 100 million years ago. Huge herds of thousands of duck-billed dinosaurs up to 30 feet long migrated seasonally a thousand kilometers or more. Following warmth and blooming vegetation, they stopped along the way only long enough to build nests and raise their young until they were able to travel with the herd. Solitary carnivores followed the herds, preying on the young and infirm. Occasionally, great volcanic eruptions from the mountains buried an entire herd of dinosaurs in ash, preserving together eggs, baby dinosaurs in nests, and adults.

A Major Change in the Tectonics of Western North America

Subduction continued along North America's western margin as the accreted terranes docked. As long as the Pacific plates collided with the west coast at high speed, the western margin of the continent was squeezed by the collisions. However, for an unknown reason, the collision suddenly slowed down about 45 million years ago. Then the squeezing decreased and the warm, thick crust of the Cordillera began to ooze out horizontally.

Only the deeper, hotter part of the crust was able to spread, however. The cooler rocks of the upper crust were brittle and unable to flow. Consequently, they fractured and faulted as they were pulled apart by the spreading lower crust. The shallow upper crust is analogous to a layer of chocolate frosting spread on the mound of honey. If the honey flowed outward, the frosting would crack into segments, which would then separate as spreading continued.

As the upper crust faulted and pulled apart in this manner, a second major episode of Cordilleran igneous activity began about 45 million years ago, and it continues today. Plutonic and volcanic rocks that formed during the last 45 million years are abundant in western North America (Fig. 9–21). The San Juan Mountains of southwestern Colorado and the older volcanic and plutonic rocks of the Cascade Range of northern California, Oregon, and Washington formed during this time.

A Second Major Tectonic Change Occurs in Western North America

Development of the San Andreas Fault

As North America moved westward, it drew closer to a portion of the mid-oceanic ridge called the **East Pacific**

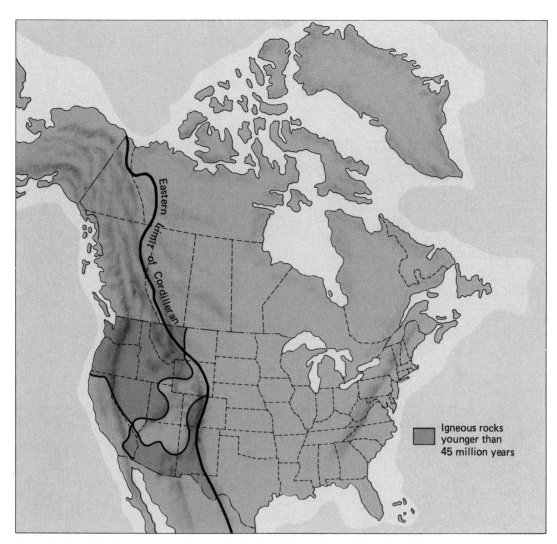

Figure 9–21 Volcanic and plutonic rocks in the United States formed from 45 million years ago to the present. On this map the igneous rocks are shown as they would appear if none had been eroded away.

rise (Fig. 9–22). By 30 million years ago, most of the oceanic crust between North America and the East Pacific rise had been subducted. The western edge of the continent reached the East Pacific rise near southern California. Notice in Figure 9–22B that the Pacific plate was moving northwestward, nearly parallel to the westward movement of the North American plate. Since these two plates were moving in nearly the same direction, they were not colliding. Therefore subduction stopped where southern California touched the Pacific plate, beginning about 30 million years ago. At present, California is subduction-free all the way to Cape Mendocino in the northern part of the state. The active volcanoes of the Cascades testify that subduction continues north of Cape Mendocino.

A new fault then developed to accommodate the small difference in direction between the two plates. It was a strike-slip fault along which the Pacific plate slid northwestward against the California coast (Fig. 9–22B). As North America continued to move westward, it overrode more and more of the East Pacific rise. The strike-slip fault grew longer and migrated inland to become the San Andreas fault (Fig. 9–22C). Eventually it became large enough to be called a transform plate boundary.

The Modern Cascade Volcanoes
Figure 9–22C shows that the San Andreas fault turns westward in northern California and runs out into the

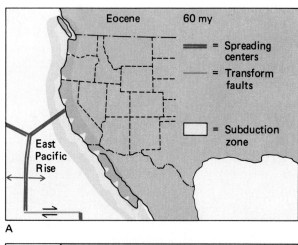

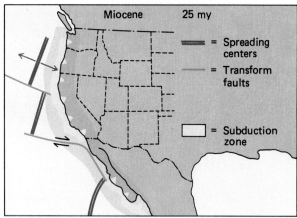

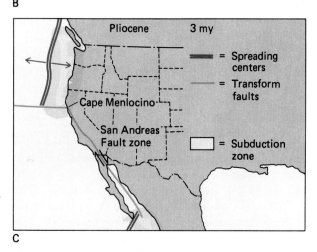

Figure 9–22   Western North America drew closer to the East Pacific rise until it hit the rise about 30 million years ago. The San Andreas fault developed where California came into contact with the rise. The fault has grown in length as more of California has hit the rise. Notice that subduction ceases where California strikes the rise and the San Andreas fault develops.

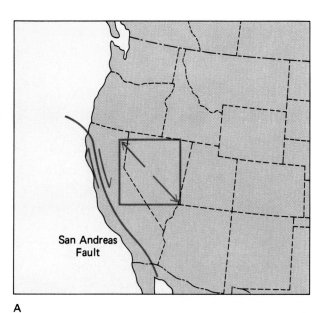

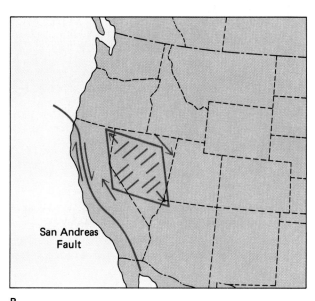

Figure 9–23   Friction along the San Andreas fault pulls the continent apart in a northwest-southeast direction (red arrows), causing normal faulting (red lines) in the Basin and Range province.

Pacific Ocean where it connects with another segment of the mid-oceanic ridge. Although the figure shows the situation as it was three million years ago, it is similar today. This segment of the mid-oceanic ridge lies just off the coasts of northern California, Oregon, Washington, and southwestern British Columbia. The oceanic plate east of the spreading center is colliding head on with the westward-moving continent. Therefore, subduction occurs today along the northwest coast of North America.

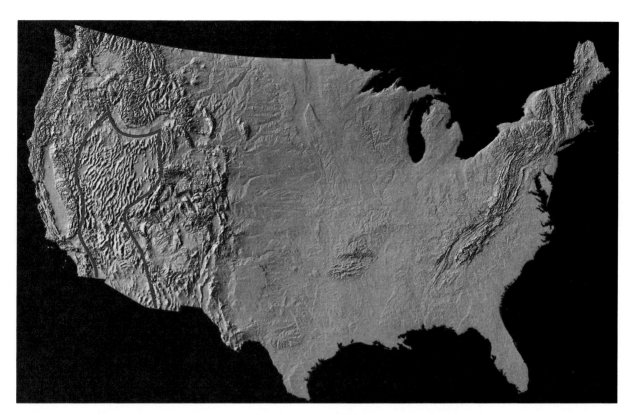

Figure 9–24   A radar image of the United States with the Basin and Range province outlined in red. Note the parallel valleys and mountain ranges caused by normal faulting.

Mounts Lassen, Shasta, Rainier, St. Helens, and Baker, as well as other volcanoes of the modern Cascades, have all formed during the past few million years. Several have erupted within the past 100 years, including Mount St. Helens in 1980. The igneous activity and volcanoes of the modern Cascade Range result from current subduction along this part of the west coast.

The Basin and Range Province

Movement along the San Andreas fault has been dragging the western margin of North America northward for the past 30 million years, since the fault first formed. Figure 9–23 shows how such a force would deform a hypothetical square of western North America into a parallelogram. The deformation would pull the crust apart in a northwest-southeast direction. That is what is happening. The crust of western North America has been stretched for the past 30 million years. As a result, many large normal faults have developed perpendicular to the extension. Large grabens have dropped downward along the faults, leaving elevated horsts between the grabens to become **block-faulted mountain ranges** (Fig. 9–24).

This region of block-faulted mountain ranges and valleys is called the **Basin and Range province**. Because the lithosphere was stretched and thinned as it was pulled apart, magma formed deep in the crust and volcanic activity accompanied the faulting. Many geologists who work in the Basin and Range believe that stretching and faulting have doubled the width of the region in an east-west direction. The tectonic forces associated with movement along the San Andreas fault continue today, and faulting and volcanism are still active in the Basin and Range.

The Colorado Plateau

A large block of western North America known as the **Colorado Plateau** (Fig. 9–25) remained strangely unaffected by the faulting and igneous activity, although it is surrounded on three sides by Basin and Range features. The entire Colorado Plateau simply rotated clockwise during Basin and Range faulting. At a later time, between five and ten million years ago, the Colorado Plateau was uplifted without much internal deformation to form a high, nearly circular plateau that extends from the Rocky

Figure 9–25 Sedimentary rocks of the Colorado Plateau in Grand Canyon, Arizona, are uplifted but show little folding.

Mountains in Colorado to Utah and includes northeastern Arizona and northwestern New Mexico.

The Columbia Plateau

The **Columbia Plateau** and its eastward extension, the Snake River Plain, make up one of the world's largest continental lava plateaus. The Columbia Plateau formed by rapid extrusion of magma from 17 to 14 million years ago. Volcanic activity occurred only a few thousand years ago in Yellowstone National Park, at the eastern end of the Snake River Plain. There the story may not be over, because active magma still underlies portions of the Park. The origin of this vast magmatic system is described in Chapter 7.

A Summary of the Cordillera

Let us quickly review the development of the Cordillera. It began to form shortly after 200 million years ago as a simple Andean-type continental margin, with the emplacement of batholiths and eruptions of volcanoes near the edge of the craton. These batholiths make up the Sierra Nevada of California and parts of the Coast Ranges of British Columbia. As subduction continued, island arcs and microcontinents from the Pacific docked against the western edge of the craton, building the margin of the continent westward. These accreted terranes became the Coast Ranges of California and parts of the Cascades of Washington and Oregon, the Coast Ranges of western Canada, and southeast Alaska. Compression resulting from the docking caused much folding and thrust faulting. These activities continued to about 45 million years ago, long after the last of the accreted terranes had docked.

A slowdown in the rate of subduction about 45 million years ago allowed the western margin of the continent to relax and spread outward. This extension caused more igneous activity, which in turn created additional plutonic and volcanic mountain ranges in the Cordillera.

Beginning about 30 million years ago, the western margin of the continent reached the East Pacific rise, and the San Andreas fault formed. Frictional drag from the San Andreas pulled the crust apart further, causing block faulting, which formed the parallel mountain ranges and valleys of the Basin and Range province. The Colorado Plateau was unaffected by most of this deformation but was uplifted as a block between ten and five million years ago.

By about five million years ago, most of the mountain ranges of the Cordilleran chain had been created by this sequence of tectonic events. Later, one additional nontectonic event occurred that sculpted the mountains of the Cordillera into their present appearance.

9.8   The Pleistocene Ice Age and the Arrival of Humans

The Spread of Glaciers

At least five major episodes of glaciation have occurred in the Earth's history. The most recent is commonly called the Pleistocene Ice Age. By two million years ago, large continental ice caps up to 3000 meters thick had formed in both the Northern and Southern Hemispheres and were rapidly spreading outward. At their greatest extent, the Pleistocene glaciers covered about one third of North America (Fig. 9–26).

The flowing ice scoured and eroded rock, soil, and other debris from the land. When they melted, the glaciers deposited huge loads of unsorted sediment, creating other landforms. Thus, glacial erosion and glacial deposition greatly modified the landscape of North America.

The Arrival of Humans in North America

About five million years ago, the precursors of modern humans evolved from a branch of apes, probably in eastern Africa. By 200,000 years ago, early humans had spread to China and Java, and by 40,000 years ago they inhabited Europe. Thus, in Pleistocene time, human precursors had spread throughout much of Africa, Asia, and Europe.

The Bering Straits are part of a shallow-water shelf lying between Alaska and Siberia. When glaciers were at their maximum size during the Pleistocene Ice Age, so much water was stored in continental glaciers that sea level fell worldwide by 100 to 200 meters. As a result, the shallow shelf underlying the Bering Straits was exposed as a low-lying, swampy isthmus connecting Asia to North America. Known as **Beringia**, this region was

(Text continues on p. 234)

Figure 9–26   The maximum extent of glaciers during the Pleistocene Ice Age.

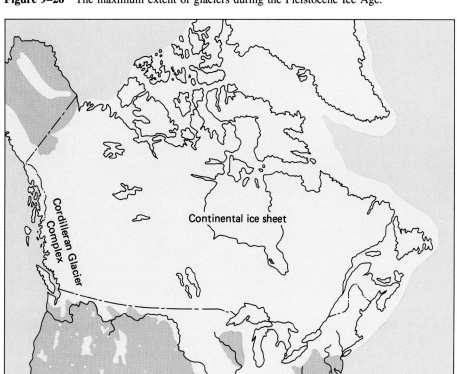

Earth Science and the Environment

Nuclear Waste Storage in North America

In a nuclear reactor, uranium nuclei split into smaller fragments called **fission products**. Many fission products are themselves radioactive. Most are also useless; therefore, it is necessary to dispose of them despite their dangerous levels of radioactivity. Many other industrial and laboratory processes generate radioactive wastes.

As an example of the nuclear waste problem, consider plutonium-239. Plutonium is one of the deadliest poisons known. "A thousandth of a gram of plutonium taken into the lungs as an invisible speck of dust will kill anyone. . . . Even a millionth of a gram is likely, eventually, to cause lung or bone cancer."* Plutonium does not occur naturally on Earth. Every atom of this element that exists on our planet has been manufactured in nuclear reactors. The half-life of plutonium-239 is 24,000 years. If 1 gram of plutonium exists today, 0.5 gram will be left after 24,000 years. After another 24,000 years, 0.25 gram will be left. The plutonium produced today will remain a problem for thousands of future generations.

The only feasible solution to the problem of radioactive wastes is to store them in a place safe from geological changes and human intervention. The U.S. Department of Energy defines a permanent repository as one that will isolate radioactive wastes for 10,000 years. In order for a location to safely keep these dangerous materials isolated for such a long time, it must meet at least three geological criteria:

1. It must be tectonically stable. That is, a radioactive waste repository must be completely safe from disruption by earthquakes, volcanic eruptions, and other tectonic disturbances.

2. It must be safe from landslides, soil creep, and other forms of mass wasting.

3. It must be dry. A safe repository must be free from seeping ground water that might corrode containers and from floods that might disrupt the site.

Let us examine three actual or potential nuclear waste repositories.

Hanford, Washington

Hanford, Washington, was chosen in 1942 as the site of the first nuclear reactor in the United States. When the Hanford reactor was built, its main purpose was to produce plutonium for atomic bombs to be used against Japan at the end of World War II. In retrospect, it is clear that the primary consideration was to build the bombs as quickly as possible, while the disposal of the nuclear wastes was given low priority. As a result, radioactive wastes were stored in steel tanks, dumped in ditches and ponds, and injected into deep wells and underground drains.

The Hanford site is situated on the Columbia River plateau, which consists of a sequence of flood basalts. Much of the basalt bedrock is fractured and interlaced with fissures and tunnels that formed as the lava cooled. Thus, the bedrock is extremely permeable to seeping ground water.

In the late 1950s, workers buried about 4000 kilograms of uranium in a large wooden box, called a crib, at the Hanford site. About a decade later, chemical wastes, including strong acids, were dumped into the same crib. The acids dissolved the uranium compounds, and the corrosive, radioactive solution leaked from the crib and seeped into the fractured basalt. It percolated downward until it encountered an impermeable layer of basalt. The solution then flowed laterally until it reached an old well drilled through the basalt layer. It then then flowed downward through the well, contaminating a local aquifer. Today, the radioactivity of water pumped from wells in the region exceeds federal drinking water standards by a factor of 1300.

The Waste Isolation Pilot Plant near Carlsbad, New Mexico

The example just cited is one of hundreds of groundwater contamination incidents that have occurred at Hanford. Obviously a safer and more reliable repository is needed. An alternative choice is a natural salt deposit. Salt (sodium chloride) is extremely soluble in water. Therefore, natural salt deposits are usually situated in dry environments. If ground water were present, the salt would have dissolved long ago.

A salt cavern near Carlsbad, New Mexico, was chosen as a repository for low-level radioactive wastes. If this site is approved, wastes will be sealed in drums, deposited in the salt cavern, and then abandoned. Because the walls of the cavern are unstable, the salt is expected to collapse, enveloping the drums and isolating them from the environment. If everything works as expected, the isolation will last for tens of thousands of years.

However, geologists have recently learned that small amounts of water seep into the cavern. Critics of the site charge that the water will dissolve the salt, creating salt brine that will corrode the drums and carry the radioactive materials into the ground water.

*John McPhee, *The Curve of Binding Energy* (New York, Ballantine Books, 1975).

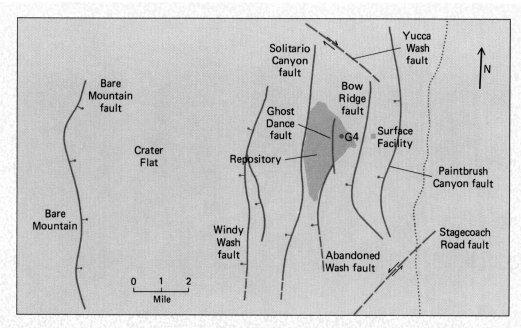

A map of faults near the proposed Yucca Mountain site. The solid red lines are normal faults with lollipops on the footwall side; the dashed lines with arrows are strike-slip faults.   *(Geotimes, January 1989)*

In October 1991, the Department of Energy (DOE), under an administrative order, overruled congressional legislation and announced it would open the Carlsbad site to store wastes from the defense industry. The State of New Mexico sued the DOE and obtained a temporary injunction barring the DOE action. As this book is being written the case remains in court.

In January 1992, an earthquake that measured 4.8 on the Richter scale occurred near the repository site. No damage was incurred as the repository is designed to withstand a magnitude 5.5 quake directly under the site.

The Yucca Mountain Repository

In December 1987, the U.S. Congress chose a site near Yucca Mountain, Nevada, about 175 kilometers from Las Vegas, as the national burial ground for all spent reactor fuel, unless sound environmental objections were found.

Critics of the Yucca Mountain site argue that, as part of the Basin and Range province, the site is tectonically active. Crustal extension, earthquakes, faulting, and volcanism continue throughout the region, and a recent history of geological activity at the Yucca Mountain site indicates ongoing tectonic events.

Bedrock at the Yucca Mountain site is welded tuff, a hard volcanic rock. The tuffs erupted from several calderas that were active from 16 to 6 million years ago. Later volcanism created the Lathrop Wells cinder cone 24 kilometers from the proposed repository. The last eruption near Lathrop Wells occurred 15,000 to 25,000 years ago.

Geologists have mapped 32 faults that formed during the past two million years adjacent to the Yucca Mountain site. The site itself is located within a structural block bounded by parallel normal faults (see figure).

The environment is desert-dry, and the water table lies 575 meters beneath the surface. Geologists estimate that even in the desert climate, ground water will percolate from the repository site to the water table sometime between 9000 and 80,000 years from now. The lower end of this estimate is close to the 10,000-year mandate for isolation.

If rocks beneath the site were fractured by tectonic activity, then contaminated ground water might disperse more rapidly than predicted. Critics point out that construction of the repository will involve blasting and drilling, and these activities could fracture underlying rock. Additionally, if the climate becomes appreciably wetter, which is possible over thousands of years, ground water flow rates may accelerate and the water table may rise.

In order to stop development of the Yucca Mountain site, the state of Nevada refused to issue air quality permits to operate drilling rigs at the repository. As this book is being written, the legal battle continues.

Supporters of the repository argue that we need nuclear power and therefore, as a society, we must accept a certain level of risk. Furthermore, the Yucca Repository is safer than the temporary storage sites now being used. Unfortunately there are no easy answers to these questions, and the debate continues.

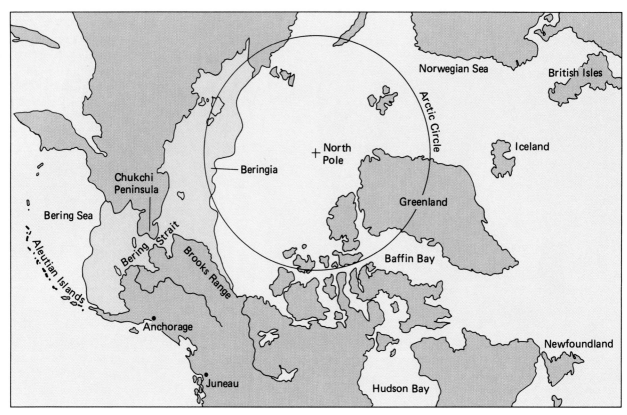

Figure 9–27 Beringia was a huge landmass, measuring 1200 by 2200 kilometers, that connected Siberia and Alaska when Pleistocene glaciers were at their greatest size.

above sea level for most of the past 150,000 years (Fig. 9–27).

Mammoths, bison, caribou, moose, musk ox, mountain sheep, and several other mammalian species that we now associate with North America migrated across Beringia to this continent within the last 150,000 years. With them, probably following as predators, came modern humans.

It is difficult to date precisely the arrival of the first wave of humans in North America. Radiocarbon ages greater than 40,000 years have been measured from charred wood in campfire rings and from burned bone fragments, establishing that the first humans had arrived by 40,000 years ago. Less reliable age measurements indicate that humans may have arrived as early as 100,000 years ago, apparently not long after the evolution of modern humans. Whenever the first humans did arrive, it is clear that the peopling of North America occurred in several waves, the most recent of which may have been only 10,000 years ago, shortly before melting glaciers inundated Beringia.

As they arrived, humans and other animals spread out from the Alaska coast, following several migration channels (Fig. 9–28). Note how far north several of the migratory routes are. Why would humans and other mammals migrate *northward* during a time when glaciers were at their maximum growth? And why would they cross to

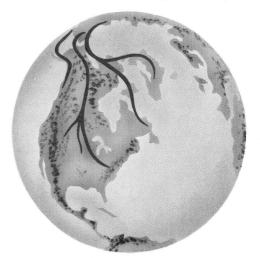

Figure 9–28 Migration routes followed by humans and other animals as they spread into North America from Beringia.

Alaska, a place known for its cold climate and glaciers even now, during a nonglacial interval? During this period of migration the ice cap was at or near its maximum size, and great glaciers covered the land to the east. However, due to local climatic patterns, ice-free corridors existed between the glaciers. One corridor extended along the northern edge of the continent, and another ran along the west coast. A third extended along the eastern side of the Cordillera, between the alpine glaciers of the mountains and the great continental ice sheet. The land in these corridors was mostly treeless, vegetated by grasses, heather, and sage. Small groves of aspen and larch grew in sheltered areas. The terrain was perfect for grazing animals. The flat plains and great animal populations made for easy hunting. The groups that continued southward arrived in the vicinity of Alberta, Canada, where they found the end of the ice and a fabulously rich new land: an endless prairie lush with rich grasses and populated by vast herds of bison, elk, deer, mountain sheep, and antelope. Bear, mountain lions, wolves, foxes, coyotes, and eagles hunted the rich herds. And now humans had come into the country.

SUMMARY

·

The geologic development of North America began at least 3.96 billion years ago. The **craton** is the oldest, central part of our continent. It consists of the **shield**, where old igneous and metamorphic rocks are exposed at the surface, and the **platform**, where they are covered by a veneer of younger sedimentary rocks. The **mountain chains** lie near the continental margins, where tectonic events younger than the rocks of the craton have created the **Appalachian**, **Innuitian**, and **Cordilleran** mountain chains. The **coastal plain** and continental shelf are made of young sedimentary rocks covering the continental margins.

The craton consists of several **provinces**, which originally were separate **microcontinents** and island arcs swept together during assembly of a **supercontinent** called **Pangaea I** between 2 and 1.8 billion years ago. This first supercontinent rifted apart a few hundred million years after its assembly, separating the North American craton from other continents. A few hundred million years after breakup of the first supercontinent, a second supercontinent called **Pangaea II** formed, with the North American craton near its center. It broke up in late Precambrian time, about 700 to 800 million years ago, isolating the North American craton again. During assembly of a third supercontinent, called **Pangaea III**, subduction began along the eastern margin of North America as Europe and Africa drew closer to the east coast. The subduction formed a mountain range made of plutonic and volcanic rocks near the east coast. About 400 million years ago, Europe, Africa, and South America began to collide with the eastern and southern margin, forming the Appalachian mountain chain. Rocks formed by this collision continue westward beneath sedimentary rocks of the coastal plain, where they can be seen in Arkansas, Oklahoma, and Texas in the Ouachita and Marathon mountains. The Innuitian mountain chain formed along the northern continental margin as the Appalachians formed along the eastern and southern margins.

During Paleozoic time, between the breakup of Pangaea II and formation of Pangaea III, high sea level flooded much of the craton, and **platform sediments** were deposited in the central portion of North America.

When Pangaea III broke up, beginning about 200 million years ago, North America once again became isolated from other continents. Subduction began along the west coast in response to westward movement of the continent, forming an Andean-type margin and beginning construction of the Cordilleran mountain chain. As subduction continued, great granite batholiths were emplaced, mountains rose, metamorphism occurred, and many island arcs and microcontinents from the Pacific Ocean basin docked, forming the **accreted terranes** of western North America. The docking of accreted terranes was accompanied by much folding and thrust faulting, which continued after docking ended, to about 45 million years ago. From 45 million years ago to the present, much of the Cordillera was pulled apart in an east-west direction, resulting in normal faulting, volcanism, and emplacement of granite plutons.

When the western margin of the continent reached the **East Pacific rise**, the **San Andreas fault** formed and extensional stress increased, causing the normal faults, grabens, and magmatic activity of the **Basin and Range province**. North of the San Andreas fault, subduction continues today and has built the volcanoes of the modern Cascades. The **Columbia Plateau** was formed by massive flood basalt flows from 17 to 14 million years ago. The **Colorado Plateau** escaped most of the folding and faulting that affected the Cordillera and was uplifted as an intact unit between five and ten million years ago.

In Pleistocene time, glaciers covered as much as one third of North America and modified landforms by both erosion and deposition. During times of maximum size of the glaciers, sea level fell enough to expose **Beringia**, a large landmass connecting Siberia to Alaska. Many species of animals, including humans, migrated to North America across this land bridge in the past 150,000 years.

KEY TERMS

•

Craton *210*	Pangaea I, II, and III *214*	Cordilleran mountain chain *222*	Block-faulted mountain ranges *229*
Shield *210*	Appalachian mountain chain *217*	Docking *223*	Basin and Range province *229*
Platform *210*	Innuitian mountain chain *219*	Accreted terrane *223*	Colorado Plateau *229*
Coastal plain *211*	Platform sedimentary rocks *220*	Cordilleran fold and thrust belt *225*	Columbia Plateau *230*
Geologic province *212*		East Pacific rise *227*	Beringia *231*
Supercontinent *212*			
Microcontinent *214*			

REVIEW QUESTIONS

•

1. What is the approximate age of the oldest rocks in North America? In what part of the continent are they found?

2. Sketch a simple map showing the outline of North America and the locations of the three main types of geologic regions that make up the continent.

3. Describe the North American craton.

4. Why are the mountain chains of North America located near the continental margins?

5. What set of geologic events would form a mountain chain near the center of a continent?

6. Explain the relationship between the continental shelves and the coastal plain of North America.

7. Explain the origins of the provinces of the North American craton. What are the main kinds of differences among the provinces?

8. Why are boundaries between the provinces of the craton commonly intensely sheared and faulted?

9. What is a microcontinent? What is a supercontinent?

10. In Paul Hoffman's model, why do supercontinents break up within a few hundred million years after they form?

11. How many supercontinents have formed in Hoffman's model? When did each exist?

12. What were the most important tectonic events that built the Appalachian mountain chain?

13. What happens to the igneous, metamorphic, and deformed sedimentary rocks of the Appalachian mountain chain in Alabama, Mississippi, and regions west of Mississippi?

14. Platform sediments overlie the craton in much of the central portion of North America. How and when did they form?

15. What is the Cordillera?

16. Describe or sketch on a map the locations of the large granite batholiths of the Cordillera. How and when did these great bodies of granite form?

17. What is an accreted terrane? Where are accreted terranes found in North America?

18. Where is the Cordilleran fold and thrust belt located? Sketch it on a map of North America.

19. Why do recent and modern volcanic activity along the west coast of North America occur only north of the San Andreas fault?

20. What are the main geologic structures of the Basin and Range province? Why are they the most common structures?

21. What is the importance of Beringia in the history of North America? How did Beringia form?

DISCUSSION QUESTIONS

•

1. The oldest known rocks in North America are gneisses. What does the fact that they are metamorphic rocks tell you about the maximum age of rocks of the craton?

2. Discuss the relationships between data and theory in the context of the geologic history of North America described in this chapter.

3. Describe and discuss Paul Hoffman's model of supercontinent cycles. Does the model seem plausible in light of the data presented in this chapter?

4. Discuss how microcontinents might have formed.

5. Discuss how and why supercontinents form.

6. The breakup of the first supercontinent was accompanied by intrusion of many granite plutons into continental crust. Most granites form in continental crust above subduction zones. Develop and discuss a model in which granite magma forms during rifting of a supercontinent.

7. Discuss the origin of the San Andreas fault and the relationships among the San Andreas fault, the Basin and Range province, and the Colorado Plateau.

8. Using a map of tectonic plate movements, give a plausible scenario for the movement of continents over the next 200 million years. Which oceans will grow larger? Which ones will shrink?

Paul Hoffman

Paul F. Hoffman was born in 1941 in Toronto, Canada. He studied at McMaster University (B.Sc. 1964) and The Johns Hopkins University (Ph.D. 1970) before joining the Geological Survey of Canada as a Research Scientist. He has also been a visiting lecturer at Franklin & Marshall College and the University of California at Santa Barbara, a Fairchild Distinguished Scholar at the California Institute of Technology, a visiting professor at the University of Texas at Dallas and the Lamont-Doherty Geological Observatory of Columbia University, and an adjunct professor at Carleton University. In 24 summers of field work in the northern Canadian shield, his work on Precambrian sedimentary basins and related orogenic belts provided evidence that the movement of tectonic plates has occurred throughout most of Earth history. Since 1985, he has been working on a synthesis of the Precambrian geology of North America.

He serves on committees of the International Lithosphere Program, the Canadian Lithoprobe Project, the Interna-

tional Commission on Stratigraphy, the Circum-Pacific Map Project, the Geological Society of America, and the Massachusetts Institute of Technology. He has made distinguished lecture tours for the

> "My current greatest interest now is with Gondwana, the ancestral supercontinent that gave birth to Africa, South America, Antarctica, India, and Australia."

American Association of Petroleum Geologists, the Canadian Institute of Mining and Metallurgy, and the Geological Association of Canada and has won best paper awards from the Society of Economic Paleontologists and Mineralogists, the American Association of Petroleum Geologists, the Canadian Society of Petroleum Geologists, and the Geological Society of Washington. A recipient of the Past-Presidents' Medal from the Geologi-

cal Association of Canada and the Bownocker Medal for Research in Earth Sciences from The Ohio State University, Hoffman was elected a Fellow of the Royal Society of Canada in 1981.

How did your interest in geology begin?

I first became interested in geology through Saturday classes for primary school students in various subjects in natural science at the Royal Ontario Museum in Toronto. At the age of 10, I joined the mineralogy and geology group at the museum. We went on field trips to collect minerals and fossils in the Paleozoic rocks around Toronto, Ontario. As a result of those field trips and mineral collecting, I decided to be a geologist by the age of 12.

I went to McMaster University in Hamilton, Canada. Initially, I went to McMaster thinking that I would go into mineralogy because mineral collecting had been my love. What I found out at the university was that studying mineralogy was mainly an indoor pursuit and involved lots of mathematics, not the kind of thing that I was interested in or very good at.

My interest in mapping began the summer of my freshman year when I worked as a field assistant on a geological mapping party in northwestern Ontario. We spent 4½ months living out of canoes and mapping a large area of the northwestern Ontario Archean terrane.

I spent every summer of the next 24 years mapping in the Northwest Territories with the Geological Survey of Canada. The outcrop exposures are much better there than in the southern Canadian shield so the geology is more challenging and rewarding. After I graduated from McMaster, I decided to do a Ph.D. thesis on the Canadian shield in the Northwest Territories. The Geological Survey of Canada funded my thesis project, which was done at The Johns Hopkins University.

•

Although you started out as a field geologist doing field mapping on a quadrangle level, you've become best known for your global scale tectonic models and interpretations. Tell us more about these.

Well, they were *large* quadrangles (10,000 square kilometers) and I've always been interested in the large-scale tectonics of the Canadian shield. My Ph.D. thesis project was basically designed to compare a Precambrian orogenic belt with the Appalachians, which is a Phanerozoic orogenic belt. Since The Johns Hopkins University is close to the Appalachians, we went on a lot of field trips there.

After reading about younger orogenic belts, I thought that the best way to begin a study of the Appalachians would be to start with the sedimentary rocks because they're the best preserved and because the layers of sediments record the evolution of the mountain belt.

In contrast, in most previous studies of the Precambrian, geologists had gone to the inner parts of these deeply eroded belts and looked at the metamorphic rocks first, which I felt were more difficult to work with. From my work as an undergraduate field assistant, I knew that several small basins of well-preserved sediments existed.

I started with a basin in the east arm of Great Slave Lake about 1000 miles north of Montana. I had the idea that the area to the north, which gave older radiometric ages, was a stable foreland and that the area to the south, which gave younger radiometric ages, was the internal part of the ancient mountain belt. In the Appalachians there is a pattern where the succession of sediments laid down on a subsiding continental margin are followed by sediments that were derived from the rising mountains formed as that continental margin collided with another continent. When I started out, I wasn't thinking in terms of plate tectonics, so I wanted to see whether there was the same pattern of sedimentation as in the Appalachians and also whether the paleocurrent indicators, the directional indicators of sediment transport, showed a reversal as they do in the Appalachians. There the early paleocurrents are directed toward the ocean, and the subsequent paleocurrents are directed toward the interior of the continent away from the rising mountain belt.

What I found in the Precambrian of the Great Slave Lake was a sedimentary succession very similar to the Appalachians and a similar reversal in the paleocurrents, but the paleocurrent directions weren't what I had expected. Instead of going toward the south initially and then switching toward the north, they were going toward the west initially and then toward the east. So my thesis postulated that there had been a Precambrian ocean and later a moun-

tain belt hidden to the west in the area of Great Slave Lake that is now covered by Devonian sediments. But I thought that relics of the belt might be exposed in the Canadian shield to the north so I predicted that it would be there. When I joined the Geological Survey in 1969, I proposed going to that more northernly area to see whether my prediction panned out. The area is called the Wopmay orogen, named after a famous bush pilot.

My prediction turned out to be correct. But that begged the question as to why in the east arm of Great Slave Lake the fold-belt trends northeast-southwest, with more intense deformation to the south than to the north. And why was it that the paleocurrents there seemed to be parallel to the structural trends rather than transverse to them, as in the Appalachians? In trying to wrestle with these problems, I stumbled on a paper by two Russian geologists outlining the idea of "aulacogens," subsided rifts that extend into the interior of continents. Thus, I proposed that the east arm of Great Slave Lake was a Precambrian aulacogen. This subsequently turned out to be a very popular model. In fact, I believed it for a number of years. But now I think that it's quite wrong. The Great Slave Lake area is in fact of collisional origin as I had originally thought, but the paleocurrents of the sedimentary fill I was looking at are related to the slightly younger belt to the west.

•

What is the difference between working for the Geological Survey and working in a university?

One of the differences of working with the Survey is that your specialty tends to be regionally defined rather than defined by discipline.

When I started working on the Wopmay orogen, I found that my compelling interest was to study the various parts of the orogen from the standpoint of stratigraphy, sedimentology, and structure to volcanology, igneous petrology, and metamorphic geology as I systematically examined its different zones. So rather than studying only the sediments of many regions as a university specialist would, I ended up studying all the different rock types and different zones of the orogen within only one general region.

•

Tell us more about your work for the Survey and your current research.

The geologic evidence in the Northwest Territories convinced me and others that plate tectonic processes were operating in the Precambrian. This became apparent to me at a very early stage; my thesis work and projects in the late 1960s showed that there was very little difference between the rocks, structure, and organization of orogenic belts of the Precambrian and those of the Phanerozoic. If Phanerozoic orogens were products of plate tectonics, it followed logically that the Precambrian ones must be also. I think that in the early 1970s, geologists were starting to come around to this view.

Anyway, I mainly worked in this one general region because it is extremely well exposed and offers a great range of structural levels. Most of my work up until 1983 was on the margins of a small microcontinent called the Slave microcontinent. I was operating in an area only about 600 by 600 kilometers studying the tectonic, volcanic, and metamorphic activity that resulted from a pair of collisions between three microcontinents, one on the east margin and one on the west

margin of the Slave. My reasoning was that to convince people of the reality of plate tectonics 2 billion years ago, I had to study orogens of that age in as much detail as Phanerozoic belts because most other Precambrian belts were very poorly known relative to Phanerozoic ones.

I had always felt that many of the other orogenic belts in the shield could also be interpreted in terms of plate tectonics and that their histories would be interrelated, but we didn't really have a good way of dating them until the late 1970s and early 1980s when Tom Krogh of the Royal Ontario Museum advanced the uranium–lead dating method. This method allowed one to date these Precambrian rocks with very high precision, and that's when the thing really started to become exciting. In about 1983, the Geological Society of America began to publish a series of volumes compiling, summarizing, and synthesizing everything that was known about the geological evolution of the North American plate. Because I had always had an interest in the Canadian shield as a whole, I asked to be the principal compiler for the Precambrian of Canada. A year later, the editor-in-chief for the North American overview volume asked me to prepare a chapter in that volume outlining the Precambrian evolution of North America as a whole. So at that point I threw myself into the North American synthesis. It was about this time that the explosion in uranium–lead age determination in the Precambrian began. We also started to get colored digital data sets for gravity and magnetic anomalies on the scale of the continent. These allowed you to see the continuation of the Precambrian structures underneath the sedimentary cover of the Great Plains.

Because a number of things were coming together at once, it made it an excellent time to take a

new look at the entire Precambrian structure of North America. What ultimately emerged from the synthesis of North America is that you can't understand North America just by looking at North America alone; you have to know what's going on in the rest of the world. North America was part of the supercontinent called Pangaea in the late Paleozoic and early Mesozoic time. Similarly, North America was joined with other continents at various times in various configurations during the Precambrian. So, it was a natural extension of the North American Precambrian synthesis to start looking at what occurred in the Precambrian on a global scale. My current greatest interest now is with Gondwana, the ancestral supercontinent that gave birth to Africa, South America, Antarctica, India, and Australia. Gondwana formed at the end of the Precambrian as a result of a series of collisions known collectively as the Pan-African orogenic episode.

•

What's your encore after describing the origin of the North American craton? What do you think you will be doing in the future?

The evolution of the North American craton shows us that it has been involved in the aggregation and dispersal of as many as six successive supercontinents over the past 2.6 billion years. If the notion of episodic supercontinents is valid, then one should see evidence on many continents of contemporaneous collisional mountain building, pointing to the aggregation of supercontinents, and contemporaneous rifting, signifying supercontinental breakup. The supercontinent ''cycle'' also predicts changes in global climate, the chemistry of ocean sediments, and changes in continental paleolatitudes that can be detected from remanent magnetism in rocks. So my next goal is to compare the Precambrian geology of all the continents. After North America, the best known Precambrian continent is Gondwana. It is not as well studied as North America, but it has the advantage of being very large, almost a supercontinent in itself. Unlike North America, which originated about 1.8 billion years ago, Gondwana was first aggregated only 0.6 billion years ago, almost at the end of the Precambrian. However, several of the older, smaller cratons within Gondwana contain collision zones that are very close in age to those in North America and to the other northern continents. Ultimately, I hope that piecing together past continental connections will lead to a better understanding of global Precambrian crustal development. Obviously, we will never fully succeed in this aim, but I suspect we will learn a lot that we can't anticipate at this time.

•

What would you tell a student interested in studying geology and possibly interested in pursuing it as a profession?

Earth science is so diverse that you can find opportunities regardless of your interests and abilities. It has changed greatly since I was a student 25 years ago. Then it was said that it was not an experimental science because of the problem of scaling phenomena of such great size and duration. Now, computerized numerical simulations of geologic processes of all scales using geo-

physical modeling are commonplace. The advent of satellite observations has given us a useful global perspective on geologic problems. There has also been a healthy breakdown of barriers between disciplines and a growing interest in the interactions between the solid Earth, biosphere, hydrosphere, atmosphere, and solar system. Concern over the future environment is encouraging Earth scientists to see their value to society beyond the traditional service to resource industries.

Despite increased technology, there is still a place for those romantics, like myself, who are attracted to the out-of-doors and the use of a vivid imagination and a fit body. Contrary to what is often heard, "low-tech" field work is intensely scientific—it requires you to constantly synthesize observations, to make conjectures that fit your observations, and to test those conjectures through predictions that can be proven wrong through new observations. In contrast, much time in

"high-tech" science is spent trying to get the equipment to work. At its best, field geology rewards the intellectual, the athletic, and the aesthetic spirits simultaneously. It also has the advantage of having enough physical discomfort to discourage the uncommitted!

•

What good is an understanding of geology to someone who doesn't plan to become a geologist?

There is intense public interest these days in various issues that are centrally related to geology, such as environmental pollution, earthquakes, landslides, climate change, resource depletion, and biological extinctions. Current issues affecting society and its future are much more interesting and meaningful if one has some knowledge of how the Earth works. Moreover, it is crucial in this age of litigation to understand the nature of scientific inquiry and the strengths and limitations of scientific evidence.

Geology provides a means of understanding geography, the oceans, volcanoes, mountain ranges, rivers, and plains, as well as the people who live there. Travel becomes a deeper experience when what one sees conjures up images of times past. Geology, the science of Earth's history, converts one's perception of the planet from a snapshot to a movie.

This interview was conducted by Graham Thompson, University of Montana, and appears in *Modern Physical Geology*, Saunders College Publishing, 1991.

Surface Processes

UNIT III

● Weathering, Soil, and Erosion

● Streams and Ground Water

● Glaciers and Wind

● Winter sunset on Leaning Tower in Yosemite Valley
(Galen Rowell/Mountain Light)

Weathering, Soil, and Erosion

Landforms are the features that make up the shapes of the Earth's land surface. Some landforms, such as a mountain chain or a plain, can be very large; others, such as a hill or stream valley, may be small. The Earth's internal processes cause tectonic activities that shape continents and build mountain chains. Those landforms are then modified by **surface processes**.

10.1 Surface Processes

Surface processes are all of the processes that work on the Earth's surface to modify landforms. They include weathering, erosion, and transport and deposition of sediment. **Weathering is the decomposition and disintegration of rocks and minerals at the Earth's surface by both mechanical and chemical processes.** It converts solid rock to gravel, sand, clay, and soil. Weathering involves little or no movement of the decomposed rocks and minerals.

Erosion is the removal of weathered rocks and minerals from the place where they formed. Flowing water, wind, glaciers, and gravity erode weathered mate-

● Moraine Lake, Alberta, Canada.

245

A

B

Figure 10–1 Surface processes develop landforms by both erosion and deposition. (A) The Colorado River created Grand Canyon of Arizona by eroding bedrock. (B) This small delta formed as the stream deposited sand and gravel near its mouth.

rial from its place of origin. Then the material may be **transported** great distances by the same agents: water, wind, ice, and gravity. Finally, those agents of transport **deposit** the sand, clay, gravel, and other material in layers at the Earth's surface.

Weathering, erosion, transport, and deposition typically occur in an orderly sequence. For example, water freezes in a crack in granite, loosening a grain of quartz. A hard rain erodes the grain and washes it into a stream. The stream then transports the quartz to the seashore and deposits it as a grain of sand on a beach.

Surface processes create landforms by wearing away mountain ranges, sculpting coastlines, and carving canyons and valleys (Fig. 10–1A). They also build landforms by depositing sediment (Fig. 10–1B). Although the processes work slowly from the perspective of human life, in geologic time they can erode an entire mountain range to a flat plain or excavate a deep canyon.

10.2 Weathering

The physical and chemical environment at the Earth's surface is corrosive to most materials. A pocketknife rusts when it is left out in the rain. For similar reasons rocks decompose naturally. Thus, over the centuries, stone cities have fallen into ruin.

If you visit the remains of ancient Greece or Rome, you can see two types of changes. First, large building stones have broken into smaller fragments (Fig. 10–2). **Mechanical weathering** is the physical disintegration of

rock into smaller pieces. For example, vegetation grows in cracks in building stones. Roots enlarge the cracks, pushing the stones aside and eventually toppling the wall. In cold climates, water expands as it freezes in cracks, fracturing rocks and reducing stone buildings to rubble. Such mechanical processes break rocks into smaller pieces, but they do not alter the chemical compositions of the rocks and minerals.

If you look closely at a building stone in an ancient city, you may see a second type of disintegration. Its face may be pitted and discolored, and once-sharp edges are rounded. In addition, the rock may be soft and earthy rather than hard and solid as it was when originally quarried. **Chemical weathering** occurs when air and water chemically attack rocks. The chemical changes are similar to rusting in that a chemically weathered rock contains different minerals, and has a different chemical composition, from those of the original rock. Certain kinds of air pollution accelerate chemical weathering. Thus, decomposed building stones can be seen in most modern industrial cities as well as in ancient cities.

Just as building stones in ancient cities decompose both mechanically and chemically, rocks in natural settings weather by both processes. Mechanical and chemical weathering reinforce one another. For example, chemical processes generally act on the surface of a solid object. Therefore, a chemical process will speed up if the surface area increases. Think of a burning log: the fire starts on the outside and works its way inward. If you want the log to burn faster, simply split it in half to

Figure 10–2 In these Roman ruins, mechanical weathering has broken large stones into fragments, and chemical weathering has rounded sharply carved edges. *(Italian State Tourist Office)*

increase its surface area. Mechanical weathering cracks rocks, exposing more surface area for chemical agents to work on (Fig. 10–3).

Mechanical Weathering

Recall that mechanical weathering does not alter the chemical nature of rocks and minerals; it simply breaks them into smaller pieces. Think of breaking a rock with a hammer: the fragments are smaller than, but otherwise identical to, the original rock. Six processes mechanically weather rocks: frost wedging, salt cracking, abrasion,

biological activity, pressure release fracturing, and thermal expansion and contraction.

Frost Wedging

Water collects in natural cracks and crevices in rocks. If the outside temperature drops below 0°C, the water may freeze. Water expands when it freezes. Thus, water freezing in a crack pushes the rock apart in a process called **frost wedging**. The ice may cement the rock together, but when it melts, rocks may tumble from a steep outcrop. If you walk through the mountains during a season when water freezes at night and thaws during the day, be careful.

Figure 10–3 When mechanical weathering processes break rocks, more surface area is available for chemical weathering.

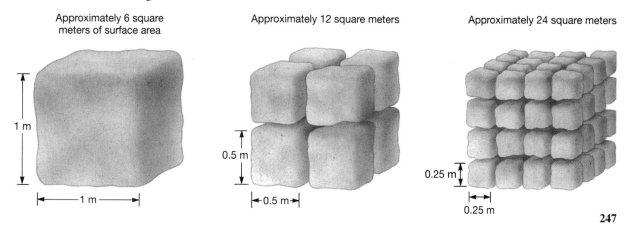

Approximately 6 square meters of surface area

1 m

1 m

Approximately 12 square meters

0.5 m

0.5 m

Approximately 24 square meters

0.25 m

0.25 m

247

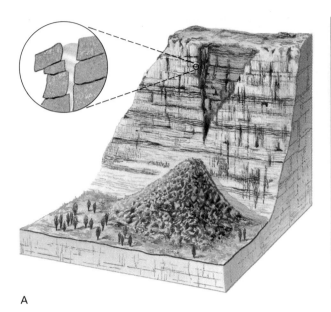

A

B

Figure 10–4 (A) Frost wedging dislodges rocks from cliffs to create talus slopes. (B) A talus slope along the San Juan River, Utah.

Rocks tumble from cliffs when the rising sun melts the ice. Mountaineers commonly travel in the early morning before rocks begin to fall.

Anyone who has spent time in the mountains has noticed large piles of broken, angular rocks at the bases of cliffs. These piles are called **talus slopes** (Fig. 10–4). The rocks in talus slopes have broken from the cliffs, mainly by frost wedging.

Figure 10–5 White salt crystallized when salty ground water evaporated from this outcrop in Grand Canyon, Arizona.

Salt Cracking

Growing salt crystals can also crack rocks. In areas where ground water is salty, salt water may seep into cracks or between mineral grains in rock. When the water evaporates, the dissolved salts crystallize (Fig. 10–5). The growing crystals exert forces great enough to widen cracks and push grains apart. Many sea cliffs are pitted by **salt cracking**. These features exist only within the reach of salt spray from the breaking waves. Salt cracking is also common in deserts, where ground water often is salty.

Abrasion

Have you ever looked at rocks in a stream or on a beach? They have been smoothed and rounded by collisions with other rocks, silt, and sand carried by the moving water. As the rocks and smaller grains collide, their sharp edges and corners wear away. The mechanical wearing and grinding of rock surfaces by friction and impact is called **abrasion**. Abrasion produces rocks with a characteristic smooth, rounded appearance (Fig. 10–6).

Like water, wind by itself is not abrasive, but it hurls dust and sand against rocks to carve unusual and beautiful landforms (Fig 10–7). Glaciers also abrade rocks. A glacier picks up particles ranging in size from clay to boulders and drags them across bedrock, abrading both the rock fragments embedded in the ice and the bedrock beneath (Fig 10–8).

Figure 10–6 Abrasion rounded these rocks in the bed of the Bitterroot River, Montana.

Figure 10–7 Windblown sand can sculpt exotic figures in bedrock, at Death Valley, California.

Figure 10–8 Rocks embedded in the base of a glacier can wear grooves in bedrock.

Figure 10–9   As this tree grew from a crack in bedrock, its roots forced the crack to widen.

Biological Activity

A plant root can widen a small crack in rock. If soil collects in a crack, a seed may fall there and grow. Roots work their way down into the crack. As the roots grow, they push the rock apart (Fig. 10–9). City dwellers can observe this effect in sidewalks, which are often cracked by tree roots.

Pressure Release Fracturing

Kilometers beneath the Earth's surface, rock is under immense pressure from the weight of overlying rocks. Imagine granite 15 kilometers deep. At that depth, pressure is about 5000 times that at the Earth's surface. Imagine further that tectonic forces push this granite up to form a mountain range. As overlying rock gradually erodes, pressure on the granite decreases and the rock expands. But since the granite is brittle, it cracks as it expands. This form of mechanical weathering is called **pressure release fracturing**.

Granite commonly fractures by **exfoliation** (Fig. 10–10), a process in which large concentric plates split off the rock like the layers of an onion. The plates may be only 10 or 20 centimeters thick at the surface, but they thicken with depth. Exfoliation fractures are usually absent below a depth of 50 to 100 meters. Thus, they appear to result from exposure of the granite at the Earth's surface.

Thermal Expansion and Contraction

Surface rocks warm and cool as air temperature changes. A rock expands when it is heated, and contracts when it cools. When the temperature of any object changes rapidly, its surface may heat or cool, and therefore expand or contract, faster than its interior. The resulting forces can fracture the object. For example, welders must heat cast iron slowly because if it is heated too rapidly, it cracks.

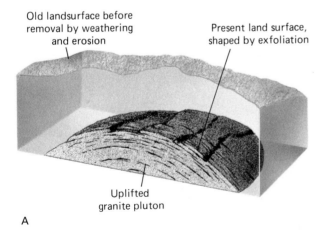

A

B

Figure 10–10   (A) Development of an exfoliation dome. The exfoliated slabs are only a few centimeters to a few meters thick. (B) Exfoliated granite near Shuteye Peak in the Sierra Nevada, California.   *(N. K. Huber, U.S. Geological Survey)*

Rock is more susceptible to fracturing than metal because a rock is a mixture of different minerals. Each mineral expands and contracts at a different rate and to a different extent with changing temperature. This varying change in volume increases the tendency of rocks to fracture when heated or cooled. If you line a campfire with certain types of rocks, such as granite, they commonly break as you cook your dinner.

In mid-latitude mountain and desert environments, daily temperatures may fluctuate between $-5°C$ and $+25°C$. Is this 30° difference sufficient to fracture rock? Geologists disagree on the answer. In one laboratory experiment, granite was subjected to repeated, rapid temperature changes of more than 100°C and did not fracture. This result implies that thermal change might not be important in mechanical weathering. On the other hand, daily heating and cooling over hundreds of thousands of years may fracture rock. Alternatively, thermal effects may only be important during occasional catastrophic events. If you walk over rock-covered soil in an area recently burned by a hot forest fire, you can easily recognize that the rocks have fractured as a result of the much greater temperature changes resulting from the fire.

Chemical Weathering

Rock seems durable when it is observed over a human life span. Return to your childhood haunts and you will see that rock outcrops in woodlands or parks have not changed. Yet, over a longer time, rocks decompose chemically at the Earth's surface. You need only visit a cemetery to see the effects of chemical weathering. Carvings on old headstones are commonly faint and poorly preserved, whereas newer inscriptions are sharp and clear.

Many natural chemicals attack rock. Oxygen, water, and carbon dioxide are the most important ones. Other corrosive compounds also occur in some streams, lakes, and underground water.

Reactions with Oxygen

Many elements react rapidly with molecular oxygen, O_2. In our everyday experience, an iron bar reacts with water and oxygen in the atmosphere to rust. Rusting is one manifestation of a more general process called **oxidation.*

About 21 percent of the Earth's atmosphere is oxygen, and as a result, oxidation is common. It usually turns useful material to waste. Wood burns (oxidizes) to form ashes, iron oxidizes to rust, and so on. Oxidation in rocks occurs when iron in minerals reacts with oxygen.

Weathering by Solution

If you put salt (halite) in water, it dissolves and the ions disperse into the water to form a **solution** (Fig. 10–11). Halite dissolves so easily that it exists only in the driest environments. Unless you live in a desert, you will never find a chunk of halite in your backyard.

Acids and Bases

In pure water, a small proportion of the water molecules break apart to form an equal number of hydrogen ions

*Oxidation is properly defined as the loss of electrons from a compound or element during a chemical reaction. In the weathering of common minerals, this usually occurs when the mineral reacts with molecular oxygen.

Figure 10–11   Halite dissolves in water because the attractions between water molecules and the sodium and chloride ions are stronger than the chemical bonds in the crystal.

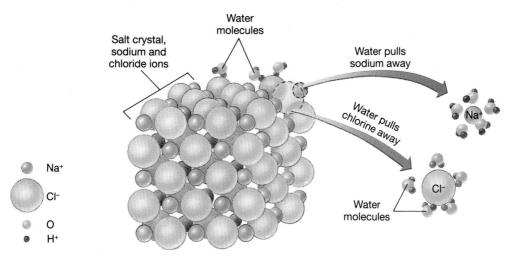

(H$^+$) and hydroxyl ions (OH$^-$).* If a chemical compound is added to increase the proportion of hydrogen ions, the resultant solution is an **acid**; if the hydroxyl ion concentration is increased, the solution is a **base**.

Acids and bases are much more corrosive than pure water because of the nature of hydrogen and hydroxyl ions. Think of an atom on the surface of a crystal. It is held in place because it is attracted to other atoms in the crystal by electrical forces. But at the same time, electrical forces in the outside environment are pulling the atom away from the crystal. The result is a tug-of-war. If the atom is bonded to the crystal more strongly than it is attracted to the outside environment, then the crystal remains intact. If outside attractions are stronger, then the crystal dissolves as its atoms are pulled away (Fig. 10–11). Acids and bases dissolve minerals faster than pure water because the electrically charged hydrogen and hydroxyl ions pull atoms from crystals more effectively than water molecules do.

Natural water is never pure. Atmospheric carbon dioxide dissolves in rainwater and reacts to form a weak acid called carbonic acid. The solution chemically attacks rocks and minerals. In addition, when water flows across or beneath the surface of the Earth, it dissolves ions from minerals. In some instances, these ions make the water acidic; in other cases, the water becomes basic. In either case, the resulting solution has a greater ability to dissolve rocks and minerals.

Chemical Weathering of Common Rocks and Minerals

Granite   Recall that continents are made up mostly of granite and similar rocks. Granite consists mainly of feldspar and quartz. When water and atmospheric gases attack granite, the feldspar crystals dissolve. Most of the dissolved ions quickly recombine to form soft, earthy clay. Commonly, the clay replaces feldspar on such a fine scale that weathered granite may look as if its feldspar crystals are unaltered. When you attempt to scratch them, however, they are soft and crumble to dust. The feldspar has converted completely to clay but preserves the shape and appearance of feldspar crystals.

As feldspar weathers to clay, the clay crystals take on water in a process called **hydration**. The clay then expands because of the addition of water, pushes against nearby mineral grains, and breaks them free. Thus, hydration, which is a form of chemical weathering, causes mechanical weathering. Some granites are so severely weathered by hydration that their mineral grains are loose

Figure 10–12   Coarse grains of quartz and feldspar accumulate directly over weathering granite.

to depths of several meters and can be pried out with a fingernail (Fig. 10–12).

However, quartz is different. Feldspar and the other rock-forming minerals except for quartz all contain soluble cations such as potassium, sodium, or calcium, in addition to silicon and oxygen. Water and acids readily dissolve these ions and therefore decompose the minerals. However, quartz is pure silica, SiO_2. Silica is relatively insoluble, and thus, quartz resists chemical weathering.

When granite weathers, the feldspar and other minerals decompose to clay. As they decompose, they release the unaltered quartz grains from the rock. Therefore, quartz accumulates as sand grains in soil. Quartz is also resistant to mechanical weathering. Therefore, when streams transport eroded soil to the seacoast, the quartz grains survive the journey and accumulate on beaches. For this reason, the sand on most beaches is quartz (Fig. 10–13).

Limestone   Limestone is a type of sedimentary rock commonly exposed at the Earth's surface. It is made up principally of the mineral calcite, which dissolves even in mild acid. Thus, limestone dissolves in rainwater. When acidic rainwater seeps into cracks in limestone, it dissolves the rock, enlarging the cracks to form subterranean

*Hydrogen ions react instantaneously and completely with water, H_2O, to form the hydronium ion, H_3O^+, but for the sake of simplicity we will consider the hydrogen ion, H^+, an independent entity.

Figure 10–13 Most beach sand is quartz.

caves and caverns. The origins of caves and related features are explored in Chapter 11.

Although rain is naturally slightly acidic, common atmospheric pollutants greatly increase its acidity. For that reason, weathering of rocks and stone buildings accelerates greatly in regions with air pollution problems. This subject is discussed in Chapter 18.

10.3 Soil

Bedrock breaks into smaller fragments as it weathers, and much of it decomposes to clay and sand. Thus, on most land surfaces, a layer of loose rock fragments mixed with clay and sand overlies bedrock. This material is called **regolith**. Scientists in different disciplines use slightly different definitions for soil. In engineering and construction, "soil" and "regolith" are interchangeable. However, soil scientists define **soil** as upper layers of regolith that

support plant growth. That is the definition we will use here.

Components of Soil

Soil is a mixture of mineral grains, organic material, water, and gas. The mineral grains include clay, silt, sand, and rock fragments. Clay grains are so small and closely packed that water does not flow through them readily. Pure clay holds water so effectively that plants growing in clayey soils suffer from lack of oxygen. In contrast, water flows easily through sandy soil. The most fertile soil is **loam**, a mixture of sand, clay, silt, and generous amounts of organic matter.

When plant or animal matter dies and falls, it retains its original shape and form until it begins to decay. Thus, if you walk through a forest or prairie, you can find bits of leaves, stems, and flowers on the surface. This material is called **litter** (Fig. 10–14). When litter decomposes

Figure 10–14 Litter is organic matter that has fallen to the ground and started to decompose, but still retains its original form.

Earth Science and the Environment

Chemical Fertilizers and Soil Deterioration

Chemical fertilizers have increased global food production and have thus helped feed the expanding human population. For this reason, modern farmers all over the world use fertilizers in ever-increasing amounts. But the long-term effects of chemical fertilizers have caused serious concern. In types of intensive agriculture where chemical fertilizers are used, crop residues such as straw are removed and soil humus is thus continuously de-

pleted. When humus is lost, soil retains less water, nutrients leach from the soil, and the agricultural system becomes totally dependent on irrigation and chemical fertilizers.

One solution to humus depletion is to use agricultural wastes or other organic matter as fertilizers. Natural fertilizers such as manure are usually more expensive than chemicals. However, many soil scientists feel that if one were to view profit and loss over a period of decades, not just one or two growing seasons, then it would be economically advantageous to use natural fertilizers.

so that you can no longer determine the origins of the individual pieces, it becomes **humus**. Humus is an essential component of most fertile soils. Scoop up some forest or rich garden soil with your hand. Soil rich in humus is light and spongy and absorbs a large quantity of water. It soaks up so much moisture that it swells after a rain and then shrinks during a dry spell. This alternate shrinking and swelling keeps the soil loose, allowing roots to grow easily. A rich layer of humus also insulates deeper soil from heat and cold and reduces water loss by evaporation.

Soil nutrients are chemical elements necessary for plants to grow and complete their life cycles. Some examples are carbon, phosphorus, nitrogen, and potassium. Another function of humus is to hold nutrients in soil and make them available to plants.

Intensive agriculture commonly destroys humus by exposing it to erosion and oxidation. Farmers then replace the lost nutrients with chemical fertilizers to maintain crop growth. But loss of humus reduces the natural ability of soil to conserve water and nutrients. Water can flow over the surface, eroding the soil. In addition, the muddy runoff carries fertilizer and pesticide residues, contaminating streams and ground water.

Soil Profiles

If you dig down through a typical soil, you can see several layers, or **soil horizons** (Fig. 10–15). The uppermost layer of a mature soil is the **O horizon**, named for its *o*rganic component. This layer is mostly litter and humus with a small proportion of minerals. The next layer down, called the **A horizon**, is a mixture of humus, sand, silt, and clay. The thicker layer including both O and A horizons is often called **topsoil**. A kilogram of average fertile topsoil

contains about 30 percent by weight organic matter including millions of living organisms. Approximately two trillion bacteria, 400 million fungi, 50 million algae, 30 million protozoa, and thousands of larger organisms such as insects, worms, and mites live in a kilogram of average topsoil.

The third layer, the **B horizon** or subsoil, is a transitional zone between topsoil and weathered parent rock below. Roots and other organic material occur in the B horizon, but the amount of organic matter is low. The lowest layer, called the **C horizon**, consists of partially weathered bedrock. It lies directly on unweathered parent rock. This zone contains very little organic matter.

Now think of what happens when rainwater falls on soil. It sinks into the O and A horizons. As it travels through the topsoil, it partially dissolves minerals and carries the dissolved ions to lower levels. This downward movement of dissolved material is called **leaching**. The A horizon is sandy because water carries clay downward leaving sand behind. Because materials are removed from the A horizon, it is called the **zone of leaching**.

Dissolved ions and clay carried downward from the A horizon accumulate in the B horizon, which is therefore called the **zone of accumulation**. This layer retains moisture because of its high clay content. Although moisture retention may be beneficial, if too much clay accumulates, a dense, waterlogged soil can develop.

Soil-Forming Factors

Why is one soil rich and another poor, or one sandy and another loamy? What factors contribute to the character of a specific soil? Five **soil-forming factors** control how soil develops as parent material weathers.

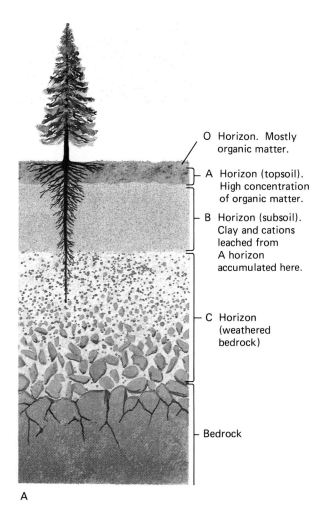

O Horizon. Mostly organic matter.

A Horizon (topsoil). High concentration of organic matter.

B Horizon (subsoil). Clay and cations leached from A horizon accumulated here.

C Horizon (weathered bedrock)

Bedrock

A

B

Figure 10–15   (A) A well-developed soil commonly shows several distinct horizons. (B) Soil horizons are often distinguished by color and texture. The dark upper layer is the A horizon; the white lower layer is the B horizon.

Parent Rock

The type of parent rock exerts a strong influence on soil. As explained above, granite contains mostly feldspar and quartz. When it weathers, its feldspar converts to clay, but the quartz grains resist chemical weathering and become sand. Thus, sandy soil commonly forms on weathering granite. In contrast, basalt contains much feldspar but no quartz, and soil formed from basalt is likely to be clay-rich but not sandy. In addition to texture, the parent material provides nutrients to soil, so nutrient abundance or deficiency depends in part on the chemical composition of the parent rock.

Time

In a geologically young soil, weathering of feldspar and other minerals may be incomplete, and so the soil is likely to be sandy. As a soil matures and more feldspar decomposes, the soil's clay content increases.

Many young soils consist mostly of slightly weathered minerals inherited from adjacent bedrock. This is particularly true in regions where glaciers have eroded old soil, exposing bare rock within the past 20,000 years. In time, however, a stream may deposit layers of sand or mud, or wind may blow in dust. The foreign minerals mix with the residual soil, changing its composition and texture.

Climate

Rain seeps downward through soil. But other factors pull the water back upward. Roots suck soil water toward the surface and water near the surface evaporates. In addition, water is electrically attracted to soil particles. If the pore size is small enough, water can be drawn upward by **capillary action**. Capillary action can be demonstrated by placing the corner of a paper towel in water and watching the water rise (Fig. 10–16).

During a rainstorm, water percolates down through the A horizon, dissolving soluble ions such as calcium, magnesium, potassium, and sodium. In arid and semiarid regions, when the water reaches the B horizon, capillary action and plant roots then draw it back up toward the

Figure 10–16 Colored water soaks upward through the pores in a paper towel by capillary action.

surface, where it may evaporate. As it evaporates or is taken up by plants, many of its dissolved ions precipitate in the B horizon, encrusting the soil with salts. A soil of this type is a **pedocal** (Fig. 10–17). Such a process often deposits enough calcium carbonate to form a hard cement called **caliche** in the soil.

In a wet climate, ground water leaches soluble ions from both the A and B horizons. The less soluble elements such as aluminum, iron, and some silicon remain behind, accumulating in the B horizon to form a soil type called a **pedalfer** (Fig. 10–17). The subsoil in a pedalfer commonly is rich in clay, which is mostly aluminum and silicon, and has the reddish color of iron oxide.

Figure 10–17 Pedocals, pedalfers, and laterites form in different climates.

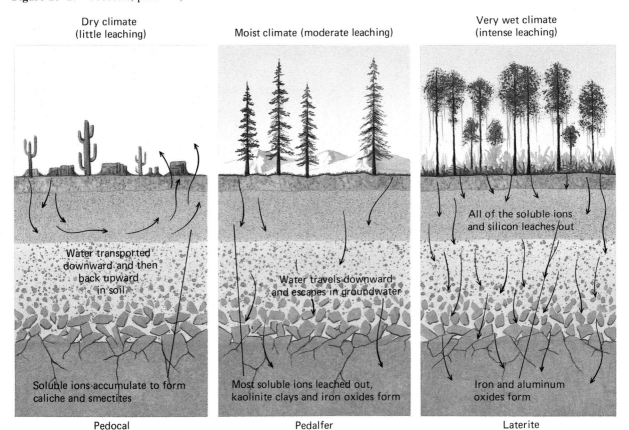

Dry climate (little leaching)

Moist climate (moderate leaching)

Very wet climate (intense leaching)

Water transported downward and then back upward in soil

Water travels downward and escapes in groundwater

All of the soluble ions and silicon leaches out

Soluble ions accumulate to form caliche and smectites

Most soluble ions leached out, kaolinite clays and iron oxides form

Iron and aluminum oxides form

Pedocal

Pedalfer

Laterite

Earth Science and the Environment

Tropical Rainforests

Most laterite soils form in the tropics, where the climate is hot and rainfall is high. Tropical rainforests support a rich and varied community of plants and animals, but paradoxically, their laterite soils are not suitable for agriculture. When litter falls to the forest floor, it decays quickly in the hot, moist environment, and growing plants absorb most of the nutrients. Any excess nutrients are leached out by the great amount of water seeping into the ground. As a result, the soil contains little humus and nutrients.

When the rainforest is cut, heavy tropical rains rapidly leach any remaining nutrients from the soil. Farmers can grow crops for a few years, but soon the small supply of nutrients is consumed and the soil becomes useless for agriculture. Because the nutrients are lost, it is then impossible for the natural rainforest vegetation to regenerate, and the abandoned soil lies bare and unprotected from erosion.

Between 1945 and 1990, more than half of the

Earth's tropical rainforests were cut, partly for timber and partly for agriculture (see figure). Today, deforestation of the tropics continues at a rate of about 7 million hectares per year, or about 13 hectares per minute, day and night, throughout the year. Only about one third of this land is capable of supporting sustained agriculture. In some regions the exposed clayey soil bakes in the hot sun to a brick-like texture, and the area becomes a desolate wasteland. This loss is especially tragic because the rainforests are a valuable renewable resource. If the nuts and fruits were harvested and trees selectively cut for fuel and lumber, the forests could provide local people with continuous economic rewards.

Beyond direct economics, many scientists fear that if vast regions of the forest are cut, the carbon dioxide–oxygen balance in our atmosphere may be significantly affected resulting in global climate change. In addition, millions of species of plants and animals in the rainforest are rapidly becoming extinct, reducing the biological diversity of our planet.

Destruction of a tropical rainforest in Peru. *(USAID)*

What kind of soil develops in a region of very high rainfall, such as a tropical rainforest? There, so much water seeps through the soil that nearly all the cations are leached away, and only very insoluble aluminum and iron minerals remain (Fig. 10–17). Soil of this type is called **laterite**. Laterites are often colored rust-red by iron oxide.

A highly aluminous laterite, called **bauxite**, is the world's main source of aluminum ore.

Rates of Growth and Decay of Organic Material

As explained previously, humus is decomposed plant material, and a thick layer of humus makes rich soil.

The most fertile soils are those of prairies and forests in temperate latitudes. There, plant growth and decay are balanced so that thick layers of humus form. In the tropics, organic material decays rapidly, so little humus accumulates. The Arctic is so cold that plant growth and decay are slow. Therefore, litter and humus form slowly and Arctic soils contain little organic matter.

Slope Angle and Aspect

Soil generally migrates downslope in response to gravity. Therefore, if all other factors are equal, soil is thinner and poorer on a hillside, and the valley floor has the deepest and richest soil.

Aspect is the orientation of a slope with respect to compass direction. Exposure of a slope to the Sun affects soil formation. For example, in the semiarid West, thick soils and dense forests cover the northern slopes of hills, but thin soils and grass dominate southern exposures. The reason for this difference is that in the Northern Hemisphere more water evaporates from the sunny southern slopes. Therefore, fewer plants grow, weathering occurs more slowly, and soil development is retarded. The moister northern slopes weather more deeply to form thicker soils.

10.4 Erosion

Natural Soil Erosion

Weathering decomposes bedrock and plants add organic material to the regolith to create soil at the Earth's surface.

Sand chokes the bed of this desert wash in southern Utah.

However, soil does not continue to accumulate and thicken throughout geologic time. If it did, the Earth would be covered by a mantle of soil hundreds or thousands of meters thick, and rocks would be unknown at the Earth's surface. Instead, flowing water, wind, and glaciers erode soil as it forms. In addition, some weathered material simply slides downhill under the influence of gravity. In fact, all forms of erosion combine to remove soil about as fast as it forms. For this reason, soil is usually only a few meters thick or less in most parts of the world.

Once soil erodes, the clay, sand, and gravel begin a long journey as they are carried downhill by the same agents that eroded them: streams, glaciers, wind, and gravity. On their journey they may come to rest in a stream bed, a sand dune, or a lake bed, but those environments are only temporary stops. Sooner or later they erode again and are carried downhill, until finally they are deposited where the land meets the sea. Here they remain and are buried by younger sediment until they lithify to form sedimentary rocks.

Natural erosion and transport of sediment by streams, glaciers, and wind are the subjects of the following two chapters. In the last three sections of this chapter, we will describe erosion by gravity.

Soil Erosion and Agriculture

In nature, soil erodes approximately as rapidly as it forms. However, improper farming, livestock grazing, and logging can accelerate erosion. Plowing removes plant cover that protects soil. Logging often removes forest cover, and the machinery breaks up the protective litter layer. Similarly, intensive grazing can strip away protective plants. Rain, wind, and gravity then erode the exposed soil easily and rapidly. Meanwhile, soil continues to form by weathering at its usual slow, natural pace. Thus, increased rates of erosion caused by agricultural practices can lead to net soil loss.

When farmers use proper conservation measures, soil can be preserved indefinitely or even improved. Some regions of Europe and China have supported continuous agriculture for centuries without soil damage. However, in recent years, marginal lands on hillsides, in tropical rainforests, and along the edges of deserts have been brought under cultivation. These regions are particularly vulnerable to soil deterioration, and today soil is being lost at an alarming rate throughout the world (Fig. 10–18).

Soil is eroding more rapidly than it is forming on about 35 percent of the world's croplands. About 23 trillion kilograms of soil (about 25 billion tons) are lost every year. The soil lost annually would fill a train of freight cars long enough to encircle our planet 150 times.

Figure 10–18   Soil erosion has formed deep gullies in this California hillside.

Table 10–1 shows the sediment loads of several major rivers. Although much of the sediment results from natural erosion, the sediment loads of major rivers have been rising, and this increase stems from excessive erosion caused by logging and improper agriculture.

In the United States, approximately one third of the topsoil that existed when the first European settlers arrived has been lost. Erosion is continuing in the United States at an average rate of about 10.5 tons per hectare* per year. However, in some regions the rate is considerably higher, and yearly losses of more than 25 tons per hectare are common. To put this number in perspective, a loss of 25 tons per hectare per year would lead to complete loss of the topsoil in about 150 years.

10.5   Erosion by Mass Wasting

Mass wasting is the movement of Earth material downslope, primarily under the influence of gravity. Look up

*One hectare equals 2.47 acres.

An expensive landslide in Hong Kong.   *(Hong Kong Government Information Services)*

toward a hill or mountain and think about the bedrock and soil near the top. Gravity constantly pulls them downward, but common sense and experience tell us that on any given day the rock and soil are not likely to slide or tumble down the slope. They are held up by their own strength and by friction. Even so, a slope can become unstable and mass wasting can occur. For example, stream erosion can undermine a rock cliff so much that it collapses. Water can add weight and lubrication to soil on a slope and cause it to slide. Mass wasting occurs naturally in any hilly or mountainous terrain. Steep slopes are especially vulnerable, and scars from recent movement of rock and soil are common in mountainous country.

TABLE   10–1

Sediment Load of Selected Major Rivers		
River	**Countries**	**Annual Sediment Load (million metric tons)**
Yellow	China	1600
Ganges	India	1455
Amazon	Brazil	363
Mississippi	United States	300
Irrawaddy	Burma	299
Kosi	India	172
Mekong	Southeast Asia	170
Nile	Egypt	111

Source: S. A. El-Swaify and E. W. Dangler, "Rainfall Erosion in the Tropics: A State of the Art," in American Society of Agronomy, *Soil Erosion and Conservation in the Tropics* (Madison, Wisc.: 1982).

In recent years, the human population has increased dramatically. As land has become overpopulated, the character of human settlements has changed. In poor countries, more and more people try to scratch out a living in mountainous areas once considered too harsh for agriculture. In rich nations, people have moved into the hills to escape congested cities. As a result, permanent settlements have grown in previously uninhabited steep terrain. Many steep slopes are naturally unstable. Others can be destabilized by construction or agriculture.

Every year, small landslides destroy homes and farmland. Occasionally, an enormous landslide buries towns, killing thousands or even tens of thousands of people. Mass wasting causes billions of dollars' worth of damage every year. In fact, the total global property damage from mass wasting in a single year approximately equals the damage caused by earthquakes in 20 years. In many

instances, losses occur because people do not recognize danger zones. In other cases, damage is caused, or at least aggravated, by improper planning and construction.

Consider three real examples of mass wasting that have affected humans:

1. A movie star builds a mansion on the edge of a picturesque California cliff. After a few years the cliff collapses and the house slides into the valley (Fig. 10–19A).

2. A ditch carrying irrigation water across a hillside in Montana leaks water into the ground. After years of seepage, the muddy soil slides downslope and piles against a house at the bottom of the hill (Fig. 10–19B).

3. In the springtime, when snow is melting, water saturates soil and bedrock on a steep mountain slope. Suddenly, an earthquake shakes the overburdened

Figure 10–19 Examples of mass wasting. (A) A few days after this photo was taken, the corner of the house hanging over the gully fell in. *(J. T. Gill, USGS)* (B) A landslide, triggered by a leaking irrigation ditch, threatens a house in Darby, Montana. (C) This landslide near Yellowstone Park buried a campground, killing 26 people. *(Donald Hyndman)*

slope, triggering a landslide that buries a campground and kills several people (Fig. 10–19C).

Why does mass wasting occur? Is it possible to avoid or predict such catastrophes to reduce property damage and loss of life?

Factors That Control Mass Wasting

Imagine that you are a geological consultant on a construction project. The developers want to construct a road at the base of a hill, and they wonder whether landslides or other forms of mass wasting are likely. What factors should you consider in your evaluation?

Steepness of the Slope

Obviously, the **steepness** of a slope is a factor in mass wasting. If frost wedging dislodges a rock from a vertical cliff, the rock will tumble to the valley below. However, a similar rock is less likely to roll down a gentle hillside.

The relationship between slope steepness and mass wasting can be illustrated by placing a block of wood on a board and slowly tilting one end of the board. When the board is nearly level, friction holds the block in place, and it does not slide. However, if you tilt the board beyond a certain critical angle, the block slides or tumbles.

Type of Rock and Orientation of Rock Layers

The block of wood is a coherent mass; either the entire block slides, or none of it moves. A hillside does not behave in the same way. Any portion of a slope can move. For example, if sedimentary rock layers dip in the same direction as a slope, then the upper layers may slide easily over the lower ones. Imagine a hill underlain by shale, sandstone, and limestone layered parallel to the slope, as shown in Figure 10–20A. If the base of the hill is undermined (Fig. 10–20B), the upper portion is left hanging and may slide along the layer of weak shale. In contrast, if the rock layers dip at an angle to the hillside,

Figure 10–20 (A) Sedimentary rock layers dip parallel to this slope. (B) If a roadcut undermines the slope, the dipping rock provides a good sliding surface, and the slope may fail. (C) Sedimentary rock layers dip at an angle to this slope. (D) The slope may remain stable even if it is undermined.

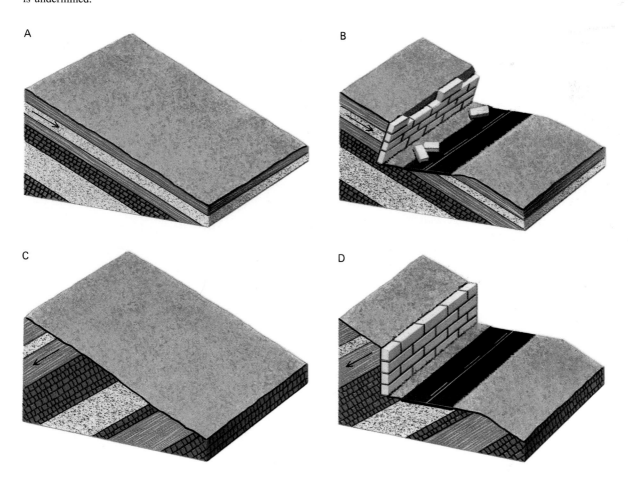

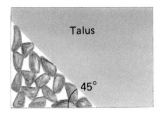 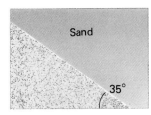

Figure 10–21 The angle of repose is the maximum slope that can be maintained by a specific material.

the slope may be stable even if it is undercut (Figs. 10–20C, D).

Several processes can undermine a slope. A stream or ocean waves can erode its base. Road cuts and other types of excavation can create instability in the same manner. Therefore, geologists and engineers must consider not only a slope's stability *before* construction, but how the project might alter its stability.

Nature of Unconsolidated Materials

The **angle of repose** is the maximum slope, or steepness, at which loose material remains stable. Recall that when chunks of rock break from a cliff, they fall to the bottom, where they collect to form a talus slope. Because rocks in talus are angular and irregular, they interlock and jam together, allowing talus to maintain a high angle of repose, up to 45°. The slope cannot be any steeper because if it were, the rock would slide. In contrast, rounded sand grains have a lower angle of repose (Fig. 10–21).

Water and Vegetation

To understand how water affects slope stability, think of a sand castle. Even a novice sand-castle builder knows that it is impossible to build steep-sided towers and walls with dry sand; the sand must be moistened first (Fig. 10–22). But if you add too much water, the castle collapses. When some soils become saturated with water, they flow downslope, just as the sand castle collapses. In addition, if water collects on an impermeable layer of clay or shale, it may lubricate overlying layers of rock or soil so that they can move easily.

Water also adds weight to a slope. The extra weight during a heavy rain can cause mass wasting. Finally, if water freezes and thaws, it can alternately bind particles together and push them apart. As the particles are pushed apart, they migrate downslope.

Roots hold soil together, and therefore a highly vegetated slope is stabler than a similar bare one. Many forested slopes that were stable for centuries have slid when the trees were removed during logging, agriculture, or construction.

Mass wasting is common in regions with low or intermittent rainfall. For example, southern California has dry summers and occasional heavy winter rainfalls. Vegetation is sparse due to summer drought and wildfires. When winter rains fall, bare hillsides often become saturated and slide. Mass wasting occurs for similar reasons during rare but intense storms in deserts.

Earthquakes and Volcanoes

Earthquakes and volcanic eruptions have triggered many devastating landslides. An earthquake can cause mass wasting by shaking the ground, causing an unstable slope to slide. A volcanic eruption may melt snow and ice near the top of a volcanic mountain. The water then soaks into the slope to release a landslide.

10.6 Types of Mass Wasting

Mass wasting can occur slowly or rapidly. Sometimes rocks fall freely down the face of a steep mountain. In other instances, material moves downslope so slowly that the movement is unnoticed by casual observers. Mass wasting falls into three categories: flow, slide, and fall (Fig. 10–23). To understand these categories, think again of building a sand castle. If you add too much water to the sand, it will flow like molasses down the face of the structure. During **flow**, loose, unconsolidated regolith moves as a viscous fluid, analogous to road tar on a hot day or fluid magma pouring down the side of a volcano.

Figure 10–22 The angle of repose depends on both the type of material and its water content. Dry sand forms low mounds, but if you moisten the sand, you can build steep, delicate towers with it.

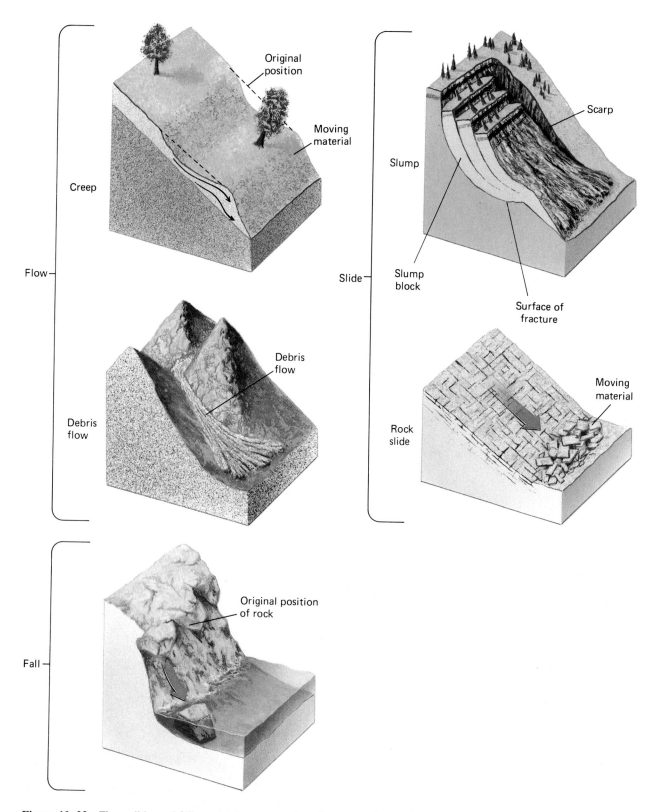

Figure 10–23   Flow, slide, and fall are the three categories of mass wasting.

● MEMORY DEVICE

"Landslide" is a general term for many different types of mass wasting and for the landforms created by those movements of soil and rock. The types of mass wasting discussed in the text, such as debris flow, mudflow, and slump, are specific kinds of landslides.

Flow can be slow; some slopes flow at a speed of about 1 centimeter per year or less. On the other hand, mud with a high water content can flow almost as rapidly as water.

If you undermine the base of a sand castle, a block of sand may break away from the rest of the castle and move downslope as a coherent unit. This type of movement is called **slide**. Slide is usually faster than flow, but it still may take several seconds for the block to slide down the face of the castle. When a hillside slides, the moving blocks or slabs remain intact. Trees might be tilted as they are carried downslope, but often they are not killed or knocked over.

If you took a huge handful of sand out of the bottom of a sand castle, the whole tower would topple. This rapid, free-falling motion is called simply **fall**. Fall is the most rapid type of mass wasting. In extreme cases, rock can fall at a speed dictated solely by the force of gravity and wind resistance.

Table 10–2 outlines the characteristics of flow, slide, and fall. Details of these three types of mass wasting are explained in the following sections.

Flow

Types of flow include creep, debris flow, earthflow, mudflow, and solifluction.

TABLE 10–2

Some Categories of Mass Wasting				
Type of Movement	**Description**	**Subcategory**	**Description**	**Comments**
Flow	Individual particles move downslope independently of one another, not as a consolidated mass. Typically occurs in loose, unconsolidated regolith.	Creep	Slow, visually imperceptible movement	Often occurs in conjunction with slump. Common in arid regions with intermittent heavy rainfall, or can be triggered by volcanic eruption
		Debris flow	More than half the particles larger than sand size; rate of movement varies from less than 1 m/year to 100 km/hr or more	
		Earthflow and mudflow	Movement of fine-grained particles with large amounts of water	
		Solifluction	Movement of waterlogged soil situated over permafrost	Can occur on very gradual slopes
Slide	Material moves as discrete blocks, can occur in regolith or bedrock.	Slump	Downward slipping of a block of earth material, usually with a backward rotation on a concave surface	Often triggers flow; trees located on slump blocks remain rooted
		Rockslide	Usually rapid movement of a newly detached segment of bedrock	
Fall	Material falls freely in air; typically occurs in bedrock.	—	—	Occurs only on steep cliffs

Figure 10–24 Creep has bent this layering in sedimentary rocks in a downslope direction.

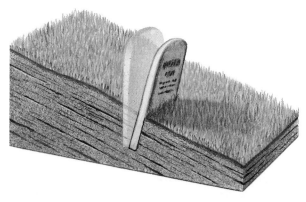

Figure 10–25 During creep, the soil surface moves more rapidly than deeper layers, so objects embedded in the soil tilt downhill.

Creep

As the name implies, **creep** is a slow downhill movement of rock or soil. Individual particles move independently of one another, not as a consolidated mass. Typically, movement occurs at a rate of about 1 centimeter per year—so slowly that the motion cannot be detected without a reference. When a slope creeps, the surface moves more rapidly than deeper layers (Fig 10–24). As a result, anything with roots or a foundation tilts downhill. For example, if you look at an old hillside cemetery, you may note that older headstones are tilted, whereas newer ones are vertical (Fig. 10–25). Over the years, soil creep has

tipped the older monuments, but the newer ones have not yet had time to tilt.

Creep also tilts fences and telephones poles, and in some instances it may tear entire buildings apart. When soil creep tilts trees, they develop curved trunks due to their natural tendency to grow upward. The result is a J-shaped appearance called pistol butt (Fig. 10–26). If you ever contemplate buying hillside land for a home site, examine the trees. If they have pistol-butt bases, the slope is probably unstable.

Creep often accelerates as heavy rain or snowmelt adds weight and lubrication to soil. In addition, movement can result from freeze-thaw cycles that occur mainly in spring and fall in temperate regions. Recall that water expands when it freezes. When wet soil freezes, it expands

Figure 10–26 If a hillside creeps as a tree grows, the tree develops pistol butt.

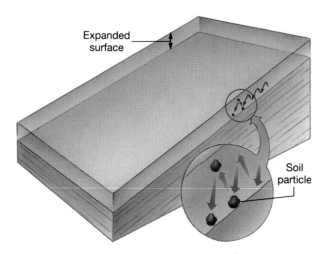

Figure 10–27 When soil freezes, it expands outward, away from the slope. But when it thaws again, the soil particles settle vertically, resulting in a net downslope migration of the soil.

outward away from the slope (Fig. 10–27). However, when the soil thaws, the particles fall vertically downward. This expansion and falling result in a net displacement downslope. The displacement in any one cycle is small, but the soil may freeze and thaw once a day for a few months each year, leading to a total movement of a centimeter or more per year.

Debris Flows, Mudflows, and Earthflows

Debris flows, mudflows, and earthflows all involve downslope flow of wet soil or regolith as a plastic or semifluid mass. Think of what can happen when heavy rain falls on an unvegetated slope, as in a desert. The rain rapidly mixes with soil to form a slurry of mud and rocks. A slurry consists of water and solid particles, but it flows as a liquid. Wet concrete is a familiar example of a slurry. It flows easily and is routinely poured or pumped from a truck.

The advancing front of a flow often forms a tongue-shaped lobe (Fig. 10–28). Slow-moving flows travel at a rate of about 1 meter per year, but others can move as fast as a car speeding along an interstate highway. The destructive potential of a flow depends on its speed and the consistency of the slurry. Flows can pick up boulders and automobiles and smash into houses, filling them with mud or even dislodging them from their foundations.

Different types of flows are characterized by the sizes of the solid particles. A **debris flow** consists of a mixture of clay, silt, sand, and rock fragments in which more than half of the particles are larger than sand. In contrast, mudflows and earthflows are predominantly sand and mud. Some **mudflows** have the consistency of wet con-

crete, and others are even more fluid. **Earthflows** have less water than mudflows and are therefore less fluid.

Flows typically occur in arid and semiarid regions when heavy rain oversaturates soil. However, leaky irrigation ditches and overwatered lawns can also oversaturate soil and cause flows.

Solifluction

In temperate regions, soil moisture freezes in winter and thaws in summer. However, in very cold regions such as the Arctic, the Antarctic, and in some mountain ranges, a layer of permanently frozen soil or subsoil called **permafrost** lies about a half meter to a few meters beneath the surface. Because ice is impermeable, summer meltwater cannot percolate downward, as it does in temperate and tropical regions, and it therefore collects on the ice layer below it. This leads to two unique characteristics in these soils.

1. Water cannot penetrate the ice layer, so it collects near the surface. As a result, even though many Arctic regions receive little annual precipitation, bogs and marshes are common.

Figure 10–28 A debris flow in the Cascade Range of Washington has a characteristic tongue-shaped lobe.

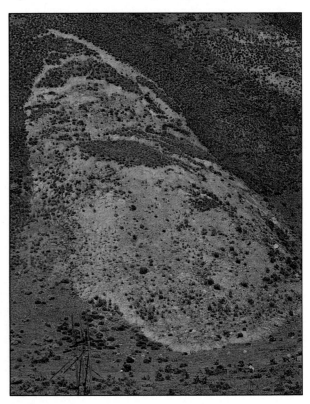

Earth Science and the Environment

Mass Wasting in the Himalayas

The Himalayas are the highest mountain chain on Earth. Located at nearly tropical latitudes, the tall peaks rise steeply from narrow valleys. In recent years the rapidly growing population in this region has increased the demand for food and fuel. The most inexpensive short-term solution to these needs has been to cut hillside forests for fuel, construct level terraces on the steep slopes, and plant grain on the terraces. Mass wasting is common where the forest has been cut and terraces built (see figure). Entire hillsides are frequently lost, and slopes that were once useful for growing timber and grazing livestock have become useless. In some cases, landslides have crushed houses and killed their inhabitants.

Mass wasting is common in the steep terrain of the Himalayas. Much terraced farmland is lost to landslides.

2. Ice, especially ice with a thin film of water on top, is extremely slippery. Therefore, Arctic regions are particularly susceptible to mass wasting of soil overlying the permafrost.

   Solifluction is mass wasting that occurs when water-saturated soil moves over permafrost (Fig. 10–29). Solifluction can occur even on very gentle slopes.

Slide

In many cases, a large block of rock or soil, sometimes an entire hillside, fractures and moves. The material does not flow as a fluid, but rather slides downslope as a coherent mass, or as several blocks that remain intact. There are two types of **slide**: slump and rockslide.

Figure 10–29 Arctic solifluction is characterized by lobes and a hummocky surface. *(R. B. Colton, USGS)*

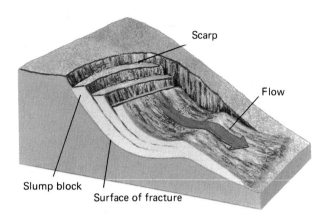

Figure 10–30 In slump, blocks of soil or rock remain intact as they move downslope.

A **slump** occurs when a gently curved fracture forms in rock or regolith. Overlying blocks of material slide downhill along the fracture, as shown in Figure 10–30. Trees remain rooted in the moving blocks. However, because the blocks rotate on the concave fracture, trees on the slumping blocks are tilted backward. Thus, you can distinguish slump from creep by the orientations of the trees. Slump tilts trees uphill, whereas creep tilts them downhill. When the blocks reach the bottom of the slope, they pile up to form a broken, jumbled, hummocky topography.

It is useful to be able to identify slump because it often recurs in the same place or on adjacent slopes. Thus a slope that shows evidence of past slump is not a good place to build a house. Figure 10–19B shows a slump that almost destroyed a home in Darby, Montana. Slump and flow often occur together. In many situations, the movement of a slump triggers a debris flow, earthflow, or mudflow farther downslope.

In the western United States, irrigation water commonly flows through unlined ditches. If the ground is porous, water seeps from the ditch into the soil and bedrock. The slump in Figure 10–19B resulted from water seeping through an irrigation ditch cut across an unstable slope. Alternatively, excessive irrigation can trigger slump. A Los Angeles man left his sprinkler on when he went away for a vacation. His property was on a hillside, and when he returned, he found that not only was his lawn gone, but his house had slid downslope as well.

During a **rockslide**, or **rock avalanche**, a fracture occurs in bedrock and the overlying rock slides downslope. Characteristically, the rock breaks into small pieces of rubble during the slide. If a rockslide occurs on a steep slope, a turbulent mass of broken rock tumbles down the hillside.

A classic example of a rock avalanche occurred in the hills above the Gros Ventre River near the town of Kelly, Wyoming. Before the slide, a layer of sandstone rested on shale, which in turn was supported by a thick bed of limestone (Fig. 10–31). The sedimentary layers dipped 15° to 20° toward the river and parallel to the slope. Over time, the Gros Ventre River had undermined the sandstone, leaving the slope unsupported above the river. Only a trigger was needed to release the hillside. Snowmelt and heavy rains provided the trigger in the spring of 1925. Water seeped into the ground, saturating the soil and bedrock and increasing their weight. The water collected on the shale, forming a slippery surface. Finally the sandstone layer broke loose and began to slide over the shale. In a few moments, approximately 38 million cubic meters of rock tumbled into the valley. The sandstone crumbled into small pieces, forming a 70-meter-high natural dam across the Gros Ventre River. (For comparison, a 20-story building is about 70 meters high.) But a dam made of rockslide debris is generally

This huge mass of broken rock fell from the mountainside in the background and destroyed the town of Frank, Alberta, Canada.

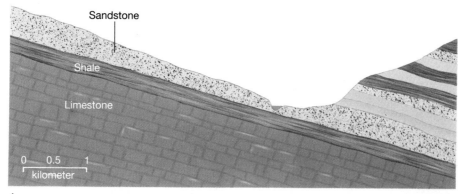

Sandstone

Shale

Limestone

0   0.5   1
kilometer

A

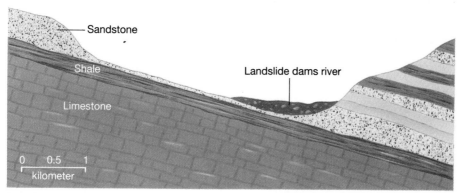

Sandstone

Shale

Limestone

Landslide dams river

0   0.5   1
kilometer

B

C

Figure 10–31   A profile of the Gros Ventre hillside (A) before and (B) after the slide. (C) About 38 million cubic meters of rock and soil broke loose and slid downhill during the Gros Ventre slide.   *(After W. C. Alden, ''Landslide and Flood at Gros Ventre, Wyoming.'' Transactions, AIME, 76 (1928), 348.*

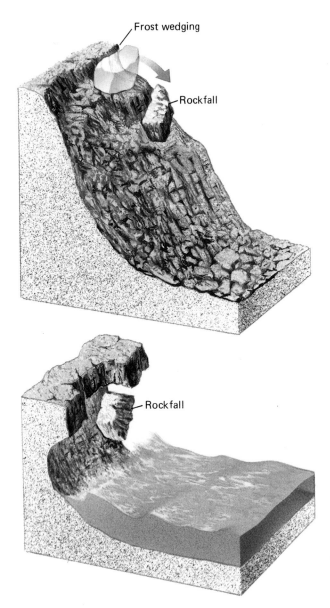

Figure 10–32 Rockfall commonly occurs in spring or fall when freezing water dislodges rocks from cliffs. Undercutting of cliffs by waves, streams, or construction can also cause rockfall.

unstable. Two years later the lake overflowed the dam, washing it out and creating a flood downstream that killed several people.

Fall

If a rock dislodges from a steep cliff, it falls rapidly under the influence of gravity. Recall from our discussion of weathering that when water freezes and thaws, the alternate expansion and contraction can dislodge rocks from cliffs and thus cause rockfall. Rockfall also occurs when a cliff is undermined. For example, if ocean waves under-

cut a steep shoreline (Fig. 10–32), rock above the waterline may tumble.

10.7 Mass Wasting Triggered by Earthquakes and Volcanoes

Mass wasting often occurs when a trigger mechanism releases an unstable slope. Water commonly triggers landslides because it adds weight, lubricates, and reduces shear strength. Earthquakes and volcanic eruptions also initiate mass wasting. In fact, in many cases, an earthquake or eruption itself causes comparatively little damage, but the resulting mass wasting is devastating. Consider the following two case histories.

Case History 1: The Madison River Slide, Montana

In August 1959, an earthquake occurred just west of Yellowstone National Park. This region is sparsely populated, and most of the buildings in the area are wood-frame structures that can withstand quakes. Thus, the earthquake itself caused little property damage and no loss of life. However, the ground motion triggered a massive rockslide from the top of Red Mountain above

Figure 10–33 The eruption of Nevado del Riuz in Colombia triggered a mudflow that buried the town of Armero, killing 20,000 people. *(Wide World Photos/Associated Press)*

the Madison River. About 30 million cubic meters of rock broke loose and slid into the valley below, burying a campground and killing 26 people. The slide's momentum carried it more than 100 meters up the mountain on the opposite side of the valley. The mass of falling rock compressed large quantities of air. Intense winds were generated as this air escaped from beneath the debris. The winds were so strong that a car was lifted off the ground and carried more than 10 meters before it crashed into nearby trees. The debris dammed the Madison River, forming a body of water behind the dam that was later named Quake Lake. Figure 10–19C shows the pile of debris and some of the damage left behind by this slide.

Case History 2: Nevado del Ruiz, Colombia

In November 1985, the volcanic mountain Nevado del Ruiz erupted in central Colombia. The eruption caused only minor damage, but heat from the ash and lava melted large quantities of ice and snow that lay on the mountain. The rushing water mixed with ash and debris on the mountainside, forming a mudflow that raced down gullies and stream valleys to the town of Armero, 48 kilometers from the mountain. Twenty thousand people were buried and killed in Armero, and additional loss of life and property damage occurred in a dozen or so other villages in nearby valleys (Fig. 10–33).

SUMMARY

•

Weathering is the decomposition and disintegration of rocks and minerals by mechanical and chemical processes. **Erosion** is the physical removal of weathered rock or soil by moving water, wind, glaciers, or gravity. After rock or soil has been eroded from the immediate environment, it may be **transported** great distances and is eventually **deposited**.

Mechanical weathering can occur by **frost wedging**, **salt cracking**, **abrasion**, **biological activity**, **pressure release**, or **thermal expansion and contraction**.

Chemical weathering occurs when chemical reactions decompose minerals. Some minerals react with oxygen to form new minerals. This process is known as **oxidation**. A few minerals are **soluble** in water. **Acids and bases** enhance solubilities of minerals. Rainwater is slightly acidic due to reactions between water and atmospheric carbon dioxide. Weathering of feldspar and other common minerals, except quartz, produces clay. Weathering releases quartz grains from rock to form sand. Limestone dissolves to form caverns as rainwater seeps through cracks in the rock.

Soil is the layer of weathered material overlying bedrock. **Sand, silt, clay,** and **humus** are common components of soil. Water **leaches** soluble ions downward through soil. Clay is also transported downward by water. The uppermost layer of soil, called the **O horizon**, consists mainly of litter and humus. The **A horizon** is the **zone of leaching**, and the **B horizon** is the **zone of accumulation**.

Five major factors control the character of soil. (1) **Parent rock** determines the composition and texture of young soils. (2) Soils change with **time**. (3) **Climate** influences soil. In dry climates, **pedocals** form. In pedocals, leached ions precipitate in the B horizon, where they accumulate and may form **caliche**. In moist climates, soluble ions are leached from the soil and **pedalfer** soils develop, with high concentrations of less-soluble aluminum and iron. **Laterite** soils form in very moist climates where leaching removes all of the soluble ions. (4) Rates of growth and decay of **organic matter** affect the characteristics of soil. (5) **Slope angle** and **aspect** affect soil development and erosion.

Mass wasting is the downhill movement of rock and soil under the influence of gravity. The stability of a slope and the severity of mass wasting depend on (1) steepness of the slope, (2) orientation and type of rock layers, (3) nature of unconsolidated materials, (4) climate and vegetation, and (5) earthquakes or volcanic eruptions.

Mass wasting falls into three categories: flow, slide, and fall. During **flow**, a mixture of rock, soil, and water moves as a viscous fluid. **Creep** is a slow type of flow that occurs at a rate of about 1 centimeter per year. A **debris flow** consists of a mixture in which more than half the particles are larger than sand. **Earthflows** and **mudflows** are mass movements of predominantly fine-grained particles mixed with water. Earthflows have less water than mudflows and are therefore less fluid. **Solifluction** is a type of flow that occurs when water-saturated soil moves over permafrost.

Slide is the movement of a coherent mass of material. **Slump** is a type of slide in which the moving mass travels on a concave surface. In a **rockslide**, a newly detached segment of bedrock slides along a tilted bedding plane or fracture.

Fall occurs when particles fall or tumble down a steep cliff.

Earthquakes and volcanic eruptions trigger devastating mass wasting. Damage to human habitation can be averted by proper planning and construction.

KEY TERMS

●

Weathering *245*
Erosion *245*
Mechanical
    weathering *246*
Chemical weathering *246*
Frost wedging *247*
Talus slope *248*
Salt cracking *248*
Abrasion *248*
Pressure release
    fracturing *250*

Exfoliation *250*
Thermal expansion *250*
Oxidation *251*
Hydration *252*
Regolith *253*
Soil *253*
Loam *253*
Litter *253*
Humus *254*
Soil nutrient *254*
Soil horizons *254*

Topsoil *254*
Leaching *254*
Capillary action *255*
Pedocal *256*
Caliche *256*
Pedalfer *256*
Laterite *257*
Mass wasting *259*
Angle of repose *262*
Flow *262*
Slide *264*

Fall *264*
Creep *265*
Debris flow *266*
Mudflow *266*
Earthflow *266*
Permafrost *266*
Solifluction *267*
Slump *268*
Rockslide *268*
Rock avalanche *268*

REVIEW QUESTIONS

●

1. Explain the differences among weathering, erosion, transportation, and deposition.

2. Explain the differences between mechanical weathering and chemical weathering. Give examples of each.

3. Explain frost wedging. What landforms are created by frost wedging?

4. Explain how thermal expansion can fracture a rock.

5. What are the components of healthy soil? What is the function of each component?

6. Characterize the four major horizons of a mature soil.

7. List the five soil-forming factors and briefly discuss each one.

8. Imagine that soil forms on granite in two regions, one wet and the other dry. Will the soil in the two regions be the same or different? Explain.

9. Explain how soils formed from granite change with time.

10. What is a laterite soil? How does it form? Why is it unsuitable for agriculture?

11. Explain how a small amount of water might increase slope stability, whereas mass wasting might occur on the same slope during heavy rainfall or rapid snowmelt.

12. Discuss the differences among flow, slide, and fall. Give examples of each.

13. Compare and contrast creep, debris flow, and mudflow.

14. Why is solifluction more likely to occur in the Arctic than in temperate or tropical regions?

15. Compare and contrast slump and rockslide.

16. Explain how trees are bent but not killed by slump. How are trees affected by rockslide?

DISCUSSION QUESTIONS

●

1. What process is responsible for each of the following observations and phenomena? Is the process a mechanical or chemical change? (a) A board is sawn in half. (b) A board is burned. (c) A cave forms when water seeps through limestone. (d) Calcite precipitates from a hot underground spring. (e) Meter-thick sheets of granite peel off a newly exposed pluton. (f) In mountains of the temperate region, rockfall is more common in the spring than in midsummer.

2. Arctic regions are cold most of the year, and summers are short. Therefore, decomposition of organic matter is slow. In the temperate regions, decay is much more rapid. How does this difference affect the fertility of the soils?

3. Have your class debate the relative merits of chemical and organic fertilization. If you live in a rural region, call your local county agricultural service (such as the Agricultural Stabilization and Conservation Service) and the Soil Conservation Service and ask their opinions. If possible, interview local farmers and record their opinions.

4. Show how time interacts with the other soil-forming factors listed in the text.

5. Explain how exfoliation might lead to mass wasting.

6. The Moon is considerably smaller than the Earth, and therefore its gravitation is less. It has no atmosphere and therefore

no rainfall. The interior of the Moon is cool, and thus it is geologically inactive. Would you expect mass wasting to be a common or an uncommon event in mountainous areas of the Moon? Defend your answer.

7. Explain how wildfires might contribute to mass wasting.

8. What types of mass wasting (if any) would be likely to occur in each of the following environments? (a) A very gradual (2 percent) slope in a heavily vegetated tropical rainforest. (b) A steep hillside composed of alternating layers of conglomerate, shale, and sandstone in a region that experiences distinct dry and rainy seasons. The dip of the rock layers is parallel to the slope. (c) A steep hillside composed of clay in a rainy environment in an active earthquake zone.

9. Identify a hillside in your city or town that might be unstable. Using as much data as you can collect, discuss the magnitude of the potential danger. Would the earth or rock movement be likely to affect human habitation?

10. Explain how the mass wasting triggered by earthquakes and volcanoes can have more serious effects than the earthquake or volcano itself. Is this always the case?

Streams and Ground Water

11

About 1.3 billion cubic kilometers of water exist at the Earth's surface. Of this huge quantity, however, 97.5 percent is salty seawater, and just under 1.8 percent is frozen into the great ice caps of Antarctica and Greenland. Only about 0.65 percent is fresh water in streams, lakes, and underground reservoirs.

All organisms that live on land need fresh water to survive. Rivers and streams serve as arteries for transportation, can be dammed to produce energy, and are used for irrigation and industry; freshwater fisheries yield protein for human consumption. In addition, although the Earth's landforms are initially created by tectonic processes, flowing water is the most important agent in modifying them.

● Havasu Creek, Grand Canyon, Arizona.

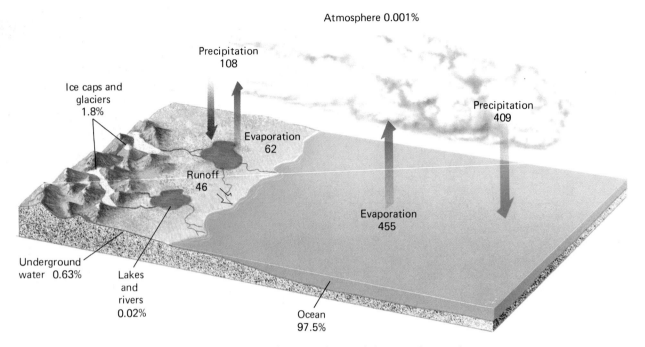

Atmosphere 0.001%

Precipitation
108

Ice caps and
glaciers
1.8%

Precipitation
409

Evaporation
62

Runoff
46

Evaporation
455

Underground
water 0.63%

Lakes
and
rivers
0.02%

Ocean
97.5%

Figure 11–1 The hydrologic cycle shows that water is constantly recycled among the sea, the atmosphere, and land. Numbers are thousands of cubic kilometers of water transferred each year. Percentages are proportions of total global water in different portions of the Earth's surface.

Fresh water is most conspicuous in streams and lakes, but about 30 times more fresh water is hidden just below the Earth's surface, saturating soil and bedrock as **ground water**. Ground water pumped from wells provides drinking water for more than half of the population of North America and is a major source of water for irrigation and industry. Ground water also creates caves and other landforms as it seeps through rock and soil.

11.1 The Water Cycle

The Earth's water moves constantly. It evaporates from the sea, falls as rain, and flows over land as it returns to the ocean. The constant circulation of water among sea, land, and the atmosphere is called the **hydrologic cycle** or the water cycle (Fig. 11–1).

Water **evaporates** from sea and land to become water vapor in the atmosphere. Most atmospheric water vapor evaporates from the oceans, and a small amount evaporates from land and inland waterways. Water also evaporates directly from plants as they breathe in a process called **transpiration**.

Precipitation includes all processes in which atmospheric moisture returns to the Earth's surface—rain, snow, hail, and sleet. **Runoff** is the water that flows back to the oceans over the surface of the land.

STREAMS

11.2 Characteristics of a Stream

When rain falls on land, much of it soaks directly into the ground, although during a hard rain some may run over the soil surface. Sooner or later, both the ground water and surface runoff seep into channels to flow over the Earth's surface. Water flowing in a channel has a variety of names, such as creek, brook, rivulet, stream, and river. To avoid confusion, the term **stream** is used for all water flowing in a channel, regardless of size. The term **river** is commonly used for any large stream fed by smaller streams called **tributaries**.

Normally a stream flows in its **channel**. The floor of the channel is called the **bed**, and the sloping edges of the channel are the **banks**. When rainfall is heavy or when snow melts rapidly, a flood may occur. During a flood, a stream overflows its banks and water covers the adjacent land, called a **flood plain**.

Stream Flow

A slow stream flows at 0.25 to 0.5 meters per second (1 to 2 kilometers per hour), whereas a steep, flooding stream may race along at about 7 meters per second (25 kilometers per hour). Three factors control stream velocity: (1) the gradient of the stream bed; (2) the flow, or discharge, of the stream; and (3) the channel shape.

Gradient

Obviously, if all other factors are equal, water will flow more rapidly down a steep slope than a gradual one. **Gradient** is the vertical drop of a stream within a certain horizontal distance. The lower Mississippi River has a shallow gradient and drops only 10 centimeters per kilometer. In contrast, a tumbling mountain stream may drop 40 meters per kilometer.

Discharge

Discharge is the volume of water flowing downstream per unit time. It is measured in cubic meters per second (m^3/sec). Discharge can be estimated by multiplying the cross-sectional area of a stream by its velocity. The velocity of a stream increases when discharge increases. Thus, streams flow faster during flood, even though the gradient is unchanged.

The largest river in the world is the Amazon, with a discharge of 150,000 m^3/sec. In contrast, the Mississippi River, the largest in North America, has a discharge of about 17,500 m^3/sec, approximately one-tenth that of the Amazon.

The discharge of a stream can change dramatically from month to month or even during a single day. For example, the Selway River, a mountain stream in Idaho, has a discharge of 100 to 130 m^3/sec during early summer,

The Selway River, Idaho.

when the mountain snowpack is melting rapidly. During the dry season in late summer, the discharge drops to about 20 m^3/sec (Fig. 11–2). In extreme cases, discharge can vary almost instantaneously. A desert streambed may be completely dry in midsummer. But a sudden thunderstorm can send a wall of water rushing violently down the streambed. After a few hours, the flow may die off to a gentle trickle, as the stream dries up again.

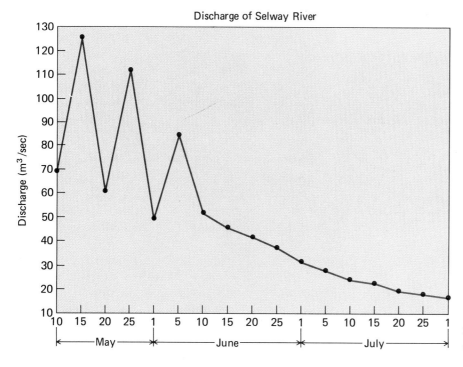

Figure 11–2   Discharge of the Selway River in northern Idaho during spring and early summer, 1988. The discharge ranges from 10 m^3/sec to 15 m^3/sec during the rest of the year.   *(U.S. Forest Service)*

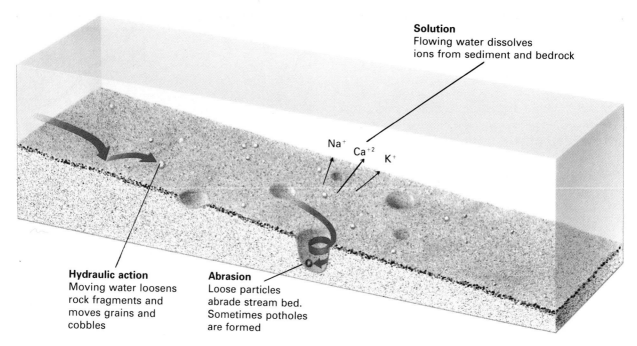

Solution
Flowing water dissolves
ions from sediment and bedrock

Na$^+$ Ca^{+2} K$^+$

Hydraulic action
Moving water loosens
rock fragments and
moves grains and
cobbles

Abrasion
Loose particles
abrade stream bed.
Sometimes potholes
are formed

Figure 11–3   A stream can weather and erode its bed and banks by hydraulic action, abrasion, and solution.

Channel Shape

Flowing water is slowed by friction between the water and the streambed and banks. Consequently, water flows more slowly near the banks than near the center of a stream. If you paddle a canoe down a straight stream

Floods on the Rogue River, Oregon, have eroded the flood plain and exposed the roots of this tree.

channel, you move faster when you stay away from the banks. The amount of friction depends on the roughness of the channel. Boulders on the banks or in the streambed increase friction and slow a stream down, whereas it can flow more rapidly if the bed and banks are smooth.

11.3 Stream Erosion

A stream erodes soil and bedrock, transports sediment, and eventually deposits the sediment. These processes shape the Earth's surface by cutting stream valleys, creating flood plains, and building deltas in oceans and lakes. Ultimately, stream erosion can level an entire mountain range.

Processes of Stream Erosion

A stream erodes its bed and banks by three different processes: hydraulic action, abrasion, and solution (Fig. 11–3).

Hydraulic action is the ability of flowing water to dislodge fragments of rock and grains of sediment. To demonstrate hydraulic action, aim a garden hose into bare dirt. In a short time the water erodes a small hole, moving soil and small pebbles. Similarly, a stream can dislodge solid particles in its bed and banks, especially when the current is moving swiftly.

Abrasion is the mechanical wearing away of rock by a stream. Water itself is not abrasive, but sand and

Figure 11–4   Circulating pebbles abrade a streambed to form potholes.

As explained in Chapter 10, many minerals dissolve in water, a process called **solution**. Streams dissolve rocks and minerals in the streambed, and then carry away the dissolved ions.

The Ability of a Stream to Erode and Transport

A rapidly flowing stream has more energy to pick up and transport sediment than a slow stream. The **competence** of a stream is a measure of the largest particle it can carry. A fast-flowing stream can carry cobbles and boulders in addition to small particles. A slow stream may carry only silt and clay.

The **capacity** of a stream is the total amount of sediment it can carry past a point in a given amount of time. Capacity is proportional to both current speed and discharge. Thus, a fast stream has a greater capacity than a slow stream of equal size, and a large stream has a greater capacity than a small one flowing with the same velocity.

Because the ability of a stream to erode and carry sediment is proportional to its velocity and discharge, most erosion and sediment transport occur every year during the few days or weeks when the stream is in flood and flowing most rapidly. Relatively little erosion and sediment transport occur during the remainder of the year. To see this effect for yourself, look at any stream during low water. It will most likely be clear, indicating little erosion or sediment transport. Look at the same stream later, when it is flooding after a heavy rain. It will prob-

other sediment carried by a stream wear away boulders and bedrock in the streambed. Thus, a stream loaded with sediment can be thought of as flowing sandpaper. As the stream-transported sediment abrades rock in the streambed, the sediment itself becomes worn and rounded. For this reason, most stream-transported particles are well rounded. Abrasion can produce **potholes** and other rounded depressions in a stream bed. If cobbles are caught in a small hollow in a rocky stream bed, the current swirls them around and around, enlarging the hollow (Fig. 11–4).

The Colorado River flowing through Grand Canyon, Arizona, becomes muddy during high water.

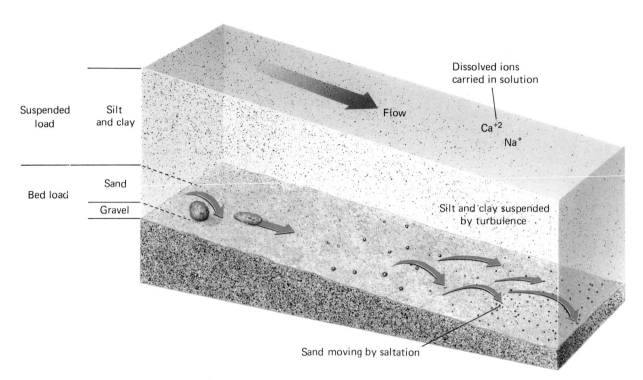

Figure 11–5 A stream transports sediment in three ways. It carries dissolved ions in solution, silt and clay in suspension, and larger particles as bed load.

ably be muddy and dark, indicating that much sediment is being eroded from the bed and banks and is being carried by the current.

11.4 Sediment Transport

After a stream erodes grains of rock or soil, the flowing water may then carry the sediment downstream. A stream transports sediment in three forms: dissolved load, suspended load, and bed load (Fig. 11–5).

Dissolved Load

A stream's **dissolved load** is the ions dissolved in the stream water. Dissolved ions travel wherever the water goes. A stream's ability to carry dissolved ions depends mainly on its discharge and its chemistry, not its velocity. Thus, dissolved ions exist even in lake and ocean waters; that is why the seas and some lakes are salty.

Suspended Load

Even a slow stream can carry fine sediment. If you shake up loamy soil in a jar of water and then let the jar sit, the sand grains settle quickly. But the smaller silt and clay particles remain suspended in the water as **suspended load**, giving it a cloudy appearance. Suspension is a mode of transport in which water turbulence keeps fine particles mixed with the water and prevents them from settling to the bottom. Clay and silt are small enough that even the slight turbulence of a slow stream keeps them in suspension. A rapidly flowing stream can carry sand in suspension.

Bed Load

Cobbles and boulders are too heavy to be carried in suspension by a stream. However, a fast current can roll or drag these large particles along the streambed as **bed load**. During a flood, when stream velocity is highest, even large boulders can be transported in this way. Sand also moves as bed load. If the stream velocity is sufficient, however, sand grains bounce along in a series of short leaps or hops. This type of movement is called **saltation** (Fig. 11–5).

The world's muddiest rivers, the Yellow River in China and the Ganges River in India, both carry more than 1.5 *billion* tons of sediment to the ocean every year. The sediment load of the Mississippi River is about 450 million tons per year. Most rivers carry the largest proportion of their sediment in suspension, a smaller proportion in solution, and the smallest proportion as bed load.

Earth Science and the Environment

Salinization

As a rough average, most of the world's large rivers contain about 110 to 120 parts per million (ppm) dissolved ions (100 ppm is equal to one hundredth of one percent). Although most rivers do not taste salty, they carry tremendous amounts of material to the oceans in this manner. When river water is used to irrigate crops, much of it evaporates, and the dissolved ions precipitate as salts in the soil. Over a period of years, the salts accumulate in a process called **salinization**. If desert or semi-desert soils are irrigated for long periods of time, they can become so salty that crops cannot grow in them.

The great civilizations of Mesopotamia, where much of Western art, science, and literature originated, were built in a desert between the Tigris and Euphrates rivers of western Asia. The agriculture of this region depended on an extensive irrigation system, one of the great achievements of early civilization. Today much of this once fertile region is barren, eroded, and desolate. Archaeologists dig up ancient irrigation canals, farming tools, and grinding stones in the desert. This ecological catastrophe was the outcome of salinization due to irrigation.

In the United States, salinization is severe in regions such as the San Joaquin Valley in California, where intensive irrigation is used to convert desert and semidesert to rich farmland. In some soils, enough calcium carbonate has precipitated from irrigation water to form a rock-hard layer called **caliche**. If the land is to be used, the caliche must be broken apart and removed at great expense with heavy machinery. Caliche development can be slowed by removing excess salty water with a drainage system. However, it is expensive to install tiles and pipes to collect and dispose of salty water. A better technique for preventing salinization may be to use less irregation water by moistening the roots of individual plants with perforated pipes, rather than sprinkling an entire field.

11.5 Sediment Deposition

A rapidly flowing stream can transport all sizes of particles, from clay to boulders. When the current slows down, it loses its ability to transport the largest particles and deposits them in the streambed. As its velocity continues to decrease, the stream deposits progressively finer and finer particles, thus **sorting** sediment according to size. A stream may slow down as it flows from mountains onto a valley floor. In this case, it deposits boulders and cobbles in the steep upper reaches of its channel, and sand and mud farther downstream.

A stream flows more rapidly when it is in flood than at low water, so its velocity diminishes as the flood wanes. In this case, it may deposit boulders and cobbles near the base of a single sediment layer, and mud in the upper part of the same layer.

Stream deposits fall into three categories. (1) **Channel deposits** form in the stream channel itself. (2) **Alluvial fan** and **delta deposits** form where the stream slows abruptly. (3) **Flood plain deposits** accumulate on the flood plain adjacent to a stream channel.

Channel Deposits

A **bar** is an elongate mound of sediment in a stream channel. Bars form when a stream is no longer able to carry the amount of sediment it has been carrying.

A stream sorts its sediment, depositing boulders of the same size together.

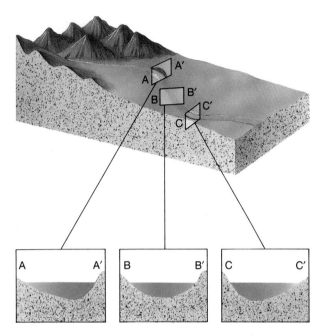

Figure 11–6   A straight stream flows most rapidly near its center, but a winding stream flows most rapidly near the outer parts of its curves. The shaded zone in each cross section shows the area with fastest flow.

Imagine a winding stream such as that in Figure 11–6. Now consider the cross section A-A′ through one of the stream bends. The water on the outside of the curve moves faster than the water on the inside. This difference occurs because the water on the outside must travel farther and must therefore move faster to maintain the flow. This is analogous to a line of musicians marching around a corner in a parade. The marchers on the outside of the turn must travel farther, and thus must move faster, than those on the inside.

A point bar forms on the inside bend of a stream.   *(Amos Turk)*

The faster water on the outside of the bend has a greater erosive ability, and therefore a stream erodes its bank on the outside of a curve. At the same time, the slower water on the inside of the bend deposits sediment. Because such deposits are located on the inside point of a bend, they are called **point bars**.

As a result of this uneven erosion and deposition, the initial bend becomes accentuated. When the current

Meanders in a mountain stream in western Montana.

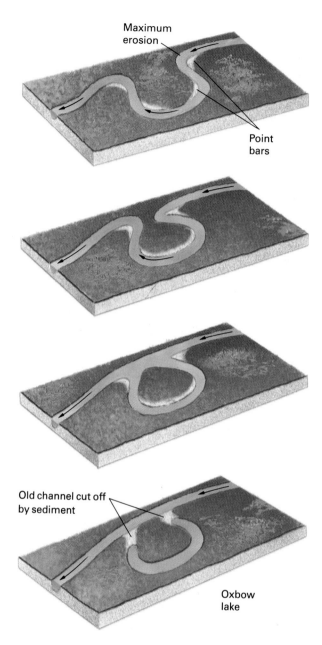

Figure 11–7 An oxbow lake forms when a stream erodes through a meander neck.

bounces off one bend, it is deflected to the other side, where a second curve develops downstream from the first. Eventually, a series of bends called **meanders** forms (Fig. 11–7). As erosion continues, the curves may become so pronounced that the outside of one meander approaches another. Given enough time, the stream will cut across the gap and create a new channel, abandoning a portion of its old bed. The old meander is isolated from the flowing water to become an **oxbow lake**.

When a stream has a greater supply of sediment than it can carry, it deposits some of the sediment in its chan-

nel. The sandy and gravelly deposits that form within stream channels are **mid-channel bars**. Bars are transient features of a stream. They may be deposited in one year and eroded in the next. If a river is used extensively for shipping, tugboat captains must be careful because bars form, disappear, and move. In industrialized nations, river channels are charted frequently, and detailed maps of the bottoms are redrawn regularly.

A **braided stream** is one that flows in many shallow, interconnecting channels (Fig. 11–8). It develops where more sediment is supplied to a stream than it can carry. The excess sediment is deposited in the channel. The accumulating sediment gradually forces the stream to overflow its banks and erode new channels. As a result, a braided stream flows simultaneously in several channels, and shifts back and forth across its flood plain as old channels fill and new ones form.

Braided streams are common in both deserts and glacial environments because both have abundant sediment. In a desert, large amounts of sediment are eroded and transported during occasional heavy rains because little or no vegetation exists to prevent erosion. Glacial erosion supplies huge amounts of sediment to streams flowing from a glacier.

Alluvial Fans and Deltas

A stream can slow down suddenly and deposit much of its sediment in two types of environments. If a stream flows from a steep mountain front onto a flat plain, its velocity decreases abruptly. Here it deposits particles of all sizes, from cobbles to fine silt, in a fan-shaped landform called an **alluvial fan**. Alluvial fans are common

Figure 11–8 The braided channel of the Chaba River in the Canadian Rockies.

Figure 11–9 Alluvial fans. *(Ward's Natural Science Establishment, Inc.)*

in many arid and semiarid mountainous regions (Fig. 11–9).

A stream also slows abruptly where it enters still water of a lake or ocean. Here its sediment settles out to form a nearly flat plain called a **delta**. Most gravel has already been deposited farther upstream, so deltas consist mostly of sand, silt, and clay. Part of the delta lies above water level, and the remainder consists of a shallow underwater plain. Deltas are commonly fan-shaped, resembling the Greek letter "delta," Δ.

Deltas are shifting, changing environments. As more sediment accumulates, old channels fill and are aban-

doned while new channels develop, as in a braided stream. As a result, the stream splits into many channels called **distributaries** (Fig. 11–10). A large delta may spread out in this manner until it covers thousands of square kilometers.

Figure 11–11 shows changes in the Mississippi delta over the past 5000 to 6000 years. During this period the main channel has changed position seven times. What will happen in the future? A river is dynamic; in its natural state it continues changing. If the Mississippi River were left alone, it would probably abandon the lower 500 kilometers of its present path and cut into the channel of the Atchafalaya River to the west. However, the mouth of the Mississippi is heavily industrialized, and it is impractical to allow the river to change its course, leaving old shipping lanes and wharves high and dry. Therefore, engineers have built great systems of levees to stabilize the channels.

Floods and Flood Plains

Large rivers flood as a result of continuous, heavy rains, often augmented by spring snowmelt in the mountains. For example, in 1937, heavy rain fell almost continuously for 25 days in the Ohio River drainage. In response, the river rose 12 meters above its normal level.

Mountain and desert streams often flood as a result of rapid snowmelt or intense thunderstorms. In 1976 a series of summer storms saturated soil and bedrock near Rocky Mountain National Park, northwest of Denver. Then a large thunderhead dropped 19 centimeters of rain in 1 hour in the headwaters of Big Thompson Canyon. The Big Thompson River flooded, filling its narrow valley with a deadly, turbulent wall of water. Some people tried to escape by driving toward the mouth of the canyon, but traffic clogged the two-lane roadway and trapped

Figure 11–10 A delta grows larger with time. *(After Ward's Natural Science Establishment, Inc., Rochester, NY).*

 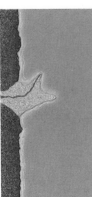 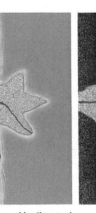 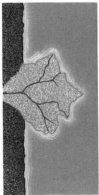 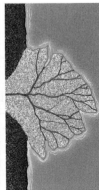

Young delta straight channel

As the delta grows the channel is diverted to one side and then the other

Distributaries form

Mature delta

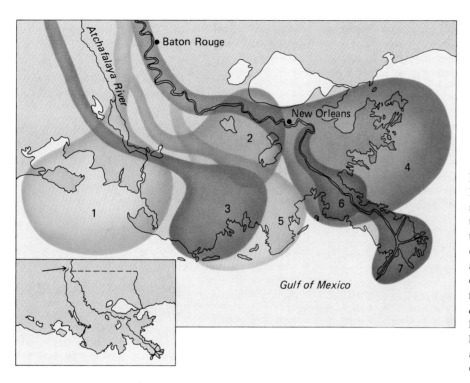

Figure 11–11 The Mississippi River has flowed into the sea by seven different channels during the past 6000 years. As a result, the modern delta is composed of seven smaller deltas formed at different times. The oldest delta is numbered 1, and the current delta is 7. *(After C. R. Kolb and J. R. Van Lopik,* Depositional Environments of the Mississippi River Deltaic Plain, *p. 22. Copyright © 1966 by the Houston Geological Society.)*

motorists in their cars, where they drowned. A few residents tried to escape by scaling the steep canyon walls; some of them were caught by the rising waters. Within a few hours, 139 people had died and five were missing. By the next day, the flood was over.

A satellite photo of the Mississippi River delta. *(NASA)*

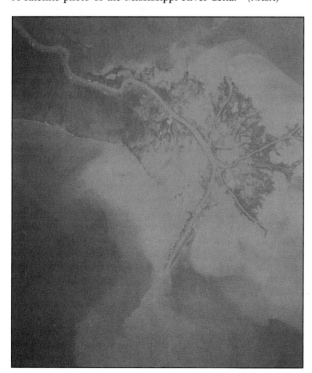

Many streams flood regularly, some every year. A given stream floods to a higher level in some years than in other years. A "100-year flood" is the largest flood that occurs in a given stream an average of once every 100 years. A 10-year flood is the largest that occurs on the average of once every 10 years. In any stream, small floods are more common than large ones. For example, a given stream may rise 7 meters above its banks during a 100-year flood, but only 2 meters during a 10-year flood.

The aftermath of the Big Thompson Canyon flood.

Thus, a 100-year flood is higher and larger, but less frequent, than a 10-year flood.

Does a "100-year flood" mean that if a large flood occurs in 1993, one of equal size will not occur until 2093? No, not at all. The 100-year cycle is a measure of probability: the chance of a 100-year flood occurring in any given year is 1 in 100. Think of a roulette wheel with 100 slots. Imagine that one is marked "F" for a 100-year flood, and 99 are marked "NF" for no flood (or just a small flood). Now you spin the wheel. The chance on *any* spin is 1 out of 100 that you will land on F. If you land on F on one spin, you might land on F the very next try, or you might spin 200 times before it happens again. It is a game of chance.

As a stream rises to flood stage, both its discharge and velocity increase. Therefore, it erodes its bed and

Figure 11–12 Natural levees forms as a flooding stream deposits sand and silt on its banks. Silt and clay accumulate on the flood plain.

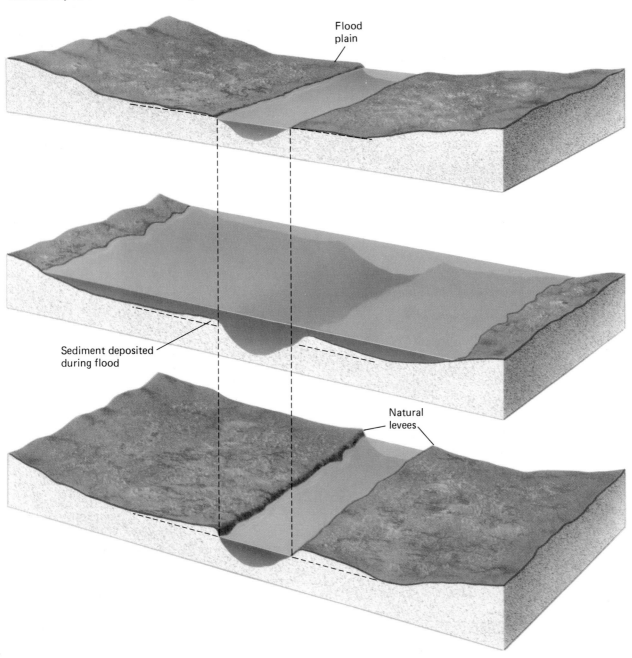

banks. When a stream floods, water flows out of its channel. The portion of a valley covered by water during a flood is called the flood plain. The current slows abruptly where water leaves the stream channel to flow onto the flood plain. Because it slows down so quickly, the current deposits sand and silt on the banks of the stream. This sediment forms ridges called **natural levees** at the margins of the channel (Fig. 11–12).

Floodwater farther out on the flood plain carries mostly clay and silt. This finer sediment accumulates on flood plains to form fertile soils.

11.6 Downcutting and Base Level

A stream creates a valley by eroding bedrock and carrying off the sediment. The flowing water erodes both downward into the streambed and laterally against the banks. Downward erosion is called **downcutting**. How deeply can a stream erode its channel? How deep can a valley become? The **base level** of a stream is the deepest level to which it can erode its bed. For most streams,* the lowest possible level of downcutting is sea level, which is called the **ultimate base level**. This concept is straightforward. Water only flows downhill. If a stream were to cut its way down to sea level, it would stop flowing and hence would no longer be able to erode its bed or banks.

In addition to ultimate base level, a stream may have a number of **local**, or **temporary, base levels**. For example, where a stream flows into a lake, the current stops and erosion ceases because the stream has reached a temporary base level. A layer of rock that resists erosion may also establish a temporary base level because it resists downcutting and flattens the stream gradient. Thus, the stream slows down and erosion decreases. The top of a waterfall is an example of a temporary base level established by resistant rock. A waterfall is a transient feature of a stream. Over geologic time, the stream erodes and weathers the resistant rock and wears down the cliff (Fig. 11–13).

If a stream has numerous temporary base levels, it erodes its bed in the steep places where flow is rapid and deposits sediment in the low-gradient stretches where it flows more slowly (Fig. 11–13B). Over time, erosion and deposition smooth out the irregularities in the gradient. The resultant **graded stream** has a smooth concave profile (Fig 11–13C). An idealized graded stream is in equilibrium with its sediment load. Its gradient, discharge, and current velocity are balanced so that the amount of sediment it *can* carry is identical to the amount supplied to it. Thus, a perfectly graded stream transports all the sediment supplied to it but does not erode its banks or

*We say "for most streams" because a few empty into valleys that lie below sea level.

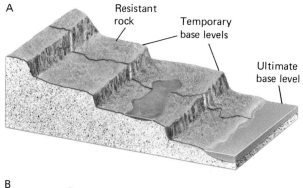

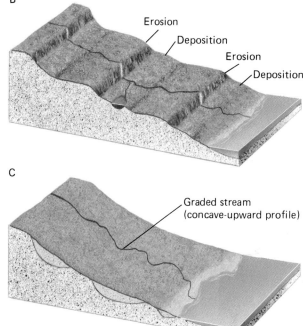

Figure 11–13 An ungraded stream (A) has many temporary base levels. With time, the stream smooths out the irregularities (B) to develop a graded profile (C).

deposit sediment. An idealized graded stream such as this does not actually exist in nature, but many streams come close.

Mountain streams usually downcut rapidly compared to the rates of lateral erosion. As a result, they cut relatively straight channels with steep-sided, V-shaped valleys. In contrast, low-gradient streams are closer to base level, and their downcutting is therefore slower. They mainly erode laterally, undercutting their valley walls, and mass wasting becomes predominant. Such streams carve out wide valleys with flat bottoms and broad flood plains. Meanders and oxbow lakes are common on their flood plains.

The V-shaped canyon of the Yellowstone River, Yellowstone National Park. *(Larry Davis)*

11.7 Drainage Basins

Only a dozen or so major rivers flow into the sea along the coastlines of the United States (Fig. 11–14). Each is fed by a number of tributaries, and in turn, the tributaries are fed by smaller streams. Adjacent river systems are separated by mountain ranges or other raised areas. These high places are called **drainage divides**. The region that is ultimately drained by a single river is called a **drainage basin**. It can be large or small and is bounded by drainage divides. For example, the Rocky Mountains separate the Colorado and Columbia drainage basins to the west from the Mississippi and Rio Grande basins to the east.

In the most common type of drainage basin, the pattern of tributaries resembles veins in a leaf. This arrangement is called a **dendritic drainage pattern** (Fig. 11–15A). Tributaries join the main stream at V-shaped junctions in dendritic drainage systems. You can determine the direction of flow because the V usually points downstream. The next time you fly in an airplane, look out the window and try to determine the directions of stream flow in the drainage systems below. Dendritic patterns occur in regions where bedrock is relatively uniform and therefore streams take the shortest route downslope.

Figure 11–14 Major drainage basins of the United States.

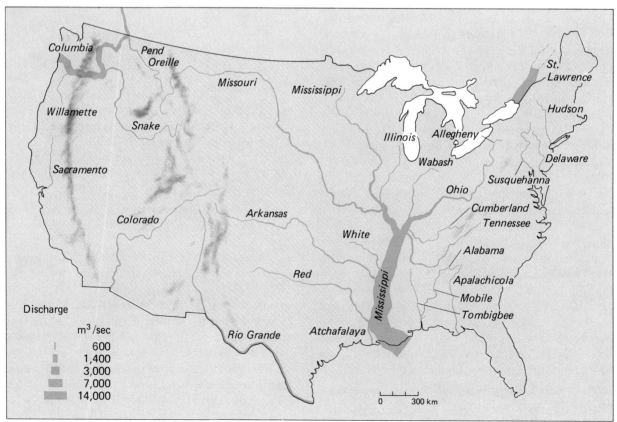

Map view

Perspective view

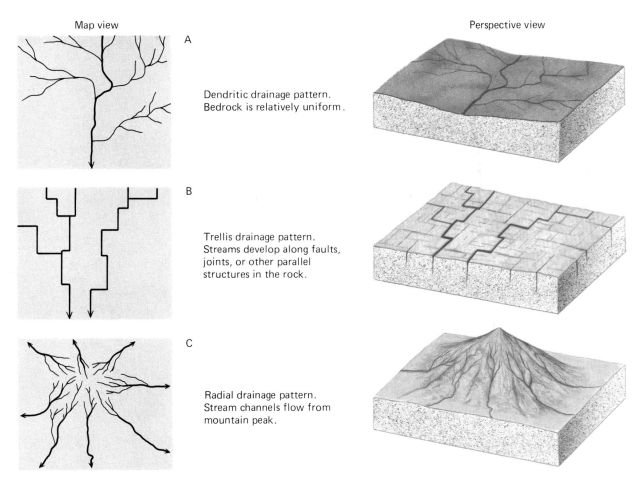

A

Dendritic drainage pattern. Bedrock is relatively uniform.

B

Trellis drainage pattern. Streams develop along faults, joints, or other parallel structures in the rock.

C

Radial drainage pattern. Stream channels flow from mountain peak.

Figure 11–15 Drainage patterns. (A) Dendritic. (B) Trellis. (C) Radial.

Bedrock is not uniform in some regions, however. For example, Figure 11–16 shows a layer of sandstone lying over granite. The rocks have been faulted and tilted. As a result, outcropping bands of easily eroded sandstone alternate with bands of resistant granite. Streams follow the softer sandstone rather than cutting across the hard granite, forming straight, parallel channels. A drainage pattern characterized by parallel channels intersected at right angles by short tributaries is called a **trellis pattern** (Fig. 11–15B). Faulting often crushes rock in the fault zone, making it relatively easy to erode. Thus, streams commonly follow fault zones. A trellis pattern can also develop where bedrock is crisscrossed by a series of faults or joints that intersect at right angles. A **radial** drainage pattern forms where a number of streams flow from a mountaintop (Fig. 11–15C).

Figure 11–16 Parallel faults and contrasting rock types may cause a parallel drainage pattern to develop.

11.8 The Evolution of Valleys and Rejuvenation

According to a model popular in the first half of this century, streams erode mountain ranges and create landforms in a particular sequence. At first, the streams cut steep V-shaped valleys. Over time, lateral erosion widens the valleys into broad flood plains. Eventually, as the streams continue to erode laterally, flood plains grow wider and the entire landscape flattens, forming a large, featureless plain called a **peneplain** (Fig. 11–17).

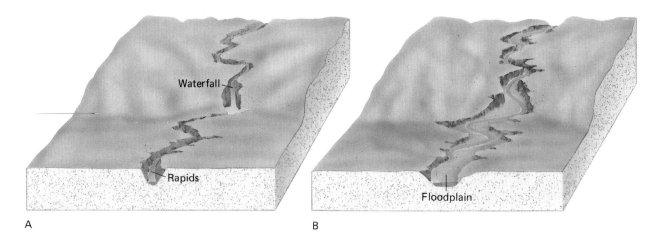

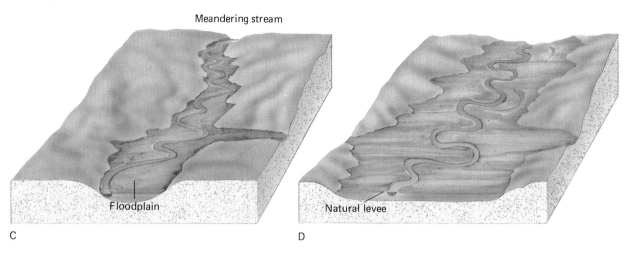

Figure 11–17   Stream erosion would widen and flatten the land to form a peneplain if no tectonic changes occurred.   *(After Ward's Natural Science Establishment, Inc., Rochester, NY).*

This model seems to make good sense based on what we have learned of stream erosion, but it is invalid because it tells only half the story. Streams do continuously erode the landscape, flattening mountains and widening flood plains. But at the same time, tectonic activity may uplift the land and interrupt the simple, idealized sequence.

Consider the Himalayas. Today, streams, glaciers, and mass wasting are eroding deep valleys. The huge amounts of sediment carried from the Himalayas to the sea by the Indus and Ganges rivers are evidence of this erosion. However, at the same time, the lithospheric plate carrying India continues to ram into Asia, uplifting the mountain chain. Thus uplift and erosion occur simultaneously and the mountain peaks are rising more rapidly than erosion is wearing them down.

A stream is **rejuvenated** when any change increases its erosive ability. Rejuvenation can result from tectonic forces uplifting the land, a drop in base level, or an increase in regional rainfall.

Incised meanders form the goosenecks of the San Juan River, Utah.   *(Tom Till)*

Incised Meanders and Stream Terraces

Imagine a stream meandering across a broad flood plain. Now suppose that tectonic forces uplift the entire plain, increasing the stream's gradient and velocity. As a result, the stream begins to cut downward into its bed. Eventually the stream cuts downward *below* its original flood plain, and if the stream keeps its meandering course as it erodes into its bed, **incised meanders** develop (Fig. 11–18). The incised stream abandons its old flood plain well above its new channel, and the abandoned flood plain is called a **stream terrace** (Fig. 11–19). Both incised meanders and stream terraces reflect rejuvenation.

In the preceding example, rejuvenation resulted from tectonic uplift, but it could equally well have resulted from a climate change. For example, if climate becomes wetter, the stream's discharge may increase. As a result, the stream may begin to downcut into its bed, forming incised meanders and terraces. Lowering of base level also causes rejuvenation. For example, a flood plain may form when a stream flows on a temporary base level supported by a layer of resistant rock. When the stream finally erodes through this resistant layer, its gradient may increase and it may start downcutting rapidly.

Streams That Flow Through Mountain Ranges

In some regions, a stream may flow right through a mountain range, ridge, or high plateau. Why doesn't it flow around the mountains, rather than cutting directly through them? Again, tectonic activity may be responsible. Imagine a stream flowing across a plain. If tectonic forces uplift the plain, but the uplift is slow enough that the stream can cut through the rising bedrock, then the stream will retain its original course through the newly formed mountains (Fig. 11–20). In this case the stream is said to be **antecedent** because it existed before the plateau rose. The Colorado River cut its way through more than 1600

Figure 11–19 Formation of terraces. (A) A stream has formed a broad flood plain. (B) Tectonic uplift or climatic change causes the stream to erode its bed. As the stream cuts downward, the old flood plain becomes a terrace above the new stream level. (C) A new flood plain forms at the lower level.

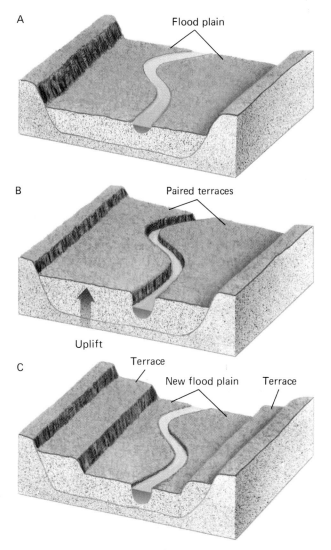

Figure 11–18 If a valley with a meandering stream is raised by tectonic uplift, the meanders may become incised into the rising flood plain.

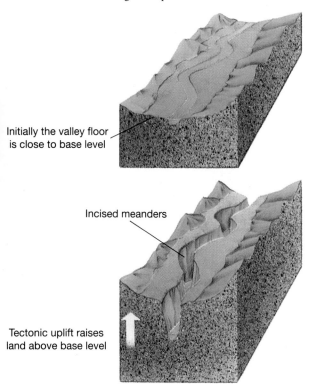

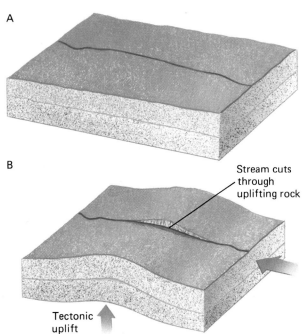

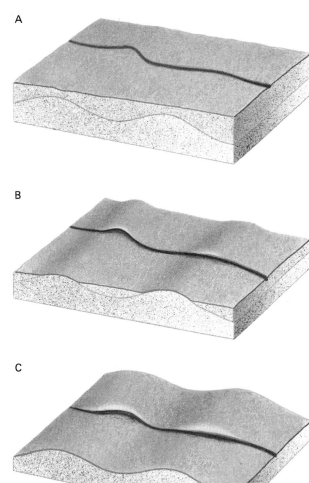

Figure 11–20 Streams that flow through mountains, case 1: an antecedent stream. A stream flows across a flat plain. Tectonic forces uplift the plain to form a ridge, but the stream erodes its bed as rapidly as the land rises. Thus, the stream is able to maintain its original path by cutting through the rising ridge.

meters of sedimentary rock, forming Grand Canyon, as the Colorado Plateau rose. Most of this erosion may have occurred within the past two to three million years, a remarkably rapid rate.

A stream also may cut through an uplift in resistant bedrock by the process of **superposition**. If a ridge of hard bedrock is covered with younger sedimentary rocks, a stream downcutting into the sedimentary rocks is unaffected by the buried feature. Eventually, however, the stream may downcut to the ridge. At this point the channel may be well established, and the stream may cut through the hard ridge rather than be diverted to one side or the other (Fig. 11–21).

Stream Piracy

Consider what may happen when two streams flow in opposite directions from a drainage divide. One of the streams may cut downward faster than the other. This can occur if one side of the range is steeper than the other, if one side receives more rainfall than the other, or if the rock on one side is softer than rock on the other. In any of these situations, the stream that is eroding more rapidly may cut its way backward into the divide. This process is called **headward erosion**. If headward erosion continues, the more deeply eroded stream may cut *through* the drain-

Figure 11–21 Streams that flow through mountains, case 2: a superposed stream. A stream flows over young sedimentary rock that has buried an ancient mountain range. As the stream erodes downward, it cuts into the old mountains, maintaining its course.

age divide. The higher stream then reverses direction and flows into the more deeply cut stream that has just penetrated the divide. Hence, one stream eventually captures the drainage of the other. This sequence of events is called **stream piracy** (Fig. 11–22).

GROUND WATER

11.9 Characteristics of Ground Water

If you drilled a hole into the ground in most places, in a few days it would fill with water to within a few meters of the surface. The water would appear even if no rain fell and no streams flowed nearby. The water that seeps

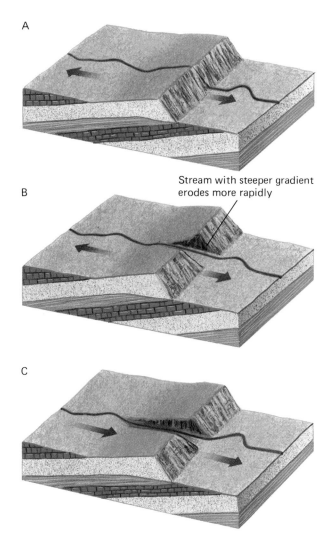

A

B

Stream with steeper gradient erodes more rapidly

C

Figure 11–22 Stream piracy. The stream on the steeper side of the ridge erodes downward more rapidly than the stream on the opposite, gentler slope. Eventually, the steeper stream cuts through the ridge to intersect the higher, gentler stream. The higher stream then reverses direction to flow into the lower one.

The springs of Vesey's Paradise flow from the walls of Grand Canyon, Arizona.

into the hole is part of the vast reservoir of subterranean ground water that saturates the Earth's crust in a zone between a few meters and a few kilometers below the surface.

Globally, 30 times more water is stored as ground water than in all streams and lakes combined. We can extract the water by digging wells and pumping it to the surface. Before the invention of advanced drilling and pumping technologies, human impact on ground-water reserves was minimal. Today, however, deep wells and high-speed pumps can extract ground water more rapidly than natural processes replace it. This situation is common in the central and western United States. It is a serious problem because ground-water reserves provide drinking water for more than half of the population of North America.

Porosity, Permeability, and Ground Water

Ground water is stored in small cracks and voids in soil and bedrock. **Porosity** is the proportion of a volume of rock or soil that consists of open spaces. Igneous and metamorphic rocks, such as granite, gneiss, and schist, have low porosities unless they are fractured. However, many sedimentary rocks can be quite porous. Sandstone can have 5 to 20 percent porosity (Fig. 11–23). The porosity of loose sediment and soil can be even greater, reaching 50 percent in sand and even 90 percent in clay.

While porosity tells us *how much* water rock or soil can retain, it tells us nothing about the *rate* at which water can flow through the pores. **Permeability** is a measure of the speed at which water can travel through porous soil or bedrock.

Well-sorted sediment

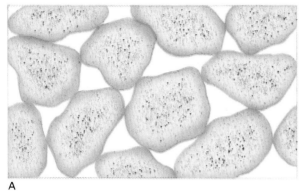

A

Poorly sorted sediment

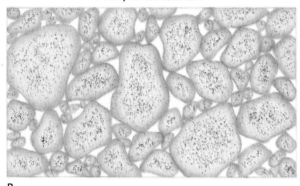

B

Sedimentary rock with cementing material between grains

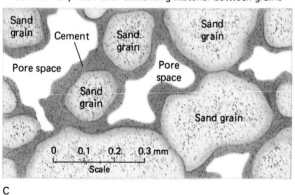

C

Figure 11–23 Different types of sediment, rock, and soil have different amounts of open pore space between grains. (A) Well-sorted sediment consists mostly of equal-size grains and thus has a high proportion of pore space. This sample has a porosity of about 30 percent. (B) In poorly sorted sediment, small grains fill the spaces among the large ones, and porosity is lower. In this drawing it is about 15 percent. (C) Cement in sedimentary rock fills pore space, lowering the porosity.

Soil and loose sediment such as sand and gravel are both porous and permeable. Thus, they can hold a lot of water, and it flows easily through them. Sandstone and conglomerate can also have high permeabilities. Although clay and shale are porous and can hold a large amount of water, the pores in these fine-grained materials are so small that water flows very slowly through them. Thus, clay and shale have low permeabilities.

The Water Table

When rain falls on dry soil, it usually soaks into the ground. Water does not descend into the crust indefinitely, however. Below a depth of a few kilometers, rock is at high enough temperature and pressure to be plastic. Here the weight of overlying rock closes the pores, and rock at this depth is both nonporous and impermeable. Water accumulates on this impermeable barrier, filling all the pores in the rock and soil above it. This completely wet layer of soil and bedrock above the barrier is called the **zone of saturation**. The **water table** is the top of the zone of saturation (Fig. 11–24). Above the water table lies the **unsaturated zone**, or **zone of aeration**. In this layer, the rock or soil may be moist but not saturated. The **soil moisture belt** is the soil layer. It holds more water than the unsaturated zone below and supplies much of the water needed by plants.

If you dig into the unsaturated zone, the hole will *not* fill with water. However, if you dig below the water table into the zone of saturation, you have dug a **well**, and **the water level in a well is at the level of the water table**.

An **aquifer** is any body of rock or regolith that can yield economically significant quantities of water. An aquifer must be both porous and saturated; that is, it must contain water. It must also be permeable so water flows into a well to replenish water that is pumped out. High-quality aquifers commonly occur in sand and gravel, sandstone, limestone, and highly fractured igneous and metamorphic rock. Think of an aquifer as a sponge that water seeps through, *not* as underground pool or stream.

Movement of Ground Water

In a few regions, underground rivers flow through caverns, but *they are the exception, not the rule*. Nearly all ground water seeps slowly through interconnected pores in bedrock and regolith. Typically, ground water flows at about 4 centimeters per day (about 15 meters per year), although flow rates may be much faster or slower. The rate depends on the permeability of regolith and on the nature of fractures in bedrock. Water can flow rapidly through large interconnected fractures.

In a temperate climate, ground water seeps *into* streams. Figure 11–25 shows a stream and water table

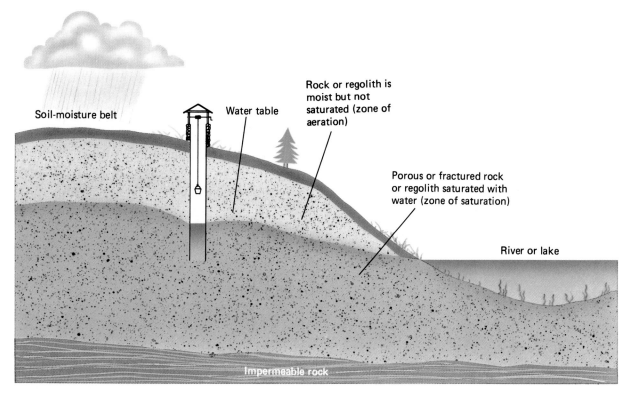

Figure 11–24   The water table is the top of the water-saturated zone near the Earth's surface. It intersects the land surface at lakes and streams and is the level of standing water in a well.

in hilly country. In general, the water table follows the contours of the land, rising and falling with the topography. Just as streams always flow downhill, ground water always seeps from areas where the water table is highest toward areas where it is lowest. Streams follow gulleys

Figure 11–25   The water table rises beneath hills and sinks beneath valleys. At any point, it also rises during a rainy season and sinks during drought.

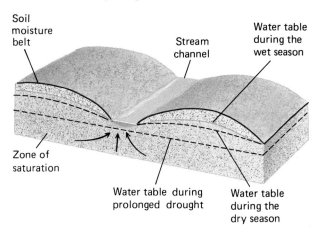

and valleys. Therefore, ground water normally flows toward streams and seeps into streambeds, which is why streams continue to flow even when rain has not fallen for weeks or months.

The water table is not fixed, but rises and falls with the seasons. During a wet season, such as spring in a temperate climate, rain seeps into the ground, and the water table rises. During a dry season, the water table falls (Fig. 11–25). That is why it is common for the water level in wells to rise and fall through the year, and with wet years and years of drought.

In contrast, in an arid climate the water table commonly lies below streambeds, and water seeps downward *from* a stream to the water table. Desert stream channels are dry most of the time. When streams do flow, however, the water often originates from wetter environments in nearby mountains, although sometimes desert storms fill the channels for short periods of time.

Springs

A **spring** is a place where ground water flows or seeps from the ground. A spring can exist wherever the water table intersects the land surface (Fig. 11–26). In some

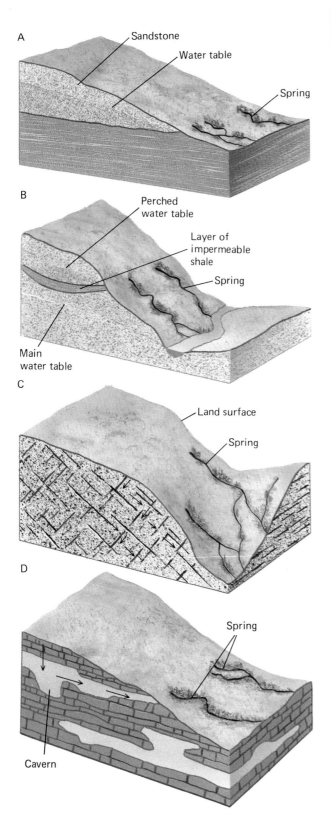

An artesian well spouts into the air. *(D. R. Crandell/USGS)*

places, a layer of impermeable rock or clay may lie above the main water table, creating a local saturated zone, the top of which is called a **perched water table**. A perched water table can also intersect the land surface to form a spring (Fig. 11–26B).

Artesian Wells

Figure 11–27 shows a layer of permeable sandstone sandwiched between two layers of impermeable shale. On the left, the strata are tilted. An inclined aquifer bounded top and bottom by layers of impermeable rock is called an **artesian aquifer**. Water in the lower part of the sandstone aquifer is under pressure from the weight of water above.

◀ **Figure 11–26** Springs form where the water table intersects the land surface. This situation can occur where (A) the land surface intersects a contact between permeable and impermeable rock layers; (B) a layer of impermeable rock or clay lies "perched" above the main water table; (C) water flows from fractures in otherwise impermeable bedrock; and (D) water flows from caverns onto the surface.

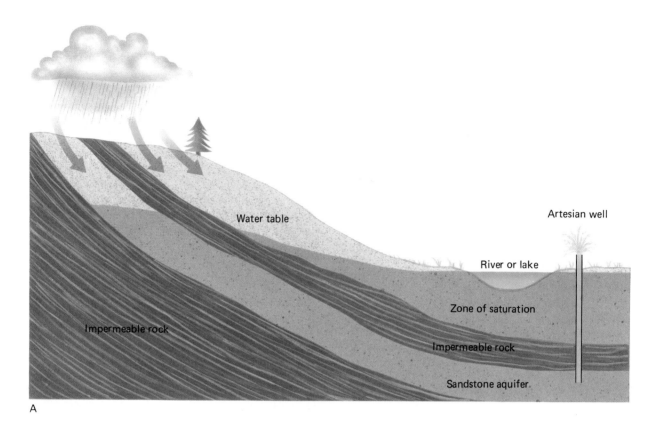

A

B

Figure 11–27 (A) An artesian aquifer forms where a tilted layer of permeable rock, such as sandstone, lies sandwiched between layers of impermeable rock, such as shale. Water rises in an artesian well without being pumped. (B) A hose with a hole shows why an artesian well flows spontaneously.

Therefore, if a well is drilled into the sandstone, water will rise without being pumped. If pressure is sufficient, the water spurts out onto the surface. A well of this kind is called an **artesian well**. As an analogy, think of a water-filled hose. If one end is held high and the lower end sealed, and you puncture the hose below the high point, water will squirt out.

11.10 Use of Ground Water

Ground water is a valuable resource for a variety of reasons.

1. It is abundant. As mentioned at the beginning of this chapter, 30 times more fresh water exists underground than in surface reservoirs.

Earth Science and the Environment

Depletion of Ground Water: Irrigation on the High Plains

Many parts of the high plains in western and midwestern North America receive scant rainfall. Early settlers simply planted crops and suffered in times of drought. In the 1930s, two sequences of events combined to change agriculture in this region. One was a great drought that destroyed crops over a wide region and exposed the soil to erosion. Dry winds blew across the land, eroding the parched soil and carrying it for hundreds and even thousands of kilometers. Thousands of families lost their farms, and the region was dubbed the Dust Bowl. The second event was the arrival of inexpensive technology. Electricity came to rural regions, and relatively cheap pumps and irrigation systems were developed. With the specter of drought fresh in people's memories and the tools to avert future calamities available, the age of modern irrigation was initiated.

The figure shows a map and cross section of the Ogallala aquifer in the central high plains. As you can see, the aquifer extends almost 900 kilometers from the Rocky Mountains eastward across the prairie, and from Texas into South Dakota. It consists of porous sandstone and conglomerate within 350 meters of the surface. The aquifer averages about 65 meters thick and contains a vast amount of water. Between 1930 and 1980, about 170,000 wells were drilled into the Ogallala aquifer and extensive irrigation systems were installed throughout Kansas, Nebraska, Oklahoma, the Texas panhandle, and parts of neighboring states.

Today, farmers and hydrologists are concerned that the Ogallala aquifer is being overexploited. They estimate that half of the water has already been removed from parts of the aquifer, and pumping rates are increasing. As explained earlier, water moves through aquifers at an average rate of about 15 meters per year. Most of the water in the Ogallala aquifer accumulated when the last Pleistocene ice sheet melted. Additional water seeps into the aquifer from the Rocky Mountains. At a flow rate of 15 meters per year, ground water takes 60,000 years to travel from the mountains to the eastern edge of the aquifer.

Under such conditions, deep ground water is, for all practical purposes, nonrenewable. Just as coal and petroleum are called fossil fuels, deep ground water is sometimes called fossil water. The removal of deep ground water is therefore analogous to mining.

If the present pattern of water use continues, wells in the Ogallala aquifer beneath the central plains will dry up early in the next century. In the mid-1980s, about 5 million hectares of land were irrigated from this aquifer. (This is an area about the size of the states of Massachusetts, Vermont, and Connecticut combined.) About 40 percent of the cattle in the United States are fed with corn and sorghum raised in this region, and large quantities of grain and cotton are grown here as well. If the aquifer is depleted and another source of irrigation water is not found, productivity in the central high plains is expected to decline by 80 percent. Farmers will go bankrupt, and food prices throughout the nation will rise.

---

2. Because ground water moves so slowly, it is stored in the Earth and remains available during dry periods.

3. In some regions, ground water flows from wet environments to arid ones, making water available in dry areas.

Most ground water that is used by humans is extracted from wells. Most successful wells are dug or drilled in valleys, close to the level of streams or lakes. If ground water is pumped continuously and rapidly, the water table can change significantly. The first disturbance occurs near the well. If water is withdrawn faster than it can flow into the well from the aquifer, a **cone of depression** forms (Fig. 11–28). When the pump is turned off, ground water flows back toward the well in a matter of days or weeks if the aquifer has good permeability, and the cone of depression disappears. On the other hand, if

water is continuously removed more rapidly than it can flow to the well through the aquifer, then the water table drops (Fig. 11–28C).

Subsidence

Subsidence, the sinking or settling of the Earth's surface, can be caused by the removal of ground water. When water is withdrawn from an aquifer, rock or soil particles may shift closer to each other, filling some of the space left by the lost water. As a result, the volume of the aquifer decreases and the overlying ground subsides. Removal of oil from petroleum reservoirs has the same effect.

Subsidence rates can reach 5 to 10 centimeters per year, depending on the rate of ground water removal and the nature of the aquifer. Some areas in the San Joaquin

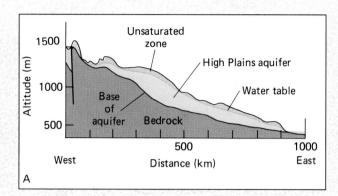

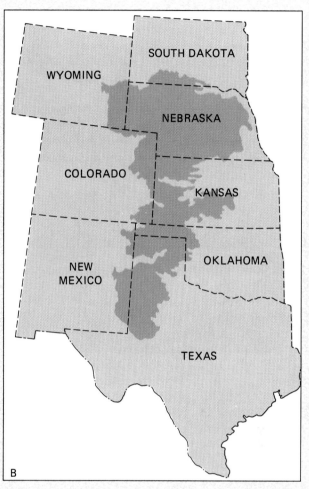

The Ogallala aquifer supplies water to much of the high plains. (A) A cross-sectional view of the aquifer shows that much of its water originates in the Rocky Mountains and flows slowly as ground water beneath the high plains. (B) A map showing the extent of the aquifer. (*After Gutentag et. al., 1984, U.S. Geological Survey Professional Paper 1400-B*)

Figure 11–28   (A) A well is drilled into an aquifer. (B) A pump draws water from the well faster than it can flow into the well through the aquifer, and a cone of depression forms. (C) If the pump continues to extract water more rapidly than it flows to the well, the water table falls.

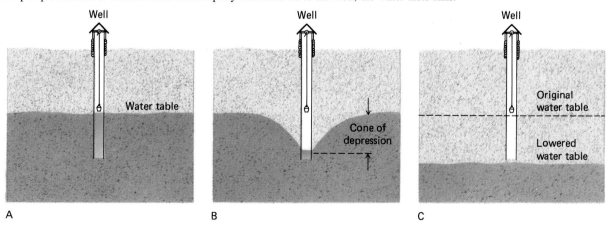

Valley of California have sunk several meters. The problem is particularly severe when it occurs beneath a city. For example, Mexico City is built on an old marsh. Over the years, as the weight of buildings and roadways has increased and much of the ground water has been removed, parts of the city have settled 1 to 2 meters. Many millions of dollars have been spent to maintain this complex city on its unstable base. Similar problems are occurring in cities in the United States, including Phoenix, Arizona, and Houston, Texas.

Unfortunately, subsidence is not a reversible process. When rock and regolith contract, their porosity is usually permanently reduced so that ground-water reserves cannot be completely recharged even if water becomes abundant again.

11.11 Ground-Water Pollution

Hooker Chemical Company's main plant was located in Niagara Falls, New York. Early in the 1940s, Hooker purchased an abandoned canal called Love Canal. During the following years, the company disposed of approximately 19,000 tons of chemical wastes from the manufacturing processes by loading them into 55-gallon steel drums and storing them in the canal. In 1953, the company covered one of the dump sites with dirt and sold the land to the Board of Education of Niagara Falls for $1. The deed of sale stated that the site was filled with "waste products resulting from the manufacturing of chemicals." The deed also specified that Hooker would no longer be responsible for the condition of the land. The site was then used for an elementary school and a playground.

The steel drums in Love Canal were exposed to moist soil from the outside and corrosive chemicals from the inside. Eventually they began to leak, and the chemicals seeped into ground water. In the spring of 1977, heavy rains raised the water table and turned the area around Love Canal into a muddy swamp. But it was no ordinary swamp; poisonous compounds from the leaking drums were mixed with the water and soil. The toxic fluid soaked the playground, seeped into basements, and saturated gardens and lawns. Children who attended the school and adults who lived nearby suffered serious illnesses. Residents and their physicians reported abnormally high incidences of epilepsy, liver malfunctions, miscarriages, skin sores, rectal bleeding, severe headaches, and birth defects. Residents of the Love Canal neighborhood brought suit against Occidental Petroleum Company (which had bought Hooker Chemical), the city of Niagara Falls, and the Niagara Falls school board. The legal battle continued for eight years. Finally, in an out-of-court settlement, residents were awarded damages according to the severity of their illnesses, from $2000 for people with minor medical problems up to $400,000 for those who claimed cancer or severe mental retardation.

The awareness generated by Love Canal and the fear that other such episodes could occur were so great that in December 1980, the U.S. Congress passed the Comprehensive Environmental Response, Compensation, and Liability Act, commonly known as Superfund. This law provides an emergency fund to clean up hazardous waste sites. The Environmental Protection Agency (EPA) has estimated that between 1950 and 1975, 5.5 billion metric tons of hazardous waste were spilled onto the land or buried in dump sites throughout the United States. A total of 20,766 sites were identified in the initial tally. The massive scale of the problem has overwhelmed the EPA and other government agencies. By early 1991, 1245 sites had been targeted for cleanup, but only 63 of these projects had been completed. About 1000 *new* sites are expected to be identified within the next decade.

Sources of Ground-Water Pollution

Sewage from septic tanks and cesspools may contaminate ground water. These sources contribute mainly bacterial and viral contamination, although chemical contamination may also occur if people flush paints, pesticide residues, or other household chemicals down the toilet.

Landfills, dumps, and hazardous waste disposal sites contribute chemical pollutants. Some industrial chemicals are poisonous even in trace amounts; others are suspected of causing cancer or birth defects. Landfills and dumps can also release bacterial and viral contamination.

Wastewater from factories, farms, and municipal sewage treatment plants is often stored in basins, pits, ponds, or lagoons. About 100,000 to 150,000 such sites exist in the United States. Many are unlined, and the soils beneath them are permeable. Therefore, the polluted water seeps downward to contaminate ground water.

Mine and mill wastes commonly contain high concentrations of metals. Some of these metals leach into the ground water, and many, such as arsenic, cobalt, lead, and other heavy metals, are toxic.

Approximately two million underground storage tanks exist in the United States. Half hold gasoline at service stations, and the remainder contain chemicals. Many of the tanks are untreated steel, and, like the 55-gallon drums at Love Canal, they eventually rust and leak.

Liquid chemical wastes are sometimes injected into deep wells below an aquifer to dispose of them without contaminating the aquifer. If the wastes are corrosive or if the pressure is high, the injection pipe may leak. The wastes may then enter ground water.

Agricultural practices also create troubling ground-water pollution. Pesticides and herbicides are sometimes heavily applied (or misapplied) to deal with infestations of insects or weeds, and they often percolate down to the ground water. Fertilizers also leach into aquifers, contributing inorganic contaminants such as nitrates and phosphates.

Solutions to the Problem of Ground-Water Pollution

Some ground-water pollution is removed by natural decay. Consider a septic system, for example. During the operation of a septic system, domestic sewage is flushed into a large tank. The solids settle to the bottom of the tank and are pumped out every five years or so. The wastewater flows out of the tank and into a series of underground trenches filled with gravel. The wastewater trickles through the gravel into soil, where the organic matter decomposes.

For purification to be effective, wastewater must move slowly through well-aerated soil or rock. Sandy soils and sandstone have the proper permeability for effective purification. If movement is slow and oxygen is available, decay organisms can feed on the sewage, decomposing it to harmless byproducts before the polluted water travels very far (Fig. 11–29A). On the other hand, imagine installing a septic tank in coarse gravel, fractured granite, or cavernous limestone. In any of these situations, the polluted water would flow rapidly through the large openings. Not enough time would elapse for the decay organisms to decompose the sewage, so wells or streams quite a distance from the source could become polluted (Fig. 11–29B).

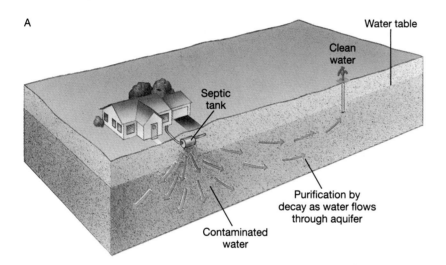

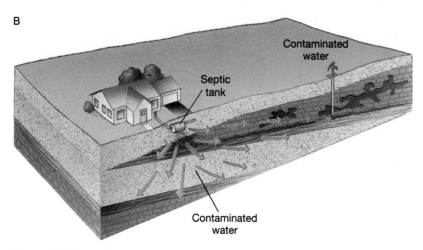

Figure 11–29 (A) Water moves slowly through permeable sandstone, giving natural processes enough time to purify it before it reaches a well. (B) But water flows through cavernous limestone too rapidly to be purified, and pollutes a nearby well.

Whereas sewage decomposes naturally, many chemicals such as pesticides, paints, and solvents are nonbiodegradable; that is, microorganisms do not decompose them. Once a nondegradable compound enters the ground water, it may persist for a long time. Recall that ground water flows slowly, so pollutants are not readily diluted or washed away. In addition, ground water does not have access to as much air as surface water does. Therefore, oxidation of persistent chemicals is less effective and less complete underground than it is in surface water.

Because natural processes are *not* effective in removing chemical contaminants from ground water, it is important to prevent contamination. Thus, underground storage containers should be lined and leak detectors installed. Landfills and waste treatment ponds should be built over a layer of impermeable clay or plastic so the contaminants are isolated from aquifers. Factories should be closely monitored so that hazardous wastes are disposed of responsibly.

Approximately 50 percent of the U.S. population depends on ground water as its primary source of drinking water. The EPA has established maximum tolerance levels for a variety of chemicals that may be present in water drawn from wells. According to the Association of Ground Water Scientists and Engineers, approximately ten million Americans drink water that does not meet EPA standards.

11.12 Caverns and Karst Topography

Just as streams erode valleys and form flood plains, ground water also creates landforms. Recall from Chapter 10 that rainwater reacts with atmospheric carbon dioxide to produce a slightly acidic solution, which is capable of dissolving the calcite that forms limestone. This reaction is reversible: the dissolved ions can precipitate to form calcite again.

Caverns

Caverns form when slightly acidic rainwater seeps into cracks in limestone bedrock, dissolving the rock and enlarging the cracks. Mammoth Caves in Kentucky and Carlsbad Caverns in New Mexico are two famous caverns formed in this way. The largest chamber in Carlsbad Caverns is taller than the U.S. Capitol building and is broad enough to accommodate 14 football fields. Most caverns form at or below the water table. If the water table drops, the chambers are opened to air.

If you entered a cavern, you would notice features obviously formed by deposition, not dissolution. Long pointed structures hang from the ceilings and rise from the floors. Collectively, all mineral deposits formed in caves by the action of water are called **speleothems**.

When a solution of water, dissolved calcite, and carbon dioxide percolates through the ground, it is under the pressure of water in the cracks above it. If a drop of this solution seeps into the ceiling of a cavern, the pressure decreases suddenly because the drop comes in contact with the air. The high humidity of the cave prevents the water from evaporating rapidly, but the lowered pressure allows some of the carbon dioxide to escape as a gas. When the carbon dioxide escapes, the drop becomes less acidic. This decrease in acidity causes some of the dis-

Figure 11–30 Stalactites and stalagmites in a limestone cavern. *(Hubbard Scientific Co.)*

solved calcite to precipitate. The net result is that as the water drips from the ceiling, calcite is slowly deposited.

Over time, beautiful and intricate speleothems form. **Stalactites** hang icicle-like from the ceiling of a cavern (Fig. 11–30). Usually, only a portion of the dissolved calcite precipitates as the drop seeps out of the ceiling. When the drop falls to the floor, it spatters, and the impact releases more carbon dioxide. The acidity of the drop decreases further, and another minute amount of calcite precipitates. Thus, **stalagmites** build from the floor upward to complement the stalactites. Because stalagmites are formed by splashing water, they tend to be broader than stalactites. As the two features continue to grow, they may eventually join to form a **column** (Fig. 11–31).

Sinkholes

If the roof of a cavern collapses, a depression called a **sinkhole** forms on the Earth's surface. A sinkhole can also form as limestone dissolves from the surface downward (Fig. 11–32). A well-documented sinkhole formed in May 1981 in Winter Park, Florida. During the initial collapse, a three-bedroom house, half a swimming pool, and six Porsches in a dealer's lot all fell into the under-

Figure 11–31 Columns in Carlsbad Caverns. *(Tom Till)*

ground cavern. Within a few days, the sinkhole was about 200 meters wide and 50 meters deep and had devoured additional structures and roadways (Fig. 11–33).

Figure 11–32 Sinkholes and caverns are characteristic of karst topography. Streams commonly disappear into sinkholes and flow through the caverns to emerge elsewhere.

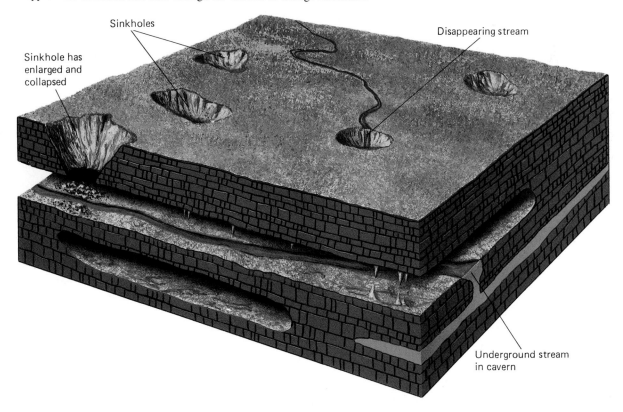

Sinkholes

Disappearing stream

Sinkhole has enlarged and collapsed

Underground stream in cavern

Figure 11–33 This sinkhole in Winter Park, Florida, collapsed suddenly in May 1981, swallowing several houses and a Porsche agency. *(Wide World Photos/Associated Press)*

Steamboat Geyser, Yellowstone National Park. *(National Park Service)*

Although sinkhole formation is a natural event, the problem can be intensified by human activities. The Winter Park sinkhole formed when the water table dropped, removing support for the ceiling of the cavern. The water table fell as a result of a severe drought augmented by excessive removal of ground water by humans.

Karst Topography

Karst topography forms in broad regions underlain by limestone and other readily soluble rocks. Caverns and sinkholes are common features of karst topography. Surface streams often pour into sinkholes and disappear into caverns. In the area around Mammoth Caves in Kentucky, streams are given names such as Sinking Creek, an indication of their fate.

The word "karst" is derived from a region in Yugoslavia where this type of topography is well developed. Karst landscapes are found in Alabama, Tennessee, Kentucky, southern Indiana, and northern and central Florida. Extensive karst landscapes also occur in China.

11.13 Hot Springs and Geysers

At numerous locations throughout the world, hot water naturally flows to the surface to produce **hot springs**. Ground water can be heated in three different ways.

1. Recall that the Earth's average geothermal gradient is approximately 25°C per kilometer in the upper portion of the crust. Therefore, if ground water descends through cracks to depths of 2 to 3 kilometers, it is heated by 50° to 75°. The hot water then rises because it is less dense than cold water. The springs at Warm Springs, Georgia, are probably heated in this manner. However, it is unusual for fissures to descend so deep into the Earth, and this type of hot spring is uncommon.

2. In regions of recent volcanism, magma, or hot igneous rock may remain near the surface and can heat ground water at relatively shallow depths. Such hot springs are common throughout western North and South America. This region has been tectonically active in the recent past and remains active today. Hot springs are reminders of recent volcanic activity. Shallow magma heats the hot springs and geysers of Yellowstone National Park.

3. Many hot springs have the odor of rotten eggs from small amounts of hydrogen sulfide (H_2S) dissolved in the hot water. These springs are heated by chemical reactions. Sulfide minerals, such as pyrite (FeS_2), react chemically with water. The reactions release heat. Hydrogen sulfide, produced by the reaction, rises with the heated ground water and gives it the strong odor.

Most hot springs bubble gently to the surface or flow from cracks in bedrock. However, a **geyser** erupts hot water and steam violently onto the Earth's surface. Geysers generally form in open cracks and channels in hot underground rock. In the first step, ground water seeps into the empty channels (Fig. 11–34). It is then heated by contact with the hot rock. Gradually, bubbles of water vapor form and start to rise, just they do in a heated teakettle. If part of the channel is constricted, the bubbles may accumulate in the narrow neck and form a temporary barrier, increasing pressure. When the pressure rises, some of the bubbles are forced upward past the constriction. On the surface, this movement appears as short bursts of steam and spurts of water. Below ground, when the bubbles rise, the pressure at the constriction is suddenly reduced. The water, which was already hot, flashes into vapor, which forces its way to the surface, blowing steam and hot water skyward. In Iceland, tourists used to make geysers blow by throwing soap into them to encourage the formation of bubbles.

The most famous geyser in North America is Old Faithful in Yellowstone Park, which erupts on the average of once every 65 minutes. Old Faithful is not as regular as people like to believe; the frequency of eruptions varies from about 30 to 95 minutes apart.

11.14 Geothermal Energy

Hot ground water can be used to generate electricity or it can be used directly to heat homes and other buildings. Energy extracted from the Earth's heat is called **geothermal energy**. The United States is the largest producer of geothermal electricity in the world, with a production capacity of 2200 megawatts. This is equivalent to the power output of two large nuclear reactors and provides about one million people with all of their electrical needs. However, this amount of energy is miniscule compared

Figure 11–34 In order for a geyser to erupt (A) ground water seeps into underground chambers and is heated by hot igneous rock. Foam constricts the geyser's neck, trapping steam and raising pressure. (B) When the pressure exceeds the strength of the blockage, the constriction blows out. Then the hot ground water flashes into vapor and the geyser erupts.

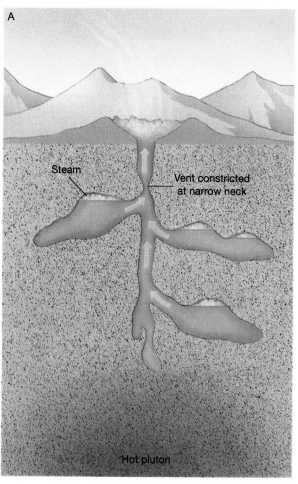

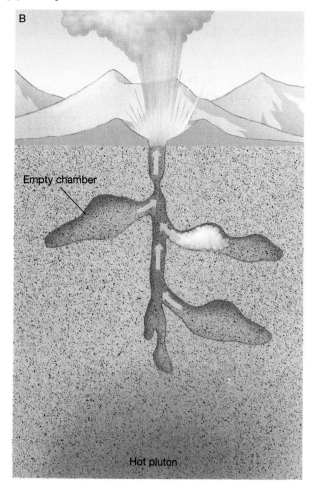

to the amount available. The United States Geological Survey estimates that the upper 4 kilometers of rock beneath the United States contains heat energy equivalent to 3000 trillion barrels of oil. This is enough energy to run the country for the next 200,000 years at current energy consumption rates.

The major problem with current methods of extracting energy from the Earth's heat is that they work only where deep ground water is naturally heated. Only a limited number of such "wet" sites exist, where abundant ground water *and* hot rock or magma occur at the same place. However, many "dry" sites exist where rising magma has heated rocks close to the surface, but little ground water is available. Wells can be drilled in these regions, and water pumped into the wells can be circulated through the hot rock and then extracted, producing steam that drives turbines to generate electricity.

The technology of extracting heat from dry sites is not free of problems. Hot rock usually lies more than 3 or 4 kilometers below the surface. It is expensive and technically difficult to drill a deep well into hot rock. In addition, the drilling and heat extraction processes use large amounts of water. Most regions of shallow, hot rock in North America are found in the west, where tectonism and igneous activity have recently taken place. Much of this region is semiarid.

Scientists and engineers are developing methods for extracting energy from dry Earth heat at a pilot project at Fenton Hill, New Mexico. They drilled two separate wells side by side (Fig. 11–35). Water is pumped down one, called the injection well, to a depth of about 4 kilometers. The pump forces the water into hot, fractured granite at the bottom of the well, and then into the extraction well, where it returns to the surface. The scientists have succeeded in pumping water into the well at 20°C and extracting it 12 hours later at 190°C.

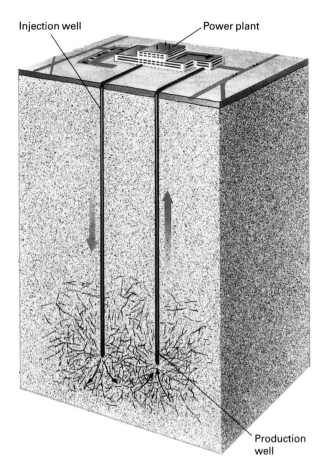

Figure 11–35 A schematic view of the Fenton Hill, New Mexico, dry geothermal energy plant.

In an era when petroleum prices are low, there is little incentive to develop expensive geothermal projects. The economic picture is expected to change within the next few decades as petroleum resources become limited and prices rise.

SUMMARY

Only about 0.65 percent of Earth's water is fresh water in streams, lakes, and underground reservoirs. The rest is salty seawater and glacial ice. **Evaporation, precipitation,** and **runoff** continuously recycle water among land, sea, and the atmosphere in the **hydrologic cycle.**

A **stream** is any body of water flowing in a channel. The floor of the channel is called the **bed,** and the sides of the channel are the **banks.** The velocity of a stream is determined by its **gradient, discharge,** and channel shape.

A stream **erodes** rock and soil, **transports** sediment downslope, and **deposits** it. The relative rates of these

three processes are determined by the velocity and discharge of the stream. A stream erodes its bed and banks by **hydraulic action, abrasion,** and **solution** and carries sediment as **dissolved load, suspended load,** and **bed load.**

A stream deposits sediment in three categories. (1) **Channel deposits,** called **bars,** form within the river channel itself. Curves in the channel are accentuated by uneven erosion and deposition, leading to formation of **point bars, meanders,** and **oxbow lakes. Braided streams** develop when more sediment is supplied to a stream than it can transport. (2) A **delta** forms at the mouth of a stream, and an **alluvial fan** forms where a fast mountain stream slows as it enters a larger valley. (3) **Flood-plain deposits** occur on the valley floor surrounding the stream. Floods create **natural levees** along the stream bank.

Ultimate base level is the lowest elevation to which a stream can erode its bed. This is usually sea level. A lake or resistant rock can form a **local**, or **temporary**, **base level**. A **graded stream**, in equilibrium with its sediment load, has a smooth, concave profile.

Downcutting, lateral erosion, and mass wasting combine to form a stream valley. Mountain streams downcut rapidly and form V-shaped valleys, whereas lower-gradient streams form wider valleys by lateral erosion and mass wasting.

A drainage basin can have **dendritic**, **trellis**, or **radial** patterns, depending on the bedrock geology.

Streams erode mountains and deposit sediment in lowlands, to flatten rugged topography. Tectonic activity usually interrupts this process, **rejuvenating** a landscape. **Incised meanders** and **stream terraces** often form where a stream downcuts into its bed as tectonic uplift occurs. In some instances, tectonic uplift can cause a stream to cut through a mountain range or resistant bedrock. **Headward erosion** can lead to **stream piracy**.

Most of the rain that falls on land seeps into soil and bedrock to become **ground water**. Ground water saturates the upper few kilometers of soil and bedrock to a level called the **water table**. **Porosity** is the proportion of rock or soil that consists of open space. **Permeability** is the speed with which fluid can move through pores in soil or bedrock. An **aquifer** is a body of rock that can yield economically significant quantities of water. Aquifers are porous and permeable.

Most ground water moves slowly, about 4 centimeters per day. **Springs** occur where the water table intersects the surface of the land. Dipping layers of permeable and impermeable rock can produce an **artesian aquifer**. If a well is drilled into an artesian aquifer, the water rises and may even spout up above the ground surface.

If water is withdrawn from a well faster than it can be replaced by the aquifer, a **cone of depression** forms. If rapid withdrawal continues, the water table falls.

Polluted ground water may be purified slowly by natural processes.

Caverns form where ground water dissolves limestone. **Speleothems** form by precipitation of calcite from ground water in caves. A **sinkhole** forms when the roof of a limestone cavern collapses. **Karst topography**, with numerous caves, sinkholes, and subterranean streams, is characteristic of limestone regions.

Hot springs develop when hot ground water rises to the surface. Ground water can be heated by (1) the geothermal gradient, (2) shallow magma or a cooling pluton, or (3) chemical reactions between ground water and sulfide minerals. Hot springs have been tapped to produce **geothermal energy**, and "dry sites" are now being explored.

KEY TERMS

●

R E V I E W   Q U E S T I O N S

●

1. In which physical state (solid, liquid, or vapor) does most of the Earth's free water exist? Which physical state accounts for the least?

2. Why can a rapidly flowing stream carry sand and cobbles, whereas a slow-moving stream can carry only silt and clay?

3. Why does the Mississippi River carry more sediment than a small mountain stream?

4. Distinguish among the three types of stream erosion: hydraulic action, solution, and abrasion.

5. List and explain three ways in which sediment can be transported by a stream. Which type of transport is independent of stream velocity? Explain.

6. Why do braided streams often develop in glacial and desert environments?

7. How are alluvial fans and deltas similar? How do they differ?

8. Give two examples of natural features that create temporary base levels. Why are they temporary?

9. Draw a profile of a graded stream and an ungraded stream.

10. Explain how a stream forms and shapes a valley.

11. In what type of terrain would you be likely to find a V-shaped valley? Where would you be likely to find a meandering stream with a broad flood plain?

12. How can a stream become rejuvenated? Give an example of a landform created by a rejuvenated stream.

13. Explain the difference between an antecedent and a superposed stream.

14. (a) Draw a cross section of soil and shallow bedrock, showing the zone of saturation, water table, and zone of aeration. (b) Explain each of the preceding terms.

15. What is an aquifer, and how does water reach it?

16. Explain why bedrock or regolith must be both porous and permeable to be an aquifer.

17. Compare the movement of ground water in an aquifer with that of water in a stream.

18. How does an artesian aquifer differ from a normal one? Why does water from an artesian well rise without being pumped?

19. Describe problems that can arise from excessive use of ground water.

20. Why is depletion of ground water likely to have longer-lasting effects than depletion of a surface reservoir?

21. Explain how land can subside when ground water is depleted. If the removal of ground water is stopped, will the land rise to its original level again? Defend your answer.

22. Explain how water from a septic tank is purified by natural processes. Explain why purification is more effective in some types of bedrock or regolith than in others.

23. What is karst topography? How can it be recognized? How does it form?

24. Explain three mechanisms for the formation of hot springs.

D I S C U S S I O N   Q U E S T I O N S

●

1. Describe the ways in which (a) a rise and (b) a fall in the average global temperature could affect the Earth's hydrologic cycle.

2. A stream is 50 meters wide at a certain point. A bridge is built across the stream, and the abutments extend into the channel, narrowing it to 40 meters. Discuss the changes that might occur as a result of this constriction.

3. Obtain a copy of a hydrograph of a local river from the county extension agent, the Coast Guard, or the Forest Service. (A hydrograph is a continuous record of discharge over time.) How does it compare with the hydrograph of the Selway River shown in Figure 11–2? The Selway is a small stream in the mountains of Idaho. Explain any differences between the two hydrographs.

4. Defend the statement that most stream erosion occurs during a relatively short time period when the stream is flooding.

5. Describe the difference between a natural levee and an artificial one. How is each formed? How effective is each in reducing floods?

6. What type of drainage pattern would you expect in the following geologic environments? (a) Platform sedimentary rocks. (b) A batholith fractured by numerous faults. (c) A flat plain with a composite volcano in the center. Explain your answers.

7. Which of the wells in Figure 11–36 would you expect to contain water? Which would you expect to be polluted? Explain your answers.

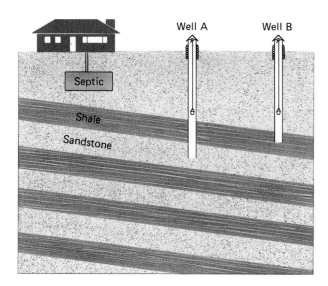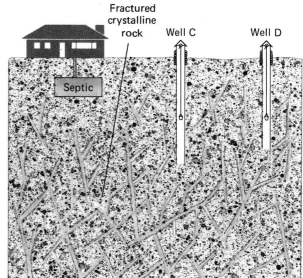

Figure 11–36

8. Imagine that you live on a hill 25 meters above a nearby stream. You drill a well 40 meters deep and do not reach water. Explain.

9. Explain why many desert streams dry up in the summer, whereas streams in humid environments do not go dry even during periods of drought.

10. The ancient civilization of Mesopotamia fell after its agricultural system collapsed due to problems resulting from failure of irrigation techniques. In his book, *Cadillac Desert*, author Marc Reisner describes the transformation of the western United States from its natural semidesert condition to its modern agricultural wealth. He maintains that this system, like others that preceded it, cannot be sustained indefinitely. Argue for or against Reisner's hypothesis.

11. Would you expect to find a cavern in granite? Would you expect to discover a cavern in shale? Defend your answers.

12. Why can't stalactites or stalagmites form when a cavern is filled with water?

13. Contrast problems of ground-water pollution in a region of karst topography with those in a sandstone-aquifer region.

Glaciers and Wind

12

One hundred thousand years ago, the world was free of ice except for the polar ice caps of Antarctica and Greenland. Then, in a period of a few thousand years, Earth's climate cooled by a few degrees. The polar ice caps grew and spread into lower latitudes as winter snow failed to melt in summer. At the same time, glaciers formed near the summits of high mountains, even near the equator. They flowed down mountain valleys and out onto nearby lowlands. By 20,000 years ago, glaciers covered one third of the Earth's continental surface.

The granite forming the beautiful cliffs and domes of Yosemite Valley solidified from magma intruded deep within the Earth's crust about 150 million years ago (Fig. 12–1). Tectonic processes then raised the granite toward the Earth's surface. Streams eroded the rising land surface, creating valleys and exposing the granite. Then, 20,000 years ago, mountain glaciers flowed down the stream valleys. The moving ice smoothed and rounded the stream-cut landforms to create the present topography of the valley.

About 15,000 years ago, Earth's climate warmed again by a few degrees. As a result, the glaciers began to

● Bugaboo glacier, British Columbia, Canada.

Figure 12–1   Glaciers sculpted the cliffs and canyons of Yosemite Valley.   *(Science Graphics/Ward's Natural Science Establishment, Inc.)*

melt rapidly. A flowing glacier picks up huge amounts of rock and soil. When it melts, it deposits that sediment. The smooth, low, rounded hills of parts of upper New York State are piles of gravel deposited by great ice sheets as they melted (Fig. 12–2).

The barren sediment deposited by the melting ice was unprotected by plants. Strong winds blowing from the dying glaciers eroded the sand and dust, blowing it about and then depositing it again. As a result, great fields of sand dunes and windblown silt formed in the wake of the glaciers.

Vast areas of windblown sand and silt are common in regions recently vacated by glaciers, but they are also common on seacoasts and deserts where plants do not protect soil from wind. Thus, the dunes of Cape Cod, the Oregon coast, and deserts of the American southwest are hills of sand deposited by wind (Fig. 12–3).

Tectonic plate movements create mountain ranges, plains, and other large-scale features of a continent. But glaciers, wind, running water, and mass wasting sculpt those features into the familiar landforms that surround us. Running water is the most effective agent of erosion because rain falls and streams flow nearly everywhere on land.

Glaciers and wind, however, create landforms in special climatic environments. Glaciers form wherever winter snowfall exceeds summer melting. Thus, glaciers and the landforms they create develop at high latitudes and high altitudes, where average temperatures are low. Wind is effective only where plants do not protect soil. Therefore, wind-created landforms commonly dominate deserts, seacoasts, and regions recently abandoned by melting glaciers.

GLACIERS

12.1 Types of Glaciers

A popular saying tells us that "no two snowflakes are alike." Surely, with the billions of snowflakes that fall every year, no one can be sure that the saying is correct. Nevertheless, falling snow consists of a nearly infinite variety of beautiful ice crystals.

Figure 12–2   Glaciers deposited sand and gravel to form these hills in New York State.   *(Science Graphics/Ward's Natural Science Establishment, Inc.)*

Figure 12–3 Wind created the sand dunes on Oregon's coast.

When snow falls on a cold day, the crystals retain their delicate, airy shapes and accumulate in a light, fluffy mass known to skiers as "powder." However, when more snow buries these fragile crystals, the weight compacts and crushes them. In spring, when the temperature rises above freezing during the day and falls below freezing at night, the edges of the crystals melt and refreeze, forming rounded grains of ice. The upper layers of snow melt in the Sun's warmth, and the water seeps downward, freezing and cementing the grains together. Thus, snow changes, or metamorphoses, with time.

In most temperate regions, winter snow melts entirely during the summer. However, in certain cold, wet environments, winter snow does not melt completely, but accumulates year after year. It becomes denser as it is buried to ever greater depths. If snow survives through one summer, it converts to a mass of loosely packed, rounded ice grains called **firn** (Fig. 12–4). Mountaineers like firn because the sharp points of their ice axes and crampons sink into it easily and hold firmly. If firn is buried deeper in the snowpack and subjected to additional thawing and freezing, it converts to **glacial ice**, a mass of interlocking ice crystals. The crystals pack together so tightly that water cannot percolate through glacial ice.

A **glacier** is a massive, long-lasting accumulation of compacted snow and ice that forms on land. A glacier forms wherever the amount of snow that falls in winter exceeds the amount that melts in summer. Glaciers in mountain regions flow slowly downhill. Glaciers on level land flow outward under their own weight, just as cold honey poured onto a tabletop slowly spreads outward.

Alpine Glaciers

Mountainous regions are generally colder and wetter than adjacent lowlands. Near the mountain summits, winter snowfall is high and summers are short and cool. These conditions create **alpine glaciers**. Alpine glaciers exist on every continent—in the Arctic and Antarctic, in temperate regions, and in the tropics. Glaciers cover the summits of Mount Kenya in Africa and Mount Cayambe in South America even though both peaks are near the equator. Some alpine glaciers in high latitudes flow great distances from the peaks onto adjacent lowlands. For

Figure 12–4 Newly fallen snow changes through several stages into glacial ice.

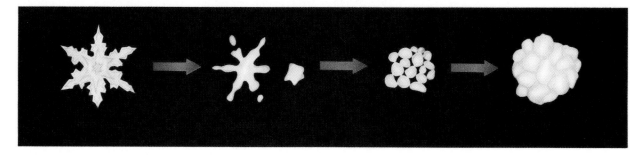

This alpine glacier flows around granite peaks in British Columbia, Canada.

example, the Kahiltna Glacier flows down the southwest side of Denali (Mount McKinley) in Alaska. It is about 65 kilometers long, 12 kilometers wide at its widest point, and about 700 meters thick. Most alpine glaciers are smaller; a few are larger.

The formation of an alpine glacier depends on both temperature and precipitation. The average annual temperature in the state of Washington is warmer than that in Montana, but alpine glaciers in Washington are larger and flow to lower elevations than those in Montana be-

The Antarctic ice sheet, Victoria Land.   *(Mugs Stump)*

cause more snow falls in Washington. Washington's mountains receive such heavy winter snowfall that even though summer melting is rapid, large quantities of snow accumulate every year. In Montana, snowfall is light enough that most of it melts in the summer, and thus, most of Montana's mountains have no glaciers.

Continental Glaciers

In Greenland and Antarctica, winters are so cold and long, and summers so short and cool, that glaciers are not confined to the mountains but cover most of the land regardless of elevation. An **ice sheet**, or **continental glacier**, is a glacier that covers an area of 50,000 square kilometers or more. The ice spreads outward in all directions under its own weight. Together, the ice sheets of Greenland and Antarctica make up 99 percent of the world's ice. The Greenland sheet is more than 2.7 kilometers thick in places and covers 1.8 million square kilometers. Yet, it is small compared with the Antarctic ice cap, which covers about 13 million square kilometers, almost 1.5 times the size of the United States. If the Antarctic ice sheet melted, the meltwater would create a river the size of the Mississippi that would last 50,000 years. In contrast, much of the Arctic is ocean, and since glaciers cannot form on oceans, no ice sheet exists at the North Pole. Instead, the upper few meters of much of the Arctic Ocean are commonly frozen.

At certain times in the past, average global temperature was lower than at present. Polar and temperate climates were colder, and possibly wetter, than now. Consequently, vast continental glaciers covered much of North America, Europe, Asia, and parts of the southern continents. These times, called the **ice ages**, are discussed later in this chapter.

12.2 Glacial Movement

Imagine that you set two poles in dry ground on opposite sides of a glacier and a third pole in the glacier, in a straight line with the other two. After a few months, the center pole would have moved downslope to form a triangle with the other two (Fig. 12–5). This simple experiment would tell us that the glacier had moved downhill. Rates of glacial movement vary depending on steepness, precipitation, and temperature. For example, in the coastal ranges of Alaska, where annual precipitation is high and average temperature relatively high (for glaciers), some glaciers move several meters a day. In contrast, in the interior of Alaska, where conditions are generally cold and dry, glaciers move only a few centimeters a day. At these rates, ice flows the length of an alpine glacier in hundreds to a few thousand years. In some

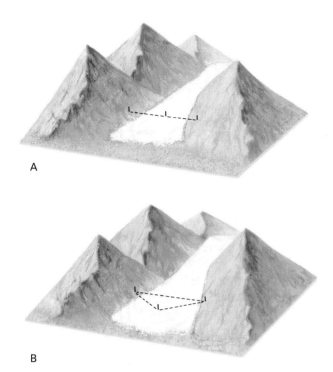

Figure 12–5 Glaciers flow. If three stakes are set in a straight line with two on land and one on a glacier (A), the stake on the ice will move (B).

instances, a glacier may **surge** at speed of 10 to 50 meters per day.

Imagine a mass of ice as large as the Kahiltna glacier flowing downslope. How does it move? A glacier moves by two mechanisms, basal slip and plastic flow.

Basal Slip

In **basal slip**, the entire glacier slides over bedrock in the same way a bar of soap would slide down a tilted board. Just as wet soap is more prone to slide than dry soap, an accumulation of water between bedrock and the base of a glacier accelerates basal slip.

Several factors can melt ice near the base of a glacier. Earth heat rises to warm the ice near bedrock. Friction from glacial movement may also generate heat. In addition, pressure at the base of a glacier can melt ice. Recall from Chapter 7 that solids melt in response to changing pressure as well as to an increase in temperature. When ice melts and converts to water, it shrinks. Pressure near the base of a glacier attempts to squeeze the ice into a smaller volume; therefore, ice at the base of a glacier can respond to pressure by melting. Thus, water may be present even when the temperature at the base of a glacier is slightly below 0°C. Additionally, during summer, ice

on the surface of a glacier may melt. Some of that meltwater can seep downward to the glacier's bottom.

Plastic Flow

Near the surface of a glacier, pressure is low and the ice is brittle, like an ice cube or ice on a frozen lake. However, at depths greater than 40 to 50 meters, pressure is sufficient to plastically deform ice. As a result, this portion of the glacier moves by **plastic flow** in addition to basal slip. In plastic flow, glacial ice deforms and flows as a very viscous fluid, rather than fracturing.

When a glacier flows over uneven bedrock, the ice follows the rock surface. The deeper, plastic part of the glacier can bend and flow over bedrock bumps, but the brittle upper zone develops fractures, or cracks, called **crevasses** (Fig. 12–6). Crevasses form in the upper 40 to

Figure 12–6 (A) Crevasses form in the upper, brittle zone of a glacier where the ice flows over uneven bedrock. (B) Crevasses in the Bugaboo Mountains of British Columbia.

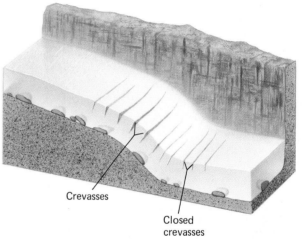

Crevasses

Closed crevasses

A

B

Earth Science and the Environment

Surging Glaciers

Case History: The Hubbard Glacier and the Town of Yakutat

Under normal conditions, the power of a glacier is masked behind its apparent stillness. It is easy to find evidence of glacial erosion, but a glacier takes a long time to scour mountains and deposit moraines. If you walk on a glacier, you are likely to hear it crack and rumble as the ice shifts and fractures, but you cannot see the ice flow. In the mid 1980s, however, the Hubbard Glacier in Alaska changed significantly in a matter of a few months.

The Hubbard glacier flows out of the St. Elias Range along the Canada-Alaska border and meets the sea at the mouth of a narrow inlet called Russell Fjord. In turn, the Hubbard is fed by a number of tributaries. One is the Valerie Glacier, which enters from the west just above the tidewater zone. Under normal conditions, these glaciers advance at a rate of about 15 centimeters per day. In the spring of 1986, however, the Valerie accelerated radically and began flowing 35 *meters* (about the length of a football field) per day. In turn, the larger Hubbard Glacier sped up to 15 meters per day. The Hubbard Glacier then advanced across the mouth of Russell Fjord, forming a dam that isolated the inner fjord from the sea.

The newly formed ice dam rose 30 meters above sea level, extended to the floor of the fjord, and was nearly half a kilometer wide. By sealing off the entrance, it converted the fjord to a salty inland lake. Fresh water flowed into the lake from nearby streams and glaciers, raising its level by about 0.5 meter per day. The fresh water diluted the seawater, and as the salinity declined, many saltwater species trapped in the isolated fjord began to die. Seals and dolphins, deprived of their food supply, starved. Residents in the nearby town of Yakutat feared that flooding would threaten the village and the rich salmon spawning grounds nearby. Then in October, approximately three months after the ice dam formed, it broke, and the fjord was once again connected to the ocean.

Glacial surges are relatively common. When an alpine glacier behaves normally, water flows along the base of the ice, promoting a constant rate of basal slip. Just prior to a glacial surge, the amount of meltwater flowing from the base of a glacier decreases greatly, indicating that the drainage system under the ice is blocked. When this occurs, water accumulates at the base of the glacier, enhancing lubrication. The glacier can then surge forward on this layer of water. Eventually, drainage becomes reestablished beneath the ice, excess water escapes, and normal glacial motion resumes.

---

50 meters, but do not continue into the plastic ice beneath. The opening of a crevasse is often covered by a weak layer of newly fallen snow. Many mountaineers have been injured or killed when they fell through such a snow bridge and tumbled into the abyss below.

The Mass Balance of a Glacier

Consider an alpine glacier starting in the high mountains and flowing into a lowland valley (Fig. 12–7). At the top of the glacier, snowfall is heavy, temperatures are low for much of the year, and avalanches carry large quantities of snow from the surrounding peaks onto the ice. In the summer some snow melts, but overall, more snow falls in winter than melts in summer. Therefore, snow piles up

from year to year. This higher-elevation part of the glacier is called the **accumulation area**. Here the glacier's surface is covered by fresh snow, which is often powdery in winter and slushy in summer.

Lower in the valley, the temperature is higher throughout the year and less snow falls. In this area, summer melting exceeds winter snowfall, so in summer all of the previous winter's snow melts, leaving a surface of old, hard glacial ice. The **firn line**, or **snowline**, is the boundary between permanent snow and seasonal snow. The firn line may shift up and down the glacier from year to year.

In the lower part of a glacier, called the **ablation area** or **zone of wastage**, more snow is lost in summer than accumulates in winter so there is a net loss of glacial

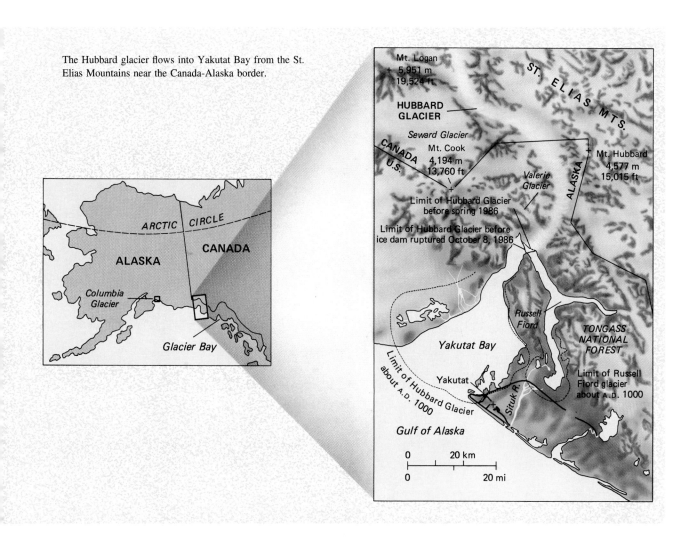

The Hubbard glacier flows into Yakutat Bay from the St. Elias Mountains near the Canada-Alaska border.

ice. Why is there any ice at all in the ablation area? Glacial ice flows downward from the accumulation area to the ablation area and continuously replenishes it.

Even farther down valley, the rate of glacial flow cannot keep pace with melting, so the glacier ends at its **terminus** (Fig. 12–8). The terminus is usually located on land. Streams or even large rivers flow out from the melting ice. In some regions, however, **tidewater glaciers** extend directly into the sea. Here, the terminus may be a steep ice cliff that drops off abruptly into the sea (Fig. 12–9). Giant chunks of ice break off, or **calve**, forming **icebergs**. The largest icebergs in the world are those that calve from the Antarctic ice shelf. Many are more than 50 meters thick, and they may be hundreds or, in rare instances, thousands of square kilometers in area.

One giant covered 30,000 square kilometers, about as large as Vancouver Island in western Canada. The tallest icebergs in the world come from tidewater glaciers in Greenland. Some rise 150 meters above sea level. Since the visible portion of an iceberg is only about 10 to 15 percent of its total mass, these bergs may be 1500 meters thick.

Glaciers grow and shrink. If average annual snowfall increases or average annual temperature drops, the accumulation area of an alpine glacier descends to a lower elevation, and the glacier grows. But what happens to the terminus? At first, no change occurs. Since the terminus is well below the firn line, winter snow in this region melts completely in summer. So, for the first few years, the terminus remains stable. However, as the accumulation area expands and thickens, more ice flows down-

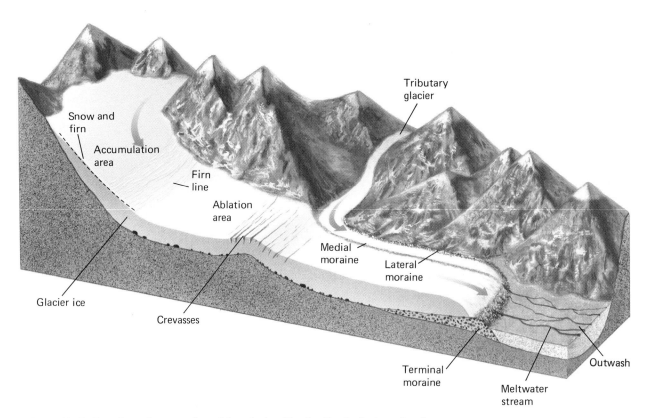

Figure 12–7 Prominent features of an alpine glacier. The firn line is the boundary between permanent snow and seasonal snow.

Figure 12–8 The terminus of an alpine glacier on Baffin Island, Canada, in midsummer. Dirty, old ice can be seen in the lower part of the glacier below the firn line, and clean snow higher up on the ice above the firn line. *(Steve Sheriff)*

Figure 12–9 A tidewater glacier flows directly into the sea, as here at Le Conte Bay, Alaska. Icebergs form when chunks of ice break away from the ice cliffs.

slope. Eventually the terminus advances farther down the valley. The lag time between a change in climate and the advance of a glacier may range from a few years to several decades, depending on the size of the glacier, its rate of motion, and the magnitude of the climate change. On the other hand, if annual snowfall decreases or the climate warms, the accumulation area shrinks and the glacier retreats. In Glacier Bay, Alaska, glaciers have retreated 60 kilometers in the past 125 years. When a glacier retreats, its ice continues to flow downhill, but the terminus melts back faster than the glacier flows downslope.

12.3 Glacial Erosion

A flowing glacier scours huge areas of bedrock and erodes landscapes. It does not merely scrape the Earth's surface as a bulldozer does, however. In addition, water seeps into cracks in the bedrock beneath a glacier and then freezes, loosening rock fragments. The flowing glacier then pries the fragments loose and incorporates them in the ice. This process is called **plucking**. The fragments range from silt-size grains to house-size boulders (Fig. 12–10).

Figure 12–10 (A) A glacier plucks rocks from bedrock and then drags them over the bedrock, abrading both the loose rocks and bedrock. (B) Plucking formed these crescent-shaped depressions in granite at Le Conte Bay, Alaska.

Ice

Water seeps into cracks, freezes and dislodges rocks which are then plucked out by glacier

Rocks are dragged along bedrock

Bedrock

A

B

Figure 12–11 Rocks embedded in the base of a glacier gouge glacial striations into bedrock.

Once a glacier picks up rocks and fine sediment, it carries them along. Ice itself is too soft to wear away bedrock. But rocks embedded in the base of the ice are extremely abrasive and scrape across bedrock like a sheet of coarse sandpaper pushed by a giant's hand. As the ice flows, rocks embedded in it gouge parallel grooves and scratches, called **glacial striations**, into the bedrock (Fig. 12–11). Striations show the direction of ice movement. They have been used to map movements of continental ice sheets that flowed over the land during the most recent ice age.

If silt or sand, rather than rock, is embedded in the base of a glacier, it abrades a smooth, shiny finish called **glacial polish** on the bedrock surface. Abrasion grinds rocks and other coarse particles into silt-size sediment

A

Figure 12–12 (A) A cirque carved near the summit of a peak in the Canadian Rockies. (B) Snow accumulates, and a glacier begins to flow downslope from the summit of a peak. (C) Glacial plucking erodes a small depression in the mountainside. (D) Continued glacial erosion and weathering enlarge the depression. When the glacier melts, it leaves a cirque carved in the side of the peak, as in the photograph.

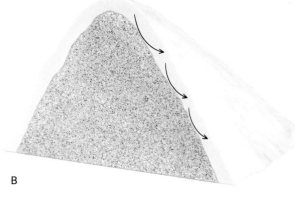

B

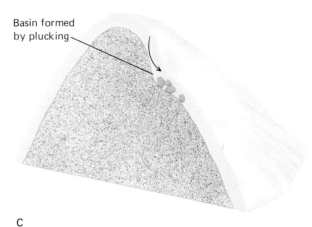

Basin formed by plucking

C

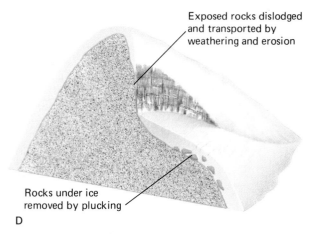

Exposed rocks dislodged and transported by weathering and erosion

Rocks under ice removed by plucking

D

called **rock flour**. Much of this fine sediment is carried away by streams flowing from the terminus of a glacier. Glacial streams are often so muddy that they are brown or gray. Sometimes, however, the suspended mud colors alpine streams and lakes turquoise, blue, or green.

Erosional Landforms Created by Alpine Glaciers

Let us take an imaginary journey through a mountain range that was glaciated in the past but is now ice-free. We will start with a helicopter ride to the summit of a high rocky peak. Perhaps the first impression from the window of the helicopter is one of sharp, jagged topography in the high country and rounded U-shaped valleys bounded by steep rock cliffs below (Fig. 12–12A).

We land on the summit and step out of the helicopter. Beneath us, a steep mountain face drops off into a horseshoe-shaped depression gouged out of the mountainside. This depression is called a **cirque**. A small glacier at the head of the cirque reminds us of the larger mass of ice that existed here in a colder, wetter time.

To understand how a glacier forms a cirque, imagine a gently rounded mountain. As snow accumulates and a glacier forms, the ice begins to flow down the mountainside (Fig. 12–12B). Abrasion and plucking erode a small depression that grows as the glacier continues to flow over many years (Fig. 12–12C). As erosion enlarges the cirque, its walls become steeper and higher. Frost wedging releases rocks from the cirque walls. The rocks fall onto the surface of the glacier, and the flowing ice carries this debris from the cirque to lower parts of the valley (Fig. 12–12D). When the glacier finally retreats and melts away, it leaves a steep-walled, rounded cirque as evidence of its existence.

As a cirque forms, the glacier often plucks a depression at its bottom. After the glacier melts, this depression fills with water, forming a small lake called a **tarn**, nestled in the cirque.

If glaciers erode three or more cirques in different sides of a peak, they may create a sharp, steep pyramid-shaped rock summit called a **horn**. The Matterhorn in the Swiss Alps is a famous example of a horn (Fig. 12–13). If two glaciers flow along opposite sides of a mountain ridge, they erode both sides of the ridge, forming a sharp, narrow **arête** between adjacent valleys (Fig. 12–14).

Looking downward from the glaciated peak, you may see a small, high valley emptying into a larger and deeper

Figure 12–13 The Matterhorn formed as three glaciers eroded cirques into the peak from three different sides. *(Swiss Tourist Board)*

Figure 12–14 An arête in the Bugaboo Mountains in British Columbia.

Figure 12–15 A hanging valley in Yosemite National Park. *(Science Graphics/Ward's Natural Science Establishment, Inc.)*

valley. A small glacial valley lying high above the floor of the main valley is called a **hanging valley** (Fig. 12–15). Perhaps a waterfall tumbles from the hanging valley into the deeper valley. To understand how a hanging valley forms, imagine these mountain valleys filled with glaciers, as they were several millennia ago (Fig. 12–16). The main glacier, flowing through the lower valley, gouged a deep trough. In contrast, the smaller tributary glacier did not scour the rock as rapidly or cut as deeply. As a result, when the ice melted, the floor of the tributary was considerably higher than the main valley floor, forming an abrupt drop where it entered the main valley. Typically, a series of hanging valleys empties into a main valley.

A mountain stream commonly erodes downward into its bed, cutting a steep-sided, V-shaped valley. A glacier, however, is not confined to a narrow stream bed, but instead fills its entire valley. As a result, it erodes outward against the valley walls as well as downward into its bed, scouring the sides as well as the bottom of its valley. In this way it forms a broad, rounded **U-shaped valley** (Fig. 12–17).

Continental glaciers erode the landscape in much the same way alpine glaciers do. The main difference is that continental glaciers are considerably thicker and not confined to valleys. Therefore, they scour the entire landscape and sometimes cover whole mountain ranges.

12.4 Glacial Deposits

In the 1800s, before geologists understood that continental glaciers covered vast parts of the land only 10,000 to 20,000 years ago, they recognized that large deposits of sand and gravel found in some places had been trans-ported from somewhere else. A theory popular at the time suggested that this material had been carried by icebergs during catastrophic floods. Owing to this mistaken mode of transport, the deposits were named "drift."

Today we know that drift was carried and deposited by glaciers. Although the word is a misnomer, it remains

A glacier transports and deposits sediment of all sizes. Baffin Island, Northwest Territories, Canada.

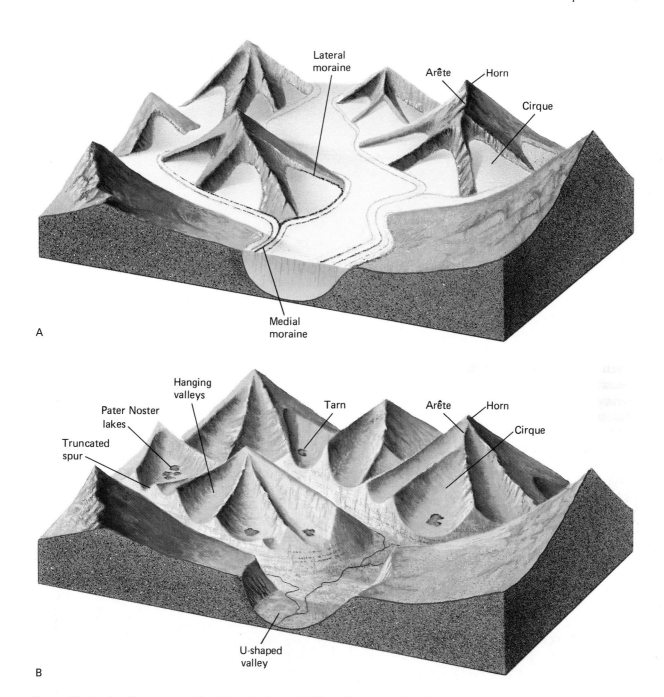

Figure 12–16   Landforms created by alpine glaciers. (A) Mountains covered by alpine glaciers. (B) The same mountains after the glaciers have melted.

in common use. Now Earth scientists define **drift** as all rock or sediment transported and deposited by a glacier. In many regions once covered by continental glaciers, tens or even hundreds of meters of drift overlie bedrock. Drift covers large regions of North America, including much of the prairie of the north-central United States and south-central Canada and the rocky hills and pastures of New England.

Drift is subdivided into two categories. **Till** is deposited directly by glacial ice. **Stratified drift** is sediment that was first carried by a glacier and then transported and deposited by a stream.

Landforms Composed of Till

Ice is so much more viscous than water that it carries particles of all sizes together. As a result, when a glacier

Figure 12–17 A U-shaped glacial valley, Purcell Mountains, British Columbia.

melts, it deposits fine clay and huge boulders together in an unsorted, unstratified, jumbled mass.

If you travel in country that was once glaciated, you occasionally find large boulders lying free on the surface. In many cases they are of a rock type different from the bedrock in the immediate vicinity. Boulders of this type are called **erratics** and were transported to their present locations by glaciers. The origins of glacial erratics can be determined by exploring the terrain in the direction from which the glacier came, until the parent rock is found. Some erratics have been carried 500 or even 1000 kilometers from their points of origin; they provide clues to the movements of past glaciers.

Moraines

A **moraine** is a mound or ridge of till deposited by a glacier. Think of a glacier as a giant conveyor belt. If you place a number of suitcases on a conveyor belt, it carries them to the end of the belt and deposits them in a jumbled pile. Similarly, a glacier carries sediment and then drops it at the terminus. If a glacier is neither advancing nor retreating, its terminus may remain in the same place for years, and the sediment piles up at the terminus in a ridge called an **end moraine** (Fig. 12–18). An end moraine that forms when a glacier is at its greatest advance, before beginning to retreat, is called a **terminal moraine**.

Now imagine what happens if the glacier retreats. If the terminus recedes at an even rate, the till disperses in a relatively thin layer over a broad area, forming **ground moraine**. Ground moraine fills old stream channels and

Figure 12–18 An end moraine is a ridge of till piled up at a glacier's terminus.

A lateral moraine deposited by a glacier flowing from Mount Sir Sanford, British Columbia, Canada.

other low spots, leveling the terrain. Often this leveling disrupts drainage patterns. Many swamps in the northern Great Lakes region are on young ground moraine.

If the glacier stops retreating and the terminus remains in the same place for a few years or more, a new end moraine called a **recessional moraine** forms in the same manner as a terminal moraine. The major differences are that the recessional moraine forms at a later date and upstream from the terminal moraine.

End moraines and ground moraines are characteristic of both alpine and continental glaciers. An end moraine deposited by a large alpine glacier may be so high that it would take an hour to climb to its top. A steep moraine can be difficult and dangerous to climb. Till is commonly loose, and large boulders are mixed randomly with rocks, cobbles, sand, and clay. A careless hiker can dislodge boulders and send them tumbling downhill.

The extents of continental glaciers of the most recent ice age can be determined by locating their terminal moraines. In North America the moraines form a broad, undulating line extending across the northern United States from Montana to New York. Enough time has passed that soil has formed on the moraines and vegetation has established itself. The hilly, wooded terrain of Ice Age National Scientific Reserve in western Wisconsin is on a terminal moraine. Part of Long Island, east of New York City, is a terminal moraine deposited in shallow water of the continental shelf.

Alpine glaciers also deposit **lateral moraines** along the sides of the ice. Recall that when an alpine glacier flows down a valley, it erodes the valley walls as well as the floor. In addition, large amounts of debris fall from the steepening valley walls onto the margins of the glacier. As a result of both processes, the edges of the glacier carry large loads of sediment. When the glacier melts, it deposits this sediment near its margins, forming lateral moraines.

If two glaciers converge to form a single larger one, sediment from the edge of each glacier is incorporated into the middle of the larger body of ice. The sediment forms a ridge of till that appears as a dark stripe on the surface of the ice downstream from the convergence (Fig. 12–19). When the glacier melts, it deposits this till as a **medial moraine** in the middle of the glacial valley.

Figure 12–19   The dark streaks in the middle of this glacier are medial moraines formed by the merging of lateral moraines where two glaciers converge.

Drumlins

A cluster of about 10,000 elongate hills, called **drumlins**, dot a region of rolling farmland in upstate New York (Fig. 12–20). Each hill looks like an upside-down spoon, or a whale swimming through the ground with its back in the air. A drumlin is typically 1 to 2 kilometers long and 15 to 50 meters high. Most drumlins are made of till, although some are stratified drift. In either case, the elongate shape forms where a glacier flows over and reshapes a mound of drift. The flowing ice streamlines the hill and elongates it parallel to the direction of flow. When the ice melts, the streamlined hill is left as a feature of the landscape.

Landforms Composed of Stratified Drift

During summer, when snow and ice melt rapidly, streams flow over the surface of a glacier. Many are so deep and wide that hikers cannot jump across them easily. A stream flowing on a glacier commonly runs into a crevasse and plunges into the interior of the glacier. Some water finds its way to the bottom of the ice and flows over bedrock or drift beneath the glacier. Eventually, all of this water flows from the terminus.

Because a glacier erodes so much sediment, a stream flowing from a glacier commonly carries large amounts of silt, sand, and gravel. The stream eventually deposits this sediment downstream from the glacier as **outwash**. Recall from Chapter 11 that a glacial stream carries such a heavy load of sediment that it often becomes braided. The outwash deposited in a mountain valley by streams flowing from an alpine glacier is called a **valley train**. If the sediment spreads out from the confines of a narrow valley into a larger valley or plain, it forms an **outwash plain**. Outwash plains are also characteristic of continental glaciers.

As a glacier melts and retreats, streams flow on top of, within, and beneath the ice. They commonly deposit small mounds of sediment called **kames**. A kame can form as a fan or delta at the margin of a melting glacier or where sediment collects in a crevasse or other depression in the ice. An **esker** is a long, snake-like ridge that forms as the bed deposit of a stream that flowed within or beneath a glacier (Fig. 12–21).

Since kames, eskers, and outwash are stream deposits, they are sorted and show sedimentary bedding. Thus, they are easily distinguished from the unsorted and unstratified till deposited directly from glacial ice.

Figure 12–20   Aerial view of drumlins in northern Saskatchewan. Each drumlin is about 40 meters high.   *(Canadian Department of Energy, Mines, and Resources, National Air Photo Library)*

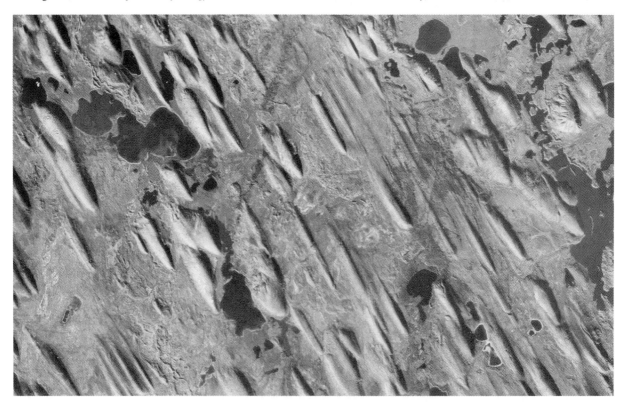

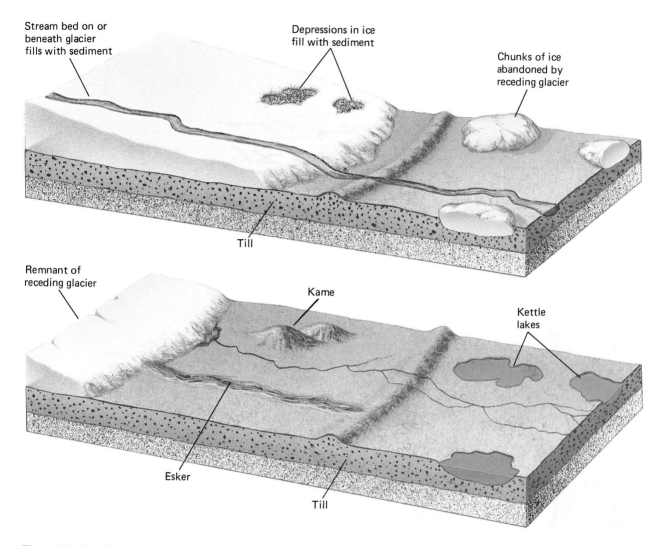

Figure 12–21   Kames, eskers, and kettle lakes are common landforms left by a retreating glacier.

A large block of ice may be stranded and buried in moraine or outwash as a glacier melts. When the block melts, it leaves a depression in the surface called a **kettle**. Kettles fill with water, forming kettle ponds or lakes. A kettle lake is as large as the stranded ice chunk. It can vary from a few tens of meters to a kilometer or so in diameter, with a typical depth of 10 meters or less.

12.5 The Ice Ages

Geologists have found moraines, drumlins, and striations in bedrock throughout northern Europe, far from the mountain glaciers. Once they understood that glaciers had formed these features, they concluded that, sometime in the past, a vast continental ice sheet must have covered northern Europe. Later, geologists found similar evidence of past glaciation on other continents. A time of extensive glacial growth, when alpine glaciers descend into lowland valleys and continental glaciers spread over higher latitudes, is called an **ice age** or a **glacial epoch**.

Geologic evidence shows that the Earth has been warm and relatively ice-free for about 90 percent of the past 2.5 billion years. However, at least five major ice ages have occurred during that time. The earliest occurred about 2.3 billion years ago. Later ice ages occurred about 600 million years ago—just before the rapid increase in plants and animals that marks the beginning of the Cambrian period—and in early and late Paleozoic times, about 500 and 250 million years ago (Fig. 12–22).

The most recent ice age occurred mainly during the Pleistocene epoch and is called the **Pleistocene Ice Age**. It began about two million years ago, and many Earth scientists think we are still in this most recent ice age. However, the Earth has not been glaciated continuously

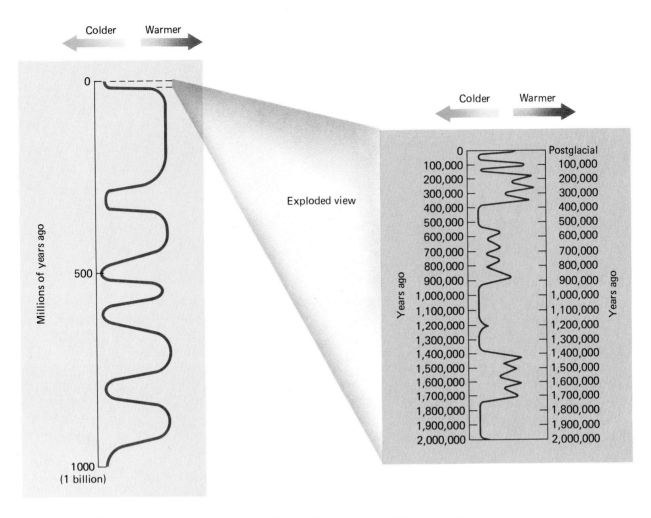

Figure 12–22 Average global temperature variations during the past one billion years. The times of lowest temperature are thought to coincide with ice ages. The right-hand scale is an expanded view to show temperature fluctuations during the Pleistocene Ice Age.

for that entire time; instead, continental ice sheets and alpine glaciers have grown and then melted away several times. Figure 12–22 also shows fluctuations in average global temperature during the Pleistocene epoch.

During the most recent advance, glaciers reached their maximum extent about 18,000 years ago. At that time, 30 percent of the Earth's land area was covered with ice (Fig. 12–23). Earth scientists often speak of events that occurred millions or billions of years ago. In contrast, 18,000 years is geologically recent. To put this time in perspective, Cro-Magnon artists were painting on cave walls in central Europe 35,000 years ago, and agricultural societies developed about 10,000 years ago.

Effects of Pleistocene Continental Glaciers

The thickest portion of the most recent North American ice sheet was located southeast of what is now Hudson

Bay in northern Canada. The ice flowed outward from that region, eroding soil and polishing and striating bedrock (Fig. 12–24). Today, vast regions of bedrock remain exposed, and the striations can be seen clearly on the ground and from the air.

The Pleistocene ice sheets created many major topographic features of North America. Their terminal moraines lie south of the Great Lakes and extend westward into Montana and eastward into southern New York. A terminal moraine forms part of Long Island, just east of New York City. Continental ice scoured and deepened the Great Lakes. The ice sheets and alpine glaciers widened and deepened old stream valleys, damming their channels with moraines to form large lakes. Today, much of the northern Great Plains is covered with ground moraine and outwash. These deposits form the fertile soil of the "breadbasket" of North America.

Recall that the lithosphere consists of several plates that float isostatically on the plastic asthenosphere. The

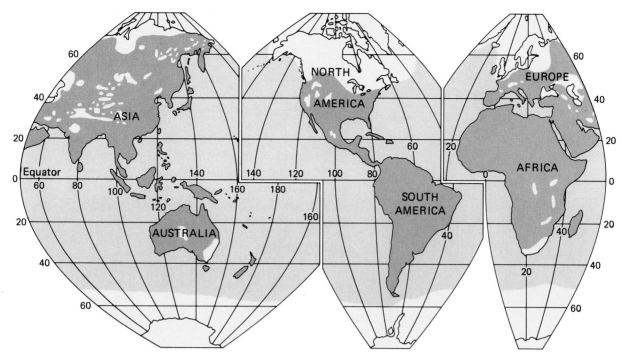

Figure 12–23 Glaciers covered 30 percent of the Earth's continents 18,000 years ago, during the most recent glacial advance.

Figure 12–24 An ice sheet covered most of northern North America 18,000 years ago. It was thickest near what is now Hudson Bay and from there flowed outward in all directions, as shown by arrows.

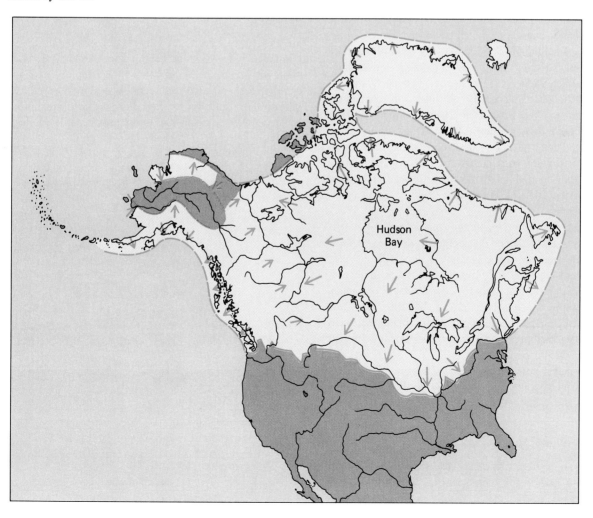

Hudson Bay

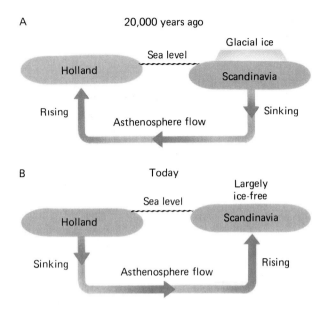

Figure 12–25 About 20,000 years ago, the weight of a continental glacier depressed Scandinavia. The displaced asthenosphere flowed laterally and raised Holland. Today the ice has melted, Scandinavia is rebounding, and Holland is sinking into the sea.

weight of a continental ice sheet 2 to 3 kilometers thick is sufficient to cause the lithosphere to sink. For example, 18,000 years ago, most of Scandinavia was buried under a continental glacier. The great weight depressed the peninsula (Fig. 12–25). As the lithosphere beneath Scandinavia settled, the displaced asthenosphere flowed southward, causing the lithosphere below Holland to rise. When the ice on Scandinavia melted, the entire process reversed. Today Scandinavia is rebounding, the asthenosphere is flowing back northward, and Holland is sinking. As their land sinks, the Dutch are spending billions of dollars on dikes to protect their country from the encroaching sea.

When glaciers grow, water is taken from the oceans and sea level falls. When glaciers melt, sea level rises. At the greatest extent of the glacial advance 18,000 years ago, global sea level fell to about 100 meters below its present elevation (Fig. 12–26). Adjacent to continents, oceanographers have found stream beds that are now below sea level and have retrieved fossilized teeth and bones of extinct mammoths from the sediment deposited by these ancient rivers.

In many coastal regions, deep, narrow inlets called **fjords** extend far into the land. Most fjords are glacially-carved valleys that were later flooded by the rising sea.

Causes of Ice Ages

Although no doubt exists that several ice ages have occurred in the past 2.5 billion years, Earth scientists still dispute the *causes* of ice ages. Recall that at least five major glacial episodes have occurred during the past 2.5 billion years. The most recent, the Pleistocene Ice Age, was characterized by several glacial advances and retreats. The causes of both the ice ages and the Pleistocene glacial fluctuations must have been related to global climate changes. Climatic variations, their causes, and their effects on glaciers are discussed in Chapter 17.

WIND

12.6 Wind Erosion

When wind blows through a forest or across a grassy prairie, the trees or grasses sway, but the vegetation protects the soil from erosion. In addition, rain commonly accompanies windstorms in temperate climates; the water dampens the soil and binds particles together, so little wind erosion occurs. In contrast, seacoasts, recently glaciated areas, and deserts commonly have little or no vegetation to protect the soil from wind. Therefore, wind erodes soil in these environments.

Deflation

One can hardly think about wind erosion without bringing to mind an image of a hide-behind-your-camel-type sandstorm in the desert. Wind often blows rapidly enough to erode and transport sediment. Wind erosion, called **deflation**, is a selective process. Because air is much less dense than water, wind can move only small particles, mainly silt and sand. (Clay particles usually stick together,

Figure 12–26 Sea level changes during the past 20,000 years. Sea level was lowest about 18,000 years ago when glaciers were largest.

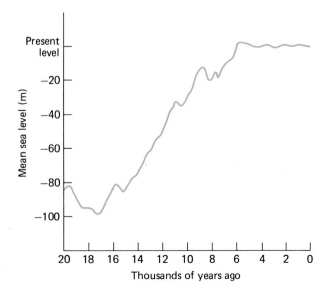

Earth Science and the Environment

Wind Erosion: The Dust Bowl

The semiarid and arid plains of the southwestern United States have experienced periodic droughts for centuries. However, the natural prairie ecosystems were resistant to these droughts. The native bushes and perennial grasses grow deep roots that collect water and nutrients from lower layers in the soil. Annual plants sprout from seed every spring, grow quickly, and then die in the fall. Annuals generally have shallow roots. During dry years, so little water is available that many of the annuals die. However, the deep-rooted perennials survive drought and protect the soil from wind erosion. In years of high rainfall, the annuals sprout quickly and prevent soil erosion and water runoff. Thus, the soil is protected in both dry and wet years.

Cultivated land is less resistant to changing weather. Generally, farmers till the land before they plant, removing the protective armor of natural vegetation. At this point, the field is vulnerable to erosion. If a drought occurs, the unprotected soil dries up and blows away. If spring rains are too heavy, the exposed soil may be eroded before the sprouts have an opportunity to grow. Many crops, such as corn, cotton, and vegetables, are cultivated in rows, and weeds growing between the rows are removed by herbicides or by further tilling. In short, much of the soil is exposed throughout the growing season. In addition, most crops are shallow-rooted annuals that cannot withstand drought.

During the early 1900s, improper farming practices led to a decline in soil fertility in the southwestern United States. Wind and water erosion had begun to damage the soil. Finally, when a prolonged drought occurred, the seeds failed to sprout, and the dry, exposed soil eroded rapidly. In 1934, a summer wind stripped the topsoil from entire counties and blew some of the dust more than 1500 kilometers eastward into the Atlantic Ocean. Altogether, 3.5 million hectares of farmland were destroyed, and productivity was reduced on an additional 30 million hectares. The same drought had little effect on parts of the prairie that had not been farmed.

During the 1920s and 1930s, windblown dust had a devastating effect on agriculture in the United States. This photograph, entitled *Buried machinery,* was taken in South Dakota on May 13, 1936. *(Reprinted with permission of the National Archives)*

and consequently wind does not move them.) Imagine soil, unprotected by vegetation, containing silt and sand mixed with larger pebbles and rocks (Fig. 12–27). When wind blows, it removes only the small particles, leaving the pebbles and rocks to form a continuous cover of stones called **desert pavement** (Fig. 12–28).

Only the smallest and lightest particles can be lifted very high or carried great distances by wind. Sand grains, which are relatively large and heavy, are usually lifted less than 1 meter in the air and are transported only a short distance. Recall from the previous chapter that a stream carries sand by **saltation**. In this process, the grains bounce along in a series of small jumps. Wind also carries sand primarily by saltation.

In contrast, wind can carry fine silt in suspension. Skiers in the Alps commonly encounter a silty surface on

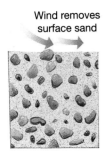

Wind removes
surface sand

Formation of desert
pavement complete–
No further wind
erosion

Figure 12–27 Wind erodes silt and sand, but leaves larger rocks behind to form desert pavement.

the snow, blown from the Sahara Desert and carried across the Mediterranean Sea.

Wind erosion commonly forms depressions called **blowouts**. In the 1930s, a series of intense, dry winds eroded large areas of the Great Plains and created the Dust Bowl. The deflation formed tens of thousands of blowouts, many of which remain today. Some are small, measuring only 1 meter deep and 2 or 3 meters across, but others are much larger (Fig. 12–29). One of the deepest blowouts in the world is the Qattara Depression in western Egypt. It is more than 100 meters deep and 10 kilometers in diameter. Ultimately, the lower limit for a blowout is

the water table. If the bottom of the depression reaches moist earth near the water table, where sand is bound together by water, wind erosion is no longer effective. If a blowout reaches the water table, a pool forms, creating an oasis.

Abrasion

Moving air by itself is not abrasive enough to erode rocks. But windblown sand and silt are abrasive and are effective agents of erosion. Since wind carries sand only a meter or less above the ground, abrasion concentrates close to the ground. If you see a tall desert pinnacle topped by a delicately perched cap, it was probably *not* created by wind erosion. The pinnacle is too high. However, if the *base* of a pinnacle is sculpted, then wind may be the responsible agent (Fig. 12–30). Cobbles and boulders lying on the surface often have faces worn flat by windblown sand. Such rocks are called **ventifacts** (Fig. 12–31).

12.7 Dunes

A **dune** is a mound or ridge of wind-deposited sand. Dunes are common in regions where glaciers have recently melted, in deserts, and along sandy coastlines.

Figure 12–28 Desert pavement is a continuous cover of stones left behind when wind blows silt and sand away.

Figure 12–29 The horse is standing behind a hummock left in the middle of a large blowout. Soil and vegetation on top of the hummock show the level of the ground surface before wind eroded the blowout. *(N. H. Darton, USGS)*

Figure 12–30 Wind abrasion selectively eroded the base of this rock because windblown sand moves mostly near the surface.

Windblown sand travels by saltation; the grains bounce along like so many Ping-Pong balls. If the wind blows over a rock or a small clump of vegetation, the downwind, or **lee**, side of the obstacle provides a small sheltered area where wind slows down. Grains of sand settle out in this protected zone. The growing mound of sand creates a larger windbreak, and more sand is deposited, forming a dune. Dunes commonly grow to heights of 30 to 100

Figure 12–31 Windblown sand sculpted the faces on this ventifact. *(W. H. Bradley, USGS)*

meters, and some giants exceed 500 meters. In places they are tens or even hundreds of kilometers long.

Most dunes are asymmetrical. Wind blows sand from the windward side of a dune, and then the sand slides down the sheltered lee side, where it accumulates. Thus, wind erodes sand from the windward side of the dune and deposits it on the lee side. For this reason, dunes migrate in the downwind direction. The leeward face of a dune is called the **slip face**. Typically the slip face slopes at about 30° to 34°, whereas the windward face slopes at about 15° (Fig. 12–32).

Migrating dunes can overrun buildings and highways. For example, U.S. Highway 95 runs across the Nevada desert. Near the town of Winnemucca, dunes advance across the highway several times a year. Highway crews must remove up to 4000 cubic meters of sand to reopen the road.

Attempts are often made to stabilize dunes in inhabited areas. One solution is to plant vegetation to reduce deflation and stop dune migration. The main problem with this approach is that desert dunes commonly form in regions that are too dry to support vegetation. Another solution is to build artificial windbreaks to create dunes in places where they do the least harm. For example, a fence stops blowing sand and forms a dune, thereby protecting areas downwind. Fencing is a temporary solution, however, because eventually the dune will cover the fence and continue to migrate. In Saudi Arabia, dunes are sometimes stabilized by covering them with tarry wastes from petroleum refining.

Figure 12–32 A dune migrates in a downwind direction as wind erodes sand from its windward side and deposits it on the slip face.

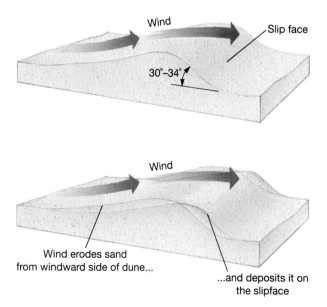

Figure 12–33 Cross-bedded sandstone in Zion National Park preserves the sedimentary bedding of ancient sand dunes.

Fossil Dunes

When dunes are buried by other sediment and eventually lithified, the resulting sandstone retains the original sedimentary structures of the dunes. For example, Figure 12–33 shows a rock face in Zion National Park in Utah. Notice the sloping sedimentary layering. This rock has not been tilted by tectonic forces. It is a lithified dune, and the dipping beds are the layering of the dune's slip face. The bedding dips in the direction in which the wind was blowing when it deposited the sand. Notice that the planes dip in various directions in different layers. These dip changes indicate changes in wind direction. The layering is an example of **cross-bedding**, described in Chapter 3.

Types of Sand Dunes

Wind speed, sand supply, and vegetation all influence the shape and orientation of a dune. If a dune forms in a rocky area with little sand and sparse vegetation, the dune's center grows higher than its edges (Fig. 12–34). When the dune migrates, the edges move faster because there is less sand to transport. The resulting **barchan dune** is crescent-shaped, with the tips of the crescents pointing downwind (Fig. 12–35A). Barchan dunes are not connected to one another, but instead migrate across the landscape independently. In a rocky desert, only a small portion of the surface is covered by barchan dunes; the remainder is mostly rock or desert pavement.

If sand is plentiful and vegetation sparse, the edges of a dune do not taper as in a barchan, and therefore they move at the same speed as the center. As a result, the dunes migrate as long, parallel ridges called **transverse dunes** (Fig. 12–35B). Transverse dunes form perpendicular to wind direction.

If vegetation is more plentiful, the plants may control dune shapes. If some disturbance destroys vegetation in a small area, deflation may create a blowout. Wind blows

Figure 12–34 A barchan dune forms where the supply of sand is limited.

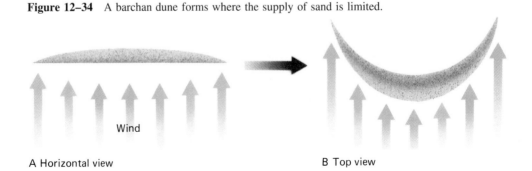

A Horizontal view

Wind

B Top view

A barchan dune, Coral Pink Sand Dunes, Utah.

sand out of the depression and deposits it nearby in a **parabolic dune** (Fig. 12–35C). Parabolic dunes are common in moist semidesert regions and along seacoasts. Note that parabolic dunes are similar in shape to barchan dunes, except that the tips of parabolic dunes point into

Transverse dunes. *(Hubbard Scientific)*

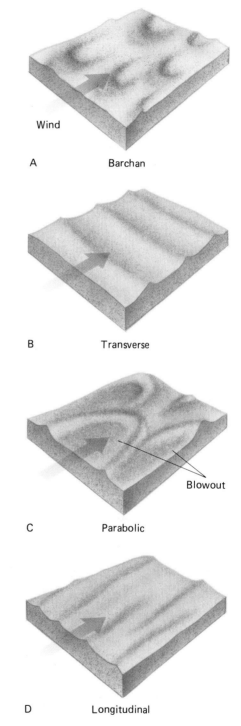

Figure 12–35 Types of sand dunes. (A) Barchan. (B) Transverse. (C) Parabolic. (D) Longitudinal.

Figure 12–36 Villagers in Askole, Pakistan, have dug caves in these vertical loess cliffs.

the wind, not downwind. The tips of a parabolic dune are often anchored by clumps of vegetation.

If wind direction is erratic but prevails from the same general quadrant of the compass, and if the supply of sand is limited, then **longitudinal dunes** form *parallel* to the wind direction (Fig. 12–35D). In portions of the Sahara Desert, longitudinal dunes are 100 to 200 meters high and as much as 100 kilometers long.

Loess

Most soils contain silt and sand. Wind-transported sand moves close to the ground by saltation and does not travel far. However, wind can carry finer silt for hundreds or even thousands of kilometers. Windblown silt can accumulate in thick deposits called **loess** (pronounced *luss*). Loess is porous and uniform, and typically lacks layering. Often the individual silt particles are angular and therefore

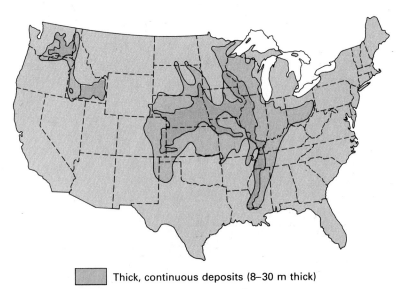

◻ Thick, continuous deposits (8–30 m thick)

◻ Thinner, intermittent deposits (1.5–8 m thick)

Figure 12–37 Loess covers much of the United States.

interlock. As a result, even though the loess is not cemented, it typically forms vertical cliffs and bluffs (Fig. 12–36).

The largest loess deposits in the world are found in central China. Here, loess beds cover 800,000 square kilometers and reach a thickness of more than 300 meters. The silt was blown from the Gobi and the Takla Makan deserts of central Asia. The particles are so cohesive that people dug caves into the loess cliffs to make their homes. However, in 1920 a great earthquake caused the cave system to collapse, burying and killing an estimated 100,000 people.

Loess is also abundant in North America. The silt accumulated during the Pleistocene Ice Age, when continental ice sheets ground bedrock into fine rock flour. Streams carried this fine sediment away from the melting glaciers and deposited it in vast outwash plains. The outwash zones were cold, windy, and devoid of life, and the silt was easily picked up and transported by wind. Loess deposits in the United States range from about 1.5 meters to 30 meters thick (Fig. 12–37). Soils formed on loess are generally fertile and make good farmland. Much of the rich soil of the central plains of the United States formed on loess.

SUMMARY

●

Glaciers and **wind** sculpt the Earth's surface and create landforms, but only under special climatic conditions.

If snow survives through one summer, it converts to granular ice crystals called **firn**. Firn can metamorphose further to form **glacial ice**. A **glacier** is a thick mass of ice that forms on land and flows plastically. Glaciers form wherever winter snowfall exceeds summer melting. **Alpine glaciers** occur in mountainous regions; a **continental glacier** covers a large part of a continent. A glacier moves by both **basal slip** and **plastic flow**. The upper 40 to 50 meters of a glacier are too brittle to flow, and large cracks called **crevasses** develop in this layer.

In the **accumulation area** of a glacier, more snow falls than melts, whereas in the **ablation area**, melting exceeds accumulation. The **snowline**, or **firn line**, is the boundary between permanent and seasonal snow. The end of a glacier is called the **terminus**.

Glaciers erode bedrock by **plucking** and by abrasion. Alpine glaciers create **cirques**, steep-sided depressions eroded into peaks; **arêtes**, thin, sharp, ridges separating two glaciated valleys; **horns**, pyramid-shaped peaks formed by the intersection of three or more cirques; and **hanging valleys**, formed by tributary glaciers, that "hang" high above the floor of a larger valley.

Glaciers scour both the bottoms and the sides of valleys, giving the valleys a characteristic U shape.

Drift is any rock or sediment transported and deposited by a glacier. The unsorted drift deposited directly by a glacier is **till**. Most glacial terrain is characterized by large mounds of till known as **moraines**. **Terminal moraines**, **ground moraines**, **recessional moraines**, **lateral moraines**, **medial moraines**, and **drumlins** are all accumulations of till deposited by a glacier. **Stratified drift** is sediment transported, sorted, and deposited by streams flowing on, under, or within a glacier. Stratified drift forms **valley trains**, **outwash plains**, **kames**, and **eskers**. A **kettle** is a depression created by the melting of a large ice block that was abandoned by a retreating glacier.

Several major ice ages occurred during the past 2.5 billion years. The most recent happened during the **Pleistocene epoch**, when both alpine and continental glaciers created many topographic features that are prominent today. Ice sheets isostatically depress continents, which later rebound when the ice melts. Sea level falls when continental ice sheets form, and rises again when the ice melts.

Wind is an important agent of erosion, transport, and deposition of sediment in environments where little vegetation is present to protect soil. Sparse vegetation is common in deserts and sandy coastlines. **Deflation** is erosion by wind. The wind selectively erodes silt and sand, leaving larger stones on the surface as **desert pavement**. Sand grains are carried short distances, at a meter or less above the ground, by **saltation**, but silt can be transported great distances at higher elevations. Wind erosion forms **blowouts**. Windblown particles are abrasive, but because the heaviest grains travel close to the surface, abrasion occurs mainly near ground level. A **ventifact** is a stone or pebble shaped by windblown sand.

A **dune** is a mound or ridge of wind-deposited sand. Most dunes are asymmetrical, with gently sloping windward sides and steeper **slip faces** on the lee sides. Dunes migrate. Types of dunes include **barchan dunes**, **transverse dunes**, **longitudinal dunes**, and **parabolic dunes**. Wind-deposited silt is called **loess**.

KEY TERMS

•

Firn *313*
Glacier *313*
Alpine glacier *313*
Ice sheet *314*
Continental glacier *314*
Ice age *314*
Basal slip *315*
Plastic flow *315*
Crevasse *315*
Accumulation area *316*
Firn line *316*
Ablation area *316*
Terminus *317*
Tidewater glacier *317*

Iceberg *317*
Plucking *319*
Glacial striation *320*
Glacial polish *320*
Rock flour *321*
Cirque *321*
Tarn *321*
Horn *321*
Arête *321*
Hanging valley *322*
Drift *323*
Till *323*
Stratified drift *323*
Erratic *324*

Moraine *324*
Terminal moraine *324*
Ground moraine *324*
Recessional moraine *325*
Lateral moraine *325*
Medial moraine *325*
Drumlin *326*
Valley train *326*
Outwash plain *326*
Kame *326*
Esker *326*
Kettle *327*
Glacial epoch *327*
Pleistocene Ice Age *327*

Deflation *330*
Desert pavement *331*
Saltation *331*
Blowout *332*
Ventifact *332*
Dune *332*
Lee *333*
Slip face *333*
Barchan dune *334*
Transverse dune *334*
Parabolic dune *335*
Longitudinal dune *336*
Loess *336*

REVIEW QUESTIONS

•

1. Describe the metamorphism of newly fallen snow to glacial ice.

2. Differentiate between an alpine glacier and a continental glacier. Where are alpine glaciers found today? Where are continental glaciers found today?

3. Distinguish between basal slip and plastic flow.

4. Why are crevasses only about 40 to 50 meters deep, even though many glaciers are much thicker?

5. Describe the surface of a glacier in the summer and in the winter in (a) the accumulation area and (b) the ablation area.

6. How do icebergs form?

7. Describe how glacial erosion can create (a) a cirque, (b) striated bedrock, and (c) smoothly polished bedrock.

8. Describe the formation of arêtes, horns, hanging valleys, and truncated spurs.

9. Distinguish among ground, recessional, terminal, lateral, and medial moraines.

10. Why are kames and eskers features of receding glaciers?

11. Describe four types of topographic features left behind by the continental ice sheets.

12. Why is wind erosion more prominent in deserts and sandy coastlines than in humid regions?

13. Describe the formation of desert pavement.

14. Describe the evolution and shape of a barchan dune, a transverse dune, a parabolic dune, and a longitudinal dune. Under what conditions does each type of dune form?

DISCUSSION QUESTIONS

•

1. Outline the changes that would occur in a glacier if (a) the average annual temperature rose and precipitation decreased; (b) the temperature remained constant but precipitation increased; and (c) the temperature decreased and precipitation remained constant.

2. In some regions of northern Canada, both summer and winter temperatures are cool enough for glaciers to form. Speculate on why continental glaciers are not forming in the region.

3. In some mountain ranges, the tops of mountain peaks are jagged and covered by rubble, whereas the lower elevations of the mountains are rubble-free. Give a plausible explanation for these observations.

4. If you found a large boulder lying in a field, how would you determine whether or not it was an erratic?

5. Imagine that you encountered some gravelly sediment. How would you determine whether it was deposited by a stream or a glacier?

6. Explain how medial moraines prove that glaciers move.

7. Compare and contrast erosion, transport, and deposition of sediment by wind, streams, mass wasting, and glaciers.

8. Imagine that someone told you that an alluvial fan had been deposited by wind. What evidence would you look for to test this statement?

9. What type of dune forms under each of the following sets of conditions? (a) Relatively high vegetation cover, sand supply, and wind strength. (b) Low vegetation and sand supply.

10. Imagine that you were looking at a satellite photograph of a distant planet. What deductions could you make if you saw numerous sand dunes?

The Oceans

UNIT

IV

Año Nuevo Point, California

Oceans and Coastlines

No oceans exist on our neighboring planets—Mercury, Venus, and Mars. No continents exist on the outer giants—Jupiter, Saturn, Uranus, and Neptune. Pluto is frozen solid. Among the planets, only Earth has both solid continents and liquid seas.

The Earth formed about 4.6 billion years ago. Water started to collect on the surface shortly afterward, and ocean basins filled to about their present level by three billion years ago. The first living organisms probably evolved in the primordial seas. Although we do not know when life began, fossils of single-celled creatures similar to modern bacteria have been found in 3.5- to 3.6-billion-year-old rocks from Australia and South Africa. These organisms lived in warm, shallow ocean water.

● A kayaker paddles along the east coast of Ellesmere Island in northern Canada.

Today oceans cover about 71 percent of the Earth's surface. Ocean currents carry heat from place to place across the globe. Without this exchange of heat, the equator would be unbearably hot, the higher latitudes would be frigid, and the temperature difference between night and day would be much greater than it is at present. Thus, the oceans profoundly affect the surface of our planet.

OCEANS

13.1 The Earth's Oceans

Why Does Earth Have Oceans?

The Earth's water comes from three sources.

1. Recall that the Earth formed when an interstellar cloud of dust and gas condensed under the influence of gravity. Over time these individual particles coalesced into a planet. The solid planet then grew hotter due to the energy generated during gravitational collapse combined with heat from radioactive decay and mete-orite bombardment. Volcanic eruptions occurred frequently, and water rose to the surface with the fiery magma. Most of this volcanic water collected on the surface during the first half-billion years of Earth's history, although volcanoes continue to emit water today.

2. As the early Earth formed, large amounts of water remaining in the inner Solar System were attracted by the Earth's gravity.

3. Comets are composed mainly of water and other light compounds. During the early history of the Solar System, comets collided with Earth, adding water to it.

These three sources of water must have been common to all planets in the inner Solar System. Why, then, has Earth remained wet while our neighbors are dry? Water exists only under a narrow range of conditions. If the temperature is too low, water freezes. Water once flowed on Mars, but today only ice is found near its surface. If the temperature is too high or if the atmospheric pressure too low, water boils. Perhaps Venus once had oceans, but today its surface temperature is so high that its water has boiled off and escaped into space.

Figure 13–1   A schematic cross section of the continents and ocean basins. The vertical axis shows elevations relative to sea level. The horizontal axis shows the relative areas of the types of topography. Thus, about 30 percent, or 150 million square kilometers, of the Earth's surface lies above sea level.

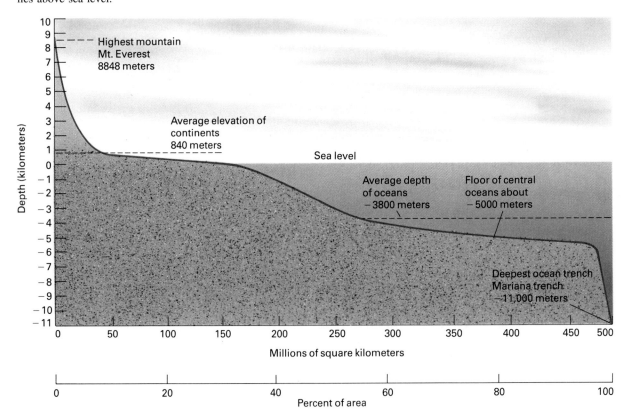

So much water exists at the Earth's surface that if Earth were a perfectly smooth sphere, it would be covered by a global ocean 2000 meters deep. Therefore, although we began this section asking "Why does the Earth have oceans?" a more appropriate question is "Why do continents rise above the seas?" Recall that the solid lithosphere floats on the plastic rock of the asthenosphere. Regions of the Earth covered with oceanic crust float at low elevations because oceanic crust is thin and dense. In contrast, continents float at higher elevations because they are thicker and less dense. Because water runs downhill, the low ocean basins fill with water and the continents protrude above sea level. The average depth of the oceans is 3800 meters, whereas the average elevation of continents is only 840 meters above sea level. The volume of the oceans is 18 times greater than that of all land above sea level (Fig. 13–1).

Geography of the Oceans

All of the Earth's oceans are connected, and water flows from one to another, so in one sense the Earth has just one global ocean. However, several distinct ocean basins exist (Fig. 13–2). Their boundaries were originally related to commerce, but today we realize that the boundaries are also geological in nature. The largest and deepest ocean is the Pacific. It covers one third of Earth's surface, more than all land combined. The Pacific Ocean averages 4000 meters deep and contains more than half of the world's water. The Atlantic Ocean has about half the surface area of the Pacific. The Indian Ocean is slightly smaller than the Atlantic. The Arctic Ocean surrounds the North Pole and extends southward to the shores of North America, Europe, and Asia. Therefore, it is bounded by land, with only a few straits and channels connecting it to the Atlantic and Pacific Oceans. The surface of the Arctic Ocean freezes in winter, and parts of it melt for a few months during summer and early fall. During summer, portions of the Arctic Ocean are ice-free while other regions are covered with a congested mass of ice floes that crash and grind against one another. The Antarctic Ocean, feared by sailors for its cold and ferocious winds, has no geological or topographical northern boundary. However, oceanographers define its boundary as the zone where warm currents from the north converge with cold antarctic water.

Figure 13–2 The oceans of the world. The "seven seas" are the North Atlantic, South Atlantic, North Pacific, South Pacific, Indian, Arctic, and Antarctic. However, these designations are related more closely to commerce than to the geology and oceanography of the ocean basins. Geologists and oceanographers recognize four major ocean basins: the Atlantic, Pacific, Indian, and Arctic.

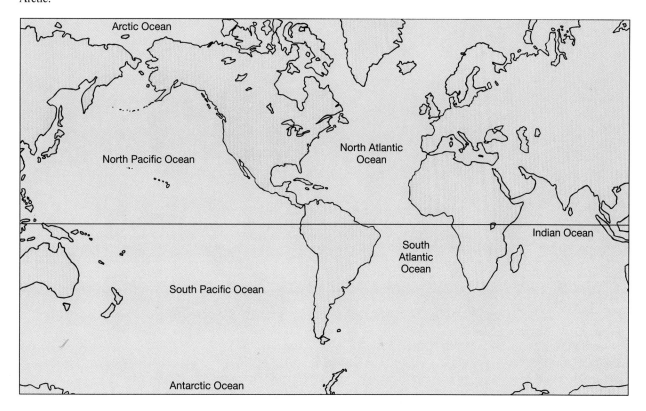

Ice floes clog portions of the Arctic Ocean even in summer.

The Southern Hemisphere is predominantly ocean, whereas the Northern Hemisphere is mostly land. Since oceans transport and store heat, this distinction has important consequences for climate. It is discussed in Chapter 17.

13.2 Seawater

Salinity

Dissolved ions make up about 3.5 percent of the weight of ocean water. Oceanographers measure **salinity** in parts per thousand (ppt). (Percent is a measure of parts per hundred.) Thus, 3.5 percent is 35 ppt. The six ions listed in Figure 13–3 make up 99 percent of the ocean's dissolved material. However, almost every element found on land is also found dissolved in seawater, albeit mainly in trace amounts. For example, seawater contains about 0.000000004 (4×10^{-9}) ppt gold. Although the concentration is small, the oceans are large and therefore contain a lot of gold. About 4.4 kilograms of gold are dissolved in each cubic kilometer of seawater. Since the oceans contain about 1.3 billion cubic kilometers of water, about 5.7 billion kilograms of gold exist in the oceans. Unfortunately, it would be hopelessly expensive to extract even a small portion of this amount. In addition to trace elements and salts, seawater also contains dissolved gases, especially oxygen and carbon dioxide.

Although the average salinity of the oceans is 35 ppt, salinity varies geographically. High rainfall near the equator dilutes seawater to 34.5 ppt. Conversely, in dry subtropical regions, where evaporation is high and precipitation low, salinity can be as high as 36 ppt (Fig. 13–4). Salinity varies more dramatically along coastlines. The Baltic Sea is a shallow ocean basin fed by many large freshwater rivers and diluted further by rain and snow. As a result, its salinity is as low as 2 ppt. In contrast, the Persian Gulf has low rainfall, high evaporation, and few large inflowing rivers, and its salinity exceeds 42 ppt.

The world's rivers carry more than 2.5 billion tons of dissolved salts to the oceans every year. Underwater volcanoes contribute additional dissolved ions. If salts are being added continuously, is the salinity of the oceans

Figure 13–3 The composition of seawater. Starting with 1 kilogram (1000 grams) of seawater, the weights of water and dissolved ions are given in grams.

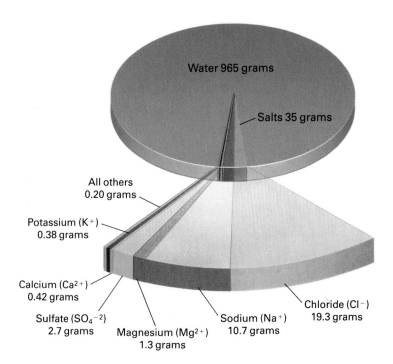

Water 965 grams

Salts 35 grams

All others 0.20 grams

Potassium (K^+) 0.38 grams

Calcium (Ca^{2+}) 0.42 grams

Sulfate (SO_4^{-2}) 2.7 grams

Magnesium (Mg^{2+}) 1.3 grams

Sodium (Na^+) 10.7 grams

Chloride (Cl^-) 19.3 grams

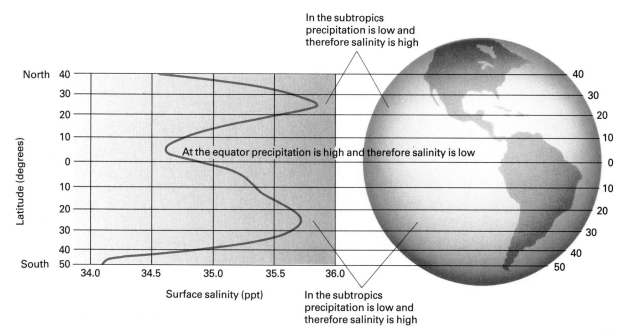

Figure 13–4 Changes in the salinity of the surface of the central oceans with latitude.

increasing? In 1900 John Joly, an Irish geologist, reasoned that the ocean basins filled with fresh water shortly after the Earth formed, and the seas became saline due to the continuous addition of salt. Using estimates of the amount of salt in river water and the volume of water flowing to the oceans each year, he calculated the total amount of salt added to the oceans per year. By comparing this number with the estimated amount of salt in the modern oceans, he calculated that the oceans, and hence the Earth, were about 90 million years old. Today we know that the Earth is 4.6 billion years old. Why were Joly's conclusions so wrong? Joly assumed that once salt was added to the seas, it remained dissolved in the seawater. In fact, vast quantities of salt are removed from seawater. When a portion of a marine basin becomes cut off from the open oceans, the water evaporates, precipitating thick sedimentary beds of salt. Additionally, large amounts of salt become incorporated into shale and other sedimentary rocks. It now appears that, throughout much of geologic time, the salinity of the oceans has been constant because salt has been removed from seawater at the same rate at which it has been added.

Temperature

Anyone who swims in lakes knows that on a hot, sunny day, surface water is warmer than water a meter or two deep. Fresh water is densest at 4°C, and salt water is densest at around 1° to 2°C. Thus, warm water floats on cold water because it is less dense.

Lakes are comfortably warm for swimming during the summer because warm water floats on the surface and does not mix with the deep, cooler water. Oceans develop a layered temperature profile as well, but because surface water is stirred by waves and currents, the warm layer in an ocean may be up to 450 meters thick. Below this upper layer temperature drops rapidly with depth. This second layer, called the **thermocline**, extends to a depth of 2 kilometers. Beneath the thermocline, ocean water is consistently around 1°C to 2°C. The cold, dense water in the ocean depths mixes very little with that of the surface layer. Thus, there are three distinct temperature zones in the ocean, as shown in Figure 13–5. This layered structure does not exist in the polar latitudes because cold surface water sinks, causing vertical mixing.

13.3 Waves

Waves form when wind blows across water. They vary from gentle ripples on a pond to destructive giants smashing against shore during a hurricane. In deep water, the size of a wave depends on (1) wind speed, (2) the length of time the wind has blown, and (3) the distance the wind has traveled without being interrupted by land. (Sailors call this factor the **fetch**.) A gentle breeze blowing at 25 kilometers per hour (16 miles per hour) for 2 to 3 hours across a bay 15 kilometers wide generates waves about 0.5 meter high. In contrast, when a large Pacific storm blows at 90 kilometers per hour (55 miles per hour)

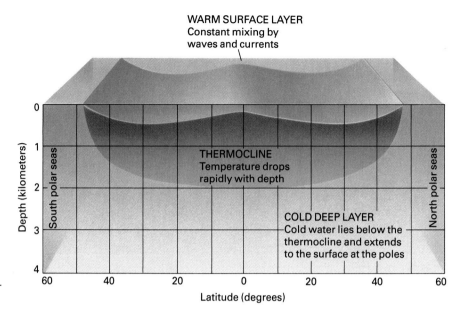

WARM SURFACE LAYER
Constant mixing by
waves and currents

Figure 13–5   A temperature profile of the oceans.

for several days and waves roll uninterrupted for 3500 kilometers, they can become 30 meters high, as high as a ship's mast.

The highest part of a wave is called the **crest**; the lowest is the **trough** (Fig. 13–6). The **wave height** is the vertical distance from the crest to the trough. We hardly ever talk about *one* wave, because waves occur in a series of crests and troughs. The **wavelength** is the distance between successive crests (or troughs). The **wave period** is the time interval between two crests as waves roll past a stationary observer.

If you make waves by throwing a rock into a calm lake, the waves travel across the water surface. However, the water does not move at the same speed or in the same direction as the waves. If you are sitting in a boat in deep water, you bob up and down and sway back and forth as the waves pass beneath you, but you do not travel along with the waves. In an ocean or lake wave, water moves in circular paths, as shown in Figure 13–7.

The movement of water continues downward below the wave trough in circles of decreasing size. At a depth equal to about one-half the wavelength, the disturbance becomes negligible. Thus, if you dive deep enough, you escape wave motion.

As a wave passes over deep water, even the lowest circles of water motion do not contact the sea floor. But when a wave approaches shore and enters shallow water, its base begins to touch bottom. Friction between the wave and the sea floor slows the wave. Consequently, the wave behind it catches up and the wavelength decreases.

Figure 13–6   Terminology used to describe waves.

Surf on the California coast.

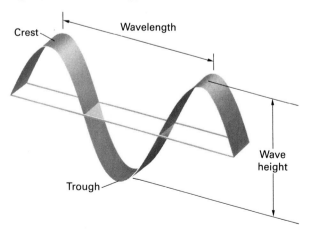

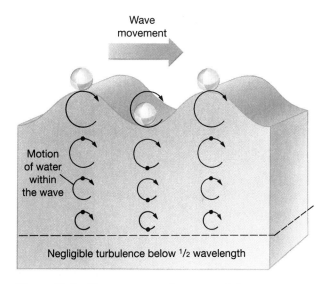

Figure 13–7 The movement of a wave and the movement of water within the wave.

In addition, friction slows the base of the wave more than the top (Fig. 13–8). As the crest rides over the trough, the wave steepens until it can no longer support itself. At this point it collapses forward, or **breaks**, as it rolls into shallower water. As an analogy, think of a skier whose skis catch on a branch hidden under the snow. The skis slow down while the person's upper body continues at constant speed. The skier pitches forward and somersaults into the snow. Once a wave breaks, its water flows toward the beach as a chaotic, turbulent mass called **surf**. A sheet of foamy water called **swash** flows onto the beach. When its energy is dissipated, the water flows back to sea as **backwash**.

Recall from Chapter 6 that a tsunami (an ocean wave produced by an underwater earthquake or volcanic eruption) may be only 1 to 3 meters high, but its wavelength may be 100 to 150 kilometers. The destructive power of a tsunami develops when the wavelength shortens and the wave height increases as the wave rolls into shallow water and breaks.

Refraction

Most waves approach shore at an angle. Thus, one end of a wave encounters shallow water first and slows down while the rest of the wave is still in deeper water. The part of the wave close to shore slows down while the remainder continues to advance at a constant speed. As a result, the wave bends. This process is called **refraction**. Consider a sled gliding down a snowy hill onto a cleared road. If the sled hits the road at an angle, the runner that hits the pavement first slows down, while the other runner, which is still on snow, continues to travel rapidly (Fig. 13–9). The result is that the sled turns abruptly. Water waves (in fact, all waves) behave in a comparable manner.

Giant Sea Waves

What is the maximum height that a wave striking shore can attain? Strong onshore winds can push the ocean surface toward shore with enough force to raise sea level temporarily. In addition, the extreme low atmospheric pressure of a large storm pulls the water upward like a giant vacuum cleaner, causing the surface to rise further. The combined effects of these two processes are called a **storm surge**.

Consider the New Jersey coast as an example. Oceanographers estimate that an intense hurricane would generate a 6-meter-high storm surge along the coast of New Jersey. Now imagine that this surge occurs at the highest possible tide level, which is almost 2 meters near Atlantic City, New Jersey. Finally, wind-driven waves rise and roll inland on the elevated water surface. Using profiles of the depth and configuration of the sea floor, oceanographers have calculated that the largest wave that could break against the beach at Atlantic City is about 12 meters.

Figure 13–8 When a wave approaches the shore, the circular motion flattens out and becomes elliptical. The wavelength shortens, and the wave steepens until it finally breaks, creating surf. The dashed line shows the shape that the wave would have had in deep water.

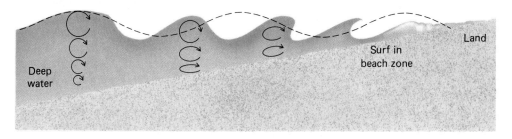

A

Sled analogy

Snowy hill

Road

B

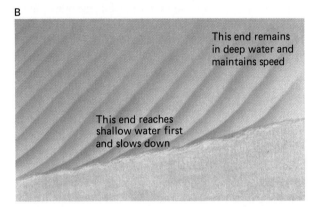

This end remains
in deep water and
maintains speed

This end reaches
shallow water first
and slows down

C

Figure 13–9 (A) A sled turns upon striking a paved road-way at an angle because one of its runners hits the roadway and slows down before the other does. (B) When a water wave strikes the shore at an angle, one end slows down, caus-ing the wave to refract, or bend. (C) Wave refraction on a lakeshore.

If the wave were higher, its base would encounter the sea floor and it would break before reaching shore.

Adding these numbers (6 + 2 + 12 meters), a wave crest of 20 meters is possible along the New Jersey coastline. Such a wave would destroy Atlantic City and

A Coast Guard rescue vessel in storm waves off the Oregon coast. *(Courtesy of U.S. Coast Guard)*

flood most of southern New Jersey. For a brief time, the hills of the Pine Barrens would be islands in a salt sea.

Of course, this is the worst case imaginable, an infrequent event. When Hurricane Hugo struck the coast of North Carolina with 220-kilometer-per-hour winds in the fall of 1989, the storm waves were 5 to 6 meters high. They destroyed nearly 1900 homes and killed 40 people as they rolled over low-lying barrier islands and inundated coastal areas. Wind and water damage totaled $3 billion.

Wave Erosion

Just as stream erosion is most intense during a flood, when a stream flows with greatest energy, most coastal erosion occurs during intense storms. A 6-meter-high storm wave strikes shore with 40 times the force of a normal 1.5-meter-high wave. A giant, 10-meter-high storm wave strikes a 10-meter-wide sea wall with four times the thrust energy of the three main orbiter engines of a space shuttle.

A wave striking a rocky cliff drives water into cracks or crevices in the rock, compressing air in the cracks. The air and water together may dislodge rock fragments or even huge boulders. Engineers built a breakwater in Wick Bay, Scotland, of car-size rocks weighing 80 to 100 tons each. The rocks were bound together with steel rods set in concrete, and the sea wall was topped by a steel-and-concrete cap weighing more than 800 tons. A large storm broke the cap and scattered the upper layer of rocks about the beach. The breakwater was rebuilt, reinforced, and strengthened, but a second storm destroyed this wall as

Destruction from Hurricane Hugo.   *(Courtesy of Wide World Photos/Associated Press)*

well. On the Oregon coast, a 60-kilogram rock was tossed over a 25-meter-high lighthouse by the impact of a storm wave. After sailing over the lighthouse, it crashed through the roof of the keeper's cottage, startling the inhabitants.

While images of flying boulders are spectacular, most wave erosion occurs gradually. Water is too soft to abrade rock, but waves carry large quantities of silt, sand, and gravel. Breaking waves roll this sediment back and forth over bedrock, acting like liquid sandpaper eroding the rock. Seawater also dissolves soluble rock and carries ions in solution.

13.4 Tides

Even the most casual observer notices that the sea surface rises and falls daily. If the water level is high at noon, it is low about 6 hours later, at 6 o'clock, and is high again near midnight. These vertical displacements are called **tides**. In most seacoasts, approximately two high tides and two low tides occur every day (Fig. 13–10).

The gravitational pulls of the Moon and Sun cause tides. Although the Moon is much smaller than the Sun, it is so much closer to the Earth that its gravitational

Figure 13–10   (A) Low tide, and (B) high tide at Digby, Nova Scotia.   *(Geological Survey of Canada—GSC 67898)*

A

B

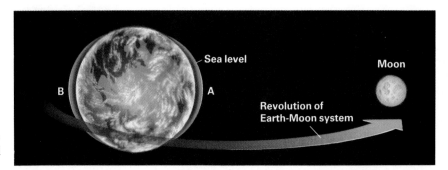

Figure 13–11 A schematic view of tide formation. Magnitudes and sizes are exaggerated for emphasis.

influence is stronger. At any time, one point on Earth (marked A in Fig. 13–11) lies directly under the Moon while all other points are farther away. Since gravitational force is greatest for objects that are closest together, the part of the ocean nearest the Moon is attracted with the strongest force. This attraction causes the ocean to bulge outward toward the Moon, resulting in a high tide in that region.

But now our simple explanation runs into trouble. As the Earth spins on its axis, a given point on the Earth passes directly under the Moon approximately once every 24 hours. Thus, our initial explanation would account for high tide if it occurred every 24 hours. But the period between successive high tides is only 12 hours. High tide occurs not only at a point on Earth directly under the Moon, but also simultaneously at a point 180° away from

Figure 13–12 The Moon moves 13.2° every day. (A) The Moon is directly above an observer on Earth. (B) One day later, the Earth has completed one complete rotation, but the Moon has traveled 13.2°. The Earth must now travel for another 53 minutes before the observer is directly under it.

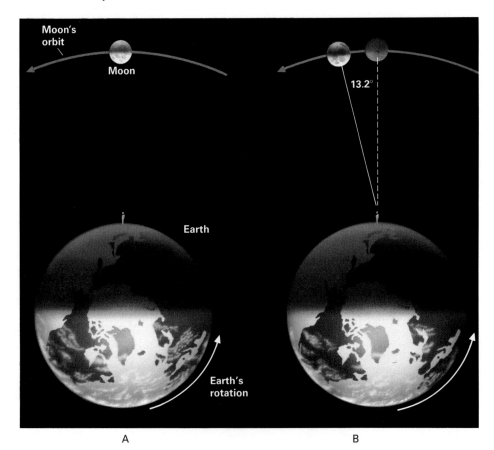

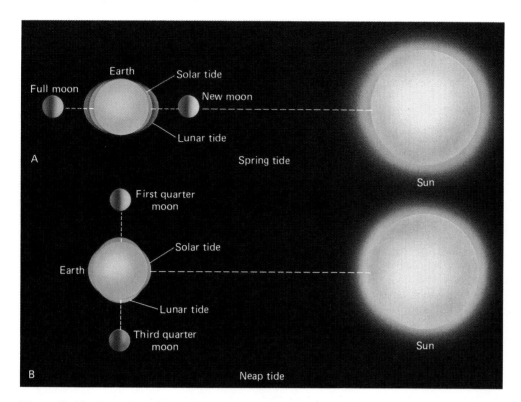

Figure 13–13 Formation of (A) spring and (B) neap tides.

the Moon. To understand this phenomenon, we must consider the Earth-Moon orbital system. Most people visualize the Moon orbiting the Earth. Although this is true, it is more accurate to say that the Earth and the Moon orbit a common center of gravity. It is as if the two were locked together like dancers spinning in each other's arms. Just as the back of a dancer's dress flies outward when she twirls around, the opposite side of the Earth's oceans bulge outward as it twirls around with the Moon.* This bulge is the high tide 180° away from the Moon (point B on Fig. 13–11). Thus, tides rise and fall twice daily.

Tides do not occur at the same time each day, but are delayed by approximately 53 minutes every 24 hours. The Earth makes one complete rotation in 24 hours, but at the same time the Moon is orbiting the Earth. After a point on Earth makes one complete rotation, it must spin for an additional 53 minutes to catch up with the orbiting Moon. This is why the Moon rises 53 minutes later each day. In the same manner, the tides are about 53 minutes later each day (Fig. 13–12).

When both the Sun and Moon are directly in line with Earth, their gravitational fields combine, creating a strong tidal bulge. During these times, called **spring tides**,

*To make the analogy complete the dancers would have to be twirling individually as they spin together in each other's arms, because both Earth and the Moon rotate on their axes.

high tides are higher than average and low tides are lower (Fig. 13–13A). If you enjoy observing the communities of plants and animals that live in the intertidal zone, low spring tides are a good time to walk the beaches. When the Sun and Moon are 90° out of alignment with Earth, each partially offsets the effect of the other and tidal differences are less. These smaller tidal differences are called **neap tides** (Fig. 13–13B).

13.5 Currents

An ocean wave moves along the sea surface, but the water in the wave travels in circles, ending up where it started. In contrast, a **current** is a continuous flow of water in a particular direction. Although river currents are familiar and easily observed, ocean currents were not recognized by early mariners.

Surface Currents

In the late 1700s, Benjamin Franklin lived in London, where he was Deputy Postmaster General for the American colonies. He noticed that mail ships took two weeks longer to sail from England to North America than did merchant ships. Franklin learned that the captains of the

(Text continues on p. 356)

Earth Science and the Environment

Energy from the Ocean

Energy from Tides

In many coastal regions, tidal currents funnel through narrow entrances into bays and estuaries. Twice a day, sea water flows into the bays with the rising tide, and twice a day it rushes outward. The energy of these currents can be harnessed if a tidal dam is built and a turbine installed.

During the 1800s, tidal power was popular in the United States, especially along the coast of Maine with its narrow bays and inlets. At that time, it was an attractive alternative to bulky steam engines that consumed large quantities of wood. However, when fossil fuels became cheap in the 1900s, almost all tidal power plants were abandoned. Today, the potential still exists, but even in the most favorable sites, tidal dams are not always economical. For example, a 250-megawatt tidal power station was built in France in 1968. Although technologically successful, this facility costs 2.5 times more per unit of energy than a comparable dam on a river.

In addition to being costly, tidal dams can create environmental problems. Tidal bays are often productive estuaries where fish come to breed, and they are also popular places for recreation. If these areas are dammed, some of the natural qualities and resources will be lost. Nevertheless, proponents of tidal projects point out that when fossil fuels become scarce, tidal energy might become attractive.

Energy from Ocean Waves

Waves strike every coastline in the world, and their combined energy is enormous. As the cost of oil rises and pollution and political availability of petroleum become more troublesome, we ask, "Is it possible and economical to harness wave energy?"

Figure 1   A schematic view of one design for harnessing wave energy. An incoming wave raises the water level inside a concrete chamber. Air inside the chamber is compressed and spins a turbine to power a generator.

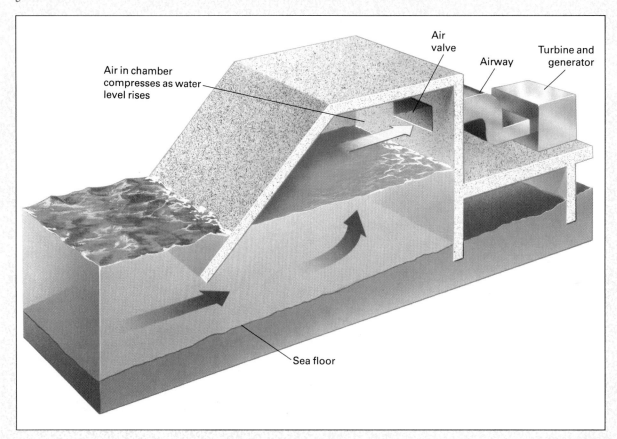

In 1990 engineers designed and built a simple wave generator in Northern Ireland, shown in Figure 1. When water from an incoming wave rises in a concrete chamber, it forces air through a narrow valve. The air spins a rotor that drives an electrical generator. The generator produces electricity for a small coastal village.

To date, the major barrier to widespread use of wave power has been an economic one. Although the wave energy along an entire coastline is enormous, the potential in any one place is small. Therefore, many structures are needed to harness large quantities of energy. At present, the structures are too expensive to be practical. However, improvements in design and rising oil prices may reverse the economic balance sheet.

Ocean Thermal Power

It is also possible to use the heat of the ocean to produce electricity. In tropical regions, the sea surface is approximately 20°C warmer than subsurface water. The warm water is hot enough to vaporize a pressurized liquid such as ammonia. The gaseous ammonia can drive a turbine just as hot steam drives a turbine in a coal-fired plant. In an ocean thermal generator, the gaseous ammonia is cooled by the subsurface water and reused, as shown in Figure 2. The efficiency of such an engine is low, but there is so much water in the oceans that a large amount of energy is available. Today, the capital costs of the ocean thermal-power plants are so high that they are uneconomical even though they produce electricity without consuming any fuel.

Figure 2 A schematic view of an ocean thermal power plant. Pressurized ammonia is boiled by the warm surface waters of tropical oceans. The ammonia gas expands against the blades of a turbine, and the spinning turbine drives a generator to produce electricity. The gases are cooled and condensed by colder subsurface water that is pumped into the power plant.

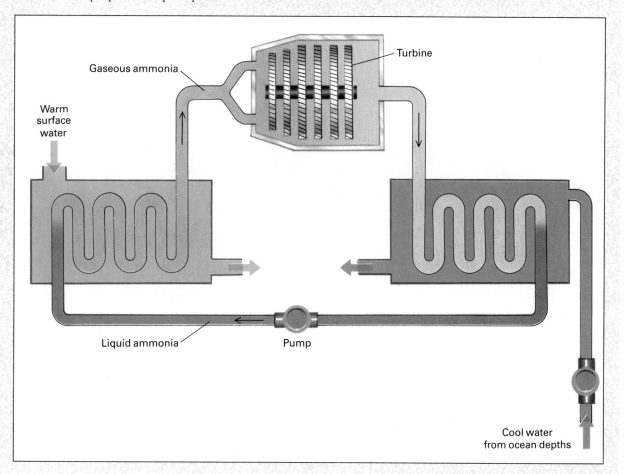

merchant ships had discovered a current flowing north-ward along the east coast of North America and then across the Atlantic to England. When sailing from Europe to North America, the merchant ships saved time by avoiding the current. The captains of the mail ships were unaware of this current and lost time sailing against it on their westward journeys. In 1769, Franklin and his cousin, Timothy Folger, a merchant captain, charted the current and named it the **Gulf Stream**.

As ship traffic increased and navigators searched for the quickest routes around the globe, they discovered other ocean currents (Fig. 13–14). Ocean currents have been described as rivers in the sea. The analogy is only partially correct; ocean currents have no well-defined banks, and they carry much more water than even the largest river. The Gulf Stream is 80 kilometers wide and 650 meters deep near the coast of Florida and moves at approximately 5 kilometers per hour, a moderate walking speed. As it moves northward and eastward, the current widens and slows; east of New York it is more than 500 kilometers wide and travels at less than 8 kilometers per day.

Another important difference between rivers and sea currents is that rivers flow in response to gravity, whereas ocean surface currents are driven primarily by wind. When wind blows across water in a constant direction for a long time, it drags surface water along with it, forming a current. In many regions the wind blows in the same

direction throughout the year, forming currents that vary little from season to season. However, in other places, changing winds cause currents to change direction. For example, in the Indian Ocean prevailing winds shift on a seasonal basis. When the winds shift, the ocean currents follow.

Ocean Currents and the Coriolis Effect

Notice in Figure 13–14 that most open ocean surface currents move in circular paths called **gyres**. Gyres rotate clockwise in the Northern Hemisphere and counterclockwise in the Southern Hemisphere.

To understand gyres, let us start with a few analogies. Imagine riding a skateboard across a smooth parking lot and trying to throw a ball to a friend who is standing a few meters to the side (Fig. 13–15). When the ball is in your hand, it is moving along with you and the skateboard. After it is thrown, the ball heads toward your friend but at the same time it continues to move in the same direction as the skateboard. As a result, the ball curves, and if your friend is not alert, the friend will not catch it. The effect is dramatic; try it. For an analogy that is closer to the rotating Earth, stand on one side of a spinning playground merry-go-round and have a friend stand in the center (Fig. 13–16). Then throw a ball to the person in the center. Again the ball veers sharply and is difficult to catch.

Figure 13–14   Major ocean currents of the world.

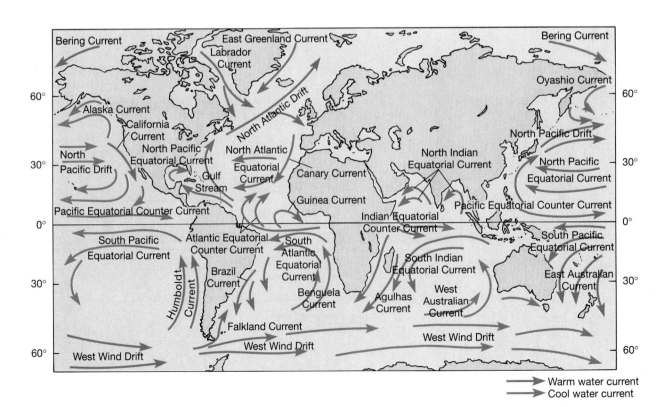

Figure 13–15   When the person on a moving skateboard throws a ball, the ball arcs in the direction of motion.

Why? When the merry-go-round makes one complete revolution, the person on the outside travels a distance equal to its circumference. Because the circumference at the center is zero, the person at the center is spinning but is not otherwise moving. Thus, when the ball is thrown, it veers in the direction of the rotation of the merry-go-round.

The spherical Earth is similar to the merry-go-round. The circumference of the Earth is greatest at the equator and decreases toward the poles. But all parts of the planet make one complete rotation every day. Since a point on the equator must travel farther than any other point on the Earth in 24 hours, the equatorial region moves faster than regions closer to the poles. At the equator, all objects

move eastward with a velocity of about 1600 kilometers per hour; at the poles there is no eastward movement at all, and the velocity is 0 kilometers per hour. Now imagine that a rocket ship is fired from the equator toward the North Pole. Before it was launched it was traveling eastward at 1600 kilometers per hour with the rotating Earth. As it takes off, it is moving both eastward and northward. But at any distance north of the equator, it is traveling eastward faster than the Earth beneath it. Thus, the rocket curves toward the east, or the right. The same kind of deflection occurs when a mass of water or air moves poleward from the equator, as shown in Figure 13–17A.

Conversely, consider an ocean current flowing southward from the Arctic Ocean toward the equator. Since it started near the North Pole, this water moves more slowly than the Earth's surface in tropical regions, and therefore it lags behind as it flows southward. The Earth spins in an easterly direction, and therefore the current veers toward the west, or to the right, as shown in Figure 13–17B.

North-south currents always veer to the right in the Northern Hemisphere. In the Southern Hemisphere currents turn toward the left for the same reason. The deflection of currents caused by the Earth's rotation is called the **Coriolis effect.** The Coriolis effect also deflects air currents, as will be discussed in Chapters 16 and 17.

Ocean currents profoundly affect the climates of large regions of the Earth. One million cubic meters of warm water flow northward in the Gulf Stream *every second,* warming coastal areas. For example, Churchill, Manitoba, is a village in the interior of Canada. It is frigid and icebound for much of the year. Polar bears regularly migrate through town. Yet it is at about the same latitude as Glasgow, Scotland, which is warmed by the Gulf Stream and therefore experiences relatively mild winters.

Figure 13–16   The Coriolis effect is demonstrated on a playground merry-go-round. The person standing on the rim is analogous to an observer at the equator, and the person standing in the center is analogous to an observer at the pole. The ball arcs in the direction of rotation of the merry-go-round.

A   Rotation of Earth toward east   B   Rotation of Earth toward east

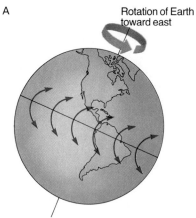 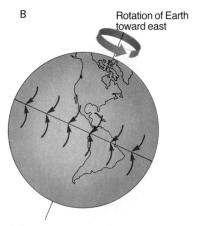

Figure 13–17 The Coriolis effect. Water moving north or south is deflected by the rotation of the Earth.

Water or air moving poleward from the Equator is traveling east faster than the land beneath it and veers to the east (turns right in the Northern Hemisphere and left in the Southern Hemisphere)

Water or air moving toward the Equator is traveling east slower than the land beneath it and veers to the west (turns right in the Northern Hemisphere and left in the Southern Hemisphere)

In general, the climate in western Europe is warmer than that at similar latitudes in other regions not heated by tropical ocean currents.

Upwelling

In Figure 13–14, note that the California current flows southward along the coast of California. In the Southern Hemisphere, the Humboldt current moves northward along the west coast of South America. Both currents are deflected westward by the Coriolis effect. As these surface currents veer away from shore, water from the deep parts of the ocean basins is drawn upward to the surface. This upward flow of water is called **upwelling**. Upwelling carries cold water from the ocean depths to the surface. In August, water on the east (Atlantic) coast of the United States is warmed by the Gulf Stream and may be a comfortable 21°C. However, on the California coast, the cool California current combines with the upwelling current to produce water that is only 15°C, and surfers and swimmers must wear wetsuits to stay in the water for long. Upwelling also brings nutrients from the deep ocean to the surface, creating rich fisheries along the coasts of California and Peru.

Deep-Sea Currents

Wind does not affect the ocean depths, and oceanographers once thought that deep ocean water was almost motionless. In her book *The Sea Around Us*, Rachael Carson wrote that the ocean depths are "a place where change comes slowly, if at all." However, in 1962 ripples and small dunes were photographed on the floor of the North Atlantic. Since flowing water forms these features,

the photographs suggested that water was moving in the ocean depths. More recently, deep-sea currents have been measured directly with flow meters, and moving sediment has been photographed with underwater television cameras.

Whereas wind drives surface currents, deep-sea currents are driven by differences in water density. When a continuous supply of dense water is available, it sinks and flows horizontally along the sea floor to form a deep-

A vertical slice through a section of the North Atlantic Ocean floor. The cross-bedding shows the existence of deep-sea currents. *(Charles Hollister, Woods Hole Oceanographic Institution)*

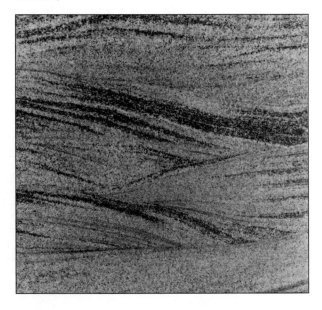

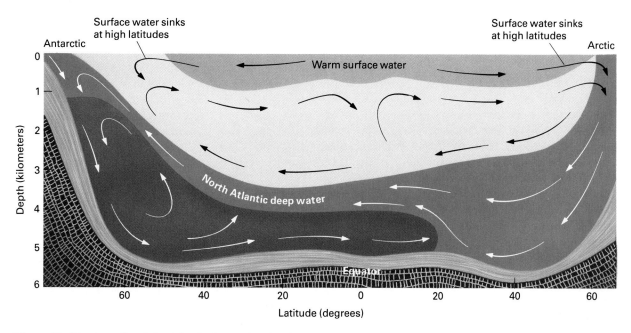

Figure 13–18 A profile of the Atlantic Ocean showing the deep-sea currents.

sea current. Two factors cause water to become dense and sink: temperature and salinity. Recall that water is densest when it is cold, close to freezing. Therefore, as tropical water moves poleward and cools, it becomes denser and sinks. Recall also that the salinity of seawater changes with location. Salty water is denser than less salty water.

The Gulf Stream originates in the subtropics, where seawater is warm and salty. When this water reaches the northern part of the Atlantic Ocean, near the tip of Greenland, it cools and sinks. This water is then deflected to the south to form the North Atlantic Deep Water, which flows along the sea floor all the way to Antarctica (Fig. 13–18). An individual water molecule that sinks near Greenland may travel for 500 to 2000 years before resurfacing half a world away in the south polar sea.

Tidal Currents

When tides rise and fall along an open coastline, water moves in and out from the shore as a broad sheet. If the flow is channeled by a bay with a narrow entrance or by islands, the moving water funnels into a **tidal current**. A tidal current is a flow of ocean water caused by tides. Tidal currents can be intense where large differences exist between high and low tides and narrow constrictions occur in the shoreline. On parts of the west coast of British Columbia, a diesel-powered fishing boat cannot make headway against tidal currents flowing between closely spaced islands. Fishermen must wait until the tide, and hence the tidal currents, reverse direction before proceeding.

Longshore Currents

Waves generate currents that flow close to shore. After a wave breaks, the water flows back toward the sea. This outward flow may concentrate in a current called a **rip current** or **undertow**, which moves toward deep water. A rip current can be strong enough to carry a hapless swimmer out to sea.

If waves regularly strike shore at an angle, a **longshore current** forms and flows parallel to the coast. Longshore currents flow in the surf zone and a little farther out to sea and may travel for tens or even hundreds of kilometers. They transport sand and other sediment parallel to coastlines (Fig. 13–19). Most sediment found on a coast is not produced by erosion at that location, but rather is carried from the interior of the continent by

Figure 13–19 Formation of a longshore current.

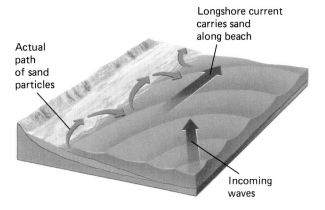

streams and deposited in deltas on the coast. It is then redistributed along the coast by longshore currents.

COASTLINES

Coastlines are among the most geologically active zones on Earth. Convergent plate boundaries occur along many continental coasts. Waves and currents weather, erode, transport, and deposit sediment continuously on all coastlines. Finally, the shallow waters adjacent to continents and oceanic islands are productive biological zones. Many organisms that live in these regions build hard shells or skeletons of calcium carbonate, the same material that composes limestone. Thus, living organisms enter into geological cycles.

13.6 Reefs

A **reef** is a wave-resistant ridge or mound built by corals, algae, and other organisms. Because these organisms need sunlight and warm, clear water to thrive, reefs develop in shallow tropical seas where no suspended clay or silt muddies the water (Fig. 13–20). As reef-building organisms die, their offspring grow on their remains. Thus, only the outer and topmost portions of a reef contain living organisms.

The South Pacific and portions of the Indian Ocean are dotted with numerous islands called atolls. An **atoll** is a circular reef that forms a ring of islands around a calm, protected body of water called a **lagoon**. Atolls vary from 1 to 130 kilometers in diameter. They are surrounded by deep water of the open sea. If corals and other reef-building organisms cannot live in deep water, how did atolls form? Charles Darwin studied the problem during his famous voyage on the *Beagle* from 1831 to 1836. He reasoned that, to make an atoll, a reef must have formed in shallow water on the flanks of a volcanic island. Eventually the island sank as the reef continued to grow upward, so that the living portion of the reef always remained in shallow water (Fig. 13–21). At first this proposal was not well received because scientists could not accept the idea that volcanic islands sank. Moreover, Darwin could not explain why an island would sink. In a sense, Darwin's theory of atoll development mirrored Wegener's continental drift theory. Both theories were well founded on observation and reasoning, but they were rejected because no plausible mechanism was offered. However, when scientists drilled into a Pacific atoll shortly after World War II and found volcanic rock beneath the reef, Darwin's original hypothesis was reconsidered. Today we know that as the lithosphere beneath a volcanic island cools, after the volcano becomes extinct it becomes denser and sinks isostatically, carrying the island down with it.

Reefs around the world have suffered severe epidemics of disease and predation within the last decade. Studies of fossils show that epidemics and even extinctions have affected reefs periodically for millions of years. However, the past few decades are different from any other interval in the Earth's history due to the phenomenal growth of human population and industry. Human activity has caused the extinction of uncounted species and has decimated the populations of many more. Therefore, we ask, are the recent epidemics among reef organisms part of

Figure 13–20 (A) An aerial view of a reef adjacent to the island of Palau in the southwestern Pacific. Note that the waves break on the reef, leaving calmer, shallow water between the reef and the island. (B) Underwater photograph of castle coral, a reef-building organism. *(Larry Davis)*

A

B

the natural cycle or a result of human activity? One possibility is that sewage provides nutrients for algae and other organisms that smother coral colonies and prevent new ones from forming. In addition, many scientists are concerned that gases released during industrial activity are increasing global temperature. This phenomenon will be discussed in Chapter 17. Reef building corals could be adversely affected by a rise in the temperature of seawater. Both of these possibilities are hard to prove. It is often difficult to locate the sources of chemical compounds in seawater. Some may originate as industrial pollutants while others form naturally. In addition, global warming remains controversial. Neither side has proven its case, and we do not know whether the corals are dying in response to natural causes or human impact.

Figure 13–21   Formation of an atoll.

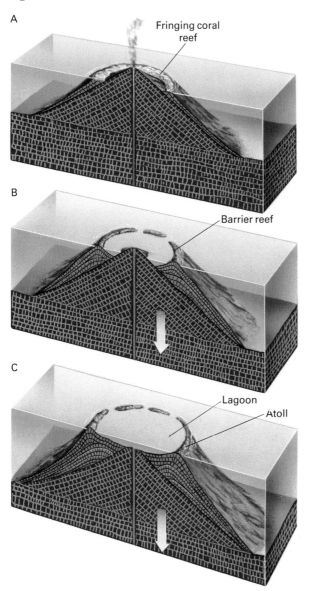

13.7 Beaches

When most of us think about going to the beach, we think of gently sloping expanses of sand. A **beach** is any strip of shoreline washed by waves and tides. Although many beaches are sandy, others are swampy, rocky, or bounded by shear cliffs (Fig. 13–22).

A beach is divided into two zones, the **foreshore** and the **backshore**. The foreshore, called the **intertidal zone**

Figure 13–22   (A) A sandy beach on the southern shore of Long Island, New York. (B) A rocky beach on the northern coast of California.

A

B

by biologists, lies between the high and low tide lines. It is alternately exposed at low tide and covered by water at high tide. The backshore is usually dry but is washed by waves during storms. These storms deposit and erode sediment and dislodge vegetation. Since many plants cannot survive periodic inundation by salt water, the backshore supports a sparse population of salt-resistant grasses. The backshore can be wide or narrow, depending on its topography and the frequency and intensity of storms. Along coasts where the elevation rises steeply, the backshore may be a narrow strip. In contrast, if the coast consists of low-lying plains where hurricanes occur regularly, the backshore may extend several kilometers inland as a zone of shifting coastal dunes partially stabilized by vegetation.

13.8 Sediment-Rich Coastlines

Many coastal landforms and the overall character of a shoreline depend on the amount of sand and other fine-grained sediment supplied to the coast. If a rich supply of sand is available, sandy beaches and other features built of sand dominate the coast.

Sources of Sediment

Three processes can supply large quantities of sand and other sediment to a coastal area: erosion, transport, and changes in sea level. Some sediment forms by erosion of bedrock or coral reefs along the coasts. However, under most conditions erosion is too slow to supply abundant sediment to a coastline.

Sediment can also be transported from some other place. Rivers carry large quantities of sand, silt, and clay to the sea and deposit it on deltas that may cover thousands of square kilometers. In some regions, glaciers deposited large quantities of till along the coast.

Longshore currents erode sediment from these sources and carry it for great distances along a coast. For example, much of the sand forming the beaches of the Carolinas and Georgia originated from the Hudson River delta between New York and New Jersey and from glacial deposits on Long Island and southern New England.

Figure 13–23 Emergent and submergent coastlines. If sea level falls or if the land rises, the sediment-laden shallow sea floor is exposed, forming sandy beaches. If coastal land sinks or sea level rises, areas that were once land are flooded. Irregular shore lines develop and beaches are commonly sediment-poor.

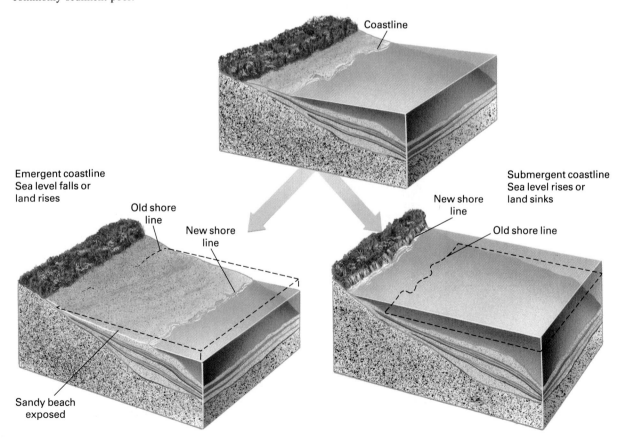

These processes deposit large quantities of sand and mud below sea level on the continental shelf. Under normal conditions, this sediment cannot be eroded and transported because it lies below the bases of waves and out of the reach of longshore currents. But if a coastline rises or sea level falls, this vast supply of sediment is exposed. This environment is called an **emergent coastline** (Fig. 13–23).

When the Pleistocene ice sheets melted, their weight was removed from land, which then rebounded isostatically all across northern North America, Europe, and Asia to form emergent coastlines. Alternatively, land can rise by local or regional tectonic uplift. An emergent coastline can also form when sea level drops.

Characteristics of Sediment-Rich Coastlines

When large quantities of sediment are available, longshore currents and waves erode, transport, and deposit the sediment along the shore. A long ridge of sand or gravel extending out from a beach is called a **spit**. As sediment continues to migrate along the coast, the spit may continue to grow. A well-developed spit may rise several meters above high-tide level and may be tens of kilometers long. If a spit grows until it blocks the entrance to a bay, it forms a **baymouth bar** (Fig. 13–24). Many seaside resorts are built on spits. Developers often ignore the fact that a spit is a transient and changing landform. The sand on a spit is continuously deposited and eroded by longshore currents and waves. If the rate of erosion exceeds that of deposition for a few years in a row, a spit can shrink or disappear completely, leading to destruction of beach homes and resorts.

Figure 13–25 An aerial view of a barrier island along the south coast of Long Island, New York.

A **barrier island** is a long, narrow, low-lying island that parallels the shoreline. It looks like a beach or spit separated from the mainland by a lagoon (Fig. 13–25). A chain of barrier islands extends along the east coast of the United States from New York all the way to Florida.

Figure 13–24 A spit forms where sediment is carried away from the shore and deposited. If a spit closes the mouth of a bay, it becomes a baymouth bar.

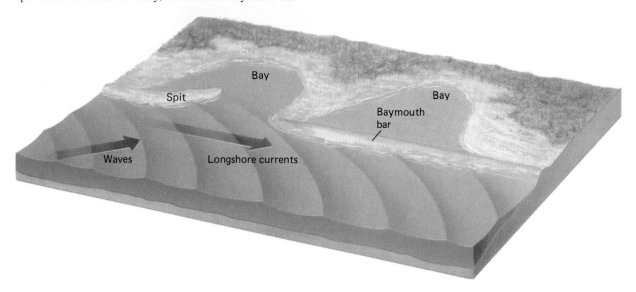

An aerial view of a spit on the south shore of Long Island.

The islands are so nearly continuous that a small boat can travel the entire coast inside the barrier island system and remain protected from the open ocean most of the time.

Barrier islands form in several ways. The two essential ingredients are a large supply of sediment and waves or currents to transport the sediment along the coast. If a coast is shallow for several kilometers outward from shore, storm waves break far out to sea. The breaking waves may carry sediment toward shore and deposit it where their energy diminishes, forming low-lying barrier islands. Barrier islands also form by deposition from longshore currents and are therefore similar to spits. If a longshore current veers out to sea, it slows down and deposits its sediment where it reaches deeper water. If this sediment is further piled up by waves, a barrier island may form. Yet another mechanism involves sea-level change. Underwater sand bars may be exposed as coastlines emerge. Alternatively, sand dunes or beaches may form barrier islands if a coast starts to submerge.

13.9 Sediment-Poor Coastlines

A **submergent coastline** develops when coastal land sinks or sea level rises. The sea then floods low-lying areas, and the shoreline moves inland. In many areas on land, bedrock is exposed or covered by only a thin layer of soil. If this type of terrain is submerged, and if rivers do not supply large amounts of sediment, the coastline is sediment-poor. Submergent coasts are commonly irregular, with many bays and headlands, because they form when the sea floods stream valleys, hills, and other topographic features. The coast of Maine, with its numerous inlets and rocky bluffs, is a submergent coastline (Fig. 13–26). Small sandy beaches form in protected coves, but most of the beaches are rocky and bordered by cliffs.

Erosion and deposition gradually straighten such an irregular coastline. As waves move inland from deeper

Figure 13–26 The Rocky coast of Maine is a submergent coastline. *(Acadia National Park)*

These expensive hotels in Miami have been built on geologically transient barrier islands. Much of the rest of the city lies on reclaimed swamps surrounded by lagoons. *(Comstock)*

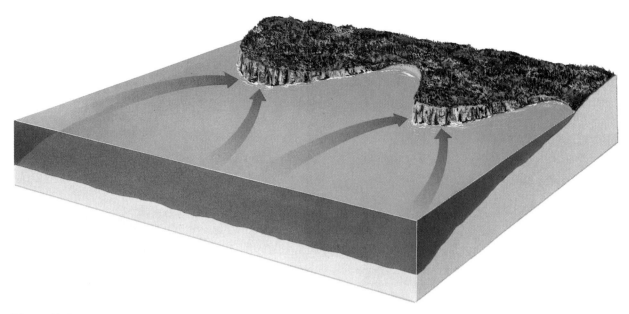

Figure 13–27   When a wave strikes a headland, the shallow water causes that portion of the wave to slow down and the wave refracts, thereby eroding the point of the headland.

water, they strike the tip of a headland first and break against the rocks, hurling sediment against the cliffs and eroding them. The part of the wave that strikes the headland first slows down. As a result, both sides refract, or bend, around the headland. The refracted waves then break against the sides of the prominence, eroding it

further (Fig. 13–27). When waves break against the headlands, most of their energy dissipates. Therefore, the waves inside the adjacent bays have lower energy, and as a result, sediment is deposited in the bays. As headlands erode and bays fill with sediment, an irregular coastline eventually becomes smooth (Fig. 13–28).

A

B

C

Figure 13–28   (A, B, C) A three-step sequence in which an irregular coastline is straightened. Erosion is greatest at the points of the headlands, and sediment is deposited inside the bays, leading to a gradual straightening of the shoreline.

Figure 13–29 A wave-cut cliff on the California coast.

A **wave-cut cliff** forms when waves erode the lower part of a rock face (Fig. 13–29). Wave erosion may undercut a cliff, and the overhanging rock eventually breaks loose and falls. As the cliff retreats, a flat **wave-cut platform** is left behind at sea level (Fig. 13–30).

If waves cut a cave into a narrow headland, the cave may eventually erode all the way through the headland, forming a scenic **sea arch.** When an arch collapses or when the inshore part of a headland erodes faster than the tip, a pillar of rock called a **sea stack** forms (Fig. 13–31). A sea stack is a temporary landform; eventually it crumbles to rubble.

Figure 13–30 A wave-cut platform on the Oregon coast.

13.10 Fjords and Estuaries

A **fjord** is a deep, long, narrow arm of the sea surrounded by high rocky cliffs or mountainous slopes (Fig. 13–32). The water in the fjord may be hundreds of meters deep, and often the cliffs drop straight into the sea. Fjords form

Figure 13–31 A sea stack on the Oregon coast.

Figure 13–32   Inugsuin Fjord on Baffin Island.

Figure 13–33   A satellite view of Chesapeake Bay. *(Chesapeake Bay Foundation)*

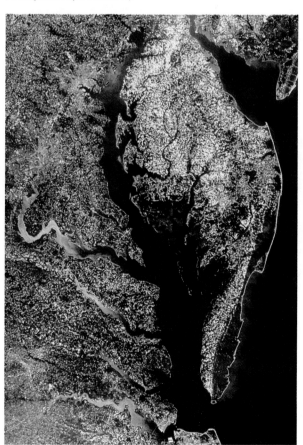

in mountainous coastal terrain where glaciers reach the sea. Whereas a stream can erode only to sea level, a glacier is massive enough and dense enough to erode bedrock below sea level. When a glacier melts, it leaves a deep fjord. Recall from Chapter 12 that when the Pleistocene ice sheets melted about 10,000 years ago, so much water was added to the oceans that sea level rose by more than 100 meters. Fjords were deepened when the rising seas flooded glacially scoured coastal valleys.

In contrast, a broad river valley submerged by rising sea level or a sinking coast is called an **estuary**. Estuaries are shallow and have gently sloping beaches. Chesapeake Bay is a major estuary along the Atlantic coast of North America (Fig. 13–33). It formed by submergence of the Susquehanna River valley. It is approximately 100 kilometers long, averages 10 to 15 kilometers wide, and contains numerous bays and inlets. Despite its great size, Chesapeake Bay is only 7 to 10 meters deep near its mouth, and its greatest depth is 50 meters.

Estuaries are the richest protein-producing environments on Earth. Streams wash nutrients into an estuary and the shallow water provides habitats for many marine organisms. Unfortunately for the organisms, estuaries also are prime sites for industrial activity. Inflowing rivers provide abundant fresh water, and despite their shallow entrances, estuaries are protected and can be developed into excellent harbors. As a result, many estuaries have become seriously polluted in recent years.

13.11 Beach Erosion and Human Settlement

The seashore has always been an attractive place for humans to live. The ocean, with its great mass of water, resists changes in temperature. Therefore coastal regions are generally cooler in summer and warmer in winter than continental interiors. The sea provides food and transportation. In addition, the seacoast is aesthetically pleasing. We enjoy the salt air and find the rhythmic pounding of surf soothing and relaxing. Vacationers sail, swim, and fish along the shore. For all these reasons, coastlines have become heavily urbanized and industrialized. In the United States, more than 50 percent of the population, 40 percent of the manufacturing plants, and 65 percent of the electrical power generators are located within 80 kilometers of the oceans or the Great Lakes.

Coastlines change constantly. Rivers transport sediment to the coast; waves and longshore currents erode the shore and transport and deposit sediment. In addition, sea level rises and falls. Thus, a fundamental conflict exists between geological change and our attempts to urbanize coastlines by building permanent structures at the water's edge.

Engineers recognize three types of solutions to the problem of shoreline erosion: (1) build "hard" barriers such as sea walls; (2) stabilize beaches by replenishing eroded sediment; and (3) abandon development on beaches and relocate homes and businesses. Unfortunately, none of these solutions is satisfactory. Hard barriers often provide little protection and sometimes even hasten erosion. Replenishment is expensive and temporary; sediment trucked in is likely to be eroded the follow-

ing season or during the next storm. Finally, relocation is expensive and politically difficult. People do not want to abandon beach-front property.

13.12 Case History: The South Shore of Long Island

Long Island extends eastward from New York City and is separated from Connecticut by Long Island Sound (Fig. 13–34). A series of narrow, low-lying barrier islands lies along the southern coast of Long Island. Longshore currents flow westward, eroding sand from the eastern tip of the island and carrying it toward Rockaway Beach and New Jersey. The sand of the beaches and barrier islands is constantly eroded and carried westward, only to be replaced by sand from the eastern end of the island.

Are the beaches of Long Island stable or unstable? The answer to that question depends on our time perspective. Over geologic time, the sand at the eastern end of the island will be exhausted and the flow of sand will cease. The entire coastline will erode and the barrier islands and beaches will disappear. This change will not occur in the near future because of the vast amount of sand still available at the east end of the island. Thus, the beaches are stable over a time perspective of years or decades. Longshore currents replace sand at the same rate at which they erode it and carry it westward, maintaining equilibrium.

If we narrow our perspective even further, and look at a Long Island beach over a season, or during a single storm, it alternately shrinks and expands. Over such short times, the rates of erosion and deposition are not equal.

Figure 13–34 A map of Long Island showing the predominant movement of longshore currents.

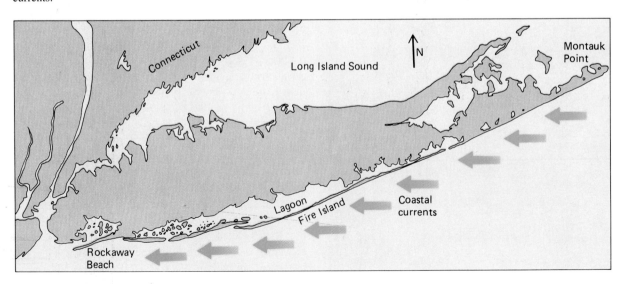

Violent winter waves and currents erode beaches, whereas sand accumulates on the beaches during the calmer summer months (Fig. 13–35). In an effort to prevent these seasonal fluctuations and protect their personal beaches, Long Island property owners have built stone barriers called **groins**. If a groin is built from the shore out into the water, it intercepts the steady flow of sand moving from the east and keeps that particular part of the beach from eroding (Fig. 13–36B). But the groin impedes the overall flow of sand. West of the groin the beach erodes as usual, but the sand is not replenished because the upstream groin traps it. As a result, beaches down-current from the groin suffer severe erosion (Fig. 13–37).

So now, land owners living down-current from a groin may decide to build another one west of their land and pass the problem farther downstream (Fig. 13–36C). The situation has a domino effect, with the net result that millions of dollars must be spent in attempts to stabilize a system that was naturally stable in its own dynamic manner.

Storms pose another dilemma. Hurricanes occasionally strike Long Island in the late summer or fall, generating storm waves that completely overrun the barrier islands and roll inland, flattening dunes and eroding beaches. When the storms are over, gentler waves and longshore currents carry sediment back to the beach and rebuild it. As the sand accumulates again, salt marshes rejuvenate and the dune grasses grow back within a few months.

These short-term fluctuations are not compatible with human activity. People build houses, resorts, and hotels

on or near the shifting sands. The owner of a home or resort hotel cannot allow the buildings to be flooded or washed away. Therefore, property owners construct large sea walls just inland of the beach. How do they affect the natural system? When a storm wave rolls across a low-lying beach, it dissipates its energy gradually as it flows over the dunes and pushes the sand inland. The beach is like a master in judo who defeats an opponent by yielding with the attack, not countering it head on. A sea wall interrupts this gradual absorption of wave energy. The waves crash violently against the barrier and erode the base of the sea wall. When the sand on which the wall is built erodes, the wall collapses. It may seem surprising that a reinforced concrete sea wall is *more* likely to be permanently destroyed than a beach of grasses and sand dunes, yet this is often the case (Fig. 13–38).

13.13 Global Warming and Sea-Level Rise

Sea level has risen and fallen repeatedly in the geologic past, and coastlines have emerged and submerged throughout the history of the planet. We have every reason to expect that coastlines will continue to change in the future. This is a sobering concept. Is sea-level change rapid and great enough to flood coastal cities and towns?

Both tectonic events and climate change lead to changes in sea level. Changes in shapes of the ocean basins or in the topography of the ocean floor may cause

Figure 13–35 In the absence of industrialization, beaches on Long Island are eroded in winter. The beach sediment is carried back inshore the following summer, and the beach is reestablished.

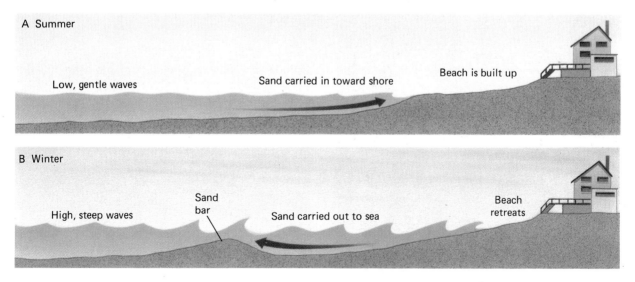

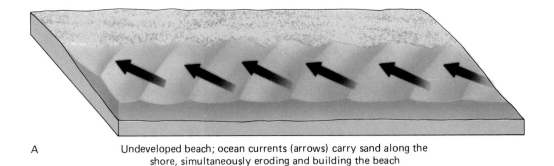

A Undeveloped beach; ocean currents (arrows) carry sand along the shore, simultaneously eroding and building the beach

B A single groin or breakwater. Sand accumulates on upstream side and is eroded downstream

C Multiple groin system

Figure 13–36 The effect of breakwaters on a beach. (A) An undeveloped beach. No net change occurs as sand is simultaneously eroded and deposited by longshore currents. (B) A single groin, or breakwater. Sand accumulates on the upstream side and is eroded downstream. (C) A multiple-groin system.

Figure 13–37 (A) Longshore currents on Long Island move from east to west. As a result, sand accumulates on the upstream (east) side of a groin, and the west side is eroded. Here waves lap against the foundation of the house along the eroded portion of the beach. (B) A close-up of the house in part (A).

A

B

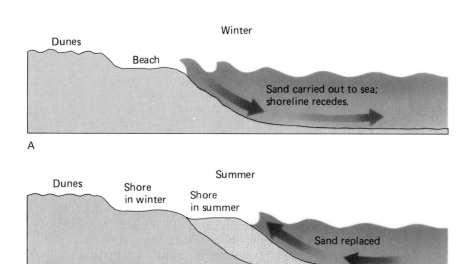

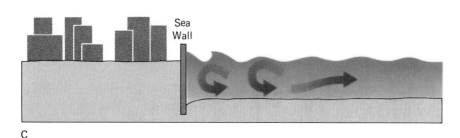

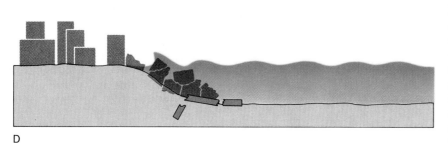

Figure 13–38 (A) In a natural system, the violent winter waves often move sand out to sea. (B) The gentler waves that occur in summer push sand toward shore and thereby rebuild the beach. (C) If a sea wall is erected, waves strike the wall and may eventually destroy it. (D) The beach is severely eroded.

sea level to rise or fall. However, these events occur over millions of years and will not affect human settlement in the near future.

During the past 40,000 years, sea level has fluctuated by 140 meters, primarily in response to growth and melting of glaciers. The rapid sea-level rise that started about 20,000 years ago began to level off about 7000 years ago. By coincidence, humans began to build cities about 7000 years ago. Thus, civilization has developed at a time when sea level has been relatively constant (Fig. 13–39).

However, small changes in sea level affect human settlement. If a shoreline consists of gently sloping land, a rise of a meter or two may move the beach a few kilometers inland, flooding thousands of square kilometers of land. During the past century, global sea level has risen by 1 to 2.5 centimeters per decade.

Many climatologists predict that industrial pollution will raise global temperature by 2° to 3°C in the next few decades. Warming can affect sea level in two ways. Portions of the huge Greenland and Antarctic ice caps may melt, and the water will flow into the ocean. Additionally, when the atmosphere warms, seawater warms and expands. This expansion can cause a rise in sea level.

Future Concerns

Many oceanographers predict that sea level will continue to rise at a rate between 1 and 2.5 centimeters per decade. In some regions, problems caused by global sea-level rise have been compounded by local effects such as tectonic sinking. In London, where the high-tide level has risen by 1 meter in the past century, a multimillion-dollar series

Pollution of the Oceans

In its article on "sewerage," the eleventh edition of the *Encyclopaedia Britannica*, published in 1910, states, "Nearly every town upon the coast turns its sewage into the sea. That the sea has a purifying effect is obvious. . . . It has been urged by competent authorities that this system is not wasteful, since the organic matter forms the food of lower organisms, which in turn are devoured by fish. Thus the sea is richer by this clean method of disposal."

This concept is based on two fallacies. As the article states, lower organisms such as bacteria eat sewage. If a small quantity of sewage is dispersed in a large volume of water, organisms can purify the water. However, sewage released along the coastline does not disperse rapidly. Instead it concentrates near the discharge sites, especially if they are located in bays or estuaries that mix slowly with deeper offshore water. When sewage concentrates in this way, it overwhelms the capacity of lower organisms to consume it. Consequently, by the late 1800s and early 1900s, coastal areas became smelly and beaches were foul and unsanitary.

The second fallacy is the belief that all pollutants are biodegradable. In reality, many types of waste are not consumed by marine organisms. Modern chemical factories synthesize large quantities of nonbiodegradable compounds, and some are toxic even in low concentrations. Thus, seawater becomes polluted when too much biodegradable waste is flushed into it or when any amount of nonbiodegradable waste is added.

Ocean pollution comes from several sources.

Sewage   When you flush your toilet in an urban area, the waste is pumped to a sewage treatment plant. At the plant, large solids are removed by filters and smaller particles collect in settling tanks. Filtration and settling remove about 60 percent of the solids and none of the dissolved waste. In most cities, the remaining material is then pumped to an aerated tank or pond where bacteria feast on the organic matter and convert it to substances that are not pollutants, such as carbon dioxide and water. After filtration and biological treatment, only 10 to 15 percent of the original organic contaminants remain. Thus, these processes represent a significant purification, but the water discharged from a sewage treatment plant is hardly fit to drink. In most urban areas, the wastewater is then dumped directly into the ocean or into rivers that flow to the sea. In the United States about 24 trillion liters of partially contaminated wastewater are discharged from sewage treatment plants every year.

In addition to body wastes, sewage contains chemical contaminants. Some come from households where people may flush old paints or solvents down the toilets. Other compounds are dumped illegally into sewage systems from industrial sources.

Industrial Wastes   In the United States, the Clean Water Act establishes guidelines intended to "restore and maintain the chemical, physical and biological integrity of the nation's waters." However, pollution control costs money. It is economically impossible to require factories and refineries to purify all of their wastewater completely and to discharge only drinkable water into rivers and coastal oceans. Therefore the law is not absolute; standards and guidelines are often changed by legislatures and courts. According to the Natural Resources Defense Council, about 20 trillion liters of industrial wastewater

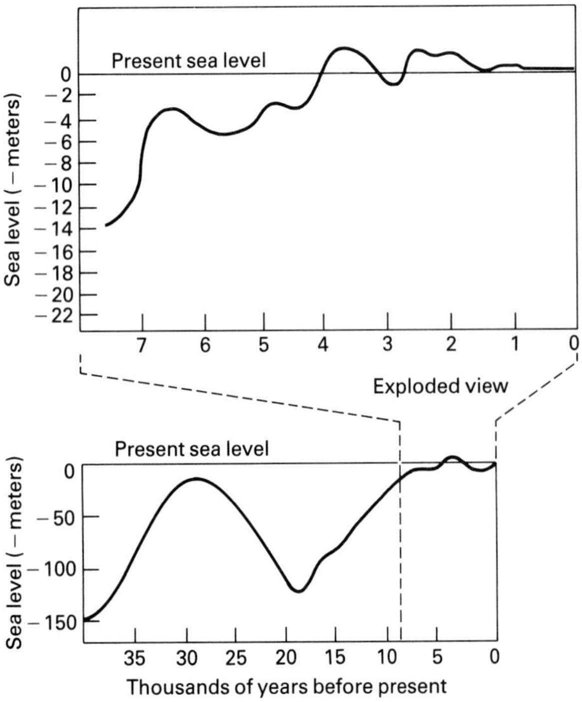

Figure 13–39   Sea level change during the past 40,000 years. The lower graph shows an exploded view of sea level change over the past 7000 years.   *(J. D. Hansom,* Coasts. *Cambridge: Cambridge University Press, 1988.)*

of storm gates was built on the Thames River, and a similar system is now planned to protect the city of Venice from further flooding. In North America, some planners are contemplating building gigantic levees to protect coastal cities, and they are visiting the Netherlands to see how the Dutch do it. The cost would exceed that of any construction project in history. People in poor lowland countries cannot afford massive costs. According to one estimate, a 3°C rise in global temperature would lead to flooding that would displace ten million people from the Nile delta and 38 million farmers from the coastal plain of Bangladesh.

are discharged yearly. Some contain toxic or carcinogenic substances.

Petroleum In March 1989, the supertanker *Exxon Valdez* ran aground on the rocks near Valdez, Alaska, spilling 40 million liters of petroleum into the waters of Prince William Sound. Other accidents have been worse. Between 100 and 700 million liters were released into the Persian Gulf during the Gulf War in 1991. Fortunately, mega-spills are relatively rare. However, hardly a week goes by without smaller oil spills occurring somewhere in the world. Over the past decade an average of 140 million liters of petroleum were spilled annually throughout the world.

Agricultural Runoff Some of the pesticides and fertilizers spread onto fields are washed into streams and then transported into the oceans.

Litter Most plastics do not decompose readily and therefore do not present a chemical hazard to the environment. However, an estimated 100,000 marine mammals and birds die each year from ingesting or becoming entangled in litter such as plastic bags, beverage can yokes, and abandoned fishing nets.

Effects of Ocean Pollution Accidental oil and chemical spills kill plankton, shellfish, fish, birds, and marine mammals. However, in any one area, a spill is usually a one-time event. After a period of a few months to a few years, most of the contaminants are diluted or washed away and the resident organisms return. On the other hand, continuous discharge from factories and sewage treatment plants prevents affected ecosystems from cleansing themselves.

The effect of chronic pollution is most pronounced in estuaries with limited access to the open oceans. As a measure of pollution and industrialization, in the Chesapeake Bay, commercial and sports fishermen caught 2.7 million kilograms of striped bass annually during the 1960s; by 1980 the catch had plummeted to 270,000 kilograms per year. The oyster harvest peaked more than 100 years ago, in 1885, and has been declining ever since. The decrease in the oyster population has been caused by overfishing, pollution, disease, and silting of the oyster beds.

Population decline is not the only measure of the effects of pollution. In 1988 the Federal Drug Administration (FDA) found "unacceptable levels" of toxins or suspected carcinogens such as pesticides, PCBs, dioxins, and benzene in 73 percent of the seafood samples tested in the United States. Pollution also affects recreational areas. In the late 1980s authorities closed many beaches in the northeastern United States when medical wastes washed up onto the sands. Even safe beaches are suspect. The Environmental Protection Agency (EPA) defines a beach as "safe" when it estimates that only 19 people out of every 1000 will become ill with intestinal problems after swimming.

Is the World Ocean Threatened by Pollution? We know that many coastal areas are polluted. But what about the central oceans? In 1990 a study released by the United Nations Conference on the Human Environment stated that the concentration of pollutants in the deep ocean basins had declined between 1972 and 1990. The decline was attributed to rising global environmental concern and resultant pollution control efforts in many countries. The open ocean is relatively clean except for floating tar and plastic litter, found mainly in shipping lanes and where ocean currents converge.

SUMMARY

●

Seawater contains about 35 parts per thousand (ppt) dissolved salts. The upper layer of the ocean, about 450 meters thick, is relatively warm. In the **thermocline**, below the warm surface layer, temperature drops rapidly with depth. Deep ocean water is consistently around 1° to 2°C.

Where waves do not interact with the sea floor, their size depends on (1) wind speed, (2) the amount of time the wind has blown, and (3) the distance the wind has traveled. The highest part of a wave is the **crest**; the lowest, the **trough**. The distance between successive crests is called the **wavelength**. **Wave height** is the vertical distance from the crest to the trough. The water in a wave moves in a circular path. When a wave nears the shore, the bottom of the wave slows and the wave **breaks**,

creating **surf**. **Refraction** is the bending of a wave as it strikes shore at an oblique angle. Ocean waves erode rock and sediment by hydraulic action, solution, and abrasion and then transport and eventually deposit sediment.

Tides are caused by gravitational pull of the Moon and Sun. Two high tides and two low tides occur every day.

A **current** is a continuous flow of water in a particular direction. **Surface currents** are driven by wind and deflected by the **Coriolis effect**. When a surface current is deflected offshore, cold, deep water rises to replace it in a process called **upwelling**. **Deep-sea currents** are driven by differences in seawater density. Cold, salty water is dense and therefore sinks and flows along the sea floor. **Tidal currents** form when the tidal flow is channeled. **Longshore currents** transport sediment along a shore.

A **beach** is a strip of shoreline washed by waves and tides. A **reef** is a wave-resistant ridge or mound built by corals and other organisms.

Most coastal sediment is transported to the sea by rivers. Glacial drift, reefs, and local erosion of rock also add sediment in certain areas. If land rises or sea level falls, the coastline migrates seaward and old beaches are abandoned above sea level, forming an **emergent coastline**. An emergent coastline is usually sediment-rich and characterized by sandy beaches, **spits**, **baymouth bars**, and **barrier islands**. In contrast, a **submergent**

coastline forms when land sinks or sea level rises. A submergent coastline is usually sediment-poor. A rocky coast with **wave-cut cliffs**, **wave-cut platforms**, **arches**, and **stacks** is common in this environment. Irregular coastlines are straightened by erosion and deposition. A **fjord** is a submerged glacial valley. An **estuary** is a submerged riverbed and flood plain.

Human intervention such as the building of **groins** may upset the natural movement of coastal sediment and alter patterns of erosion and deposition on beaches. Sea level has risen over the past century and may continue to rise into the next.

KEY TERMS

●

Salinity *346*
Thermocline *347*
Crest *348*
Trough *348*
Wave height *348*
Wavelength *348*
Wave period *348*
Surf *349*
Swash *349*
Backwash *349*
Refraction *349*

Storm surge *349*
Tide *351*
Spring tide *353*
Neap tide *353*
Current *353*
Gulf Stream *356*
Gyre *356*
Coriolis effect *357*
Upwelling *358*
Tidal current *359*
Rip current *359*

Longshore current *359*
Reef *360*
Atoll *360*
Lagoon *360*
Beach *361*
Foreshore *361*
Backshore *361*
Intertidal zone *361*
Emergent coastline *363*
Spit *363*

Baymouth bar *363*
Barrier island *363*
Submergent coastline *364*
Wave-cut cliff *366*
Wave-cut platform *366*
Sea arch *366*
Sea stack *366*
Fjord *366*
Estuary *367*
Groin *369*

REVIEW QUESTIONS

●

1. How did oceans form during the early history of the Earth?

2. Why does the Earth have continents and oceans rather than one global ocean that covers the entire planet?

3. Name the major ocean basins and give their locations.

4. Using Figure 13–1, what percentage of the Earth's area (a) lies above an elevation of 840 meters, (b) lies 4 kilometers or more below sea level, (c) lies 5 kilometers or more below sea level?

5. Compare the salinity of the open ocean at the equator and in dry subtropical regions.

6. What factors affect the salinity of seawater near shore?

7. Describe the temperature profile of the open oceans.

8. List the three factors that determine the size of a wave.

9. Draw a picture of a wave and label the crest, the trough, the wavelength, and the wave height.

10. Describe the motion of both the surface and the deeper layers of water as a wave passes.

11. Explain how surf forms.

12. What is refraction? How does it affect coastal erosion?

13. How is a tidal current different from a longshore current?

14. Explain why two high tides occur every day even though the Moon lies directly above any portion of the Earth only once a day.

15. Explain the difference between spring and neap tides.

16. Explain the Coriolis effect.

17. What drives surface currents, deep-sea currents, tidal currents, and longshore currents?

18. Explain the formation of an atoll.

19. List three different sources of coastal sediment.

20. What is an emergent coastline, and how does it form? Are emergent coastlines sediment-rich or sediment-poor? Why?

21. Compare and contrast a beach, a barrier island, and a spit.

22. Explain how coastal processes straighten an irregular coastline.

23. Describe some dominant features of a sediment-poor coastline.

24. What is a groin? How does it affect the beach in its immediate vicinity? How does it affect the entire shoreline?

25. Explain how warming of the atmosphere could lead to a rise in sea level.

DISCUSSION QUESTIONS

•

1. Earthquake waves were discussed in Chapter 10. Compare and contrast earthquake waves with water waves.

2. How can a ship survive 10-meter-high storm waves, whereas a beach house can be smashed by waves of the same size?

3. Explain why very large waves cannot strike a beach directly in shallow coastal waters.

4. During World War II, few maps of the underwater profile of shorelines existed. When planning amphibious attacks on the Pacific islands, the Allied commanders needed to know near-shore water depths. Explain how this information could be deduced from aerial photographs of breaking waves and surf.

5. Imagine that an oil spill occurs from a tanker accident. Discuss the effects of mid-ocean currents, deep-sea currents, longshore currents, storm waves, and tides on the dispersal of the oil.

6. Compare and contrast coastal erosion with stream erosion.

7. In Section 13.9 we explained how erosion and deposition tend to smooth out an irregular coastline by eroding headlands and depositing sediment in bays. If coastlines are affected in this manner, why haven't they all been smoothed out in the 4.6-billion-year history of the Earth?

8. Is Chesapeake Bay an emergent or submergent coastline? Briefly outline a plausible sequence of events that could have led to the formation of this coastline.

9. Prepare a three-way debate. Have one side argue that the government should support the construction of groins. Have the second side argue that the government should prohibit the construction of groins. Have the third position defend the argument that groins should be permitted, but not supported.

10. Prepare another debate on whether government funding should be used to repair storm damage to property on barrier islands.

11. Solutions to environmental problems can be divided into two general categories: social solutions and technical solutions. Social solutions involve changes in attitudes and life styles but do not generally require expensive industrial or technological processes. Technical solutions do not mandate social adjustments, but require advanced and often expensive engineering. Describe a social solution and a technical solution to the problem of coastline changes in Long Island. Which approach do you feel would be more effective?

12. Imagine that sea level rises by 25 centimeters in the next century. How far inland would the shoreline advance if the beach (a) consisted of a vertical cliff, (b) sloped steeply at a 45-degree angle, (c) sloped gently at a 5-degree angle, (d) were almost flat and rose only 1 degree?

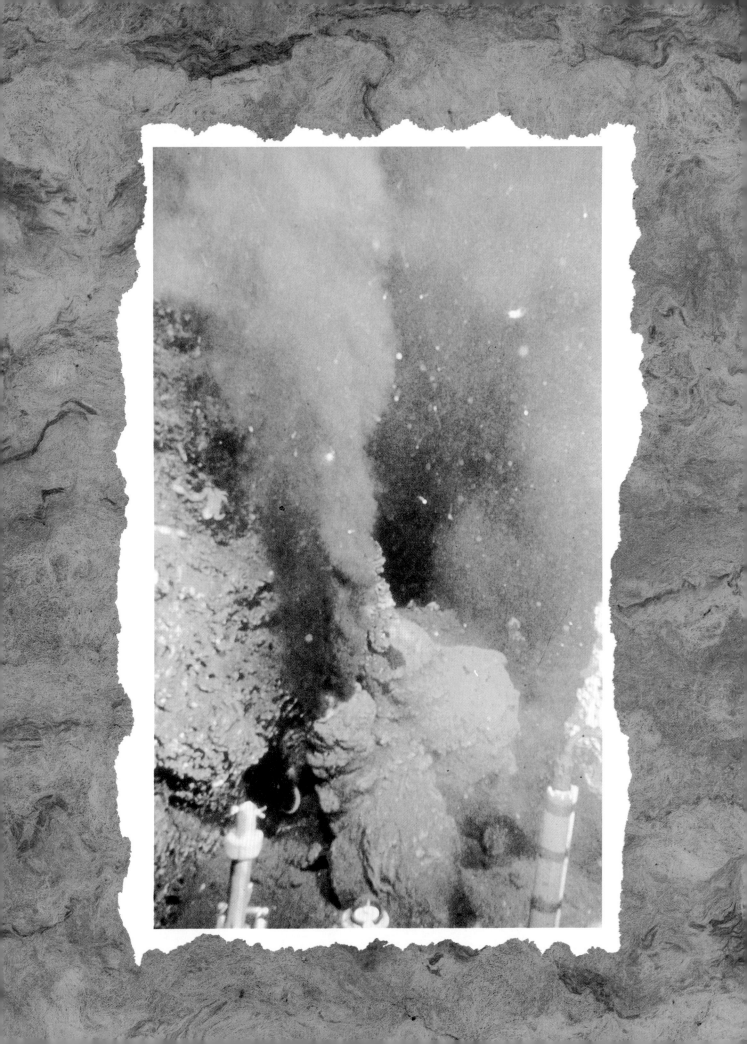

The Geology of the Ocean Floor

I f you were to ask most people to describe the difference between continents and oceans, they might show surprise at a question with such an apparently simple answer and reply, "Why, obviously oceans are water and continents are land!" This is true, of course; but to geologists, a more important distinction exists. They would explain that the rocks making up oceanic crust are very different from those of continental crust. The accumulation of seawater in the world's ocean basins is a *result* of that difference.

Oceanic crust is basalt. In contrast, continental crust is granite. Granite is less dense than basalt. In addition to the difference in density, continental crust is thicker than oceanic crust. The entire lithosphere, including both continents and oceans, floats on the plastic asthenosphere. Continents float to high elevations because they are thicker and less dense than oceanic crust. Oceanic crust floats at low elevations because it is thin and dense. Water naturally collects in the huge depressions formed by oceanic crust.

● A submarine volcanic vent spewing out mineral-laden hot water on the Pacific Ocean floor. *(Dudley Foster / Woods Hole Oceanographic Institution)*

Figure 14–1   After the sediment core is retrieved from the sea floor, it is carefully labeled and cut into sections. Technicians split the sectioned core lengthwise to produce half cylinders, which are then studied by scientists.   *(Ocean Drilling Program, Texas A&M University)*

The sizes of all ocean basins change over geologic time because new oceanic crust forms at spreading centers such as the Mid-Atlantic ridge, and old sea floor is consumed at subduction zones. At present, the Atlantic Ocean is growing while the Pacific is shrinking. These changes in size and distribution of the oceans affect the movement of currents and the transport of heat across the globe. Therefore, such tectonic changes affect climate and life on Earth.

14.1 Studying the Ocean Floor

Geologists and oceanographers have always asked questions such as "What does the sea floor look like?" "What types of sediment or rock make up oceanic crust?" Seventy-five years ago, scientists had better maps of the Moon than of the sea floor. The Moon is clearly visible in the night sky and we can view its surface with a telescope. The sea floor, on the other hand, lies at an average depth of nearly 5 kilometers and in places is more than 10 kilometers deep. Today a variety of techniques are used to study the sea floor, including sampling, remote sensing, and direct observation from deep-diving submarines.

Sampling Techniques

Several devices collect samples of sediment and rock from the ocean floor. A **rock dredge** is an open-mouthed steel net dragged along the sea floor behind a research

ship. The dredge breaks rocks from submarine outcrops and hauls them to the surface. Sediment near the surface of the sea floor can be sampled by a **coring device**, a weighted, hollow steel pipe lowered on a cable from a research vessel. The weight drives the pipe into the soft sediment, which is forced into the pipe. The sediment **core** is retrieved from the pipe after it is winched back to the surface. If the core is taken and removed from the pipe carefully, even the most delicate sedimentary layering is preserved (Fig. 14–1).

Both sediment and rock samples are retrieved by **sea-floor drilling** methods that were developed for oil exploration and recovery. Large drill rigs are mounted on offshore platforms and on research vessels (Fig. 14–2). The drill cuts cylindrical cores from both sediment and rock, which are then brought to the surface for study.

Remote Sensing

Remote sensing methods do not require direct physical contact with the ocean floor, and for some studies this approach is both effective and economical. The **echo sounder** is the principal tool for mapping sea-floor topog-

Figure 14–2   The *Joides Resolution*, a deep-sea drilling ship.   *(Ocean Drilling Program, Texas A&M University)*

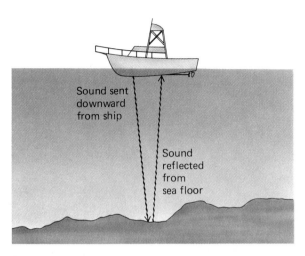

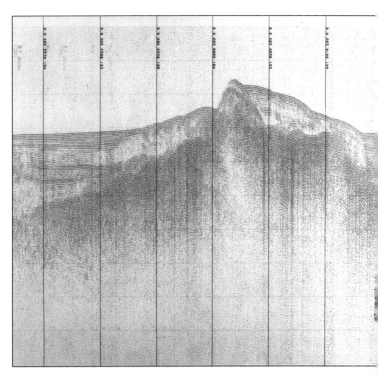

Figure 14–3   (A) Measuring the topography of the sea floor with an echo sounder. A sound signal generated by the echo sounder bounces off the sea floor and back up to the ship, where its travel time is recorded. (B) A seismic profiler record of sediment layers and basaltic ocean crust in the Sea of Japan.   *(Ocean Drilling Program, Texas A&M University)*

raphy. It emits a sound signal from a research ship and then records the signal after it bounces off the sea floor and travels back up to the ship (Fig. 14–3A). The water depth is calculated from the time required for the sound to make the round trip. The ship steers a carefully navigated course with the echo sounder operating continuously, and in this way a topographic profile of the ocean floor is constructed. Many such profiles are then combined to produce a topographical map of the sea floor. The **seismic profiler** works in the same way but uses a higher-energy signal. In addition to bouncing off the sea floor, some of the sound waves penetrate its surface and reflect off layers within the underlying sediment and rock. This gives a picture of the layering and structure of sediment and rocks of oceanic crust well as sea-floor topography (Fig. 14–3B).

In the early 1970s, small, specially designed research submarines were developed jointly by French and American scientists. The submarines carry researchers to the ocean floor to view, photograph, and sample sea-floor rocks and sediment directly (Fig. 14–4).

14.2 The Mid-Oceanic Ridge

At the turn of the century, oceanographers believed, on the basis of little or no evidence, that the sea floor was a broad, level, featureless plain that had remained unchanged since the formation of the Earth. This view

changed radically just before and during World War II. The extensive use of submarines during the war made it essential to have topographic maps of the sea floor. Those maps, made with early versions of the echo sounder, were kept secret by the military. When they became available to the public after peace was restored, scientists were

Figure 14–4   *Alvin*, a research submarine capable of diving to portions of the ocean floor. Scientists on board control robot arms and grapples to sample sea-floor rocks and sediment.   *(Red Catanach, Woods Hole Oceanographic Institution)*

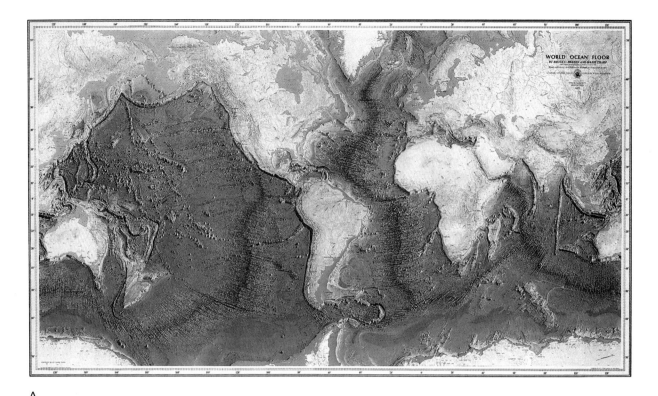

A

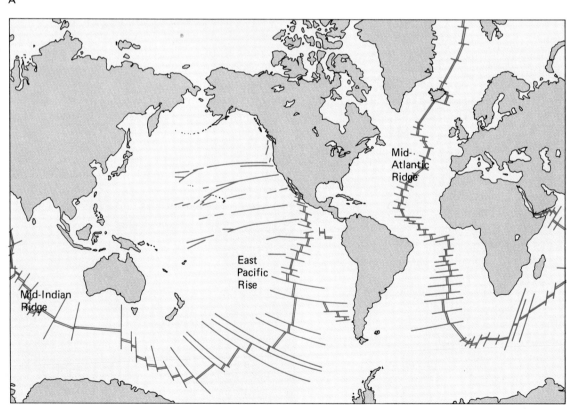

B

Figure 14–5 (A) Topography of the ocean floor. *(Copyright © Marie Tharp)* (B) A map of the ocean floor showing the mid-oceanic ridge in red. Double lines indicate the ridge axis; single lines indicate transform faults. Comparing the two maps, note that the major deep-sea mountain ranges shown in (A) form the mid-oceanic ridge system shown in (B).

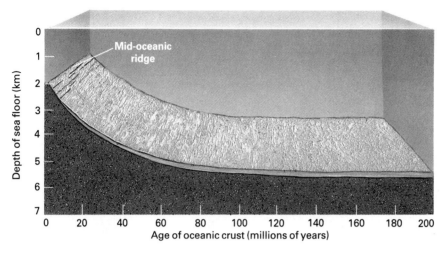

Figure 14–6 The relationship between average depth and age of the sea floor. At the mid-oceanic ridge, new lithosphere is thermally expanded. It ages and cools as it moves away from the ridge and consequently sinks to greater depths. The central portion of the sea floor lies at a depth of about 5 kilometers.

surprised to learn that the ocean floor has at least as much topographic diversity and relief as the continents (Fig. 14–5). Great undersea mountain ranges, broad plains, isolated high peaks, and deep valleys make up a varied and fascinating landscape beneath the oceans. Plate tectonics theory now enables us to explain the origins of these features.

The **mid-oceanic ridge** is a continuous submarine mountain chain that encircles the globe (Fig. 14–5B). It is the largest mountain system on Earth, with a total length of more than 80,000 kilometers and a width of more than 1500 kilometers in most places. The ridge rises an average of 3 kilometers above the surrounding deep-sea floor. It covers more than 20 percent of the Earth's surface, nearly as much as all continents combined.

This submarine mountain chain is completely unlike mountain ranges found on continents. Continental ranges are usually composed of granite and folded and faulted sedimentary and metamorphic rocks. In contrast, the mid-oceanic ridge is made up entirely of undeformed basalt.

A **rift valley** 1 to 2 kilometers deep and several kilometers wide splits the ridge crest. In 1974 French and American scientists used a small research submarine to dive into the rift valley in the Atlantic Ocean. They saw gaping vertical cracks up to 3 meters wide on the floor of the rift. Nearby were submarine basalt flows so young that they were not covered by sediment. The cracks form when oceanic crust separates at the ridge axis. Basalt magma then rises through the cracks and flows onto the floor of the rift valley. This basalt becomes new oceanic crust as two lithospheric plates spread outward from the ridge axis. Thus, the rift valley is the boundary between two diverging plates. Scientists also saw cliffs hundreds of meters high formed by faulting as the two lithospheric plates separated.

The mid-oceanic ridge rises high above the surrounding sea floor because new lithosphere forming at the ridge axis is hot and therefore of relatively low density. Its

low density causes it to float higher than older and cooler lithosphere on both sides of the ridge. Because the new lithosphere cools as it spreads away from the ridge, it becomes denser. Therefore, it sinks to lower elevations, forming the deeper sea floor adjacent to the ridge (Fig. 14–6).

Shallow earthquakes and high heat flow occur at the mid-oceanic ridge (heat flow is the amount of heat flowing outward from the Earth's surface). The earthquakes result from fracturing and faulting of oceanic crust as the two plates separate at the spreading center (Fig. 14–7). Blocks of new oceanic crust drop downward along the faults, forming the rift valley in the center of the ridge. The heat flow at the ridge is several times as great as that measured

Figure 14–7 A cross-sectional view of the central rift valley of the Mid-Atlantic ridge. The pulling apart of two lithospheric plates at the rift causes fracturing, faulting, and numerous shallow earthquakes. The rift valley itself is formed by blocks of oceanic crust that drop down along the faults as the crust separates.

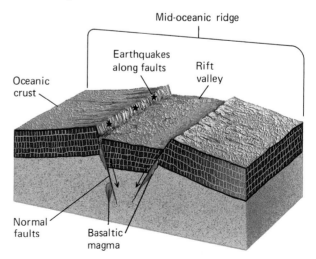

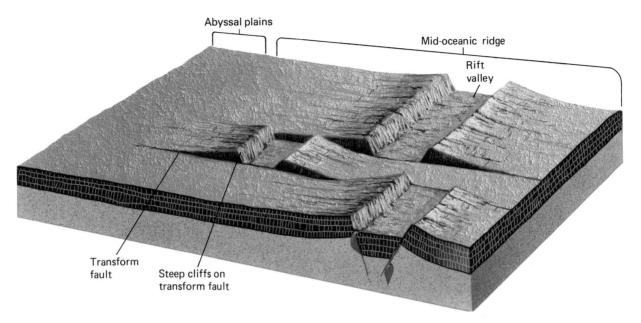

Abyssal plains

Mid-oceanic ridge

Rift valley

Transform fault

Steep cliffs on transform fault

Figure 14–8 Transform faults offset segments of a mid-oceanic ridge. Because of the offsetting, adjacent segments of the ridge are separated by steep cliffs that may be up to 3 or 4 kilometers high. Note the nearly level surface, the abyssal plain, that lies far from the axis of the ridge.

in other parts of the ocean basins because the spreading lithosphere is stretched thin under the ridge and the hot asthenosphere bulges toward the surface. Rising magma carries additional heat upward.

In describing the mid-oceanic ridge, we have established an image of a mountain range bisected by a deep rift valley, snaking its way beneath the world's major oceans. Closer examination shows that the rift valley and the ridge are not really continuous, but rather are cut and offset by numerous fractures called **transform faults** (Fig. 14–8).

Transform faults extend through the entire thickness of the lithosphere. They develop because different segments of a lithospheric plate move at different rates. If one portion of a plate moves faster or slower than an adjoining portion, transform faults develop to accommodate the differences in motion. In some cases, a transform fault can grow so large that it forms a plate boundary. In such an instance, the fault becomes a transform plate boundary. The San Andreas fault in California is a transform plate boundary.

14.3 Sediment and Rocks of the Deep-Sea Floor

The Earth is 4.6 billion years old, and rocks as old as 4.1 to 4.2 billion years have been found on continents.

However, no oceanic crust is older than about 200 million years. Oceanic crust is so young because it forms continuously at spreading centers and recycles into the mantle at subduction zones. Thus, oceanic crust is youngest at the mid-oceanic ridge and becomes older with increasing distance from the ridge. In contrast, once continental crust forms, it cannot return into the mantle because of its low density.

Seismic profiling and sea-floor drilling show that oceanic crust varies from about 7 to 10 kilometers thick. It consists of three layers. The lower two are basalt and the upper is sediment (Fig. 14–9).

Basaltic Oceanic Crust

Layer 3, 5 to 7 kilometers thick, is the deepest and thickest layer of oceanic crust. It directly overlies the mantle. The upper part consists of vertical basalt dikes. They form as magma oozing toward the surface freezes in the cracks of the rift valley. The lower portion of Layer 3 consists of horizontally layered bodies of gabbro, the coarse-grained equivalent of basalt. These sills form as magma cools beneath the basalt dikes.

Layer 2 lies above Layer 3 and is about 1.5 kilometers thick. It consists mostly of **pillow basalt**, which forms as basalt magma oozes onto the sea floor through the cracks in the rift valley. Contact with cold seawater causes the molten lava to contract into pillow-shaped spheroids (Fig. 14–10).

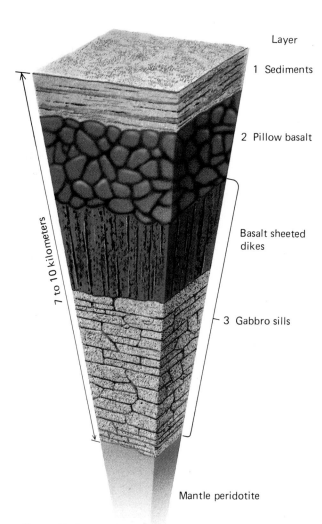

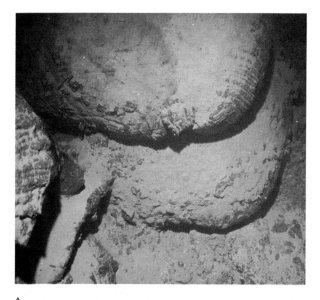

Figure 14–9 The three layers of oceanic crust. Layer 1 consists of terrigenous and pelagic sediments. Layer 2 is pillow basalt. Layer 3 consists of an upper part made up of sheeted dikes and a lower part composed of intrusive gabbro. Below Layer 3 is peridotite, thought to be the top of the upper mantle.

The pillow lavas, basalt dikes, and gabbro sills form only at the mid-oceanic ridge. However, these rocks make up the foundation of all oceanic crust because all oceanic crust forms at the ridge axis and then spreads outward.

Ocean-Floor Sediment

The uppermost layer of oceanic crust, called **Layer 1**, consists of two different types of sediment: terrigenous and pelagic. **Terrigenous sediment** is derived directly from land. It is composed of sand, silt, and clay eroded from the continents and deposited on the ocean floor near the continents by submarine currents. **Pelagic sediment**, on the other hand, is found even in deep basins far from continents. It is a gray and red-brown mixture of clay and

Figure 14–10 (A) Sea-floor pillow basalt in the Cayman trough. *(Woods Hole Oceanographic Institution)* (B) Occasionally, slices of the sea floor are shoved up onto the continents by tectonic activity. These pillow basalts were formed underwater but today are exposed in western Oregon.

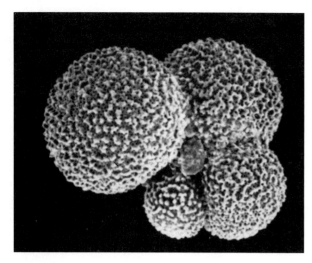

A scanning electron microscope photo of pelagic foraminifera, the tiny organisms that live by floating near the surface of the oceans.   *(Ocean Drilling Program, Texas A&M University)*

the remains of tiny plants and animals that live in the surface waters of the oceans. When these organisms die, their remains slowly settle to the ocean floor. Much of the clay in the mixture was transported from continents by wind.

Pelagic sediment accumulates at a rate of about 2.5 millimeters every 1000 years. At the axis of the mid-oceanic ridge where oceanic crust is newly formed, there is no sediment. The sediment layer becomes thicker with increasing distance from the ridge because the sea floor becomes older with increasing distance from the ridge axis (Fig. 14–11). Close to shore, pelagic sediment gradually merges with the much thicker layers of terrigenous sediment, which can be 3 or 4 kilometers thick.

Parts of the ocean floor beyond the mid-oceanic ridge are flat, level, featureless submarine surfaces called the **abyssal plains** (Fig. 14–8). They are the flattest surfaces on Earth. Seismic profiling shows that the surface of the basaltic crust is rough and jagged throughout the ocean. On the abyssal plains, however, this rugged profile is buried by sediment. Thus, the extraordinarily level sur-

Figure 14–11   The thickness of pelagic sediments increases with distance away from the mid-oceanic ridge.

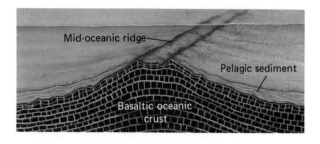

faces of the abyssal plains result from accumulation of mud in the deep ocean. The mud fills in the hills and valleys of the basaltic crust and creates a smooth surface. If you were to remove all of the sediment, you would see the same rugged topography that you see near the mid-oceanic ridge.

14.4   Geological Resources on the Ocean Floor

Manganese Nodules

About 25 to 50 percent of the Pacific Ocean floor is covered with golf ball– to bowling ball–size **manganese nodules** (Fig. 14–12). The average nodule contains 20 to 30 percent manganese, 6 percent iron, about 1 percent each of copper and nickel, and lesser amounts of other metals. (Much of the remaining 60 to 70 percent consists of anions chemically bonded to the metal ions.) The metal ions may be introduced into seawater from continuing volcanic activity at the mid-oceanic ridge. The dissolved ions then precipitate to form the nodules by chemical reaction between seawater and pelagic sediment.

The nodules grow by about ten layers of atoms per year, which amounts to 3 millimeters per million years. Curiously, they are found only on the surface of, but never within, the sediment on the ocean floor. Since sediment accumulates much faster than the nodules grow, why doesn't it bury the nodules as they form? Photographs show that animals churn up sea-floor sediment. Worms burrow into it and other animals pile sediment against the nodules to build protective shelters. Some geologists suggest that these activities constantly lift the nodules onto the surface.

Figure 14–12   Manganese nodules covering the sea floor on the Blake Plateau east of Florida. Each nodule is from a few centimeters to about 30 centimeters in diameter, so this photograph covers a few hundred square meters.   *(Frank Manheim, Woods Hole Oceanographic Institution)*

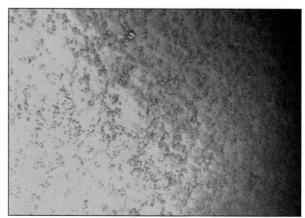

A trillion or more tons of manganese nodules lie on the sea floor. They contain several valuable industrial metals that could be harvested without drilling or blasting. One can imagine a scenario in which exploration is performed using undersea television cameras and the nodules are collected by a vacuum diver, a scoop, or a robot. But, despite its tremendous wealth, the ocean floor is not the easiest environment in which to operate complex machinery, so at present, exploitation of manganese nodules is not profitable.

Ore Deposits Formed by Submarine Volcanic Activity

In volcanically active regions of the sea floor—near the mid-oceanic ridge and submarine volcanoes—warm rock heats seawater as it seeps into cracks in oceanic crust. The warm water then circulates through the fractured basalt. In this way, hot, salty seawater migrates through vast volumes of rock. The rate at which seawater circulates through oceanic crust is equivalent to the entire combined volume of the oceans passing through once every eight million years! Recall from Chapter 3 that hot brine dissolves metals from rock. In this way, the hot seawater picks up metals as it circulates through oceanic crust. Then, as it rises through the upper layers of oceanic crust, it cools, and chemical conditions change. As a result, huge deposits of iron, copper, lead, and zinc precipitate within sea-floor sediment and upper layers of basalt. The metal-bearing solutions can be seen as jets of black water, called **black smokers**, spouting from fractures in the mid-oceanic ridge (Fig. 14–13). The black color is caused by precipitation of fine-grained metal sulfide minerals as the solutions cool upon contact with seawater.

Submarine metal deposits formed in this manner can be highly concentrated. On land an ore deposit containing 2.5 percent zinc is rich enough to mine commercially, some undersea zinc deposits contain as much as 55 percent zinc. The cost of operating machinery beneath the sea is so great that such undersea deposits are currently not profitable to mine. However, in some places ore deposits formed underwater have been uplifted by tectonic forces. The ancient Romans mined copper, lead, and zinc ores of this type in the Apennines Mountains of Italy. This geological wealth contributed to their political and military ascendancy. In modern times, rich deposits of lead, zinc, and silver formed in underwater volcanic environments are now mined in Australia, North America, and other continents.

In addition to these metal deposits, geologists have discovered vast undersea petroleum reserves along continental margins. These resources are discussed in Chapter 22.

Figure 14–13 A black smoker rising from the East Pacific rise. The "smoke" is hot, mineral-rich seawater that has circulated through oceanic crust, dissolving sulfur, iron, zinc, copper, and other metals. It is heated by the hot basalt near a spreading center. When the hot solution meets cold seawater, the metals precipitate. The hot, nutrient-rich water sustains thriving plant and animal communities. *(Dudley Foster, Woods Hole Oceanographic Institution)*

14.5   Continental Margins

Continental margins are regions where continental crust meets oceanic crust. Two principal types of continental margins exist. A **passive continental margin** is characterized by a firm connection between continental and oceanic crust. Little tectonic activity occurs at this type of boundary. Continental margins on both sides of the Atlantic Ocean are passive margins.

In contrast, an **active continental margin** is characterized by subduction of an oceanic lithospheric plate beneath a continental plate. At an active continental margin, subduction typically occurs at or very close to the edge of the continent, and the subducting plate descends at an angle for hundreds of kilometers beneath the continent. The west coast of South America is an example of an active margin. Each type of continental margin has its own characteristic features.

Passive Continental Margins

Consider the passive margin of eastern North America. Recall from Chapter 9 that, about 200 million years ago, all of the Earth's continents were joined, forming the supercontinent Pangaea. Shortly thereafter, Pangaea began to break up into the continents as we know them today. The breakup may have been caused by a plume of hot mantle rock rising beneath the supercontinent. As

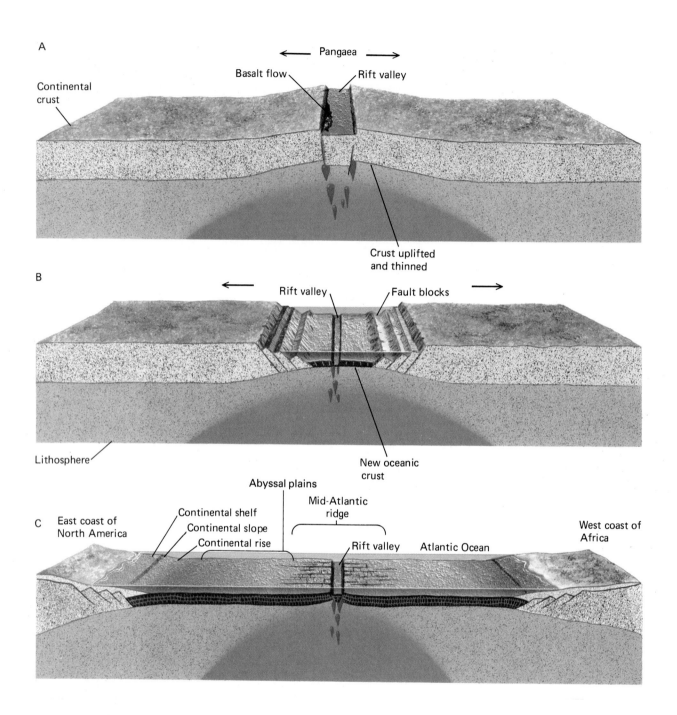

Figure 14–14   A model for the development of the passive continental margin of eastern North America. The Atlantic Ocean basin formed as a result of the rifting of Pangaea. (A) Pangaea is elevated over a rising mantle plume. (B) The crust thins due to faulting and erosion as a rift valley forms. Pangaea tears apart, and rising basalt magma forms new oceanic crust between the two halves of continental crust. (C) As the new Atlantic Ocean basin widens, sediment from the continents accumulates to form a broad continental shelf-slope-rise complex and buries the faults.

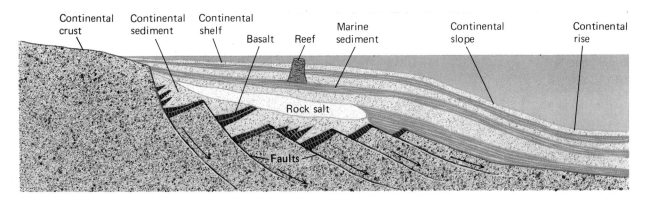

Figure 14–15 A passive continental margin is characterized by a broad continental shelf, slope, and rise formed by accumulation of sediment eroded from the continent. In some areas, salt deposits form. In tropical areas, reefs may also grow on the continental shelf.

Pangaea was heated and pulled apart, its crust stretched and thinned (Fig. 14–14A). Eventually the lithosphere fractured and began to separate where the crust was thinnest. Basalt magma rose through the cracks and flowed out onto the splitting continent. Continued eruption of basalt formed new oceanic crust between the separating fragments of Pangaea as they drifted apart (Fig. 14–14B). However, no further tectonic activity occurred along the ocean-continent boundary, hence the term passive continental margin.

Continental Shelf

Streams and rivers carry sediment from land to the sea and deposit it on coastal deltas. Then longshore currents redistribute the sediment along the coast. As sediment accumulated on the east coast of North America, it built a shallow, gently sloping submarine surface on the submerged edge of the continent. This surface is a **continental shelf** (Fig. 14–15). The average inclination of a continental shelf is about 0.1°, or approximately 1 meter drop in 600 meters of horizontal distance. Its depth increases gradually from the shoreline to about 200 meters at the outer shelf edge. A continental shelf on a passive margin is often a large feature. The continental shelf off the coast of southeastern Canada is about 500 kilometers wide. Parts of the shelves of Siberia and northwestern Europe are even wider.

Most continental shelves are covered by young sediment carried to the continental margin by modern rivers. Thick layers of shale, limestone, and sandstone lie beneath the younger sediment. Many of these older sedimentary rocks contain oil. Some of the world's richest offshore petroleum reserves occur in the North Sea between England and Scandinavia, in the Gulf of Mexico, and in the Beaufort Sea on the northern coast of Alaska

and western Canada. In recent years, extensive exploration and development of offshore petroleum reserves have taken place on continental shelves. Deep drilling for oil has revealed that beneath these older sedimentary rocks of the continental shelves lie granitic continental crustal rocks, confirming that the continental shelves are truly parts of the continents despite the fact that they are covered by seawater.

Continental Slope and Rise

At the outer edge of a continental shelf, the sea floor suddenly steepens from about 0.1° to 4° or 5° as the sea floor depth increases from 200 meters to about 5 kilometers. This steep region of the sea floor averages

An offshore drilling platform producing oil in the North Sea between England and Scandinavia. *(Oryx Energy Co., Dallas, Texas)*

Earth Science and the Environment

Life in the Sea

On land, most photosynthesis is conducted by multicellular plants such as mosses, ferns, grasses, and trees. In turn, large animals such as cows, deer, elephants, and bison consume the plants. In contrast, most of the photosynthesis and consumption in the ocean are carried out by small organisms called **plankton**. Many plankton are microscopic; others are up to a few centimeters long. **Phytoplankton** are plankton that conduct photosynthesis like land-based plants. Therefore, they are the ultimate source of food for aquatic animals. When you look across the surface of the ocean, you do not notice the phytoplankton, but they are so abundant that they supply about 50 percent of the oxygen in our atmosphere. **Zooplankton** are tiny animals that feed on the phytoplankton. The larger and more familiar marine plants such as seaweed and animals such as fish, sharks, whales, and dolphins play a relatively small role in oceanic photosynthesis and consumption.

Coastal waters above the continental shelf are generally 200 meters or less deep. In some regions this shallow zone extends up to 500 kilometers from the shore; in others the zone is much narrower. In contrast, the central oceans average about 5 kilometers deep.

Coastal areas support large populations of marine organisms. In addition, many deep-sea fish spawn in shallow water within a kilometer or two of shore. These shallow zones are hospitable to life because they have (1) easy access to the deep sea, (2) lower salinity than the open ocean, (3) a high concentration of nutrients originating from land and sea, (4) shelter, and (5) abundant plant life rooted to the sea floor in addition to the phytoplankton floating on the surface. As a result, about 99 percent of the marine fish caught every year are harvested from the shallow waters of the continental shelves. In many areas these fisheries are threatened by pollution.

The productivity of the central oceans is limited by the fact that light is available only at the surface, whereas gravity pulls nutrients downward toward the sea floor. As a result, key nutrients, especially nitrates and phosphates, are scarce in upper layers of the ocean where photosynthesis can occur. Productivity is so low that the central oceans have been likened to a great desert.

For many years, people assumed that little life exists in the depths of the open oceans because they are too dark for photosynthesis. However, improved diving and sampling techniques have revealed fascinating communities on the deep ocean floor. It is now clear that the ocean floor, at all depths, supports populations of both large animals and smaller organisms such as bacteria. These organisms eat the carcasses and wastes of plants and animals that sink from the surface layers.

---

about 50 kilometers wide and is called the **continental slope** (Fig. 14–15). It is formed by sediment much like that of the shelf. Its steeper angle is primarily due to rapid thinning of underlying continental crust as it approaches the junction with oceanic crust. The sedimentary layering is disrupted in places because some of the sediment has slumped and slid down the incline.

The steepness of a continental slope usually decreases as it gradually merges with the deep ocean floor. This region, called the **continental rise**, consists of an apron of sediment that was transported across the continental shelf and down the slope and is the terrigenous sediment described in Section 14.3. It came to rest on the deep ocean floor at the foot of the slope. The continental rise averages a few hundred kilometers wide. Typically, it joins the deep-sea floor at a depth of about 5 kilometers.

In essence, then, the continental shelf-slope-rise complex on a passive continental margin is a topographic surface formed by accumulation of sediment near the continental margin. The sediment is derived by erosion of the continent and smooths and conceals topography at the junction of continental and oceanic crust.

Submarine Canyons and Abyssal Fans

Deep, V-shaped, steep-walled valleys called **submarine canyons** are eroded into continental shelves and slopes. They look like submarine stream valleys. These canyons typically start on the outer edge of a continental shelf and continue across the slope to the rise (Fig. 14–16). At their lower ends submarine canyons commonly lead into **abyssal fans** (sometimes called **submarine fans**), large, fan-shaped accumulations of sediment deposited on the continental rise.

Most geologists now agree that submarine canyons are cut by **turbidity currents**. A turbidity current develops when loose, wet sediment resting on the continental shelf or slope starts to slip downslope, drawn by gravity. The movement may be triggered by an earthquake or simply by oversteepening of the slope as sediment accumulates. When the sediment starts to move, it mixes with water. Since the mixture of sediment and water is fluid and denser than water alone, it begins to flow downslope as a turbulent, chaotic avalanche over the surface of the shelf and slope. A turbidity current can travel at speeds

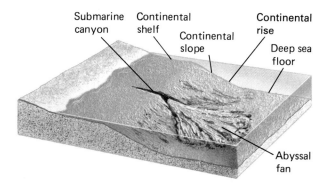

Figure 14–16   A submarine canyon and an abyssal fan.

greater than 100 kilometers per hour and for distances up to 700 kilometers. Sediment-laden water traveling at such speeds has tremendous erosive power. Once a turbidity current erodes a small channel into the continental shelf

and slope, subsequent currents tend to use the same channel, just as intermittent surface streams use the same channel year after year. Over time, the currents erode a deep submarine canyon into the continental margin. Turbidity currents slow down when they reach the level deep-sea floor beyond the continental slope. The sediment accumulates there to form an abyssal fan.

Active Continental Margins

An active continental margin forms where an oceanic plate subducts beneath a continental plate. A long, narrow, steep-sided depression called a **trench** forms on the sea floor where the oceanic plate bends downward and begins to subduct (Fig. 14–17). The deepest place on Earth is in the Mariana trench, north of New Guinea in the southwestern Pacific, with a maximum depth of nearly 11 kilometers below sea level. Depths of 8 to 10 kilome-

Figure 14–17   An active continental margin is characterized by subduction of an oceanic lithospheric plate beneath a continental plate.

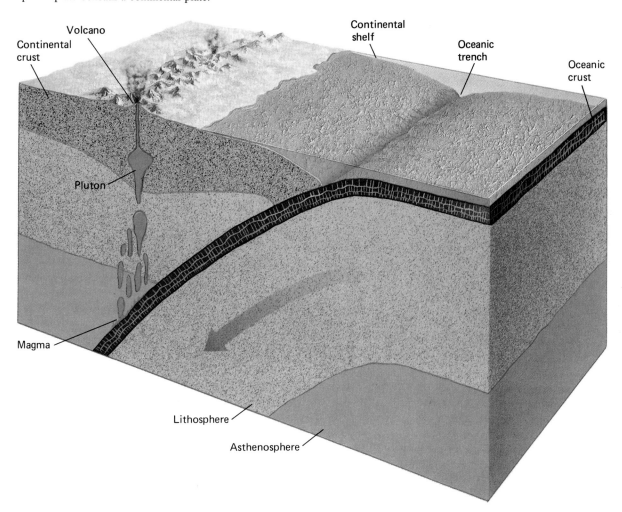

ters are not uncommon in other trenches. Trenches form wherever subduction occurs, at an active continental margin or in the middle of an ocean basin where two oceanic plates converge.

An active continental margin commonly has a much narrower continental shelf and a considerably steeper continental slope than does a passive margin. The continental rise is absent because sediment flows into the trench instead of accumulating on the ocean floor. The landward wall of the trench forms the continental slope of an active margin. It typically has an inclination of 4° or 5° in its upper part and steepens to 15° or more near its bottom.

Subduction at an active continental margin causes earthquakes, mountain building, and volcanic eruptions. The west coast of South America is an example of such an environment.

14.6 Island Arcs

In many parts of the Pacific Ocean and elsewhere, two oceanic plates are colliding. When oceanic plates collide, one plate subducts beneath the other, diving into the mantle and forming a mid-oceanic trench (Fig. 14–18). As we learned in Chapter 5, when a plate subducts, portions of it melt, forming magma. In an ocean-ocean collision, the magma rises and erupts on the sea floor to form submarine volcanoes next to the trench. The volcanoes eventually grow to become a chain of islands, called an **island arc**. The western Aleutian Islands are an example

Figure 14–18 An island arc forms at a convergent boundary between two oceanic lithospheric plates. One of the plates subducts, generating magma that rises to form the islands.

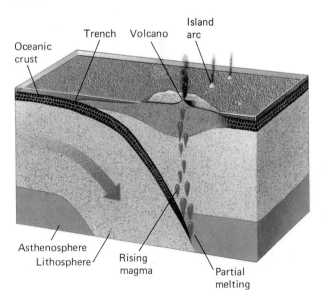

of an island arc. Many others are found in the southwestern Pacific.

14.7 Seamounts, Oceanic Islands, and Aseismic Ridges

A **seamount** is a submarine mountain that rises 1 kilometer or more above the surrounding sea floor. An **oceanic island** is a seamount that protrudes above sea level. Both seamounts and oceanic islands are common in all ocean basins, but they are particularly abundant in the southwestern Pacific Ocean. Seamounts and oceanic islands occur most commonly as isolated peaks on the ocean floor. However, some occur in chains of mountains called **aseismic ridges**. Aseismic means "no earthquake activity," as opposed to the mid-oceanic ridge and island arcs where earthquakes are common. The Hawaiian Island–Emperor Seamount Chain is an example of an aseismic ridge. Dredge samples show that seamounts, like oceanic islands and the ocean floor itself, are made of basalt.

Seamounts and oceanic islands are submarine volcanoes probably formed at hot spots above mantle plumes. Isolated seamounts and islands must have formed over plumes that persisted only for a short time. In contrast, aseismic ridges, such as the Hawaiian Island–Emperor Seamount Chain, formed over long-lasting plumes. In this case the lithospheric plate on which the volcanoes formed migrated over the plume as the magma continued to rise. Each volcano formed directly over the plume and then became extinct as the moving plate carried it away from the plume. As a result, the seamounts and islands become progressively younger toward one end of the chain (Fig. 14–19).

After a volcanic island or seamount forms, it begins to sink. Three factors contribute to the sinking.

1. If the hot spot feeding the volcanic eruptions cools and stops supplying magma, the lithosphere beneath the island cools, becomes denser, and contracts. Alternatively, the island migrates away from the hot spot if the lithospheric plate moves. This also causes cooling, contraction, and sinking of the island.

2. The weight of the newly formed volcano causes isostatic sinking.

3. Erosion lowers the top of the volcano.

These three factors commonly result in gradual transformation of a volcanic island to a seamount (Fig. 14–20). If the Pacific Ocean plate continues to move at its present rate, the island of Hawaii may sink beneath the sea within 10 to 15 million years. We learned in the last chapter that a reef often forms around an oceanic island. When the island sinks, the reef continues to grow and forms an atoll.

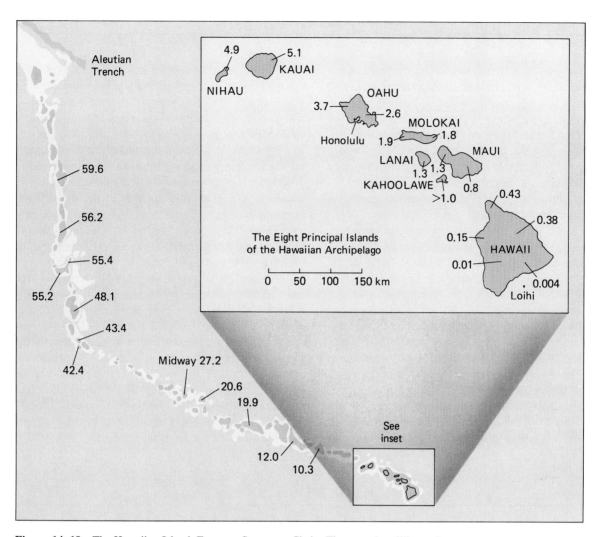

Figure 14–19 The Hawaiian Island–Emperor Seamount Chain. The ages, in millions of years, are for the oldest volcanic rocks found on each island or seamount.

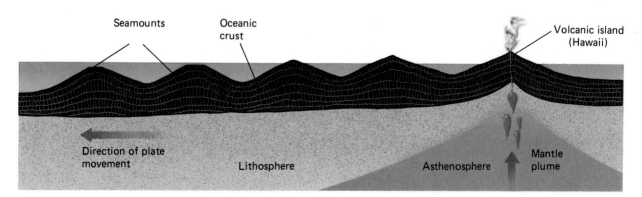

Figure 14–20 A schematic profile of the progressive sinking of volcanoes of the Hawaiian Island–Emperor Seamount Chain away from the mantle plume. As the sea floor sinks, the islands sink and eventually drop below sea level.

SUMMARY

•

The major geologic difference between a continent and an ocean basin is that the continent is composed of relatively thick, low-density granite, whereas oceanic crust is mostly thin, dense basalt. Thin, dense oceanic crust lies at low topographic levels and forms ocean basins. Because of the great depth and remoteness of the ocean floor and oceanic crust, our knowledge of them comes mainly from **sampling** and **remote sensing**.

The **mid-oceanic ridge** is a submarine mountain chain that extends through all of the Earth's major ocean basins. A **rift valley** runs down the center of the ridge, and the ridge and rift valley are both offset by numerous **transform faults**. The mid-oceanic ridge forms at the center of lithospheric spreading where new oceanic crust is added to the sea floor.

Abyssal plains are flat areas of the deep-sea floor where the rugged topography of the basaltic oceanic crust is covered by deep-sea sediment. Oceanic crust varies from about 7 to 10 kilometers thick and consists of three layers. The top layer is sediment, which varies from zero to a few kilometers thick. Beneath this lies about 1.5 kilometers of **pillow basalt**. The deepest layer of oceanic crust is from 5 to 7 kilometers thick and consists of basalt dikes on top of gabbro sills. The base of this layer is the boundary between oceanic crust and mantle. The age of sea-floor rocks increases regularly away from the mid-oceanic ridge. Almost no oceanic crust is older than about 200 million years because it recycles into the mantle at subduction zones.

A **passive continental margin** includes a **continental shelf**, a **slope**, and a **rise** formed by accumulation of sediment in the region where continental crust joins oceanic crust. **Submarine canyons**, eroded by **turbidity currents**, notch continental margins and commonly lead into **abyssal fans** where the turbidity currents deposit sediments on the continental rise. An **active continental margin**, where oceanic crust subducts beneath the margin of a continent, usually includes a narrow continental shelf and a continental slope that steepens rapidly into a trench. A **trench** is an elongate trough in the ocean floor formed where oceanic crust dives downward at a subduction zone. Trenches are the deepest parts of ocean basins.

Island arcs are common features of some ocean basins, particularly the southwestern Pacific. They are chains of volcanoes formed at subduction zones where two oceanic plates collide. **Seamounts, oceanic islands**, and **aseismic ridges** also form in oceanic crust as a result of volcanic activity over mantle plumes.

KEY TERMS

•

Rock dredge *378*	Terrigenous	Active continental	Turbidity current *388*
Coring device *378*	sediment *383*	margin *385*	Trench *389*
Echo sounder *378*	Pelagic sediment *383*	Continental shelf *387*	Island arc *390*
Seismic profiler *379*	Abyssal plains *384*	Continental slope *388*	Seamount *390*
Mid-oceanic ridge *381*	Manganese nodules *384*	Continental rise *388*	Oceanic island *390*
Rift valley *381*	Black smoker *385*	Submarine canyon *388*	Aseismic ridge *390*
Transform fault *382*	Passive continental	Abyssal fan (submarine	
Pillow basalt *382*	margin *385*	fan) *388*	

REVIEW QUESTIONS

•

1. Describe the main differences between oceans and continents.

2. Sketch a cross section of the mid-oceanic ridge, including the rift valley.

3. Describe the dimensions of the mid-oceanic ridge.

4. Explain why the mid-oceanic ridge is topographically elevated above the surrounding ocean floor. Why does its elevation gradually decrease away from the ridge axis?

5. Explain the origin of the rift valley in the center of the mid-oceanic ridge.

6. Why is heat flow unusually high at the mid-oceanic ridge?

7. Why are the abyssal plains characterized by such low relief?

8. Sketch a cross section of oceanic crust from a deep-sea basin. Label, describe, and indicate the approximate thickness of each layer.

9. Describe the two main types of sea-floor sediment. What is the origin of each type?

10. Compare the ages of oceanic crust with the ages of continental rocks. Why are they so different?

11. Sketch a cross section of both an active continental margin and a passive continental margin. Label the features of each. Give approximate depths.

12. Explain why a continental shelf is made up of a foundation of granitic crust, whereas the deep ocean floor is composed of basalt.

13. Why does an active continental margin typically have a steeper continental slope than a passive margin? Why does an active margin typically have no continental rise?

14. Explain the relationships among submarine canyons, abyssal fans, and turbidity currents.

15. Why are turbidity currents often associated with earthquakes or with large floods in major rivers?

16. Explain the origins of and differences between seamounts and aseismic ridges.

17. Why do oceanic islands sink after they form?

DISCUSSION QUESTIONS

•

1. How and why does an oceanic trench form?

2. The east coast of South America has a wide continental shelf, whereas the west coast has a very narrow shelf. Discuss and explain this contrast.

3. Seismic data indicate that continental crust thins where it joins oceanic crust at a passive continental margin, such as on the east coast of North America. Other than that, we know relatively little about the nature of the junction between the two types of crust. Speculate on the nature of that junction. Consider rock types, geologic structures, ages of rocks, and other features of the junction.

4. Discuss the topography of the Earth in an imaginary scenario in which all conditions are identical to present ones except that there is no water. In contrast, what would be the effect if there were enough water to cover all of the Earth's surface?

Cindy Lee Van Dover

Cindy Lee Van Dover received a B.S. in environmental science from Rutgers University (1977), followed by an M.S. in ecology (1985) from the University of California, Los Angeles. In 1980 she published the first of three articles on decapod crustaceans from the Indian River region of Florida. Then, in 1982, she participated in the Oasis Expedition to hydrothermal vents in the East Pacific rise. This was only the beginning of her relationship with the deep ocean and specifically hydrothermal vent communities. She took part in several other expeditions—aboard the research vessel *Atlantis II* and, as a scientist, aboard the submersible *ALVIN*—to places such as the Rose Garden and Gorda ridge.

In 1989 Van Dover completed a Ph.D. in biological oceanography in the Massachusetts Institute of Technology and Woods Hole Oceanographic Institution Joint Program. Since then she has held the position of postdoctoral investigator with the Biology Department at Woods Hole Oceanographic Institution in Woods Hole, Massachusetts. Also in 1989, Van Dover began her pilot train-

ing, and by 1990 she had piloted *ALVIN* to a variety of underwater locations throughout the Pacific Ocean. Later that year, she participated in a joint United

> As an ecologist, it seemed important to get a broad perspective of what the sea floor and hydrothermal vents are like.

States–Soviet Union dive aboard the *MIR-2* submarine in Monterey Bay, off Baja California, Mexico, and in the nearby Guaymas basin. At the controls of *ALVIN*, Van Dover explores hydrothermal vents with dexterity and thoroughness, combining her expert knowledge of sea life with her Navy-certified piloting skills. Most recently, she has employed data gathered from *ALVIN* expeditions to study the effects of sludge from deep-sea waste dumping on the biota of hydrothermal vent communities.

Van Dover is a member of the Deep Submersible Pilots' Association and the American Geophysical Union. In addition

to her considerable scholarly contributions, she has published several nontechnical articles on the undersea world as seen from her unusual perspective. In 1988 she was honored as one of *Ms.* magazine's "women of the year." Her commitment to science and her pioneering spirit have melded to help unlock the mysteries of the deep ocean, one of the few remaining earthly frontiers.

•

Where were you born and where did you attend school?

I was born and grew up in New Jersey, about 5 miles from the coast, in Eatontown; I went to a regional high school there, which happened to have a summer marine biology program and excellent teachers. That's how I got into marine science to begin with. As an undergraduate, I went to Rutgers University. After that, I spent some years working up and down the East Coast at various marine labs, as a technician. Eventually, I traveled west to earn my master's degree at UCLA. Then, back once more on the East Coast, I worked as a technician at the Marine Biological Lab in Woods Hole before starting work on my Ph.D. in the Woods Hole Oceanographic Institution–Massachusetts Institute of Technology Joint Program in biological oceanography.

•

So, from an early time in your career, you've been involved in marine biology or biological oceanography.

I always liked animals—especially invertebrates, because they come in so many different forms. Bugs and spiders I still think are creepy, an inheritance from my mother, but the animals that live in water are wonderful; some look more like plants than animals, others have fantastic features—feather-like appendages or multiple pairs of tiny eyes. Marine animals can be the stuff of science fiction, and I wanted to know more about them.

Then, too, I always wanted to be on a boat when I was little, even though I would get terribly seasick. I don't know where the sense of the romance of the sea comes from, but I certainly had it. Partly the animals fascinated me. But I admit that the adventure, the sense of independence or strength, the chance to prove myself against the sea lies somewhere behind my wanting to sail.

•

What did you do your Ph.D. on?

My Ph.D. was on the ecology of deep-sea hydrothermal vents. I started off my dissertation work studying food webs at vents—who eats whom. Geologists had just discovered hydrothermal vents on the Mid-Atlantic ridge, back in 1985. Thousands of shrimp swarm over the sulfide chimneys there. One scientist described them as looking like maggots swarming on a hunk of rotten meat—a not very poetic, but very accurate, description. My advisor was given some specimens; he passed them on to me: "Find out

what they are doing, how they are making a living."

As part of my dissertation work, I joined an *ALVIN* expedition to vent sites off the coast of Washington. I had learned that the chief scientist was planning to use an electronic camera developed here in Woods Hole. At the local bar on a Friday afternoon, I talked with the engineers who built the camera and learned that it ought to be able to detect the sort of light we supposed might be generated at a high-temperature vent. The chief scientist was willing to configure the camera appropriately, position it in front of a black smoker, turn out the submersible lights, and collect a time exposure of the ambient light. He and one of his colleagues were able to do this successfully, capturing a spectacular image of a glow emitted by the hot water.

•

Can you tell us some more about what these black smokers are, and something about the animal or plant life associated with them?

Black smokers occur along mid-ocean ridge spreading centers, which are amazing places. Along the spreading centers, the Earth's plates are pulling apart and new ocean crust is forming. It is a very black, high-contrast environment with incredible terrain, basically a linear volcano; pillowed and ponded lavas cover the sea floor. It is a very fluid-looking, dynamic landscape, with lava in frozen pools and swirls, buckles and fractures. Fissures there are often big enough for the submersible to drive into; exposed along the walls are the histories of eruptions, flows on top of flows.

Where the sea floor is moving apart, there is a lot of earthquake activity, a lot of cracking of the basalt. Seawater fills those cracks and actually reaches, if not the magma

chamber, then very, very hot rock. It is heated, reacts with the rock, and becomes modified. The heated water is buoyant and rises up through the crust to exit at the sea floor with a very interesting chemistry—enriched in metal sulfides and depleted in magnesium and oxygen. It is the chemical characteristics of the vent water that support microorganisms, the bacteria at the base of the food web. Life at vents is based *not* on photosynthesis, like we are used to in shallow-water and terrestrial systems, but on chemosynthesis. Instead of using light energy, the microorganisms use energy in reduced compounds to fix carbon. Instead of plants, the bacteria are the "grasses" of the hydrothermal vent community. The bacterias live both as free-living organisms suspended in the water column and on surfaces of rocks, sulfides, and animals; they also live as endosymbionts in tube worms and bivalves, supplying nutrients to their invertebrate hosts.

•

Can you describe the macrofauna around a black smoker a little bit more? What, in addition to tube worms, lives there?

In some places tube worms are the dominant organism. They are beautiful; they have bright red plumes and very white tubes. Living with them are a variety of crabs and other crustaceans. There can also be large golden-brown mussels and giant white clams with blood-red bodies occurring in large numbers at the base of black smokers, in the cracks and crevices where warm, sulfide-laden vent fluids are flowing.

•

How did you originally become interested in learning to pilot the *ALVIN*?

I think I published something like a dozen papers on the biology or ecology of vent fauna, but I'd only dived once to one vent. I was writing about animals I never saw alive, in their natural setting. I would go on lots of cruises, out at sea for a month or more each time, yet felt very lucky to get one dive. A simple look around me showed me who dove the most: the pilots. They could dive every other day, every three days, depending on the number of pilots. If you are a scientist, you dive only at the specific site you wrote the proposal for—you can't go all over the globe, diving anywhere. But the pilots get to go to all the different sites. As an ecologist, it seemed important to get a broad perspective of what the sea floor and hydrothermal vents are like. So that was the scientific motivation. I also wanted to do it just because of the adventure of it.

•

How did you first get involved with the *ALVIN* group?

ALVIN is Navy-certified, which means they have to get recertified at intervals and prove that they are operating a safe sub. Traditionally, repairs on *ALVIN* were supervised by a shop manager, and he knew how to maintain the thing. There was little in the way of procedures written down. But the Navy insisted they put together a maintenance manual. I came along at the right time with the talent to write and organize a document—I'd just finished my Ph.D. dissertation as proof of that ability. I began as a volunteer and was eventually hired.

•

What sort of training do you need to drive, or pilot, the submarine?

The training is all on the job, and when I trained, it was all done independently; the schematics and instrument manuals were in files and I was encouraged to work my way

through them all. In order to qualify, you have to pass a qualifying dive with the chief pilot, proving you can competently operate the sub. You also have to pass a series of qualifying boards, with the two most important being an Engineering Board and a Navy Board. During the Engineering Board, you stand in front of the engineers who designed and built the submarine. They have you draw and explain the schematics of every major system in the submarine. It was, at my exam, an unfriendly board. I passed. The Navy Board was mostly related to safety procedures and was a thoroughly professional exchange.

•

Have you ever had to troubleshoot beneath the surface?

All the time. We constantly work

with untested scientific gear that needs troubleshooting. Some of the failures are more scary than others. You learn there are certain failures you cannot continue a dive with; when one of those happens, you really pay attention.

As an example, once I had a 10K leak indication, which meant that I had water in one of the outside electrical boxes. These boxes are oil-filled; water in a box could short out the system and cause serious damage. When I saw that 10K indication, my pulse shot up rapidly. You see, at first you don't know where the 10K leak is. It could be in a battery box—bad news; these are 120-volt batteries and they don't like salt water. There are a series of toggle switches to go through that will locate where the problem is. You go through them one at a time, systematically, fast. It

is tense. You know once you find out where the problem is, you'll know what to do. Still, it's tense. In this particular case, the problem was in the variable ballast system. I just shut that system down and continued the dive. On deck we discovered that a single drop of water had closed the circuit in the leak detector. But there was no way of knowing the problem was minor underwater. The moment of a failure is always a challenge; you have to be on top of things, know your submarine; you have to know those schematics to know what to do next.

•

What do you see as your professional direction in the next decade?

I hope to continue to be a research scientist and to continue to study the ecology of the sea floor. It is an important field if we are to appreciate the world in which we live—not the small proportion that we ourselves occupy and modify, but the vast ocean environment that buffers our world and makes it livable. Continuing work on hydrothermal vents is an exciting prospect, pure scientific adventure, where the rules have not even begun to be defined. I'm also involved in some of the sewage sludge disposal issues off New Jersey. I'm interested in trying to trace the entry of sewage sludge into the benthic community of the deep sea. Dumping in deep water was initially started with the belief that none of the discharged materials would be detectable on the sea floor—the particles were so fine, they would be diluted and dispersed in the surface waters. My advisor and some of his colleagues thought maybe that wasn't the case and obtained funding to investigate. My contribution to the project is to use stable isotopes as tracers of the sewage

Tubeworms and other organisms living on sulfide chimneys at the Juan de Fuca ridge. *(Cindy Lee Van Dover)*

sludge. The technique requires an hypothesis that the organisms have two potential, isotopically distinct food sources. In this case, the sewage sludge has a very distinctive sulfur isotopic composition compared to normal sources of deep-sea organic sulfur. We collected benthic invertebrates—sea stars, sea urchins, etc.—from within the dump site and from a reference area upstream along the same depth contour. From the isotopic data for one species, an urchin, there was a strong sewage-sludge signal in the dump-site specimens. In fact, as much as 25 to 30 percent of the organic sulfur in dump-site urchins may be derived from sewage sludge.

•

What do you think that means? Is it good or bad?

Well, that's a little hard to say. It is a perturbation. In a way, it is good in that the sewage sludge provides a source of nutrition for the urchins in a food-limited environment. But the urchins may pay a price in accumu-

lating contaminants. There is the chance for magnification in the food web. It might also select for opportunistic species and change the structure of the benthic community. What strikes me as significant is the fact that the original thought was "Oh, we don't need to worry about this; it can't possibly get there." But sewage sludge is reaching the sea floor. If we can still be so naive about what we are doing to the environment, there is a lesson: we need to assess what we are doing a little more carefully, we need to know more about the deep sea.

•

What would you tell a student who might be interested in further investigating or even professionally pursuing a field of biological oceanography or marine biology, your field?

If you make it in this field, you will have a life that is exciting and adventurous. I believe there truly are unimaginable discoveries waiting for us in the oceans. But you only get there through hard work and creative thinking. You don't do this for money or prestige or glory. Each scientist has a philosophy, a way of viewing the world and tackling questions. My approach is probably more unorthodox than most, and I pay my dues for that. I don't hesitate to cut across boundaries, to venture into fields that I know nothing about in order to get at some aspect of my work. Becoming an *ALVIN* pilot is a good example of that; I had no technical skills appropriate for that job. My first cruise with the *ALVIN* group, I sailed as an electrician, of all things! But I learned, and learned fast. Have the confidence in yourself to go after what you want; have the consciousness of your abilities to know how far you can go.

The Atmosphere

UNIT

V

● Rainbow over the Potala Palace in Tibet *(Galen Rowell/ Mountain Light)*

The Earth's Atmosphere

15

I f you landed on the Moon without a space suit, you would not live long. The Moon has no atmosphere and therefore no air to breathe. The temperature on the sunny side hovers around 200°C, higher than that of boiling water. On the dark side the temperature is − 120°C, much lower than the lowest temperature ever recorded on Earth.

The Earth's atmosphere contains oxygen, essential to both plants and animals, and carbon dioxide needed by plants. The gases that envelop us insulate the Earth and distribute the Sun's heat, so the surface is neither too hot nor too cold for life. Clouds form from water dissolved in the atmosphere, and rain falls from clouds. In addition, the atmosphere filters out much of the Sun's ultraviolet radiation, which can destroy living tissue. The atmosphere carries sound; without air we would live in silence. Without an atmosphere, airplanes and birds could not fly, wind would not transport pollen and seeds, the sky would be black rather than blue, and no reds, purples, or pinks would color the sunset.

● Towering clouds over the Atlantic Ocean between Africa and South America. The yellow glow in the lower right is a dust storm from the Sahara. *(NASA)*

When you wake up in the morning and look out the window, you may remark, "It is a nice day," or "What a nasty gray day." Your remarks refer to atmospheric conditions such as temperature, wind, cloudiness, humidity, and precipitation. If you live in a city you might also note how smoggy or clear the air is. The atmosphere changes constantly. Some changes occur from day to day or even hour to hour. **Weather** is the state of the atmosphere at a given place and time. Temperature, precipitation, cloudiness, humidity, and wind are all components of weather.

Climate is a composite of weather patterns from season to season, averaged over many years. Miami and Los Angeles have warm climates: summers are hot and even the winters are warm. In contrast, New York and Chicago have cooler climates with winter snow. Seattle experiences moderate temperatures with foggy, cloudy winters. In this and the following two chapters we will discuss the composition and structure of the modern atmosphere, weather, and climate. We will also look at past changes in the atmosphere and consider its future.

15.1 Evolution of the Atmosphere

All planets and moons of the Solar System evolved from the same mixture of dust and gas, but today, no two have similar atmospheres. Mercury and our Moon have no

atmosphere. The atmosphere of Venus is mainly carbon dioxide and that of Jupiter mostly hydrogen and helium.

The Earth's present-day atmosphere is dramatically different from its original one. As the Earth initially coalesced into a rocky sphere, its atmosphere consisted mainly of hydrogen with smaller amounts of helium and other gases. Most of the hydrogen and helium escaped into space shortly after the Earth formed, 4.6 billion years ago. A second atmosphere then formed as volcanoes gave off gases trapped within the planet, and the Earth's gravitation attracted other gases from outer space. By studying the atmosphere of our two nearest neighbors, Venus and Mars, and by analyzing the compositions of old rocks on the Earth's surface, scientists have deduced that this second atmosphere consisted of carbon dioxide (CO_2), nitrogen (N_2), and water vapor (H_2O), with smaller amounts of methane (CH_4), ammonia (NH_3), hydrogen (H_2), and carbon monoxide (CO). Oxygen was present in trace quantities only. This atmosphere formed within the first 500 million years of Earth history. Most modern organisms would rapidly suffocate and die in such an environment.

Recall from Chapter 4 that the earliest fossils date back to about 3.5 billion years ago, although life probably originated earlier. Early life forms evolved the ability to conduct photosynthesis, combining carbon dioxide and water in the presence of sunlight to form glucose (sugar) and oxygen (Fig. 15–1). They used glucose for food, and

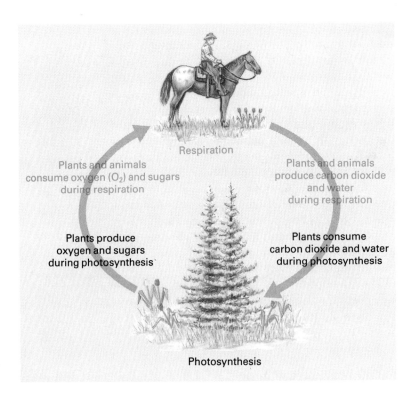

Respiration

Plants and animals consume oxygen (O_2) and sugars during respiration

Plants and animals produce carbon dioxide and water during respiration

Plants produce oxygen and sugars during photosynthesis

Plants consume carbon dioxide and water during photosynthesis

Photosynthesis

Figure 15–1 Exchange of gases and nutrients among plants and animals.

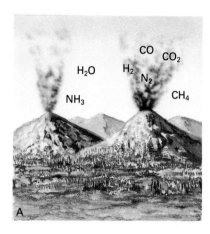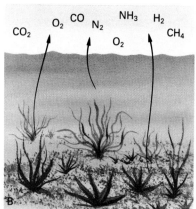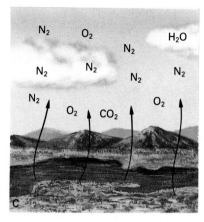

Figure 15–2   Evolution of the atmosphere. (A) The primitive atmosphere contained carbon dioxide, nitrogen, and other gases. (B) As plants evolved, the composition of the atmosphere began to change. Oxygen, released during photosynthesis, began to accumulate. (C) The modern atmosphere is composed mainly of nitrogen and oxygen, with smaller concentrations of water, carbon dioxide, and other gases. The ratio of oxygen to carbon dioxide is maintained by dynamic exchange among plants and animals.

released the oxygen into the atmosphere. In this way, photosynthesizers convert solar energy to chemical energy. As early organisms evolved and multiplied, they released more oxygen into the atmosphere.

For the next three billion years, single-celled organisms dominated the Earth, and the oxygen content of the atmosphere slowly increased. Then about 570 million years ago, at the beginning of the Cambrian period, multicellular plants and animals suddenly evolved and thrived. According to one theory, multicellular organisms could not have evolved earlier because the oxygen concentration was too low to support them. The Earth's atmosphere reached its present oxygen level of about 21 percent around 450 million years ago, 100 million years after the sudden bloom of macroscopic plants and animals (Fig. 15–2).

Fires burn rapidly if oxygen is abundant, so if its concentration were to increase even by a few percent, fires would burn uncontrollably across the planet. If the oxygen concentration were to decrease appreciably, most modern plants and animals would not survive. If the carbon dioxide concentration were to increase by a small amount, the average temperature of the Earth would rise because carbon dioxide traps heat.

Are we merely fortunate to have inherited such a nearly perfect atmosphere? The answer is no; luck is not responsible for the compatibility of Earth's atmosphere with its organisms. Organisms did not simply adapt to an existing atmosphere; they partially created it by photosynthesis and respiration. Thus, not only is our delicate oxygen–carbon dioxide balance biologically maintained, but

the very presence of oxygen in our atmosphere can be explained only by biological activity. If all life on Earth were to cease, the atmosphere would revert to its primitive, oxygen-poor composition and become poisonous to modern plants and animals.

15.2   The Composition of the Atmosphere

Air is mostly gas, with small quantities of water droplets and dust. The gaseous composition of dry air is roughly 78 percent nitrogen (N_2), 21 percent oxygen (O_2), and 1 percent other gases (Fig. 15–3, Table 15–1). Nitrogen,

Figure 15–3   The gaseous composition of natural dry air. Refer to Table 15–1 for a list of the 1 percent other gases.

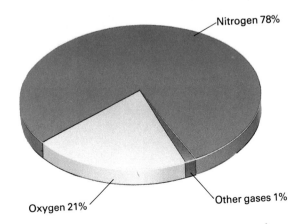

TABLE 15–1

Gaseous Composition of Natural Dry Air*	Concentration
Gas	*Percent*
Nitrogen, N_2	78.09
Oxygen, O_2	20.94
Inert gases, mostly argon, with much smaller concentrations of neon, helium, krypton, and xenon	0.93
Carbon dioxide, CO_2	0.03
Methane, CH_4, a natural part of the carbon cycle	0.0001
Hydrogen, H_2	0.00005
Oxides of nitrogen, mostly N_2O and NO_2, both produced by solar radiation and by lightning	0.00005
Carbon monoxide, CO, from oxidation of methane and other natural sources	0.00003
Ozone, O_3, produced by solar radiation and by lightning	Trace

*Natural dry air is defined as air without water or industrial pollutants. Carbon dioxide, methane, oxides of nitrogen, carbon dioxide, and ozone are all components of natural air, but they are also industrial pollutants. Therefore, the concentrations of these gases may vary, especially in urban areas. Pollution and its consequences are discussed in Chapter 18.

the most abundant gas, does not react readily with other substances. Oxygen, on the other hand, reacts chemically as fires burn, iron rusts, and plants and animals respire.

In addition to the gases listed in Table 15–1, air contains water vapor, water droplets, and dust. The types and quantities of these components vary with both location and altitude. Most natural air contains some water vapor. In a hot, steamy jungle, air may contain 5 percent water vapor by weight, whereas in a desert or cold polar region, only a small fraction of a percent may be present.

If you sit in a house on a sunny day, you may see a sunbeam passing through a window. The visible beam is light reflected from tiny specks of suspended dust. Clay, salt, pollen, bacteria, viruses, bits of cloth, hair, and skin are all components of dust. People travel to the seaside to enjoy the "salt air." Visitors to the Great Smoky Mountains in Tennessee view the bluish, hazy air formed by sunlight reflecting from pollen and other dust particles.

Within the past century, humans have altered the chemical composition of the atmosphere by releasing

This research balloon is 240 meters (800 feet) high, yet it is constructed of plastic no thicker than an average dry-cleaning bag. Instruments carried aloft by it can measure chemical reactions that lead to the depletion of the ozone layer about 30 to 45 kilometers above the Earth's surface. *(National Center for Atmospheric Research)*

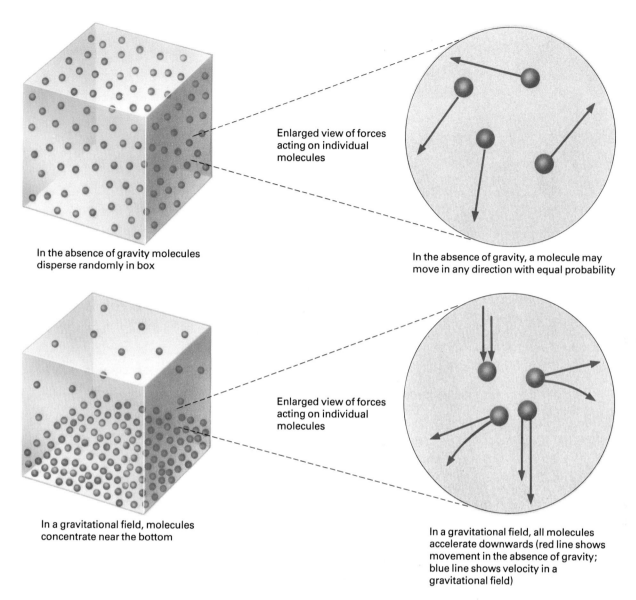

Enlarged view of forces acting on individual molecules

In the absence of gravity molecules disperse randomly in box

In the absence of gravity, a molecule may move in any direction with equal probability

Enlarged view of forces acting on individual molecules

In a gravitational field, molecules concentrate near the bottom

In a gravitational field, all molecules accelerate downwards (red line shows movement in the absence of gravity; blue line shows velocity in a gravitational field)

Figure 15–4 Gas molecules under the influence of gravity in the absence of wind and air currents.

large quantities of pollutants. Some are gases such as sulfur dioxide and nitric oxide; others, such as soot and smoke, are dust particles.

15.3 Atmospheric Pressure

Gas molecules in the atmosphere move rapidly and chaotically, frequently colliding with one another. Gravity pulls these molecules downward (Fig. 15–4). As a result, the atmosphere is densest near the Earth's surface and grows less dense with increasing altitude. Anyone who has ever climbed a high mountain has experienced atmospheric thinning with height. At 3000 meters (about 10,000 feet), even a person in good physical condition notices that

breathing is more difficult than at sea level. At 4500 meters (about 15,000 feet), a person's actions slow considerably, and above 6000 meters (about 20,000 feet), climbers move surprisingly slowly and breathe with difficulty.

If the double pan balance shown in Figure 15–5 sits on a table it reads zero—not because air is weightless, but because air pushes down equally on both pans. In order to weigh the atmosphere, you must compare the weight of a column of air with the weight of a vacuum. In practice it is impossible to place one pan of a scale in a vacuum chamber. However, the same effect can be achieved by placing the bottom of an open glass tube in a dish of a liquid such as mercury, sealing the top, and

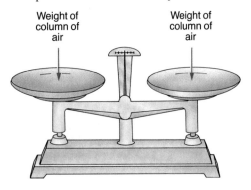

Figure 15–5 You cannot measure the weight of the atmosphere with a double balance because the weight of air is equal on both pans.

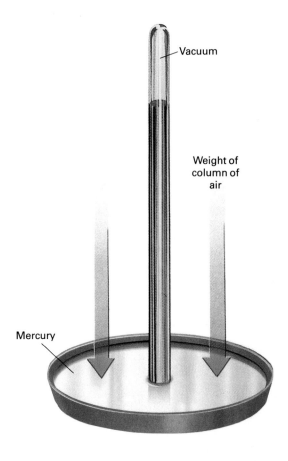

Figure 15–6 A barometer. In a barometer, the weight of the column of air on the mercury in the dish is *not* balanced by any air pressure from within the tube, because the upper region is evacuated. Instead, the air pressure is balanced by the weight of the mercury in the tube. As a result, the height of the mercury in the column is a measure of the outside air pressure.

evacuating the air from the tube. The mercury rises in the tube because the weight of the atmosphere pushes downward on the mercury in the dish but there is no air in the vacuum tube (Fig. 15–6). At sea level mercury rises approximately 76 centimeters, or 760 millimeters (about 30 inches), into a vacuum tube. The pressure exerted by the air is the **barometric pressure**, and a device that measures barometric pressure is a **barometer**. Barometric pressure is expressed in a variety of different units. Weather forecasters commonly express pressure in inches or millimeters of mercury, referring to the height of a column of mercury in a barometer. Another common unit is the **millibar**, which is 1/1000 of a bar. (A **bar** is a pressure unit defined in terms of force per unit area.) Figure 15–7 gives the conversion between millibars and inches of mercury. A mercury barometer is a cumbersome device nearly a meter tall, and mercury vapor is poisonous. A safer and more portable instrument for measuring pressure, called an **aneroid barometer**, consists of a partially evacuated metal chamber connected to a pointer. When atmospheric pressure increases, it compresses the chamber and the pointer moves in one direction. When pressure decreases, the chamber expands, directing the pointer the other way (Fig. 15–8).

A barometer measures air pressure. Two different factors cause air pressure to change. One is changing weather. At any given location, air pressure changes from day to day or even hour to hour. On a typical stormy day at sea level, pressure may be 980 millibars (28.94 inches), although it has been known to drop to 900 millibars (26.58 inches) or less during a hurricane. In contrast, during a

period of clear, dry weather a typical high pressure reading may be 1025 millibars (30.27 inches). (These variations are discussed in the following chapter.)

In addition, pressure decreases with altitude. Mountain climbers and airplane pilots routinely use an **altimeter**, a barometer with a scale calibrated in units of elevation rather than pressure. Figure 15–9 is a graph of changes in atmospheric pressure with elevation. Close to Earth's surface, pressure decreases sharply with altitude, and at high elevations the rate of decrease tapers off. If you stand on the summit of a mountain 5600 meters

Figure 15–7 Conversion between millibars and inches of mercury.

											Standard sea level pressure							
PRESSURE																		

Millibars 956 960 964 968 972 976 980 984 988 992 996 1000 1008 1016 1024 1032 1040 1048 1056

Inches 28.2 28.4 28.6 28.8 29.0 29.2 29.4 29.6 29.8 30.0 30.2 30.4 30.6 30.8 31.0 31.2

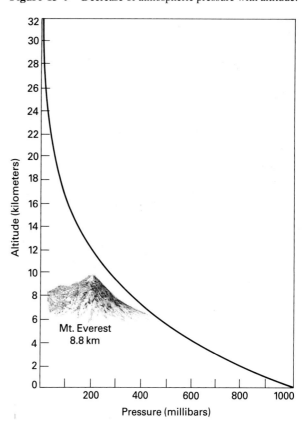

Focus On

The Upper Fringe of the Atmosphere

How high is the top of the atmosphere? Figure 15–16 shows that atmospheric pressure gradually diminishes with altitude, but the line never reaches zero and nothing in the graph defines an upper boundary. One definition of the top of the Earth's atmosphere is that it is where the pressure is about equal to the background pressure in the Solar System.

Most of the gases in space between the planets come from the Sun. The outer fringe of the Sun is the **corona**, which can be observed as a beautiful halo during a full solar eclipse. This region is extremely hot, about 2,000,000°C. Ordinary matter cannot exist at 2,000,000°C. Electrons are stripped from atoms, and gases become a collection of bare nuclei in a sea of electrons. Many of these nuclei and electrons are so hot and

are moving so fast that they escape the Sun's gravity and fly off into space. This stream of charged particles is called the **solar wind**. It blows past the Earth and other planets and flies outward toward the far reaches of the Solar System. When you think about the solar wind, do not think of a gentle breeze blowing against your face. At the Earth's surface, the atmosphere contains approximately 10^{19} (10 followed by 19 zeros) particles per cubic centimeter, and what you feel when an Earth wind blows is the effect of these particles striking your cheek. However, the solar wind contains only five particles per cubic centimeter and certainly could not be felt. The outer boundary of the Earth's atmosphere is approximately 9600 kilometers (6000 miles) high where atmospheric density approaches five particles per cubic centimeter, the density of the solar wind.

(18,400 feet) high, half of the atmosphere lies below you and you must survive on half as much oxygen for every breath you take. As a reference, the summit of Denali, the highest mountain in North America, is about 6200 meters high. If you ascended in a balloon to 16 kilometers (10 miles) above sea level, you would be above 90 percent of the atmosphere and would need an oxygen mask to survive. At an elevation of 100 kilometers, pressure is

only 0.00003 that of sea level, approaching the vacuum of outer space. There is no absolute upper boundary to the atmosphere.

Figure 15–9 Decrease of atmospheric pressure with altitude.

Figure 15–8 A mechanical barometer. When the atmospheric pressure increases, the chamber is compressed, deflecting the dial one way. When the pressure decreases, the chamber expands, deflecting the dial the other way.

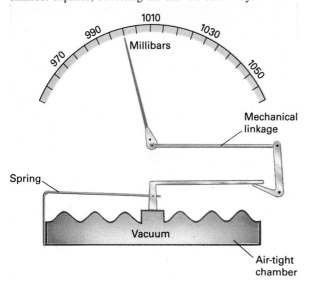

15.4 Solar Radiation

From the Sun, Earth would appear as a tiny, star-size speck. Solar energy streams out from the Sun in all directions, and Earth intercepts only 1/2,000,000,000 (one two-billionth) of the total output. However, even this tiny fraction warms Earth's surface and makes it habitable.

The space between Earth and the Sun is nearly empty. How does sunlight travel through a vacuum? In the late 1600s and early 1700s, light was poorly understood. Isaac Newton postulated that light consists of streams of particles, which he called "packets" of light. Two other physicists, Robert Hooke and Christian Huygens, argued that light travels in waves. Today we know that Hook and Huygens were correct—light behaves as a wave; but Newton was also right—light acts as if it is composed of particles. But how can light be both a wave and a particle at the same time? In a sense this is an unfair question because light is fundamentally different from familiar household and garden objects. Light is unique; it behaves as a wave and a particle simultaneously. There is no fundamental reason why light should only be like an ocean wave or only like a speeding bullet.

Particles of light are called **photons**. In a vacuum, photons travel only at one speed, the speed of light—never faster and never slower. The speed of light is 3 × 10^8 meters/second, or 186,000 miles/second. At that rate a photon covers the 150 million kilometers between Sun and Earth in about 8 minutes. Photons are unlike ordinary

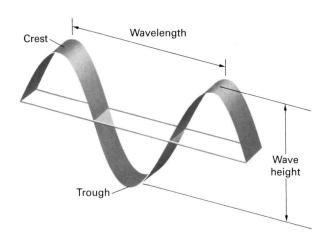

Figure 15–10 Terminology to describe waves.

matter in that they appear when they are emitted and disappear when they are absorbed.

Light also behaves as an electrical and magnetic wave, called **electromagnetic radiation**. Its **wavelength** is the distance between successive crests (Fig. 15–10). The **frequency** of a wave is the number of complete cycles, from crest to crest, that pass by any point in a second. (Think of how *frequently* the waves pass by.) Electromagnetic radiation occurs in a wide range of wavelengths. The **electromagnetic spectrum** is the continuum of radiation of different wavelengths (Fig. 15–11). At one

Figure 15–11 The electromagnetic spectrum. The wave shown above the number scale is not to scale. In reality the wavelength varies by a factor of 10^{22}, and this difference cannot be drawn.

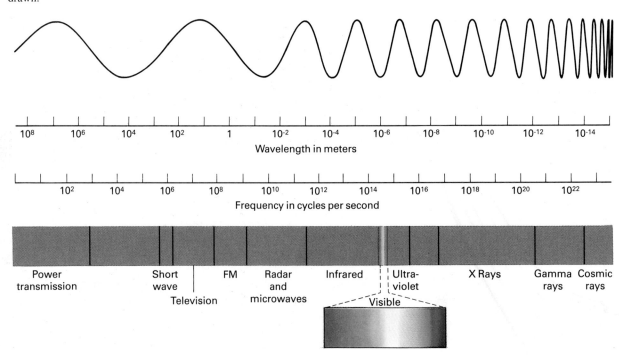

end of the spectrum, radiation given off by the current in an ordinary household wire has a long wavelength (5 million meters or 5000 kilometers) and low frequency (60 cycles/second). At the other end, cosmic rays from outer space have a short wavelength (about one trillionth of a centimeter, or 10^{-14} meter, and very high frequency (10^{22} cycles/second). Visible light is a tiny portion (about one millionth of one percent) of the electromagnetic spectrum.

Emission

Imagine that you have an iron bar at room temperature and place it in a hot flame. When the bar is hot enough it begins to glow dull red. The heat from the flame has excited electrons in the iron bar, and the excited electrons then emit electromagnetic radiation. If you heat the bar further, it gradually changes color until it becomes white. Figure 15–11 shows that color is a measure of wavelength. All objects emit radiation at some wavelength. The demonstration with the iron bar shows that the wavelength of emitted light is determined by its temperature. A hot bar emits high-energy radiation, which has high frequency and short wavelength. An iron bar at room temperature emits low-energy radiation, which has relatively long wavelength and lower frequency. This radiation carries so little energy that it does not activate the sensors of our eyes and is invisible infrared radiation.

Transmission

Radiation travels through a vacuum in straight lines. It can travel through space for trillions upon trillions of kilometers without any change in wavelength or loss of energy.

Radiation is also **transmitted** directly through transparent media. Thus, visible light passes through a window or through clear air. When light is transmitted through a perfectly transparent medium, no change in wavelength or loss of energy occurs. All materials are transparent to some wavelengths and opaque (not transparent) to others. For example, skin and muscle are opaque to visible light but transparent to X-rays. The walls of your house are opaque to visible light but transparent to radio waves, which is why you can listen to music from a radio playing in another room.

Absorption

A black cast-iron frying pan is not transparent to visible light. When light strikes its surface, a small amount is reflected, but most is **absorbed**. When an object absorbs radiation, the photons disappear and convert to another form of energy. If you go outside on a sunny day, your face absorbs energy and feels warm. Similarly, rock and soil absorb radiation, and a sunlit cliff can be warm on a winter afternoon. Thus, radiant energy converts to heat.

The Sun's surface temperature is about 6000°C. Due to this high temperature, the Sun emits relatively high-energy radiation, primarily in the ultraviolet and visible portions of the spectrum. Recall that high-energy radiation has short wavelength and high frequency. When this radiation strikes Earth, it is absorbed by rock and soil. After the radiant energy is absorbed, the rock and soil re-emit it. But the Earth's surface is much cooler than the Sun's. Therefore the Earth emits low-energy infrared radiation, which has relatively long wavelength and low frequency. Thus, the Earth absorbs high-energy visible light and emits low-energy, invisible infrared radiation.

Reflection and Refraction

Radiation **reflects**, or bounces back, from many surfaces. We are familiar with the images reflected by a mirror or the surface of a still lake. Even some dull-looking objects

A reflection in a still mountain lake.

Earth Science and the Environment

Solar Cells

When solar radiation is absorbed by a material, the photons disappear, but the energy is not lost; it converts to another form. In 1954, a research team at Bell Laboratories discovered that the energy from light-activated electrons of certain materials can be converted directly to electricity. A device that produces electricity directly from sunlight is called a **solar cell**. Although the invention of solar cells was exciting, the first devices were prohibitively expensive. In the 1950s, cells cost approximately $2000 per watt of output. Thus, a capital investment of $200,000 was required to generate enough electricity to light one 100-watt bulb. Nevertheless, the technology was used in exotic applications such as spacecraft, where power requirements were low and cost was relatively unimportant. The first practical use of solar cells was to power the second United States satellite, Vanguard I, launched in 1958.

Over the past 25 years, dramatic improvements have been made in the technology for producing solar cells, and prices have plummeted (see figure). In 1991, solar cells cost about $6 per watt. At that price you could provide electricity for an average home with an investment of about $10,000, assuming energy-efficient lighting and appliances were used and some other source of energy were used for cooking and heating. If you borrowed the $10,000 at 10 percent interest, your payments for electricity would be $83 per month, or about three times as much as it costs to buy the equivalent amount of electricity from the power company. Thus, solar cells remain uneconomical for most homes.

However, solar cells have obvious advantages. They can generate electricity without need for transmission lines. Furthermore, the cells require no maintenance. They therefore become economical when electricity is needed in remote places, such as light buoys anchored at sea and telephones and other communication systems located far from electric lines. In addition, if a house is situated more than a kilometer from the nearest power line, it is cheaper to install solar cells than to string a new line. In 1990 about 10,000 buildings in the United States were powered exclusively by solar cells. The technology has also blossomed in many other countries. About 60 percent of the population of the world has no access to commercial electricity. Portable generators are expensive, require maintenance that is often unavailable, and burn expensive fuel. In one report, health officials in India stated that solar-powered refrigerators were urgently needed to store heat-sensitive medicines and vaccines.

Some people argue that the way we measure energy costs is incomplete. Conventional electric generators degrade the environment. Coal and oil-fired generators pollute the air. The problems of nuclear waste disposal and decommissioning of obsolete nuclear power plants have not yet been solved. Dams for hydroelectricity disrupt aquatic ecosystems and destroy recreational and scenic areas. These factors are not added to the cost of electricity but must be paid by society. If these hidden costs were added to the cost of commercial electricity, solar cells would become more attractive.

If the price of fossil fuels rises and the price of solar cells continues to fall, solar generation of electricity will become economical. Then we will be able to envision a new energy age in which people obtain electric power by bolting solar panels on their rooftops, and centralized solar power plants are built in deserts.

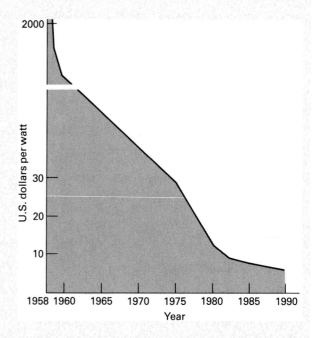

The cost of photovoltaic cells from 1958 to 1990.

are efficient reflectors. Light is bouncing back to your eye from the white paper of this page, although very little is reflected from the black letters.

Recall that both ocean waves and seismic waves bend, or **refract**. Because radiation also behaves as a wave, it too refracts. If you place a pencil in a dish partially filled with water, the pencil appears bent. The pencil actually remains straight, but our eyes are fooled because light refracts as it travels across the boundary between air and water.

Scattering

On a clear day the Sun shines directly through windows on the south side of a building, but if you look through a north-facing window, you cannot see the Sun. Even so, light enters through the window, and the sky outside is blue. If sunlight were only transmitted directly, a shaded, north-facing room would be dark and the sky outside the window would be black. Atmospheric gases, water droplets, and dust particles **scatter** sunlight in all directions, as shown in Figure 15–12. Light that has been scattered is called **diffused** light. Short-wavelength violet and blue light scatter more than longer-wavelength red light. This scattered sunlight illuminates a room on the shady side of a house.

If Earth had no atmosphere, sunlight would travel directly to the surface and the Sun would look white in a black sky. However, Earth's atmosphere scatters the violet and blue wavelengths throughout the sky, and they

Red colors in a sunset over Bristol Bay, Alaska.

turn the sky blue. The yellow color of the Sun is a mix of the colors left over after the blues are removed by scattering. At noon, sunlight passes through the atmosphere by the shortest possible path (Fig. 15–13). At sunrise and sunset, the Sun is low in the sky, and sunlight travels a greater distance through the atmosphere to reach the Earth. During this journey, the blues scatter into space,

Figure 15–12 When incoming radiation is scattered, the direction changes but the wavelength does not.

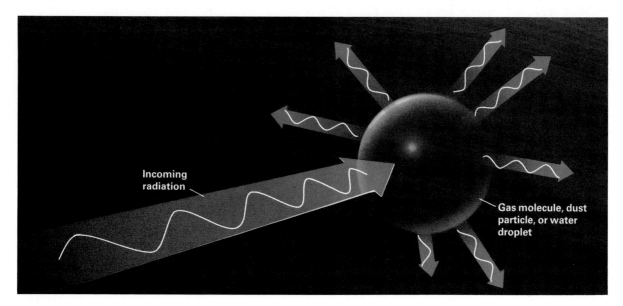

Incoming radiation

Gas molecule, dust particle, or water droplet

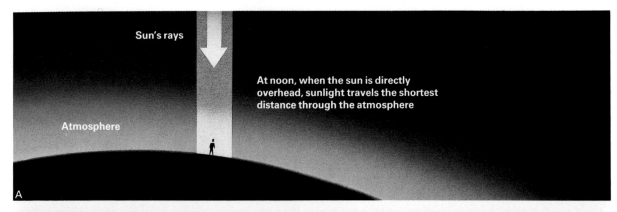

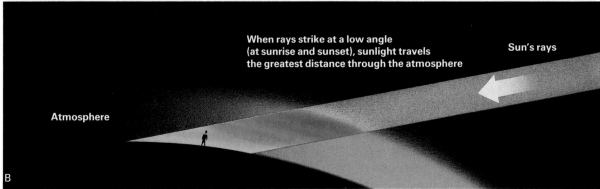

Figure 15–13 (A) When the Sun is directly overhead, sunlight passes the shortest distance through the atmosphere. Violet and blue wavelengths, which are most easily scattered, turn the sky blue. (B) At dawn and dusk, sunlight travels a greater distance through the atmosphere to reach the Earth. Red light, which is scattered the least, passes through to the Earth, and the sky appears red.

and red light, which is scattered the least, passes through to the Earth. As a result, the sky is often red at dawn and dusk. If the air is filled with dust or water droplets, all the wavelengths scatter and the sky becomes white or gray.

15.5 The Radiation Balance

With this background, let us examine the fate of sunlight as it reaches Earth (Fig. 15–14).

1. About 8 percent of incoming solar radiation is scattered in the atmosphere and diffuses back into space. As a result, it never reaches the ground.

2. Nineteen percent is absorbed by the atmosphere before it reaches the ground.

3. Twenty-six percent is reflected back into space. Clouds reflect most of this amount (23 percent), and the ground reflects the rest (3 percent). Some surfaces are better reflectors than others. **Albedo** is the reflectivity of a surface. Snowfields and glaciers have high albedos and reflect 80 to 90 percent of sunlight. On the other

hand, city buildings and dark pavement have albedos of only 10 to 20 percent (Table 15–2). If the Earth's albedo were to rise by growth of glaciers or increase in cloud cover, the surface of our planet would cool; alternatively, a decrease in albedo might cause warming.

4. Forty-seven percent of sunlight passes through the atmosphere and is absorbed by soil, rock, and the oceans. This energy warms the Earth's surface.

TABLE 15–2

Albedos of Natural Surfaces	
Surface	**Percent Sunlight Reflected**
Snow	80–90
Sand	30–60
Clouds (average)	50–55
Forests and farmlands	5–25
Cities	10–20
Water (average)	8

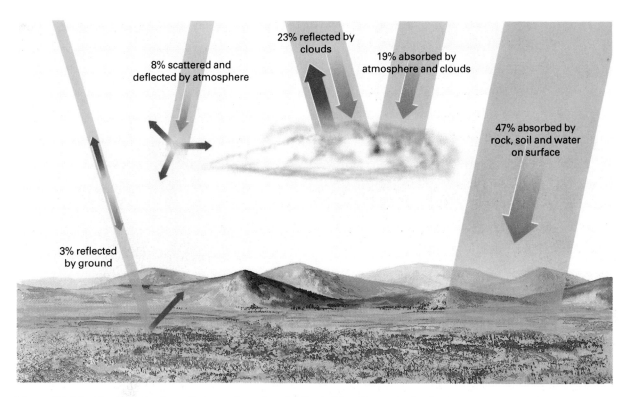

Figure 15–14   The fate of solar radiation as it reaches the Earth. The 47 percent that is absorbed on the surface is eventually reradiated as long-wavelength radiation, as shown in Figure 15–16.

If the Earth absorbs radiant energy from the Sun, why doesn't it get hotter and hotter until the oceans boil and the rocks melt? The answer is that the Earth re-emits all the energy it absorbs. As explained above, most solar energy that reaches the Earth is short-wavelength visible and ultraviolet radiation. This radiation is absorbed at the Earth's surface and is then re-emitted, mostly as long-wavelength, invisible infrared radiation, sometimes called "heat rays." Some of this heat escapes directly into space, but some is absorbed by the atmosphere. Thus, the atmosphere traps heat radiating from Earth and acts as an insulating blanket.

If Earth had no atmosphere, radiant heat loss would be so rapid that the Earth's surface would cool drastically at night. Recall that the temperature on the dark side of the Moon is −120°C. The Earth remains warm at night because the atmosphere absorbs and retains much of the radiation emitted by the ground. If the atmosphere were to absorb even more of the long-wavelength radiant heat from the Earth, the atmosphere and Earth's surface would become warmer. This warming process is called the **greenhouse effect** (Fig. 15–15).

Some molecules in the atmosphere absorb infrared radiation and others do not. Oxygen and nitrogen, which together make up almost 99 percent of dry air at ground level, do not absorb infrared radiation. Water, carbon dioxide, methane, and other gases do. Water plays the major role in absorbing heat because it is so abundant. Carbon dioxide is also important because its abundance in the atmosphere can vary due to several natural and industrial processes. Methane has become important because large quantities are released by industry and agriculture. The role of pollutants in altering global temperature is discussed in Chapter 18.

15.6   Temperature Changes with Elevation

The temperature of the atmosphere changes with altitude (Fig. 15–16). The layer of air closest to the Earth, the layer we live in, is the **troposphere**. Virtually all of the water vapor and clouds exist in this layer, and almost all weather occurs here. The Earth's surface absorbs solar energy, and thus the surface of the planet is warm. But, as explained above, continents and oceans also radiate heat, and some of this energy is absorbed in the atmosphere. At higher elevations in the troposphere, the atmosphere is thinner and absorbs less energy, and therefore temperature decreases. Thus, mountaintops are generally

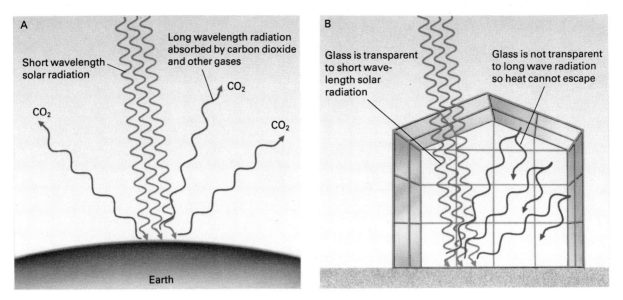

Figure 15–15 (A) The greenhouse effect can be viewed as a three-step process. 1. The Earth absorbs short-wavelength solar radiation. 2. The Earth reradiates the energy in the form of long-wavelength (heat) rays. 3. Molecules in the atmosphere absorb some of the long-wavelength radiation. (B) The analogy of the greenhouse. Glass is transparent to short-wavelength solar radiation. However, it is opaque to longer-wavelength radiation emitted from the soil. (*Note:* The comparison between the atmosphere and a greenhouse is only partially correct. Both the glass in a greenhouse and the Earth's atmosphere are transparent to incoming short-wavelength and partially opaque to emitted long-wavelength radiation. However, the glass in a greenhouse is also a physical barrier that prevents heat loss through air movement, whereas the atmosphere is not. Despite this inaccuracy, the term "greenhouse effect" is commonly used.)

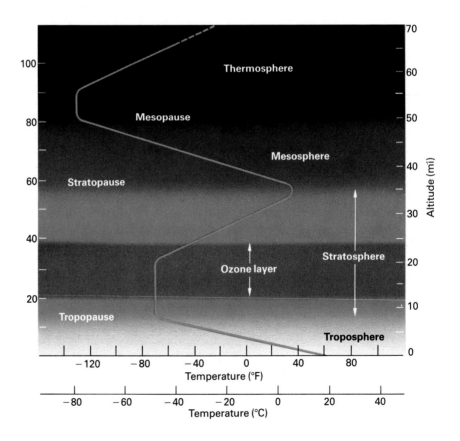

Figure 15–16 Temperature change with altitude and the major layers of the atmosphere.

Glaciers and ice cliffs on Mount Kilimanjaro, Africa, near the Equator. *(Dougal McCarty)*

colder than valley floors, and pilots flying at high altitudes must heat their cabins.

The top of the troposphere is at an altitude of about 17 kilometers (10 miles) at the equator, although it is lower at the poles; this ceiling is the **tropopause**. At the tropopause the steady decline in temperature with altitude ceases abruptly. Cold air from the upper troposphere is too dense to rise above the tropopause. As a result, little mixing occurs between the troposphere and the layer above it, called the **stratosphere**.

In the stratosphere, temperature remains constant to 35 kilometers and then *increases* with altitude until, at about 50 kilometers, it is as warm as that at the Earth's surface. This reversal in the temperature profile occurs because the troposphere and stratosphere are heated by different mechanisms. As already explained, the troposphere is heated primarily from below, by the Earth. The stratosphere, on the other hand, is heated primarily from above, by solar radiation.

Oxygen molecules in the stratosphere absorb energetic ultraviolet rays from the Sun. The radiant energy breaks the oxygen molecules apart, releasing free oxygen atoms. The oxygen atoms then recombine to form ozone, (O_3). Ozone absorbs ultraviolet light more efficiently than oxygen does, warming the upper stratosphere. Ultraviolet light is energetic enough to affect organisms. Small quantities give us a suntan, but large doses cause skin cancer, inhibit the growth of many plants, and otherwise harm living tissue. The ozone in the upper atmosphere protects life on Earth by absorbing much of this high-energy radiation before it reaches Earth's surface.

Ozone concentration declines in the upper portion of the stratosphere, and therefore at about 55 kilometers above the Earth, temperature once more begins to decline rapidly with elevation. This boundary between rising and falling temperature is the **stratopause**, the ceiling of the stratosphere. The second zone of declining temperature is the **mesosphere**. Little radiation is absorbed in the mesosphere, and the thin air is extremely cold. Starting at about 80 kilometers above the Earth, the temperature again remains constant and then rises rapidly in the **thermosphere**. Here high-energy X-rays and ultraviolet radiation from the Sun are absorbed. High-energy reactions strip electrons from atoms and molecules to produce ions. The temperature in the upper portion of the thermosphere is just below freezing, not extremely cold by surface standards.

Even though the temperature in the thermosphere is no colder than that on a winter day in Chicago, if you ascended into the thermosphere in a balloon you would rapidly freeze to death. Recall from Section 15.3 that at an elevation of 100 kilometers, atmospheric pressure is only 0.00003 that of sea level. The heat contained in a parcel of air is proportional to both the temperature and

(Text continues on p. 418)

415

Earth Science and the Environment

Depletion of the Ozone Layer

As explained in the text, ozone in the stratosphere absorbs ultraviolet (UV) radiation and thereby protects life on Earth. UV radiation from the Sun tans our skin. Heavier doses of UV cause burns and increase the chance of skin cancer. If more UV light were to reach the Earth's surface, the risks of such damages would increase. Plants, too, might be adversely affected. Some food crops, such as tomatoes and peas, grow more slowly if subjected to high doses of UV light.

Some air pollutants rise into the stratosphere where they destroy ozone. Organic compounds contain-ing chlorine and fluorine, called **chlorofluorocarbons**, or **CFCs**, are major contributors to the destruction of the ozone layer.* CFCs are used as cooling agents in almost all refrigerators and air conditioners, as propellants in some aerosol cans, as cleaning solvents used during the manufacture of weapons, and in plastic foams in products such as coffee cups and some building insulation. Nitrogen oxides from autos and airplanes also react in the stratosphere to destroy ozone.

In the 1950s, when CFCs were first developed, they were thought to be nonreactive and hence not harmful to the environment. However, in the mid-1970s laboratory studies showed that CFCs destroy ozone. These studies

*These compounds are also referred to as **Freons**, a Du Pont trade name.

The three-step reaction sequence that leads to the destruction of the ozone layer from CFCs. *Step 1:* Molecules of CFCs rise into the stratosphere. When struck by ultraviolet radiation, a molecule breaks apart, releasing a chlorine atom. *Step 2:* A chlorine atom reacts with ozone, O_3, to destroy the ozone molecule and release oxygen, O_2. The extra oxygen atom attaches itself to the chlorine to produce ClO. *Step 3:* The ClO sheds its oxygen to produce another free chlorine atom. Thus chlorine is not used up in the reaction, and one chlorine atom reacts over and over again to destroy many ozone molecules.

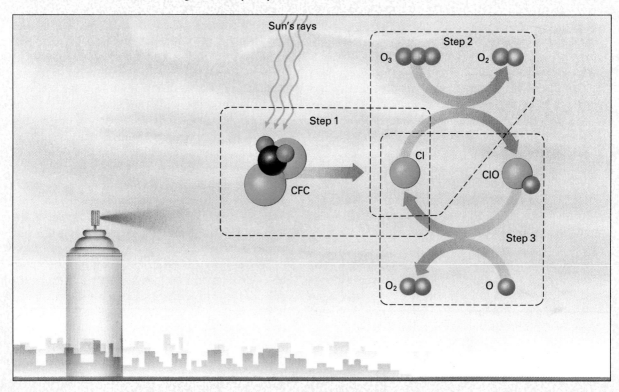

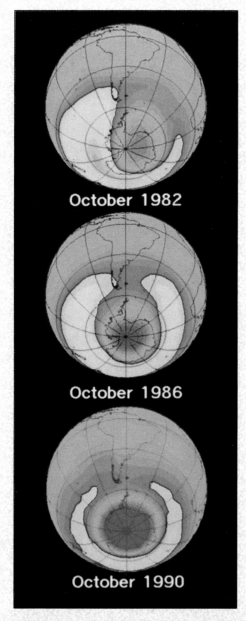

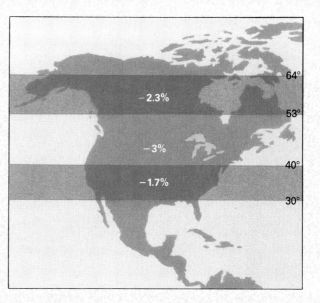

Stratospheric ozone depletion in the Northern Hemisphere. *(EPA)*

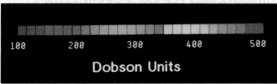

Growth of the ozone hole from 1982 to 1990. The scale is in Dobson units, which measure stratospheric ozone. Purple indicates the least ozone (greatest depletion). *(NASA)*

were soon supported by measurements of actual ozone depletion. In the 1980s scientists observed an unusually low ozone concentration over Antarctica, called the **ozone hole**. During the mid-1980s the ozone hole appeared in some years, but not in others. However, severe ozone depletion was recorded for three consecutive years; 1989, 1990, and 1991. After the 1991 data were recorded, Charles Jackman, a research scientist at NASA, speculated, "Three years in a row of low ozone leads one to wonder that maybe most years will be low in the future." In 1992, a second ozone depletion zone was recorded over New England.

These data persuaded the industrial nations of the world to limit the use of CFCs. In 1978 the United States, Canada, and the Scandinavian countries banned the use of CFCs as propellants in aerosol cans. In 1987, 24 nations signed a stricter treaty that held production of CFCs at 1986 levels, starting in 1989. Between 1989 and 1999, production in these countries will be cut in stages until it is at half its 1986 level.

CFC molecules take a few decades to rise from the Earth's surface into the stratosphere. Once aloft, they persist for about 100 to 300 years before decomposing. Therefore, even if production were curtailed tomorrow, the problem would continue for a century or more.

Focus On

Latitude and Longitude

If someone handed you a perfectly smooth ball with a dot on it and asked you to describe the location of the dot, you would be at a loss to do so because all positions on the surface of a sphere are equal. How, then, can locations on a spherical Earth be described? Even if we ignore irregularities, continents, and oceans, the Earth has points of reference because it rotates on its axis and has a magnetic field that nearly (but not exactly) coincides with the axis of rotation. The North and South Poles lie on the rotational axis. If we consider the North Pole to lie at the top (there is no astronomical reason to do so), then lines of **latitude** form imaginary horizontal rings around the Earth. Mathematicians measure distance on a sphere in degrees. Using this system, the equator is defined as 0°

latitude, the North Pole is 90° north latitude, and the South Pole is 90° south latitude.

No natural east-west reference exists, so a line running through Greenwich, England, was arbitrarily chosen as the 0° line. The planet was then divided by lines of **longitude**, also measured in degrees. On a globe with the poles at the top and bottom, lines of longitude run vertically. To a navigator they measure east-west distance from Greenwich, England.

The system is easy to use. Minneapolis–St. Paul lies at 45° north latitude and 93° west longitude. The latitude tells us that the city lies halfway between the equator (0°) and the North Pole (90°). Since a circle has 360°, the longitude tells us that Minneapolis is about one quarter of the way around the world in a westward direction from Greenwich, England.

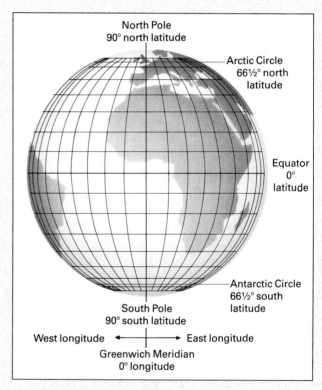

North Pole
90° north latitude

Arctic Circle
66½° north
latitude

Equator
0°
latitude

Antarctic Circle
66½° south
latitude

South Pole
90° south latitude

West longitude ⟷ East longitude

Greenwich Meridian
0° longitude

Latitude and longitude.

the number of molecules. In such a rarefied atmosphere, so few molecules exist that little heat is retained.

15.7 Temperature Changes with Latitude and Season

Figure 15–16 shows that the temperature at the Earth's surface is about 20°C (68°F). Of course, this is a global

average; surface temperature fluctuates with both location and season. If you travel extensively, you may have experienced bitter-cold arctic air or the oppressive heat of the tropics.

Temperature Changes with Latitude

The region near the equator is warm throughout the year, whereas polar regions are cold and ice-bound even in

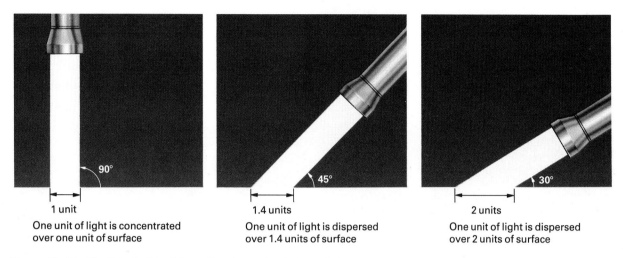

90°

1 unit

One unit of light is concentrated
over one unit of surface

45°

1.4 units

One unit of light is dispersed
over 1.4 units of surface

30°

2 units

One unit of light is dispersed
over 2 units of surface

Figure 15–17 If a light is shined from directly overhead, the radiation is concentrated on the surface. However, if the source of the light (or the surface) is tilted, the light spreads over a larger area and the radiation is dispersed.

summer. To understand this temperature difference, consider first what happens if you hold a flashlight above a flat board. If the light is held directly overhead and the beam shines straight down, a small area is brightly lit. If the flashlight is held at an angle to the board, a larger area is illuminated. However, because the same amount of light is spread over a larger area, the intensity is reduced (Fig. 15–17).

Now consider what happens when the Sun shines on the spherical Earth. The region directly beneath the Sun, analogous to the part of the flat board under a direct light, receives the most concentrated radiation. The rest of the globe receives light at an angle and therefore receives less intense radiation. Thus, radiation is less concentrated at a higher latitude (Fig. 15–18). Because the equator receives the most concentrated solar energy, it is generally warm throughout the year. Average temperature becomes progressively cooler poleward (north and south of the equator).

The Seasons

Earth circles the Sun in a planar orbit while simultaneously spinning on its axis. This axis is tilted at 23.5° from a line drawn perpendicular to the orbital plane.

The Earth revolves around the Sun once a year. As shown in Figure 15–19, the North Pole tilts toward the Sun in summer and away from it in winter. June 21 is the **summer solstice** in the Northern Hemisphere because at this time the North Pole leans the full 23.5° toward the Sun. As a result, sunlight strikes the Earth from directly overhead at a latitude 23.5° north of the equator. This

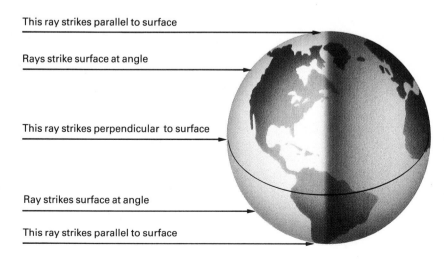

This ray strikes parallel to surface

Rays strike surface at angle

This ray strikes perpendicular to surface

Ray strikes surface at angle

This ray strikes parallel to surface

Figure 15–18 If the Earth were not tilted, the Sun's rays would always strike perpendicular to the Earth's surface at the equator and parallel at the poles. Therefore, the equatorial regions would receive the most intense solar radiation, and the polar regions would receive hardly any.

The midnight Sun. This photograph was taken exactly at midnight in July, at 70° north latitude in the Canadian Arctic.

latitude is the **Tropic of Cancer**. If you stood on the Tropic of Cancer at noon on June 21, you would cast no shadow. June is warm in the Northern Hemisphere for two reasons:

1. Because the Sun is high in the sky, sunlight is more concentrated than in winter.

2. When the North Pole is tilted toward the Sun, it receives 24 hours of daylight. Polar regions are called "lands of the midnight sun" because the Sun never sets in the summertime. Below the Arctic Circle the Sun sets in the summer, but the days are always longer than in winter (Table 15–3).

While it is summer in the Northern Hemisphere, the South Pole tilts away from the Sun and the Southern Hemisphere receives low-intensity sunlight and has short days. June 21 is the first day of winter in the Southern Hemisphere. Six months later, on December 22, the seasons are reversed. The North Pole tilts away from the Sun, giving rise to the **winter solstice** in the Northern Hemisphere, while it is summer in the Southern Hemisphere. On this day sunlight strikes the Earth directly overhead at the **Tropic of Capricorn**, latitude 23.5° south. At the North Pole the Sun never rises and it is continuously dark, while the South Pole is bathed in continuous daylight.

On March 21 and September 23, the Earth's axis is at right angles to the Sun, and sunlight shines directly on the equator. If you stood at the equator at noon on either of these two dates, you would cast no shadow. But north or south of the equator, a person casts a shadow even at

Figure 15–19 A schematic view of the Earth's orbit showing the progression of the seasons.

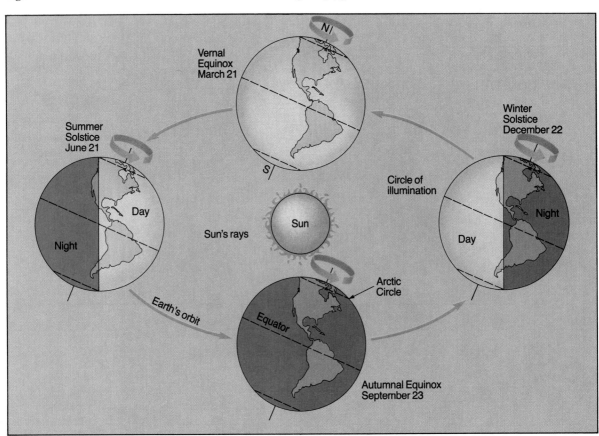

T A B L E   15 – 3

		Hours of Sunlight		
Latitude	Geographic Reference	Summer Solstice	Winter Solstice	Equinoxes
0°	Equator	12 h	12 h	12 h
30°	New Orleans	13 h 56 min	10 h 04 min	12 h
40°	Denver	14 h 52 min	9 h 08 min	12 h
50°	Vancouver	16 h 18 min	7 h 42 min	12 h
90°	North Pole	6 mo	0 00	12 h

noon. In the Northern Hemisphere, March 21 is the first day of spring and September 21 is the first day of autumn, whereas the seasons are reversed in the Southern Hemisphere. On the first days of spring and autumn, every portion of the globe receives 12 hours of direct sunlight and 12 hours of darkness. For this reason, March 21 and September 21 are called the **equinoxes**, meaning equal nights.

All areas of the globe receive the same total number of hours of sunlight every year. The North and South Poles receive direct sunlight in dramatic opposition, six months of continuous light and six months of continuous darkness, whereas at the equator each day and night is close to 12 hours long throughout the year. Although the poles receive the same number of sunlight hours as the equatorial regions, the sunlight reaches the poles at a much lower angle and therefore delivers much less total energy.

15.8 Temperature Changes Due to Heat Transport and Storage

Even though all locations at a given latitude receive equal amounts of solar radiation, temperature is *not* constant with latitude. Figure 15–20 is a map of the world with lines, called **isotherms**, connecting areas of the same

Figure 15–20 Global temperatures in January.

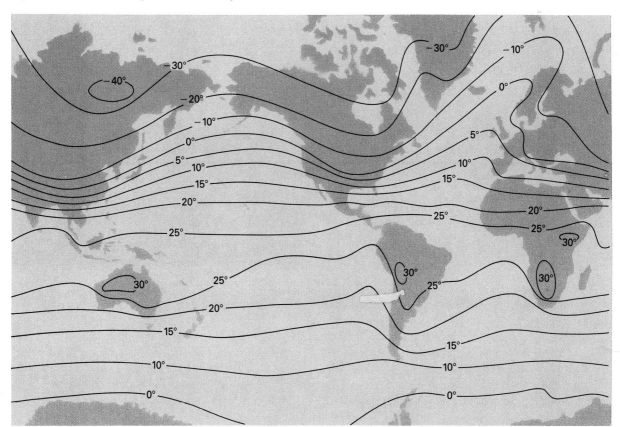

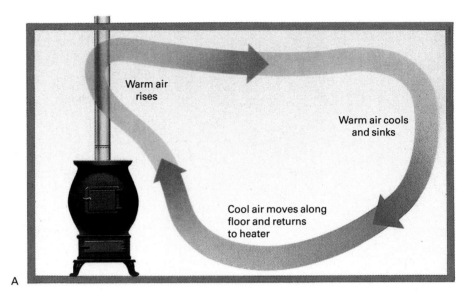

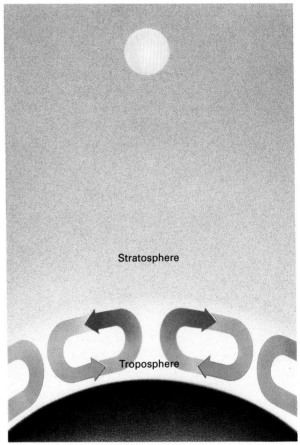

Figure 15–21 (A) Convection currents created by a heater in one corner of a room. (B) Convection currents created when the Sun heats the Earth. In this case the ceiling is the boundary between the troposphere and the stratosphere. This ceiling exists because the lower stratosphere is warmer than the upper troposphere.

mean temperature in January. Note that the isotherms loop and dip across lines of latitude. For example, the 0°C line runs through Seattle, dips southward across the center of the United States, and then swings northward to northern Norway. Such variations occur because winds and ocean currents transport heat from one region of the Earth to another, and heat storage varies with location.

Heat Transport

If you place a metal frying pan on the stove, the handle gets hot even though it is not in contact with the burner. Heat from the burner is absorbed by the frying pan and then conducted through the metal. **Conduction** is the transport of heat by atomic or molecular motion. When

A hang-glider pilot spirals upward on a column of rising air. *(Jeff Elgart, United States Hang Gliding Association)*

the frying pan is heated, the atoms just above the burner move more rapidly. They then collide with their neighbors and transfer energy to them. Like a falling row of dominoes, energy is passed from one atom to another until the handle becomes hot. Air, in contrast, is a poor conductor, and energy is not transported from one region of the globe to another in this manner.

To understand how air transports heat, imagine that a heater is placed in one corner of a cold room. The heated air in the corner expands, becoming less dense. This light, hot air rises to the ceiling. It flows along the ceiling, cools, falls, and returns to the stove, where it is reheated (Fig. 15–21A). **Convection** is the transport of heat by the movement of currents. Convection readily occurs in liquids and gases. Recall from Chapter 13 that ocean currents transport large quantities of heat northward or southward, thereby altering climate. For example, the Gulf Stream carries tropical water northward and warms the west coast of Europe.

Similar currents occur in the atmosphere. If air in one region is heated above the temperature of surrounding air, this warm air becomes less dense and rises. As the warm air rises, cooler, denser air in another portion of the atmosphere sinks. Air then flows along the surface to complete the cycle (Fig 15–21B). In meteorology, this horizontal motion is called **advection**, whereas convection is reserved for vertical air flow. The steady winds that blow across the tropical oceans, a tornado that ravages a city in the Midwest, and a thunderstorm that drops rain and hail on your Sunday picnic are all caused by convective and advective processes in the atmosphere. We will develop our understanding of weather and climate in Chapters 16 and 17.

Changes of State

Given the proper temperature and pressure, almost all substances can exist in three states: solid, liquid, and gas. However, at the Earth's surface, many substances commonly exist in only one state. In our experience, rock is almost always solid and oxygen is almost always a gas. Water plays an important role in heat transfer and storage because it commonly exists in all three states—as solid ice, as a liquid, and as gaseous water vapor.

Large amounts of energy are required to melt ice and to vaporize water. **Latent heat** (stored heat) is the energy released or absorbed when a substance changes from one state to another. About 80 calories are required to melt a gram of ice at 0°C. As a comparison, 100 calories are needed to heat the same amount of water from freezing to boiling (0°C to 100°C). Another 540 to 600 calories are needed to evaporate a gram of water.*

The energy transfers also work in reverse. **Condensation** is the conversion of water vapor to liquid. When 1 gram of water vapor condenses, 540 to 600 calories are

*The heat of vaporization varies with temperature and pressure; at 100°C, the value is 539.5 cal/g.

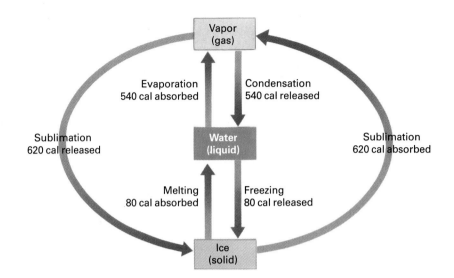

Figure 15–22   Changes of state for water. (Calories are given per gram at 0°C and 100°C. The values vary with temperature.)

released. When 1 gram of water freezes, 80 calories are released (Fig. 15–22).

The energy absorbed and released during freezing, melting, evaporation, and condensation of water is important in the atmospheric energy balance. For example, in the northern latitudes March is usually colder than September, even though equal amounts of sunlight are received in both months. However, in March much of the incident heat is absorbed by melting snow. Evaporation cools seacoasts, and hurricanes are driven in part by the

energy released when massive quantities of water vapor condense.

Heat Storage

If you place a pan of water and a rock outside on a hot summer day, the rock becomes hotter than the water. Both have received identical quantities of solar radiation. Why is the rock hotter?

Figure 15–23   Monthly temperatures for Portland, Oregon, and Eau Claire, Wisconsin. Both lie at the same latitude, but Portland is influenced by proximity to the Pacific Ocean, whereas Eau Claire is situated in the center of the continent. As a result, Portland has warmer winters, cooler summers, and less temperature difference between seasons.

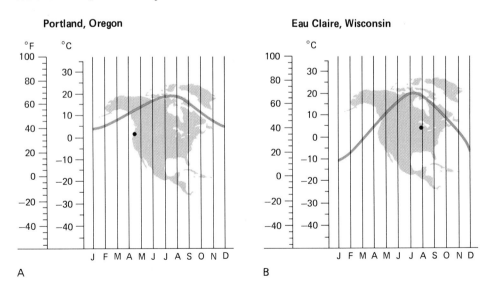

1. **Specific heat** is the amount of energy needed to raise the temperature of 1 gram of material by 1°C. Specific heat is different for every substance, and water has an unusually high specific heat. Thus, if water and rock absorb equal amounts of energy, the rock becomes hotter than the water.

2. Rock absorbs heat only at its surface, and the heat travels slowly through the solid. As a result, heat concentrates at the surface, thereby raising its temperature. Heat is dispersed much more evenly through water for two reasons. First, solar radiation penetrates several meters below the surface of the water, warming it to this depth. Second, water is a fluid, and fluids transport heat by convection.

3. Water evaporates, and since evaporation is a cooling process, water loses more heat than rock.

Think of the consequences of the temperature difference between rock and water. On a hot summer day you may burn your feet walking across dry sand or rock, but the surface of a lake or ocean is never burning hot. Suppose that both the ocean and the adjacent coastline are at the same temperature in spring. As summer approaches, both land and sea receive equal amounts of solar energy. But the land becomes hotter, just as in our demonstration the rock became hotter than the water. Along the seacoast the land is cooled by sea breezes. The interior of a continent is not cooled in this manner and is generally hotter than the coast. In winter the opposite effect occurs, and inland areas are generally colder than the coastal regions. Thus, coastal areas are commonly cooler in summer and warmer in winter than continental interiors (Fig. 15–23). This effect is shown by the 0°C temperature line in Figure 15–20, which dips southward through the continents and rises to higher latitudes over the oceans. The coldest temperatures recorded in the Northern Hemisphere occurred in central Siberia and not at the North Pole, because Siberia is landlocked, whereas the North Pole lies in the middle of the Arctic Ocean.* In summer, however, Siberia is considerably warmer than the North Pole. In fact, the average temperature in some places in Siberia ranges from −50°C in winter to +20°C in summer, about the greatest range in the world.

*Even though the Arctic Ocean is covered with ice, water lies only a few meters below the surface and it still influences climate.

SUMMARY

•

Weather is current atmospheric conditions, and **climate** is a composite of the weather conditions averaged over years.

The Earth's atmosphere has been modified by organisms since they first evolved. Most oxygen in the modern atmosphere was generated by plants, and the composition of the modern atmosphere is partly maintained by organisms. Today, dry air is roughly 78 percent nitrogen (N_2), 21 percent oxygen (O_2), and 1 percent other gases. Air also contains water vapor, dust, liquid droplets, and industrial pollutants.

Atmospheric pressure is the weight of the atmosphere per unit area. Pressure decreases continuously with altitude.

Light is a form of **electromagnetic radiation** and exhibits properties of both waves and particles simultaneously. When radiation encounters matter, it may be **transmitted**, **reflected**, **absorbed**, or **scattered**. Earth receives energy in the form of high-energy, short-wavelength solar radiation. Rock and soil re-emit low-energy, long-wavelength radiation. The **greenhouse effect** is the warming of the atmosphere due to the absorption of this long-wavelength radiation by water, carbon dioxide, methane, and other atmospheric gases. About 8 percent of solar radiation is scattered back into space, 19 percent is absorbed in the atmosphere, 26 percent reflected, and 47 percent absorbed. However, the absorbed radiation is eventually re-emitted back into space.

In the **troposphere**, temperature decreases with altitude. The temperature rises in the **stratosphere** because ozone absorbs radiation. The temperature decreases again in the **mesosphere**, and then in the uppermost layer, the **thermosphere**, temperature increases as high-energy radiation is absorbed.

The general temperature gradient from the equator to the poles results from the decreasing intensity of solar radiation from the equator to the poles. Changes of seasons are caused by the tilt of the Earth's axis relative to the Earth-Sun plane.

Winds and ocean currents transfer heat from one region of the globe to another by convection and **advection**. Large quantities of heat are absorbed or emitted when water freezes, melts, vaporizes, or condenses. Oceans also affect weather because water has a high **specific heat**; it stores heat in summer and releases it in winter.

KEY TERMS

•

Weather *402*	Electromagnetic	Troposphere *413*	Equinox *421*
Climate *402*	spectrum *408*	Tropopause *415*	Isotherm *421*
Barometric pressure *406*	Emission *409*	Stratosphere *415*	Conduction *422*
Barometer *406*	Transmission *409*	Stratopause *415*	Convection *423*
Millibar *406*	Absorption *409*	Mesosphere *415*	Advection *423*
Aneroid barometer *406*	Reflection *409*	Thermosphere *415*	Latent heat *423*
Altimeter *406*	Refraction *411*	Summer solstice *419*	Condensation *423*
Photons *408*	Scattering *411*	Tropic of Cancer *420*	Specific heat *425*
Electromagnetic	Albedo *412*	Winter solstice *420*	
radiation *408*	Greenhouse effect *413*	Tropic of Capricorn *420*	

REVIEW QUESTIONS

•

1. How did the primitive atmosphere differ from our atmosphere today? How did the modern atmosphere evolve?

2. Explain how plants maintain an atmosphere that can support animal life.

3. List the two most abundant gases in the atmosphere. List three other, less abundant gases. List three non-gaseous components of natural air.

4. Draw a graph of the change in pressure with altitude. Explain why the pressure changes as you have shown.

5. What is a barometer, and how does it work?

6. What happens to light when it is transmitted, reflected, absorbed, and scattered? Give an example of each of these phenomena.

7. Why is the sky blue? Why are sunsets and sunrises red?

8. What is the fate of the solar radiation that reaches the Earth?

9. List the layers of the atmosphere. Discuss the temperature changes within each.

10. Describe the difference between weather and climate. Which is more predictable?

11. Explain how the tilt of the Earth's axis affects climate in the temperate and polar regions.

12. Discuss temperature and lengths of days at the poles, the mid-latitudes, and the equator at the following times of year: June 21, December 21, and the equinoxes.

13. How is a convection current generated? Discuss the effect of wind on world climate.

14. Discuss the effect of the oceans on world climate. Are the coastal regions always warmer than inland areas? Explain.

DISCUSSION QUESTIONS

•

1. Explain why humans could not survive in the Earth's primitive atmosphere, yet life as we know it could not have evolved in the present atmosphere.

2. Imagine that enough matter vanished from the Earth's core so that the Earth decreased in mass to half its present mass. In what ways would the atmosphere change? Would normal pressure at sea level be affected? Would the thickness of the atmosphere change? Explain.

3. An astronaut on a space walk must wear protective clothing as a shield against the Sun's rays, but the same person is likely to relax in a bathing suit in sunlight down on Earth. Explain.

4. Why must climbers wear dark glasses to protect their eyes while they are on high mountains? Why do they get sunburned even when the temperature is below freezing?

5. As the winter ends, the snow generally melts first around trees, twigs, and rocks. Explain why the line of melting radiates outward from these objects, and the snow in open areas melts last.

6. In central Alaska the sky is often red at noon in December. Explain why the sky is red, not blue, at this time.

7. Refer to Figure 15–15A. (a) What percent of the incident solar energy is absorbed by the Earth? (b) Is the Earth

growing warmer or colder, or is the global temperature fairly constant? Explain.

8. If we lived on the surface of a flat Earth, would different regions experience different climates, or similar ones? Assume that the flat Earth is tilted 23.5° degrees with respect to the plane of its orbit.

9. If the North Pole receives the same number of hours of sunlight per year as do the equatorial regions, why is it so much colder than the equator?

10. Would a large inland lake be likely to affect the climate of the land surrounding it? Deep lakes seldom freeze com-

pletely in winter, whereas shallow ones do. Would a deep lake have a greater or a lesser effect on weather than a shallow one?

11. Neither oxygen nor nitrogen is an efficient absorber of visible or infrared radiation. Since these two gases together make up 99 percent of the atmosphere, the atmosphere is largely transparent to visible and infrared radiation. Predict what would happen if oxygen and nitrogen absorbed visible radiation. What would happen if these two gases absorbed infrared radiation?

Weather

The Sun shines almost as brightly today as it did yesterday, but the air temperature may not be the same. For example, in Havre, Montana, the temperature plummeted to −30°C (−22°F) during the last week of October 1991 and then soared to +10°C (+50°F) a week later. Similar changes occur elsewhere. If you live in New York City, you may notice that a wind from the northwest brings cool air. When the wind shifts and blows from the southeast, the temperature rises dramatically and a storm develops as warmer maritime air flows into the city. In this chapter we continue our study of the atmosphere by examining weather—the current state of the atmosphere and its daily and weekly fluctuations.

16.1 Moisture in Air

Humidity

When water boils on a stove, a steamy mist rises above the pan, and then the mist disappears into the air. Of course, the water molecules have not been lost. In the

● Towering cumulus cloud. *(Arjen and Jerrine Verkaik, Skyart)*

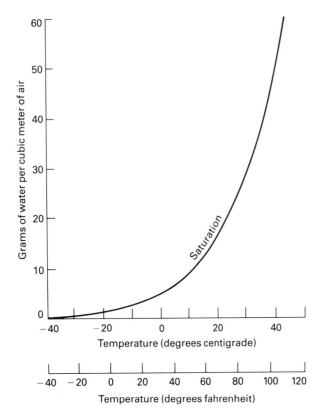

Figure 16–1 Maximum quantity of water vapor in air as a function of temperature.

pan, water is liquid, and in the mist above, the water exists as tiny droplets. These droplets then evaporate, and the water vapor mixes with air and becomes invisible.

All air contains some water vapor. **Humidity** is the amount of water vapor in air. **Absolute humidity** is the mass of water vapor in a given volume of air, expressed in grams per cubic meter (g/m³).

Air can hold only a certain amount of water vapor. Warm air can hold more water vapor than cold air. For example, air at 25°C can hold 23 g/m³ of water vapor, but at 12°C, it can hold only half that quantity, 11.5 g/m³ (Fig. 16–1). **Relative humidity** is the amount of water vapor in air relative to the maximum it can hold at a given temperature. It is expressed as a percentage:

$$\text{Relative humidity (\%)} = \frac{\begin{array}{c}\text{actual quantity of water} \\ \text{per unit of air}\end{array}}{\begin{array}{c}\text{maximum quantity at} \\ \text{the same temperature}\end{array}} \times 100$$

If air contains half as much water vapor as it can hold, its relative humidity is 50 percent. Suppose that air at 25°C contains 11.5 g/m³ of water vapor. Since air at that temperature can hold 23 g/m³, it is carrying half of its maximum, and the relative humidity is 11.5 g/23 g ×

100 = 50 percent. Now let us take some of this air and cool it without adding or removing any water vapor. Since cold air holds less water vapor than warm air, the relative humidity increases even though the *amount* of water vapor remains constant. If the air cools to 12°C and it still contains 11.5 g/m³, the relative humidity reaches 100 percent because air at that temperature can hold only 11.5 g/m³. When relative humidity reaches 100 percent, the air is **saturated**. If saturated air cools further, water vapor condenses into liquid droplets.

Three atmospheric processes cool air and cause condensation: (1) Air loses heat by radiation. (2) Air can cool by contact with a cool surface such as water, ice, rock, soil, or vegetation. (3) Air cools when it rises. Dew, frost, and some types of fog form by radiative and contact cooling. However, rain is almost always caused by cooling that occurs when air rises.

Dew and Frost

You can observe condensation on a cool surface with a simple demonstration. Heat water on a stove until it boils and hold a cool drinking glass in the clear air just above the steam. Water droplets will condense on the surface of the glass. The glass cools the hot, moist air. As the air cools, vapor condenses into liquid droplets. The same effect occurs in a house on a cold day. Water droplets or ice crystals appear on windows as warm, moist indoor air cools on the glass (Fig. 16–2).

In some regions, the air on a typical summer evening is warm and humid. After the Sun sets, plants, houses, windows, and most other objects lose heat by radiation and therefore become cool. During the night, water vapor condenses on the cool objects. This condensation is called **dew**. The **dew point** is the temperature at which relative humidity reaches 100 percent and air becomes saturated

Figure 16–2 Ice crystals on a window on a frosty morning.

A cloud is a visible concentration of water droplets or ice crystals in air.

with moisture. At that point condensation occurs. If the dew point is below freezing, **frost** forms. Thus, frost is not frozen dew, but ice crystals formed directly from vapor.

Supersaturation and Supercooling

If air contains dust, smoke, or salt particles or if it is in contact with the ground, then water vapor condenses on solid surfaces when the relative humidity reaches 100 percent. In most regions of the lower atmosphere, suspended solid particles are abundant and vapor condenses at or just below the dew point. However, under certain conditions condensation occurs so slowly that for all practical purposes it does not happen. Air becomes **supersaturated** when its relative humidity rises above 100 percent; that is, when it has cooled below its dew point but water remains as vapor. Supersaturation occurs only in clear, dust-free air high in the atmosphere.

Similarly, liquid water does not always freeze at its freezing point. Small droplets can remain liquid in a cloud even when the temperature is as cold as −10°C. Such water is **supercooled**. Supercooled droplets do not freeze because there are no solid particles to condense onto.

16.2 Clouds and Precipitation

A cloud is a visible concentration of water droplets or ice crystals in air. The basic requirements for cloud formation are a supply of humid air and a means of cooling it so that water vapor condenses to form droplets or ice crys-

tals. Most clouds form well above the Earth's surface and so are not cooled by direct contact with the ground. Almost all cloud formation and precipitation occur when air rises.

Why Does Air Cool When It Rises?

Work and heat are both forms of energy. Work can be converted to heat or heat can be converted to work, but energy is never lost. If you pump up a bicycle tire you are performing work to compress the air. This energy is not lost; much of it converts to heat. Therefore, both the pump and the newly filled tire feel warm. Conversely, if you puncture a tire, the air rushes out. It must perform work to expand, so the air rushing from a punctured tire cools. Variations in temperature caused by compression and expansion of gas are called **adiabatic temperature changes**. Adiabatic means without gain or loss of heat.

Adiabatic temperature changes are different from temperature changes caused by addition or removal of heat. You put a kettle of water on the stove and heat is added, so the water gets hot; you put the kettle in the refrigerator and heat is removed, so the water cools. In contrast, during adiabatic warming, air warms up because work is done on it, not because heat is added. During adiabatic cooling, air cools because it performs work, not because heat is removed.

As explained in Chapter 15, air pressure decreases with elevation. When surface air rises into lower-pressure zones, it expands, just as a balloon would expand if you placed it in a partial vacuum. Rising air performs work in order to expand, and therefore it cools adiabatically.

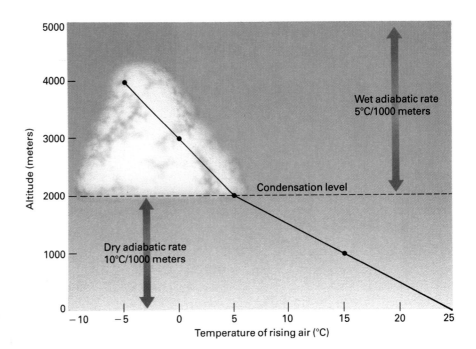

Figure 16–3 Dry and wet adiabatic lapse rates.

Dry air cools by 10°C for every 1000 meters it rises (5.5°F/1000 ft). This cooling rate is called the **dry adiabatic lapse rate**. Thus, if dry air were to rise from sea level to 9000 meters (about the height of Mount Everest), it would cool by 90°C (162°F).

Almost all air contains at least some water vapor. As moist air rises and cools adiabatically, its temperature eventually decreases to the dew point. At the dew point moisture condenses to form droplets, and therefore a cloud forms. But recall from Chapter 15 that latent heat is released when vapor condenses. As the air continues to rise through the cloud, it now is affected by two opposing processes. It continues to cool adiabatically, but at the same time it is heated by the latent heat released during condensation. The net result is that the air continues to cool, but more slowly than at the dry adiabatic lapse rate. The **wet adiabatic lapse rate** is the cooling rate after condensation has begun. It is 5°C/1000 m (2.7°F/1000 ft), or half the dry adiabatic lapse rate (Fig. 16–3). Thus, once clouds start to form, rising air no longer cools as rapidly as it did lower in the atmosphere.

Formation of Clouds

On some days clouds hang low over the land and obscure nearby hills. At other times clouds float high in the sky, well above the mountain peaks. What factors determine the height and shape of a cloud?

Recall that upper-level air in the troposphere is generally cooler than air near the Earth's surface because upper-level air is heated less by energy radiating from the Earth.

This cooling with elevation is called the **normal lapse rate**. The normal lapse rate affects air that is neither rising nor falling. It is 6°C/1000 m (3.3°F/1000 ft) and thus is less than the dry adiabatic lapse rate.

In Figure 16–4A, the normal lapse rate is shown on the vertical axis. In this diagram, the Earth's surface temperature is 21°C. Now suppose air in one locality is heated to 25°C. As a result, it expands and rises. As it rises, it cools adiabatically. When the rising air cools to the same temperature as the surrounding air, it is no longer buoyant and stops rising. In Figure 16–4B the initial surface temperature is lower, so the air rises higher.

Figure 16–3 shows the elevation at which condensation occurs and clouds form. Figure 16–4 shows the maximum height to which air will rise. If air stops rising before it cools to its dew point, no clouds form. Clouds form only when air continues to rise after it has cooled below its dew point.

Types of Clouds

Cirrus clouds are wispy clouds that look like the ends of a person's hair blowing in the wind or feathers floating across the sky (from Latin: wisp of hair). Cirrus clouds form at high altitudes, 6000 to 15,000 meters (20,000 to 50,000 feet). The air is so cold at these elevations that cirrus clouds are composed of ice crystals rather than water droplets. High winds aloft blow them out into long, gently curved streamers (Fig. 16–5).

Stratus clouds are horizontally layered, sheet-like clouds (from Latin: layer). They form when condensation

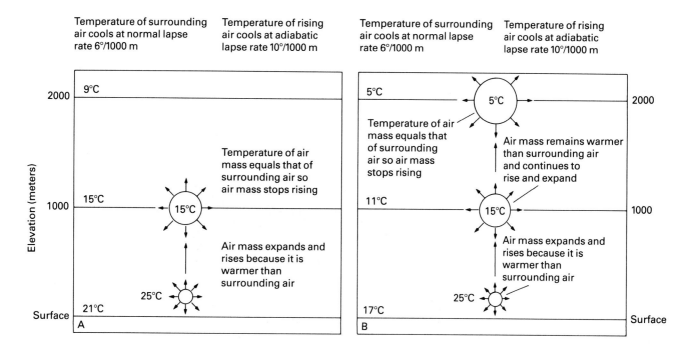

Figure 16–4 Warm air rises. As it rises, it expands and cools adiabatically. When the air mass cools to the temperature of the surrounding air, it stops rising. A and B show the heights to which the warm air would rise, given two different starting conditions.

occurs at the same elevation at which air stops rising. In Figure 16–4A, the warm air cools enough to stop rising at 1000 meters. If condensation also occurs at 1000 meters, the clouds spread out into a broad sheet. Stratus

clouds form the dark, dull gray, overcast skies that may persist for days and bring steady rain (Fig. 16–6).

Cumulus clouds are column-like clouds with a flat bottom and a billowy top (from Latin: heap or pile). The largest cumulus clouds measure more than 10 kilometers from bottom to top. On a hot summer day the top of a cumulus cloud may rise in cauliflower-like masses high above its base. The base of the cloud forms at the altitude at which the rising air cools to its dew point and condensations starts. However, in this situation the rising air remains warmer than the surrounding air and therefore continues to rise. As it rises, more vapor condenses, forming the billowing columns (Fig. 16–7).

Other types of clouds are named by combining these three basic terms (Fig. 16–8). **Stratocumulus clouds** are low, sheet-like clouds with some vertical structure. The term **nimbo** refers to a cloud that precipitates. Thus, a **cumulonimbus cloud** is a towering rain cloud. If you see one, you should seek shelter, because cumulonimbus clouds commonly produce intense rain, thunder, and lightning. A **nimbostratus cloud** is a stratus cloud from which rain or snow falls. Other prefixes are also added. Alti- is derived from the Latin root *altus*, meaning high. An **altostratus cloud** is simply a high stratus cloud.

(Text continues on p. 436)

● M E M O R Y   D E V I C E

Names of Clouds

Name	From Latin Root Meaning	Other Words or Meanings from the Same Root
Cirrus	A filament like a wisp of hair	In biology, "cirrus" is used to describe any slender appendage such as a tentacle, tendril, foot, or arm.
Stratus	To spread out into a layer	In geology, a stratum is a single bed (or layer) of sedimentary rock, and stratigraphy is the study of bedded sedimentary rocks.
Alti	High	Altitude
Nimbus	Rain	No other common usages

Figure 16–5 Cirrus are high, wispy clouds.

Figure 16–6 Stratus are low clouds spread out across the sky like a continuous blanket.

Figure 16–7 Cumulus are column-like clouds with flat bottoms.

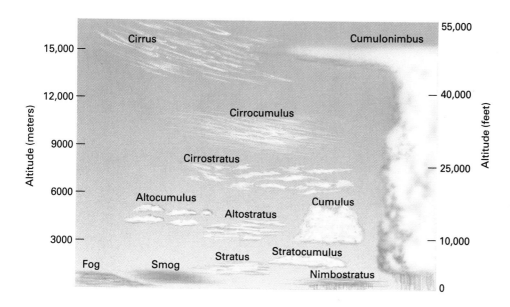

Figure 16–8 The major types of clouds.

Cirrocumulus are high clouds with some vertical structure caused by rising air.

Altocumulus are middle-elevation clouds that often form in symmetrical waves.

Cumulonimbus are towering rain clouds that often bring thunder and lightning.

Figure 16–9 Radiation fog is seen as a morning mist in this field in Idaho.

Fog

Fog is a cloud that forms at or very close to ground level. Several different mechanisms create fog. **Advection fog** occurs when warm moist air from the sea blows onto cooler land. The air cools, and water vapor condenses at ground level. San Francisco, Seattle, and Vancouver all experience foggy winters as moist air from the Pacific Ocean is cooled first by the cold California current and then by land. The foggiest location in the United States is Cape Disappointment, Washington, where visibility is obscured by fog 29 percent of the time. **Radiation fog** occurs when Earth and air cool by radiation loss during the night (Fig. 16–9). Often this cool, dense foggy air settles into valleys. If you are driving late at night in hilly terrain, beware, because a sudden dip in the roadway may lead you into a thick fog where visibility is low. **Evaporation fog** occurs when air is cooled by evaporation from a body of water, commonly a lake or river. Evaporation fogs are common in late fall and early winter, when the air has become cool but the water is still warm. The water evaporates, but the vapor cools and condenses almost immediately upon contact with the cold air. **Upslope fog** occurs when air cools as it rises along a land surface. Upslope fogs occur both on gradually sloping plains and on steep mountains. For example, the Great Plains rise from sea level at the Mississippi delta to 1500 meters (5000 feet) at the Rocky Mountain front. When humid air moves northwest from the Gulf of Mexico toward the Rockies, it rises and cools adiabatically. This cooling may form upslope fogs. The rapid rise at the mountain front also forms fogs.

Types of Precipitation

Rain

Why does rain fall from some clouds, whereas other clouds float across a blue sky on a sunny day and produce no rain? The droplets in a cloud are small, about 0.01 millimeter in diameter (about 1/7 the diameter of a human hair). In still air, such a droplet would require 48 hours to fall from a cloud 1000 meters above the Earth. But, these tiny droplets never reach the Earth because they evaporate first.

In dense stratus clouds, the droplets become so concentrated that they collide and stick together. If the droplets grow large enough, they fall as **drizzle** (0.1 to 0.5 millimeter in diameter) or light rain (0.5 to 2 millimeters in diameter). About one million cloud droplets must combine to form an average-size raindrop.

If you have ever been caught in a thunderstorm, you may remember raindrops large enough to be painful as they struck your face or hands. Recall that a cumulus cloud forms from rising air and that its top may be several kilometers above its base. Upper portions of the cloud may be so cold that the water forms tiny ice crystals rather than water droplets. Water vapor may condense directly onto the crystals so that they grow large and heavy enough

Earth Science and the Environment

Cloud Seeding

A cloud contains moisture whether precipitation occurs or not. During droughts, it is frustrating to watch clouds pass overhead but receive no rain. In the 1940s, meteorologists realized that precipitation could be induced artificially by creating surfaces to enhance the growth of water droplets or ice crystals. Many clouds are so high that air temperature is below freezing and the water droplets are supercooled. In the first cloud-seeding experiments, scientists poured pellets of dry ice from an airplane into a supercooled cloud. (Dry ice is solid carbon dioxide, $-78°C$.) The dry ice froze some of the droplets. The surfaces of the ice crystals formed nuclei for additional deposition from neighboring supercooled droplets. The crystals grew rapidly until they were heavy enough to fall. If the temperature in the lower atmosphere was sufficiently warm, the crystals melted to form raindrops.

Because dry ice is expensive, further experiments were done, which showed that a warm crystal would induce ice crystal formation as long as it had a shape similar to that of an ice crystal. The most effective crystal was silver iodide. Thus, if silver iodide is sprinkled onto a supercooled cloud, ice deposits on the crystal.

Cloud seeding does not produce moisture, but generally shifts rainfall from one location to another. One region's gain is then another region's loss. Furthermore, control over the redistribution of rain is not always precise. As a result, conflicts of interest and political problems may arise. Thus, ranchers in Colorado objected when ski areas used cloud seeding to increase winter snowfall. They argued that the ski industry was stealing their water. Another problem arises because silver iodide is poisonous and if highly concentrated can harm plants and animals.

to fall. Condensation continues as the crystal falls through the towering cloud, and the crystal grows. If the lower atmosphere is warm enough, the ice melts before it reaches the surface. Raindrops formed in this manner may be 3 to 5 millimeters in diameter, large enough to hurt when they hit.

Snow, Sleet, and Glaze

As explained above, when the temperature in a cloud is below freezing, the cloud is composed of ice crystals rather than water droplets. If the temperature near the ground is also below freezing, the crystals remain frozen and fall as **snow** (Fig. 16–10). In contrast, if raindrops form in a warm cloud and fall through a layer of cold air at lower elevation, the drops freeze and fall as small spheres of ice called **sleet**. Sometimes the freezing zone near the ground is so thin that raindrops do not have time to freeze before they reach the Earth. However, when they land on subfreezing surfaces, they form a coating of ice called **glaze**. Glaze can be heavy enough to break tree limbs and electrical transmission lines. It also coats highways with a dangerous icy veneer.

Hail

Occasionally, precipitation takes the form of very large ice globules called **hail**. Hailstones vary from 5 millimeters in diameter to a record giant 14 centimeters in diame-

Figure 16–10 Windblown snow in British Columbia.

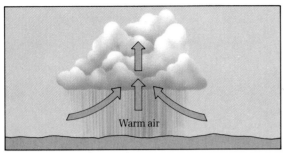

A   Convection–convergence

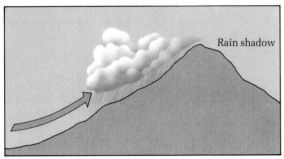

B   Orographic lifting

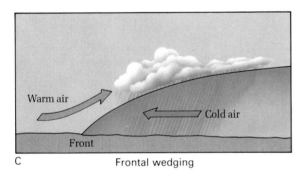

C   Frontal wedging

Figure 16–11 Three mechanisms cause air to rise: (A) convection-convergence, (B) orographic lifting, and (C) frontal wedging.

ter, weighing 765 grams (more than 1.5 pounds), that fell in Kansas. A 500 gram (1-pound) hailstone crashing to Earth at 160 kilometers (100 miles) per hour can shatter windows, dent car roofs, and kill people and livestock. Even small hailstones are destructive enough to damage crops. Hail always falls from a cumulonimbus cloud. Because cumulonimbus clouds form in columns with distinct boundaries, hailstorms occur in local, well-defined areas. Thus, one farmer may lose an entire crop while a neighbor is unaffected.

A hailstone consists of concentric shells of ice like the layers of an onion. Two mechanisms have been proposed for their formation. In one, ice crystals alternately

fall and rise through the cloud. When the crystals first start falling, they grow in the lower portion of the cloud. If the rising air is strong enough, it blows the ice particles back upward. As they rise, more moisture collects, causing the particles to grow even larger. An individual particle may rise and fall several times until it is so large that it drops out of the cloud. In another proposed mechanism, hailstones form in a single pass through the cloud. During their descent, supercooled water freezes onto the ice crystals. The layered structure occurs because different temperatures and amounts of supercooled water exist in different portions of the cloud. Therefore, each layer is deposited in a different portion of the cloud.

Why Does Air Rise?

In Section 16.1 we established that three atmospheric processes cool air and cause condensation: radiation, contact, and rising. Rising air causes almost all cloud formation and precipitation. Now let us explore the three mechanisms that cause air to rise: convection-convergence, orographic lifting, and frontal wedging (Fig. 16–11). Each mechanism is defined below and discussed in the remainder of the chapter.

> **Convection-convergence** If one portion of the atmosphere becomes warmer than surrounding air, the warm air expands and rises. Surface air then rushes in to replace the rising air and converges with it.
>
> **Orographic lifting** When air blows against a mountainside, it is forced to rise.
>
> **Frontal wedging** When a mass of warm air comes in contact with cool air, the warm air rises over the cool air.

16.3 Pressure and Wind

Warm air rises because it is less dense than cooler air. Warm rising air causes convection and convergence. Rising air exerts less downward force than still air. Consequently, atmospheric pressure is relatively low beneath a rising air mass, forming a **low pressure** region. Air rises slowly above a typical low pressure region, about 1 kilometer per day.

If air in the upper atmosphere cools, it becomes denser than the air beneath it and it sinks. The sinking air exerts a greater downward force, forming a **high-pressure** region (Fig. 16–12). Vertical air flow in both high and low pressure regions is accompanied by horizontal airflow. **Wind** is the horizontal movement of air in response to differences in air pressure. Wind always flows

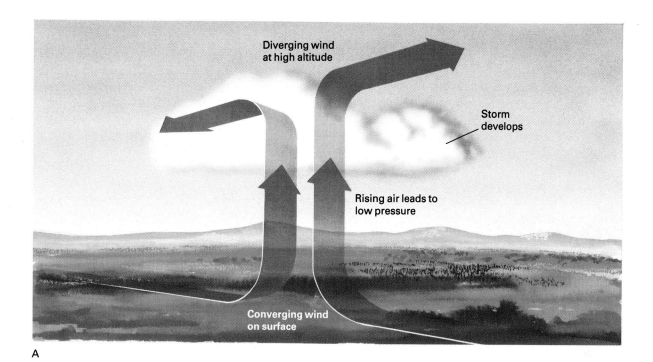

A

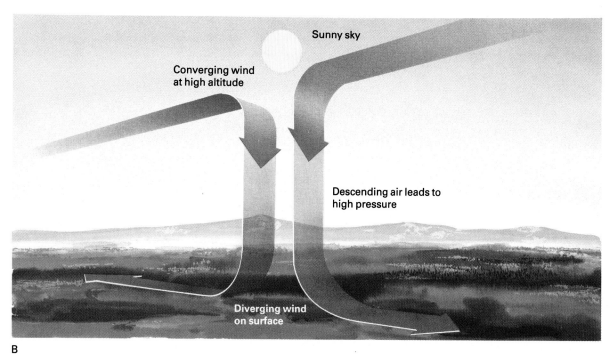

B

Figure 16–12 Generalized relationships among air pressure, air movement, and weather. Both high- and low-pressure zones generate winds. Rising air forms low pressure and clouds, whereas falling air leads to high pressure and sunny skies.

away from a region of high pressure and toward a low-pressure region. Ultimately, all wind is caused by unequal heating of the Earth's surface.

We can now understand a few simple relationships between barometric pressure and weather. When air is heated, it expands and rises, creating low pressure. The rising air cools adiabatically. If the cooling is sufficient, clouds form and rain or snow may fall. Thus, low barometric pressure is an indication that precipitation may soon follow. Alternatively, when air is cooled, it falls, and the barometric pressure rises. This falling air is compressed and heated adiabatically. Since warm air can hold more moisture than cold air, clouds generally do not form over high-pressure regions. Instead, the warm air absorbs moisture from the Earth. Thus, high pressure generally accompanies fair, dry weather.

Pressure Gradient

Wind blows in response to *differences* in pressure. Imagine that you are sitting in a room and the air is still. Now you open a can of vacuum-packed coffee and hear the hissing as air rushes into the can. The pressure in the room is higher than that inside the coffee can, and wind blows from the room into the can. But if you blow up a balloon, the air inside the balloon is at higher pressure than the air in the room. When the balloon is punctured, wind blows from the high-pressure zone of the balloon into the lower-pressure zone of the room (Fig. 16–13).

Wind speed is determined by the magnitude of the pressure difference over distance, called the **pressure gradient**. Thus, wind blows rapidly if a large pressure difference exists over a short distance. A large, or steep, pressure gradient is analogous to a steep hill. Just as a ball rolls quickly down a steep hill, wind flows rapidly across a steep pressure gradient. To create a pressure-gradient map, air pressure is measured at hundreds of different weather stations. Points of equal pressure are connected by lines called **isobars**. A steep pressure gradient is shown by closely spaced isobars, whereas a weak pressure gradient is illustrated by isobars that are spaced farther apart (Fig. 16–14). Pressure-gradients change daily, or sometimes hourly, as high- and low-pressure zones move. Thus, maps are updated frequently.

Coriolis Effect

Recall from Chapter 13 that ocean currents are deflected by the spin of the Earth and that this deflection is called the **Coriolis effect**. Wind is similarly deflected by the Earth's spin. In the Northern Hemisphere wind is deflected toward the right, and in the Southern Hemisphere, to the left (Fig 16–15).

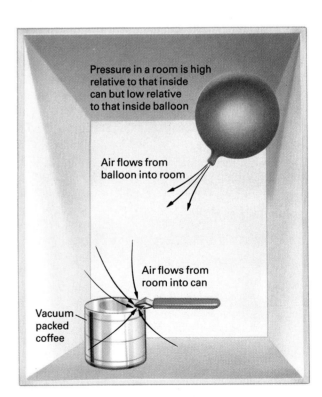

Figure 16–13 Winds blow in response to differences in pressure.

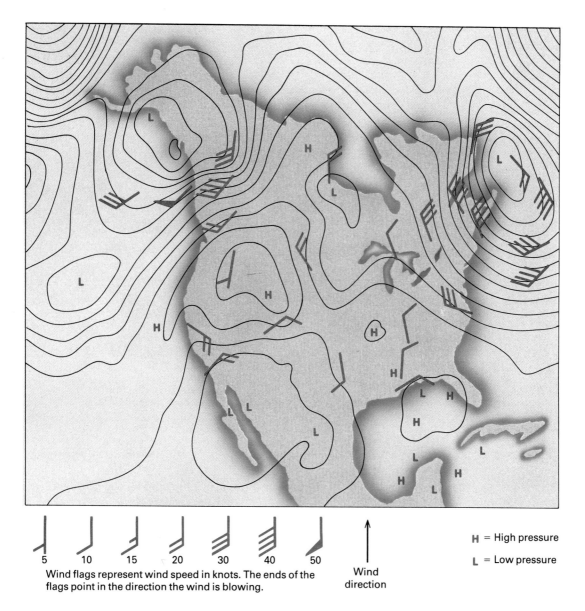

Wind flags represent wind speed in knots. The ends of the flags point in the direction the wind is blowing.

Wind direction

H = High pressure

L = Low pressure

Figure 16–14 Pressure map and winds at 5000 feet in North America on February 3, 1992. High altitude data are shown because the winds are not affected by surface topography and thus the effect of pressure gradient is well illustrated. Note that in the northeast and northwest, steep pressure gradients cause high winds that spiral counterclockwise into the low pressures. More widely spaced isobars around high pressure zones in the central United States cause less severe winds. *(NOAA weather station, Missoula, MT)*

Friction

Rising and falling air generates wind both along the Earth's surface and at higher elevations. As surface wind flows along the ground it is affected by **friction** with the Earth's surface, whereas high-altitude wind is not. As a result, wind speed normally increases with elevation. This

fact was first noted during World War II. On November 24, 1944, U.S. bombers were approaching Tokyo for the first mass bombing of the Japanese capital. Flying between 8000 and 10,000 meters (27,000 to 33,000 feet), the pilots suddenly found themselves roaring past landmarks 140 kilometers (90 miles) per hour faster than the theoretical top speed of their airplanes! Amid the

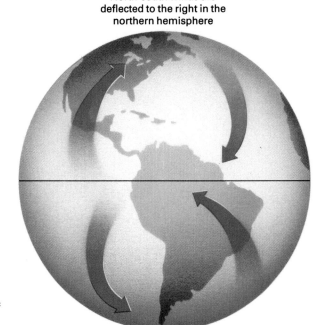

North-south winds are deflected to the right in the northern hemisphere

North-south winds are deflected to the left in the southern hemisphere

Figure 16–15 North-south winds are deflected by the Coriolis effect, to the right in the Northern Hemisphere and to the left in the Southern Hemisphere.

confusion, most of the bombs missed their targets, and the mission was a military failure. However, this experience introduced meteorologists to **jet streams**, narrow bands of high-altitude wind. The jet stream in the Northern Hemisphere flows from west to east at speeds between 120 and 240 kilometers per hour (75 and 150 miles/hr). As a comparison, surface winds attain such velocities only in hurricanes and tornadoes. Airplane pilots traveling from Los Angeles to New York fly with the jet stream to gain speed and save fuel, whereas pilots moving from east to west try to avoid it.

Cyclones and Anticyclones

Three major factors affect wind: pressure gradient, spin of the Earth, and friction. Winds blow from high-pressure to low-pressure regions. As air flows it is deflected by the Coriolis effect and slowed by friction. Figure 16–16A shows the movement of air in the Northern Hemisphere as it converges toward a low-pressure area. If the Earth did not spin, wind would flow directly across the isobars, as shown by the black pressure-gradient arrows. However, the Earth does spin, and the Coriolis deflection is shown by the red arrows. The generalized wind flow follows the small black arrows, as shown by the broad

magenta arrow. Note that the generalized wind pattern spirals counterclockwise.

A low-pressure region with its accompanying surface wind is called a **cyclone**.* The opposite mechanism forms an **anticyclone** around a high-pressure region. When descending air reaches the surface, it spreads out in all directions. The diverging winds are deflected, forming a pinwheel pattern that moves away from the center (Fig. 16–16B).

16.4 Wind and Mountains: Rain-Shadow Deserts

Air is forced to rise when it flows over a mountain range. This process is called **orographic lifting**. As the air rises, it cools adiabatically, and water vapor may condense into clouds that produce rain or snow. These conditions create abundant precipitation on the windward side and the crest of the range. When the air passes over the crest onto the leeward (downwind) side, it sinks (Fig. 16–17). This air

---

*In this usage, cyclone means a system of rotating winds and is not used to describe the violent storms otherwise known as hurricanes or typhoons.

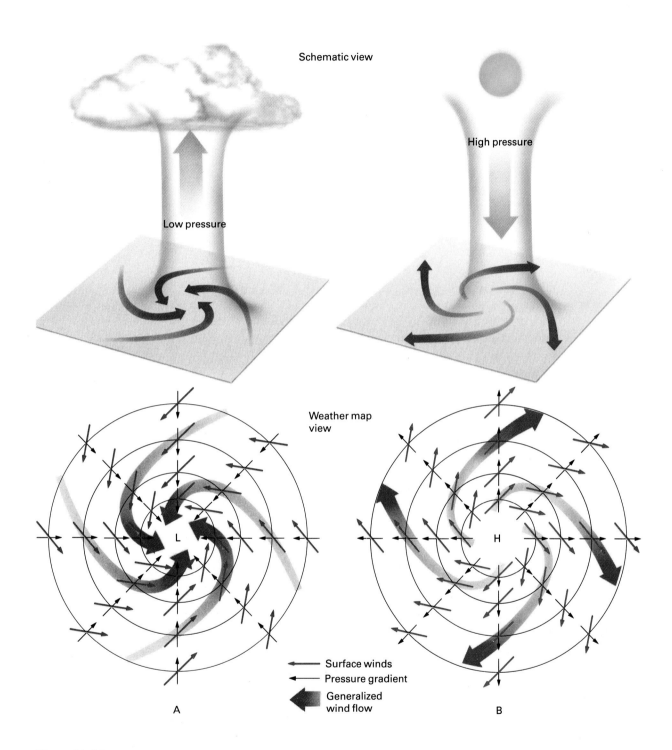

Figure 16–16 Schematic views and weather-map views of (A) cyclones and (B) anticyclones.

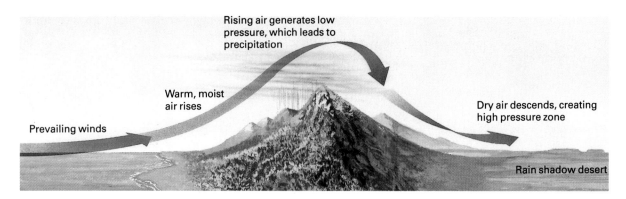

Figure 16–17   Formation of a rain-shadow desert. Warm moist air from the ocean rises. As it rises, it cools, and water vapor condenses to form rain. The dry, descending air on the lee side absorbs moisture, forming a desert.

has already lost much of its moisture. In addition, it warms adiabatically as it falls, absorbing moisture and creating a **rain-shadow desert** on the leeward side of the range. Figure 16–18 shows rainfall distribution in California. Note that the leeward valleys receive much less moisture than the mountains to the west.

16.5 Fronts and Frontal Weather

An **air mass** is a large body of air with approximately constant temperature and humidity at any given altitude.

Typically, an air mass is 1600 kilometers or more across and several kilometers thick. Because air acquires characteristics from the Earth's surface, an air mass is classified by its place of origin and is described in terms of its temperature and moisture. Temperature can be either **polar** (cold) or **tropical** (warm). **Maritime air** originates over water and has high moisture content, whereas **continental air** has low moisture content (Fig. 16–19, Table 16–1).

Air masses move and collide. The boundary between two colliding air masses is a **front**. The term was first used during World War I because weather systems were

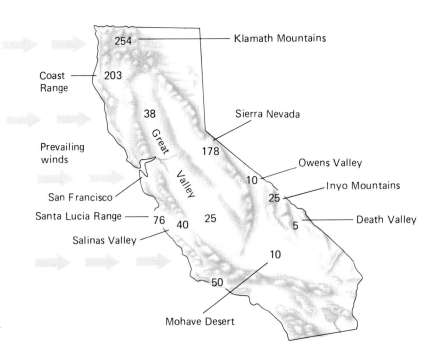

Figure 16–18   Rainfall patterns in the state of California. Note that rain-shadow deserts lie east of the mountain ranges. Rainfall is reported in centimeters per year.

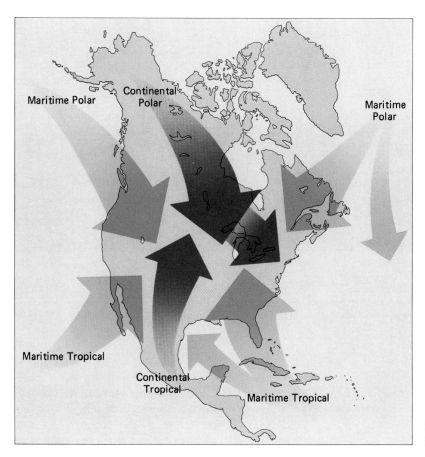

Figure 16–19 Air masses classified by their source regions.

considered analogous to armies that advance and clash along battle lines. When two air masses collide, at first each retains its identity, just as a drop of ink remains concentrated after it falls into a glass of water. If left alone for a few minutes, however, the ink diffuses in the water (Fig. 16–20). In the same manner, two air masses eventually mix to become one homogeneous mass. Even so, during the collision and initial mixing, air is forced to rise, which often results in cloudiness and precipitation. Frontal weather patterns are determined by the types of

TABLE 16–1

Classification of Air Masses		
Classification according to latitude (temperature):		
Polar (P) air masses originate in high latitudes and are cold.		
Tropical (T) air masses originate in low latitudes and are warm.		
Classification according to moisture content:		
Continental (c) air masses originate over land and are dry.		
Maritime (m) air masses originate over water and are moist.		
Symbol	**Name**	**Characteristics**
mP	Maritime polar	Moist and cold
cP	Continental polar	Dry and cold
mT	Maritime tropical	Moist and warm
cT	Continental tropical	Dry and warm

Figure 16–20 The Earth's dynamic atmosphere is enormously more complex that a static glass of water. Nevertheless this sequence illustrates the dispersal of one fluid into another. Warm air mixes with cold air just as this drop of blue ink disperses throughout the glass of water. In the atmosphere, swirling winds and precipitation often accompany the collision of air masses just as the ink forms patterns in the water.

air masses that collide and their relative speeds and directions. The symbols commonly used on weather maps to describe frontal systems are shown in Figure 16–21.

Warm Fronts

If a warmer air mass travels toward a colder one and collides with it, a **warm front** forms. Because warm air is less dense than cold air, the warm air rises over the cold air as the warm air mass continues to move (Fig. 16–22). The rising warm air generates an extensive area of cloudiness and precipitation. Notice that the boundary

Figure 16–21 Symbols commonly used on weather maps. Note that the words "warm" and "cold" are relative only. Air that is situated over the central plains of Montana at a temperature of 0°C may be warm relative to polar air above northern Canada but cold relative to a 20°C air mass over the southeastern United States.

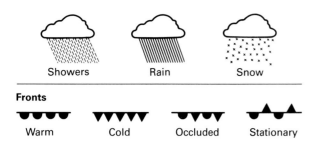

between the two air masses slopes gently. The warm air rises slowly along this boundary, forming stratus and nimbostratus clouds that become lower and thicker as the front approaches. Due to the gradual lifting, precipitation is generally light, but it is widespread and long-lasting. As shown in Figure 16–22, a characteristic sequence of clouds precedes a warm front. About 1000 kilometers from the front, air has risen so high and dropped so much of its moisture that wispy cirrus clouds float across a blue sky. Closer to the front, lower, thicker clouds appear, and precipitation generally starts about 300 kilometers from the frontal boundary. Although cirrus clouds are fair-weather clouds, they are often harbingers of a storm pushing in behind them.

Cold Fronts

A **cold front** forms when colder air overtakes and displaces warmer air. The faster-moving cold air is too dense to rise over the warm air. Instead, it distorts into a blunt wedge or bulge and pushes under the warmer air (Fig. 16–23). Note that the leading edge of a cold front is much steeper than that of a warm front. The steep contact between the two air masses causes the warm air to rise rapidly, creating a narrow band of violent weather. The rapidly rising air often forms cumulus and cumulonimbus clouds. The storm system may be only 25 to 100 kilometers wide, but within this zone downpours, thunderstorms, and violent winds are common.

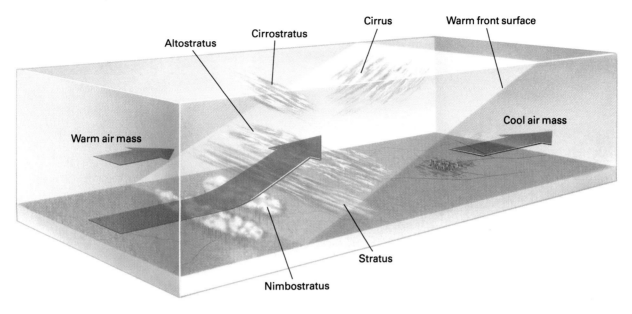

Figure 16–22 A profile of a warm front.

Occluded Front

An **occluded front** forms when a warm air mass is trapped between two colder air masses (Fig. 16–24). The warm air is lifted completely off the ground. Precipitation occurs along both frontal boundaries, combining the narrow band of heavy precipitation of a cold front with the wider band of lighter precipitation of a warm front. The net result is a large zone of inclement weather.

Stationary Front

A **stationary front** occurs when two air masses come in contact but neither is moving. Under these conditions, the

Figure 16–23 A profile of a cold front.

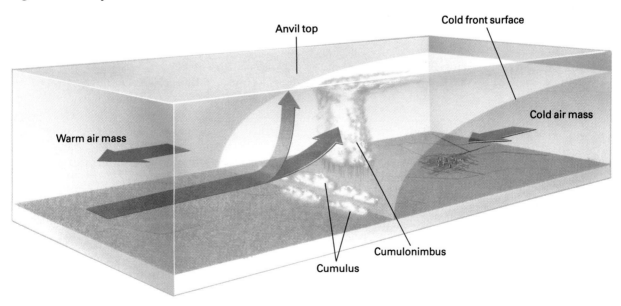

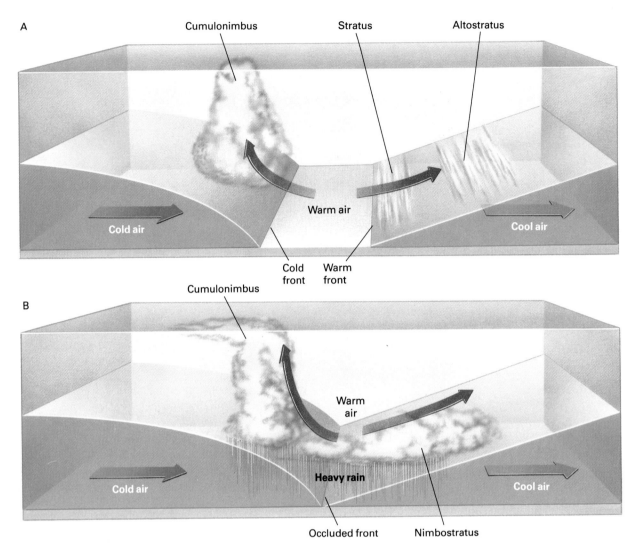

A

Cumulonimbus   Stratus   Altostratus

Warm air

Cold air   Cool air

Cold   Warm
front   front

B

Cumulonimbus

Warm
air

Heavy rain

Cold air   Cool air

Occluded front   Nimbostratus

Figure 16–24 A profile of a storm caused by an occluded front.

front can remain over an area for several days. Warm air rises, forming conditions similar to those in a warm front. As a result, rain, drizzle, and fog may occur.

The Life Cycle of a Middle-Latitude Cyclone

Most low-pressure cyclones in the middle latitudes of the Northern Hemisphere develop along fronts between polar and tropical air masses. Consider first a typical sequence involving a stationary front between a continental polar air mass and a maritime tropical air mass (Fig 16–25A). Cold winds flow parallel to one another, as shown. (Prevailing wind directions are explained in Chapter 17.) In Figure 16–25B, the cold air mass has started to push southward. As movement begins, warm air is lifted off

the ground. Then some small disturbance deforms the straight edge of the front. This disturbance may be a topographic feature, such as a mountain range, or it may be induced by air flow from a local storm or a temperature variation over a lake or basin. Once the bulge forms, the winds on both sides are deflected to strike the front at an angle. Thus, a warm front forms to the east and a cold front forms to the west.

Rising warm air then forms a low-pressure region at the apex of the bulge between the two fronts (Fig. 16–25C). Air begins to circulate counterclockwise around the bulge, as explained in Section 16.3. To the west the cold front advances southward, and to the east the warm front advances northward. In a well-developed cyclone, air spirals counterclockwise around the low-pressure zone and rain or snow falls. Over a period of one to three days,

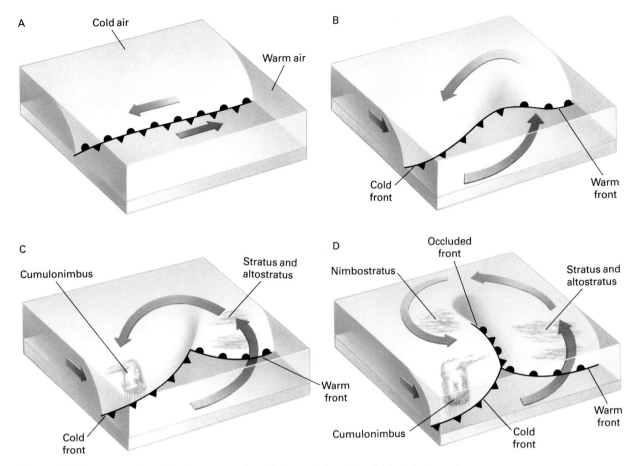

Figure 16–25   A three-dimensional representation of the evolution of a middle-latitude cyclone.

the air rushing into the low-pressure region equalizes pressure differences and the storm dissipates.

16.6 Sea and Land Breezes

Anyone who has lived near an ocean or large lake has encountered winds blowing from water to land and from land to water. Sea and land breezes are caused by uneven heating and cooling of land and water (Fig. 16–26). Recall that land surfaces heat up faster than adjacent bodies of water, and cool more quickly. If land and sea are nearly the same temperature on a summer morning, during the day the land warms and heats the air above it. Hot air then rises over the land, producing a local low-pressure area. Cooler air from the sea flows inland to replace the rising air. Thus, on a hot, sunny day, winds generally blow from the sea onto land. The rising air is good for flying kites or hang gliding, but often brings afternoon thunderstorms.

At night the reverse process occurs. The land cools faster than the sea and descending air creates a high pressure over the land. Then the winds reverse, and breezes blow from the shore out toward the sea.

Monsoons

A **monsoon** is a seasonal wind caused by uneven heating and cooling of continents and oceans. Just as sea and land breezes reverse direction with day and night, monsoons reverse direction with the passing seasons. In the summertime the continents become warmer than the sea. Warm air rises over land, creating a large low-pressure area and drawing moisture-laden maritime air inland. When the moist air rises as it flows over the land, clouds form and heavy monsoon rains fall. In winter the process is reversed. The land cools below the sea temperature, and air rises over the ocean and descends over land, producing dry continental high pressure. The prevailing winds blow from land to sea. More than half of the inhabitants of the Earth depend on monsoons because the predictable heavy summer rains bring water to the fields of Africa and Asia. If the monsoons fail to arrive, crops cannot grow and people starve.

Figure 16–26   Sea breezes blow inland during the day, and land breezes blow out to sea at night.

Figure 16–27   Thunderstorm frequency in the United States.   *(NOAA)*

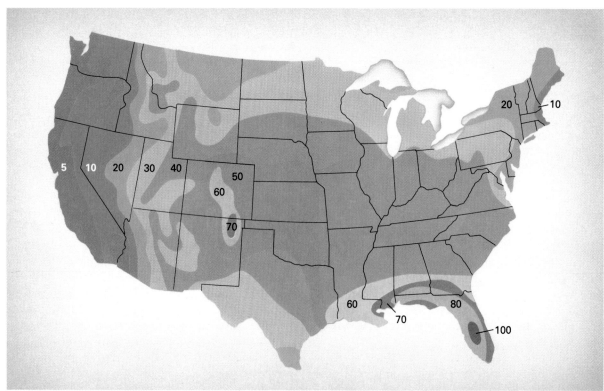

16.7 Thunderstorms

An estimated 16 million thunderstorms occur every year, and at any given moment about 2000 thunderstorms are in progress over different parts of the Earth (Fig. 16–27). A single bolt of lightning discharges several hundred million volts and for a few seconds produces as much power as a nuclear power plant. It heats the surrounding air to 10,000 °C or more, hotter than the surface of the Sun.

Despite their violence, thunderstorms are local systems, often too small to be included on national weather maps. A typical thunderstorm forms and then dissipates in a few hours and covers from about ten to a few hundred square kilometers. It is not unusual to stand on a hilltop in the sunshine and watch rain squalls and lightning a few kilometers away. All thunderstorms develop when warm moist air rises, forming cumulus clouds that develop into towering cumulonimbus clouds. Different conditions cause these local regions of rising air.

1. **Sea breeze convergence**. Central Florida is the most active thunderstorm region in the United States. As the subtropical Sun heats the Florida peninsula, rising air draws converging moist air from both the east and west coasts.

2. **Convection**. Thunderstorms also form in continental interiors during the spring or summer, when afternoon sunshine heats the ground and generates cells of rising moist air.

3. **Orographic rise**. Moist air rising as it flows over hills and mountain ranges may generate thunderstorms.

Lightning in the Uinta Mountains, Utah. *(Jeffery Vanuga, Earth Images)*

A typical thunderstorm occurs in three stages. In the initial, **cumulus stage**, rising air condenses, forming a cumulus cloud (Fig. 16–28A). As the cloud forms, the condensing vapor releases latent heat. In a typical thunderstorm, 400,000 tons of water vapor condense within the cloud, and the energy released by the condensation is equivalent to the explosion of 12 atomic bombs the size of the one dropped on Hiroshima during World War II. Large droplets or ice crystals develop within the cloud at

Figure 16–28 Three stages in the evolution of a thunderstorm.

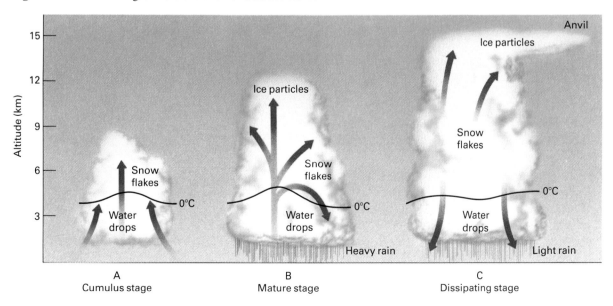

A towering thunderhead building over the Colorado Rockies.

this stage, but the rising air keeps them in suspension, and no precipitation falls to the ground.

Eventually, water droplets or hailstones become so heavy that updrafts can no longer support them, and they fall as rain or hail. During this stage, warm air continues to rise and may attain velocities of over 300 kilometers per hour in the central and upper portions of the cloud. The cloud may double its height in minutes. At the same time, ice falling through the cloud chills the lower regions and this cool air sinks, creating a downdraft. Thus, air currents rise and fall simultaneously within the same cloud. These conditions, known as **wind shear**, are dangerous for aircraft, and pilots avoid large thunderheads (Fig. 16–28B).

As explained earlier, rainfall from a cumulonimbus cloud can be unusually heavy. In one extreme example in 1976, a stationary complex of thunderstorms dropped 25 centimeters of rain over Big Thompson Canyon on the eastern edge of the Colorado Rockies in about 4 hours. The river flooded the narrow canyon, killing 139 people.

The mature stage of a thunderstorm, with rain or hail and lightning, usually lasts for about 15 to 30 minutes and no longer than an hour. The cool downdraft reduces the temperature in the lower regions of the cloud. As the temperature drops, convection currents weaken and warm, moist air is no longer drawn into the cloud (Fig. 16–28C). Once the water supply is cut off, condensation ceases, and the storm loses energy. Within minutes the rapid vertical air motion dies and the storm dissipates. Although a single thundercloud dissipates rapidly, new thunderheads can build in the same region, causing disasters such as that at Big Thompson Canyon.

Lightning is an electrical spark that jumps from a cloud to ground or to another cloud. It occurs when the buildup of static electricity is great enough to overcome the insulating properties of air. If you walk across a carpet on a dry day, the friction between your feet and the rug shears electrons off the atoms on the rug. The electrons migrate into your body and concentrate there. If you then touch a metal doorknob, a spark consisting of many electrons jumps from your finger to the metal knob.

In 1752 Benjamin Franklin showed that lightning is an electrical spark. In the nearly 250 years since Franklin, atmospheric physicists have been unable to explain the exact mechanism of lightning.* Charges separate within cumulonimbus clouds and build until a bolt of lightning jumps from the cloud. According to one theory, charge separation occurs in much the same way as when you walk across a carpet. Raindrops and ice particles collide as the heavier particles fall past smaller suspended water droplets and ice crystals in the cloud. Friction transfers electrons to the heavier particles. Another theory suggests that cosmic rays, bombarding the cloud from outer space, produce ions at the top of the cloud. Other ions form on the ground as winds blow over sharp objects at the Earth's surface. Convection transports these ions through the cloud (Fig. 16–29). Perhaps neither theory is entirely accurate and some combination of the mechanisms is responsible for lightning.

16.8 Tornadoes and Tropical Cyclones

Tornadoes and tropical cyclones are both characterized by intense low-pressure centers. Strong winds follow the steep pressure gradients and spiral inward toward their central columns of rising air.

---

*Earle R. Williams, "The Electrification of Thunderstorms," *Scientific American*, November 1988: 88ff.

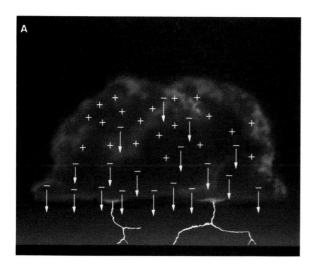
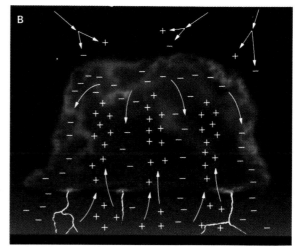

Figure 16–29   Two theories for the development of lightning. (A) Friction from falling droplets generates charge separation. (B) Charged particles are produced from above by cosmic rays and below by interactions with the ground. The particles are then distributed by convection currents. Theory A ignores convection, and theory B ignores precipitation. Since both processes occur within the cloud, the full explanation must include elements of both theories.

Tornadoes

A **tornado** is a small, short-lived, funnel-shaped storm that protrudes from the base of a cumulonimbus cloud (Fig. 16–30). The base of the funnel can be as small as 2 meters or as wide as 3 kilometers in diameter. Some tornadoes remain suspended in air; others touch the ground. After a tornado touches ground, it may travel a few meters to a few hundred kilometers across the surface. The entire funnel travels from 40 to 65 kilometers per hour, but the spiraling winds within the funnel are much more intense. Few direct measurements have been made of pressure and wind speed inside a tornado. However, we know that a steep pressure gradient develops because

Figure 16–30   Tornadoes. (A) Seymour, Texas. (B) Union City, Oklahoma.   *(NOAA, National Severe Storms Laboratory)*

a large pressure difference occurs over a very short distance. Meteorologists estimate that winds in tornadoes may reach 500 kilometers per hour or greater. These winds rush into the narrow low-pressure zone and then spiral upward. After a few seconds to a few hours, the tornado lifts off the ground and dissipates.

Tornadoes are by far the most violent of all storms. In 1910, one tornado lifted a team of horses and then deposited them, unhurt, several hundred meters away. They were lucky. About 120 Americans are killed every year by these storms, and property damage runs in the millions of dollars. Tornado winds can lift the roof off a house and then flatten the walls. Flying debris kills people and livestock caught in the open. Even so, the total destruction from tornadoes is not as great as that from hurricanes because the path of a tornado is narrow and its duration short.

Although tornadoes can occur anywhere in the world, 75 percent of the world's twisters concentrate in the Great Plains, east of the Rocky Mountains. They frequently form in the spring or early summer. At that time, continental polar (dry, cold) air from Canada comes in contact with maritime tropical (warm, moist) air from the Gulf of Mexico. As explained previously, thunderstorms develop under these conditions. Meteorologists cannot explain why most thunderstorms dissipate harmlessly and a few develop tornadoes. However, one fact is apparent: tornadoes are most likely to occur when large differences in temperature and moisture exist between the two air masses and the boundary between them is sharp.

The probability that any particular place will be struck by a tornado is small. Nevertheless, Codell, Kansas, was struck three years in a row—in 1916, 1917, and 1918—and each time the disaster occurred on May 20! In one particularly intense period in 1974, 148 tornadoes occurred in 2 days in 13 states.

Tropical Cyclones

A **tropical cyclone** (called a hurricane in North America and the Caribbean, a typhoon in the western Pacific, and a cyclone in the Indian Ocean) is less intense than a tornado but much larger and longer-lived (Table 16–2). Tropical cyclones are circular storms that average 600 kilometers in diameter and persist for days or even weeks (Fig. 16–31). Intense low pressure in the center of a hurricane generates wind that varies from 100 to more than 300 kilometers per hour. As these cyclones rage over tropical oceans, they create storm surges and huge waves that flood coastal regions. The most destructive hurricane in the United States struck Galveston, Texas, in September 1900. Eight thousand people died and millions of dollars of damage occurred. One reason the death toll was so high was that the population was caught unaware. A tropical cyclone has a sharp boundary, and even a few hundred kilometers outside that boundary, fluffy white clouds may be floating about in a blue sky. Today, in the United States, hurricanes are detected by satellite and people are evacuated from coastal areas well in advance of oncoming storms. Warning cannot eliminate property damage, however. When Hurricane Hugo struck the North Carolina coast in 1989 with 220-kilometer-per-hour winds, waves flooded coastal areas, destroying 1900 homes and killing 40 people. Property damage totaled $3 billion.

In the spring of 1991, more than 100,000 people were killed when a cyclone sent 7-meter waves across the heavily populated, low-lying coast of Bangladesh. Although meteorologists had been tracking the storm, communication was so poor and transportation facilities so inadequate that people were not evacuated in time.

Tropical cyclones form only over warm oceans, never over cold oceans or land. Thus, moist warm air is

TABLE 16-2

Comparison of Tornadoes and Tropical Cyclones		
	Range	
Feature	**Tornado**	**Tropical Cyclone**
Diameter	2–3 km	400–800 km
Path length (distance traveled across terrain)	A few meters to hundreds of kilometers	A few hundred to a few thousand kilometers
Duration	A few seconds to a few hours	A few days to a week
Wind speed	300–800 km/hr	120–250 km/hr
Speed of motion	0–70 km/hr	20–30 km/hr
Pressure fall	20–200 mb	20–60 mb

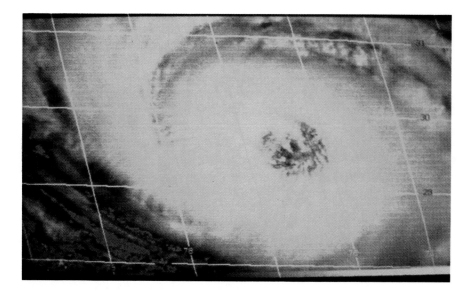

Figure 16–31 A satellite image of Hurricane Hugo as it approached the South Carolina coast in September 1989. *(NOAA, National Hurricane Service)*

crucial to development of this type of storm. Recall from Section 16.5 that a mid-latitude cyclone develops when a small disturbance produces a bulge in a previously linear front. A similar mechanism initiates a tropical cyclone. In the summer, the Sun warms the subtropical air. As the hot air rises, a vast belt of low pressure circles the globe. In addition, many local low-pressure disturbances move across the subtropical oceans at this time of year. If a local disturbance intersects with the global subtropical low, it creates a bulge in the isobars. Winds are deflected by the bulge and begin to spiral toward a growing low. As moisture-laden winds spiral inward, warm moist air rises. Water vapor condenses from the rising air, and the heat released warms the air further, which causes even more air to rise. The low pressure becomes more intense, strong winds blow inward to replace the rising air, and more precipitation occurs. But in turn the additional pre-

cipitation gives off more heat, which continues to add energy to the storm. At the center of the storm, called the **eye**, air that has been rushing inward spirals upward. Thus, the horizontal wind speed in the eye is reduced to near zero. Survivors who have been in the eye of a hurricane report an eerie calm. Rain stops, and the Sun may even shine weakly through scattered clouds, but this is only a momentary reprieve. A typical eye is only 20 kilometers in diameter, and after it passes the hurricane rages again in full intensity.

Once a hurricane develops, it is powered by continuing condensation and pushed by prevailing winds. It dissipates after it reaches land or passes over colder water. Condensing water vapor in a single tropical cyclone releases as much energy as that produced by all the electric generators in the United States in six months.

SUMMARY

•

Absolute humidity is the mass of water vapor in a given volume of air. **Relative humidity** is the amount of water vapor in air compared to the amount the air could hold at that temperature. **Condensation** occurs when moist air cools below its **dew point**. When warm air rises, it performs work and therefore cools **adiabatically**.

The characteristics of a cloud depend on the height to which air rises and the elevation at which condensation occurs. The three fundamental types of clouds are **cirrus**, **stratus**, and **cumulus**. Precipitation occurs when small

water droplets coalesce until they become large enough to fall.

When air is heated, it expands and rises, creating **low pressure**; but, the air cools adiabatically as it rises. The cooling creates clouds and rain or snow. Thus, low barometric pressure is an indication that precipitation may soon follow. Alternatively, when cool air falls, creating **high pressure**, it is compressed and heats up adiabatically. Since warm air can hold more moisture than cold air, clouds generally do not form under high-pressure conditions. Thus, rising barometric pressure generally precedes fair weather.

Uneven heating of the Earth's surface causes pressure differences and wind. Wind speed is determined by the **pressure gradient**. **Cyclones** and **anticyclones** are winds that spiral into low-pressure zones and away from high-pressure zones.

Air cools adiabatically when it rises over a mountain range, causing precipitation. A **rain-shadow desert** forms where air sinks down the leeward side of a mountain.

When two air masses collide, the warmer air rises along the front, forming clouds and often precipitation.

Sea breezes and **monsoons** arise because ocean temperature changes slowly in response to daily and seasonal changes in solar radiation, whereas land temperature changes quickly. A **thunderstorm** is a small, short-lived storm from a cumulonimbus cloud. Lightning occurs when charged particles separate within the cloud. A **tornado** is a small, short-lived, funnel-shaped storm that protrudes out of the bottom of a cumulonimbus cloud. A **tropical cyclone** is a larger, longer-lived storm that forms over warm oceans and is powered by the energy released when water vapor condenses to rain.

KEY TERMS

•

Humidity *430*	Normal lapse rate *432*	Upslope fog *436*	Rain-shadow desert *444*
Absolute humidity *430*	Cirrus cloud *432*	Convection-	Air mass *444*
Relative humidity *430*	Stratus cloud *432*	convergence *438*	Front *444*
Dew point *430*	Cumulus cloud *433*	Orographic lifting *438*	Occluded front *447*
Supersaturation *431*	Stratocumulus cloud *433*	Frontal wedging *438*	Stationary front *447*
Supercooling *431*	Nimbo *433*	Pressure gradient *440*	Monsoon *449*
Adiabatic temperature	Cumulonimbus cloud *433*	Isobar *440*	Wind shear *452*
changes *431*	Nimbostratus cloud *433*	Coriolis effect *440*	Tornado *453*
Dry adiabatic lapse	Altostratus cloud *433*	Jet stream *442*	Tropical cyclone *454*
rate *432*	Advection fog *436*	Cyclone *442*	
Wet adiabatic lapse	Radiation fog *436*	Anticyclone *442*	
rate *432*	Evaporation fog *436*	Orographic lifting *442*	

REVIEW QUESTIONS

•

1. Describe the difference between absolute humidity and relative humidity. How can the relative humidity change while the absolute humidity remains constant?

2. List three atmospheric processes that cool air.

3. What is the dew point? How does dew form?

4. List a set of atmospheric conditions that might produce supersaturation or supercooling.

5. What are the differences among dew, frost, and fog?

6. Explain why air cools as it rises.

7. What is an adiabatic temperature change? How does it differ from a nonadiabatic temperature change?

8. Compare and contrast the dry adiabatic lapse rate, the wet adiabatic lapse rate, and the normal lapse rate.

9. Discuss the factors that determine the height at which a cloud forms.

10. Describe cirrus, stratus, and cumulus clouds. Include their shapes and the type of precipitation to be expected from each.

11. How does rain form? How do droplets falling from a stratus cloud differ from those formed in a cumulus cloud?

12. Compare and contrast snow, sleet, and hail.

13. List three mechanisms that cause air to rise.

14. Why does low pressure often lead to rain? Why does high pressure often bring sunny skies?

15. If the wind is blowing southward in the Northern Hemisphere, will the Earth's spin cause it to veer east or west? If

the wind is moving southward in the Southern Hemisphere, which way will it veer? Explain your answer.

16. Compare and contrast sea and land breezes with monsoons.

17. What is a rain-shadow desert? How does it form?

18. How do warm and cold fronts form, and what types of weather are caused by each?

19. Describe the three stages in the life of a thunderstorm.

20. Briefly explain how lightning forms.

21. Compare and contrast tornadoes with tropical cyclones. How does each form, and how does each affect human settlements?

DISCUSSION QUESTIONS

•

1. Using Figure 16–1, estimate the maximum absolute humidity at 0°C, 10°C, 20°C, and 40°C. Estimate the quantity of water in air, at 50 percent relative humidity, at each of the above temperatures.

2. Explain why frost forms on the inside of a refrigerator (assuming it is an old-fashioned one and not a modern frost-free unit). Would more frost tend to form in (a) summer or winter, and (b) in a dry desert region or a humid one? Explain.

3. Which of the following conditions produces frost? Which produces dew? Explain. (a) A constant temperature throughout the day. (b) A warm summer day followed by a cool night. (c) A cool fall afternoon followed by freezing temperature at night.

4. Using the data in Figures 16–1, 16–3, and 16–4, predict the elevation at which clouds will form (if they form at all) under the following conditions. (a) Surface temperature is 20°C. An air mass with a temperature of 30°C carries 10 grams of water per cubic meter. (b) Surface temperature is 20°C. An air mass with a temperature of 20°C carries 10 grams of water per cubic meter. (c) Surface temperature is 15°C. An air mass with a temperature of 30°C carries 10 grams of water per cubic meter. (d) Surface temperature

is 10°C. An air mass with a temperature of 15°C carries 5 grams of water per cubic meter.

5. Draw a chart showing temperatures at cloud level and at ground level that will cause condensing water vapor to fall as rain, snow, sleet, and glaze.

6. What is the energy source that powers the wind?

7. Are sea breezes more likely to be strong on an overcast day or on a bright sunny one? Explain.

8. What is an air mass? Describe what would likely happen if a polar air mass collided with a humid subtropical air mass.

9. Study the weather map in today's newspaper, and predict the weather two days from now in your area, Salt Lake City, Chicago, and New York City. Defend your prediction. Check the paper in two days to see if you were right or wrong.

10. Using Figure 16–27, estimate the number of days per year that thunderstorms occur in your area. How does this number compare with the highest and lowest frequency values for other regions in the United States? Which of the three mechanisms listed in the text are responsible for most of the thunderstorms in your area? Defend your answer.

Climate

17

T oday most of North America has cool or cold winters, warm summers, and seasonal variations in precipitation. However, these conditions are geologically recent. In the past, North America has experienced both tropical warmth and the arctic cold of the ice ages. One hundred million years ago dinosaurs wallowed in hot, humid swamps, and only 25,000 years ago woolly mammoths roamed the edges of giant glaciers that covered one third of the continent.

If temperature and precipitation were to change again, conditions for agriculture and patterns of fuel consumption would also change, with profound implications for our already crowded planet. In addition, climate affects sea level, glaciation, and other factors that define the character of the Earth's surface and its plants and animals. In this chapter we introduce global climate zones and then ask two questions:

1. What natural factors cause climate change?

2. What human factors cause climate change? Can industrial development cause climate change, and if so, what kinds of change?

● A rainbow over the Bering Sea. *(Terry Domico, Earth Images)*

17.1 Global Winds and Precipitation Zones

In 1735 a British meteorologist, George Hadley, reasoned that global wind systems are generated solely by the temperature difference between the equator and the poles. Hadley inferred that air is heated at the equator, rises, and then flows poleward at high elevations. At the poles, it cools, sinks, and flows back toward the equator along the Earth's surface (Fig. 17–1). This pattern has since been called a single convection cell model of atmospheric circulation. However, when meteorologists began to map global weather patterns shortly after World War I, they found high- and low-pressure systems migrating across the mid-latitudes like writhing snakes (Fig. 17–2). A Norwegian scientist, Carl Rossby, concluded that the single cell model could not explain these migrating storms, because it did not account for the effects of the Earth's rotation.

In the 1950s global wind systems were modeled experimentally by climatologists at the University of Chicago. They mounted a circular pan on a variable-speed turntable and placed a heating element around the rim and a cooling coil at the center (Fig. 17–3). The pan represented the Earth, with the rim analogous to the equator and the center analogous to the poles. They filled the pan with water and added dye in order to trace currents.

When the pan was stationary, water rose at the heated edge, traveled across the surface, sank at the cooled center, and returned to the edge along the bottom, thus forming a Hadley cell. When the pan rotated slowly, this

A woodcut to accompany Samuel Taylor Coleridge's poem, *The Rhyme of the Ancient Mariner*. The woodcut illustrates the thirst and despair of sailors attempting to cross the doldrums, a region with no prevailing winds.
(C. Dord collection)

current was deflected by the Coriolis effect, but the single-cell pattern was retained. However, when the scientists increased the rotational speed, the cell broke apart. Midway between the edge and the center—the area representing the Earth's middle latitudes—the current diverged into whirls and eddies that looked very much like the real storms observed in the middle latitudes.

Our modern understanding of global climate is derived from this experiment combined with mathematical analysis and thousands of direct measurements of natural wind patterns. As already explained, the Sun shines most directly at or near the equator and warms the air near the Earth's surface. The warm air gathers moisture from the equatorial oceans. The warm, moist rising air forms a vast region of low pressure near the equator, with little horizontal air flow. As the rising air cools adiabatically, the water vapor condenses and falls as rain. Therefore, local squalls and thunderstorms are common, but steady winds are rare. This hot, still region was a serious barrier in the age of sailing ships. Mariners called the equatorial

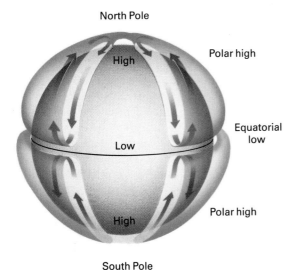

Figure 17–1 The Hadley one-cell model for global circulation patterns, proposed in 1735. It has since been replaced by the three-cell model (Fig. 17–5).

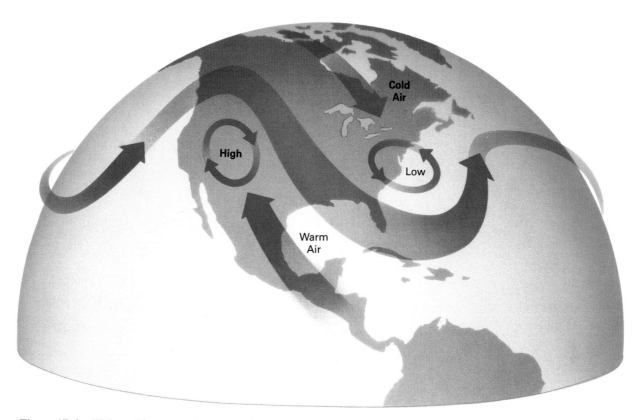

Figure 17–2   High- and low-pressure systems that snake their way across the Northern Hemisphere cannot be explained by the Hadley one-cell model.

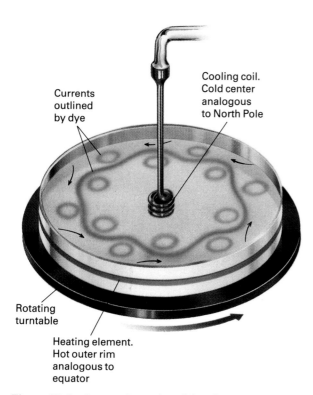

Figure 17–3   An experimental model to demonstrate the Earth's winds.

region the **doldrums**, and literature is alive with descriptions of their despair at being unable to move across the vast, windless seas. On land, the frequent rains near the equatorial low-pressure zone nurture lush tropical rainforests.

The air rising at the equator splits to flow north and south at high altitudes. However, these high-altitude winds do not continue to flow due north and south as Hadley predicted, because they are deflected by the Coriolis effect. Thus, their poleward movement is interrupted. In both the Northern and the Southern Hemispheres, this air veers until it flows due east, at about 30° north and south latitudes (Fig. 17–4). It then cools enough to sink to the surface, creating subtropical high-pressure zones at 30° north and south latitudes. The sinking air warms adiabatically, absorbing water and forming clear blue skies. At the center of the high-pressure area the air moves vertically and not horizontally, and therefore few steady surface winds blow. This calm high pressure belt circling the globe is called the **horse latitudes**. The region was so named during the 1500s and 1600s because sailing ships were becalmed for long periods of time. Horses transported as cargo on the ships often died of thirst and hunger. The warm, dry descending air in this high-

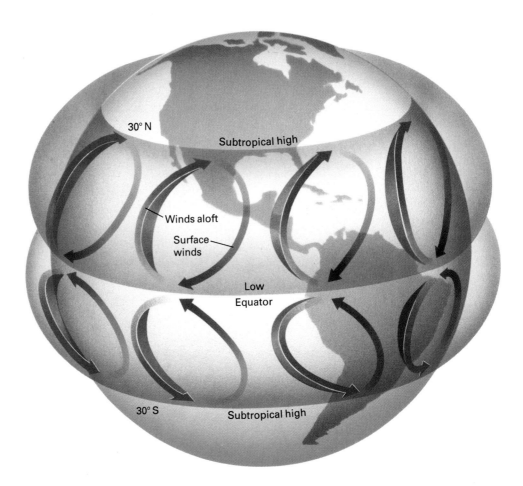

Figure 17–4   Air rising at the equator moves poleward at high elevations, falls at about 30°
north and south latitudes, and returns to the equator, forming trade winds. (Winds at higher latitudes are shown in Fig. 17–5.)

pressure zone forms many of the world's great deserts,
including the Sahara in North Africa, the Kalahari in
South Africa, and the Australian interior desert.

Descending air at the horse latitudes splits and flows
over the Earth's surface in two directions, toward the
equator and toward the poles. The surface winds moving
toward the equator are deflected by the Coriolis effect, so
they blow from the northeast in the Northern Hemisphere
and from the southeast in the Southern Hemisphere. In
the days of sailing ships, sailors transporting goods for
trade depended on these reliable winds and hence called
them the **trade winds**. The winds moving toward the
poles are also deflected by the Coriolis effect. The predominant winds in the mid-latitudes are called the **prevailing westerlies**. They flow from the southwest in the
Northern Hemisphere and from the northwest in the
Southern Hemisphere (Fig 17–5A).

The poles are cold year-round. The cold polar air
sinks, creating yet another band of high pressure. The
sinking air flows over the surface toward lower latitudes.
In the Northern Hemisphere these surface winds are deflected by the Coriolis effect to form the **polar easterlies**.
The polar easterlies and prevailing westerlies converge at
about 60° latitude. Air is forced upward at the convergence, forming a low-pressure boundary zone called the
polar front.

Recall that the Hadley model for global winds depicts
one convection cell in each hemisphere. The more accurate modern model, called the **three-cell model**, depicts
three convection cells in each hemisphere. The cells are
bordered by alternating bands of high and low pressure
(Fig 17–5B). In the three-cell model, global winds are
generated by heat-driven convection currents, and then
their direction is altered by the Earth's rotation.

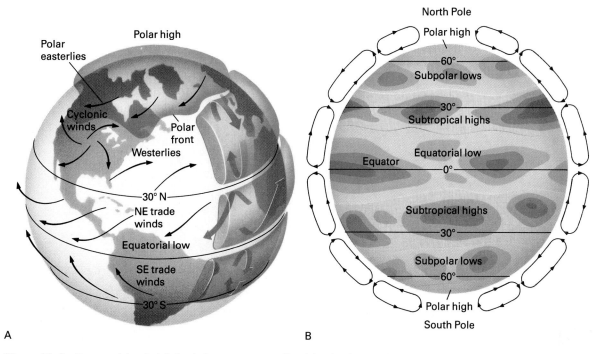

Figure 17–5 Two models of global wind patterns as predicted by the three-cell theory. (A) An artist's rendition showing the three cells and the directions of surface winds. (B) High- and low-pressure belts with the major convection currents shown on the edges of the sphere.

Figure 17–5B shows stationary linear boundaries between the cells. In reality these boundaries migrate north and south with the seasons. They are also distorted by surface topography and local air movement. For example, in the Northern Hemisphere, cyclones and anticyclones develop along the polar front as explained in Chapter 16 and shown in (Fig. 17–6). These storms bring alternating rain and sunshine, which are favorable for agriculture.

Thus, the great wheat belts of the United States, Canada, and Russia all lie between 30° and 60° north latitude.

Recall from Chapter 16 that a jet stream is a narrow band of fast-moving high-altitude air. Jet streams form at boundaries between the Earth's major climate cells as high-altitude air is deflected by the rotation of the Earth. The **subtropical jet stream** flows between the trade winds and the westerlies, and the **polar jet stream** forms

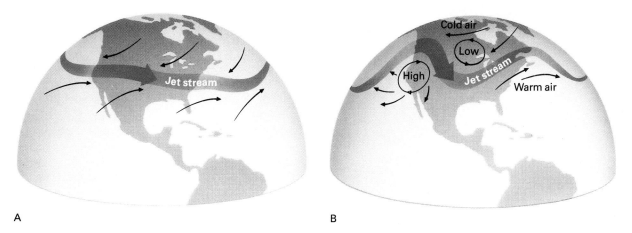

Figure 17–6 The polar front and jet stream.

A fertile prairie in Alberta, Canada.

along the polar front. When you watch a weather forecast on TV, the meteorologist commonly shows the movement and direction of the polar jet stream as it snakes across North America. Storms commonly occur along this line. We can now understand this relationship. The jet stream marks the boundary between cold polar air and the warm, moist westerly flow that originates in the subtropics. Storms develop where the two converge.

17.2 Factors That Determine Climate

In this and previous chapters, we have studied the factors that control temperature and precipitation and therefore determine climate. In this section we will briefly review these factors (Fig. 17–7).

1. Latitude Solar radiation is most intense at the equator and decreases at higher latitudes. Therefore, temperature is generally highest at the equator and lowest toward the poles. Seasonal temperature changes are also more pronounced at higher latitudes than at the equator (reference: Section 15.7).

2. Pressure and Wind Belts The three-cell model defines global bands of high and low pressure and prevailing winds between them. Examples include the equatorial low (the doldrums), the subtropical high (the horse latitudes), the trade winds, the mid-latitude westerlies, the polar front, and the polar easterlies (reference: Section 17.1).

3. Altitude Air is cooler at higher elevations, so mountainous regions and high plateaus are almost always colder than adjacent lowlands. Thus, glaciers exist on Mount Kenya in Africa and Mount Cayambe in South America even though both lie directly on the equator (reference: Section 15.6).

4. The Oceans Oceans affect coastal climates in two ways.

1. Currents transport heat northward or southward (reference: Section 13.5).
2. Even in the absence of currents, the ocean affects both the temperature and precipitation over adjacent land. Coastlines are generally warmer in winter and cooler in summer than continental interiors (reference: Section 15.8). In addition, in regions where warm oceans lie adjacent to cooler land, humid maritime air blows inland, leading to abundant precipitation (reference: Section 16.6). However, not all coastal areas are hu-

Major factors that determine climate

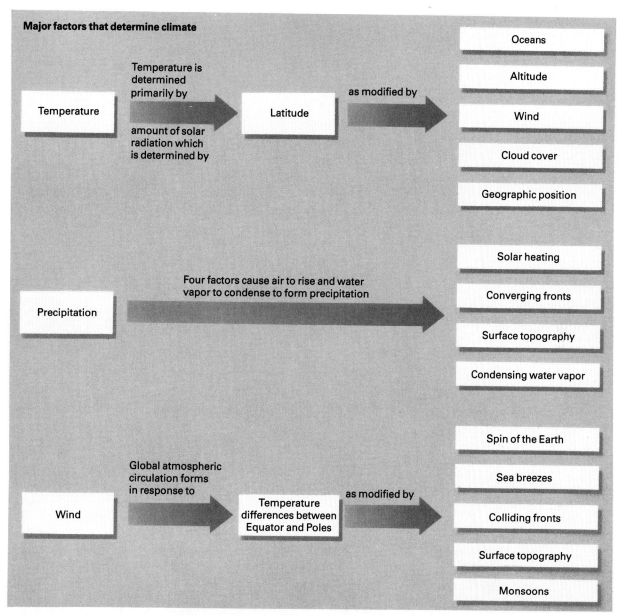

Figure 17–7   Major factors that determine climate.

mid. The Atacama Desert, which is perhaps the driest region on the Earth, lies along the Pacific coast of South America. This desert forms as cold air lying above the cold Pacific Ocean current sinks over the coast of Chile and Peru.

5. Mountain Ranges   Mountains alter the movement of air and generate local climate zones. In general, air rises on the windward side of a range, causing abundant precipitation. After it has passed over the mountains, it

sinks and warms adiabatically, forming a rain-shadow desert on the downwind side of the mountains (reference: Section 16.4).

6. Geographic Position   So many different factors affect the Earth's climate that it is impossible to write a few general rules predicting a specific area's climate based on its latitude, altitude, and distance from the ocean. As just explained, many coastal areas receive abundant precipitation, whereas in other places deserts exist along the shore.

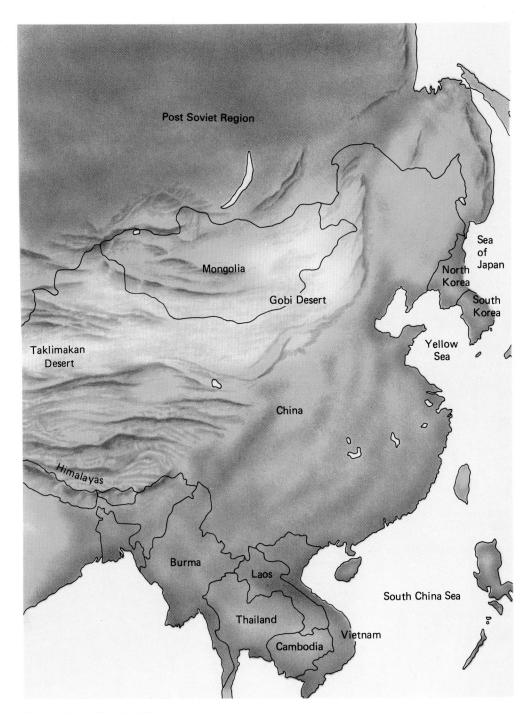

Figure 17–8   The Gobi Desert.

As another example, the influence of a mountain chain on climate depends, in part, on prevailing winds. The Gobi Desert in central Asia lies at about 40° north latitude, and its eastern edge is a little more than 400 kilometers from the Yellow Sea (Fig. 17–8). As a comparison, Pittsburgh, Pennsylvania, lies at about the same latitude and is 400 kilometers from the Atlantic Ocean. If latitude and distance from the ocean were the only factors, the two regions would have similar climates. In fact, the Gobi is a desert and western Pennsylvania receives enough rainfall to support forests and rich farmlands. The Gobi is bounded by the Himalayas to the south and the Urals to the west, which shadow it from the prevailing winds. In contrast, winds carry abundant moisture to western Pennsylvania from the Gulf of Mexico, the Great Lakes, and the Atlantic Ocean.

17.3 Climate Zones of the Earth

The Earth's major climate zones are classified primarily by temperature and precipitation. But an area with both wet and dry seasons has a different climate from one with moderate rainfall all year long, even though the two areas may have identical total annual precipitation. Therefore, climatic zones are also classified on the basis of seasonal variations in temperature and precipitation (Fig. 17–9). The **Köppen climate classification**, used by climatologists throughout the world, defines five principal groups.

A   Humid tropical   Every month is warm, with a mean temperature over 18°C (64°F). The temperature difference between day and night is greater than the difference between December and June averages. There is enough moisture to support abundant plant communities.

B   Dry   B climates have a chronic water deficiency; in most months evaporation exceeds precipitation.

C   Humid middle latitudes with mild winters   C climates have distinct winter and summer seasons and enough moisture to support abundant plant communities. In C climates the winters are mild, with the average temperature in the coldest month above −3°C (27°F).

D   Humid middle latitudes with severe winters   D climates are similar to C climates with distinct summer and winter seasons. However, D climates are colder, with the average temperature in the coldest month below −3°C (27°F).

E   Polar   Winters are extremely cold and even the summers are cool, with the average temperature in the warmest month below 10°C (50°F).

Figure 17–9 A model climograph. This graph shows the average monthly temperature (curved line) and average monthly precipitation for Nashville, Tennessee. Other features of the climograph are highlighted by the labels.

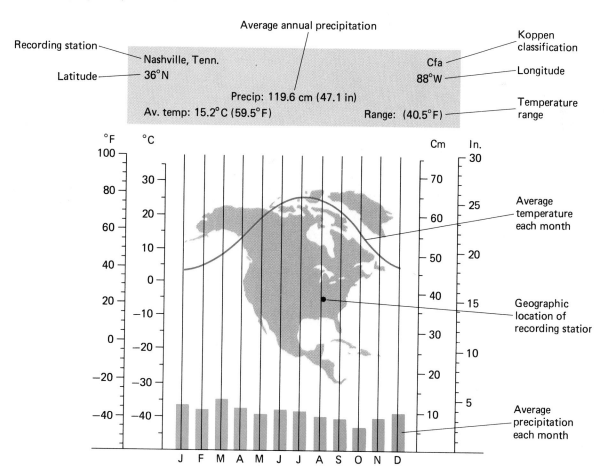

Earth Science and the Environment

Expansion and Contraction of Deserts

Most deserts are surrounded by semiarid regions that receive more moisture than a true desert but less than forests and grasslands. Semiarid lands support small populations of plants and animals. The Sahara is the largest desert on the planet. South of the Sahara lies a semiarid belt called the Sahel. Reports issued in the 1970s and 1980s claimed that the Sahara was advancing steadily southward into the Sahel at a rate of 5 kilometers per year. Scientists argued that human mismanagement of the land was causing the desert advance. However this interpretation has recently been questioned by a new study that indicates that the Sahara has not grown continuously, but has alternately expanded and contracted. The desert boundary is clearly visible on satellite photographs as a line of vegetation. Researchers plotted changes in the desert by studying photographs taken between 1980 and 1990. As shown in the figure, the desert expands (blue line) when rainfall declines (red line).

Although land mismanagement may not increase the size of the Sahara, human activity—for example, overgrazing, extensive cultivation, irrigation, and devegetation for firewood—can reduce the productivity of semiarid lands. One hundred years ago, nomads moved across the Sahara and the Sahel with little regard for national boundaries, traveling with the seasons and abandoning an area after it had been grazed for a short period of time. This constant movement prevented overgrazing. In addition, population levels of the nomadic tribes were stable and low. In recent years, however, their life style has changed. As medical attention and sanitation have improved, the population has expanded dramatically. In addition, enforcement of national borders and civil strife in some countries have curtailed nomadism. As populations have increased and the social structures changed, increasing pressure has been put on the land to produce food for more people from smaller areas.

Cattle eat the most nutritious grasses and ignore noxious weeds or inedible shrubs. If an area is lightly grazed, grasses remain healthy and are able to compete successfully with weeds and shrubs for water, nutrients, and space. However, if the grasses are overgrazed, their growth can be so seriously disrupted that they cannot reseed. This loss of grasses leads to several problems. Less edible plants become dominant, and the value of the range decreases. Loss of grasses leaves the soil susceptible to erosion. When soil is devoid of vegetation and baked in the Sun, it becomes so impermeable that water evaporates before it soaks in. Therefore, the water table

drops. Because the deep-rooted bushes depend on underground water, lowering of the water table ultimately kills these plants. The process is accelerated as the grazing cattle pack the earth with their hooves, blocking the natural seepage of air and water through the soil. Firewood gatherers cut the woody shrubs, further destroying the vegetation.

Crops can be grown in deserts if they are irrigated, but extensive irrigation can destroy the land. Recall that all river water is slightly salty. If this water is used to irrigate fields, some of the water evaporates, leaving the salts behind. Over time, the salt content of the soil may accumulate until the soil becomes too salty to support plants.

Even in the United States, overgrazing of semiarid range land has created a problem. Some critics argue that government policies favor short-term profits for ranchers rather than long-term stability of public range land, and that millions of hectares are being destroyed needlessly.

Expansion and contraction of the Sahara Desert from 1980 to 1990. Note that the Sahara has expanded (blue line) when rainfall has decreased (red line). Thus, rainfall, and not land management, appears to be the prime factor governing the size of the Sahara. *(Compton J. Tucker, Harold E. Dregne, and Wilbur W. Newcomb, "Expansion and Contraction of the Sahara Desert from 1980 to 1990,"* Science 253 *(July 19, 1991): 299ff)*

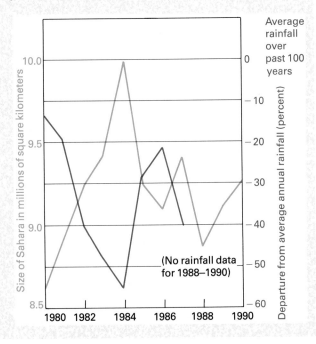

(No rainfall data for 1988–1990)

Subclassifications of these five groups are shown in Table 17–1.

Although climates are defined by temperature and precipitation, you can often estimate climate types from a photograph of an area. Visual classification is possible because specific types of plants grow in particular climatic environments; for instance, cactus grows in the desert, and trees grow only if moisture is abundant. A **biome** is a community of plants living in a large geographic area characterized by a particular climate. Following is a brief discussion of the major climate zones and their biomes (Fig. 17–10):

(Text continues on p. 472)

TABLE 17–1

World Climates					
Symbol					
1st	*2nd*	*3rd*	**Description**	**Climate Type**	**Vegetation**
A			**Humid tropical; no winter**		
	f		Wet year-round	Af = tropical wet	Rainforest
	m		Wet with short dry season	Am = tropical monsoon	Rainforest
	w		Winter dry season	Aw = tropical wet and dry	Grassland savannah
B			**Dry; evaporation greater than precipitation**		
	S			Steppe	
		h	Semiarid	BSh = tropical steppe	Steppe grassland
		k	Cool and dry continental interior or rain shadow location	BSk = mid-latitude steppe	Steppe grassland
	W		Desert		
		h	Warm	BWh = tropical desert	Desert
		k	Cool and dry; continental interior or rain-shadow location	BWk = mid-latitude desert	Desert
C			**Humid mid-latitude; mild winter**		
	f	a	East coast: warm, wet summers with westerly winds and winter cyclonic storms	Cfa = humid subtropical	Forest
	f	b	Cool summer and mild winter; maritime, westerlies, and cyclonic storms	Cfb = marine west coast	Forest to temperate rainforest
	s	a	Dry summer and wet winter with cyclonic storms	Csa = dry summer, wet winter, subtropical	Mediterranean
D			**Humid mid-latitude; severe winter**		
	f	a	Mid-latituide continental warm summer, cold winter	Dfa = humid continental; warm summer, cold winter	Prairie and forest
	f	b	High mid-latitude continental	Dfb = humid continental; cool summer, cold winter	Prairie and forest
	f	c & d	High-latitude continental	Dfc/d = subarctic	Boreal forest taiga
E			**Polar with no warmth in summer**		

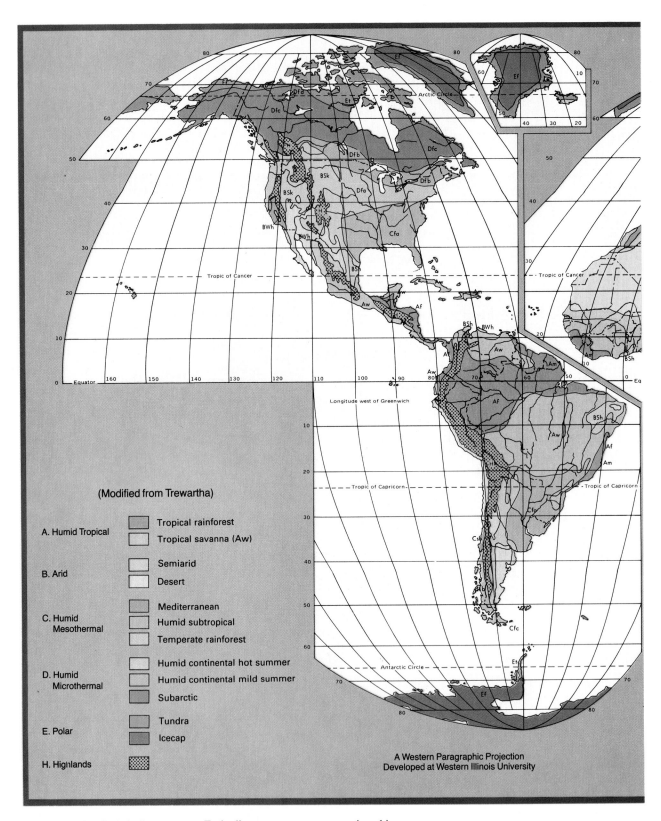

Figure 17–10   Global climate zones. Each climate zone supports a unique biome.

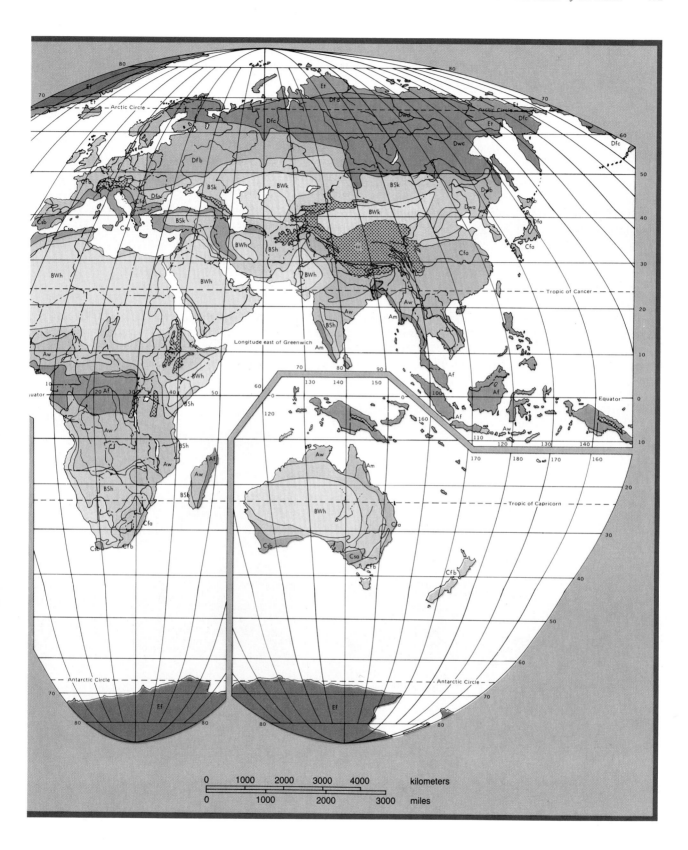

A   Humid Tropical Climates

Tropical Climates with Abundant Rainfall

The large low-pressure zone near the equator causes abundant rainfall, often exceeding 400 centimeters per year (160 in/yr), that supports **tropical rainforests** (Fig. 17–11). The dominant plants in a tropical rainforest are tall trees with slender trunks. These trees branch only near the top, covering the forest with a dense canopy of leaves. The canopy blocks out most of the light, so as little as 0.1 percent of the sunlight reaches the forest floor, which consequently has relatively few plants. The ground in a tropical forest is soggy, the tree trunks are wet, and water drips everywhere.

Tropical Climates with Distinct Wet and Dry Seasons

Tropical monsoon (Fig. 17–12) and **tropical savanna** (Fig. 17–13) both have large seasonal variations in rainfall, but a monsoon climate has greater total precipitation and greater monthly variation. Precipitation is great enough in tropical monsoon biomes to support rainforests. The seasonal precipitation is also ideal for agriculture.

A tropical savanna is a grassland with scattered small trees and shrubs. Such grasslands extend over large areas, often in the interiors of continents, where rainfall is insufficient to support forests or where forest growth is prevented by recurrent fires. Savannas are most extensive in Africa, where they support a rich collection of grazing animals such as zebras, wildebeest, and gazelles. Grasses sprout and grow with the seasonal rains, and the great African herds migrate with the foliage.

B   Dry Climates

In dry zones where the annual precipitation varies from 25 to 35 centimeters per year (10 to 15 in/yr), the climate is **semiarid** and **grasslands** predominate. If the rainfall is less than 25 centimeters per year (10 in/yr), **deserts**

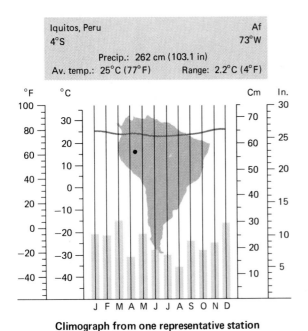

Climograph from one representative station

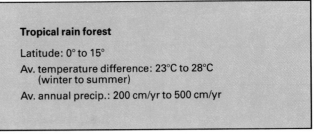

General statistics for climate type

Figure 17–11   Tropical rainforest.

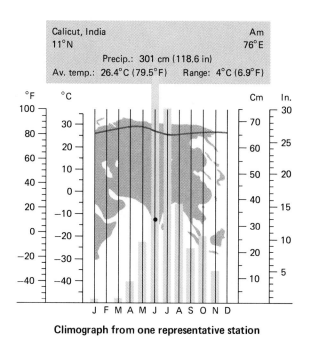

Tropical monsoon

Latitude: 5° to 30°

Av. temperature difference: 20°C to 30°C
(winter to summer)

Av. annual precip.: 150 cm/yr to 400 cm/yr

Climograph from one representative station

General statistics for climate type

Figure 17–12   Tropical monsoon.

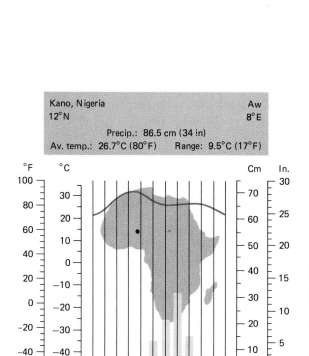

Tropical savanna

Latitude: 5° to 25°

Av. temperature difference: 20°C to 30°C
(winter to summer)

Av. annual precip.: 100 cm/yr to 180 cm/yr

Climograph from one representative station

General statistics for climate type

Figure 17–13   Tropical savanna.

Earth Science and the Environment

Tropical Rainforests and Climate

At present the world's tropical rainforests are being logged and cleared to make way for farms. According to the World Resources Institute, approximately one third of the original tropical rainforest has already been cleared, and deforestation continues at a rate of about 7 million hectares per year, or about 0.6 percent of the total rainforest each year. This rate amounts to about 13 hectares per minute, day and night, throughout the year.

A mature tropical forest influences climate because water evaporation from leaf surfaces cools the atmosphere. In turn, the cooling induces condensation, which leads to rain that sustains the forest. In large areas that have been deforested, local climate has changed. For example, in Panama, rainfall in areas deforested 50 years ago has decreased by 20 percent compared with nearby unlogged regions. If rainfall declines enough, it becomes impossible to reestablish the forest after it is cut. In addition, when the forest is cut and the vegetation burned or allowed to

rot, large quantities of carbon dioxide are released, adding to the greenhouse effect. Conservationists are also concerned that millions of species will become extinct if deforestation continues.

Destruction of a tropical rainforest in Costa Rica.

form and support only sparse vegetation (Fig. 17–14). The world's largest deserts lie along the 30° latitude high-pressure zones, although rain-shadow and coastal deserts exist in other latitudes.

C Humid Mid-Latitude Climates with Mild Winters

Humid Subtropics

The southeastern United States is a **humid subtropical** region (Fig. 17–15). During the summer, conditions can be as hot and humid as in the tropics, and rain and thundershowers are common. However, during the winter, arctic air pushes southward, forming cyclonic storms. Although the average monthly temperature seldom falls below 7°C (44°F), cold fronts occasionally bring frost and snow. Precipitation is relatively constant year-round due to convection-driven thunderstorms in summer and cyclonic storms in winter. This zone supports both trees with needles (conifers) and trees with broad leaves (deciduous) as well as valuable crops such as vegetables, cotton, tobacco, and citrus fruits.

Mediterranean Climate

The **Mediterranean climate** is characterized by dry summers, rainy winters, and moderate temperature (Fig. 17–16). These conditions occur on the west coasts of all continents between latitudes 30° and 40°. In summer the subtropical high migrates to higher latitudes, producing to near-desert conditions with clear skies as much as 90 percent of the time. In winter the prevailing westerlies bring warm, moist air from the ocean, leading to fog and rain. Thus, 75 percent or more of the annual rainfall occurs in winter. Although redwoods, the largest trees on Earth, grow in specific environments in central California, the summer heat and drought of Mediterranean climates generally retard the growth of large trees. Instead, shrubs with scattered trees dominate. Fires occur frequently during the dry summers and spread rapidly through the dense shrubbery. During the dry fall of 1991, a particularly devastating fire swept though the hills of Oakland and Berkeley, California, destroying 3354 homes worth $1.5 billion and killing 24 people. If the vegetation is destroyed by fire, landslides often occur when the winter rains return. Torrential rains in the winter of 1992 brought extensive flooding to southern California.

(Text continues on p. 477)

Lima, Peru   BWh
12°S   77°W
Precip.: 4 cm (1.6 in)
Av. temp.: 20°C (68°F)   Range: 9°C (15.5°F)

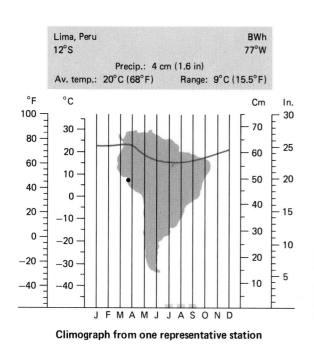

Climograph from one representative station

Figure 17–14   Desert.

Desert

Latitude: Variable

Av. temperature difference: Variable
   (winter to summer)

Av. annual precip.: Less than 25 cm/yr

General statistics for climate type

New Orleans, La.   Cfa
30°N   90°W
Precip.: 146 cm (57.4 in)
Av. temp.: 21°C (69.5°F)   Range: 16°C (28.5°F)

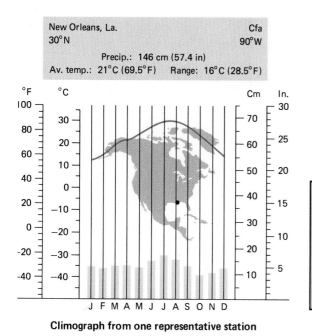

Climograph from one representative station

Figure 17–15   Humid subtropics.

Humid subtropics

Latitude: 15° to 40°

Av. temperature difference: 7°C to 32°C
   (winter to summer)

Av. annual precip.: 60 cm/yr to 250 cm/yr

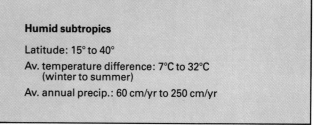

General statistics for climate type

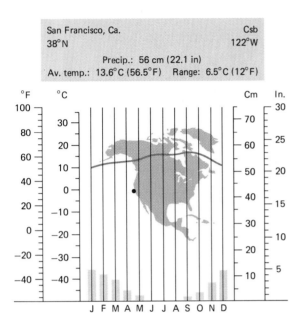

San Francisco, Ca.   Csb
38°N   122°W

Precip.: 56 cm (22.1 in)
Av. temp.: 13.6°C (56.5°F)   Range: 6.5°C (12°F)

Climograph from one representative station

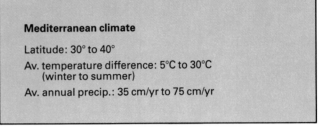

Mediterranean climate

Latitude: 30° to 40°

Av. temperature difference: 5°C to 30°C
(winter to summer)

Av. annual precip.: 35 cm/yr to 75 cm/yr

General statistics for climate type

Figure 17–16   Mediterranean climate.

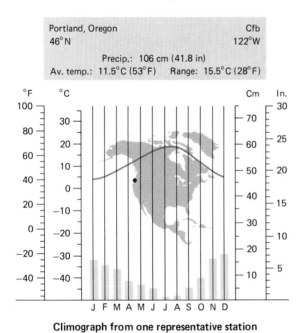

Portland, Oregon   Cfb
46°N   122°W

Precip.: 106 cm (41.8 in)
Av. temp.: 11.5°C (53°F)   Range: 15.5°C (28°F)

Climograph from one representative station

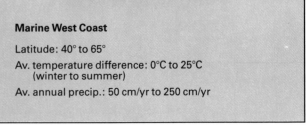

Marine West Coast

Latitude: 40° to 65°

Av. temperature difference: 0°C to 25°C
(winter to summer)

Av. annual precip.: 50 cm/yr to 250 cm/yr

General statistics for climate type

Figure 17–17   Marine West Coast.

Marine West Coast

Marine west coast climate zones border the Mediterranean zones and extend poleward to 65° (Fig. 17–17). They are influenced by warm ocean currents that moderate temperature and bring abundant precipitation. Thus, summers are cool and winters warm, and the temperature difference between the seasons is small. Seattle, Portland, and other northwestern coastal cities experience rain and drizzle for days at a time, especially during the winter, when the warm, moist maritime air from the Pacific flows first over cool currents close to shore and then over cool land surfaces. The total rainfall varies from moderate, 50 centimeters per year (20 in/yr) to wet, 250 centimeters per year (100 in/yr). The wettest climates occur where mountains interrupt the maritime air. **Temperate rainforests** grow where rainfall is greater than 100 centimeters per year (40 in/yr) and is constant throughout the year. Temperate rainforests are common along the northwest coast of North America, from Oregon to Alaska.

D   Humid Mid-Latitude Climate with Severe Winters

Continental interiors in the mid-latitudes are characterized by hot summers and cold winters (Fig. 17–18). Thus, in the northern Great Plains the temperature can drop to −40°C (−40°F) in winter and soar to 38°C (100°F) in summer. Even in a given season the temperature may vary greatly as the polar front moves northward or southward. Thus, in winter the northern continental United States may experience arctic cold one day and rain a few days later. If rainfall is sufficient, this climate supports abundant coniferous forests, whereas grasslands dominate the drier regions. Millions of bison once roamed the continental grasslands of North America, and today wheat and other grains grow from horizon to horizon. The northernmost portion of the D climate zone is the subarctic, which supports the **taiga** biome. Taiga consist of conifers that survive extremely cold winters.

E   Polar

In the Arctic, winters are harsh and long, and the temperature remains above freezing only during a short summer. Trees cannot survive, and low-lying plants such as mosses, grasses, flowers, and a few small bushes cover the land. This biome is called **tundra** (Fig. 17–19).

17.4   Urban Climates

If you ride a bicycle from the center of a city toward the countryside, you may notice that the air gradually feels cooler and more refreshing as you leave the city streets and enter the green fields or hills of the outlying area. This feeling is not entirely psychological; the climate of

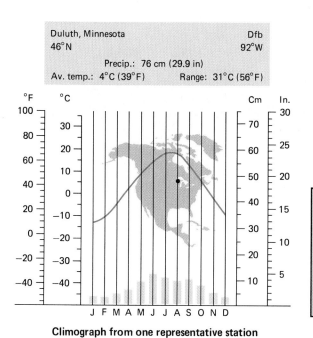

Climograph from one representative station

Figure 17–18   Mid-latitude with severe winters.

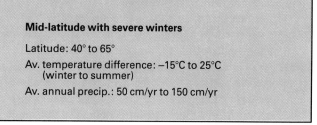

Mid-latitude with severe winters

Latitude: 40° to 65°

Av. temperature difference: −15°C to 25°C (winter to summer)

Av. annual precip.: 50 cm/yr to 150 cm/yr

General statistics for climate type

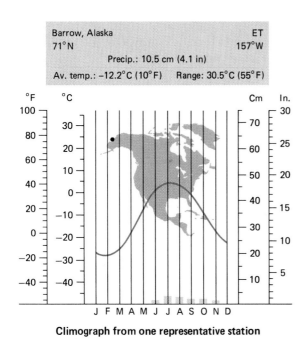

Barrow, Alaska   ET
71°N   157°W
Precip.: 10.5 cm (4.1 in)
Av. temp.: –12.2°C (10°F)   Range: 30.5°C (55°F)

Climograph from one representative station

Arctic Tundra

Latitude: 65° to 80°

Av. temperature difference: –35° to + 10°C
   (winter to summer)

Av. annual precip.: generally low, 10 cm/yr to 35 cm/yr

General statistics for climate type

Figure 17–19 Arctic tundra.

cities is measurably different from that of the surrounding rural regions (Table 17–2).

As shown in Figure 17–20, the average winter minimum temperature in the center of Washington, D.C., is more than 3°C (5.4°F) warmer than that in outlying areas. This difference is called the **urban heat island effect**. Stone and concrete buildings and asphalt roadways absorb solar radiation. Cities are also warmer, because little surface water exists and as a result little evaporative cooling occurs. In contrast, in the countryside, water collects in the soil and evaporates for days after a storm. Roots draw water from deeper in the soil, and this water evaporates from leaf surfaces. Urban environments are also warmed by the heat released when fuels are burned. In New York City in winter, the combined heat output of all the vehicles, buildings, factories, and electrical generators is 2.5 times the solar energy reaching the ground. The heat that is absorbed or produced within the city is retained because buildings have enough mass to store it. This heat keeps cities warmer at night than the surrounding countryside. In addition, the tall buildings block winds that might otherwise disperse the warm air. Finally, air pollutants absorb long-wave radiation (heat rays) emitted from the ground and produce a local greenhouse effect.

TABLE 17–2

Average Changes in Climatic Elements Caused by Urbanization	
Element	**Comparison with Rural Environment**
Cloudiness:	
Cover	5–10% more
Fog—winter	100% more
Fog—summer	30% more
Precipitation, total	5–10% more
Relative humidity:	
Winter	2% lower
Summer	8% lower
Radiation:	
Total	15–20% less
Direct sunshine	5–15% less
Temperature:	
Annual mean	0.5%–1.0°C higher
Winter minimum (average)	1.0–3.0°C higher
Wind speed:	
Annual mean	20–30°C lower
Extreme gusts	10–20°C lower
Calms	5–20°C higher

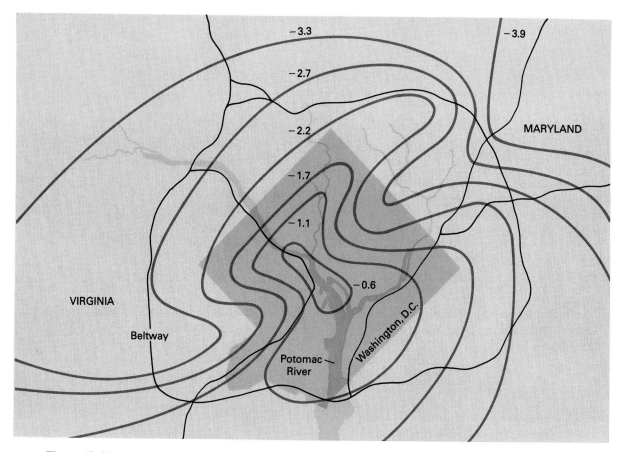

Figure 17–20   The urban heat island effect. The average minimum temperature in and around Washington, D.C., during the winter.   *(From C. A. Woollum, Weatherwise 17 (1964): 264)*

As warm air rises over a city, a local low-pressure zone develops, and rainfall is generally greater over the city than in the surrounding areas (Fig. 17–21). Water condenses on dust particles, which are abundant in polluted urban air. Weather systems collide with the city buildings and linger, much as they do on the windward side of mountains. Thus, a front that might pass quickly over rural farmland remains longer over a city and releases more precipitation.

In 1600, less than 1 percent of the global population lived in cities. By 1950, 30 percent of the world's population was urban, and by 1990 that ratio had grown to 50

Figure 17–21   Warm air rising over a city creates low-pressure zone. As a result, precipitation is greater in the city than over the surrounding countryside.

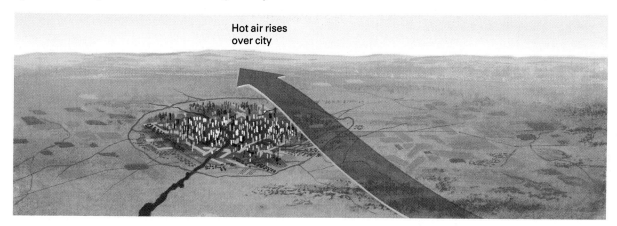

Hot air rises over city

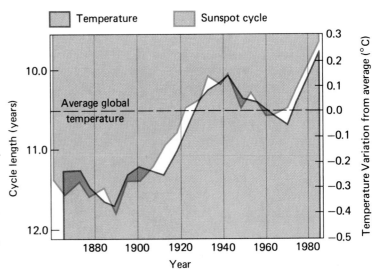

Figure 17–22 The correlation between time interval between maximum sunspot abundance (blue curve) and temperature (brown curve). Global temperature is expressed as change from average annual temperature in the years 1951 to 1980. *(Source: E. Friis-Christensen and K. Lassen, "Length of the Solar Cycle: An Indicator of Solar Activity Closely Associated with Climate."* Science 254 *(Nov. 1, 1991): 698ff)*

percent. Therefore, although urban climate change may not affect global climate, it does affect the lives of the many people living in cities.

17.5 Climate Change

Five natural factors alter global climate: changes in solar radiation, tectonic events, changes in the Earth's orbit around the Sun, catastrophic events such as volcanic eruptions and meteorite impacts, and changes in atmospheric composition.

Changes in Solar Radiation

In ancient times, astronomers assumed that the Sun did not change and would not change through time. This belief was first questioned by the Renaissance astronomer and physicist Galileo, who observed alternately appearing and disappearing dark spots on the Sun, called sunspots. In more recent times, astronomers have studied both sunspots and total solar output. Sunspots are easy to see with a telescope, but accurate measurements of total solar output are difficult because variations in the atmosphere obscure observations. In 1980 the Solar Maximum Mission spacecraft was launched to study the Sun from space, where no such interference exists.

Solar output dropped by about 0.1 percent between 1980 and 1985 and then increased by late 1987. However, the Earth's climate warmed steadily in this period. Therefore no obvious relationship exists between short-term fluctuations in solar output and climate. It is more difficult to assess long-term correlation between solar output and climate because accurate data have been available only since 1980.

Several statistical correlations have been made between sunspot frequency and climate. A study, published in 1991, shows a close correspondence between length of sunspot cycles and global temperature (Fig. 17–22). However, the statistical correlation does not prove that sunspot cycles cause climate change, and the explanation of the correlation remains under debate.

Tectonic Activity

According to one theory, continental climates were cool and ice ages occurred during periods when the continents clustered close to the poles. For example, during the Permian period (about 250 million years ago), all the continents were gathered near the South Pole into one giant supercontinent called Pangaea. Glaciation was extensive at this time. As Pangaea broke apart and the continents moved toward the equator, the Permian Ice Age ended (Fig. 17–23).

Other ice ages may have resulted from mountain building. Global cooling during the past 40 million years has coincided with the formation of the Himalayas and the North American Cordillera.* Scientists have used computer simulations to predict changes in climate that would result from uplift of these mountain chains. Their mathematical models mimicked the observed cooling.

Since mountain building and continental migration require tens of millions of years, tectonic events are only responsible for long-term climate change.

Changes in the Earth's Orbit Around the Sun

Recall that major ice ages have occurred at least five times in the Earth's history. During the most recent, the Pleistocene Ice Age, glaciers advanced and retreated sev-

*William F. Ruddiman and John Kutzbach, "Plateau Uplift and Climatic Change," *Scientific American*, March 1991: 66ff.

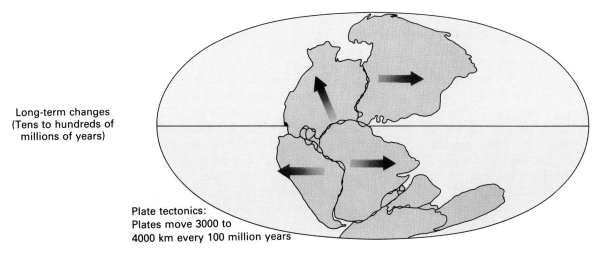

Long-term changes
(Tens to hundreds of
millions of years)

Plate tectonics:
Plates move 3000 to
4000 km every 100 million years

Figure 17–23 Tectonic movement may explain temperature changes for long-range cycles, measured in tens to hundreds of millions of years.

eral times. Although tectonic plate movement could have caused the five ice ages, it is too slow to account for the periodic advances and retreats of ice during the Pleistocene Ice Age.

One explanation for the rapid Pleistocene advances and retreats is that the Earth's temperature changes in response to variations in the Earth's orbit. Three types of orbital changes occur (Fig 17–24):

1. The shape of the Earth's orbit around the Sun changes on about a 100,000-year cycle. This is known as **eccentricity**.

2. The Earth's axis is currently tilted at about 23.5° with respect to a line perpendicular to the plane of its orbit around the Sun. The **tilt** oscillates by 1.5° on about a 41,000-year cycle.

3. The Earth's axis, which now points directly toward the north star, circles like the axis of a wobbling top, completing a full cycle every 23,000 years. This movement is called **precession**.

Orbital changes do not appreciably affect the total solar radiation received by the Earth. However, they do affect the distribution of solar energy with respect to latitude and season. Seasonal changes in sunlight reaching the Earth can cause an onset of glaciation by affecting summer temperature. At higher latitudes, snow falls and ice forms during the winter, even during interglacial pe-

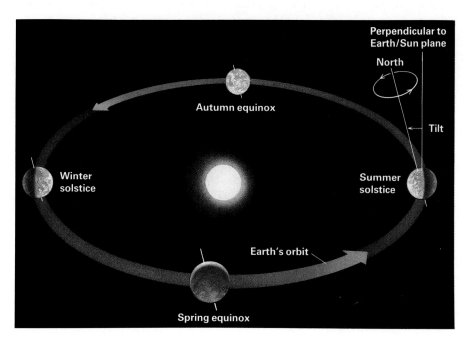

Figure 17–24 Orbital variations may explain the temperature oscillations and glacial advances and retreats during the Pleistocene epoch. Orbital variations occur over time spans of tens of thousands of years.

riods. If summers are hot, the snow and ice melt, but if summers are cool, snow and ice persist, leading to glaciation.

Early in the twentieth century, a Yugoslavian astronomer, Milutin Milankovitch, calculated the combined effects on the Earth's climate of all three variations. His calculations showed that the three regular orbital variations should combine periodically to generate alternating cool and warm climates in the higher latitudes. Moreover, the timing of the calculated climate changes coincided with the known timing of Pleistocene glacial advances and retreats.

Milankovitch's work is supported by other studies. Certain microorganisms that live near the surface of the ocean are sensitive to temperature. When the organisms die, their remains fall to the sea floor. Certain characteristics of these fossils preserve records of seawater temperatures when the organisms were alive. Therefore, it is possible to graph variations in global temperature over the past several million years by studying the fossils preserved in deep-sea sediment. A close correlation was observed between temperature changes recorded by examining the fossils and Milankovitch's calculated changes.*

Catastrophic Events Such as Meteorite Impacts and Volcanic Eruptions

As explained in Chapter 4, 65 million years ago one quarter of all known animal species and an uncounted number of plants suddenly became extinct. Included in the list were the dinosaurs as well as many less conspicuous species of clams, fish, single-celled plankton, and insects. In 1977 Walter and Luis Alvarez suggested that a giant meteorite struck the Earth 65 million years ago. The impact vaporized the meteorite and surrounding rock and started enormous fires. The vaporized rock and soot from the fires rose into the upper atmosphere and were carried by winds over the entire globe. This thick, dark cloud blocked the Sun. Plants died, surface waters froze, and a large percentage of fauna and flora on the planet froze and starved to death.

An alternative, but related, theory is that volcanic ash blasted into the upper atmosphere by repeated volcanic eruptions caused global cooling and may have been responsible for the dinosaurs' demise. The largest historical eruption occurred in 1815 when Mount Tambora in the southwestern Pacific Ocean exploded, ejecting approximately 100 times more ash into the atmosphere than did the 1980 eruption of Mount St. Helens. The following year, 1816, was one of the coldest years in history and has been called "the year without a summer" and "eighteen

hundred and froze to death." Crop failures, compounded by the devastation caused by the Napoleonic Wars, led to widespread famine in Europe.

The most recent volcanic event with potential to alter climate was the 1991 eruption of Mount Pinatubo in the Philippines. Beginning on June 16, 1991, Pinatubo ejected approximately 40 to 50 million tons of sulfur dioxide into the stratosphere (Fig. 17–25). Sulfur dioxide dissolves in water droplets and reacts to form sulfuric acid aerosols, which reflect sunlight. Scientists estimate that sunlight in the tropics will be reduced by 7 to 15 percent for a year after the eruption. In the absence of other factors, this reduction would cause global cooling. However, 1991 was the second warmest year on record. Some climatologists cite the greenhouse effect and a periodic warming of parts of the Pacific Ocean, called El Niño, as the cause of the high global temperature in 1991. They are carefully monitoring global climate in 1992 to see what will happen.

Changes in Atmospheric Composition —The Greenhouse Effect

As you learned in Chapter 15, the composition of the atmosphere has changed dramatically over the 4.6-billion-year history of our planet. Some molecules in the atmosphere absorb infrared radiation (heat rays), and others do not. If the concentration of infrared-absorbing molecules increases, then the atmosphere warms. This warming is called the **greenhouse effect**. Although water vapor is the most abundant infrared-absorbing gas, carbon dioxide has received more attention because it is released by human activity and has increased during the past century.

Only a small fraction of the total carbon near the Earth's surface is present in the atmosphere as carbon dioxide at any one time. Large quantities exist (1) dissolved in seawater, (2) in limestone on the Earth's surface, (3) in plant and animal tissue, and (4) in rocks of the lower crust and upper mantle (Fig. 17–26).

Seawater   When seawater is heated, dissolved carbon dioxide escapes into the atmosphere. In turn, the increased atmospheric carbon dioxide causes greenhouse warming, which further heats the oceans causing more carbon dioxide to escape. The warmth also evaporates seawater. The water vapor absorbs additional infrared radiation, which causes even more warming. Such a feedback mechanism can escalate. A runaway greenhouse effect on Venus has raised its surface temperature to 450°C, about as hot as the inside of a self-cleaning oven.

Limestone   Most limestone forms when marine organisms absorb calcium and carbon dioxide and convert them to solid calcium carbonate. The reverse process occurs as well. When limestone dissolves, the carbon dioxide is

*Hays, J. D., John Imbrie, and N. J. Shackleton, "Variations in the Earth's Orbit: Pacemaker of the Ice Ages," *Science* 194 (1976): 1121.

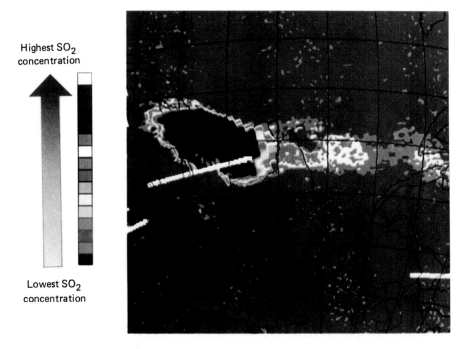

Highest SO₂
concentration

Lowest SO₂
concentration

Figure 17–25 A false color image of the sulfur dioxide cloud from the eruption of Mount Pinatubo. The volcano is at the lower right side of the photograph. The image was taken on June 18, 1991, two days after the initial eruption. The cloud had already drifted 8000 kilometers. The white lines represent missing data, and the light blue specks scattered throughout the image are caused by "noise" in the experimental data. *(NASA)*

Figure 17–26 A simplified carbon cycle. Carbon passes through the processes indicated by the red arrows much more rapidly than through those indicated by the yellow arrows.

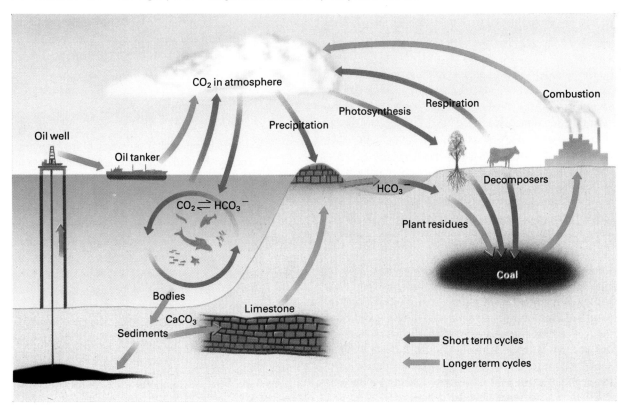

Earth Science and the Environment

Threshold Effects

No linear correlation exists between quantity of pollution emitted into the air and the effects on humans and global climate. Some pollutants are absorbed by rain-water, dissolve in the oceans, or are removed by other processes. Others accumulate. In some cases a compound may accumulate and not alter atmospheric conditions until a **threshold** concentration is reached. Once the threshold is crossed, a small additional amount may cause a large change. As an example of a threshold effect, consider the potential rise in sea level from melting of glaciers.

Ice melts at 0°C. Imagine that during the summer, the average temperature on a continental glacier is −2°C. If the temperature increases by 1.5°C, nothing much happens because the temperature rises to only −0.5°C and the glacier remains frozen. If you did not know the melting point of ice and observed that a temperature rise of 1.5°C had little effect, you might make the extrapolation that another 1.5°C temperature rise would also cause

little change. However, this prediction would be wrong, because it ignores the fact that the first temperature rise placed the environment at a threshold. If the warming trend continued, another 0.5°C rise would melt the ice. Then sea level would rise, coastal cities would be flooded, and climates might change. Anyone who tries to extrapolate future environmental problems must ask, "Are we already poised at the edge of a threshold?" In a recent publication by the Worldwatch Institute, the authors warn:

> With many of the natural systems now at risk, however, thresholds are not well defined, systemic responses to threshold crossings are not well understood, and the consequences of those crossings are largely incalculable. . . .
>
> Any system pushed out of equilibrium behaves in unpredictable ways. Small external pressures may be sufficient to cause dramatic changes. Stresses may become self-reinforcing, rapidly increasing the system's instability.*

*Lester Brown et al., *State of the World 1987* (New York: W. W. Norton and Co., 1987).

---

released as a gas. Thus marine organisms remove carbon dioxide from the atmosphere, and weathering releases the carbon dioxide back into the atmosphere. Increasing temperature leads to higher rates of limestone weathering adding to the feedback effect described in the previous paragraph.

Living Tissue During photosynthesis, carbon dioxide gas combines with other elements to form sugars that are energy-rich organic molecules. These sugars break down during respiration, releasing carbon dioxide gas. Large amounts of carbon are stored in plant communities such as tropical rainforests. Increased plant growth removes carbon dioxide from the atmosphere. On the other hand, carbon dioxide is released when plants die and rot or burn. Additional carbon is stored in fossil fuels, which formed from living organisms. Carbon dioxide is released when these fuels are burned.

Rocks in the Lower Crust and Upper Mantle Volcanic eruptions release carbon dioxide from the lower crust and upper mantle into the atmosphere. Volcanic activity is high during periods of rapid seafloor spreading and subduction. Thus, large quantities of carbon dioxide are released when lithospheric plates are moving rapidly. Some geologists also contend that carbonate rock and sediment are carried into the mantle during subduction, although this inference is under debate.

Life as we know it exists only in a narrow temperature range. Our atmosphere distributes and absorbs heat to make the Earth habitable. However, atmospheric composition is delicately balanced by many processes. If oceanographic, geological, or biological conditions changed, climate would also change, with profound implications for the Earth's plants and animals.

17.6 Climate Change and Human Activity

James Watt began manufacturing practical steam engines in 1775, and by the mid-1800s the industrial revolution was well under way. During the past century and a half, human population has increased dramatically, and people have altered the character of the planet by building cities, clearing forests, and burning fuels. Carbon dioxide, methane, chlorofluorocarbons, and several other air pollutants absorb infrared radiation and are collectively called the greenhouse gases (Fig. 17–27).* Carbon dioxide is the most abundant greenhouse gas because it is released whenever fuels are burned. Global atmospheric carbon dioxide increased from about 290 parts per million (ppm) in 1870 to 350 ppm in 1990. Half of this increase has

*Although water vapor absorbs infrared radiation, it is not considered a greenhouse gas because its concentration is not altered appreciably by human activity.

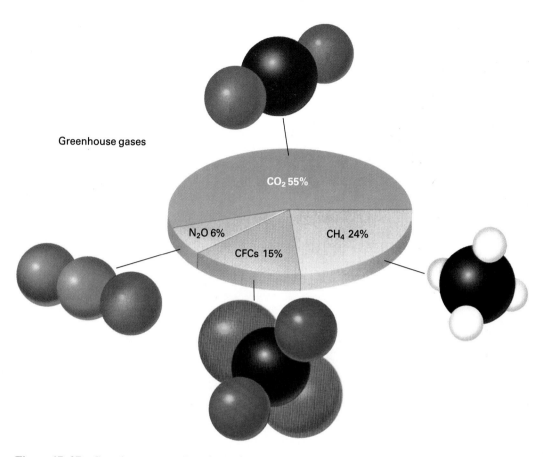

Figure 17–27   Greenhouse gases. Drawings of molecular models of the gases are included.

occurred in the past 30 years (Fig. 17–28). However, the other greenhouse gases are significant and their concentrations have also been increasing.

Between 1880 and 1991, global temperature has oscillated, with an overall upward trend. 1990 was the warmest year in the past century and every year between 1980 and 1991 was warmer than average (Fig. 17–29). As documented in Section 17.5, climate variation has occurred throughout the history of the planet. Compared to past changes, a temperature rise of a few tenths of a degree over a few decades is minor and could be attributed to any of a number of mechanisms. Yet many scientists

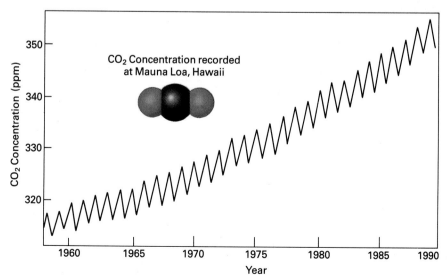

Figure 17–28   Increase in atmospheric carbon dioxide concentration from 1960 to 1990. The small oscillations in the curve are caused by seasonal changes in plant respiration and photosynthesis. *(Mauna Loa Observatory)*

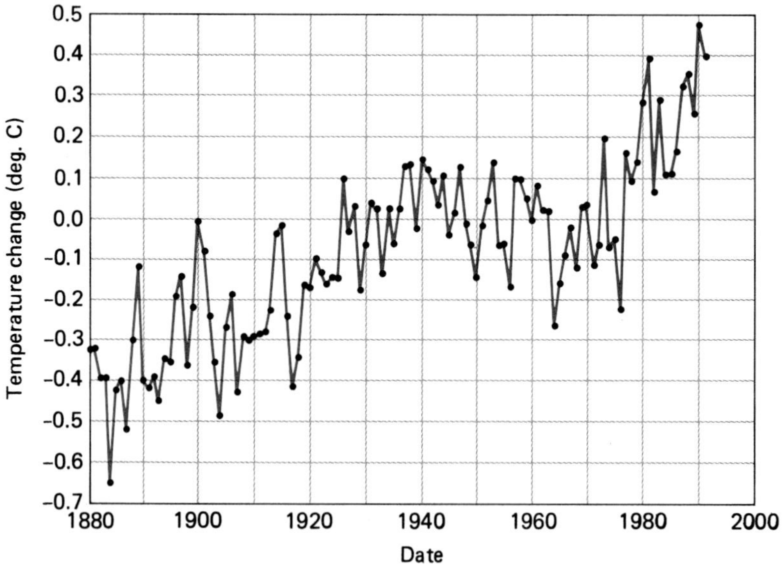

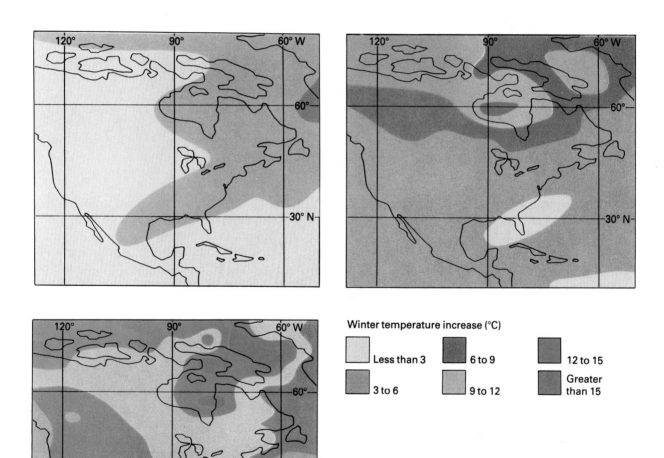

Figure 17–29 Average global temperature from 1880 to 1991. The zero line represents the average from 1951 to 1980 and plus or minus values represent deviations from the average. *(NASA/Goddard Space Center)*

Winter temperature increase (°C)

Less than 3
3 to 6
6 to 9
9 to 12
12 to 15
Greater than 15

Figure 17–30 Three estimates for increase in global temperature in response to a doubling of atmospheric carbon dioxide. Note that the estimates vary considerably, but all three agree that warming will occur. *(A) National Center for Atmospheric Research, (B) NASA/Goddard Institute for Space Studies, (C) United Kingdom Meteorological Office.*

are concerned because it is possible that human activities are altering climate, and these alterations may, in turn, affect the human condition and the survival of other species.

In recent years scientists have been debating the magnitude of the greenhouse effect and possible solutions to it. In this debate, it is important to evaluate the reliability of the data, and to distinguish fact from inference.

The absorptive properties of greenhouse gases can be measured precisely in the laboratory, and their concentrations in the atmosphere have been measured accurately in field stations for many years. Thus, scientists are certain that greenhouse gases absorb infrared radiation and that their concentrations in the atmosphere have increased during the past century.

However, temperature data such as those shown in Figure 17–29 are open to dispute. Most weather stations are located in or near cities. But recall that urban areas are generally warmer than surrounding countryside. Therefore, the urban heat island effect could bias the data. It is simple to move recording stations away from cities and thus make *present and future* data more reliable, but past data remain suspect. Another objection to using past data to interpret global temperature trends is that not enough recording stations exist in the oceans, the Southern Hemisphere, the Arctic, the Antarctic, and other remote locations. Researchers are aware of these problems and try to compensate for them. Still another problem in interpreting global temperature data is that the observed

warming trend could be caused by some factor other than the greenhouse effect.

Thus, no one can prove that human activity is causing greenhouse warming. Two opposing viewpoints have emerged. One side argues that since we cannot prove that greenhouse warming is occurring, we should study the problem further but delay action. The counter argument is that people frequently act on the basis of incomplete proof. If a mechanic told you that your brakes were faulty and likely to fail within the next 1000 miles, you would recognize this as an opinion, not a fact. Yet, would you wait for proof, or replace the brakes now?

Three different predictions of the magnitude of global warming are presented in Figure 17–30. Despite the large variability in the predictions, all three models project warming. If global temperature were to increase, productive agricultural areas could suffer droughts, and abundant rain might fall in locations that are now desert (Fig. 17–31). Global warming could disrupt natural ecosystems, leading to an increased rate of species extinction. The Antarctic and Greenland ice caps could melt, causing a rise in sea level.

Most air pollutants are byproducts of incomplete combustion of fuels and can be controlled or eliminated if we are willing to spend the money to install pollution control devices. However, carbon dioxide is released *whenever* fossil fuels are burned. It is produced both in the smoky, sooty fire built by a Nepali family to cook dinner and in the cleanest experimental engine in the

Figure 17–31 One model for possible changes in rainfall patterns from global warming. The green depicts wetter areas; the brown shows areas that will become drier. Note that if this model is correct, the Southwest, parts of the central plains, the Southeastern United States will become drier and other regions will become wetter. *(Gerald A. Meehl and Warren M. Washington, Journal of Atmospheric Sciences, 45, no. 9, (May 1988)*

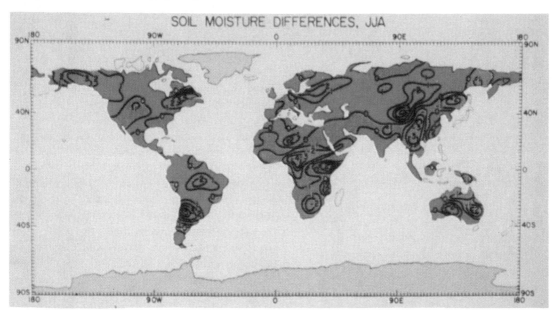

General Motors research laboratories. Therefore, the only way to reduce carbon dioxide emissions is to burn less fuel. One approach is to make do with less. If you turn down the thermostat in winter and wear a sweater, or turn the air conditioner off in summer, or ride your bicycle to work, you burn less fuel and release less carbon dioxide. Another approach is to increase the efficiency of cars, buildings, appliances, and factories. A car capable of traveling 40 miles per gallon of gas gets you to work as quickly as one that travels 20 miles per gallon, but the more efficient vehicle releases half as much carbon dioxide. A third approach is to use alternative forms of energy. Solar, hydroelectric, wind, geothermal, and nuclear energy sources do not operate by combustion and do not release greenhouse gases.

In 1991, representatives of 130 nations met under the auspices of the United Nations to discuss an international treaty to reduce emissions of greenhouse gases. Eighteen European countries and Japan have agreed to stabilize or reduce carbon dioxide output. The U.S. position is ambiguous. Administration officials have agreed, in principle, to stabilize output of greenhouse gases at 1987 levels, but have offered no legislation aimed at achieving the goal.

How much will it cost to reduce emissions of greenhouse gases? According to the U.S. Office of Technology Assessment (OTA), it would cost nothing to make small reductions in carbon dioxide emissions because the cost of conservation measures would be balanced by the money saved on fuel. A more ambitious program to reduce emission of greenhouse gases by 20 to 30 percent would cost between $20 and $150 billion per year, or about $75 to $550 per person per year. However, critics of the OTA calculations argue that fuel savings would be so great that a 20 percent reduction in carbon dioxide would *save* $200 billion per year in the United States, or $750 per person per year.

To make the debate over climate change and human activity even more complex, greenhouse gases are not the only pollutants capable of altering climate. The recent article in *Science* argues that dust, smoke, soot, and liquid droplets released during the combustion of fossil fuels have enveloped the earth in a slightly opaque shield.* These air pollutant aerosols reflect sunlight and are expected to cool the Earth about as much as greenhouse gases will warm it. Thus, possibly, the two effects will cancel. This premise was published just as *Earth Science and the Environment* was going to press and we can expect lively debate on the subject.

*R. J. Charlson, S. E. Schwartz, J. M. Hales, R. D. Cess, J. A. Coakley, Jr., J. E. Hanson, and D. J. Hofmann, "Climate Forcing by Anthropogenic Aerosols," *Science* 255 (1992): 423.

SUMMARY

•

The **three-cell model** of global wind circulation shows three cells of global air flow bordered by alternating bands of high and low pressure.

Tropical Cell   Warm air rises near the equator, forming a low-pressure region called the **doldrums**. This air flows north and south at high altitudes until it cools and falls at 30° latitude, forming a high-pressure region. The falling air splits. A portion flows back toward the equator along the surface, forming the **trade winds** and completing the tropical cell.

Mid-Latitude Cell   The remaining air that falls at 30° latitude flows poleward along the surface. In the Northern Hemisphere, this air is deflected to form the **westerlies**. Air rises at a low-pressure region near 60° latitude and returns at high altitude to complete the cell.

Polar Cell   A second region of high pressure arises over the poles, where the descending air spreads outward to form the polar easterlies. The **polar front** forms where the polar easterlies and the prevailing westerlies converge. The jet stream blows at high altitude along the polar front and along the boundaries between cells at the horse latitudes.

The climate in any particular region is determined by the seasonal cycles of solar radiation, temperature, wind, cloudiness, humidity, and precipitation. In turn, these atmospheric conditions are controlled by latitude, altitude, the oceans, pressure zones and winds, and mountain ranges.

Climate zones are classified according to annual temperature, annual precipitation, and variability in either of these factors from month to month or season to season. A **biome** is a community of plants that live in a large geographic area characterized by a particular climate. Biomes are determined mainly by climate.

The air over cities is generally warmer and wetter than that over the surrounding countryside.

Five natural factors alter global climate: changes in solar radiation, changes in the Earth's orbit around the Sun, catastrophic events such as volcanic eruptions and meteorite impacts, changes caused by tectonic events, and changes in atmospheric composition.

Increased concentrations of carbon dioxide, methane, and other gases absorb infrared radiation from the Earth and may thus cause a warming known as the **greenhouse effect**.

KEY TERMS

•

REVIEW QUESTIONS

•

1. Explain the one-cell model and the three-cell model for global circulation. Discuss limitations of each.

2. Describe the trade winds. Why are they so predictable?

3. Why is the doldrum region relatively calm and rainy? Why are the horse latitudes calm and dry?

4. Describe the polar front and the jet stream. How do they affect weather?

5. How is climate affected by latitude, altitude, oceans, and mountain ranges?

6. Discuss the effect of the oceans on world climate. Are coastal regions always warmer than inland areas? Explain.

7. List the Earth's climate zones and the biomes associated with each.

8. How does urban climate differ from that of the surrounding countryside? What factors are responsible for these differences?

9. List the five major natural factors that cause climates to change. Describe the mechanisms of each, and briefly estimate the time required for each factor to affect climate.

10. Describe the greenhouse effect.

11. Discuss the difficulties and costs involved in reducing carbon dioxide emissions.

DISCUSSION QUESTIONS

•

1. Explain why a wind moving from south to north in the Northern Hemisphere is deflected by the Coriolis effect until it is moving due east.

2. Would the exact location of the doldrum low-pressure area be likely to change from month to month? From year to year? Explain.

3. Sailors traveling in the Northern Hemisphere expect to incur predictable winds from the northeast between about 5°N and 30°N latitudes. Should airplane pilots expect northeast trade winds while flying at high altitudes in the same region? Explain.

4. Why doesn't the air that is heated at the equator continue to rise indefinitely?

5. Predict the climate at the following locations: (a) at 45°N latitude, 200 kilometers from the ocean, on the leeward side of a mountain range; (b) at 45°N latitude, on a coastline influenced by warm currents; (c) on the equator in a continental interior; (d) at 30°N latitude in a continental interior; (e) at 30°N latitude, influenced by warm currents.

6. What climate zone do you live in? What biome thrives in this climate?

7. Discuss how the urban heat island effect alters (a) global agriculture, (b) personal comfort, and (c) energy consumption.

8. Which of the following gases can contribute to the greenhouse effect by absorbing infrared radiation? If the gas is an absorber, identify its possible sources. (a) Carbon dioxide; (b) oxygen; (c) methane; (d) nitrogen; (e) oxides of nitrogen; (f) chlorofluorocarbons.

9. Explain why a small change in average global temperatures can have environmental, economic, and political effects.

10. If you were running for Congress, what proposals, if any, would you make for reducing carbon emissions?

11. Two hundred million years ago, all the continents were collected into a single land mass. Speculate on how this concentration of continents would affect global climate.

Air Pollution

18

In 1948 Donora was an industrial town of about 14,000, located 50 kilometers south of Pittsburgh, Pennsylvania. One large factory manufactured structural steel and wire, and another produced zinc and sulfuric acid. Anyone familiar with the hill country of Pennsylvania knows that foggy days are common there, especially in the fall. During the last week of October 1948, an unusually dense fog settled over the town and the air remained calm for several days. But it was no ordinary fog; the moisture mixed with pollutants emitted by the two factories. After four days the air became so thick that people could not see well enough to drive, even at noon with their lights on. Gradually at first, and then in increasing numbers, residents sought medical attention for nausea, shortness of breath, and constrictions in the throat and chest. On the morning of the fifth day, a retired steel worker died of respiratory problems, and several deaths followed in rapid succession. Within a week 20 people

● Air pollution from a paper mill in Prince Rupert, British Columbia.

491

had died and about half of the residents of the town were seriously ill.

The incident at Donora shocked people into realizing that air pollution can be harmful and even deadly, and that factories should be shut down when the air becomes too unhealthful. During the past 45 years, federal and state governments have enacted laws to control air pollutants and to assure us that Donora-type incidents will not recur. However, a sense of unease remains. You may look outside and see a brownish haze obscuring nearby hills. The morning news states that pollution levels are higher than the law allows. The daily newspaper carries reports on acid rain and destruction of the ozone layer. You ask yourself, "Are current legislation and enforcement sufficient? How healthful is the air I breathe?"

This chapter discusses the sources of air pollution and its effects on human health.

18.1 Sources of Air Pollution

Gases Released When Fuels Are Burned

Air pollution from human activity is as old as our ability to start a fire. But large-scale air pollution from industry is a relatively recent development. Since the industrial revolution began in about 1750, the major fuels in the developed areas of the world have been coal and petroleum. Coal is largely carbon, which, when burned completely, produces **carbon dioxide**, CO_2. Petroleum is a mixture of **hydrocarbons**, compounds composed of carbon and hydrogen. When hydrocarbons burn completely, they produce carbon dioxide and water. Neither is poisonous, but both are greenhouse gases. If fuels were composed purely of compounds of carbon and hydrogen, and if they always burned completely, air pollution from burning of fossil fuels would pose little direct threat to our health. However, fossil fuels contain impurities, and combustion is usually incomplete. As a result, other products form.

Products of incomplete combustion include hydrocarbons such as **benzene** and **methane**. Benzene is a carcinogen (a compound that causes cancer), and methane is another greenhouse gas. Combustion of fossil fuels releases many other hydrocarbon pollutants including **carbon monoxide**, CO, which is colorless and odorless yet very toxic.

Additional problems arise because coal and petroleum contain impurities that generate other kinds of pollution when they are burned. Small amounts of sulfur are present in coal and, to a lesser extent, in petroleum. When these fuels burn, the sulfur forms **oxides**, mainly **sulfur dioxide**, SO_2, and **sulfur trioxide**, SO_3. High sulfur dioxide concentrations have been associated with major air pollution disasters of the type that occurred in Donora. Today the primary global source of sulfur dioxide pollution is coal-fired electric generators.

Nitrogen, like sulfur, is common in living tissue and therefore is found in all fossil fuels. This nitrogen, together with a small amount of atmospheric nitrogen, reacts when coal or petroleum is burned. The products are oxides of nitrogen, mostly **nitrogen oxide**, NO, and **nitrogen dioxide**, NO_2. Nitrogen dioxide is a reddish brown gas with a strong odor. It therefore contributes to the "browning" and odor of some polluted urban atmospheres. Automobile exhaust is the primary source of nitrogen oxide pollution.

Mud from lakes and marshes in northern Minnesota contains toxophene, a pesticide used on cotton in the south.

Volatiles

A **volatile** compound is one that evaporates readily and therefore easily escapes into the atmosphere. Whenever chemicals are manufactured or petroleum is refined, some volatile byproducts escape into the atmosphere. When metals are extracted from ores, gases such as sulfur dioxide are released. When pesticides are sprayed onto fields and orchards, some of the spray is carried off by wind. When you paint your house, what happens to the solvents, the volatile parts of the paint? They evaporate into the air. As a result of all these processes, tens of thousands of different volatile compounds are present in polluted air. Some are harmless, others are poisonous, and many have not been studied.

Northern Minnesota is far from any industrial center. Yet, in 1987, when scientists dredged mud from a lake in a pristine wolf and moose refuge, they found a pesticide called toxaphene. Toxaphene had been used primarily on cotton fields in the South, and had been banned in the United States in 1982. The only logical explanation for its presence in northern Minnesota was that it had been carried in by wind (Fig. 18–1).

Particulate Matter

Whenever you build a fire, smoke rises above the flames. Smoke consists of **particles**.* In air pollution terminology, a particle is any pollutant larger than a molecule. Molecules never settle out of still air, but particles do.

*The terms ''particle'' and ''particulate'' are sometimes used interchangeably.

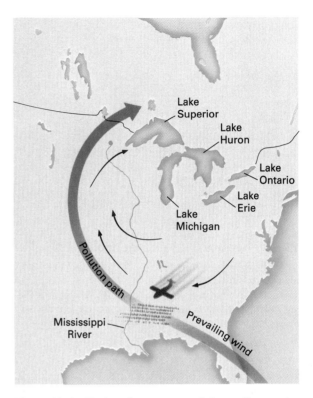

Figure 18–1 Toxic pollutants are carried vast distances by the wind. Pesticides from the gulf states were found in a moose and wolf refuge in Minnesota and in lakes in southern Canada.

The settling rate may be so slow, however, that even slight winds keep particles aloft (Fig. 18–2). Particles called soot consist mostly of carbon and are released whenever fossil fuel is burned. Soot is carcinogenic. Coal

Particles sampled from coal-fired (A) and oil-fired (B) power plants. These photographs were taken with a scanning electron microscope. *(Atmospheric Sciences Research Center, State University of New York at Albany)*

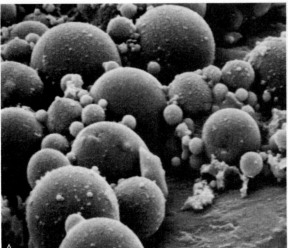
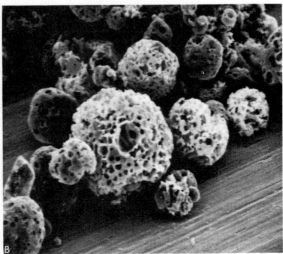

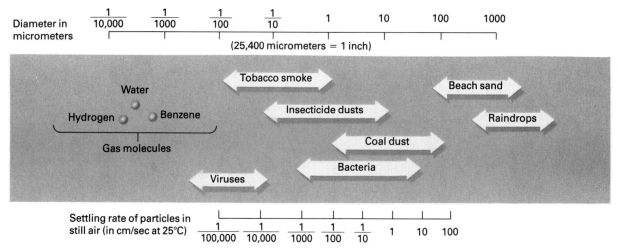

Figure 18–2 Small particles in air. Note that the size scale, shown in micrometers (mu), is not linear but progresses by factors to ten. The largest diameter shown, 1000 mu, is equal to 1 mm; it is the size of a large grain of sand or a small raindrop. The lower scale tells how rapidly or slowly particles settle in still air. Very small particles can be kept aloft indefinitely by wind currents.

always contains noncombustible minerals that accumulated when the coal formed in the muddy bottoms of ancient swamps. When the coal burns, some of these minerals escape from the chimney as **fly ash**, which settles out as gritty dust. When metals are mined, the drilling, blasting, and digging raise dust, and this, too, adds to the total load of particulates. Some particles settle out quickly, but others are transported great distances by wind.

Secondary Air Pollutants

Gases and particulates released during combustion and manufacturing are called **primary air pollutants** (Fig. 18–3). **Secondary air pollutants** are not released directly into the air by any industrial process, but are generated by reactions within the atmosphere. Two examples are smog and acid precipitation.

Figure 18–3 (A) Sources of air pollution in the United States. (B) Types of pollutants in the United States. (Note that carbon dioxide is not on this chart. Even though carbon dioxide affects the environment as a greenhouse gas, it is not generally listed as a pollutant because it is not toxic.) *(Source: EPA)*

Sources of pollution in the United States

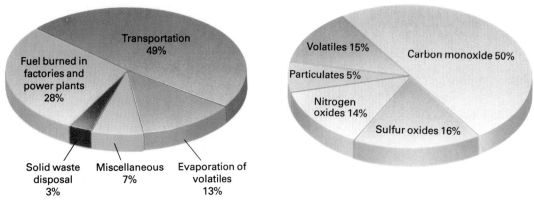

A   B

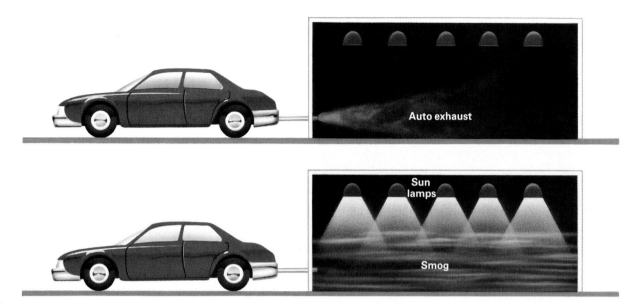

Figure 18–4   Smog is produced when automobile exhaust is exposed to sunlight.

18.2 Motor Vehicles and Smog

By the early 1900s, many industrial cities were heavily polluted. The major sources of pollution were no mystery. The burning of coal was number one. Other sources, such as steel mills and copper smelters, were readily identifiable. The most abundant air pollutants were sulfur dioxide, soot, and fly ash. When pollution was heavy, the air was dark. Black dust collected on window sills and shirt collars, and new-fallen snow did not stay white for long.

Imagine that your great grandfather had entered the exciting new business of making moving pictures. Old-time photographic film was "slow" and required lots of sunlight, so he would hardly have moved to Pittsburgh. Southern California, with its warm, sunny climate and little need for coal, was preferable. Thus, a region of Los Angeles called Hollywood became the center of the movie industry. Its population boomed, and after World War II automobiles became about as numerous as people. Then the quality of the atmosphere began to deteriorate in a strange way. Four effects were noted. (1) A brownish haze called **smog** settled over the city; (2) people felt irritation in their eyes and throats; (3) vegetable crops became damaged; and (4) the sidewalls of rubber tires began to crack.

In the early 1900s air pollution experts worked mostly in the industrialized regions of the East Coast and the Midwest. When they were called to diagnose the problem, they looked for the sources of air pollution they knew well, especially sulfur dioxide. But they could not find much sulfur dioxide in the air, and the smog was nothing like the pollution they were familiar with. Then, in the early 1950s, a California chemist, A. J. Haagen-Smit, proved that Los Angeles smog forms when automobile exhaust is exposed to sunlight.

Haagen-Smit piped automobile exhaust into a sealed room equipped with sunlamps (Fig. 18–4). The room contained plants and pieces of rubber. It was fitted with mask-like windows that permitted people to stick their faces in and smell the air. With the automobile exhaust in the room but the sunlamps off, the room smelled like automobile exhaust, not like Los Angeles smog. But when the sunlamps were turned on, smog developed. After a time, the air brought tears to the eyes, the plants showed typical smog damage, and the rubber cracked.

Further research showed that incompletely burned compounds in automobile exhaust react with nitrogen oxides in the presence of sunlight and atmospheric oxygen to form **ozone**. The ozone then reacts further with automobile exhaust to form smog (Fig. 18–5).

We read about the harmful effects of excessive ozone in the air over cities such as Los Angeles. We also read about harmful effects caused by depletion of the ozone layer in the upper atmosphere. So is ozone a pollutant, or a beneficial component of the atmosphere that we want to preserve? The answer is that it is both, depending on *where* it is found. Ozone exists in both the lower atmosphere (the troposphere) and the upper atmosphere (the stratosphere). Ozone in the troposphere reacts with

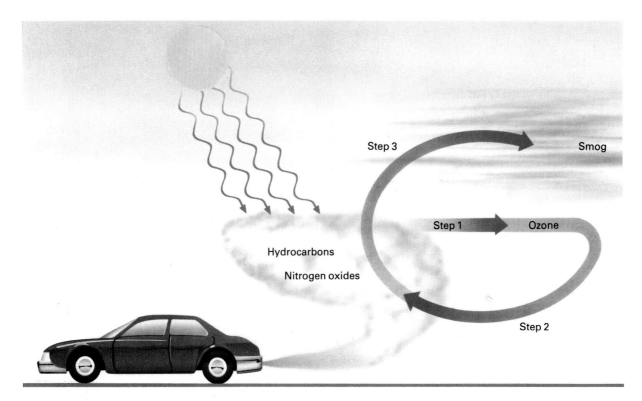

Figure 18–5 Smog forms in a multistep process that includes the following. *Step 1:* Automobile exhaust reacts with air in the presence of sunlight to form ozone. *Step 2:* Ozone reacts with automobile exhaust to form *Step 3:* Smog.

automobile exhaust to produce smog. As a result, ozone in the lower atmosphere is a pollutant. Ozone in the stratosphere absorbs ultraviolet radiation. Therefore, stratospheric ozone is a shield that protects life on Earth. (The depletion of stratospheric ozone and its effects were discussed in Chapter 15.)

18.3 Acid Precipitation

Recall that sulfur dioxide (SO_2) and oxides of nitrogen (NO and NO_2) are released when coal and petroleum are burned. Sulfur dioxide reacts in moist air to produce sulfuric acid. Similarly, oxides of nitrogen react in air to produce nitric acid. Both sulfuric and nitric acids are strong acids. They dissolve in water droplets in the atmosphere and fall to Earth as **acid precipitation** (Fig. 18–6).

Acidity is measured on the **pH scale**. A solution with a pH of 7 is neutral, neither acidic nor basic. Numbers lower than 7 represent acidic solutions, and numbers higher than 7 represent basic ones. For example, soapy water is basic and has a pH of about 10, while vinegar is an acid with a pH of 2.4.

Under normal conditions, atmospheric carbon dioxide dissolves in rainwater and reacts to form a weak acid. Thus, the pH of "pure" rainwater is a little less than 6. But in recent years industrial emissions of nitrogen and sulfur oxides have increased the acidity of rainfall in many parts of the world (Fig. 18–7). Much acid rainfall is between pH 4 and pH 5, but more severe episodes occur occasionally. A rainstorm in Baltimore in 1981 had a pH of 2.7, which is about as acidic as vinegar, and a fog in southern California in 1986 reached a pH of 1.7, which approaches the acidity of hydrochloric acid solutions used to clean toilets.

18.4 Meteorology of Air Pollution

To assess the extent to which your health is affected by any specific air pollutant—say, sulfur dioxide—you must be concerned with both the amount that you inhale and the effects of that pollutant on your body. The total quantity of sulfur dioxide in the Earth's atmosphere and the concentration in the smokestack of a coal-fired electric generator do not affect you directly because you do not breathe all the world's air, nor do you stick your head in the

Earth Science and the Environment

Air Pollution in Los Angeles

For many, the "California dream" of sunny skies, open spaces, and opportunity has been replaced by long hours of commuting on smoggy, crowded freeways. Every year 1.7 million cars and light trucks are sold in California, more than are sold annually in all but seven nations in the world. The Los Angeles metropolitan area has 14 million residents, 8 million automobiles, and about 25,000 industrial and commercial emission sources, mostly petroleum refineries, factories, and chemical plants. About 9000 tons of air pollutants escape into the air *every day.*

Air pollutants frequently concentrate during atmospheric inversions. Abundant sunlight favors smog formation. As a result, Los Angeles leads the United States in air pollution. In 1988, one or more EPA air quality standards were violated on 219 days—or about two thirds of the time—somewhere in the Los Angeles basin. On the smoggiest days, ozone readings were three times the allowable level. High ozone levels cause respiratory and reproductive problems in humans and are responsible for $300 to $500 million in crop damage in southern California. Laboratory and epidemiological studies predict that chemically contaminated air will cause cancer in about 30,000 of California's 27 million people.

In the late 1980s the people of southern California adopted a sweeping anti-pollution law that enforces stricter standards than those mandated by federal law. The objective is to reduce air pollution in the L.A. area by 70 percent by the year 2000. This goal can be achieved both by reducing the emissions of internal combustion engines and by reducing the number of miles driven.

Most of the inexpensive pollution control mechanisms have already been added to new automobiles, and additional emission control will be increasingly expensive. Nevertheless, the law mandates a 50 percent emission reduction by 1994–96 and further reduction after that. Emission could be reduced immediately by cutting down the miles driven, through car pooling and the use of public transportation. By the mid-1990s, electric cars will start to carry commuter traffic. Electric cars emit no exhaust directly, but much of the electricity that powers them is derived from fossil-fuel-burning generators that do pollute. Taking the power plant emissions into account, electric cars emit about one tenth as much pollution as internal combustion cars do.

Implementation of the entire plan will cost up to $15 billion a year over the next 10 to 20 years. However, a 1987 survey showed that 73 percent of the state's citizens were willing to pay dramatically increased costs for better air, and in recent years many of California's voter initiatives dealing with improving the environment have been approved.

Los Angeles smog.

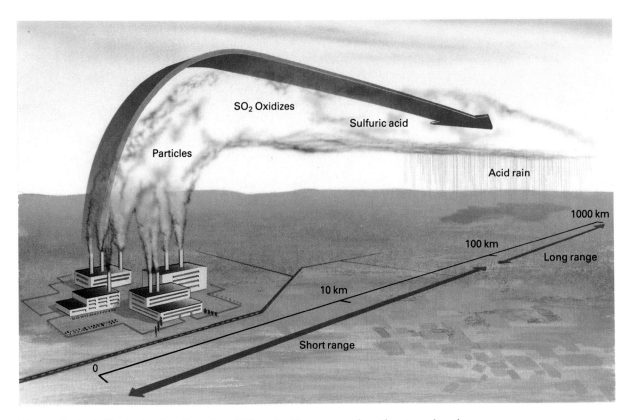

Figure 18–6 Acid rain develops from the addition of sulfur compounds to the atmosphere by industrial smokestacks.

Figure 18–7 Increases in acid precipitation in the eastern United States between 1955 and 1984. *(Source: World Resources Institute)*

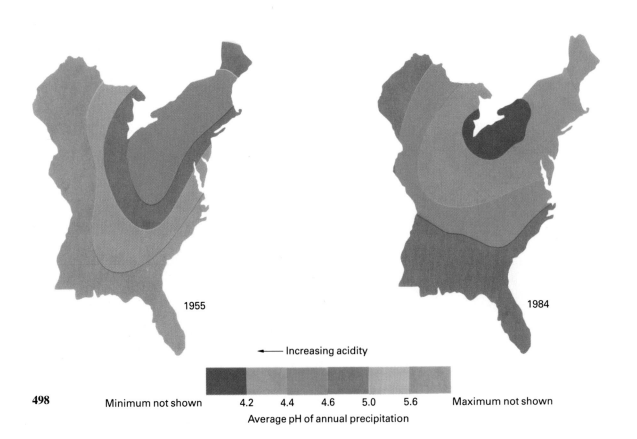

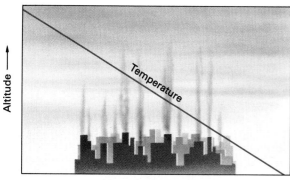

Figure 18–8 (A) Under normal atmospheric conditions, the temperature decreases continuously with altitude (red line). Warm polluted air rises through the cooler high-elevation air, and the pollutants disperse. (B) Smoke from this burning garbage dump disperses because air temperature decreases with altitude.

smokestack. Therefore, you must consider how pollutants are transported and diluted in the atmosphere.

Exhaust from an industrial stack or automobile tail pipe mixes with the surrounding air rapidly if the atmosphere is turbulent, or slowly if the air is calm. Rapid, extensive mixing dilutes the pollutants so that the concentration you breathe is lower. When little mixing occurs, the pollutant concentration remains high.

Under normal conditions, the Sun warms the Earth, which in turn warms the lower atmosphere. The warm air rises to mix with the cooler air above it, creating turbu-

lence and diluting pollutants near the ground (Fig. 18–8). At night, however, the ground radiates heat and cools. Thus, the ground and the air close to the ground become cooler than the air at higher elevations (Fig. 18–9). This condition, called an **atmospheric inversion** or **temperature inversion**, is stable because the cool air is denser than warm air; therefore, air does not mix. The stagnant layer of cool air next to the ground may be ten or a few hundred meters thick.

Frequently, as morning sunlight heats the ground, air above the ground is warmed enough to rise and break up

Figure 18–9 (A) During an inversion, temperature increases with altitude then decreases at higher elevations (red line). Polluted air cannot rise above the layer of warm air and remains trapped near the ground. The effect is intensified if the city lies in a valley that confines air movement. (B) A temperature inversion holds effluent from this paper mill close to the ground, forming a smoggy haze.

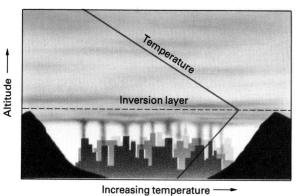

the inversion layer. However, under certain conditions the inversion may persist for days. A large, slow-moving high pressure may sweep warm air in at high altitude and hold it in place above the colder air near the ground. The episode at Donora was caused by such a high pressure. Inversions are also common along coastlines and large lakes when surface air is cooled by the water. The Los Angeles basin is a densely populated region surrounded by mountains. Maritime air, cooled by cold ocean currents, blows in from the Pacific Ocean. In addition, subtropical highs bring stable, stagnant air. The cool maritime air is trapped beneath the warm subtropical air. It cannot move eastward because of the mountains. It cannot rise because it is too dense. Pollutants from the city's automobiles and factories then concentrate until the stagnant air becomes unhealthful to breathe.

Two types of weather changes can break up an inversion. If the Sun heats the Earth's surface sufficiently, the cool air near the ground warms and rises, dispersing the pollutants. Alternatively, if a storm moves in, it causes vertical mixing that breaks up the inversion layer.

18.5 Effects of Air Pollution on Human Health

You step outside and notice a line of brownish haze, the air smells foul, and a nasty cough has lingered for the past week. You wonder if there is a connection between your persistent cough and that brown haze.

As explained at the beginning of the chapter, when air pollution concentrates enough, people sicken and die. But what about the effects of moderately polluted air?

Two types of studies are used to determine the health effects of air pollution: laboratory tests and epidemiological surveys.

Laboratory Tests

It is unethical to perform laboratory tests on people, so scientists use rats. However, it is too expensive to use millions of rats so scientists substitute fewer rats and higher doses. Suppose that a large proportion of an experimental rat population develops cancer when exposed to 10,000 times as much compound X (in proportion to body size) as you would inhale by breathing the air around your house. What does this result mean? (1) It may be valid to **extrapolate** the data and state that a smaller amount of compound X would cause *a few* rats to develop cancer. However, such extrapolation may not be valid. For example, very high doses of ordinary table salt cause body malfunctions, whereas low doses are essential to survival. (2) It may be valid to infer that, after correction for differences in body weight, the chances of a tumor in a rat and in a human are equal. But such an assumption

may be incorrect, because rats and people are different species. As a result of these uncertainties, laboratory studies are never conclusive.

Epidemiology

Epidemiology is the study of the distribution and determination of health and its disorders. One way to test a suspected pollutant is to separate a population into two groups: one that is exposed to it, and one that is not. A classic example of this type of analysis is the study of cigarette smokers. Since a clear distinction exists between people who smoke and those who do not, it is easy to examine the medical records (especially death certificates) of large numbers of people in these two groups. The knowledge that cigarette smoking is harmful has come from many studies of this kind.

Determining the effects of a widely dispersed air pollutant is more difficult. It is impossible to find a group of people who have had zero exposure to such a pollutant, so there is no way to compare the health records of those who are exposed to the pollutant with those who are not. The most widely used approach is to compare people with high levels of exposure to those with lower levels of exposure.

Workers in some industries are exposed to high concentrations of specific pollutants. In another classic epidemiological study, one quarter of the male employees working in a coal-tar dye factory in the United States between 1912 and 1962 contracted bladder cancer. This observation led to the conclusion that several compounds in coal-tar dyes are carcinogenic.

Epidemiologists also compare populations in polluted areas with those who live in less polluted environments. For example, 13 major chemical plants in the Kanawha Valley in West Virginia release a variety of toxic air pollutants. Between 1968 and 1977 the incidence of respiratory cancer in the region was more than 20 percent above the national average.

A third type of epidemiological study compares diseases in a specific region during periods of high and low air pollution contamination. In August 1986, a strike shut down the Geneva steel plant near Salt Lake City, Utah, for 13 months. When the mill shut down, the concentration of particulates in the air decreased, and the number of children hospitalized for bronchitis, asthma, pneumonia, and pleurisy declined by 35 percent. Thus, an epidemiological connection exists between particulate air pollution in Salt Lake City and respiratory diseases in children.

During 1987, 1.2 million tons of toxic air pollutants were released in the United States. Some of the major pollutants and their health effects follow (Fig. 18–10):

Sulfur oxides   Impair lung function, aggravating diseases such as asthma and emphysema.

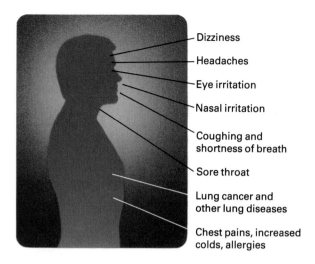

- Dizziness
- Headaches
- Eye irritation
- Nasal irritation
- Coughing and shortness of breath
- Sore throat
- Lung cancer and other lung diseases
- Chest pains, increased colds, allergies

Figure 18–10   Effects of air pollution on the human body.

Nitrogen oxides	Impair lung function and increase susceptibility to lung diseases. They also affect heart and liver and have been shown to increase vulnerability to viral infections such as influenza.
Carbon monoxide	Interferes with body's ability to absorb oxygen. In high doses, carbon monoxide leads to death through brain damage. In low doses, symptoms include impaired perception and thinking, drowsiness, and headaches. May threaten fetuses in pregnant women.
Ozone	Irritates the membranes of the respiratory system, causing loss of lung function. Also increases susceptibility to lung and heart diseases and is a suspected carcinogen.
Particulates	Irritate the bronchia and lung and can cause lung scarring, decreased lung function, and lung cancer.
Volatiles	Tens of thousands of different chemicals are released into the air every year. Many volatiles have been shown to be carcinogenic; others cause birth defects; some are directly poisonous. Most have not been studied.

18.6   Effects of Air Pollution on Vegetation and Materials

Air pollution causes widespread damage to trees, fruits, vegetables, and ornamental flowers. The most dramatic early example of plant damage was the total destruction of vegetation by sulfur dioxide in areas near smelters (Fig. 18–11). Air pollutants damage plants in several ways. Smog impedes the growth of leafy plants such as spinach, lettuce, alfalfa, and tobacco. California navel

---

● MEMORY DEVICE

The abbreviation ppm stands for parts per million. If a dissolved ion such as calcium is present in water at a concentration of 1 ppm, there is 1 kilogram of calcium for every 1 million kilograms of water. Similarly, ppb stands for parts per billion.

Is 1 ppb a large concentration or a small one? One billion pennies laid out rim to rim in a straight line would extend almost halfway around the Earth's equator. If one of these pennies were a calcium ion, it would indeed be hard to find. This example makes 1 ppb sound small. Now think of it in another way. If the calcium ion concentration were 1 ppb, there would be 15,000,000,000,000 calcium ions in every cubic centimeter of water. This example makes 1 ppb sound large. Do not let either example fool you. For some substances, a concentration of 1 ppb is inconsequential; for others, it is highly significant.

orange trees exposed to 0.0625 parts per million (ppm)* of nitrogen dioxide for 290 days yielded 57 percent less fruit than those in unpolluted air. To put these numbers in perspective, the maximum allowable outdoor concentration of nitrogen dioxide under the Clean Air Act is 0.05 ppm. In 1988 this level was exceeded once a month

*See box for a more complete explanation of parts per million.

Figure 18–11   An eroded hillside near an abandoned copper smelter in Anaconda, Montana. This hillside was once covered by forest. The trees were cut for fuel for the smelter. The soil is so heavily polluted that revegetation has not occurred.

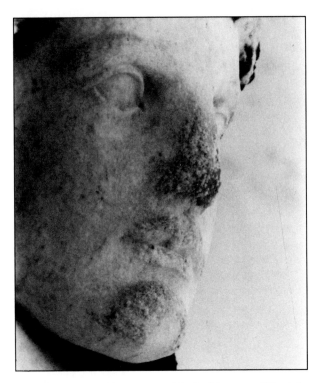

Facial features on this statue outside the Museum of Natural History in Chicago have been weathered by acid rain.   *(Wide World Photos)*

Germany were unhealthy. A year later, 34 percent of the trees were affected, and by 1985 more than half of the trees in West German forests were dying. Experiments indicate that damage caused by acid precipitation, combined with the effects of other air pollutants, caused the deaths of the trees. Forest decline has caused the loss of about $10 billion worth of timber in Germany alone. But the loss goes beyond economics. Forests provide recreation for people and habitat for many species; they regulate water and protect the soil. Tree death is not as rampant in the United States, but the U.S. Forest Service has reported that pines in the southeast grew 20 to 30 percent more slowly between 1972 and 1982 than they did between 1961 and 1971, when rain was less acidic.

Acid precipitation corrodes metal and rock. Recent air pollution has eaten away at ancient carvings, doing more damage in the past 20 years than occurred by natural weathering in the previous millennium. In some places, government officials have removed statues from their outdoor settings and sealed them in glass cases in museums. In the United States the cost of deterioration of buildings and materials from air pollution is estimated at several billion dollars per year.

in Los Angeles. Air pollution leads to an estimated $3 to $4 billion in crop damage every year in the United States.

In the late 1970s European foresters noticed that the needles and leaves of many trees were yellowing and that trees were dying. In the 1980s, tree deaths increased dramatically. In 1982, about 8 percent of the trees in West

18.7 Aesthetic Insults

A view of distant mountains through clear, fresh air is pleasing, and a brown haze is offensive. Unpleasant aesthetic effects cannot be neatly separated from other effects of air pollution. An acrid haze is perceived not only as a visual annoyance but also as a sign of possible physical harm, just as the taste of spoiled food tells you that you

Figure 18–12 Particulate air pollution levels that affect an individual during a typical working day. Note that the indoor pollution was significantly higher than outdoor pollution even in a busy, congested city.   *(After J. L. Repace and A. H. Lowery, "Tobacco Smoke, Ventilation, and Indoor Air Quality," Science 208 [1982]: 894–914).*

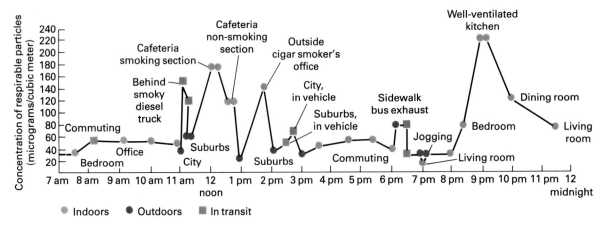

will experience food poisoning. Thus, pollution causes anxiety, and anxiety may depress our appetites or rob us of sleep, which in turn can be directly harmful.

18.8   Indoor Air

Many people spend little time outdoors, going only from home to car and from car to factory or office. Indoor air contains pollutants from outdoors that are drawn in by fans or leak through windows and doors, as well as pollutants that are generated indoors. Because of this double burden, indoor air is generally more polluted than outdoor air (Fig. 18–12). The largest indoor air pollutant is cigarette smoke. Other indoor sources include cooking vapors, aerosol sprays, house dust, dog and cat hair, body odors, and gases emitted by the building itself. Some poorly manufactured insulation decomposes in moist air, generating **formaldehyde**, a toxic gas. In addition, many stone, brick, and concrete structures, as well as most soils, contain tiny concentrations of radioactive materials that

release **radon**, a radioactive gas. In very tightly sealed houses, radon gas can build up to hazardous concentrations.

Measures that conserve heat by reducing air leaks usually increase indoor pollution by slowing down the escape of pollutants to the outdoors. However, you can protect the quality of your indoor air by several means, shown in Table 18–1.

18.9   Legislation and Public Policy in Air Pollution Control

The first air pollution laws were enacted in the fourteenth century in Great Britain. During the reign of Edward II (1307–27), a man was tortured for filling the air with a "pestilential odor" by burning coal. Later, less brutal regulation taxed and restricted the use of coal in congested areas such as London.

The major air pollution legislation in the United States is the **Clean Air Act**. As first written in 1963,

T A B L E 1 8 – 1

Measures to Protect Your Own Indoor Air Quality		
What You Can Do	**Advantages**	**Disadvantages**
Don't stop *all* air leaks—allow for a reasonable amount of ventilation.	Outdoor air replaces some indoor air, forcing pollutants outside.	Either (a) your fuel bills will be higher or (b) you will have to put up with cooler temperatures in winter and less air conditioning in summer.
Do not allow smoking indoors.	Less indoor pollution.	Inconvenient for smokers.
Insulate only with a high-grade material such as fibrous glass	Does not decompose readily.	Cannot be blown behind walls in existing homes.
After using paint, lacquers, glues, and so on, ventilate thoroughly until all odors are gone.	Removes large amounts of temporary indoor pollution.	Fuel bills will temporarily be higher.
After clothes are dry-cleaned, air them outdoors for a day before bringing them inside.	Much of the residual dry-cleaning solvent (trichlorethylene or perchlorethylene) evaporates.	Inconvenient; risk of rain.
If you use a wood stove, learn how to operate it so that it produces a minimal amount of smoke.	Reduces indoor pollution by particulate matter.	None.
If your building is so tightly sealed that some forced ventilation is needed, bring air in through an air-to-air heat exchanger.	Saves heating and air-conditioning costs.	None except for initial cost.
Use an indoor air purifier.	Saves energy compared with outdoor ventilation.	Some electrostatic indoor air purifiers generate ozone, which is toxic. Air cleaners based on adsorbers are effective, but the adsorber must be replaced when it becomes saturated.

this law was ineffective, so it has been amended and strengthened three times, in 1970, 1977, and 1990. As it stands today, the law addresses air pollution in two ways.

1. *Emission from individual sources.* The Clean Air Act limits the quantities of air pollutants that may be emitted by any source. Thus, it directs automobile manufacturers to meet standards for tail pipe emissions. Smokestack emissions from factories and power plants are also regulated. Executives who violate the law are subject to a fine of up to $25,000 *per day* and/ or imprisonment for up to one year.

2. *National Ambient Air Quality Standards.* The Environmental Protection Agency (EPA) is directed to monitor the purity of the air in general. Even if each factory, power plant, and automobile complies with the law, if the total pollution level exceeds National Ambient Air Quality Standards, then the EPA must set stricter controls.

The 1990 amendment focused on four major areas:

(a) **Sulfur dioxide and acid rain.** The new law requires that the total sulfur dioxide emissions be reduced to half the 1980 levels.

(b) **Smog.** Stricter guidelines have been established to reduce nitrogen oxides, ozone, and particulates and thereby reduce smog.

(c) In the 1970s, Congress directed the EPA to identify **volatile toxic pollutants** and enforce standards that would provide "an ample margin of safety to protect the public health from such hazardous air pollutants." During the following seven years, the EPA identified several hundred toxic volatile air pollutants but established standards for only seven. In the 1990 amendment, Congress specified 191 toxic volatiles to be regulated by the EPA.

(d) The new legislation also accelerates controls on **ozone-destroying chemicals** used in household and industrial appliances. To accomplish this goal, it sets stricter guidelines for disposal and recycling of chlorofluorocarbons, which deplete stratospheric ozone.

In view of all of these legislative safeguards, why do we still see and smell air pollution in many places? Consider a factory that emits air pollutants in violation of regulations. The company may claim that it needs more time to develop, test, and install the control systems required to reduce its emissions. EPA officials have been reluctant to close down an otherwise lawful business while such negotiations are in progress. Even after grace periods are exhausted, the EPA often grants additional extensions.

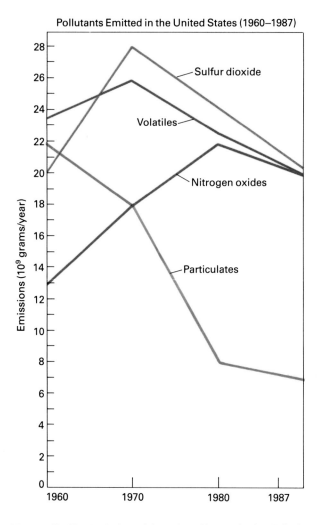

Figure 18–13 Emission of four air pollutants in the United States from 1960 to 1987. The Clean Air Act was first enacted in 1963. However, emissions of three of the four pollutants listed here increased during the 1960s. Stricter laws written in 1970 and 1977 led to a decline of emissions of all four pollutants. This graph does not show local concentrations in heavily congested areas such as Los Angeles. One purpose of the 1990 amendment is to reduce emissions in these areas. *(Source: EPA document 450/4-88-022)*

As a result, guidelines set by the Clean Air Act still have not been fully enforced (Fig. 18–13). According to the 1977 law, health standards were to be met by December 31, 1987. As the deadline approached, the law was violated regularly. The air in 60 cities was more polluted than the law allows at least 30 days per year. In Los Angeles, health standards were violated on an average of 219 days a year—two days out of three (Table 18–2). Rather than enforce the law, Congress extended the deadline on several occasions. In 1990, roughly half of the population of the United States breathed air at least once a week that did not meet National Ambient Air Quality Standards.

TABLE 18-2

Days per Year That Cities Failed the Clean-Air Standard	
Los Angeles metropolitan area	219
Bakersfield, Calif.	44
Fresno, Calif.	24
New York metropolitan area	17
Sacramento	16
Chicago	13
San Diego	12
Houston metropolitan area	12
Baltimore	11

18.10 The Cost of Pollution Control

Cars and factories can be designed to operate virtually pollution-free if we are willing to pay the price. In general, pollution control becomes more expensive as higher proportions of pollutants are removed. Limited pollution control can be relatively inexpensive, but an essentially pollution-free environment is very costly (Fig 18-14).

How much pollution control should we pay for? Some people suggest that pollution control should be applied only when there is a positive *economic* return on the investment. This approach, known as **cost-benefit**

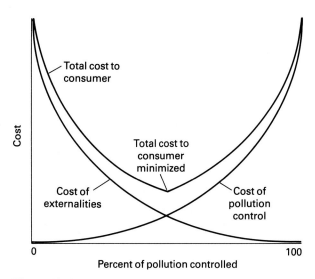

Figure 18-15 The relationship between the cost of pollution control and the cost of externalities. The actual numbers vary with the situation.

analysis, balances the monetary cost of pollution control against the monetary benefits of having the pollutants removed.

Imagine, first, that no pollution control devices were added to factories and vehicles, as shown on the left side of Figure 18-15. People would live in polluted environments similar to that in Donora in the 1940s. The cost of living in such a polluted environment includes medical bills, lost work, and damage to materials, crops, and livestock. It also includes indirect costs such as reduction in tourism and lowered land values if people no longer want to visit or live in a polluted area (Fig. 18-16). All of these costs are called *externalities*.

Pollution control costs money, but money is saved when pollution is reduced because the cost of externalities decreases. Note from Figure 18-15 that people pay both the cost of pollution control and the cost of externalities.

It is relatively inexpensive to eliminate 50 percent of the pollution and when this is done, the cost of externalities drops dramatically. Thus the total cost to the consumer decreases. As more efficient pollution control devices are added and more pollution is removed, the control cost rises rapidly and the cost of externalities decreases only slightly. In this scenario, shown by the right side of the graph, people benefit by living in a cleaner environment, but they must pay for it because the total cost to the consumer rises.

So, how much pollution control is desirable? Some people suggest we should minimize the total cost and thus accept considerable pollution. Others argue that money

Figure 18-14 The cost of reducing automobile emissions from 1985 levels. Notice that a 20 percent additional reduction in emissions would be relatively inexpensive, but after that further reductions would become quite costly. *(Source: Alan J. Krupnick and Paul R. Portney, "Controlling Urban Air Pollution: A Benefit-Cost Assessment," Science 252 [April 26, 1991]: 522ff)*

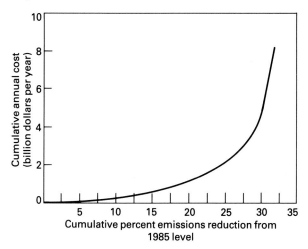

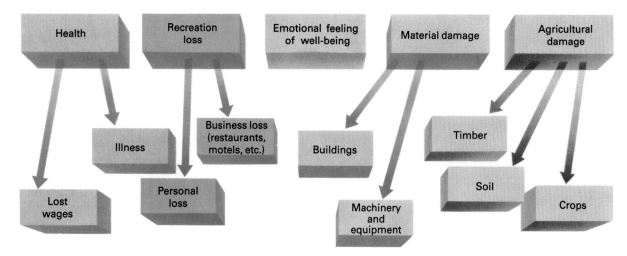

Figure 18–16   Some external and human costs of pollution.

is not the only measure of well-being, and that we would be happier if we paid more to live in a cleaner environment.

When Congress passed the latest amendments to the Clean Air Act in 1990, two scientists estimated that the cost of meeting the new ozone standards will be about $8.8 to $12.8 billion per year in the United States.* In comparison, the total benefits were estimated to be $250 million to $1 billion. Therefore, these authors concluded that the cost of compliance with the Clean Air Act is greater than the benefits.

Not everyone agrees with these conclusions. Both the cost of living in a polluted environment and the cost of pollution control are estimates. For example, how do you place dollar values on effects that are not bought and sold, such as the annoyance of a vile odor or a cough that persists but is not severe enough to warrant medical attention. In an attempt to quantify these effects, they interviewed both healthy and sick people and asked, "How much would you be willing to pay not to be sick?" Based on these interviews, they estimated that the average person would pay $25 to prevent an asthma attack, $20 not to be bedridden for a day from a respiratory disease, and $5 not to have repeated coughing fits during the day.

Many critics argue that the report places too low a value on the cost of being sick. Others argue that the costs of pollution control were overestimated because the authors ignored shifts in behavior. For example, most of the inexpensive pollution control devices have already been added to automobiles, and further emission reductions will be expensive. However, if automobiles become

too expensive, people may shift to other, less polluting ways of getting to work: bicycles, public transportation, or electric cars. Many people could work at home and be connected to their offices electronically. All of these approaches would reduce pollution inexpensively.

18.11 Warfare and the Atmosphere

Conventional Warfare

Since prehistoric times armies have set their enemies' possessions on fire. The fires lit during battle burn out of control and produce huge amounts of smoke. Air pollution from the Persian Gulf War was particularly intense because the Kuwaiti oil fields were set ablaze. When the smoke first billowed into the atmosphere, some scientists predicted it would cause global cooling. However, the particles settled fairly quickly or were washed out by rain. Therefore, although the plume covered thousands of square kilometers with an oily, toxic cloud, it was not nearly large enough to affect global climate. Nevertheless, the fires consumed 4.6 million barrels of oil per day, about equal to the United States' daily oil import, and released 1.9 million tons of carbon dioxide into the atmosphere every day, about equal to 2 percent of the daily worldwide emissions of carbon dioxide. The last burning well was extinguished on November 6, 1991.

Nuclear Warfare

In October 1983, an international conference of 100 atmospheric physicists and biologists from the United States, western Europe, and the Soviet Union studied the atmospheric effects that could result from nuclear warfare.

*Alan J. Krupnick and Paul R. Portney, "Controlling Urban Air Pollution: A Benefit Cost Assessment," *Science* 252 (April 26, 1991): 522ff.

western Europe, and the Soviet Union studied the atmospheric effects that could result from nuclear warfare. They concluded that:

1. Multiple ground-level nuclear blasts would lift huge amounts of pulverized soil into the stratosphere. This dust would be accompanied by soot from fires lit by the blasts. The dust and soot would block out 95 percent of normal solar radiation. Temperatures in the Northern Hemisphere would plummet to $-25°C$ ($-13°F$), even if the war occurred during the summer. The freeze that would result has been named **nuclear winter**. Crops and forests would die, and billions of humans as well as an uncountable number of animals would freeze and starve.

2. Large quantities of chemical compounds are integrated throughout our society in the form of raw chemicals or products used in industry, commerce, agriculture, and the home. In the event of a nuclear war, many would be blasted into the atmosphere and converted into vapors or particles that would be dispersed over the entire globe in the form of acid and other chemical precipitation.

3. The heat of the blasts would convert large quantities of atmospheric nitrogen to nitrogen oxides. In turn, these compounds would destroy much of the ozone layer, exposing survivors to high levels of ultraviolet radiation.

Twenty prominent scientists summarized the findings of the conference:

> The extinction of a large fraction of the Earth's animals, plants, and microorganisms seems possible. The population size of *Homo sapiens* conceivably could be reduced to prehistoric levels or below, and extinction of the human species itself cannot be excluded.*

*R. P. Turco, O. B. Toon, T. P. Ackerman, J. B. Pollack, and Carl Sagan, "Nuclear Winter: Global Consequences of Multiple Nuclear Explosions," *Science* 222 (1983): 1283ff. Also see Paul Erlich et al., "Long-Term Biological Consequences of Nuclear War," *Science* 222 (1983): 129ff.

SUMMARY

•

The major gases released when fossil fuels are burned are carbon dioxide, hydrocarbons, carbon monoxide, sulfur oxides, and nitrogen oxides. **Volatiles** are released by a wide range of activities and include many toxins and carcinogens. **Particles** are released when fuels are burned and by other industrial processes.

Sunlight interacts with automobile exhaust to form smog. Nitrogen and sulfur oxides react in the atmosphere to produce **acid precipitation**. Acid precipitation has caused extensive damage to forests, agricultural crops, masonry, metals, and other materials.

Under normal meteorological conditions, air temperature drops steadily with altitude. However, if the upper air is warmer than the air beneath it, the atmosphere near the Earth's surface stagnates and pollutants are trapped under the warmer air. This condition is known as **atmospheric inversion** or **temperature inversion**.

Air pollution damages human health, plants, and materials. Indoor air carries additional burdens from such sources as gas cooking stoves, tobacco smoke, leaky wood stoves, formaldehyde from resins and adhesives, and radon from soils and masonry.

The **Clean Air Act** regulates emissions from individual sources and sets National Ambient Air Quality Standards. Nevertheless, the air remains polluted in many cities. **Cost-benefit analysis** balances the cost of pollution control against the cost of living in a polluted environment.

The smoke from the Persian Gulf War did not affect global climate, but dust raised by nuclear bombs in a full-scale war could obstruct sunlight to a degree that could cause **nuclear winter**.

KEY TERMS

•

Carbon dioxide *492*
Hydrocarbon *492*
Carbon monoxide *492*
Sulfur oxides *492*
Nitrogen oxides *492*

Volatile *493*
Particle *493*
Primary air pollutant *494*
Secondary air pollutant *494*

Smog *495*
Ozone *495*
Acid precipitation *446*
Atmospheric inversion *499*

Temperature inversion *499*
Epidemiology *500*
Clean Air Act *503*
Cost-benefit analysis *505*
Nuclear winter *507*

REVIEW QUESTIONS

•

1. List the major pollutants released when fuels are burned. Explain how each is released.

2. What are volatiles and particulates, and how are they released?

3. Describe the difference between primary and secondary pollutants.

4. What is smog, and how does it differ from automobile exhaust?

5. Describe the experiments that led to our understanding of smog formation.

6. What is acid rain, and how does it form? Is acid rain a primary or a secondary air pollutant?

7. What is an atmospheric inversion, and why may it persist for several days?

8. Describe both the utility and limitations of laboratory studies and epidemiological studies in determining the health effects of air pollution.

9. List the health effects of sulfur dioxide, nitrogen oxides, carbon monoxide, ozone, and particulates.

10. Why is indoor air generally more polluted than outdoor air?

11. Discuss the major provisions of the Clean Air Act.

12. What is cost-benefit analysis? How is it applied to air pollution control?

DISCUSSION QUESTIONS

•

1. If coal were burned completely, would it produce air pollution?

2. Is the air pollution generated by coal-fired steam locomotives equivalent to that generated by diesel engines?

3. State which of the following processes are potential sources of gaseous air pollutants, particulate air pollutants, both, or neither. (a) Gravel is screened to separate sand, small stones, and large stones into different piles. (b) A factory stores drums of liquid chemicals outdoors. Some of the drums are not tightly closed, and others have rusted and are leaking. The exposed liquids evaporate. (c) A waterfall drives a turbine, which makes electricity. (d) Automobile bodies in an assembly plant are sprayed with a coating consisting of pigments dispersed in a solvent. The automobile bodies then move through an oven that drives off the solvent. (e) A garbage dump catches fire.

4. Is the speed of settling of particles in the air directly proportional to their diameters? (If the diameter is multiplied by 10, is the settling speed 10 times faster?) Justify your answer with data from Figure 18–2. Would a settling chamber be an effective device for removing particulate air pollution? Explain.

5. How can automobile exhaust contribute to the acidity of rainwater? (Assume that the automobile uses sulfur-free gasoline.) What happens in the cylinders? What happens in the outdoor atmosphere?

6. Gasoline vapor plus ultraviolet lamps do not produce the same smog symptoms as do automobile exhaust plus ultraviolet lamps. What is missing from gasoline vapor that helps to produce smog?

7. How can sulfur in coal contribute to the acidity of rainwater? What happens in the furnace when the coal is burned? What happens in the outdoor atmosphere?

8. Sulfur dioxide emitted from a stack is responsible for both acid dusts and acid rain. Which is more likely to fall to Earth closer to the stack? Explain.

9. Is the environmental damage caused by acid rain proportional to the quantity of acid deposited, or are other factors important as well?

10. Which is the more stable condition—cool air above warm air, or warm air above cool air? Which of these conditions is called an atmospheric inversion?

11. Describe the meteorological conditions most conducive to the rapid dispersal of pollutants. Describe those that are least conducive.

12. Under some conditions, two inversion layers may exist at the same time in the same vertical atmospheric structure. Draw a diagram of temperature versus height that shows one inversion layer between the ground and 200 meters, and another aloft, between 1000 and 1200 meters, while the temperature variations between them approximate equilibrium conditions.

13. Argue for or against this statement: "Since tall chimneys do not collect or destroy pollutants, all they do is protect the nearby areas at the expense of more distant places."

14. Imagine that you are performing a study to determine whether a rise in pollution is due to natural causes or industrial activity. Which of the following experiments would you rely on? Defend your choices. (a) Compare the pollution level with that of previous years, when population

and industrial activity were less. (b) Compare effects during weekdays, when industrial activity is higher, with those on weekends, when it is low. (c) Compare effects during different seasons of the year. (d) Compare effects just before and just after the switch from daylight saving time to see whether there is a sharp 1-hour shift in the data. (e) Compare effects in different areas where population and industrial activities differ.

15. A report of air pollutant concentrations shows high concentrations of sulfur dioxide, particulate matter, and nitrogen dioxide at the top of a 200-meter stack of a power plant. If you were the health officer or mayor, what action, if any, would you recommend? Would you require any additional information? If so, describe the data you would request.

16. In Section 18.10 we cited a study that reported that people would be willing to pay $25 to prevent an asthma attack, $20 not to be bedridden for a day from a respiratory disease, and $5 not to have repeated coughing fits for a day. What dollar value would you place on freedom from each of these health disorders? Discuss your values with those of your classmates. If your values differ from those reported in the cited study, how would the differences affect the cost-benefit analysis?

Stephen Schneider

Stephen Henry Schneider was born in 1945 in New York City. He received his B.S. and M.S. in mechanical engineering and his Ph.D. in mechanical engineering and plasma physics at Columbia University. Following a brief stint with NASA's Goddard Space Center, Dr. Schneider in 1972 signed on with the National Center for Atmospheric Research in Boulder, Colorado, and from 1987 to 1992 served as the head of the Interdisciplinary Climate Systems Section there. In 1992 he became a professor of biological sciences and international studies at Stanford University.

The author of several books on climate and climate change, and coauthor of many others, Dr. Schneider has focused his research on climate modeling and the forecasting of the implications of climate change on our environment. He applies his comprehensive knowledge of the field as an editor of the international journal *Climatic Change* and of *The Encyclopedia of Climate and Weather*. He also serves as a science advisor and editorial board member for many other organizations. As a result of his nearly 100 articles in publica-

tions as diverse as *Scientific American* and *Good Housekeeping*, and his appearances on several television programs, in 1991 he was given the American Association for Advancement of Science/Westing-

> Climate theory is not a perfect replica of nature; it's a model.

house Award for Public Understanding of Science, and in 1992, a MacArthur Foundation Fellow for creativity.

Dr. Schneider has been engaged as an expert speaker at many universities, including the University of Virginia, Stanford University, University of Wisconsin–Madison, and Northwestern University. He has also served as consultant to environmental and government agencies around the world on energy policies and the implications of nuclear weapons. Included as one of *Science Digest*'s "One Hundred Outstanding Young Scientists in

America" in 1984, Dr. Schneider has proven worthy of the honor through his many contributions to both the scientific and global communities.

Where were you born and raised?

I grew up on the south shore of Long Island in a town called Woodmere. What I remember enjoying a lot about Long Island was going to a square-mile acre of woods, where I would run around and just enjoy streams and nature. One day a hurricane came by, and I went back to the forest, and half the trees were knocked down and it all had been disturbed. Even at the age of nine, I realized that ecology and climate and soils were all connected systems. Later on, I suppose, this was an emotional driver for getting involved in Earth sciences.

At what point did you become interested in, and then commit yourself to, atmospheric research?

I liked racing cars when I was in high school. So I went to Columbia University's engineering school to learn how to build the fastest race car. I ended up a mechanical engineering student studying fluid mechanics, then called engineering physics. So my initial practical notion to build race cars was completely dashed by my own choice to go into more theoretical parts of engineering.

My interest in atmospheric research began around 1970, when I attended the first Earth Day celebrations at Columbia. I was a little bored working on a plasma physics problem. I wanted to do something environmentally useful. At one of the Earth Day presentations, somebody (I think it was Barry Commoner) said, ''What if pollution could change the climate—it could either heat it up if it's greenhouse gases or cool it down if it's sulphur injections,'' and I didn't believe it.

There was an atmospheric sciences course taught at Columbia by Ichtiaque Rasool. He went over the difference between Mars, Earth, and Venus. He said Venus is very hot with its very thick atmosphere, a super greenhouse effect. Mars is very cold with a very thin atmosphere, a weak greenhouse effect. Earth is right in the middle; water is what makes us different, and pollution could, in fact, dirty the greenhouse window. This was fascinating to me. Rasool said, ''I will give you your postdoctoral fellowship if you'll leave plasma physics and convert to atmospheric science—to mathematically modeling the climate.'' So I took up Rasool's offer and became a post-doc at the Goddard Institute for Space Studies, a NASA laboratory at Columbia.

•

What did you do next?

In 1972 I went to the National Center for Atmospheric Research (NCAR) to help them start what we dubbed the Climate Project, and began doing climate research in Boulder. Climate involves the integration of materials from many disciplines—from oceanography, ecology, geography, meteorology, chemistry, and physics. Critics argued that such an integration was premature because each of these subtopics is not yet understood to the satisfaction of the practitioners. And my answer is that the world has to have the answer whether the disciplines are ready or not, so why don't we take halting steps to try to see how the system is integrated and connected—what we now call Earth systems science?

•

Much of your work in atmospheric research is based on computer modeling of the atmosphere. Can you tell us how that works?

Climate theory is not a perfect replica of nature; it's a model. And it's not a physical model in the laboratory, because you cannot make a physical model in a laboratory that includes enough of the important complexity. For example, the single most important component of the Earth's climate—the evaporation of water at the surface, and the recondensation in clouds—can't be done meaningfully in the lab. So the lab is very limited. Thus, the lab that you have to have for your model, literally, is a computer Earth. It's an Earth that we can pollute—it's an Earth we can modify by just changing something in the computer, and then resimulate the climate under these new conditions.

•

Tell us more about your modeling technique.

You break the atmosphere up into a bunch of boxes, or what we call grid squares. You break it up into a latitude-longitude grid of $4\frac{1}{2}$ degrees latitude and $7\frac{1}{2}$ degrees longitude. There are 20,000 of these grid squares around the world if this grid is piled into ten vertical layers.

Now, if you have a box 500 by 500 kilometers—say, the size of the state of Colorado typically—you can't, obviously, resolve an individual cloud. Nobody's ever seen a cloud the size of Colorado. So we've got a problem, because clouds are the venetian blinds of the earth. They control, more than any other elements, the amount of solar energy absorbed, which is how much the planet is heated and the amount and distribution of infrared radiative energy that escapes back to space—the so-called greenhouse effect. They are more important radiatively than water vapor, carbon dioxide, chlorofluorocarbons, methane, and all those things that we argue about for the human-induced greenhouse effect.

Yet our models cannot explicitly resolve clouds, because they are too small. They are not 500 by 500 kilometers square. They are 5 by 5, maybe, or less. So what do we do? Some people say, ''Well, your models are no darn good. Throw them away.'' But you don't need to know every detail in order to make a prediction about how something works. I mean, you don't have to predict what happens in every play to know that the 49ers would beat Stanford in a football game.

So the question, then, is: How can we get the *average effects* of clouds at the grid scale, even though we're not explicitly calculating individual clouds? We use a technique called parameterization, which is a short contraction for parametric representation.

•

Can you give an example of this so we can get an idea about the strengths and weaknesses of models?

Say the model produces humidity, temperature, and wind at the grid box. We know the relative humidity of, say, Colorado even if we don't know how many clouds there are. So we can make a rule, a parametric representation which says if the air is humid it's more likely to be cloudy than if the air is dry. Every farmer knows that. You don't need a Ph.D. in atmospheric science to know that. So then you have a parametric form which says cloudiness is equal to a number, a parameter, times the relative humidity.

But it's more sophisticated than that. Is it more likely to be cloudy if the air is rising or sinking? Think about it for a second. If air is rising, it would be drawing in the humid air from below. If air is sinking it's starting from a high place, which is dry, and coming down. Plus it's heating when it's going down, which tends to cause evaporation. And when it's rising, the air is expanding, which means it's cooling, which means it would tend to make condensation. So then you can say, "All right, my cloud parameterization will get more sophisticated. The cloudinesss in the grid box is equal to a parameter times the relative humidity plus another parameter times the vertical velocity, both of which are calculated at the grid box average." The point is, you can get more and more sophisticated without resolving the explicit nature of the details. You are able to get the averages.

The problem is, how well are we doing this? And that's why there's an endless debate among the practitioners as to whether it's been done well, or medium well, or terribly. And how do we validate it? The Earth warms up somewhere between $1\frac{1}{2}$ and $4\frac{1}{2}°C$ for almost all of the models that have been run for carbon dioxide doubling. Now, there are some that are a little hotter and some that aren't quite as hot. But the bulk of them fall in that range. So there's an uncertainty factor of 3 (between $1\frac{1}{2}$ and $4\frac{1}{2}$). So our

models could be off by a factor of 3 global average temperatures. And this factor of 3 is largely because of differences in models' cloudiness parameterizations.

•

We're talking about clouds being the critical factor in changing the atmospheric temperature in one way or another. Why, then, are we all so interested in carbon dioxide and possibly other greenhouse gases?

The reason we worry about CO_2 is, we know beyond a doubt that if you double CO_2 you're going to trap something like 4 watts of energy over every square meter of Earth. Now, since the industrial revolution, we've added 25 percent more CO_2 and doubled the methane; that's known beyond doubt. The point is that because we've added all these chemicals, we've trapped about 3 extra watts of energy in the last 100 years in the Earth's surface layers. That's not the debate. The debate is should 3 watts of energy warm up the Earth a quarter of a degree, 2 degrees, 4 degrees? And in order to answer that question, you've got to then say, "Now, what happens to the natural system? Do the trees get darker and greener, which would accelerate the warming? Or do they get slightly lighter and brighter, which would retard the warming? What do clouds do?" All these so-called "feedback processes" are endlessly debated and cause that uncertainty factor of 3 or so that we argue about.

The point is, there's positive and negative feedback in nature and in a physical and chemical world. And the sum of the positive and negatives, as to who wins, is not known. That's the debate. So the fact that clouds are a very large factor doesn't mean that that factor will necessarily be a positive or a negative feedback. We simply don't

know. If you double CO_2, as is forecast sometime in the next century, from growing populations using fossil fuel at a higher and higher rate and continuation of deforestation, then what's going to happen is, we're going to trap another couple of watts of energy.

The debate, then, is translating that into temperature change. Is it going to end up just a half a degree to a degree more warming? Or are we going to get 3 or 4 degrees? In other words, are we going to have a mild change or a catastrophic change—because that's the range. And the debate goes on between mild and catastrophic without any clear and obvious answer right now, because we do not have either the theory or the measurements to validate the overall effect on the globe to much better than a factor of 3.

•

We know, based on your work and that of many other people, that CO_2 concentration in the atmosphere has increased. We know the absorptive qualities of CO_2. Basically it boils down to the fact that we know that increased CO_2 in the atmosphere is going to lead to global warming. Tell us more about the rate and magnitude of global warming.

The rate at which nature changes is on the order of 5°C. That's how much colder an ice age is than a so-called interglacial period, on a global average basis. Say it takes nature 5000 years to end the ice age, then we end up with a rate that's about 1°C per millennium as natural average rate of change.

The conservative forecast for global warming is a degree a century, or ten times faster than the natural average rate of change. The radical forecast says we could warm up 10 degrees in a century. That's a hundred times faster than natural rates.

Pick a middle number, and it's something like 2 to 5 degrees' warming projected to occur over the next century. That would be something like 20 to 50 times faster than natural rates of change.

•

Why do you think there's such difficulty in producing a social, political, and human response to this problem even though the logical conclusion is that if the global climate warms significantly in a short time, the effects will be more negative than positive?

It's not like chlorofluorocarbons and the ozone hole, because we all admit we can do without the spray cans and fluorocarbon refrigerators and change what's in the air conditioners in our cars. It might cost a few percent more, but very few people are willing to risk skin cancer and disruption of nature for a few chemicals that are substitutable.

When we're talking about global warming, we're talking about methane produced by agriculture, coal mining, natural gas, and landfills. We're talking about CO_2 produced by coal, oil, and gas, which the Third World is expecting to use to power their industrial revolution just the way the western countries did in the Victorian era. And people do not want to hear that these mainstays of economic growth that permit growing populations to increase standards of living have side effects that might be dangerous.

So the problem is, it's not easy to get political agreement to slow down emissions, because it could be painful. Thus, people use the honest and legitimate scientific debate as an excuse to wait and see. They'll say, "Well, we're really not sure."

No honest scientist says that we know the answer. The degree of uncertainty ranges from mild to catastrophic, and we don't know where in this range the actual outcome is going to happen. And we're not going to know in the next 5, 10, or probably even 20 years. Whether to take the chance is not a scientific question per se, but a personal value judgment about which you fear more—investing *present* resources as the hedge against potentially catastrophic change, even though that change may not be so bad if you're lucky, or *not* investing present resources to find that you might be unlucky and, in not trying to slow it down, you have gotten a really whopping big dose that will be impossible to stop without irreversible damage.

•

So what should we do?

Study after study has shown that, depending upon how much we're willing to invest to buy these new materials, we can cut somewhere between 10 and 40 percent of our greenhouse emissions by replacing existing inefficient technology, and do it at below zero net cost. In other words, the amount of money you'll save in reduced energy costs actually will pay for itself without even counting the free extras, like reducing acid rain and the threat of disrupted ecosystems.

So my personal opinion is, why don't we do those things first that are free-standing and make sense anyway? Get rid of the fluorocarbons, because they not only trap 25 percent of the heat, but they also help cause ozone depletion. That's the easiest and the first thing to do. Let's use energy efficiently, because not only does it reduce global warming, but it also reduces acid rain and reduces dependency on foreign supplies, which are expensive and sometimes militarily dangerous to protect. It reduces local air pollution. It also makes our products more competitive in the long run. All of these things have to take place, and can, and are, but we need to push them harder. We can also pursue vigorously research and development on non-fossil-fueled energy systems. And then in the time frame of 5 to 15 years, while we're slowing down the rate at which we're making the system change, and therefore slowing down the rate at which nature will have to adapt, we can buy time to have the scientists determine what is likely to happen. To use the excuse that the thing is now very uncertain and therefore we shouldn't act is to say we should never have insurance, the police, or the military.

•

What advice would you give to students taking Earth science?

What students need to recognize is, we're not talking about the environment *or* the economy. What we're talking about is, there are trade-offs between the two, but that if we use our brains and hands cleverly we can find ways to have the economy grow in a much less environmentally destructive way.

Problems are increasingly crosscutting. Environmental problems are only one example. Health problems are another example. And the way we're organized is not set up to deal with problems that crosscut. In universities we learn disciplines; in governments we deal with departments. Real issues, like global warming, involve solutions that have a little bit in population, forestry, agriculture, energy, and foreign policy. In order to manage that problem, each group has to do a little, and *it has to be coordinated.* And therefore an Earth systems science approach, such as this book is trying to take by integrating the disciplines, is the only way to be in tune with the way the world's real problems are.

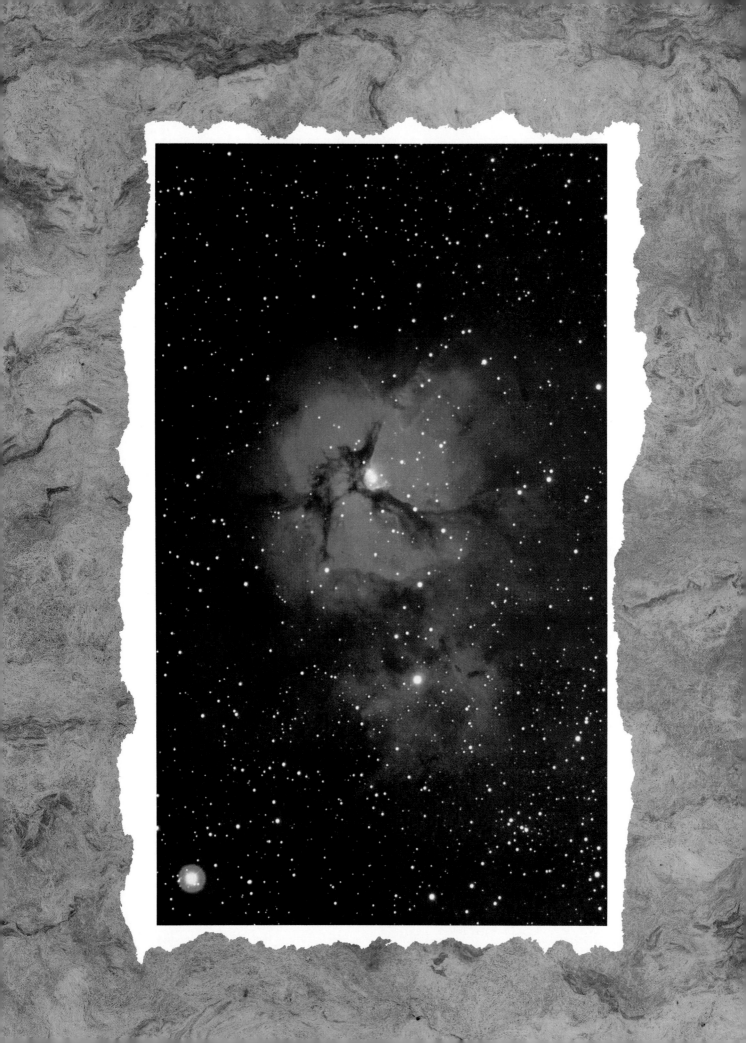

Astronomy

- Motion in the Heavens

- Planets and Moons

- The Sun, Stars, and Space

● Trifid Nebula, in the constellation Sagittarius. This colorful nebula consists of glowing clouds of hydrogen and helium. Dark lanes are opaque regions of dust and gas. This distant nebula is approximately 30 light years in size.   *(National Optical Astronomy Observatories)*

Motion in the Heavens

19

Ancient astronomers used cycles of the Sun, Moon, and stars to tell time and to mark the seasons. However, these objects are so far away that, until recently, astronomers knew very little about them. The ancient Greeks believed that the Earth is stationary and that the Sun rises and sets because it orbits our planet daily. Today we realize that the Earth's rotation causes the cycle of night and day.

The Sun shines brightly in the sky and warms the Earth, whereas the stars twinkle feebly in the darkness. Modern astronomers understand that the Sun and stars are similar and that the Sun is merely much closer, but the ancients had little concept of astronomical distances or sizes and believed that the Sun was unique.

In this and the following two chapters we will learn how astronomers deduced the motions of celestial bodies and what we have learned about their compositions and structures.

19.1 The Motions of Heavenly Bodies

Even a casual observer notices that both the Sun and Moon rise in the east and set in the west. In the mid-

● An astronaut in orbit around the Earth. *(NASA)*

Focus On

The Constellations

The oldest written record of the constellations appears in a 2000 B.C. Sumerian manuscript. Other accounts occur in old manuscripts and legends from numerous cultures. Ancient astronomers and religious leaders imagined that constellations represented animals such as Leo, the lion, or gods such as Aquarius, the Sumerian deity who pours the waters of immortality onto the Earth. The constellations have no particular significance to modern astronomy other than as a convenience to naming parts of the sky map. For example, Orion, the hunter, is easily recognized by the three bright stars of his sword belt (Fig. 1A,B). The brightest star in Orion is Rigel, which forms his left kneecap. If you look at Rigel through a telescope or a good set of binoculars, you will note that it is bluish. In contrast, Betelgeuse (pronounced beetle juice), which defines Orion's right shoulder, is red. As we will learn in Chapter 20, color is an indication of the temperature of a star, and the temperature is an indication of its size and age. Blue stars such as Rigel are quite different from red ones such as Betelgeuse. In addition, the two stars are not even close to one another: Rigel is almost twice as far from Earth as Betelgeuse is. The only relationship between the two is that they appear close in the night sky.

Figure 1 A seventeenth-century star atlas showing the constellation Orion. *(J. M. Pasachoff and the Chapin Library)*

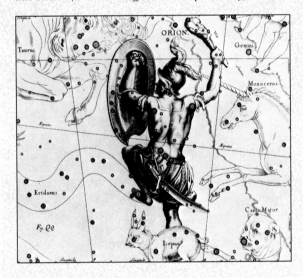

Figure 2 A view of Orion through a modern telescope. The three stars in a line in the center of the photograph represent his belt; the bright stars below it, his sword. *(Harvard College Observatory)*

latitudes, summer days are long and the Sun rises high in the sky. Winter days are shorter and the Sun does not rise as high. In contrast to the yearly cycle of seasons, the Moon completes its cycle once a month. It is full and round one night then darkens slowly each night until it is just a thin crescent. It "disappears" completely after two weeks, then reappears as a thin sliver that grows again, completing the entire cycle in about 29.5 days.

If you had been a cowboy in the 1800s, the foreman might have told you to guard the herd at night until the "dipper spills water." This phrase refers to two features of the night sky. First, stars remain in fixed positions relative to one another. This fact led the ancients to identify groups of stars, which they called **constellations**. Second, the stars revolve about a fixed point in the sky marked by the Pole Star, or North Star. In the Northern Hemisphere, the Pole Star remains motionless and all other stars revolve around it (Fig. 19–1). This motion provided the clock for the cowboy on night watch, because the dipper alternately holds and spills water as it revolves around the Pole Star.

Constellations appear and disappear with the seasons (Fig. 19–2). For example, the Egyptians noted that Sirius, the brightest star in the sky, became visible at dawn just

Figure 19–1 A time exposure of the night sky showing the rotation of the stars around the Pole Star, which appears to be motionless.
(Yerkes Observatory)

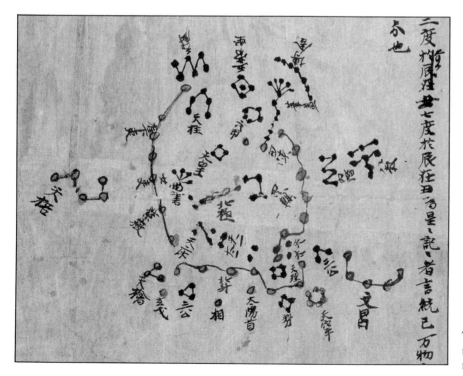

The oldest existing portable star map is in a tenth-century Chinese manuscript. *(British Library)*

519

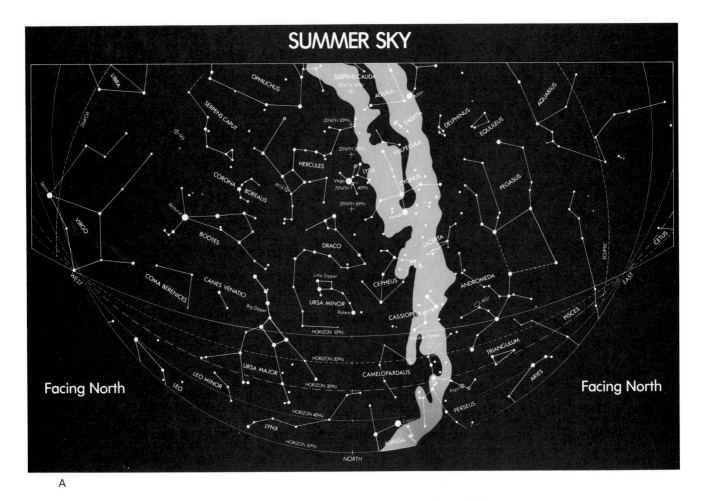

A

Figure 19–2   (A) The summer sky, viewed facing north. (B) The winter sky, viewed facing north. In both, the lightly shaded area is the Milky Way.   *(Wil Tiron)*

before the Nile began to flood. Farmers would therefore plant crops when this star first appeared, with the assurance that the high water soon would irrigate their fields.

A few objects in the sky change position with respect to the stars. Ancient Greeks called these objects **planets**, from the word meaning "wanderers." For most of the year, planets appear to drift eastward with respect to the stars, but sometimes they seem to reverse direction and drift westward. This apparent reverse movement is called **retrograde motion** (Fig. 19–3).

Figure 19–3   A planetarium simulation of the movement of Mars from August 1, 1990, through April 1, 1991. Notice that Mars appears to reverse direction, forming a retrograde loop. This motion was difficult to explain with the geocentric model.   *(Geoff Chester, Albert Einstein Planetarium, Smithsonian Institution)*

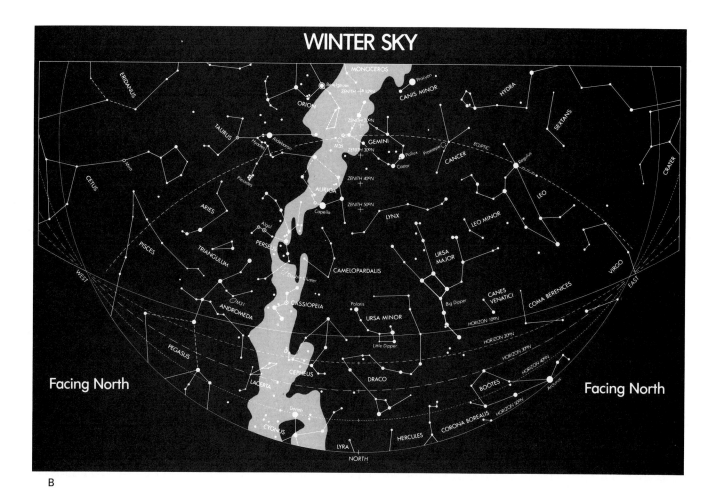

WINTER SKY

Facing North

Facing North

B

19.2 Aristotle and the Earth-Centered Universe

The ancient Greek philosopher and scientist Aristotle proposed a **geocentric**, or Earth-centered, Universe. In this model, the Earth is stationary and positioned at the center of the Universe. A series of concentric **celestial spheres** made of transparent crystal surrounds the Earth. The Sun, Moon, planets, and stars are imbedded in the spheres like jewels (Fig. 19–4). At any one time you can see only a portion of each sphere, but as the sphere revolves around the Earth, different objects appear in turn.

Today we all know that this geocentric model is incorrect. The Earth is not stationary, and it is not at the center of the Universe. In fact, the Universe has no center. The planets orbit around the Sun; the Sun and planets orbit the center of the galaxy; and the galaxy, with its 100 billion stars, is flying through space. In a brief summary

of the history of science, we usually omit mistakes because in retrospect they are insignificant dead-end paths. However, Aristotle's incorrect model is important because the debate that it provoked was the beginning of modern science.

Two observations supported the geocentric model. First, Aristotle concluded that the Earth must be stationary because people have no sensation of motion. People tend to fall off the back of a chariot when horses start to gallop, but they do not fall off the Earth, so Aristotle reasoned that the Earth must not be moving. As we will learn in Section 19.3, Galileo showed this reasoning to be wrong.

Aristotle's second observation was based on **parallax**, the apparent change in position of an object due to the change in position of the observer. To understand parallax, imagine that you were drifting on a raft in an unknown ocean. If there were no landmarks, it would be impossible to tell whether you were moving or stationary. Now suppose that two islands came into sight, one close to your raft and the other farther away. If you were

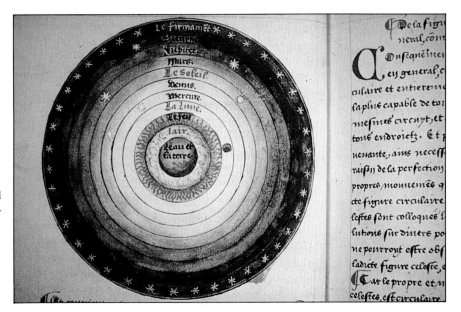

Figure 19–4   Aristotle's cosmology. Water and Earth lie in the center, surrounded by air and fire. Beyond these four basic elements, the Moon, Mercury, Venus, the Sun, Mars, Jupiter, Saturn, and the stars lie in concentric celestial spheres.   *(Houghton Library, Harvard University)*

stationary, the two islands would remain in the same positions relative to each other. But if you moved, they would appear to shift positions relative to one another, even though they were really stationary. As another example, the fence posts in Figure 19–5 are stationary. Yet they appear to line up when the photographer is in one position (A) and seem offset when the photographer steps to the side (B). The ancients correctly reasoned that if the Earth moved around the Sun, the stars should change position relative to one another, as shown in Figure

Figure 19–5   The concept of parallax is illustrated by two photographs of a fence on the Montana prairie. (A) The photographer is nearly in line with the fence, and the distant posts block one another so that we see no spaces between them. (B) When the photographer moves, the posts appear to shift position. Now we can see spaces between the more distant posts. The same effect has been observed in astronomical studies. As the Earth revolves about the Sun, the stars appear to shift, position relative to one another.

A   B

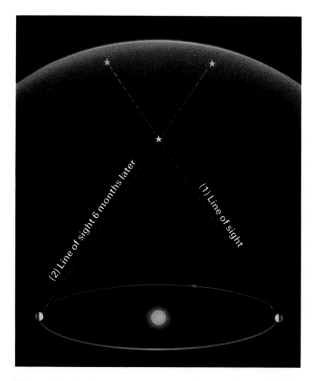

Figure 19–6 A nearby star appears to change position with respect to the distant stars as the Earth orbits around the Sun. This drawing is greatly exaggerated: in reality, the distance to the nearest stars is so much greater than the diameter of the Earth's orbit that the parallax angle is only a small fraction of a degree.

19–6. Since they observed no parallax shift, they concluded that the Earth must be stationary. Their mistake arose not out of faulty reasoning, but out of inaccurate measurements. The stars are so far away that their parallax shift is too small to be detected with the naked eye.

Aristotle incorporated philosophical concepts, in addition to observation, into scientific theory. He argued that the gods would create only perfection in the heavens and that a sphere is a perfectly symmetrical shape. Therefore, the celestial spheres must be a natural expression of the will of the gods. In addition, the Sun, Moon, planets, and stars must also be unblemished spheres. As we shall see, this incorporation of philosophy into science retarded debate and allowed dogma to rule over logic.

Aristotle's theory, although incorrect, did explain the motions of the Sun, Moon, and stars (Fig 19–7). However, it failed to explain the retrograde motion of the planets. In about A.D. 150, Claudius Ptolemy modified the celestial sphere model to incorporate retrograde motion. In Ptolemy's model, each planet moves in small circles as it follows its larger orbit around the Earth (Fig. 19–8). Ptolemy's sophisticated mathematics accurately described planetary motion, and therefore his model was

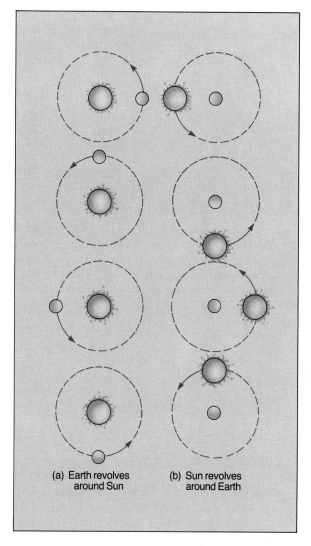

Figure 19–7 Series (a) shows the Earth revolving around the Sun, and series (b) shows the Sun revolving around the Earth. Now lay some thin paper over series (a) and trace the outlines of the Sun and Earth, but not the arrows or orbits. Lay this tracing over series (b) and notice that they match exactly, after you shift the paper to make sure the Sun and the Earth superimpose. Conclusion: There is no difference between the Earth revolving around the Sun and the reverse, provided you do not refer to anything else, such as the outline of these diagrams or another star.

accepted. Still, he retained Aristotle's erroneous idea that the Sun and the planets orbit a stationary Earth.

19.3 The Renaissance and the Heliocentric Solar System

Aristotle's and Ptolemy's ideas remained essentially unchallenged for 1400 years. Then, in the 120 years from 1530 to 1650, several Renaissance scholars changed our

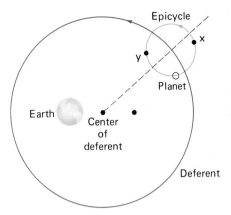

Figure 19–8   Ptolemy's explanation of retrograde motion. Each planet revolves in a small orbit (the epicycle) around the larger orbit (the deferent). When the planet is in position X, it appears to be moving eastward. When it is in position Y, it appears to have reversed direction and to be moving westward. Ptolemy did not realize that the planets move in elliptical orbits. To compensate for this error, he placed the Earth away from the center of the deferent.

understanding of motion in the Solar System and, in the process, revolutionized scientific thought.

Copernicus

In 1530 a Polish astronomer and cleric, Nicolaus Copernicus, proposed that the Sun, not the Earth, is the center of the Solar System and that the Earth is a planet, like the other "wanderers" in the sky. Copernicus based his hypothesis on the philosophical premise that the Universe must operate by the simplest possible laws. Copernicus believed Ptolemy's model was too complex and that the motions of the heavenly bodies could be explained more concisely by a **heliocentric** model with the Sun at the center of the Solar System.

Nicolaus Copernicus.   *(Marek Demianski, Torun Museum)*

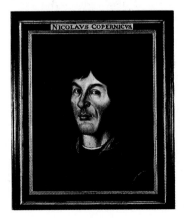

Figure 19–9 shows how the heliocentric model explains retrograde motion. Assume that initially the Earth is in position 1 and Mars is in position 1′. An observer on Earth looks past Mars (as shown by the white line) and records its position relative to more distant stars. Mars appears to be in position 1″ in the night sky. After a few weeks the Earth has moved to position 2, Mars has moved to 2′, and Mars's position relative to the stars is indicated by 2″. In the Copernican model, Earth moves faster than Mars because Earth is closer to the Sun. Therefore, Earth catches up to and eventually passes Mars, as shown by positions 3, 3′ and 4, 4′. During this passage, Mars appears to turn around and move backward, although, of course, this appearance is merely an illusion against the backdrop of the stars. Mars appears to reverse direction again through position 5 and 6, thus completing one cycle of retrograde motion. In reality, Mars never reverses direction, it just appears to behave in this way as Earth catches up to it and then passes it.

Brahe and Kepler

In the late 1500s, Tycho Brahe, a Danish astronomer, accurately mapped the positions and motions of all known bodies in the Solar System. His maps enabled him to predict where any planet would be seen at any time in the near future, but he never developed a theory to explain their motions. Brahe drank himself to death at a party in 1601, and his student Johannes Kepler inherited the vast amount of data Brahe had collected.

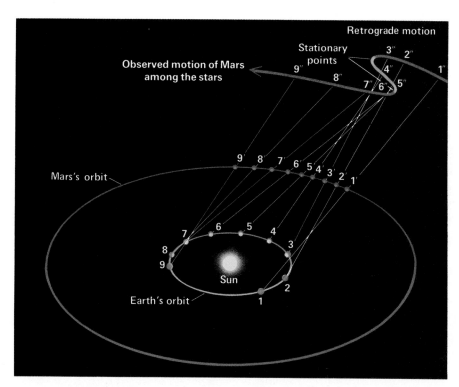

Figure 19–9 Retrograde motion as explained within the heliocentric model.

Kepler calculated that the planets moved in elliptical orbits, not circular ones, and he derived a set of rules and formulas to describe their paths. The rules are still used today. However, Kepler never answered the important question "Why do the planets move in orbits around the Sun rather than flying off into space in straight lines?"

Galileo

Galileo was an Italian mathematician, astronomer, and physicist who made so many contributions to science that he is often called the father of modern science. Perhaps most significant was his realization that the laws of nature must be understood through observation, experimentation, and mathematical analysis. This concept freed scientists from the confines of Aristotelian thought.

Recall that Aristotle's geocentric theory was based on two observations and one philosophical premise: (1) If the Earth were moving, people should fall off but they do not do so. (2) Aristotle was unable to observe parallax shift of the stars. (3) The gods would create only an unblemished, symmetrical Universe. Galileo's experiments and observations showed that Aristotle's model was incorrect and led him to support Copernicus's heliocentric model.

Galileo studied the motion of balls rolling across a smooth marble floor, and legend tells us he also dropped objects from the Leaning Tower of Pisa. He organized the results of these experiments into laws of motion that

were later expanded and quantified by Isaac Newton. One of these laws states that "an object at rest remains at rest and an object in uniform motion remains in uniform motion until forced to change." This corresponds to Newton's First Law of Motion, the law of **inertia**. Inertia is the tendency of an object to resist a change in motion. If the Earth were in uniform motion, a person on its surface would be in uniform motion along with it. The person would therefore travel with the Earth and would not fall off and be left behind, as Aristotle had assumed. In fact, the person could not even feel the motion.

Galileo Galilei. *(Tintoretto, National Maritime Museum)*

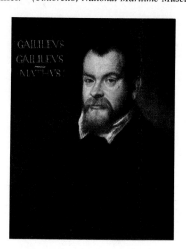

Galileo built his first telescope in 1609, one year after a much smaller one was invented by a Dutch spectacle maker. True to his careful experimental technique, he studied objects on Earth first to be sure that his new instrument enlarged them but did not distort them. Turning his telescope to the heavens he learned that the hazy white line across the sky called the Milky Way was not a cloud of light as Aristotle had proposed, but a vast collection of individual stars. Next, he trained his telescope at the Moon and saw hills, mountains, giant craters, and broad, flat regions on its surface, which he thought were seas and therefore named *maria* (Latin for "seas") (Fig. 19–10). Looking at the Sun, Galileo recorded dark regions, called sunspots, that appeared and then vanished.

Although these discoveries had no direct bearing on the controversy of a geocentric versus a heliocentric Universe, they were important because they led Galileo to question Aristotle's views. The prevailing scientific opinion at the time was that if the Milky Way were a collection of stars, Aristotle would have known about it. Furthermore, Galileo's observations of the Sun and the Moon did not agree with Aristotle's philosophical assumptions that the heavenly bodies were perfectly homogeneous and unblemished. Galileo reasoned that if Aristotle were wrong about the structures of the Milky Way, the Sun, and the Moon, perhaps he was also wrong about the motions of these celestial bodies.

When Galileo studied Jupiter, he saw four moons orbiting the giant planet. (Today we know that Jupiter has 16 moons, but only four are large enough to have been seen with Galileo's telescope.) According to the geocentric model, *every* celestial body orbits the Earth.

Figure 19–10 A comparison of Galileo's drawing of the Moon with a modern photograph. Notice how accurately he drew topographical features (A, B, C, D, and E). *(J. M. Pasachoff after Ewen Whitaker, Lunar and Planetary Laboratory, University of Arizona)*

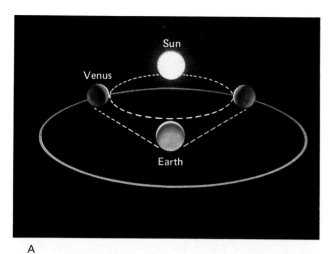

A

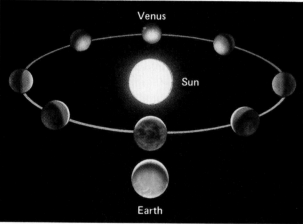

B

Figure 19–11 (A) In Ptolemy's theory, Venus could never move farther from the Sun than is shown by the dotted lines. Therefore, it would always appear as a crescent. (B) In the heliocentric theory, Venus passes through phases like the Moon. Galileo observed phases of Venus through a telescope and concluded that the planets must orbit the Sun.

However, Jupiter's moons clearly orbited Jupiter, not the Earth. The contradiction increased Galileo's doubt about Aristotelian theories.

Finally, he observed that the planet Venus passes through phases as the Moon does. Such cyclical phases could not be readily explained in the geocentric model, even with Ptolemy's modifications (Fig. 19–11). As a result of his observations, Galileo proposed that the Sun is the center of the Solar System and that the planets orbit around it. The Roman Catholic Church rejected Galileo's model and in 1633 ordered him to recant his conclusions. He bowed to the pressure rather than undergo imprison-

ment and torture. Despite his recantations, he died blind, poor, and under arrest.

Isaac Newton and the Glue of the Universe

Galileo successfully described the motions in the Solar System, but he never addressed the question "Why do the planets orbit the Sun rather than flying off in straight lines into space?" Aristotle had observed that an arrow shot from a bow flies in a straight line, but celestial bodies move in curved paths. He reasoned that arrows have an essence that compels them to move in straight lines, and planets and stars have an essence that compels them to move in circles. In the Renaissance, however, this answer was no longer acceptable.

Isaac Newton was born in 1643, the year Galileo died. During his lifetime Newton made important contributions to physics and developed calculus. A popular legend tells that Newton was sitting under an apple tree one day when an apple fell on his head and—presto—he discovered gravity. Of course, people knew that unsupported objects fall to the ground long before Newton was born. But, he was the first to recognize that gravity is a universal force that governs all objects, including a falling apple, a flying arrow, and an orbiting planet.

According to the laws of motion introduced by Galileo and expanded by Newton, a moving body travels in a straight path unless it is acted on by an outside force. Just as a ball rolls in a straight line unless it is forced to change direction, a planet also moves in a straight line unless a force is exerted on it. The gravitational attraction between the Sun and a planet forces the planet to change direction and move in an elliptical orbit. Gravity is the glue of the Universe and affects the motions of all celestial bodies.

Sir Isaac Newton. *(National Portrait Gallery, London)*

Focus On

What Orbits What?

The heliocentric model of the Solar System depicts a central Sun with nine planets orbiting around it. This model is not entirely correct, however. Recall from our discussion of the tides that the Moon does not simply orbit the Earth; instead, the Earth and the Moon orbit one another like dancers in each other's arms. Similarly, the planets and the Sun mutually orbit one another.

Imagine that you tie a Ping-Pong ball to a string and spin it around your head. You are pulling the Ping-Pong ball, but at the same time the ball pulls on you. Since you are much more massive, the ball's influence on you is

barely noticeable. However, if you spun a bowling ball around your head, it would pull on you enough to cause your body to wobble.

The Sun is much more massive than the planets, so our Solar System is analogous to a person swinging nine Ping-Pong balls. For all practical purposes we can say that the planets orbit the Sun and we can forget that, in reality, the Sun also orbits the planets. However, if the planets were more massive, their effects on the Sun would be more noticeable. One way to search for planets in outer space is to look for wobbling stars. The wobbling could be caused by a large planet orbiting the star.

19.4 Modern Understanding of the Motions of Heavenly Bodies

By about 1700 the motions of objects in space were well understood. Astronomers knew that the Sun is the center of our Solar System and that planets revolve around it in elliptical orbits. In addition to revolving around the Sun, the planets simultaneously spin on their axes. The Earth spins approximately 365 times for each complete orbit

around the Sun. Each complete rotation of the Earth represents one day. As the Earth rotates about its axis, the Sun, Moon, and stars appear to move across the sky from east to west. We explained in Chapter 15 that the Earth's axis is tilted and that this tilt combined with the Earth's orbit around the Sun produces the seasons.

Recall from Section 19.1 that different stars and constellations are visible during different seasons. The Earth's revolution around the Sun causes this seasonal

Figure 19–12 The night sky changes with the seasons because the Earth is continuously changing position as it orbits the Sun.

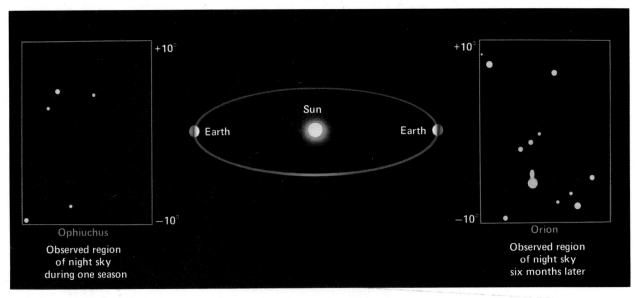

change in our view of the night sky (Figure 19–12). Part of the sky is visible on a winter night when the Earth is on one side of the Sun, and a different part is visible on a summer night six months later.

More recent measurements have revealed several additional types of planetary motion. As the Earth rotates, its axis wobbles. In addition, the Moon's gravity pulls the Earth slightly out of its orbit. Although we normally draw the Earth's orbit around the Sun as a line, the Moon's effect actually causes the Earth to spiral as it circles the Sun. At the same time, the entire Solar System is moving toward the star Vega. The stars in our neighborhood revolve around the center of the Milky Way galaxy, and the galaxy itself speeds along, carrying our Sun and planets through the void of intergalactic space.

Motion of the Moon

The gravitational attraction between the Moon and the Earth not only holds the two in orbit around each other, but affects the surfaces and interiors of both. We learned that the Moon's gravitation causes tides on Earth. In turn, the Earth's gravitation pulls on the Moon sufficiently to cause it to bulge. This bulge is attracted by Earth's gravity, so it always faces the Earth. As a result, the Moon rotates on its axis at the same rate at which it orbits the Earth. Thus, we always see the same lunar surface, and the other side was invisible to us until the Space Age.

Almost everyone has observed that the Moon appears to change shape over a month's time. It is said to be a **full moon** when it appears circular. A few evenings later, part of the disk is darkened. As the days progress, the dark portion grows until only a tiny curved sliver, called a **crescent moon**, is left, and finally, about 15 days after the Moon was full, it is totally dark and nearly impossible to see from Earth. This dark phase is called the **new moon**. (The authors of a popular astronomy text suggest that the "new" moon should properly be called the "no" moon because for all practical purposes it is invisible.)* Shortly after the new moon, the thin crescent reappears; it grows until the Moon is full again, after a total cycle of about 29.5 Earth days.

When the Moon is full, it rises approximately at sunset. On each successive evening it rises about 53 minutes later, so that in 7 days it rises in the middle of the night. In about a month the cycle is complete, and the next full moon rises on schedule in the early evening.

*Abell, Morrison, and Wolff, *Exploration of the Universe*, 6th ed. Philadelphia: Saunders College Publishing, 1991.

Phases of the Moon.   *(The Observatories of the Carnegie Institution)*

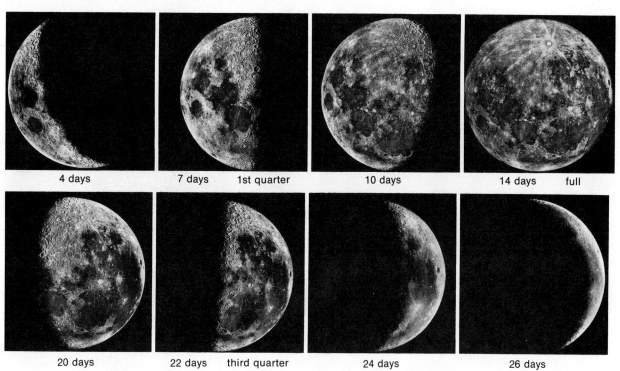

| 4 days | 7 days | 1st quarter | 10 days | 14 days | full |

| 20 days | 22 days | third quarter | 24 days | 26 days |

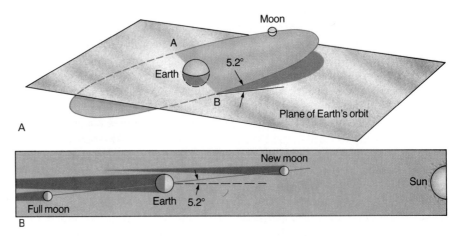

Figure 19–13 (A) The Sun and the Earth lie in one plane, and the Moon's orbit around the Earth lies in another. (B) A sideways look at the Sun, Moon, and Earth shows that most of the time the Moon's shadow misses the Earth and the Earth's shadow misses the Moon. (Scales are exaggerated for emphasis.)

Figure 19–14 Phases of the Moon. The upper part of the drawing shows the Earth-Moon orbital system viewed over the course of a month. The lower portion shows what the Moon looks like from Earth at each phase.

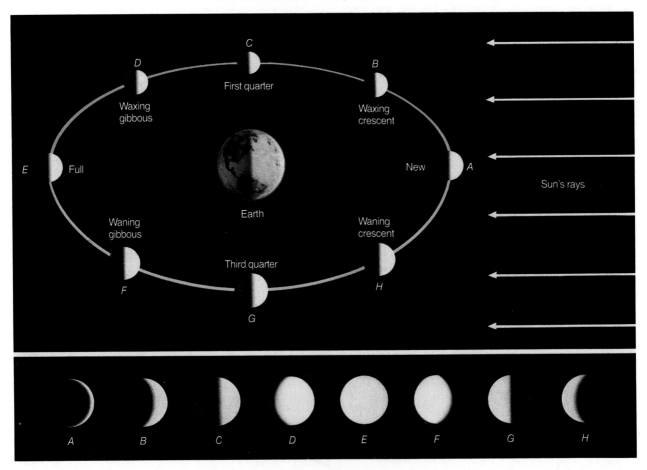

To understand the phases of the Moon, we must first realize that the Moon does not emit its own light but reflects light from the Sun. Thus, the phase of Moon depends on the relative positions of the Sun, Moon, and Earth. The Earth revolves around the Sun in a planar orbit, as if the two were positioned on a tabletop. The Moon's orbit around Earth is tilted with respect to this plane, so that generally the Moon is not in the same plane as the Earth and Sun (Fig. 19–13A). Therefore, the Earth's shadow does not normally fall on the Moon and the Moon's shadow does not normally fall on Earth.

As the Moon orbits Earth, the side of the Moon facing the Sun is fully illuminated. When the Moon is on the opposite side of the Earth from the Sun, the entire sunlit area is visible, and the Moon is full (Fig. 19–13B). However, if the Moon is between the Earth and the Sun, the sunlit side is facing away from us. In this position the Moon is new because the surface that faces the Earth is not illuminated and therefore hard to see. Midway between these two extremes, when the Moon is located 90 degrees from the line between the Earth and the Sun,

half of the illuminated side is visible, and the Moon is quartered. With each complete revolution around the Earth, the Moon passes through one complete cycle, from new to first quarter to full to third quarter and back to new again (Fig. 19–14).

But why does the Moon rise at a different time each day? It travels in a complete circle about the Earth about once a month. Imagine that today the Moon rises at 6:00 PM. Twenty-four hours later the Earth will have rotated once, but in the meantime the Moon will have moved a short distance across the sky; therefore, the Earth must rotate an additional 13.2° in order to catch up with the Moon (Fig 19–15). Thus, the timing of the moonrise depends on the Moon's position above a point on the Earth, and this also changes in a monthly cyclical pattern.

Eclipses of the Sun and the Moon

Recall that the plane of the Moon's orbit is tilted with respect to that of the Earth's orbit about the Sun, so

Figure 19–15 The Moon moves 13.2° every day. (A) The Moon is directly above an observer on Earth. (B) One day later the Earth has completed one rotation, but the Moon has traveled 13.2°. The Earth must now travel for another 53 minutes before the observer is directly under the Moon again.

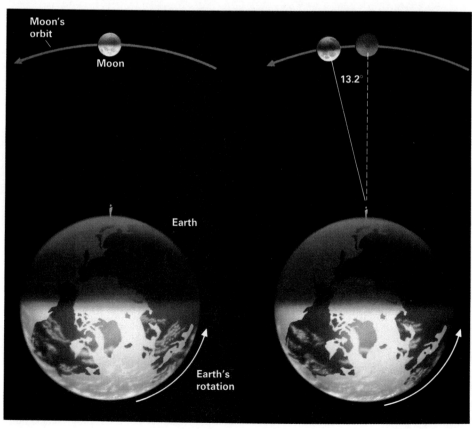

A   B

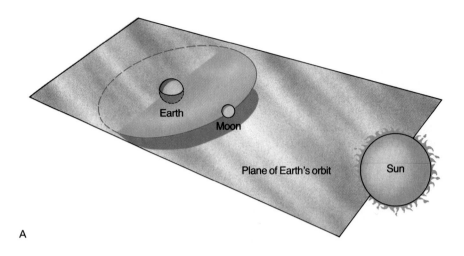

A

B

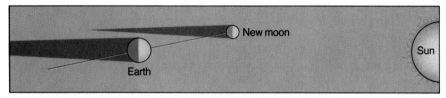

C

Figure 19–16 Eclipses of the Sun and Moon. The Sun and the Earth lie in one plane, and the Moon's orbit around the Earth lies in another. (Scales are exaggerated for emphasis.) (A) Normally the Moon lies out of the plane of the Earth-Sun orbit. (B) During the new moon, the Moon's shadow misses the Earth. (C) During the full moon, the Earth's shadow misses the Moon. (D) However, if the Moon passes through the Earth-Sun plane when the three bodies are aligned properly, then an eclipse occurs. (E) An eclipse of the Sun occurs when the Moon is directly between the Sun and the Earth, and the Moon's shadow is cast on the Earth. (F) An eclipse of the Moon occurs when the Earth's shadow is cast on the Moon.

normally the Moon lies slightly out of the plane of the Earth-Sun orbit (Fig. 19–16A). As a result, at a new moon the Moon's shadow misses the Earth (Fig 19–16B), and at a full moon the Earth's shadow misses the Moon (Fig. 19–16C).

However, on rare occasions, the Moon passes directly between the Earth and Sun (Fig 19–16D). When this happens, the Moon's shadow falls on the Earth, producing a **solar eclipse** (Fig. 19–16E). As the Moon slides in front of the Sun, an unnatural darkness descends, and the Earth becomes still and quiet. Birds return to their nests and stop singing. While the eclipse is total, the Moon blocks out the entire surface of the Sun, but the

outer solar atmosphere, or **corona**, normally invisible owing to the Sun's brilliance, appears as a halo around the black Moon (Fig. 19–17). Due to the relative distances between Sun, Moon, and Earth and their respective sizes, the Moon's shadow is only a narrow band on the Earth (Fig 19–18). The band where the Sun is totally eclipsed, called the **umbra**, is never wider than 275 kilometers. In the **penumbra**, a wider band outside the umbra, only a portion of the Sun is eclipsed. During a partial eclipse of the Sun, the sky loses some of its brilliance but does not become dark. Viewed through a dark filter such as a welder's mask, a semicircular shadow cuts across the Sun (Fig 19–19).

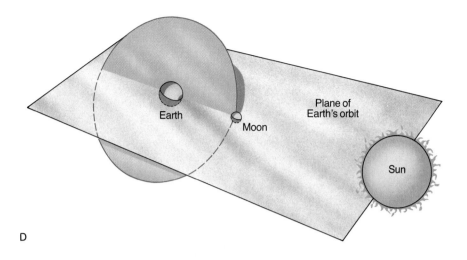

D

E

F

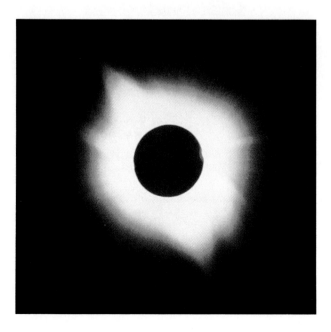

Figure 19–17 The solar corona as seen during the July 11, 1991, total eclipse of the Sun, photographed from La Paz, Baja California, Mexico. *(Stephen J. Edberg)*

Figure 19–18   A total solar eclipse is viewed in the narrow band, called the umbra, formed by the projection of the Moon's shadow on the Earth. The penumbra is the wider band where a partial eclipse is visible. (Drawing is not to scale.)

When the Earth lies directly between the Sun and Moon, the Earth's shadow falls on the Moon and the Moon temporarily darkens to produce a **lunar eclipse** (Fig. 19–16F). Lunar eclipses are more common and last longer than solar eclipses because the Earth is larger than the Moon, and therefore its shadow is more likely to cover the entire lunar surface. A lunar eclipse can last a few hours and occurs only during a full moon.

19.5 Modern Astronomy

Once astronomers understood the relative motions of the Sun, Moon, and planets, they began to ask questions about the nature and composition of these bodies, and to probe more deeply into space to study stars and other objects in the Universe. How do we gather data about such distant objects?

If you stand outside at night and look at a speck of light in the sky, the information you receive is limited by several factors. For one, your eye detects only visible

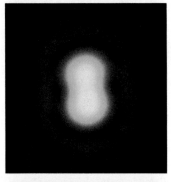

Figure 19–19   A view of the Sun during a partial eclipse. *(Jay Pasachoff)*

Figure 19–20   Resolution is the degree to which details are distinguishable in an image. The top photograph is poorly resolved and appears to show a single object, but with increasing resolution (middle and bottom) we clearly see two distinct objects.   *(Chris Jones, Union College)*

Newtonian

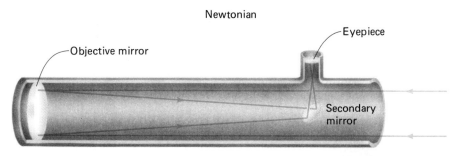

Figure 19–21   A Newtonian reflecting telescope. Incoming light (right) is collected and focused by a curved objective mirror. The light is then reflected to the eyepiece by a secondary mirror.

light, which is only one millionth of 1 percent of the electromagnetic spectrum. Thus, more than 99. 99 percent of the spectrum is invisible to the naked eye. In addition, the naked eye collects little light, and you may not see faint or distant objects at all. Your eye also has poor **resolution**; it may see one dot when two exist (Fig. 19–20). Finally, the light you see has been distorted by Earth's atmosphere. Modern astronomers attempt to overcome these difficulties with telescopes and other instruments.

Optical telescopes

The first telescopes, built in the early 1600s, collected and focused visible light and therefore are called **optical telescopes**. Most modern optical telescopes are **reflecting telescopes**. They collect light with a large curved mirror and reflect it to an eyepiece (Fig. 19–21). The reflecting mirror at the Palomar Observatory is 508 centimeters (200 inches) in diameter. It has 2.8 million times as much surface area as your pupil and collects 2.8 million times as much light. In order to create a detectable image from a weak signal, it collects light for several hours with a photographic plate or electronic detector at the eyepiece.

Astronomers have made many great discoveries with telescopes such as the one on Palomar Mountain, but telescopes of this type are limited. A mirror much larger than 600 centimeters sags under its own weight. In order to collect more light, the most recent telescopes use an array of smaller mirrors. When completed, the Keck telescope on Mount Mauna Kea in Hawaii will contain 36 individual mirrors aligned and focused by computer (Fig. 19–22). The total mirror area will be four times larger than the one at Palomar.

Problems caused by interference of the Earth's atmosphere can be solved by lifting a telescope into space. In 1990 the Hubble Space Telescope was launched into orbit

around the Earth (Fig 19-23). Unfortunately, its main mirror is flawed, so the images it sends to Earth are not as clear as the designers had hoped. Nevertheless the Hubble has begun to collect useful data, and its images are superior to those obtained by any telescopes on Earth.

Figure 19–22   The Keck telescope began observations in late 1990, when 9 of the 36 hexagonal mirrors were in place, as shown in the photograph. The remaining 27 mirrors were installed in the winter of 1991–92.   *(California Association for Research in Astronomy)*

Figure 19–23 The Hubble space telescope in orbit. *(NASA)*

Telescopes Using Other Wavelengths

Visible light is only a small portion of the electromagnetic spectrum. The wavelengths of electromagnetic radiation emitted by a star are determined by several factors, including the types of nuclear reactions that occur in the star, its chemical composition, and its temperature. In recent years, astronomers have enhanced our knowledge of stars and other objects in space by studying many different wavelengths, from low-energy radio and infrared signals to high-energy gamma and X-rays (Fig. 19–24).

Emission and Absorption Spectra

If light passes through a prism, it separates into a **spectrum**, an ordered array of colors* (Fig. 19–25). A rainbow is such an array, with white sunlight separated into its individual colors. Each color is formed by a band of wavelengths.

As light passes from the hot interior of a star through the cooler outer layers, some wavelengths are selectively

*In modern instruments light is dispersed by a diffraction grating, not a prism, but the effect is the same.

Figure 19–24 An aerial view of the Very Large Array (VLA), a radio telescope consisting of 27 mobile antennas spread out over 36 kilometers. *(National Radio Astronomy Observatory)*

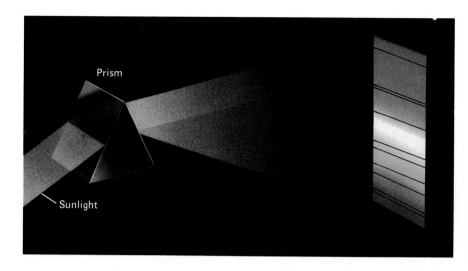

Figure 19–25   A prism disperses a beam of sunlight into a spectrum of the component wavelengths.

absorbed by atoms in the star's outer atmosphere. Therefore, in a spectrum of starlight, lines of darkness cross the band of colors. This is called an **absorption spectrum** (Fig. 19–26). Each dark line represents a wavelength that is absorbed by atoms of a particular element. Thus, an absorption spectrum enables us to determine the chemical composition of a star. This type of analysis is so effective that astronomers discovered helium in the Sun 27 years before chemists detected it on Earth. Because an atom's spectrum changes with temperature and pressure, spectra can also be used to determine surface temperatures and pressures of stars.

Whenever an atom absorbs radiation, it must eventually re-emit it. **Emission spectra** are often hard to detect against the bright background of starlight, but they can be seen along the outer edges of some stars and in large clouds of dust and gas in space.

Spacecraft

On October 4, 1957, the U.S.S.R. launched the first spacecraft, Sputnik I. A few months later America launched its space probe. Since then, dozens of spacecraft have studied our neighbors in space. Some of the most notable missions include six manned lunar landings, 25 orbital or landing missions to Venus, two successful landings on Mars, and four probes to the outer planets. In addition, Earth satellites study both the Earth and distant objects. The data from these spacecraft will be presented in the following two chapters.

Figure 19–26   A copy of the first solar absorption spectrum, taken in 1811. The dark lines are the absorption lines.   *(Deutsches Museum)*

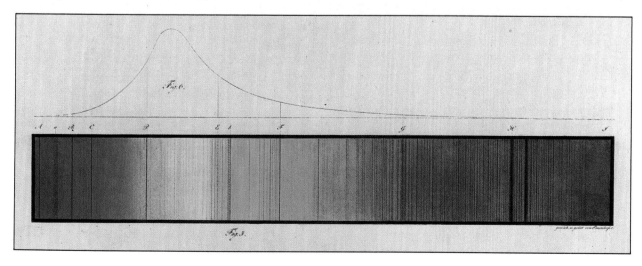

SUMMARY

•

Aristotle proposed a **geocentric**, or Earth-centered, Universe in which a stationary, central Earth is surrounded by **celestial spheres** that contain the Sun, the Moon, the planets, and the stars. Ptolemy modified the geocentric model to explain the **retrograde motion** of the planets. In Ptolemy's model each planet moves in small circles within its larger orbit. Copernicus believed that the Universe should operate in the simplest manner possible and showed that a **heliocentric** Solar System best explains planetary motion. Kepler described the elliptical orbits of the planets mathematically. Galileo used observation and experimentation to discredit the geocentric model and prove that the planets revolve around the Sun. Newton proved that gravity holds the planets in elliptical orbits.

The revolution of the Moon around the Earth causes the phases of the Moon. A **lunar eclipse** occurs when the Earth lies directly between the Sun and the Moon. A **solar eclipse** occurs when the Moon lies directly between the Sun and Earth.

Objects in space are studied with both optical telescopes and telescopes sensitive to invisible wavelengths. Atomic **spectra** provide information about the chemical compositions and temperatures of stars and other objects. Instruments carried aloft by spacecraft eliminate interference by the Earth's atmosphere and allow close inspection of objects in the Solar System.

KEY TERMS

•

Constellation   *578*
Planet   *520*
Retrograde motion   *520*
Geocentric   *521*
Celestial sphere   *521*

Parallax   *521*
Rotation   *524*
Revolution   *524*
Heliocentric   *524*

Full moon   *529*
Crescent moon   *529*
New moon   *529*
Solar eclipse   *532*

Corona   *532*
Umbra   *532*
Penumbra   *532*
Lunar eclipse   *534*
Sprectum   *536*

REVIEW QUESTIONS

•

1. What is a constellation? What is its astronomical significance?

2. Describe the apparent motion of the stars, as seen by an observer on Earth, over the course of a single night and over the course of a year.

3. List and explain Aristotle's observations and the reasoning he used to support his geocentric model of the Solar System.

4. What is retrograde motion? How is it explained in Ptolemy's and Copernicus's models?

5. Explain how Galileo's studies of physics contributed to his rejection of Aristotle's geocentric model.

6. Explain how Galileo's observations of the Milky Way, Sun, and Moon led him to question Aristotle's geocentric model.

7. What evidence convinced Galileo that the Earth revolves around the Sun?

8. What was the astronomical significance of Newton's studies of gravity?

9. How is the Moon positioned with respect to the Earth and the Sun when it is (a) full? (b) new? (c) crescent?

10. Why does the Moon rise approximately 50 minutes later each day than it did the previous day?

11. Draw a picture of the Sun, Earth, and Moon as they will be during an eclipse of the Sun. Draw a picture of the Sun, Earth, and Moon as they are aligned during an eclipse of the Moon. Explain how these positions produce eclipses.

DISCUSSION QUESTIONS

1. The Earth simultaneously rotates on its axis and revolves around the Sun. Which of these two motions is responsible for each of the following observations? (a) The Sun rises in the east and sets in the west. (b) Different constellations appear in the summer and winter. (c) On any given night the stars in the Northern Hemisphere appear to revolve around the Pole Star.

2. The Moon revolves around the Earth while the Earth rotates on its axis. Which of these two motions is responsible for each of the following observations? (a) The Moon rises in the east and sets in the west. (b) The Moon rises approximately 50 minutes later each day. (c) The Moon passes through monthly phases.

3. If you were the editor of a modern scientific journal, how would you respond to Aristotle's arguments for the geocentric model of the Universe? Defend your position.

4. Suppose that you are driving a car and the speedometer reads 80 km/hr. A person sitting next to you and looking at the speedometer at the same time may read a different value, perhaps 70 km/hr. Explain how two people can look at the same instrument at the same time and read different values.

5. With modern telescopes, astronomers can determine how far away some stars are by observing the parallactic shift as the Earth travels around the Sun. Using this technique, would it be easier to estimate distances to nearby stars or to distant stars, or would it be equally difficult for all stars? Defend your answer.

6. Describe the lunar cycle if the Sun, Earth, and Moon lay in a single plane.

7. Many societies use a lunar calendar instead of a solar one. In a lunar calendar, each month represents one full cycle of the Moon. How many days are there in a lunar month? If there are 12 months in a lunar year, is a lunar year longer or shorter than a solar year? Explain.

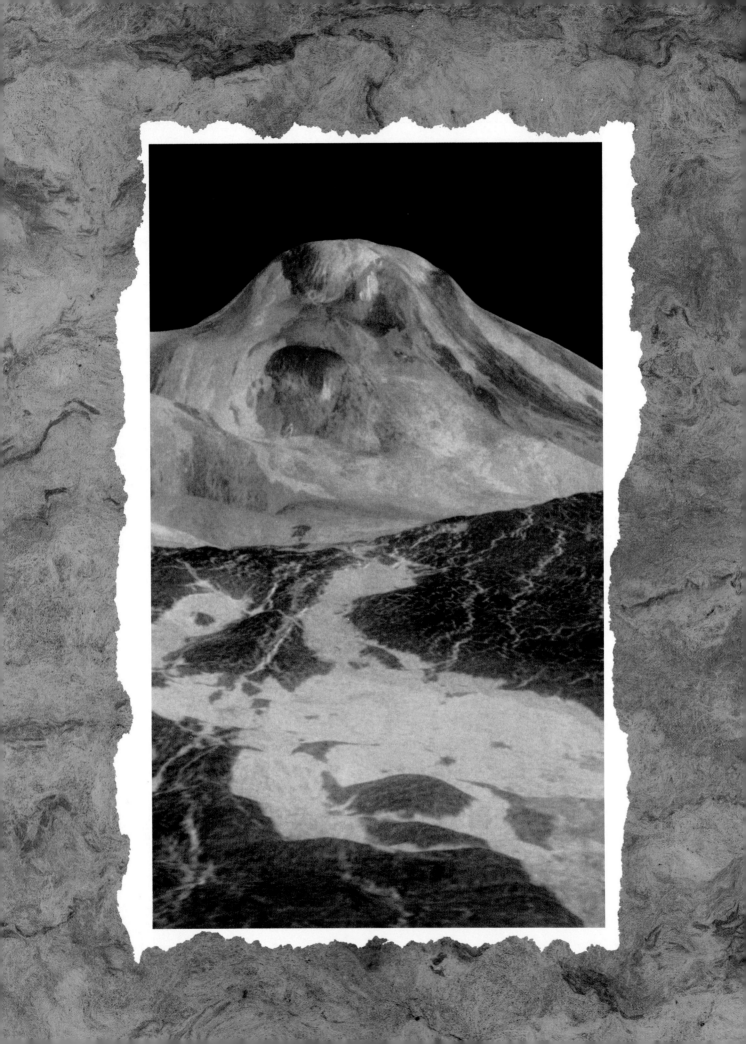

Planets and
Their Moons

20

T wenty-five years ago scientists knew more about the compositions and structures of distant stars than they knew about the planets and moons in our own Solar System. As explained in Chapter 19, stars emit their own light. Spectral analysis of starlight provides information about composition, temperature, and other important parameters. Planets and moons, on the other hand, emit no light. They merely reflect sunlight, so spectral analysis provides little information about them. Our closest neighbor in space, Venus, is particularly difficult to study because it is surrounded by dense clouds.

Thus, much of our knowledge about the planets and moons has been obtained from spacecraft launched within the past few decades. Several important missions are in space now, and the data they are relaying to Earth will modify some of the information presented in this chapter.

20.1 The Origin of the Solar System

Recall from Chapter 1 that the Solar System formed about five billion years ago from a cloud of dust and gas rotating

● The volcano, Maat Mons, on Venus. *(NASA/JPL)*

slowly in space. More than 90 percent of the matter in the cloud gravitated toward its center to form the proto-sun, the earliest form of the Sun. The protosun was heated by the gravitational collapse of the cloud, but it was not a true star because it did not yet generate energy by nuclear fusion. Heat from the protosun warmed the inner region of the disk-shaped cloud. Then, as the gravitational collapse became nearly complete, the disk cooled. Gases condensed to form small aggregates, much as raindrops or snowflakes form when moist air cools.

Pressure inside the protosun then became great enough for fusion to begin, and the modern Sun was born. At about the same time, matter surrounding the Sun continued to coalesce, forming planets. Solar radiation boiled off most of the light elements, such as hydrogen and helium, from the inner planets, leaving mainly rock and metal behind. Thus, Mercury, Venus, Earth, and Mars are mostly solid spheres and are called the **terrestrial planets** (Fig. 20–1). The outer **Jovian planets**, Jupiter, Saturn, Uranus, and Neptune, are far enough from the Sun to have retained large amounts of hydrogen, helium, and other light elements. All have relatively small solid cores surrounded by swirling liquids and gaseous atmospheres. The outermost planet, Pluto, is anomalous and is discussed further in Section 20–9.

Figure 20–1   The terrestrial planets and the larger moons of the Solar System at the same scale.   *(NASA)*

20.2   Mercury: A Planet the Size of the Moon

Mercury has a radius of 2400 kilometers, less than four-tenths that of Earth. It is the closest planet to the Sun and therefore orbits faster than any other planet. Each Mercurial year is only 88 Earth days long.* Mercury rotates slowly on its axis, completing only three rotations for each two revolutions around the Sun. Since it is so close to the Sun and its days are so long, temperature on its sunny side reaches 450°C, as hot as the inside of a self-cleaning oven and hot enough to melt tin or lead. In contrast, temperature on its dark side drops to −175°C, nearly cold enough to liquefy gaseous oxygen if any existed on the planet. The lack of an atmosphere is partly responsible for these extreme temperatures because there is no wind to carry heat from one region to another.

Little was known about the surface of Mercury before the spring of 1974, when the spacecraft *Mariner 10* passed within a few hundred kilometers of the planet and began relaying information to Earth. The first photographs revealed a cratered surface similar to that of the Moon (Fig. 20–2). The craters on both Mercury and the Moon formed during intense meteorite bombardment early in the history of the Solar System. Why are Mercury and the Moon

*Unless otherwise noted, all indications of time in this chapter are given in Earth days and years.

so heavily cratered, and Earth is not? Earth was also pockmarked by the same episode of meteorite impacts. If you could go back about four billion years, its surface would appear similar to that of Mercury today. However, tectonic activity and erosion have erased the Earth's craters. The preservation of those features on Mercury tells us that tectonic activity and erosion have not occurred on Mercury during the past four billion years.

Flat plains on Mercury are probably vast lava flows that formed early in its history when the interior of the planet was hot enough to produce magma. However, Mercury is so small that its interior has cooled and tectonic activity has ceased. It is so close to the Sun that its atmosphere has boiled off into space, so no wind, rain, or flowing water has eroded its surface.

In 1991, radar images of Mercury revealed highly reflective regions at the poles. One possible explanation is that the poles consist of ice. If polar ice caps exist, then scientists will have to modify their inferences of temperatures on Mercury.

20.3   Venus: The Greenhouse Planet

Of all the planets in our Solar System, **Venus** most closely resembles Earth in size, density, and distance from the Sun. Therefore, astronomers once thought that the envi-

Figure 20–2 A close-up of Mercury. The photograph shows an area 580 kilometers from side to side. *(NASA)*

ronment on Venus might be similar to that on Earth and that life might be found there. Until recently it was impossible to study the surface of Venus, for it is obscured by a thick, dense atmosphere with an opaque cloud cover (Fig. 20–3). Today our knowledge of Venus has been greatly increased by data obtained from spacecraft. It now seems certain that no life exists on Venus because its environment is too harsh. Its surface is nearly as hot as that of Mercury, hot enough to destroy the complex organic molecules necessary for life.

The Atmosphere of Venus

Although Venus is slightly smaller than Earth and its gravitational force is less, its atmosphere is 90 times denser than the Earth's. Thus, atmospheric pressure at the surface of Venus is equal to the pressure 1000 meters beneath the sea on our planet. The Venusian atmosphere is more than 97 percent carbon dioxide, with small amounts of nitrogen, helium, neon, sulfur dioxide, and other gases. Drops of concentrated sulfuric acid rain fall from sulfurous clouds.

Why is Venus's atmosphere so different from ours? We can imagine that early in the history of the Solar System, Venus and Earth were similar to one another, because they formed from the same cloud of dust and gas. A space traveler would have viewed the two planets as sisters. Rivers may have flowed over the surface of

Venus, and oceans may have filled its low-lying basins. On Earth, some limestone precipitates chemically. The same process may have occurred on Venus, forming beds of limestone beneath Venusian seas.

Figure 20–3 The solid surface of Venus is obscured by a turbulent cloud cover. *(NASA/JPL)*

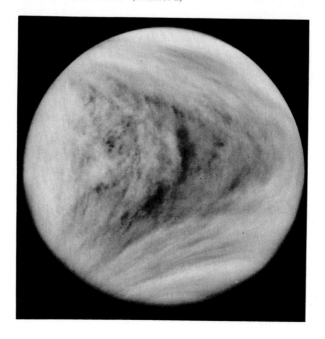

TABLE 20–1

	Distance from Sun (Millions of Kilometers)	Radius (Compared to Radius of Earth = 1)	Mass (Compared to Mass of Earth = 1)	Density (Compared to Density of Water = 1)	Composition of Planet	Density of Atmosphere (Compared to Earth's Atmosphere = 1)	Number of Satellites
Planet							
Terrestrial Planets							
Mercury	58	0.38	0.06	5.4		One billionth	0
Venus	108	0.95	0.82	5.2	Rocky with	90	0
Earth	150	1	1	5.5	metallic core	1	1
Mars	229	0.53	0.11	3.9		0.01	2
Jovian Planets							
Jupiter	778	11.2	318	1.3	Liquid hydrogen surface with liquid metalic mantle and solid core	Dense and turbulent	16
Saturn	1420	9.4	94	0.7			19
Uranus	2860	4.0	15	1.3 }	Hydrogen and helium outer layers with solid core	Similar to Jupiter except that some compounds that are gases on Jupiter are frozen on the outer planets	15
Neptune	4490	3.9	17	1.7 }			8
Most Distant Planet							
Pluto	5910	0.17	0.0025	2.0	Rock and ice	0.00001	1

Comparison of the Nine Major Planets

However, there is one important difference between the two planets. Venus is closer to the Sun than Earth is, and it therefore receives more solar heat. This small difference in solar radiation led to chemical changes in the atmosphere of Venus. These changes made Venus very hot. Recall from Chapter 15 that carbon dioxide gas and water vapor both absorb infrared radiation and thereby warm the surface of a planet in a process called the greenhouse effect. Carbon dioxide exists in many forms: as a gas in the atmosphere, dissolved in water, and combined with other compounds to form limestone. Water commonly exists as a solid (ice), liquid, or gas.

Three processes may have combined to produce a runaway greenhouse effect on Venus.

1. Carbon dioxide dissolved in Venusian seawater escaped into the atmosphere when the water was heated.

2. The heating also accelerated chemical weathering of limestone, which released more carbon dioxide.

3. Finally, as the primordial Venusian oceans warmed,

more water evaporated from them. Since both carbon dioxide and water vapor absorb infrared energy, greenhouse warming occurred.

As the temperature increased, all three of these processes speeded up. Most of the water reacted with sulfur dioxide to form sulfuric acid. Thus, the atmosphere of Venus became hot, acidic, and rich in carbon dioxide. Greenhouse warming heated the surface of the planet until eventually it reached its current temperature of about 450°C.

The Surface and Geology of Venus

Astronomers use radar to penetrate the Venusian atmosphere and produce photo-like images of its surface. The most spectacular images have come from the *Magellan* spacecraft, which began relaying data in September 1990. At this writing, only 20 percent of the mission has been completed and a fraction of the data analyzed. Therefore, both the data and the conclusions discussed here are preliminary.

Recall from our discussion of Mercury that dense swarms of meteorites bombarded the Solar System early in its history. Scientists have sampled rocks from cratered regions of the Moon and dated them radiometrically. Thus, they have established a calendar of major episodes of meteorite bombardment in the inner Solar System. The largest swarms of meteorites bombarded the terrestrial planets shortly after they formed 4.6 billion years ago and again between 4.2 and 3.9 billion years ago. With this calendar astronomers can deduce the age of the surface of a planet or moon by counting the density of craters. Few craters exist on Venus, indicating that its surface was reshaped after the major meteorite bombardments. Volcanic eruptions caused much of this resurfacing between 100 million to 1 billion years ago. Today Venus has several large volcanic mountains, and relatively recent lava flows cover much of its surface (Fig. 20–4).

Most volcanoes on Earth occur at tectonic plate boundaries, so the discovery of volcanoes on Venus led planetary geologists to look for evidence of plate tectonic activity there. Radar images from *Magellan* and earlier *Pioneer* spacecraft show that 60 percent of Venus's surface is covered by a flat plain. Two large and several smaller mountain chains rise from the plain (Fig. 20–5).

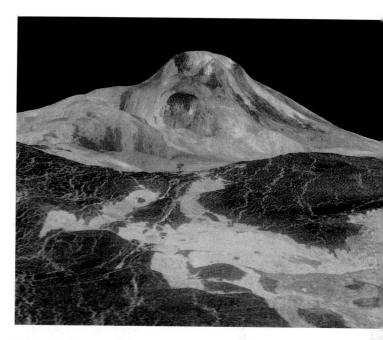

Figure 20–4   The volcano, Maat Mons, on Venus, with large lava flows in the foreground. This image was produced from radar data recorded by *Magellan* spacecraft. Simulated color is based on color images supplied by Soviet spacecraft.   *(NASA/JPL)*

Figure 20–5   A map of Venus produced from radar images from the *Pioneer Venus Orbiter.* The lowland plains are shown in blue, and the highlands are shown in yellow and red-brown. (The colors are assigned arbitrarily.)   *(NASA/Ames Research Center)*

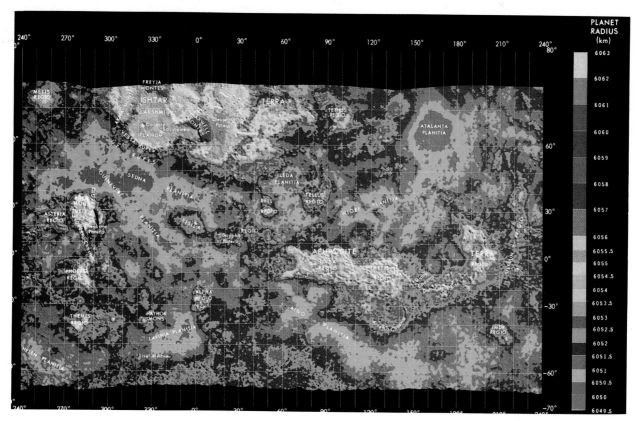

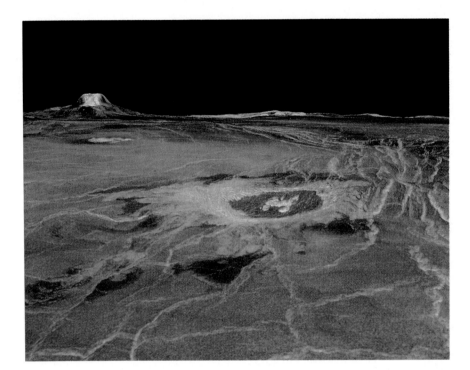

Figure 20–6   A portion of the surface of Venus showing a volcano, Gula Mons, in the background and a large impact crater in the foreground. Large crustal fractures appear near the crater. This image was produced from radar data recorded by *Magellan* spacecraft. Simulated color is based on color images supplied by Soviet spacecraft.   *(NASA/JPL)*

The tallest mountain is 11 kilometers high, 2 kilometers higher than Mount Everest. The images also show large crustal fractures and deep canyons (Fig. 20–6). If Earth-like horizontal motion of tectonic plates caused these features, then spreading centers and subduction zones would exist on Venus. After studying low-resolution images from *Pioneer* spacecraft, James Head and Larry Crumpler of Brown University predicted that an Africa-size highland, Aphrodite Terra, might be a spreading center, similar to the mid-oceanic ridges on Earth. However, when the higher resolution *Magellan* data were scrutinized, no steep scarps, transform faults, or other evidence of crustal spreading was observed. Furthermore, nothing like oceanic trenches on our own planet has been seen.

Other planetary geologists have suggested that up-wellings of hot rock from the planet's interior dominate tectonic activity on Venus. In some regions the rising rock has melted and erupted from volcanoes in a manner similar to processes that formed the Hawaiian Islands. In other regions, the hot Venusian mantle plumes have lifted the crust to form nonvolcanic mountain ranges.

Some geologists have suggested the term "**blob tectonics**" to describe the geology of Venus. Tectonics is the shaping and deformation of a planetary surface by internal processes. Clearly, Venus has experienced tectonic activity. However, its primary process seems to be rising and sinking of the surface rather than horizontal motions of lithospheric plates. (Think of blobs of rock rising and falling.)

Recall from Chapter 5 that mantle plumes on Earth may initiate rifting of the lithosphere and formation of a spreading center. Why has a similar process not occurred on Venus? Its surface temperature may be so high that the surface rocks are much more plastic than those on Earth. Therefore, rock stretches and lithosphere-deep cracks do not form. It is also possible that since Venus is smaller than Earth, its mantle is cooler and it has a thicker lithosphere. The internal heat is sufficient to cause vertical movement of the lithosphere but not to cause plates to move horizontally.

20.4 The Moon: Our Nearest Neighbor

Earth is the third planet from the Sun; it has been the subject of all the foregoing chapters of this book. Most planets have small orbiting satellites called **moons**. The Earth's Moon is close enough so that we can see some of its surface features with the naked eye. In the 1600s Galileo studied the Moon with the aid of a telescope and mapped its mountain ranges, craters, and plains. Galileo thought that the plains were oceans and called them seas, or **maria**. The word maria is still used today, although we now know that these regions are dry, barren, flat expanses of volcanic rock (Fig. 20–7). The first close-up photographs of the Moon were taken by a Russian orbiter in 1959. A decade later the United States landed the first of six manned *Apollo* spacecraft on the Moon's surface.

A

B

Figure 20–7 (A) The Moon as photographed from the *Apollo* spacecraft at a distance of 18,000 kilometers. Note the cratered regions and the flat lava flows. (B) A close-up of a portion of the heavily cratered surface of the Moon. Notice that smaller craters lie within the larger ones. The larger craters formed first. *(NASA)*

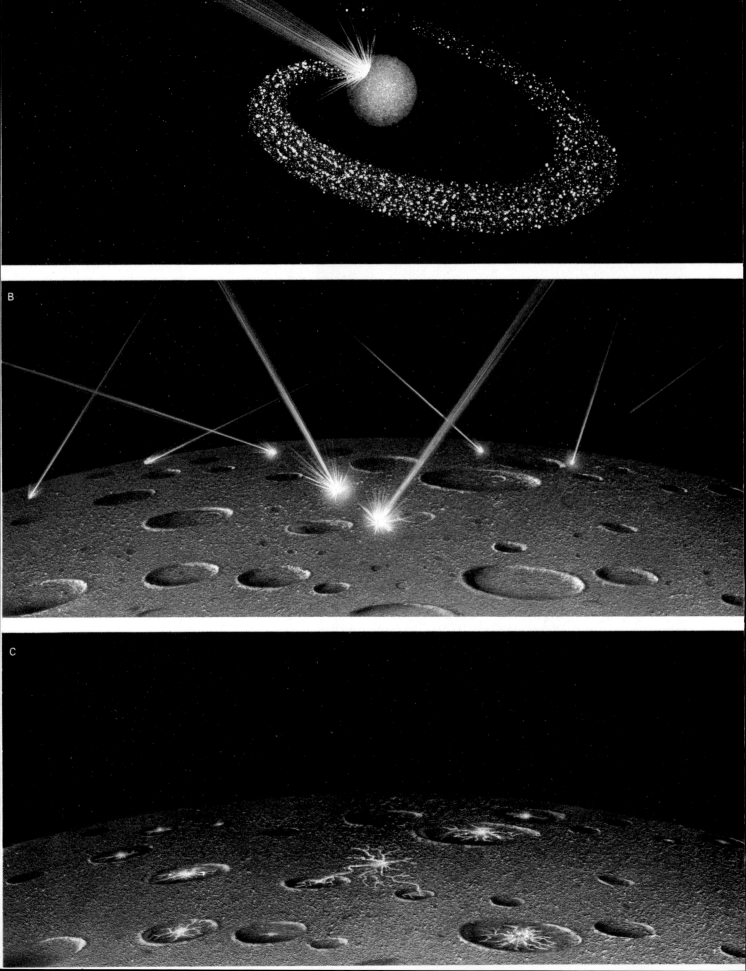

The *Apollo 11* lander returning to the command module after the first humans landed on the Moon. In the background the Earth rises over the flat lunar plain.   *(NASA)*

The Apollo program was designed to answer several questions about the Moon: How did it form? What is its geologic history? Was it once hot and molten like the Earth? If so, does it still have a molten core, and is it tectonically active?

Formation of the Moon

The question "How did the Moon form?" continues to be debated. According to the most popular theory, a giant object, perhaps a mini-planet, smashed into the Earth shortly after it formed. The collision blasted parts of this object and portions of the Earth into space. The fragments began to orbit the Earth and eventually coalesced, forming the Moon (Fig. 20–8A).

Perhaps the single most significant discovery of the Apollo program was that much of the Moon's surface consists of igneous rocks. The maria are mainly basalt flows. The highland rocks are predominantly anorthosite, a feldspar-rich igneous rock not common on Earth. Additionally, in both the maria and the highlands, the country rock was crushed by meteorite impacts and then the fragments were welded together by lava. Since igneous rocks form only from magma, it is clear that portions of the Moon were once hot and liquid.

How did the Moon become hot enough to melt? The Earth was heated initially from collisions among gases, dust, and larger particles as they collapsed under the influence of gravity. This process is called **gravitational coalescence.** Later, radioactive decay and intense meteorite bombardment heated the Earth further. But what about the Moon? Geologists have calculated that it would have taken 800 million years for radioactive decay to melt Moon rock. Radiometric age dating shows that the oldest lunar igneous rocks formed when the Moon was a mere 200 million years old. Thus, since there was insufficient time for the rock to have been melted by radioactivity, gravitational coalescence and meteorite bombardment must have been the main causes of early melting of the Moon.

History of the Moon

The Moon formed about 4.5 billion years ago shortly after the Earth formed. Soon thereafter, meteorite bombardment melted its outermost 100 kilometers. Then the bombardment diminished and the surface cooled. The igneous rocks of the lunar highlands are about 4.4 billion years old, indicating that parts of the Moon's surface must have become solid by that time. Swarms of meteorites bombarded the Moon again between 4.2 and 3.9 billion years ago (Fig. 20–8B). Billions of meteorites, some as large as the state of Rhode Island smashed into the surface, blasting huge craters and raising the Moon's temper-

Figure 20–8   A brief history of the Moon. (A) According to one theory, the Moon was formed about 4.5 billion years ago when a Mars-size object struck the Earth, sending a cloud of vaporized rock into orbit. The vaporized rock rapidly coalesced to form the Moon. (B) During its first half-billion years, intense meteorite bombardment cratered the Moon's surface. (C) The dark, flat maria formed about 3.8 billion years ago as lava flows spread across portions of the surface. Today all volcanic activity has ceased.

ature again. At the same time, radioactive decay heated the lunar interior.

By about 3.8 billion years ago, most of the Moon's interior was molten. The denser elements, such as iron and nickel, gravitated toward the center, and the less dense elements floated toward the surface. Thus, the Moon, like the Earth, has a metallic core surrounded by silicate rocks of lower density. Some of the molten rock erupted onto the Moon's surface. The lunar maria formed when lava filled circular meteorite craters (Fig. 20–8C). This volcanic activity lasted approximately 700 million years and ended about 3.1 billion years ago.

The Earth must have shared a similar history until this time, but the Moon is so much smaller that it soon cooled and has remained geologically quiet and inactive for the past three billion years. Seismographs left on the lunar surface by *Apollo* astronauts indicate that the energy released by moonquakes is only one-billionth to one-trillionth as much as that released by earthquakes on our own planet. Seismic data indicate that the core of the Moon is probably molten, or at least hot enough to be soft and plastic. However, the cool, solid upper mantle and crust are too thick for seismic and tectonic activity. Meteorite bombardment of the Moon has continued throughout this period of tectonic dormancy, but never again has there been such intense rains of meteorites as occurred early in its history.

20.5 Mars: A Search for Extraterrestrial Life

In the 11 years between 1965 and 1976, a total of 12 United States and Russian spacecraft visited **Mars**. They included two *Viking* craft that landed on the surface of the planet and collected and analyzed samples of Martian soil.

The Geology of Mars

In many ways the geology of Mars is similar to that of Venus. It has old (heavily cratered) uplands and younger (lightly cratered) lowland plains (Fig. 20–9). Lava flows much like those on Venus and the Moon cover the plains. The 10-kilometer high Tharsis bulge is the largest plateau, crowned by Olympus Mons, the largest volcano in the Solar System (Fig. 20–10). Olympus Mons is nearly three times as high as Mount Everest, with a diameter of 500 kilometers and a height of 25 kilometers. Tremendous parallel cracks split the crust adjacent to the Tharsis bulge (Fig. 20–11). If this bulge lay near a tectonic plate boundary, we would expect to see folding or offsetting of the cracks. However, the cracks are neither folded nor offset.

Figure 20–9 A view of Mars showing mountains and lowland plains with several impact craters. *(NASA/JPL)*

Therefore, geologists think that a rising mantle plume may have formed the Tharsis bulge and its volcanoes. On Earth, the lithosphere continuously slides over a mantle plume, forming a chain of volcanoes such as the Hawaiian Islands. On Mars, however, the lithosphere is stationary, allowing continuous volcanic activity in one location. The parallel cracks may be the result of stretching as the crust uplifted.

The Search for Life

One of the most exciting aspects of the exploration of Mars has been the search for life. Before the space age, photographs from earthbound telescopes showed white Martian polar regions that shrink in Martian summer and expand in winter (Fig. 20–12). Some observers thought that the white regions were ice caps. If this hypothesis were correct, and if the ice melts in summer, water must be available. Astronomers also observed that, each spring, large regions near the Martian equator darken, only to become light again in winter. Many people speculated that annual blooms of vegetation caused these color changes.

However, speculations about life on Mars have been refuted by data from orbiters and landers. High-resolution images of the Martian surface show no forests or grassy prairies. The seasonal dark patches that appear in equatorial regions each spring are not blooms of vegetation, but

Figure 20–10 Olympus Mons, the largest volcano on Mars and probably the largest in the Solar System. *(NASA/JPL)*

Figure 20–11 Crustal fractures in the Tharsis region of Mars. Two large volcanoes appear at right. *(NASA/JPL)*

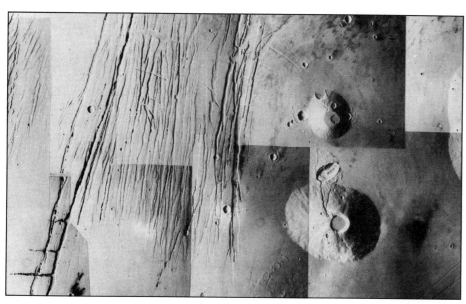

Focus On

Extraterrestrial Life

For many years, people have asked whether or not Earth is unique in its ability to support life. Early searches for extraterrestrial life looked to the nearby planets, Venus and Mars. But, as we have seen, Venus is too hot, and Mars, although potentially more hospitable, seems to be completely void of traces of life. Today astronomers are expanding their search for life toward more distant regions of the Solar System and other regions of our galaxy.

Some biologists have speculated that primitive life may have evolved in the atmosphere of Jupiter. The outer surface of the Jovian atmosphere is too cold ($-140°C$) for organisms to survive, and the interior is too hot. But somewhere in between, the temperature must be favorable. The Jovian atmosphere is composed primarily of hydrogen, helium, ammonia, methane, water, and hydrogen sulfide. These compounds contain the major elements needed to build living tissue. Lightning storms occur on Jupiter, and these electrical discharges could provide the energy needed to synthesize amino acids, the basic building blocks of proteins. In short, all the ingredients are present, and it would be thrilling, but not totally surprising, to find microorganisms floating about in the clouds of the giant planet.

If living organisms are not found on any of the planets, they might occur on a moon. Perhaps biological evolution has occurred under the ice on Europa or in the methane seas of Titan.

One can only guess about life in other solar systems. An average galaxy contains roughly 100 billion stars, and millions of galaxies exist in the Universe. So many stars exist that is reasonable to believe that planetary systems have formed around some. In recent years, this statistical inference has been supported by direct evidence. Recall from Chapter 1 that astronomers have photographed a disk of particles surrounding the star **Beta Pictoris** that appears to be a solar system in the process of forming. Furthermore, the motion of several other stars indicates that they may be orbited by planets.

We live in a time when science fiction is popular, and most science fiction authors write about extraterrestrial life forms. However, in the world of science it is important to separate what is known from what is not known. The facts are summarized in the following table.

What We Do Know

1. Life exists on Earth.

2. Some of the molecules essential to life as we know it (such as amino acids) have been formed in the laboratory by adding energy to a mixture of simple inorganic compounds. Some of these same essential substances have been found in meteorites. In addition, some organic molecules have been detected in interstellar space.

3. We can formulate reasonable hypotheses for the formation of planets. From these formulations, it is reasonable to suppose that other stars besides the Sun have planets. Actual evidence exists in one or two cases.

4. There are very, very many stars.

What We Have Not Found

1. No spacecraft has found evidence of extraterrestrial life in the Solar System.

2. The transformation of simple chemical compounds essential to life into an actual living organism has never been observed.

3. If advanced civilizations exist outside the Solar System, we have heard nothing from them, although we are listening carefully.

Conclusions

1. We cannot rule out the existence of extraterrestrial life. The data indicate that generation of organic molecules is an ordinary chemical process in the Universe. Many scientists deduce that, under the right conditions and if enough time is available, these molecules can combine to form living organisms. The Universe has been in existence for billions of years, and many billions of stars exist. Therefore, one argument concludes that the formation of living organisms—and even their evolution into intelligent beings—is so probable that it almost surely must have happened in many parts of the Universe.

2. On the other hand, we cannot rule out the possibility that life is unique to Earth and exists nowhere else. Remember, there is no direct proof whatsoever of living entities anywhere but on our planet.

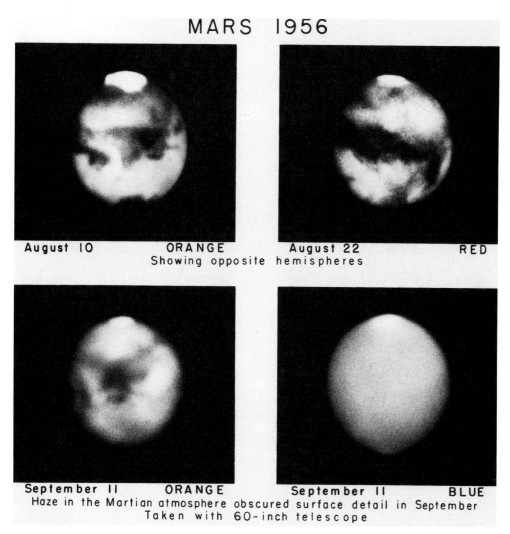

Figure 20–12 Seasonal changes in the size of the polar ice cap and the dark equatorial markings on Mars, as seen by ground-based telescopes. *(Lowell Observatory)*

rather great, dry dust storms powered by seasonal winds (Fig. 20–13). When the winds subside, bright dust particles settle onto the land surface, causing it to appear light-colored. The atmospheric density at Mars' surface is less then 1 percent that of Earth. Such a light atmosphere can produce high-velocity winds, but the winds cannot move much dust. Therefore, very little dust actually moves during these storms. In fact, cameras on the surface of the planet have barely detected any loss of visibility even during intense storms. The darkening appears significant only when seen from above the planet.

The winter ice caps, once thought to be the source of spring floods, are mostly frozen carbon dioxide, commonly called dry ice. Water ice is present on Mars, but it lies beneath the surface and never melts. At some time in the distant past, however, rivers must have flowed across

Mars's surface, eroding stream beds and canyons, for these features can be observed today (Fig. 20–14). Other photographs indicate that seas or lakes once covered parts of the Martian surface and massive floods transported sediment. But no rain has fallen for millions or hundreds of millions of years, and now Mars is a dry planet.

The Atmosphere and Climate of Mars

In trying to deduce the history of the Martian atmosphere and climate, it is helpful to compare Mars with Venus and Earth. Recall that Venus is closer to the Sun than Earth is, and thus has received more solar radiation. This radiation led to greenhouse warming and increasing temperature on Venus. Most of the planet's carbon dioxide escaped into the atmosphere.

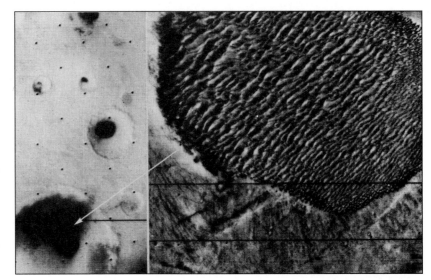

Figure 20–13 One of the mysterious "dark spots" on Mars, as seen by a ground-based telescope, is visible at the left. A close-up of the same spot taken by *Mariner 9* (right) clearly shows that the region is not vegetated, but covered with sand dunes. *(NASA/JPL)*

Figure 20–14 Streambeds on the surface of Mars were carved long ago by running water. *(Viking image from, NASA/JPL)*

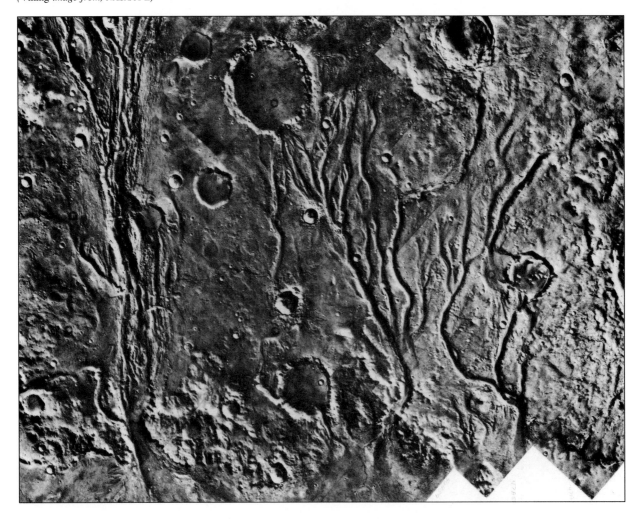

Earth receives less sunlight than Venus and therefore started out with a cooler surface. Oceans remained liquid, and additional water was frozen in glaciers. Carbon is distributed among the atmosphere, the biosphere, and rocks. This distribution has prevented excess carbon dioxide from accumulating in the atmosphere and has protected Earth from the runaway greenhouse process that has heated Venus's surface.

We see ancient volcanoes on Mars and we therefore assume that past volcanic eruptions must have released carbon dioxide and water vapor into the atmosphere, causing greenhouse warming. Clouds formed, rain fell, and streams flowed over the surface. But then water vapor and carbon dioxide froze into the ice caps and the planet became cooler. Today atmospheric pressure at the surface is only 0.006 that at the surface of Earth. Why did this reverse greenhouse effect occur?

Mars is half again as far from the Sun as Earth is. With less solar energy, it was initially cooler. The low temperature prevented water from evaporating, or froze water vapor that had been released during volcanic eruptions. With less water vapor, the atmosphere's insulating properties were reduced. As a result, radiation escaped into space and the planet's surface cooled. When the temperature became low enough, some of the gaseous carbon dioxide also froze into the ice caps.

In addition, Mars is smaller than Earth, so its gravitational field is weaker and some of its original water and carbon dioxide escaped into space. In short, mainly because of differences in size and distance from the Sun, Venus boiled, life evolved on Earth under moderate temperatures, and Mars froze.

Although Mars is inhospitable today, some scientists speculate that life may have evolved there during its more temperate past. If so, it is possible that fossils exist on Mars. Supporters of this hypothesis point out that Martian volcanoes must have released water, carbon dioxide, carbon monoxide, oxygen, hydrogen, nitrogen, and sulfur. On Earth these gases combined to form organic molecules—the building blocks of life. Thus, water and the other necessary compounds for the evolution of life probably once existed on Mars. This conclusion does not tell us that life did evolve, just that the components were present at one time.

The *Viking* landers searched for microorganisms in the soil and found none. However, they analyzed only a minuscule portion of the Martian surface. Many scientists suggest that the next spacecraft should land on an abandoned lake bed or stream channel, where fossils and organic matter are more likely to be found.

20.6 Jupiter: A Star That Failed

The Giant Planets

As explained in Section 20.1, Mercury, Venus, Earth, and Mars are the terrestrial planets. They are relatively small, with rocky outer layers, and all orbit close to the Sun.

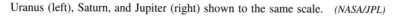

Uranus (left), Saturn, and Jupiter (right) shown to the same scale. *(NASA/JPL)*

Despite their differences, the four have similar chemical compositions. In contrast, the giant outer planets—Jupiter, Saturn, Uranus, and Neptune—are considerably different from the terrestrial group.

Visualize once again the primordial dust cloud that condensed to form the Solar System. The protosun and all the protoplanets were composed mostly of hydrogen and helium. As gravity pulled hydrogen toward the center of the Sun, the pressure became so great that hydrogen began to combine, or fuse, to form helium. Hydrogen fusion is still the source of the Sun's energy. The terrestrial planets were so close to the Sun, and their gravitational fields so weak, that most of their hydrogen, helium, and other light gases escaped and boiled off into space or were blown away by energetic particles streaming off from the Sun. In contrast, the protoplanets in the outer reaches of the Solar System were so far from the Sun that they remained cool. In addition, they were more massive and thus had stronger gravitational fields. As a result, they retained most of their gases. They may even have grown as they captured gases that escaped from the terrestrial planets. Thus, today each of the Jovian planets has a dense gaseous atmosphere, a very large liquid interior, and a much smaller solid core.

Structure and Composition of Jupiter

Jupiter retained most of its original hydrogen and helium. Therefore, its chemical composition is much like that of the Sun. However, Jupiter was not massive enough to generate fusion temperatures so it never became a star. Jupiter is 71,000 kilometers in radius. Its outer 12,000 kilometers are a vast sea of cold, liquid, molecular hydrogen, H_2. It has no hard, solid, rocky crust on which an astronaut could land or walk. A middle layer between the outer sea of liquid hydrogen and the core is also composed of hydrogen (Fig. 20–15). In this layer, temperatures are as high as 30,000°C and pressures are as great as 100 million times the Earth's atmospheric pressure at sea level. Under these extreme conditions, hydrogen molecules dissociate to form atoms. The pressure forces the atoms together so tightly that the electrons separate from their nuclei. These free electrons travel throughout the tightly packed nuclei much as electrons travel freely among metal atoms. As a result, the hydrogen conducts electricity and is called **liquid metallic hydrogen**. Movement within this fluid conductor generates a magnetic field ten times stronger than that of Earth.

Jupiter's core is a sphere about ten to twenty times as massive as the Earth and is probably composed of metals and rock surrounded by lighter elements such as carbon, nitrogen, and oxygen.

Above the liquid layers, the Jovian atmosphere is a mixture of gases, liquid droplets, and crystals consisting mainly of hydrogen, with smaller amounts of helium, ammonia, methane, water, hydrogen sulfide, and other compounds. The atmosphere is turbulent; great storms and changing weather patterns can be seen even from earthbound telescopes (Fig. 20–16). Powerful winds blow at speeds up to 500 kilometers per hour, in alternating bands from east to west and from west to east. Most storms form, distort, and dissipate within a few hours or days, but some are surprisingly stable over long periods of time.

More than 300 years ago, two European astronomers reported seeing a Great Red Spot on the surface of Jupiter. Although its shape and color change from year to year, the spot remains intact to this day (Fig. 20–17). The **Great Red Spot** is a giant hurricane-like storm. If the entire Earth's crust were peeled off like an orange rind and laid flat, it would fit entirely within the Great Red

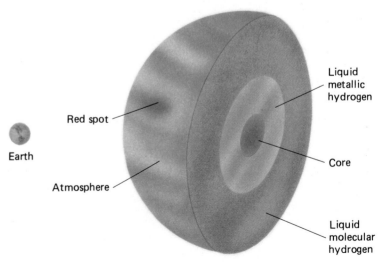

Figure 20–15   The internal structure of Jupiter, with the Earth drawn to scale on the left.

Figure 20–16 The colorful, turbulent complex cloud system of Jupiter, photographed by *Voyager*. The sphere on the left, in front of the great Red Spot, is Io; Europa lies to the right against a white oval. *(NASA/JPL)*

Figure 20–17 Jupiter from a distance of 5 million kilometers, with its Great Red Spot (upper left) and the turbulent region adjacent to it. The smallest details that can be seen in this photograph are about 95 kilometers across. *(NASA)*

Spot. No one knows exactly why the Great Red Spot has persisted over all these years. Hurricanes on Earth dissipate after a week or so, yet this storm on Jupiter has persisted for centuries.

The Moons of Jupiter

In 1610 Galileo discovered four tiny specks of light near Jupiter. He noted that they orbit Jupiter and reasoned that they are satellites of the giant planet. Astronomers have now found a total of 16 moons orbiting Jupiter. The four discovered by Galileo are the most widely studied. In addition to the moons, a rocky ring, similar to the rings of Saturn, lies inside the orbit of the closest moon.

Io

The innermost moon of Jupiter, **Io**, is about the size of the Earth's Moon and slightly denser. Since it is too small to have retained heat generated during its formation and by radioactive decay, many astronomers expected to see a cold, lifeless, cratered, Moon-like surface. Nothing could be further from the truth. *Voyager* photos showed huge masses of gas and rock erupting to a height of

Figure 20–18 A volcanic explosion on Io can be seen on the horizon. Solid material is rising to an altitude of about 200 kilometers. *(Voyager I photograph from NASA)*

200 kilometers above Io's surface (Fig. 20–18). These pictures were the first evidence of active extraterrestrial volcanism.

A few weeks before the *Voyager* photographs were transmitted to Earth, two scientists suggested on theoretical grounds that gravitational forces might heat Io's interior enough to make the planet tectonically active. Recall that the gravitational field of our own Moon causes the rise and fall of ocean tides on Earth. At the same time, the Earth's gravity distorts lunar rock. Thus, the Earth's gravitation may be partly responsible for moon quakes. Jupiter is 300 times more massive than Earth, so its gravitational effects on Io are correspondingly greater. In addition, the three nearby satellites, Europa, Ganymede, and Callisto, are large enough to exert significant gravitational forces on Io, but these forces pull in different directions from that of Jupiter. Apparently, this combination of opposing and changing tidal effects causes so much frictional heating that volcanic activity is nearly continuous on Io. The frequent lava flows have obliterated all ancient landforms, giving Io a smooth and nearly crater-free surface.

Europa

Calculations show that the second closest of the Galilean moons, **Europa**, should also be subject to powerful gravitational forces, but since Europa is farther from Jupiter, the effects are less. Although no active volcanoes have

been observed on Europa, its surface appears smooth and relatively crater-free. Meteorites must have bombarded Europa as they did all other bodies in the Solar System. Therefore, the smooth surface indicates that old impact craters have been eliminated by tectonic or erosional processes. The interior of Europa is rocky, but its surface is covered with ice crisscrossed with a complex array of streaks, which may be fractures or faults (Fig. 20–19).

Ganymede and Callisto

Both **Ganymede** and **Callisto** are probably composed of a rocky core about the size of the Earth's Moon, surrounded by a thick, icy crust. Their surface temperatures are so low that the ice is brittle and behaves much like rock. Two different types of terrain have been observed on Ganymede: one is densely cratered, and the other contains fewer craters and is crisscrossed by grooves and ridges (Fig. 20–20). The meteorite storms that swept through the Solar System four billion years ago formed the cratered regions. The smooth regions developed when the crust cracked and water from the warm interior flowed over the surface and froze, much as lava flowed over the surfaces of the terrestrial planets. The grooves and ridges

Figure 20–19 Europa, the smallest Galilean satellite, as imaged by *Voyager I,* from a distance of about 2 million kilometers. The bright areas are probably ice; the dark areas may be rocky. Note the long linear structures that crisscross the surface. Some of them are more than 1000 kilometers long and about 200 or 300 kilometers wide. They may be fractures or faults. *(NASA)*

Figure 20–20 Ganymede is Jupiter's largest satellite, with a radius of approximately 2600 kilometers, about 1.5 times that of our Moon. Ganymede is probably composed of a mixture of rock and ice. The long white filaments radiating from the white areas may have been caused by meteorite impacts. *(NASA/JPL)*

may be tectonic plate boundaries and mountain ranges formed by tectonic activity.

Callisto, the outermost of the moons discovered by Galileo, is heavily cratered, indicating that its surface is very old. Its craters are shaped differently from those on either Ganymede or the Earth's Moon. Perhaps they have been modified by ice flowing slowly across its surface.

20.7 Saturn: The Ringed Giant

Saturn, the second largest planet, is similar to Jupiter. It has the lowest density of all the planets—so low, in fact, that the entire planet would float on water if there were a basin large enough to hold it. Such a low density implies that it, too, must be composed primarily of hydrogen and helium with only a small core of rock and metal. Saturn's atmosphere is similar to that of Jupiter. Dense clouds and great storm systems cover the planet.

The Rings of Saturn

The most distinctive feature of Saturn is its spectacular array of rings visible from Earth even through a small telescope. Photographs from space probes show seven major rings, each containing thousands of smaller ringlets (Figs. 20–21 and 20–22). The entire ring system is only 10 to 25 meters thick, less than the length of a football field. However, the ring system is extremely wide. The innermost ring is only 7000 kilometers from Saturn's surface, whereas the outer edge of the most distant ring is 432,000 kilometers from the planet. Thus, it measures 425,000 kilometers from its inner to its outer edge. If you were to make a scale model of the entire ring system the thickness of a phonograph record, it would be 30 kilometers in diameter.

The rings are composed of dust, rock, and ice. The particles in the outer rings are only a few ten-thousandths

Figure 20–21 Saturn and its ring system. *(NASA/JPL)*

Figure 20–22 A close-up view of Saturn's A rings obtained from *Voyager II.* *(NASA/JPL)*

of a centimeter in diameter (about the size of a clay particle), but those in the innermost rings are chunks a few meters across. Each piece orbits the planet independently; some move faster, some more slowly, in a continuous chaotic parade.

Saturn's rings may be fragments of a moon that never formed, or remnants of one that formed and was then ripped apart by Saturn's gravitational field. Imagine what happens to a small satellite orbiting a larger planet. Gravitation between the Earth and the Moon cause tides on Earth and seismic rumblings on the Moon, whereas forces between Jupiter and Io are great enough to heat and melt the rock of Io. If a moon were close enough to its planet, the tidal effects would be greater than the gravitational attraction holding the moon together. Thus, the rings of Saturn may be the debris of one or more moons that got too close to the planet.

Titan

In addition to the rings, 17 moons orbit Saturn. **Titan** is the largest. It is unique in that it is the only moon with an appreciable atmosphere. This atmosphere is primarily nitrogen with a few percent methane, CH_4. (Methane is commonly used on Earth as a fuel and is the major component of natural gas.) The average temperature on the surface of Titan is $-180°C$, and the atmospheric pressure is 1.5 times greater than that on the Earth's surface. These conditions are close to the point where methane can exist as a solid, liquid, or vapor. Therefore, small changes in temperature or pressure on Titan could

cause its atmosphere to freeze, melt, vaporize, or condense. This situation is analogous to the environment of Earth, where water can exist as liquid, gas, or solid and frequently changes among those three states. Thus, methane clouds float in Titan's atmosphere, and methane seas, lakes, rivers, and ice caps may also exist.

Why is Titan the only satellite to have an atmosphere? Its large size and low temperature prevent gases from escaping into space. In addition, volcanoes may emit gases into the atmosphere, replacing those that escape.

Organic compounds do not decompose at the low temperature and in the inert atmosphere of Titan. Therefore, its surface is likely be to covered by a thick layer of tarry goo. It is possible that a similar layer also collected on the early Earth and later reacted to form life.

20.8 Uranus and Neptune: Distant Giants

Uranus and Neptune were unknown to the ancients. They are so distant that even today they can be seen only poorly from Earth. In 1977 the *Voyager II* spacecraft was launched to study the Jovian planets. It flew by Jupiter in 1979 and Saturn in 1981. It encountered Uranus in 1986 and then Neptune in 1989. The journey from Earth to Neptune covered 7.1 billion kilometers and took 12 years. The craft passed within 4800 kilometers of Neptune's cloud tops, only 33 kilometers away from the planned path. The strength of the radio signals received from *Voyager* measured a ten-quadrillionth of a watt ($1/10^{16}$), and it took 38 radio antennas on four continents to absorb enough radio energy to interpret the signals.

Composition of Uranus and Neptune

All four Jovian planets have dense atmospheres, liquid surfaces and interiors, and solid mineral cores, but Uranus and Neptune are denser because they have higher proportions of heavier elements. Both of these outer giants are enveloped in thick atmospheres composed primarily of hydrogen and helium with smaller amounts of compounds of carbon, nitrogen, and oxygen. Their outer layers are molecular hydrogen, but neither is massive enough to generate an interior of metallic hydrogen. Instead, their interiors are composed of methane, ammonia, and water, and the cores are probably a mixture of rock and metals.

Magnetic Fields of Uranus and Neptune

Voyager recorded that the magnetic field of Uranus is tilted 58° from its axis. This was unexpected, because our explanation of the origin of the Earth's magnetic field

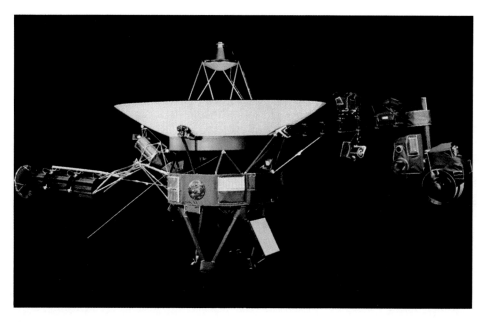

The *Voyager* planetary spacecraft.   *(NASA/JPL)*

suggests that the magnetic fields of all planets should be roughly aligned with the spin axis. One explanation was that *Voyager* just happened to pass Uranus during a magnetic field reversal. However, when *Voyager* reached Neptune, scientists learned that its magnetic field is tilted 50° from its axis. Since the probability of catching two planets during magnetic reversals is extremely low, another explanation must exist. At present no satisfactory theory has been developed.

Neptune's Violent Weather

While calm weather prevails on Uranus, Neptune is remarkably stormy. Planetary meteorologists have observed winds of at least 1100 kilometers per hour, rising and falling clouds, and a cyclonic storm system called the Great Dark Spot, similar to Jupiter's Great Red Spot (Fig. 20–23). Neptune receives only 5 percent of the solar energy that Jupiter receives, so solar heating cannot be

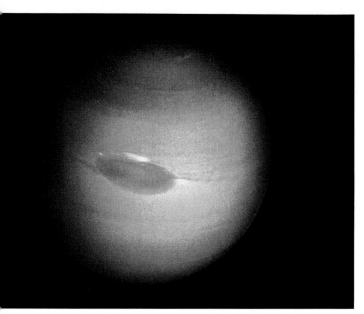

A

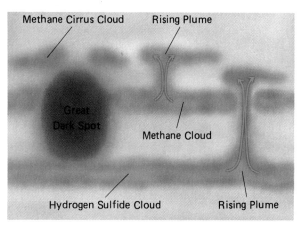

B

Figure 20–23   (A) An image of Neptune from *Voyager II*, taken through colored filters. Note the Great Dark Spot near the equator.   *(NASA/JPL)*   (B) A cross section of the cloud cover of Neptune.

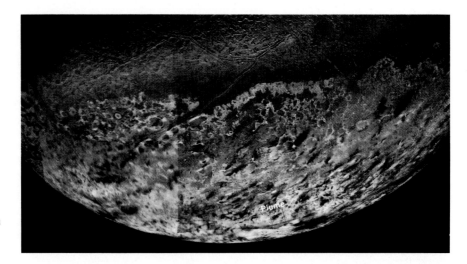

Figure 20–24 The south polar region of Triton. The small plume marked in the lower portion of the photograph is caused by a rapid vaporization of frozen nitrogen as the surface is heated by the Sun. *(NASA/JPL)*

the energy source for the violent winds. According to one fascinating but controversial theory, under the intense pressure near the core, methane decomposes into carbon and hydrogen and the carbon then crystallizes into diamond. The heat released during the formation of diamond is carried toward the surface by convection currents and then powers the winds.

Rings and Moons

A ring system and 15 moons orbit Uranus. Several of the moons are small and irregular, indicating that they may be debris from a collision with a smaller planet or moon.

Neptune has a ring system and eight moons. The largest moon is Triton, which is composed of about 75 percent rock and 25 percent ice. Like many other planets and moons, it has impact craters, mountains, and flat, crater-free plains (Fig. 20–24). While the maria on the Earth's Moon are covered by lava, those on Triton are composed of ice or frozen methane.

20.9 Pluto: The Ice Dwarf

Pluto is the outermost of the known planets and has never been visited by spacecraft. Our highest-resolution

photographs of Pluto are of poor quality compared with those of other planets (Fig. 20–25). Yet astronomers have deduced the properties of this planet from a variety of data.

In 1978 a satellite, Charon, was discovered orbiting Pluto. Using mathematical laws derived by Kepler and Newton in the 1600s, it is possible to calculate the relative masses of a planet and its satellite if the radius of the satellite's orbit and the time required for one complete revolution of the satellite are known. When Pluto and Charon orbit each other, they periodically block one another from view. By precisely measuring these appearances and disappearances, astronomers calculate that Pluto is the smallest planet in the Solar System, smaller than the Earth's Moon. It has a mass only 1/500 that of Earth.

Once the diameter and mass are known, it is easy to calculate the density, which is 2 g/cm³. The Earth, composed of a metallic core and rocky mantle and crust, has a density of 5.5 g/cm³, and ice has a density of a little less than 1 g/cm³. Since Pluto's density is between these two values, we infer that Pluto is a mixture of rock and ice. Infrared measurements show that its surface temperature is about −220°C. Spectral analysis of Pluto's bright

Figure 20–25 The best ground-based image of Pluto and Charon (left) and a similar view from the Hubble space telescope (right). *(Canada-France-Hawaii Telescope and NASA/STSCI)*

Earth Science and the Environment

Impacts of Asteroids with the Earth

Although scientists continue to debate whether an asteroid impact led to the demise of the dinosaurs, significant impacts do occur and have been an important part of the geologic history of the Earth, the Moon, and the planets. One has only to look at the pockmarked surface of the Moon to see evidence of such impacts. Impact craters are less abundant on Earth only because most have been obliterated by erosion and tectonic activity.

Eugene Shoemaker, a geologist for the U.S. Geological Survey, estimates that about 1000 asteroids greater than 1 kilometer in diameter travel in orbits that cross the Earth's orbit. Any one of these could be sucked in by the Earth's gravity to collide with our planet. Large asteroids, with diameters greater than 10 kilometers, strike the Earth on the average of once every 40 million years; those with 1- to 10-kilometer diameters strike about once in 250,000 years. Collisions with smaller objects occur even more frequently.

On February 12, 1947, people in western Siberia saw a fireball "as bright as the Sun" when a small asteroid or swarm of meteorites crashed into the Earth. More than 100 craters formed, and 23 tons of meteorite fragments were recovered in the area. In October 1989, an asteroid with a diameter of 100 to 400 meters flew by within about 1.1 million kilometers of Earth. In January 1991, a smaller one passed within 170,000 kilometers of the Earth, half the distance from the Earth to the Moon.

The effect of a collision depends on the size of the asteroid and the location of impact. If the asteroid that missed us in 1989 had collided, it would have crashed with energy equivalent to that of a 1000-megaton nuclear bomb. If it had crashed on land, everything flammable within a radius of 100 kilometers would have been ignited instantly, and trees and structures within 250 kilometers would have been leveled by the shock waves.

How concerned should we be about the dangers of such an impact? Concern is dictated partly by your own personality, but the probability that an asteroid impact will affect you during your lifetime is greater then the chance of winning the state lottery if you only buy one ticket.

---

surface show that it contains frozen methane. Its atmosphere is thin and composed mainly of carbon monoxide and nitrogen with some methane.

Pluto is similar in size and density to Neptune's moon, Triton. Their similarity has led some astronomers to postulate that both Pluto and Charon were once moons of Neptune and were pulled out of their orbits by a close encounter with another object. Other astronomers have suggested that many ice dwarfs exist in the outer Solar System. They are now searching for similar bodies with both optical and infrared telescopes.

20.10 Asteroids, Meteoroids, and Comets

Asteroids

Eighteenth-century astronomers noted that the dimensions of the planetary orbits increase in a regular pattern, starting with Mercury's orbit, the smallest, and continuing with the orbits of Venus, Earth, and Mars. But a gap in the pattern exists between Mars and Jupiter. The astronomers predicted that another planet might be found in the "open space" between Mars and Jupiter. Instead, they found tens of thousands of smaller bodies orbiting the Sun in a wide ring. These bodies are called **asteroids**. The largest asteroid has a diameter of 770 kilometers. Three others are about half that size, and most are far smaller. The orbit of an asteroid is not permanent, like that of a planet. If an asteroid passes too close to a nearby planet, it falls onto the planet's surface. However, if an asteroid passes near a planet without getting too close, the planet's gravity pulls the asteroid out of its current orbit and deflects it into a new orbit around the Sun. Thus, an asteroid may change its orbit frequently and erratically.

Meteoroids

Imagine tens of thousands of asteroids racing through the Solar System in changing paths. It is not surprising that many of them collide with each other. After a collision, the asteroids often break apart, forming small fragments and pieces of dust called **meteoroids**. These meteoroids leave the collision site in widely divergent directions. Some cross the orbit of the Earth and are pulled inward by our gravitational field. As the meteoroid falls to Earth, friction with the atmosphere heats the meteoroid until it glows. To our eyes it is a fiery streak in the sky, which we call a **meteor** or, colloquially, a **shooting star**. Most meteoroids are barely larger than a grain of sand when they enter the atmosphere and completely vaporize before they reach the Earth. Larger ones, however, may reach

Figure 20–26   A meteorite believed to be a fragment from the asteroid Vesta.   *(NASA)*

the Earth's surface before they vaporize completely. A fallen meteoroid is called a **meteorite** (Fig. 20–26).

Meteorites may provide a window into the past by reflecting the primordial composition of the Solar System. About 90 percent of all meteorites are **stony meteorites**, composed primarily of rock similar to that of the Earth's mantle. The remaining 10 percent are **metallic** and consist mainly of iron and nickel, the elements that make up the Earth's core. The 90:10 ratio of stony to metallic meteorites comes quite close to the 80:20 volume ratio of the mantle to the core in the Earth.

Most stony meteorites contain small, round grains about 1 millimeter in diameter, called **chondrules**, that are composed largely of olivine and pyroxene. Many chondrules also contain fairly complex organic molecules including amino acids, the building blocks of proteins. These molecules are of the type synthesized by living organisms, and form the molecular framework for life. How can we explain organic compounds coming from the cold vacuum of space? The answer must be that organic molecules form by inorganic (that is, nonliving) processes as well as by organic ones. Organic molecules have also been detected in dust clouds deep in interstellar space.

Figure 20–27   Halley's Comet.   *(Akira Fujii)*

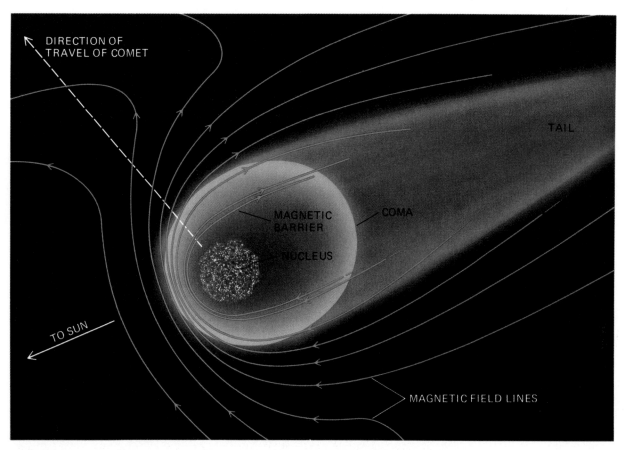

Figure 20–28 A schematic view of a comet. The size of the nucleus has been enlarged several thousand times to show detail. When the comet interacts with the solar wind, magnetic field lines are generated as shown. Ions produced from gases streaming away from the nucleus are trapped within the field, creating the characteristically shaped tail.

Comets

Occasionally, a glowing object appears in the sky. It travels slowly around the Sun in an elongated elliptical orbit and then disappears into space (Fig. 20–27). Such an object is called a **comet**, after the Greek word for "long-haired." Despite their fiery appearance, comets are cold, and their light is reflected sunlight.

Comets originate in the outer reaches of the Solar System, and much of the time they travel through the cold void far from the Sun, even beyond the orbit of Pluto. A comet is composed of ice mixed with bits of silicate rock, metals, and frozen crystals of methane, ammonia, carbon dioxide, carbon monoxide, and other compounds.

When a comet is far out in space, millions of kilometers from the Sun, it has no tail at all, but is simply a ball. As it approaches the Sun, its surface warms and some of it vaporizes. A light "wind" of radiation and ions streaming from the Sun, called the **solar wind**, blows some of the lighter particles away from the sphere to form a long tail. At this time the comet consists of a dense, solid **nucleus**, a bright outer sheath called a **coma**, and a long **tail** (Fig. 20–28). Comet tails more than 140 million kilometers long (almost as long as the distance from the Earth to the Sun) have been observed. As a comet orbits the Sun, the solar wind constantly blows the tail so that it always extends away from the Sun. There is very little matter in a comet tail. By terrestrial standards, it would represent a good, cold laboratory vacuum, yet viewed from a celestial perspective it looks like a hot, dense, fiery arrow.

Halley's comet passed through the inner Solar System between 1985 and 1987 and was studied by six spacecraft and as well as by several ground-based observatories. Its nucleus is a peanut-shaped mass approximately 16 by 8 by 8 kilometers, about the same size and shape as Manhattan Island. The cold, relatively dense coma of Halley's comet had a radius of about 4500 kilometers when it passed by.

SUMMARY

•

Mercury is the smallest planet and the closest to the Sun. It rotates slowly on its axis and therefore experiences extremes of temperature. Its surface is heavily cratered from meteorite bombardment that occurred early in the history of the Solar System. **Venus** has a hot, dense atmosphere as a result of greenhouse warming. Its surface shows signs of recent geologic activity but no Earth-like plate tectonic activity. The Earth's **Moon** probably formed from the debris of a collision between an asteroid and the Earth. The Moon was heated by the energy released during condensation of the debris, by radioactive decay, and by meteorite bombardment. Evidence of ancient volcanism exists, but the Moon is cold and inactive today. **Mars** is a dry, cold planet with a thin atmosphere, but its surface bears signs of erosion and tectonic activity.

Jupiter, **Saturn**, **Uranus**, and **Neptune** are all large planets with low densities. Jupiter and Saturn have dense atmospheres, surfaces of liquid hydrogen, inner zones of liquid metallic hydrogen, and cores of rock and metal. The largest moons of Jupiter are **Io**, which is heated by gravitational forces; **Europa**, which is smooth and ice covered; and **Ganymede** and **Callisto**, which are large spheres of rock and ice. The rings of **Saturn** are made up of many small particles of dust, rock, and ice. They formed from a moon that was fragmented or from rock and ice that never coalesced to form a moon. **Titan**, the largest moon of Saturn, has an atmosphere and may be tectonically active. **Uranus** and **Neptune** have higher proportions of rock and ice than Jupiter and Saturn, and their magnetic fields are not in line with their axes of rotation. **Pluto**, the most distant planet, has a low density and is a small icy planet.

Asteroids are small planet-like bodies. A **meteorite** is a fallen **meteoroid**, a piece of matter from interplanetary space. Most meteorites are stony, and some contain organic molecules. About 10 percent of all meteorites are metallic. When a **comet** is in the inner part of the Solar System, it consists of a small, dense **nucleus** composed of ice, rock, and carbonaceous material; an outer sheath or **coma** composed of gases, water vapor, and dust; and a long **tail** made up of particles blown outward or left behind.

KEY TERMS

•

Terrestrial planets   *542*
Jovian planets   *542*
Mercury   *542*
Venus   *542*
Maria   *546*
Mars   *550*
Jupiter   *556*

Liquid metallic
   hydrogen   *556*
Great Red Spot   *556*
Io   *557*
Europa   *558*
Ganymede   *558*
Callisto   *558*

Saturn   *559*
Titan   *560*
Uranus   *560*
Neptune   *560*
Pluto   *562*
Asteroid   *563*
Meteoroid   *563*

Meteor   *563*
Meteorite   *564*
Chondrule   *564*
Comet   *565*
Coma   *565*
Tail   *565*

REVIEW QUESTIONS

•

1. List the nine planets in order of distance from the Sun, and distinguish between the terrestrial and Jovian planets.

2. Give a brief description of the planet Mercury. Include its atmosphere, surface temperature, surface features, and speed of rotation about its axis.

3. Why are very few meteorite craters visible on Venus?

4. Compare and contrast the surface topography of Mercury, Venus, the Moon, Earth, and Mars.

5. Compare and contrast the atmospheres of Mercury, Venus, the Moon, Earth, and Mars.

6. The text states that "Venus boiled, life evolved on Earth under moderate temperatures, and Mars froze." Discuss the evolution of the climates on these three planets.

7. What leads us to believe that the Moon was hot at one time in its history? How was the Moon heated?

8. Discuss the evidence indicating that the Martian atmosphere was once considerably different than it is today.

9. Describe the composition of the planet Jupiter. How does it differ from that of Earth?

10. Explain why the mass of Jupiter was an important factor in determining its present composition and structure.

11. Compare and contrast the four Galilean moons of Jupiter.

12. Compare and contrast Saturn with Jupiter.

13. Compare and contrast Titan with the Earth.

14. Compare and contrast Pluto with Earth and Jupiter.

15. Is a comet really hot, dense, and fiery? If so, what is the energy source? If not, why do comets look like burning masses of gas?

DISCUSSION QUESTIONS

●

1. If Mercury rotated once every 24 hours as the Earth does, would you expect daytime temperatures on that planet to be higher or lower than they are? Defend your answer.

2. Suppose the oldest igneous rocks on the Moon formed when the Moon was 800 million years old. What conclusions would we then draw about the geologic history of the Moon? Could we answer the question "Was the Moon heated by internal radioactivity or by external bombardment?" Defend your answer.

3. Explain how we can learn about Earth's early history by studying the Moon.

4. At one time, Venus and Earth probably had similar climates, except that Venus was about 20°C warmer. If you could somehow cool the surface of Venus by 20°C, would conditions on that planet likely become similar to those on Earth? Explain.

5. Speculate on how a mantle plume could have generated faults on the surface of Venus even though it did not lead to rifting of tectonic plates.

6. Stephen Saunders, project scientist for NASA's *Magellan* mission, wrote, "Venus has been shaped by processes fundamentally similar to those that have taken place on Earth, but often with dramatically different results." Give examples to support or refute this statement.

7. Refer to Figures 20–9 and 20–14. Imagine that you know nothing about Mars except that it is a planet, and these two photographs were taken of its surface. What information can you deduce from these pictures alone? Defend your conclusions.

8. Imagine that three new planets were found between Earth and Mars. What could you tell about the geologic history of each, given the following limited data? (a) Planet X: The entire surface of this planet is covered with sedimentary rock. (b) Planet Y: This planet's atmosphere contains large quantities of water, ammonia, methane, and hydrogen sulfide. (c) Planet Z: About one third of this planet is covered with numerous impact craters. Smaller craters can be seen within the largest ones. Another third of the surface is much smoother and scattered, with a few small craters. The remainder of the planet has no visible craters but is marked by great topographic relief, including mountain ranges and smooth plains, but no canyons or river channels.

9. In his novel *2010: Odyssey II*, Arthur C. Clarke tells of a group of astronauts who traveled to the vicinity of Jupiter to retrieve a damaged spacecraft. At the end of the novel, Jupiter undergoes rapid changes and becomes a star. Is such a scenario plausible? Why did Mr. Clarke choose Jupiter as the planet to undergo such a change? Would any other planet have been as believable?

10. About four billion years from now, the Sun will probably grow significantly larger and hotter. How will this affect the composition and structure of Jupiter?

11. Would you expect to find gases in the ring system of Saturn? Why or why not?

12. Write a short science fiction story about space travel within the Solar System. You may make the plot fantastic and fictitious, but place the characters in scientifically plausible settings.

Stars, Space, and Galaxies

21

Imagine a world with no clocks or calendars. The movement of the Sun across the sky would mark a passing day, the phases of the Moon would define a month, and the cyclical appearance and disappearance of stars would herald the change of seasons. In preindustrial societies, the cycles of the Sun, Moon, stars, and planets were used to measure time, to signal seasons for planting and harvesting, and to announce the beginning of religious festivals.

Before the invention of the telescope, astronomers in many cultures charted the motions of the celestial bodies. However, they knew little about the natures of the stars, moons, and planets—their sizes, temperatures, compositions, and distances from the Earth. Myths were used to explain the unexplainable, and many stars and planets were thought to represent gods or animals that lived in the heavens.

Today, what we know about the celestial bodies is even more fantastic than ancient myths. A few planets are solid and rocky like the Earth, whereas others are

● Lagoon Nebula in the constellation Sagittarius. Stars are forming within this diffuse cloud of dust and gas.

(National Optical Astronomy Observatories)

569

masses of swirling liquids and gases. The Sun and stars are intensely hot spheres of gas. Occasionally a star explodes, sending streams of gas into space. After the explosion, the remains of the star may become so small and dense that they could rest on Manhattan Island, and a teaspoon of the material weighs 1 billion tons. When a truly massive star collapses, it shrinks until it is so dense that it sucks light rays into its unseeable center.

We live in an exciting period in astronomy. Increasingly sophisticated telescopes and spacecraft enable us to peer deeper into space with finer resolution than ever before. Many old mysteries are being solved, but at the same time, many new ones are appearing.

21.1 Measuring Distances in Space

Distance on Earth is measured in centimeters, meters, or kilometers. However, even a kilometer is so small that it is rarely used to express distance in space. The speed of light in a vacuum is a universal constant. Light travels at 3.0×10^8 m/sec, never faster and never slower. One **light-year** is the distance traveled by light in a year, 9.5 trillion (9.5×10^{12}) kilometers. The closest star to our Solar System is 38×10^{12} (38 trillion) kilometers, or 4 light-years, away. But an object 4 light-years away is a close neighbor in space. Some objects are *15 billion light-years* from Earth.

Recall from Chapter 19 that astronomers detected the Earth's motion around the Sun by observing the parallax shift of distant stars. The same approach can be used to calculate the distance to a star. The **parallax angle** is

measured by observing the apparent shift in position of a star at a six-month interval, after the Earth has completed half a revolution around the Sun (Fig 21–1). The angles even to the closest stars are small and are expressed in **arc seconds**; 1 arc second is equal to 1/3600 of a degree. If you were to view a coin the size of a quarter from a distance of 5 kilometers, it would appear to have a diameter of 1 arc second. A common unit of distance, the **parsec**, is the distance to a star with a parallax angle of 1 arc second. One parsec is equal to 3.2 light-years or 3.1 $\times 10^{13}$ kilometers.

21.2 The Sun

The Sun is an average-size star, but it is so much closer than any other star that we can observe it in more detail. Therefore, let us begin our study of stars with a look at the Sun. Recall that Galileo observed dark spots on the Sun's surface. His critics disagreed even though they refused to look through his telescope. They argued on philosophical grounds that if Galileo questioned the perfection of the Sun he was also questioning the perfection of God. We now know that Galileo was right: sunspots exist, and the Sun is complex and heterogeneous.

The Sun's diameter is 1.4 million kilometers (109 Earth diameters). It is almost entirely composed of two elements, hydrogen and helium. Hydrogen accounts for 92 percent of the Sun's atoms, and helium is second in abundance at 7.8 percent. All the remaining elements make up only 0.2 percent. The Sun's structure is shown in Figure 21–2.

Figure 21–1   Measurement of parallax angle.

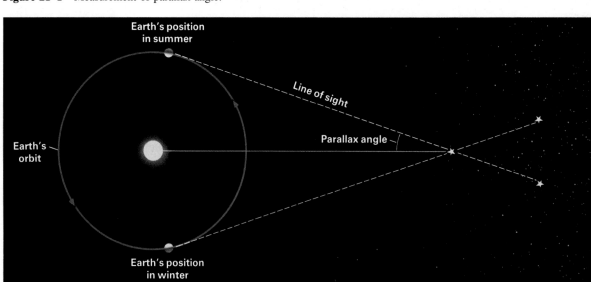

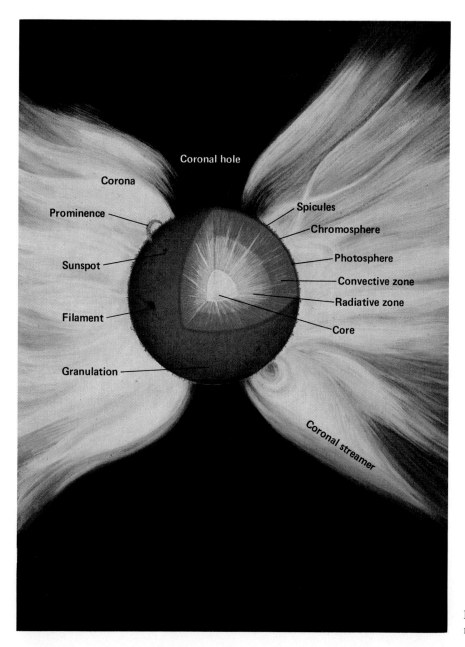

Coronal hole

Corona

Prominence

Sunspot

Filament

Granulation

Spicules

Chromosphere

Photosphere

Convective zone

Radiative zone

Core

Coronal streamer

Figure 21–2 A cut-away schematic of the Sun. *(Jay Pasachoff)*

The Core

The **core** of the Sun is extremely hot, more than 15,000,000°C, and its density is 150 times that of water.* These conditions strip electrons away from their atoms and force hydrogen nuclei to fuse into the heavier element helium. This nuclear fusion is similar to reactions that occur in a hydrogen bomb. If a sphere of hydrogen the size of a pinhead were to fuse completely, it would release as much energy as is released by burning several thousand tons of coal. The Sun's core is 140,000 kilometers in diameter and contains enough hydrogen to fuel this reaction for another five billion years. When the hydrogen in the core is used up, the Sun will change drastically, as described in Section 21.3.

The energy generated during fusion is released mostly as electromagnetic radiation. However, the interior of the Sun is so dense that radiation cannot escape directly. Instead, it is absorbed, then re-emitted and reabsorbed, many times as it works its way toward the surface.

---

*Astronomers generally report temperature on the Kelvin scale, which is based on fundamental thermodynamic properties. We use the Celsius scale here because it is more familiar, and at high temperatures the difference between the two is negligible.

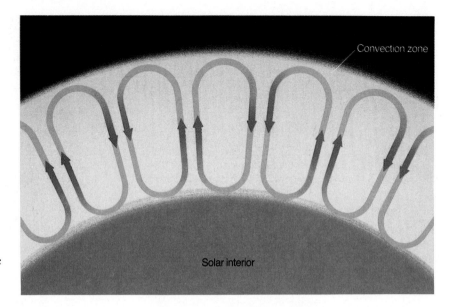

Figure 21–3 A close-up schematic showing convection currents in the Sun's outer layer.

Finally, the radiation is almost completely absorbed in a particularly opaque region close to the Sun's outer layers. Energy is carried outward from there by convection (Fig. 21–3).

The Photosphere

The Sun's visible surface is called the **photosphere**, a word derived from the Greek root *photos*, which means light. Even though the entire Sun is gaseous, the photosphere is called the solar atmosphere because it is cool and diffuse compared to the core. It is a thin surface veneer, only 400 kilometers deep. The pressure within the photosphere is about 1/100 that of the Earth's atmosphere at sea level, and its average temperature is 6000°C, about the same as that in the Earth's core. Fusion does

not occur at this relatively low temperature. Thus, the Sun's core heats the photosphere, and the sunlight we see from Earth comes from this thin, glowing atmosphere of hydrogen and helium.

The photosphere has a granular structure, each grain being about 1000 kilometers across. The granules are caused by convection currents that carry energy to the surface. If you could watch a close-up videotape of the Sun, the granules would appear and disappear like bubbles in a pot of boiling water. The bright yellow dots in Figure 21–4 are formed by hot rising gas, and the darker reddish regions are cooler areas of descending gas.

Large dark spots called **sunspots**, observed by Galileo more than 350 years ago, also appear regularly on the Sun's surface. A single sunspot may be small and last for

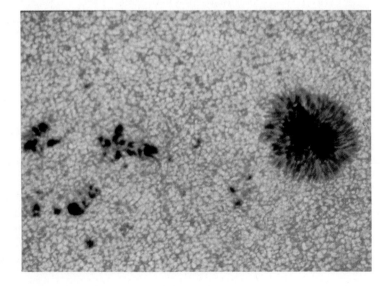

Figure 21–4 Granular structure and sunspots on the surface of the Sun. The large black structures are sunspots. Each small, bright yellow dot is a rising column of hot gas about 1000 kilometers in diameter. The darker, reddish regions between the dots are cooler, descending gases. *(National Solar Observatory/National Optical Astronomy Observatory)*

only a few days, or it may be as large as 150,000 kilometers in diameter and remain visible for months. Because a sunspot is 1000°C cooler than the surrounding area, it radiates less energy and hence appears dark compared to the rest of the photosphere. Since heat normally flows from a hot body to a cooler one, a large, cool region on the Sun's surface should quickly be heated and disappear. Yet some sunspots persist for long times. Astronomers believe that the Sun's magnetic fields restrict solar turbulence and inhibit warming of sunspots.

The magnetic fields associated with sunspots also produce flares of hot gas by accelerating gases to velocities in excess of 900 km/hr and sending shock waves smashing through the solar atmosphere. Particles emitted by the flares interrupt radio communication and cause northern lights on Earth. Truly the Sun is not a static, homogeneous, symmetrical sphere as the Greek philosophers envisioned it.

The Sun's Outer Layers

The photosphere is not bounded by a sharp outer surface. A turbulent, diffuse, gaseous layer called the **chromosphere** lies above the photosphere and is about 2000 to 3000 kilometers thick. The chromosphere can be seen with the naked eye only when the Moon blocks out the photosphere during a solar eclipse. Then, it appears as a narrow red fringe at the edge of the blocked-out Sun. Jets of gas called **spicules** shoot upward from the chromosphere, looking like flames from a burning log. An average spicule is about 700 kilometers across and 7000 kilometers high, and lasts about 5 to 15 minutes.

An even more diffuse region called the **corona** lies beyond the chromosphere. During a full solar eclipse it appears as a beautiful halo around the Sun (Fig. 21–5). Its density is one-billionth that of the atmosphere at the Earth's surface. In a physics laboratory, that would be a good vacuum.

The corona is extremely hot, about 2,000,000°C. How does the photosphere, at 6000°C, heat the corona to 2,000,000°C? According to one theory, twisting magnetic fields carry energy upward to the corona. The energy causes particles in the corona to move very fast, generating the high temperature. **Prominences** are red, flame-like jets of gas that rise out of the corona and travel as much as 1 million kilometers into space (Fig. 21–6). Some prominences are held aloft for weeks or months by the Sun's magnetic fields.

The high temperature in the corona strips electrons from their atoms, reducing hydrogen and helium to bare nuclei in a sea of electrons. These nuclei and electrons are moving so rapidly that some fly off into space, forming

Figure 21–5 During a solar eclipse, the photosphere is blocked by the Moon. The thin red streaks beyond the moon's outline are portions of the chromosphere, and the large white zone is the corona. *(NASA)*

the **solar wind**. It surrounds the Earth and extends outward toward the far reaches of the Solar System.

21.3 Stars: The Main Sequence

From our view on Earth, the Sun is the biggest and brightest object in the sky. However, it is not the largest object in the Universe, nor does it emit the most energy. The Sun looks so big and bright because it is the closest star. When astronomers began to study stars in the early 1900s, one of their first efforts was to catalogue them by their brightness. The **apparent magnitude** of a star is its brightness as seen from Earth. A star can appear luminous either because it really *is* bright or because it is close. The **absolute magnitude** is how bright a star would appear if it were 32 light-years (10 parsecs) away. Thus, astronomers correct for distance by using absolute magnitude. Absolute magnitude is sometimes expressed in solar luminosities, a relative scale in which luminosity of the Sun equals one.

A second visible property of a star is its color. If you study the sky with a telescope, or even with the naked eye, you can see that different stars have different colors; some are reddish, others white or blue (Fig. 21–7). As explained in Chapter 19, the color of starlight is a measure of its temperature.

After astronomers had catalogued many stars, they began to search for similarities and differences among them. One obvious question was: Why are the absolute magnitudes of some stars greater than those of others?

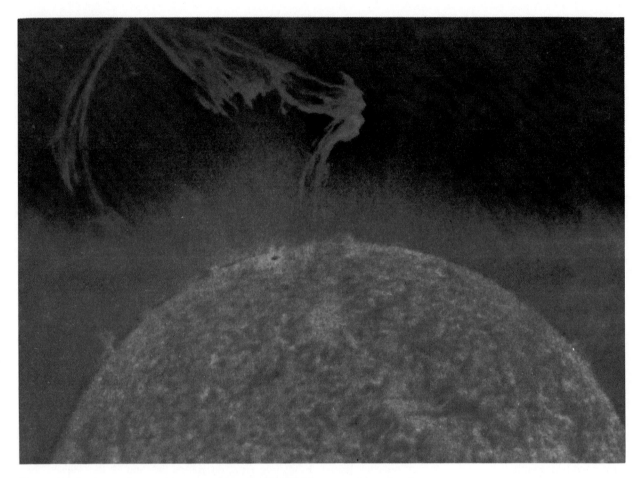

Figure 21–6 A prominence rising approximately 505,000 kilometers above the photosphere. *(National Solar Observatory/National Optical Astronomy Observatory)*

Figure 21–7 A time exposure of the constellation Orion. The stars appear as streaks due to the movement of the Earth. The varied colors are caused by differences in temperature. *(National Optical Astronomy Observatory)*

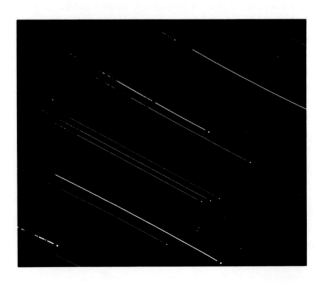

Between 1911 and 1913, Ejnar Hertzsprung and Henry Russell made graphs of absolute magnitudes versus temperature. Figure 21–8 is a **Hertzsprung-Russell**, or **H-R, diagram**.

By convention, in an H-R diagram luminosity increases toward the top and temperature increases toward the left. Note that most of the stars fall along a line from upper left to lower right. This band is called the **main sequence**. About 90 percent of all stars lie within it. As you can see from the graph, luminosity increases with temperature in the main sequence.

All main sequence stars are composed primarily of hydrogen and helium, although most also contain smaller amounts of other gases. All such stars are fueled by hydrogen fusion. **The major reason for differences in temperature and luminosity among main sequence stars is that some are more massive than others.** The force of gravity is stronger in massive stars than in less massive stars. Therefore, hydrogen fusion is more rapid and intense in massive stars, so they are hotter and more luminous (upper left of the H-R diagram). The least mas-

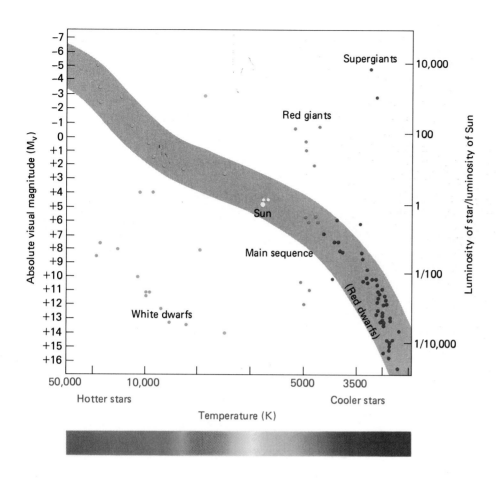

Figure 21–8   An H-R diagram for selected stars. The color bands coincide with changes in temperature. Notice that the Sun is in the yellow band.

sive stars, shown on the lower right, are cool and less luminous. Notice that our Sun is about halfway along the main sequence, in the yellow band.

To explain the stars that do not lie on the main sequence, we must first consider the life and death of a star.

21.4   The Life and Death of a Star

A star exists for billions or tens of billions of years, and we can never hope to observe one long enough to watch its birth, life, and death. We can, however, observe young, middle-aged, and old stars and thus piece together the story of stellar evolution. It is as if an alien came to Earth for one day to observe the life of a human being. It would be impossible to observe the birth, growth, and death of a single person, but the alien could observe babies, children, middle-aged people, and old people and thus infer the stages of human life.

A star originates when a large mass of dust and gas condenses under the influence of gravity to form a

protostar. As gravity pulls its atoms inward, the protostar heats up until its core becomes hot enough for nuclear fusion to begin. When fusion starts, the protostar becomes a star. Fusion generates photons and heats atomic particles. Both the photons and the hot atomic particles accelerate outward against the force of gravity. Thus, two opposing forces occur in a star. Gravity pulls particles inward, but at the same time fusion energy drives them outward (Fig. 21–9). The balance between these two forces determines the size and density of a star. At equilibrium, an average-size star has a dense core surrounded by a less dense shell.

Stars the Size of Our Sun

A young star such as our Sun is composed almost entirely of hydrogen and helium. The hydrogen nuclei fuse to form more helium. The helium nuclei, however, do not fuse in an average mature star. For fusion to occur, two nuclei must be pushed very close together. But nuclei are positively charged, and two positive charges repel each other. This mutual repulsion can be overcome if the nuclei

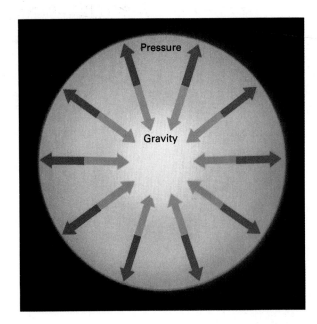

Figure 21–9 The size and density of a star are determined by two opposing forces. Gravity pulls particles inward while thermal and radiation pressures force them outward.

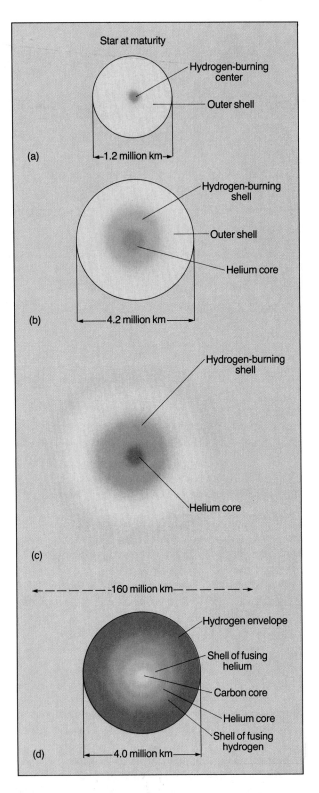

Figure 21–10 The aging of a star the size of our Sun. (The four parts cannot be drawn to scale because the Sun's diameter varies from 1.2 million kilometers to 160 million kilometers. Also, in all cases the core has been drawn larger than scale to show detail.) *(After Jay Pasachoff,* Journey Through the Universe, *Saunders College Publishing, 1992)*

are moving very rapidly. Since higher temperature causes nuclei to move more rapidly, fusion occurs only when the system is very hot. A hydrogen nucleus contains only one proton. Therefore, in hydrogen the repulsion of one proton by another is overcome at the temperature of a young star. A helium nucleus, on the other hand, contains two protons, so in order for helium fusion to occur, a much greater repulsion must be overcome. As a result, higher temperature is required to make helium nuclei fuse.

The life cycle of a star the size of our Sun is shown in Figure 21–10. The Sun evolved from a cloud of dust and gas in space and is now midway through its mature phase as a main-sequence star. It has been shining for five billion years and will continue to shine much as it is today for another five billion years (Fig. 21–10A). During this entire period, the Sun produces energy by hydrogen fusion and remains on the main sequence.

After about ten billion years, the outer shell of a star such as our Sun still contains large quantities of hydrogen, but most of the hydrogen in the core has converted to helium (Fig. 21–10B). The star's behavior now changes drastically. Since the hydrogen in the core is used up, little nuclear energy is produced and the core cools. As it cools, the outward flow of particles and energy decreases. Then the core starts to contract under the force of gravity. This gravitational contraction causes the core to grow hotter. It seems a paradox that when the nuclear reactions decrease, the core becomes hotter, but that is exactly what

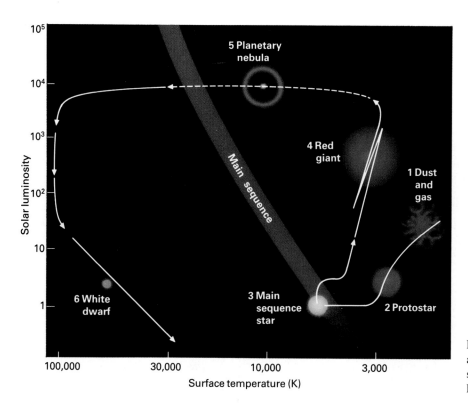

Figure 21–11   The life cycle of a star as massive as our Sun, shown by its movement along the H-R diagram.

happens. To repeat: **when hydrogen fusion ends in the core, gravity compresses the core and its temperature rises.** The increased temperature initiates hydrogen fusion in the outer shell, and the star releases hundreds of times as much energy as it did when it was mature. This intense energy output now causes the outer parts of the star to expand and become brighter. Thus, **when the core contracts, the outer shell undergoes hydrogen fusion and expands.** The star has become a **red giant** (Fig 21–10C). A red giant is hundreds of times larger than an ordinary star. Its core is hotter, but its surface is so large that heat escapes and the surface cools. This cool surface emits red light because red wavelengths have the lowest energy of the visible spectrum. These sudden changes in energy production and size move the star off the main sequence (Fig. 21–11).

Five billion years from now, the hydrogen in our Sun's core will be exhausted and the Sun will expand into a red giant. It will engulf Mercury, Venus, and Earth (Fig. 21–12). Perhaps the heat will blow much of Jupiter's atmosphere away, exposing a rocky surface.

The core of a red giant condenses under the influence of gravity and gets hotter until its temperature reaches 100,000,000°C. At this temperature, helium nuclei begin to fuse to form carbon nuclei. When helium fusion starts, radiant energy pushes outward once again, and the core expands. The star cools, its outer layers contract, and it enters a second stable phase. Gradually, as more helium fuses to carbon, the carbon accumulates in the core just

Figure 21–12   The Sun will engulf the Earth when it expands to its red-giant stage.   *(After Jay Pasachoff,* Journey Through the Universe, *Saunders College Publishing, 1992)*

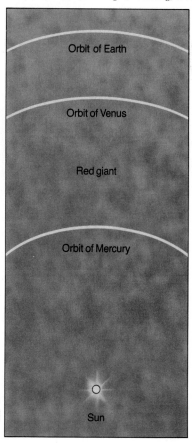

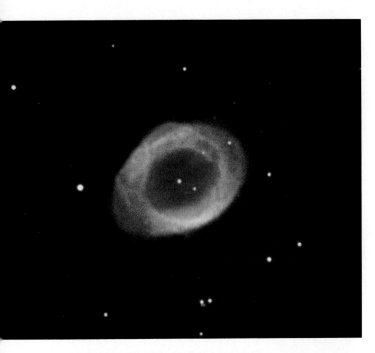

as helium did during the earlier life of the star (Fig. 21–10D). When the helium is used up, fusion ceases again and the carbon core contracts. This gravitational contraction causes the core to heat up again.

What happens next depends on the star's initial mass. Astronomers measure the mass of a star relative to that of the Sun. **One solar mass** is the mass of the Sun. In a star with one solar mass, contraction of the carbon core is not intense enough to raise its temperature sufficiently to initiate fusion of the carbon nuclei. In some stars the size of the Sun, gravitational contraction of the carbon core releases enough energy to blow a ring of gas, like a giant smoke ring, out into space. This ring is called a **planetary nebula** (Fig. 21–13). Meanwhile, the material remaining in the star contracts until atoms are squeezed so tightly together that only the strength of the electrons, called the **degeneracy pressure**, prevents further compression. A dying star as massive as our Sun will eventually shrink until its diameter is approximately that of the Earth. Such a shrunken star no longer produces energy, and it therefore glows solely from the residual heat produced during past eras. The star has become a **white dwarf** (Fig. 21–14). It will continue to cool slowly over

Figure 21–13   The Ring Nebula, the first planetary nebula discovered, is 5000 light-years from Earth.   *(National Optical Astronomy Observatories)*

Figure 21–14   The evolution and life cycles of stars of different masses   *(After Jay Pasachoff, Journey Through the Universe, Saunders College Publishing, 1992)*

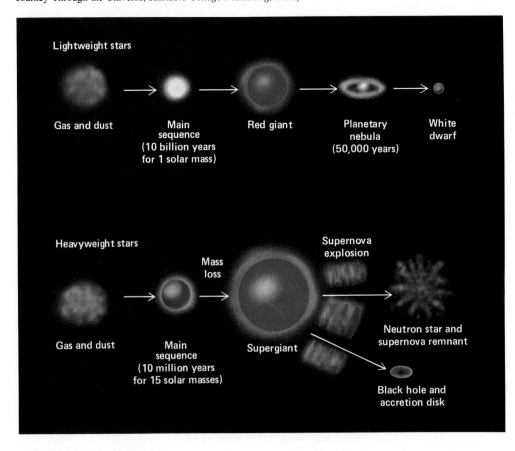

tens of billions of years, but it will never change size again. No further nuclear reactions will occur. Its gravitational force will not be strong enough to overcome the degeneracy pressure, so it will never contract further.

Large Stars

Some stars do not die so gently. If the star is larger than 1.44 solar masses, a white dwarf does not form. Instead, as the helium fusion ends, gravitational contraction produces enough heat to fuse carbon. Renewed fusion produces increasingly heavier elements in a stepwise process until iron forms. Iron is different from the lighter elements. When hydrogen or helium nuclei fuse, energy is released. In contrast, iron fusion absorbs energy and thus cools a star. When this happens, the thermal pressure that forced the stellar gases outward diminishes, and the star collapses under the influence of gravity. This collapse releases large amounts of heat. Within a few seconds—a fantastically short time in the life of a star—the star's temperature reaches trillions of degrees and the star explodes to create a **supernova**. For a brief period, a single star shines as brightly as hundreds of billions of normal stars and emits as much energy as an entire galaxy. To observers on Earth, it appears as though a new brilliant star suddenly materializes in the sky, only to become dim and disappear to the naked eye within a few months.

On February 24, 1987, Ian Shelton, an astronomer, was carrying out research unrelated to supernovas. When he developed one of his photographic plates, he saw a bright star where no star should have been (Fig. 21–15). He walked outside, looked into the sky, and saw the star with his naked eye. This was the first supernova explosion visible to the naked eye since 1604, five years before the invention of the telescope.

A supernova explosion is violent enough to send shock waves racing through the atmosphere of the star, fragmenting atomic nuclei and shooting subatomic particles in all directions. Many of the nuclear particles collide with sufficient energy both to fuse and to split apart. These processes form all the known elements heavier than iron. Thus, in studying the evolution of stars, scientists learned how the heavy elements originated.

First and Second Generation Stars

Hydrogen was the first element to form when our Universe was young. Originally, all the stars in the Universe were composed of nearly pure hydrogen. These old stars are called **population II stars**, and a few still exist. Within the cores of population II stars, hydrogen fused to helium, helium to carbon, and carbon to heavier elements, up to iron. Elements heavier than iron formed during supernova explosions of massive population II stars. The abundance of heavy elements in a population II star is proportional to its mass and age. However, population II stars have much lower abundances of heavy elements than population I stars.

Both large and small stars blast vast clouds of gas and dust, called **nebulae**, into space when they die. Eventually, the nebulae condense once again into new stars, called **population I stars**. Population I stars begin their life with primordial hydrogen mixed with heavy elements inherited from population II stars and from supernova explosions. The Sun is a population I star that condensed from a nebula containing many elements. Thus, our Solar System was born from the debris of a dying star. Solar systems containing terrestrial planets and living organisms could never form around population II stars because these stars have few heavy elements. Thus, in studying

Figure 21–15 The supernova explosion of 1987. The sky (A) before and (B) after the explosion. *(Anglo-Australian Telescope Board)*

A

B

the life cycle of stars, we also learn about the origins of the elements that make up the Earth and all its living organisms.

21.5 Neutron Stars, Pulsars, and Black Holes

Neutron Stars and Pulsars

In a supernova explosion, most of the matter in a star is blasted into a nebula, but a substantial fraction remains behind, compressed into a tight sphere. In the 1930s, scientists developed a theory to explain what happens within this sphere. If it is between 2 and 3 solar masses, the gravitational force is so intense that the star cannot resist further compression the way a white dwarf does. Instead, the electrons and protons are squeezed together to form neutrons:

$$\text{electrons} + \text{protons} \longrightarrow \text{neutrons}$$

The neutrons then resist further compression and remain tightly packed. This ball of compressed neutrons is called a **neutron star**. A neutron star is extremely dense—approximately 10^{13} kg/cm^3. If the entire Earth were as dense, it would fit inside a football stadium (Fig. 21–16). The first neutron star was discovered by accident in 1967, thirty years after the existence of neutron stars was

Figure 21–16   A comparison of the densities of stellar material from different sources.

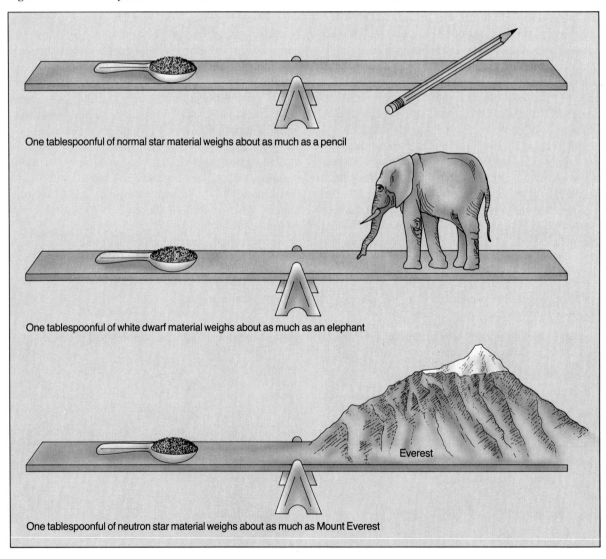

One tablespoonful of normal star material weighs about as much as a pencil

One tablespoonful of white dwarf material weighs about as much as an elephant

Everest

One tablespoonful of neutron star material weighs about as much as Mount Everest

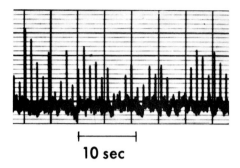

Figure 21–17 Pulsar signals as recorded by a radio telescope.

postulated on theoretical grounds. Jocelyn Bell Burnell, a graduate student at the time, was studying radio emissions from distant galaxies. In one part of the sky she detected a radio signal that switched on and off with a frequency of about one pulse every 1.33 seconds (Fig. 21–17). If such a signal were fed into the speaker of a conventional radio, you would hear a beep, beep, beep, beep evenly spaced, with one beep every 1.33 seconds. Many radio signals arrive at Earth from outer space, but the emissions Burnell heard were unusual because they were sharp, regular, and spaced a little over a second apart. At first, astronomers considered the possibility that they might be a signal from intelligent life, so they called the signals LGM, for "Little Green Men." But when Burnell found a similar pulsating source in a different region of the sky, scientists ruled out the possibility that two life forms in different parts of the Universe would both send similar signals. Once it was established that the signals did not originate from intelligent beings, their sources were called **pulsars**. But naming the source did not help to describe it.

The first step toward identifying pulsars was to estimate their sizes. Not all parts of an object in space are equidistant from Earth (Fig. 21–18). If a large sphere emits a sharp burst of energy from its entire surface, some of the radio waves start their journey closer to Earth than others and therefore arrive sooner. A person on Earth listening to the radio noise hears not a sharp beep but a more prolonged beeeeeep, because it takes a while for all the radio waves to arrive. Alternatively, a signal from a

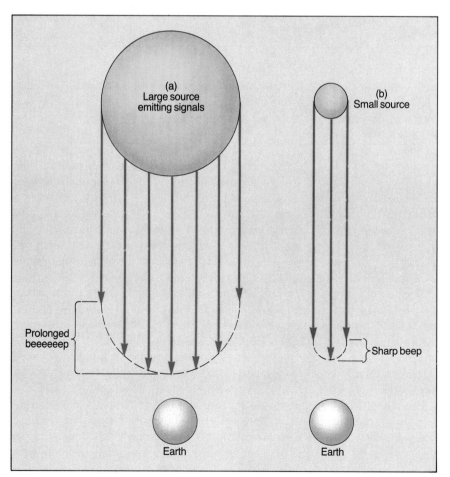

Figure 21–18 A sharp signal from a larger sphere (A) arrives over a longer time interval than a sharp signal from a smaller sphere (B).

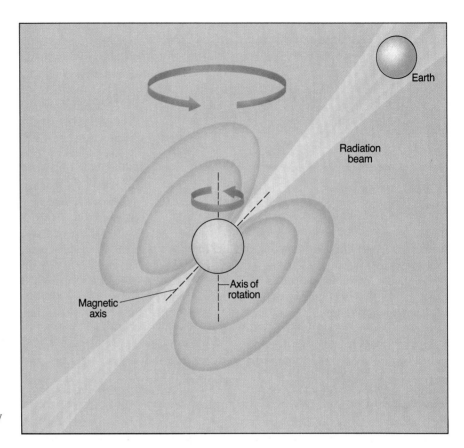

Figure 21–19 The radiation beam of a pulsar is detected only when it sweeps across the Earth.

small source is much sharper. Pulsar signals are sharp, indicating that the source must be unusually small for an energetic object in space—30 kilometers in diameter or less. The smallest star previously recorded was a white dwarf 16,000 kilometers in diameter. Scientists then reasoned that the pulsar detected by Burnell might be the long-searched-for neutron star.

Astronomers suggested that the radio signals are emitted by an electromagnetic storm on the surface of a neutron star or pulsar. According to this theory, as the star rotates, so does the storm center. A receiver on Earth detects one beep per revolution of the star, just as a lookout on a ship sees a lighthouse beam flash periodically as the beacon rotates (Fig. 21–19). Since a pulse is received once every 1.33 seconds, the pulsar must rotate that rapidly. White dwarfs can again be ruled out since they are too big to rotate so rapidly. But neutron stars are small enough to rotate that fast.

Astronomers searched for an example to test the theory. They focused a radio telescope on the Crab Nebula, where a supernova explosion occurred during the Middle Ages. A pulsar was found exactly where the supernova had occurred about 950 years ago—precisely where a neutron star should be.

To summarize, when a small star dies, a white dwarf forms. A neutron star, or pulsar, forms when a larger star

dies. But, what happens after a supernova explosion when a very large star dies?

Black Holes

If a star with more than 5 solar masses explodes, the sphere remaining after the supernova explosion is thought to be more massive than 2 to 3 solar masses. When a sphere this massive contracts, the neutrons are not sufficiently strong to resist the gravitational force. Then the star shrinks to a size much smaller than a neutron star and becomes a **black hole**. Such a collapse is impossible to imagine in earthly terms. A tremendous mass, perhaps a trillion, trillion, trillion kilograms of matter, shrinks to the size of a pinhead, then continues to shrink to the size of an atom, and then even smaller. Eventually it collapses to a point of infinite density.

Such a point of mass creates an extremely intense gravitational field. According to Einstein's theory of relativity, gravity affects photons. Thus, starlight bends as it passes the Sun. If an object is massive and dense enough, its gravitational field becomes so intense that light and other radiant energy cannot escape. So, just as you cannot throw a ball from the Earth to space because it falls back down, light cannot escape from a black hole because it is pulled back downward. Since no light can escape such an object, it is invisible; hence the name black hole. If you were to shine a flashlight beam, a radar beam, or any kind of radiation at a black hole, the energy would be absorbed. The beam could never be reflected back to your eyes; therefore, you would never see it again. It would be as though the beam just vanished into space. Similarly, if a spaceship flew too close to a black hole, it would be sucked in. No engine could possibly be powerful enough to accelerate the rocket back out, for no object can travel faster than the speed of light.

The search for a black hole is therefore even more difficult than the search for a neutron star. How do you find an object that is invisible and can emit no energy? In short, how do you find a hole in space? Although it is theoretically impossible to see a black hole, it is possible to observe the effects of its gravitational field. Many stars exist in pairs or small clusters. If two stars are close together, they orbit around each other. Even if one becomes a black hole, the two still orbit about each other, but one is visible and the other invisible. The visible one appears to be orbiting an imaginary partner. Astronomers have studied several stars that seem to orbit in this unusual manner. In at least one case, the invisible member of the pair is of at least 3 solar masses. Since a normal star of 3 solar masses would be visible, the invisible partner may be a black hole. However, the simple observation that a star moves around an invisible companion does not prove that a black hole exists.

If a star were orbiting a black hole, great masses of gas from the star would be sucked into the black hole, to disappear forever (Fig. 21–20). As this matter started to

fall into the hole, it would accelerate, just as a meteorite accelerates as it falls toward Earth. The gravitational field of a black hole is so intense that particles drawn into it would collide against each other with enough energy to emit X-rays. Thus, just as a falling meteorite glows white-hot as it approaches the Earth, anything tumbling into a black hole would glow even more energetically; that is, it would emit X-rays. These X-rays might then be detected here on Earth. But, you may ask, if light cannot escape from a black hole, how can the X-rays escape? The answer is that the X-rays are produced and escape from just outside the black hole. It is as if matter being sucked into a black hole sends off one final message before being pulled into the void from which no message can ever be sent.

An important experiment, therefore, was to focus an X-ray telescope on portions of the sky where a star appeared to orbit an invisible partner. Such telescopes must be located aboard space satellites, since X-rays do not penetrate the Earth's atmosphere. Karl Schwarzschild first predicted the existence of black holes in 1916. About 70 years later, in the 1980s, an orbiting X-ray telescope detected X-ray sources adjacent to stars that appeared to orbit an unseen partner. These findings have convinced most astronomers that black holes exist.

What happens to the matter that falls into a black hole? No one knows, but according to one hypothesis, a black hole is a doorway into another universe. Matter that falls into a black hole leaves our Universe and disappears into another. Although this idea has received considerable attention in the press, it remains highly speculative.

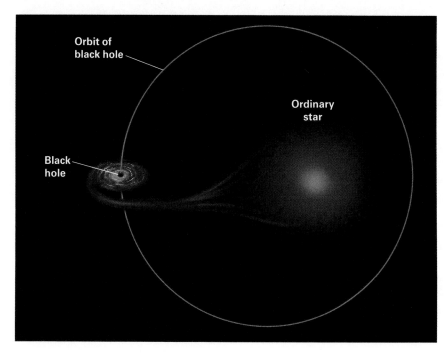

Figure 21–20 An artist's interpretation of a black hole and an ordinary star in orbit around one another. Notice that gases from the star are sucked into the black hole.

Figure 21–21 The spiral galaxy NGC 2997. *(Anglo-Australian Telescope Board)*

21.6 Galaxies

The Milky Way

Stars are not scattered throughout the Universe, but rather concentrate in groups. A **galaxy** is a concentration of billions of stars held together by their mutual gravitation. In a **spiral galaxy**, the stars are arranged in a thin disk with arms radiating outward from a spherical center, or **nucleus** (Fig. 21–21). The stars in the outer arms rotate around the nucleus like a giant pinwheel. A typical spiral galaxy contains 100 billion stars.

Our Sun lies in a spiral galaxy called the **Milky Way** (Fig. 21–22). The Milky Way's disk is 2000 light-years thick and nearly 100,000 light-years in diameter. Because the disk is thin, like a phonograph record, an observer on

Earth sees relatively few stars perpendicular to its plane. Thus, most of the night sky contains a diffuse scattering of stars with large expanses of black space between them. However, if you look *into* the plane of the disk, you see a dense band of stars from horizon to horizon (Fig. 21–23). The entire disk rotates about its center once every 200 million years, so in the 4.6-billion-year history of the Earth we have completed 23 rotations.

Large clouds of dust and gas, called **nebulae**, exist between the stars in a galaxy. The densest nebulae are 10^{-13} as dense as the Earth's atmosphere at sea level, and less dense ones are only 10^{-18} as dense as our atmosphere! A nebula consists mainly of atoms of hydrogen and helium, with traces of heavier atoms and some simple molecules. About 70 different kinds of molecules and molecu-

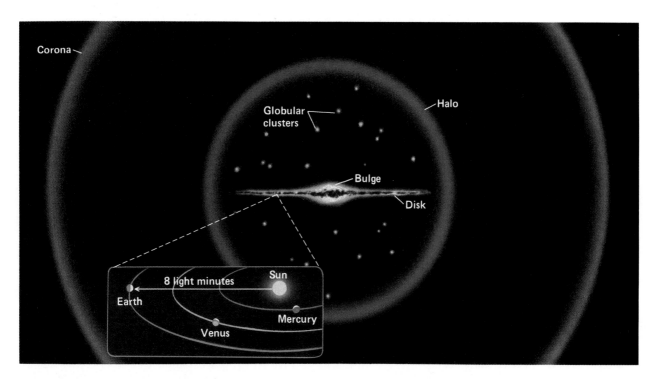

Figure 21–22   An artist's drawing of the Milky Way galaxy. The galactic disk is nearly 100,000 light-years in diameter and 200 light-years thick. As shown in the drawing, the distance from the Sun to Earth is only 8 light-minutes.   *(After Jay Pasachoff,* Journey Through the Universe, *Saunders College Publishing, 1992)*

Figure 21–23   The Milky Way as viewed from the Earth.

Figure 21–24 The Horsehead Nebula in Orion. The bright white lights are stars. The reddish glows are emission nebulae, and the horsehead-shaped dark area is an absorption nebula in front of an emission nebula. *(Royal Observatory, Edinburgh/Anglo-Australian Telescope Board)*

lar fragments have been detected in nebulae. Most are simple, such as water or ammonia.

An **emission nebula** is so hot that its atoms and molecules glow like a neon light. A dark region, shaped like a horse's head, occupies the middle of one of the red emission nebulae in Figure 21–24. This dark region is an **absorption nebula**. The gases here are cool and absorb light from the emission nebula behind it. Nebulae form from gases emitted by exploding stars. Many nebulae may combine to form a cloud 200 light years across with a mass 100,000 times that of our Sun.

A spherical **galactic halo** of dust and gas surrounds the Milky Way's galactic disk. This halo is so large that even though it is extremely diffuse, as much as 90 percent of the mass of the galaxy may be located in it. Many dim and relatively old stars exist within this halo. Some are concentrated in groups of 10,000 to 1,000,000 stars. Each group is called a **globular cluster**. The galactic halo and globular clusters are probably remnants of the original protogalaxy that condensed to form the Milky Way. The spherical structure of the halo suggests that the entire galaxy was once spherical.

Photographs of other spiral galaxies show that the galactic nucleus shines much more brightly than the disk. The concentration of stars in the galactic nucleus is perhaps one million times greater than in the outer disk. If you could visit a planet orbiting one of these stars, you would never experience night, for the stars would light the planet from all directions. However, stable solar systems could not exist in this region; during near collisions between stars gravitational forces would rip planets from their orbits.

It is impossible to see into the nucleus of our own galaxy because interstellar clouds lie in the way. Looking into the nucleus is like trying to see a ship on a foggy day. However, just as sailors use radar to penetrate the fog, astronomers study the center of the galaxy by analyzing radio, infrared, and X-ray emissions that reach us through the clouds. These studies provide a picture of the nucleus of the Milky Way.

A ring of dust and gas orbits the galactic nucleus (Fig. 21–25). Infrared measurements show that this cloud is so hot that it must be heated by an energy source equal to 10 to 80 *million* times the output of our Sun. The ring has a hole in its center like a doughnut. Relatively recently, 10,000 to 100,000 years ago, a giant explosion blew out the center of a larger cloud and created the hole. X-ray and gamma-ray emissions tell us that matter is now accelerating inward, back into the center (Fig. 21–26). What exists in the nucleus to create such disturbances? According to one theory, many new stars are forming simultaneously in the nucleus. A star emits large amounts of energy in the initial stages of formation, so if many stars were evolving, they could create the observed energy. Alternatively, the center of the galaxy could be a massive black hole or a group of smaller black holes. According to this theory, early in the history of the galaxy, many massive, short-lived stars formed in the galactic center. They have already passed through their life cycles, exploded, and collapsed to form black holes. Recall that a black hole does not emit energy, but it draws matter into itself, and the matter emits radiation as it accelerates into the hole.

Other Galaxies

In the late 1700s, a French astronomer, Charles Messier, was studying comets. In an effort to recognize comets

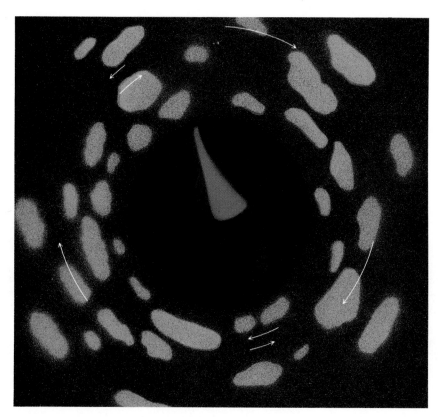

Figure 21–25 A schematic drawing of the center of the Milky Way galaxy. The orange represents streams of ionized gas falling into the center. Surrounding this region is a ring of dust and clouds of gas, shown in blue.

Figure 21–26 Radio emissions from the center of the galaxy. The bright region in the center is a small but powerful radio source. The red, orange, and blue streamers are produced by gases falling toward the center. *(K. Y. Lo, University of Illinois)*

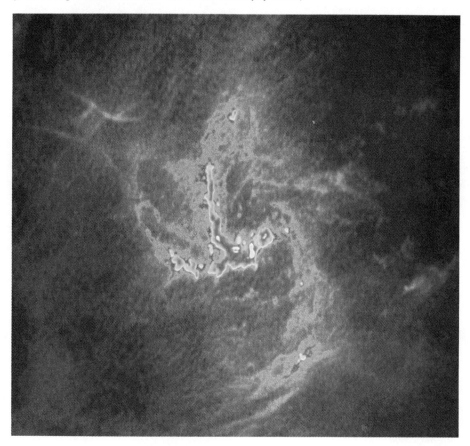

Figure 21–27 An elliptical galaxy, M 87. *(National Optical Astronomy Observatory)*

before they came close to the Sun, he recorded about 100 fuzzy objects in the sky that clearly were not stars. When these objects were studied in the 1850s with more powerful telescopes, many were observed to have spiral structures like pinwheels. They were not comets, but what were they, and how far away were they? These questions were not answered until 1924 when Edwin Hubble determined that they were farther away than even the most distant known stars. In order for us to see them at all, they must be much more luminous than a star. Hubble concluded that each object was a galaxy composed of billions of stars. Today we recognize that galaxies and clusters of galaxies form the basic structure of the Universe.

Most galaxies are elliptical. The **elliptical galaxy** shown in Figure 21–27 is nearly circular; others are elongate. A smaller number of galaxies, including the Milky Way, are **spiral** galaxies, with the characteristic pinwheel

Figure 21–28 A barred spiral galaxy, NGC 1365. *(European Southern Observatory)*

shape. About 30 percent of the spiral galaxies are **barred spirals**, having a straight bar of stars, gas, and dust extending across the nucleus (Fig 21–28). A few percent of all galaxies are **irregular** and show no obvious pattern.

Galactic Motion

Have you ever stood by a train track and listened to a train speed by, blowing its whistle? As it approaches, the pitch of the whistle sounds higher than usual, and after it passes, the pitch lowers. The same effect can be duplica-

Figure 21–29 The Doppler effect. When a signal is emitted by a stationary source (bottom drawing), observers in any direction detect the same frequency. When a signal is emitted by a moving source (B and C in upper drawing), an observer positioned in the direction of movement (toward the left in this drawing) detects waves squeezed close together and therefore with higher frequency than those detected from the same source when it is stationary. If the source is traveling away from an observer, the signal is shifted to lower frequency. *(Jay Pasachoff)*

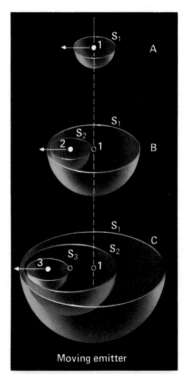

Moving emitter

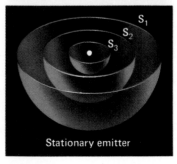

Stationary emitter

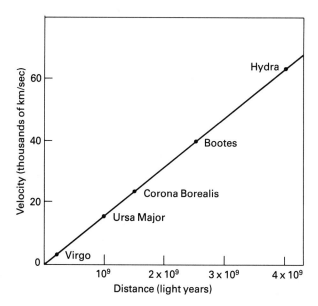

Figure 21–30 The Hubble diagram for the galaxies Virgo, Ursa Major, Corona Borealis, Bootes, and Hydra.

ted with an electric razor. If you turn on an electric razor, hold it still, and listen to it, it produces a constant sound. Now move it quickly past your ear, and listen to the change in pitch. This change, called the **Doppler effect**, was first explained for both light and sound waves by the Austrian physicist Johann Christian Doppler in 1842. Three years later, a Dutch meteorologist, Christopher Heinrich Buys-ballot, mounted an orchestra of trumpeters on an open railroad car and measured the frequency shifts as the musicians rode past him through the Dutch countryside.

A stationary object remains in the center of the circular waves it generates. The waves from a moving object crowd each other in the direction of the object's motion. The object, in effect, is catching up with its own waves (Fig. 21–29). If the object is moving toward you, you receive more waves per second (higher frequency) than you would if it were stationary, and if it is moving away from you, you receive fewer waves per second (lower frequency).

In the same way, the frequency of light waves changes with relative motion. The Doppler effect causes light from an object moving away from Earth to reach us at a lower frequency than it had when it was emitted. Lower-frequency light is closer to the red end of the spectrum. Thus, a Doppler shift to lower frequency is called a **red shift**. Alternatively, light from an object traveling toward us reaches us at higher frequency and is **blue shifted**. Using these principles, astronomers measure the relative velocities of objects billions of light-years away.

In 1929, five years after he had described galaxies, Hubble noted that light from almost every galaxy is red

shifted. Therefore, all the galaxies are flying away from us and from each other; the Universe is expanding. Moreover, he observed that the most distant galaxies are moving outward at the greatest speeds, whereas the closer ones are receding more slowly. This relationship is known as **Hubble's Law**, (Fig. 21–30). Using Hubble's Law, the distance from Earth to a galaxy can be calculated by measuring the galaxy's red shift.

21.7 Quasars

In 1960 astronomers studied many objects that looked like stars but emitted energy as radio waves. These objects were perplexing because normal stars, such as our Sun, emit mostly visible and ultraviolet light. During the following decade, astronomers discovered many other star-like objects emitting energy at frequencies unlike those of normal stars. Two of the objects showed very large red shifts. Such objects are now called **quasars** (Fig. 21–31).

Figure 21–31 Quasar 3C 275.1 is the brightest object in the center of this photograph. The nucleus is surrounded by an elliptical gas cloud. This quasar is about 2 billion light-years from Earth. The smaller objects are normal galaxies. *(National Optical Astronomy Observatory)*

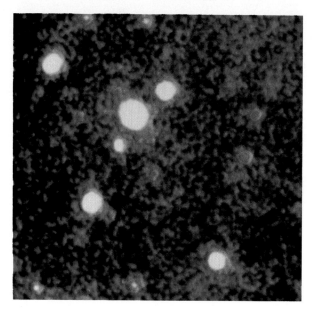

Recall that a large red shift indicates that an object is moving away from Earth at a high speed. If quasars obey Hubble's Law, then they are very far away.* The most distant quasar recorded is 15 billion light-years from Earth. In order to be visible when they are so far away, quasars must emit tremendous amounts of energy, 10 to 100 times more than an entire galaxy!

Quasars emit erratic bursts of energy. Recall from our discussion of pulsars that astronomers can estimate the size of a pulsar by the sharpness of a short burst of energy emitted from it. Similar techniques indicate that quasars are about 4 light-years across, about the distance from the Earth to the nearest star.

*Some astronomers contend that there is some other explanation for the large red shift and that quasars are closer than calculations based on Hubble's Law indicate.

Thus, a quasar is much smaller than a galaxy but emits much more energy. Furthermore, it emits energy over a wide range of wavelengths. Most quasars are very distant from Earth. One common theory for their origin is that a massive black hole, perhaps of a billion solar masses, lies at the center of a quasar, and the quasar emits energy as dust and gas accelerate into the hole.

Looking Backward into Time

If an object is 15 billion light-years away, the light we see today started its journey 15 billion years ago. Therefore, we see what was happening in the past, *but not what is happening today*. If the object blew up and disappeared 14 billion years ago, we will not know about it for another billion years! Thus, when we look at close objects we see

Figure 21–32 A three-dimensional drawing of a portion of the Universe. Notice that the galaxies are distributed unevenly. *(Harvard Smithsonian Center for Astrophysics)*

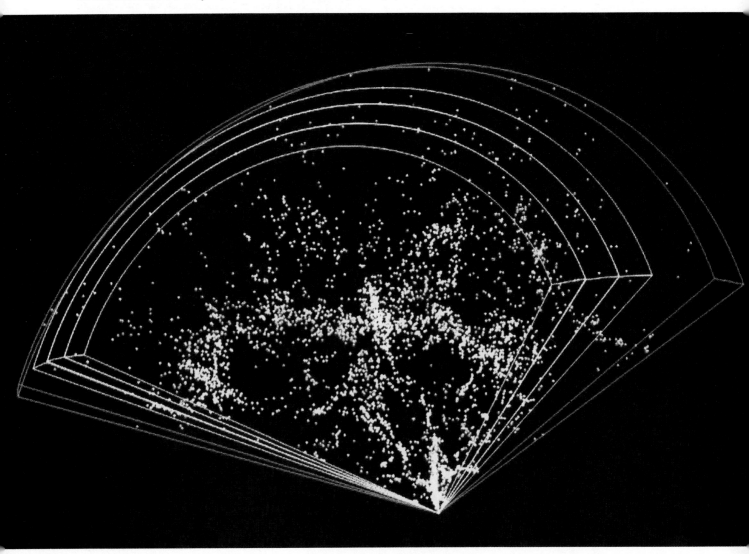

what is happening now, but when we look at distant objects we see what happened in the past. One goal of building more powerful telescopes is to study more distant objects and therefore probe farther back in time.

Recall that the Universe is estimated to be 15 to 20 billion years old. The oldest quasar is estimated to be 15 billion years old, and therefore it must have formed when the Universe was young. Moreover, quasars contain heavy elements, and heavy elements form only in the explosions of dying stars. Therefore, quasars are second-generation structures; they formed after earlier stars were born, evolved, exploded, and died. Hence, stars must have formed shortly after the Universe itself formed. If we can understand how and why quasars formed, we will increase our understanding of the evolution and structure of the Universe.

21.8 The Evolution and Structure of the Universe

In Chapter 1 we discussed the **big bang theory** and the origin of the Universe. Recall that, according to this theory, an initial explosion created space and time and ejected radiation and matter. In 1964, Arno Penzias and Robert Wilson at Bell Laboratories detected very faint microwave radiation that appeared to be uniformly distributed throughout space. This **cosmic background radiation** began traveling through space when the Universe was only a few thousand years old and has been called the echo of the big bang. They interpreted the apparent uniform distribution of this cosmic background radiation to mean that the Universe was homogeneous in its infancy.

However, the modern Universe is clearly not homogeneous. Matter is concentrated into stars, stars are clumped into galaxies, and galaxies are grouped into **clusters**. A large cluster contains tens of thousands of galaxies. Even the clusters group into **superclusters**. Most of the space between the clusters and superclusters contains no galaxies at all (Fig. 21–32).

The discrepancy between the apparent homogeneous cosmic background radiation and the nonhomogeneous Universe disturbed cosmologists for nearly three decades,

and was interpreted by some to reveal a basic flaw in the big bang theory. In 1989, the Cosmic Background Explorer (COBE) satellite was launched to make more precise measurements of the background radiation. In 1992, astrophysicists reported that cosmic background radiation actually varies by tiny amounts from one region of the Universe to another. Thus, the primordial Universe was not homogeneous as Penzias and Wilson had claimed. Matter and energy concentrated into clumps during the earliest infancy of the Universe, perhaps in the first billionth of a second. George Smoot, a researcher on the project, called these variations "the imprints of tiny ripples in the fabric of space-time put there by the primeval explosion."

From the first instant of the big bang, the Universe has been affected by two opposing influences. Matter is expanding as a result of the big bang, but at the same time gravity is pulling it inward. You can study the effect of gravity on an object by studying its motion. Thus, you can calculate the Earth's mass by observing the flight of a ball thrown into the air, and you can calculate the relative masses of the Sun and the Earth by measuring the Earth's orbit. Observations of stellar and galactic motions yield a surprising result. Gravitational forces acting on the Universe are much greater than can be accounted for by all the known stars, dust clouds, and galaxies. By some estimates, matter that we see is only 10 percent of the mass of the Universe. The other 90 percent is invisible and is called **dark matter**. What is it? Suggestions include planet-like objects that do not radiate energy, and are therefore invisible, and black holes that cannot radiate energy. One suggestion is that dark matter is made up of a type of particle that we have not yet detected.

Often, many seemingly unrelated mysteries exist in a field of science. Then, when one part of the puzzle is solved, many of the other mysteries suddenly become explainable. In the early 1990s, many questions regarding the structure and evolution of the Universe are unanswered. If we learned what dark matter is and how it is distributed, we might understand how the visible Universe became heterogeneous. Then we might understand how both quasars and modern galaxies formed.

SUMMARY

·

Distances beyond the Solar System are measured in **light-years** and **parsecs**.

The central **core** of the Sun has a temperature of 15,000,000°C and a density 150 times that of water. Hydrogen fusion occurs in this region. The visible surface of the Sun, called the **photosphere**, is only 6000°C and has a pressure of $\frac{1}{10}$ of an atmosphere. **Sunspots** are

magnetic storms on the surface of the Sun. The outer layers of the Sun, called the **chromosphere** and the **corona**, are turbulent and diffuse. The **solar wind** is a stream of particles flowing from the Sun.

A **Hertzsprung Russell**, or **H-R, diagram** is a plot of luminosity versus stellar temperature. Most stars fall within the **main sequence**. All main-sequence stars are composed primarily of hydrogen and generate energy by hydrogen fusion.

Stars form from condensing clouds of dust and gas. Hydrogen nuclei in the core of a mature star fuse to form helium, with the release of large amounts of energy. The star's size and density are maintained by a balance of radiation pushing outward and gravity pulling inward. When the hydrogen in the core is exhausted and hydrogen fusion ends in the core, gravity compresses the core and the temperature rises. Fusion then begins in the outer shell, and the star expands to become a **red giant**. In the following stage, helium nuclei fuse to form carbon. After helium fusion in an average-size star ends, the star releases a **planetary nebula** and then shrinks to become a **white dwarf**. In a larger star, enough heat is produced to fuse heavier elements; the iron-rich core then explodes to become a **supernova**. The remnant can contract to become a **neutron star** or, if the mass is great enough, a **black hole**.

Our Sun lies in the **Milky Way**, a typical **spiral galaxy**. In addition to stars in the main disk, a galaxy contains diffuse clouds of dust and gas between the stars, and a **galactic halo** and **globular clusters** of stars surrounding the disk. The nucleus of the galaxy is dense and emits large amounts of energy. **Nebulae** are large clouds of gas in interstellar space. Some are hot enough to emit light; others are cooler and absorb light.

Galaxies and clusters of galaxies form the basic structure of the Universe. Most commonly, galaxies are **elliptical**, although other shapes, such as **spiral** and **barred spiral**, also exist. The relative velocities of stars and galaxies are measured using the **Doppler effect**. **Hubble's Law** states that the most distant galaxies are moving away from Earth at the greatest speeds, while closer ones are receding more slowly.

Quasars are very far away, emit as much as 100 times the energy of an entire galaxy, and are small compared with a galaxy. A black hole may lie in the center of a quasar. We can study the evolution of the Universe by viewing distant objects because the radiation we receive from them is old.

The **big bang theory** describes a homogenous Universe immediately after its formation. Today, however, the Universe is heterogeneous; galaxies are clustered in groups, and within each galaxy matter is concentrated into stars.

KEY TERMS

Light-year *570*
Parsec *570*
Photosphere *572*
Sunspot *573*
Chromosphere *573*
Corona *573*
Protostar *575*
Red giant *577*

Solar mass *578*
Planetary nebula *578*
Degeneracy pressure *578*
White dwarf *578*
Supernova *579*
Neutron star *580*
Pulsar *581*

Black hole *582*
Galaxy *584*
Spiral galaxy *584*
Milky Way *584*
Nebula *584*
Galactic halo *586*
Globular cluster *586*

Doppler effect *589*
Red shift *589*
Blue shift *589*
Hubble's Law *589*
Quasar *589*
Supercluster *591*
Dark matter *591*

REVIEW QUESTIONS

1. Explain how the chemical composition, the temperature, and the velocity of a distant star can be determined.

2. Discuss some similarities and differences between a radio signal and one in the visible or ultraviolet region. What information can be obtained from each?

3. What information about a star can be gained from absorption and emission spectra?

4. What is a light-year? How far is it, in kilometers?

5. Draw a cutaway diagram of the Sun, labeling the core, the photosphere, sunspots, granules, the chromosphere, and the corona. Label the temperature and the relative density of each region.

6. What is the fundamental source of energy within the Sun? Is the Sun's chemical composition constant, or is it continuously changing?

7. How does energy travel from the core of the Sun to the surface? How does it travel from the surface of the Sun to the Earth?

8. Compare and contrast the core of the Sun with its surface.

9. Compare and contrast the life cycles of stars the size of our Sun, four times as massive as our Sun, and 20 times as massive as our Sun. At what points do the life cycles differ?

10. What is a supernova? Do all stars eventually explode as supernovas? Explain. What is a neutron star? Do all supernovas lead to the formation of neutron stars?

11. What is a black hole? How does it form, and how can we detect it?

12. Draw a picture of the Milky Way galaxy. Label the disk, the core, the halo, and globular clusters. Draw the approximate position of the Earth.

13. What evidence indicates that the center of our galaxy was once the scene of a violent explosion?

14. List four characteristics of quasars that distinguish them from ordinary stars.

DISCUSSION QUESTIONS

•

1. Could information be gained by building large microphones to detect the sounds of giant explosions in space?

2. What is the Doppler effect? Could the waves from a rowboat traveling through water exhibit a Doppler effect? Compare the wave frequency in front of the boat with the frequency behind it.

3. Many of the stars in our galaxy exhibit spectra that are blue shifted. Does this invalidate Hubble's Law? Explain.

4. What information (if any) could be gained by studying the absorption and emission spectra of the Moon and the planets.

5. Could a telescope be built to study a quasar as it exists today?

6. Hydrogen burns rapidly and explosively in air. Is this chemical combustion of hydrogen an important process within a star? Why or why not?

7. Explain why the density of the gases near the surface of the Sun is less than the density of the gases near the surface of the Earth, even though the gravitational force of the Sun is much greater.

8. What could you tell about the past history of a star if you knew that its core was composed primarily of carbon? Explain.

9. Compare and contrast a white dwarf with a red giant. Can a single star ever be both a red giant and a white dwarf during its lifetime?

10. Would you be likely to find life on planets that are orbiting around a star composed solely of hydrogen and helium? Explain.

11. Certain stars that lie above and below the plane of the Milky Way contain fewer heavy elements than does our own Sun. From this information alone, what can you tell about the history of these stars?

12. Explain why the pulsar signals detected by Jocelyn Bell Burnell could not have originated from (a) an ordinary star, (b) an unknown planet in our Solar System, (c) a distant galaxy, or (d) a large magnetic storm on a nearby star.

13. Could a black hole be hidden in our Solar System? Explain.

14. Would black holes represent a hazard to a rocket ship traveling to distant stars? Could the crew of such a rocket detect a black hole well in advance and avoid an encounter?

15. Is a quasar similar to a star, a galaxy, or neither? Discuss.

16. Arrange the following different environments in order of increasing densities: (a) intergalactic space, (b) the core of the Sun, (c) the corona of the Sun, (d) the region of space between planets in our Solar System, (e) galactic space, (f) the Earth's atmosphere at sea level.

17. List several objects or particles that might make up dark matter. Explain why each is invisible in the environment where it might exist.

Natural Resources

UNIT VII

- Fuels and Ore

● Drilling for petroleum in Wyoming. *(John Barstow)*

Geologic Resources

22

ENERGY RESOURCES

22.1 Fossil Fuels

22.2 Fossil Fuel Sources and Availability

22.3 Environmental Problems Resulting from
Production of Fossil Fuels

22.4 Nuclear Energy and Uranium Reserves

MINERAL RESOURCES

22.5 Ore

22.6 How Ore Deposits Form

22.7 Future Availability of Mineral Resources

22.8 Nonmetallic Resources

Many animals use tools. Apes use sticks as back scratchers and occasionally as clubs. Woodpecker finches on the Galápagos Islands of South America use small twigs to prod insects from trees. No other species uses tools to the extent that humans do, however, and no other animal controls fire.

Tools and fire are essential to civilization. Prehistoric people made wooden tools, and wood was the major fuel of antiquity. Prehistoric people also used geologic resources such as flint and obsidian to make weapons and hide scrapers. Later, people learned to mine and refine metals and to extract coal and petroleum from the Earth, and today we depend on those resources.

Modern geologic resources fall into two categories.

1. *Energy Resources* Petroleum, coal, and natural gas are called **fossil fuels** because they formed from the

● An oil drilling platform. *(John Barstow)*

remains of plants and animals that lived in the geologic past. Uranium is the basic fuel for nuclear reactors.

2. *Mineral Resources*   About 30 metals are commercially important. Some, such as iron, lead, copper, aluminum, silver, and gold, are familiar to all of us. Others, such as vanadium, titanium, and tellurium, are less well known but are vital to industry.

Many nonmetals are also mined for a variety of uses. Sand and gravel are mined for road building and the manufacture of concrete. Limestone is quarried to produce cement. Granite, limestone, marble, and other rocks are quarried for use as building stone. Phosphorus and potassium are essential fertilizers. Sulfur is used widely in the chemical industry.

ENERGY RESOURCES

22.1 Fossil Fuels

Coal

In forests and grasslands, dead plants fall to the ground. In most environments, oxygen and water mix with this plant litter. The oxygen supports organisms that decompose the litter. The water flows downward through the soil and removes the decay products.

In some swamps, however, plants grow and die so rapidly that plant litter accumulates faster than it decomposes. Newly fallen vegetation covers the older, partially rotted material, so atmospheric oxygen cannot penetrate into the deeper layers. Therefore, decomposition stops before it is complete, leaving brown, partially decayed plant matter called **peat**. Commonly, peat is then buried by mud deposited in the swamp. Burial compresses the peat, squeezing out the water and other volatiles. Plant matter is composed mainly of carbon, hydrogen, and oxygen. During burial, most of the hydrogen and oxygen escape as gases and the proportion of carbon increases. The result is **coal**, a combustible rock composed mainly of carbon (Fig. 22–1). Several types of coal are distinguished by their hardness, carbon content, and heat value (Table 22–1).

Petroleum

Organic matter eroded from land is carried to the oceans by streams and deposited with mud in shallow coastal waters. As plants and animals living in the sea die and settle to the bottom, they add more organic matter to the mud. This organic-rich mud is then buried by younger

Figure 22–1   Formation of coal deposits.

A Litter falls to floor of stagnant swamp

B Debris accumulates, barrier forms, decay is incomplete

C Sediment accumulates, organic matter is converted to peat

D Peat is lithified to coal

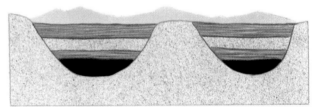

			Other Volatiles and Noncombustible		
Type	Color	Water (%)	Compounds (%)	Carbon (%)	Heat Value (Btu/pound)
Peat	Brown	75	10	15	3,000–5,000
Lignite	Dark brown	45	20	35	7,000
Bituminous (soft coal)	Black	5–15	20–30	55–75	12,000
Anthracite (hard coal)	Black	4	1	95	14,000

sediment. Burial increases temperature and pressure. These conditions convert the mud to shale and the organic material to liquid petroleum (Fig. 22–2). Typically, petroleum forms in the range 50 to 100°C. (As a comparison, lukewarm tea is about 50°C, and water boils at 100°C.)

Initially, the liquid oil is dispersed throughout the shale. But dispersed oil in shale does not constitute a reservoir. A commercial petroleum **reservoir** forms when oil flows from the shale and concentrates in other rock. A commercial reservoir develops in three phases.

Formation of Source Rock

Organic material accumulates in clay-rich mud. When the mud is buried and heated, the solid organic matter converts to liquid oil at about the same time the mud converts to shale, which is called the **source rock**. Oil cannot be recovered from shale because shale is relatively **impermeable**; that is, liquids do not flow through it rapidly.

Oil Migration

If conditions are favorable, petroleum is forced out of the source rock and **migrates** to a nearby layer of sandstone or limestone with open spaces called pores between the grains. If liquids can flow readily through porous rock, the rock is said to be **permeable**.

Concentration of Oil in a Trap

Petroleum, being less dense than water and much less dense than rock, rises through permeable rock. An **oil trap** is any barrier to the upward migration of oil or gas. When oil or gas seeps into the rock below a trap, it accumulates there and forms a reservoir. Most reservoirs form in permeable sandstone or limestone. Many types of traps accumulate oil. In one common type, a dome of impermeable **cap rock**, such as shale, covers the perme-

A

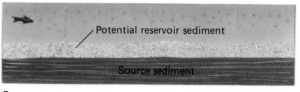

B

C

Figure 22–2 The first stages in the development of petroleum reserves.

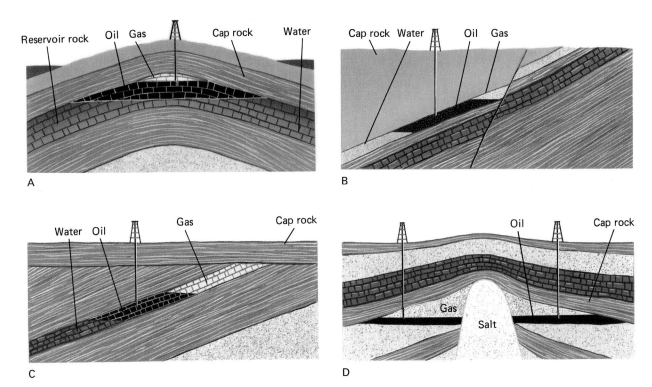

Figure 22–3 Formation of oil traps. (A) Petroleum rises into a dome in the rock. (B) A trap is created when rock fractures and slips along the fracture, moving impermeable shale above the oil reservoir. (C) Horizontally bedded shale overlies inclined, petroleum-bearing sandstones. (D) A salt dome. Layers of salt, emplaced at great depth, rise as pressure accumulates. This rising bubble of salt deforms the overlying strata.

able reservoir rock (Fig. 22–3A, D). The cap rock prevents the petroleum from rising farther. In other types of traps, rock is faulted or layered so that impermeable caps cover the reservoir (Fig. 22–3B, C).

It is important to emphasize that a petroleum reservoir is not an underground pool or lake of oil. Instead, it is permeable, porous rock saturated with oil, more like an oil-soaked sponge than a bottle of oil.

The oil pumped from the ground, called **crude oil**, is a gooey, viscous, dark liquid made up of thousands of different chemical compounds. Oil refineries convert crude oil to propane, gasoline, jet fuel, heating oil, motor oil, road tar, and other useful materials. Some of the compounds in oil are used to manufacture plastics, medicines, and many other products.

Natural Gas

Natural gas, or methane (CH_4), forms when crude oil is heated above 100 to 150°C during burial. Many oil wells contain natural gas that floats above the heavier liquid

petroleum. In other instances, the lighter, more mobile gas escapes and is trapped elsewhere in a separate reservoir.

22.2 Fossil Fuel Sources and Availability

Coal is forming today in some swamps. Oil and gas are forming in organic-rich muddy sediments in many marine basins and coastal areas. However, because the processes are extremely slow—much slower than current rates of consumption—fossil fuels will eventually be depleted. To estimate the time remaining before humans will exhaust Earth's fossil fuels, we must estimate the amount left in the ground and future rates of consumption.

One method of estimating future fuel consumption is to use past patterns to predict future trends. Figure 22–4 shows energy consumption in the United States from 1949 through 1989. Notice that the curves are not smooth. Both total energy consumption and petroleum consumption increased steadily from 1948 to 1973, dou-

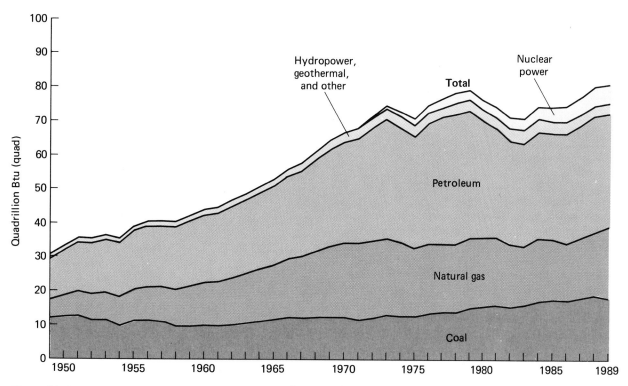

Figure 22–4   Energy consumption in the United States from 1949 through 1989.   *(Annual Energy Review 1989, Energy Information Administration.)*

bling from 1948 to 1970. If oil consumption were to continue to double every 22 years, 660 years from now we would consume as much petroleum annually as there is water in the world's oceans. Calculations of this sort assure us that we cannot go on doubling our fossil fuel consumption every 22 years.

However, the rapid increase in oil consumption between 1948 and 1970 was not sustained. In the mid-1970s, oil-producing countries began to realize how valuable and limited their fuels were, so they raised prices. As prices went up, consumers used less fuel. People drove fewer miles and purchased smaller, more fuel-efficient automobiles. They added insulation to buildings and saved energy in many other ways. Conservation resulted in a decline in energy use between 1973 and 1975. Then, from 1975 to 1990, energy consumption fluctuated but generally increased.

Petroleum

Many predictions have been made regarding the future availability of petroleum. Although they differ as to the exact time limit, they all share the conclusion that some-

time in the next 10 to 35 years, petroleum supply will be insufficient to meet demand. The United States produced 145 billion barrels of petroleum in the 100 years from the start of commercial drilling to 1990. About 27 billion barrels of recoverable oil remain in known fields.* How much oil lies undiscovered? The discovery rate of new wells has declined in recent years. For every meter of drilling, we now find half as much oil as we did in the 1950s. According to a report published in 1989 by the United States Geological Survey, geologists expect that an additional 55 billion barrels of oil will be discovered in the United States. Thus, as of 1989, known plus estimated reserves in the United States equaled approximately 82 billion barrels of oil. At the current consumption rate of 5.4 billion barrels per year, this is a 15-year supply. Currently, imports account for about 45 percent of the petroleum consumed in the United States. If the flow of oil is not impeded by political instability or war, and that percentage remains constant, imports will extend the domestic supply 35 years, or to about the year 2025.

Annual Energy Review 1990, Energy Information Administration.

(Text continues on p. 606)

Earth Science and the Environment

Energy Strategies for the United States

In the United States, petroleum consumption has increased by a factor of 2.5 since 1950. In contrast, domestic oil production peaked in 1970 and has declined for the last two decades. These two opposing trends have led to a steady increase in oil imports (Fig. 1). Today we import about 45 percent of our petroleum at an annual cost of $25 billion. This cash outflow has a negative impact on our national economy.

In addition, much of the imported oil is purchased from the Middle East, which is vulnerable to political disruptions. Twice in the 1970s, the major oil-producing nations reduced oil production and raised the price, causing economic problems in oil-importing nations. In 1990 Iraq invaded Kuwait and threw global petroleum markets into turmoil. The price of petroleum skyrocketed from below

$20 per barrel to $40 per barrel, and then fell again after the brief Gulf War.

In 1989, before the Gulf War, the United States spent between $15 billion and $54 billion to safeguard petroleum supplies and shipping lanes in the Middle East. This wide range in estimates reflects difficulties in assigning costs to military operations. The Gulf War in 1990–91 added another $30 billion to the cost of protection. Using the lowest estimates, these figures add about $23.50 to the cost of each barrel of petroleum. Thus, the cost of defending oil imports more than doubled the cost of imported oil. Such military expenditures are not added to the cost of gasoline at the retail level, but they must be paid for nevertheless.

In short, the United States no longer has an abundant, low-cost energy resource. . . . Indeed, the U.S.

Figure 1 The sources of oil used in the United States. The left side of the graph shows what has happened in the past. The right side shows what will happen in the future if current trends continue. Notice that imports grew from near zero in 1950 to about 45 percent of consumption in 1990 due to increased consumption and decreased internal production. If these trends continue, our dependence on imports will rise steadily. *(From "Improving Technology: Modeling Energy Futures for the National Energy Strategy," Energy Information Administration paper SR/NES/90-01, 1991.)*

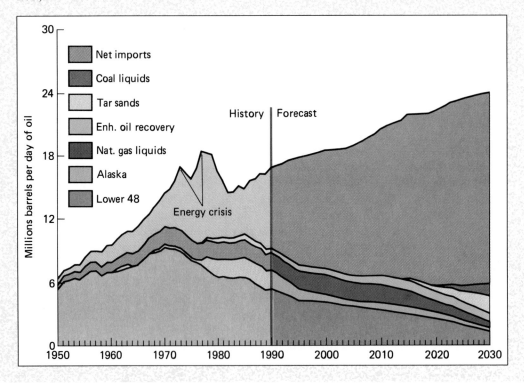

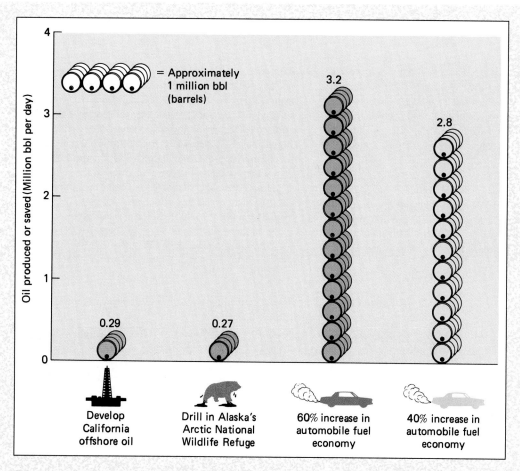

Figure 2 Petroleum savings from conservation versus drilling in the Arctic National Wildlife Refuge (ANWR) and in California offshore fields. *(From Union of Concerned Scientists, Nucleus 12, no. 3 [1990])*

energy picture is growing closer to Europe and Japan, which lack cheap domestic energy sources and have to work hard to function amidst high energy costs. The U.S. has a long way to go considering that it uses around twice as much oil per person as European nations do.*

Most people agree that the current situation is undesirable, but disagree about solutions. Three options are possible: (1) drill for more oil in the United States, (2) exploit other energy sources, and (3) conserve energy.

Drill for More Oil in the United States Existing U.S. oil fields are being depleted, production is declining, and discovery rates per foot of drilling have decreased by half since 1950. But what about finding new oil fields? Let us examine one case in particular.

In 1980 Congress set aside an 18-million-acre wildlife refuge in northeastern Alaska called the Arctic National Wildlife Refuge (ANWR). After the conclusion of the Gulf War in 1991, President Bush advocated drilling for oil in the refuge. The argument for this proposal is that our life style, economy, and national security depend on petroleum, and therefore it is imperative to drill for domestic oil wherever it exists. The argument against drilling in ANWR is that the refuge is one of the finest wildlife sanctuaries in the world and that the government must protect what little is left of such places (Fig. 2). The U.S. Department of the Interior estimates that 3.2 billion barrels of oil lie beneath ANWR. Under normal operating conditions, the oil reserves would be extracted over 30

(Continues on p. 604)

*Flavin, Christopher, "Conquering U.S. Oil Independence," *Worldwatch*, January–February, 1991.

Earth Science and the Environment (continued)

years. Production would peak at 500,000 barrels per day and average about 270,000 barrels per day. Today we import about 8 million barrels per day. Therefore, the average contribution from ANWR would result in a 3 percent reduction of imports. Calculations show that production of other reserves would result in similarly small reductions of imports. Thus, we cannot obtain petroleum independence simply by exploiting fragile environments (Fig. 2). In 1991 Congress voted to maintain protection of ANWR and prohibit drilling.

Exploit Non–Fossil Fuel Energy Sources Several non-fossil fuel energy sources exist. **Solar energy** can be harnessed in several ways. Many buildings are designed and oriented to collect sunlight. In others, **solar collectors** trap the Sun's radiation and use it to heat water. **Solar**

cells convert sunlight directly to electricity. **Biomass energy** is the energy obtained by using plants or plant products as fuels. In 1991, more homes in the United States were heated with wood than with electricity generated by nuclear power plants. More than half of the household trash in the United States and Canada is paper, which can be used as fuel. The largest electrical generator that operates solely on trash is located in Rotterdam, Netherlands. It generates 550 megawatts, enough to supply the domestic consumption of 250,000 people. **Hydroelectric energy** is produced when falling water is directed through a turbine to produce electricity. About 5 percent of the world's energy is supplied by hydroelectric generators (Fig. 3). In 1990, 13,000 **wind** generators produced electricity in the United States (Fig. 4). Their total capacity was 1000 megawatts, about the output of a large nuclear

Figure 3 The Glen Canyon dam. Water retained by the dam is stored for irrigation, and water falling through the dam produces electricity.

Figure 4   When the West was being pioneered, windmills like this one were used to pump water. Modern windmills produce electricity.

reactor. As discussed in Chapter 11, the heat of the Earth's interior can be used to produce **geothermal energy**. At present, geothermal resources satisfy only a tiny portion of human energy needs, but there is considerable potential to be exploited.

In 1990, 7.6 percent of total energy consumed in the United States came from nuclear fission, 3.5 percent from hydroelectric power, and lesser amounts from solar, geothermal, wind, and biomass sources. Although these numbers are small compared to energy provided by fossil fuels, greater potential exists. We have already discussed the problems and promise of nuclear energy. The state of California has established a program to exploit renewable energy sources, including wind, solar, hydroelectric, and geothermal power, for more than 40 percent of its electricity.

Although these renewable energy sources produce electricity, none can be used directly in automobiles. Eventually transportation may be dominated by electric cars, trains, and trolleys, but the changes will not occur quickly or cheaply.

Conservation   A new nuclear power plant or mass transit system can be built in about ten years if no major prob-

lems arise. If you and your neighbor carpool to work rather than commute separately, you can cut your transportation fuel consumption and air pollution in half starting tomorrow. Conservation is the cheapest and quickest way to reduce dependence on foreign oil. Two conservation strategies exist. **Social solutions** involve personal choices and sacrifices. If you choose to carpool rather than drive your own car, you inconvenience yourself by coordinating your schedule with your neighbor, driving or walking to a meeting place, and waiting if your friend is late. However, if everyone carpooled or used mass transit, highways would be less congested and there would be fewer traffic jams and more parking places. **Technical solutions** involve using using more efficient machinery. For example, a car that runs 40 miles on a gallon of gasoline can carry you as quickly and comfortably as an old gas hog that gets 12 miles per gallon. In 1989 people in the United States used 14.5 million barrels of oil per day *less* than they would have if they had retained the technology of 1973. Further improvements are possible. A 40 percent increase in average automobile fuel economy would save 2.8 million barrels of oil a day, about one third of the current import level.

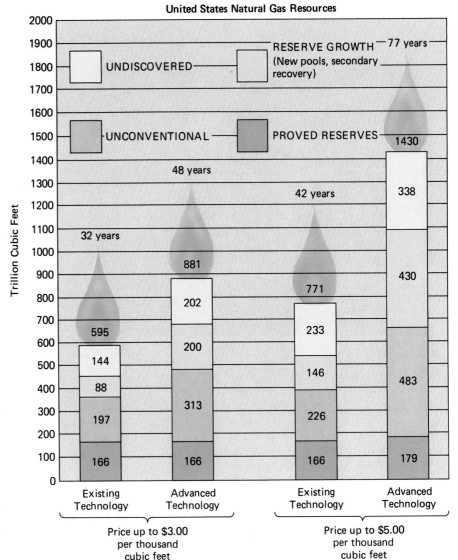

United States Natural Gas Resources

Figure 22-5 Future availability of natural gas, given four different possibilities of price and extraction technology. Above each column is the number of years our gas reserves will last, given the conditions outlined in the column. All predictions assume that consumption rates remain constant. *(American Association of Petroleum as reported in Geotimes, November 1989)*

People in their twenties in the early part of the 1990s will see, within their lifetimes, a restructuring of our society resulting from a permanent petroleum shortage.

Humans and their immediate ancestors have lived for about two million years. Thus, the "petroleum age" will appear as a short episode in the history of our species.

Natural Gas

Natural gas reserves in the United States will last from 32 to 77 years if consumption remains constant (Fig. 22-5). The large range reflects uncertainties regarding advances in extraction technology. One thing is certain, however: consumption is increasing. Natural gas is ex-

tracted as a nearly pure compound; it releases no sulfur and other pollutants when it burns. As a result, gas is becoming increasingly popular and will be even more desirable if pollution laws become stricter. If natural gas consumption increases, reserves will be depleted faster than predicted in Figure 22-5.

Coal

Large reserves of coal exist in many parts of the world. Figure 22-6 shows that widespread availability of this fuel is expected until at least the year 2200. However, coal cannot be used directly in conventional automobiles, in most home furnaces, or in many industries. One solu-

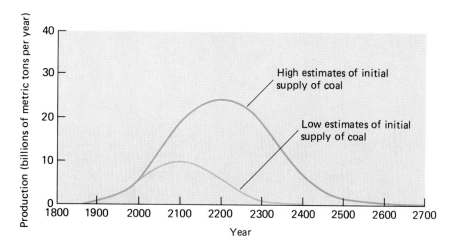

Figure 22-6 Past and predicted world coal production based on two different estimates of initial supply. *(Adapted from M. King Hubbert)*

tion to this problem is the conversion of coal to liquid and gaseous fuels, which has been accomplished both in laboratories and in industry. This process is not used commercially simply because it is too costly to compete with petroleum. However, when petroleum becomes scarce and prices rise, conversion of coal to liquid and gas will probably occur on a large scale.

Secondary Recovery from Oil Wells

On the average, more than half of the oil in a reservoir is left behind after a well has "gone dry." Conventional pumping cannot recover this oil. In the United States, more than 300 billion barrels of this type of oil remain in oil fields. One technique for recovering it is to force superheated steam into old wells at high pressure. The steam heats the oil and increases its ability to flow into the well. Of course, a great deal of energy is needed to heat the steam, so this type of extraction is not always cost-effective or energy-efficient. Another process involves pumping detergent into the reservoir. The detergent carries oil to the well. The petroleum is then recovered and the detergent recycled. Not all of the 300 billion barrels can be recovered by these techniques, but if the price of fuel rises, such **secondary recovery** of oil will become economical.

Oil Shale

Some shales and other sedimentary rocks contain large amounts of a waxy, solid organic substance called **kerogen**, the precursor of liquid petroleum. Kerogen-bearing rock is called **oil shale**. If oil shale is mined and heated properly, the kerogen converts to petroleum. In the United States, oil shales contain the energy equivalent of 2 to 5 trillion barrels of petroleum, enough to fuel the nation for

400 to 900 years at current rates of consumption (Fig. 22–7). However, low-grade oil shales require more energy to mine and convert the kerogen to petroleum than is generated by burning the oil, so they will probably never be used for fuel. Oil refined from higher-grade oil shales in the United States would supply this country for about 75 years if consumption remained at current levels. Worldwide deposits are not as rich, so the global situation is not as promising.

When oil prices skyrocketed from $22 per barrel in 1978 to $45 per barrel in 1981, major oil companies began to explore the development of oil shale, and built experimental recovery plants. However, when prices plummeted a few years later, most of this activity came to a halt. In 1991, the cost of crude oil was about $20 per barrel, and gasoline cost a little over $1 per gallon at the pump. We can expect renewed interest in oil shale when the wholesale price of oil doubles.

22.3 Environmental Problems Resulting from Production of Fossil Fuels

All steps in fossil fuel extraction and refining are potential causes of environmental problems.

Coal Mining

Coal is mined in both underground tunnels and surface mines. Underground **tunnel mines** do not directly disturb the surface of the land. However, many abandoned mines collapse, and occasionally houses have fallen into the holes. In addition, the bedrock dug out to get at the coal must go somewhere. This material, called **mine spoil**, is often deposited as banks or mounds of loose, unvegetated

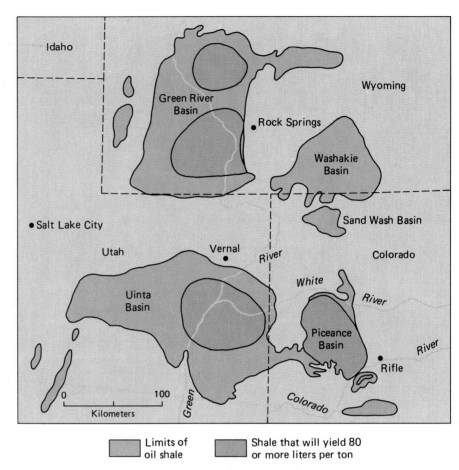

Figure 22–7 The locations of major oil shale fields in the United States. *(Adapted from U.S. Geological Survey Circular 2223, 1965)*

An underground tunnel mine.

rock near a mine. Mine spoil erodes easily, and the muddy runoff silts nearby streams and destroys aquatic habitats.

Tunnel mines are often more expensive and less efficient than **surface mines** (Fig. 22–8). To dig a surface mine, huge power shovels scrape off soil and bedrock to expose the coal. The mine spoil is piled up on the surface. In the United States, The Surface Mining Control and Reclamation Act requires that mining companies restore mined land so that it is useful for the same purposes as before the mining began. In addition, a tax is levied to reclaim land that was mined and destroyed before the law was enacted. However, many citizens claim that the government has not adequately enforced the law and that 6000 surface mine sites have not been restored in compliance with the law (Fig. 22–9).

Most coal contains sulfur. When the coal is exposed, some of this sulfur reacts with water and air to produce sulfuric acid (H_2SO_4). If pollution control is inadequate, the sulfuric acid runs into streams and ground water below the mine or mill. This type of pollution is called **acid**

A   B

Figure 22–8   (A) An aerial view of a coal strip mine, the Decker Mine in Montana.   *(E. N. Hinrichs/USGS)* (B) The Navajo coal strip mine in New Mexico.   *(H. E. Malde/USGS).*

Figure 22–9   Areas in the United States where economically attractive coal deposits are found. Lightly shaded areas indicate coal fields where some surface mining has already occurred. Dark shading indicates coal fields that may be surface mined in the future.

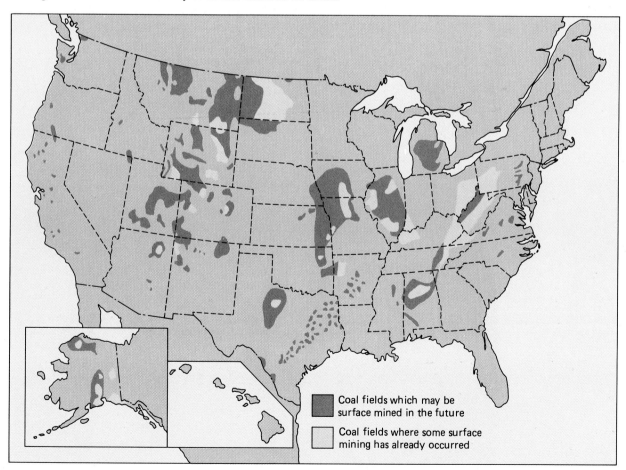

Coal fields which may be surface mined in the future

Coal fields where some surface mining has already occurred

After an oil well has been drilled, the expensive drill platform is removed and replaced by a pumper, such as this one on the Wyoming prairie.

mine drainage. In addition, when sulfur-rich coal burns, the sulfur reacts to form hydrogen sulfide and sulfuric acid, which in turn create acid precipitation.

Petroleum Extraction, Transport, and Refining

Relatively little environmental damage occurs when oil wells are drilled in temperate and arid regions. An oil well occupies only a few hundred square meters of land. Wells are even drilled in people's backyards, with minimal environmental problems. As these relatively accessible reserves are being depleted, however, oil companies have begun to extract petroleum from more fragile environments, such as the ocean floor and the Arctic tundra.

Rich offshore oil reserves exist on continental shelves in many parts of the world, including the coast of southern California, the Gulf of Mexico, and the North Sea in Europe. To obtain the oil, engineers must first build platforms on pilings driven into the ocean floor. Drilling rigs are then mounted on these steel islands (Fig. 22–10). Despite great care, accidents occur during drilling and extraction of oil. Broken pipes, excess pressure, and difficulty in capping a new well have all led to blowouts,

Figure 22–10 An offshore oil drilling platform. *(Schlumberger Inc.)*

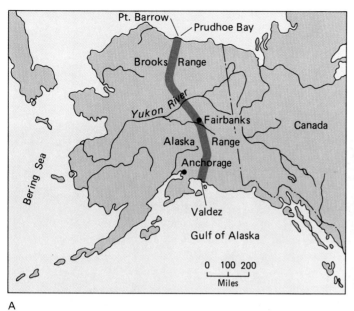

A   B

Figure 22–11   (A) The route of the Alaska Pipeline. (B) The Alaska Pipeline is built in zigzag
fashion so it will not rupture when the permafrost soil expands and contracts.   *(Alyeska Pipeline
Company)*

spills, and oil fires. When these accidents occur on shore,
they can destroy small areas of farmland but do not affect
entire ecosystems. When accidents occur at sea, however,
millions of barrels of oil can be dispersed throughout
the waters, poisoning marine life and disrupting ocean
ecosystems. Significant oil spills have occurred in virtually all offshore drilling areas.

Petroleum from Alaska's northern coast is piped
through the Alaska pipeline to Valdez on the southern
coast of Alaska (Fig. 22–11). There it is pumped into
tankers that carry it to refineries. In March 1989, the
supertanker *Exxon Valdez* was navigating around floating
ice when it ran aground near Valdez, spilling 260,000
barrels of petroleum into Prince William Sound (Fig.
22–12). The thick, black crude oil covered hundreds of
kilometers of coastline and killed thousands of birds and
sea mammals. It temporarily disrupted commercially important fisheries and damaged the local tourism industry.
It is unclear how long it will take for the ecosystem to
return to normal.

Crude petroleum must be **refined** to produce gasoline, propane, diesel fuel, motor oil, and chemicals. During refining, the crude oil is heated under pressure to
break apart its large molecules and form useful products
such as high-octane gasoline. A large refinery covers
hundreds of acres and processes thousands of barrels of
petroleum every day (Fig. 22–13). Each stage in the
refining process is a potential source of air pollution.

Figure 22–12   The *Exxon Valdez* after it ran aground and
began to spill oil. The slick appears off the ship's bow.
(Wide World Photos)

Figure 22–13 An oil refinery.

While all refineries use expensive pollution control equipment, these devices are never completely effective. As a result, toxic and carcinogenic chemicals escape into the atmosphere.

Mining and Refining of Oil Shale

Oil shale can be mined either in surface mines or by underground methods. In either case, environmental disruptions similar to those created by coal mining are routinely created. But additional problems specific to shale development also occur. When oil shale is broken into small pieces and exposed to the atmosphere, it absorbs water and swells. Thus, the volume of rock actually increases even though oil is removed from it. (As an analogy, take a potato, slice it into thin slices, and fry the slices to make potato chips. Now place the potato chips in a pile. Which occupies more volume, the pile of potato chips or the potato?) After the kerogen is removed, the expanded shale must be dumped somewhere, and the piles of rock are unsightly and easily eroded.

Another problem inherent in shale development is water consumption. Approximately 2 barrels of water are needed to produce each barrel of oil from shale. Oil shale occurs most abundantly in western North America where the climate is semiarid. In this region, water is also needed for agriculture, domestic consumption, and industry. Proponents of energy development claim that when oil prices rise and oil shale becomes economical, new dams can be built to conserve water, and industries can move elsewhere. Moreover, most of the ranchers in the region use

water inefficiently and much of the land is marginally productive. Therefore, according to this argument, it might be in the national interest to use that water instead for oil shale development. But the law allocates water rights according to history of use. The original ranchers and farmers passed down the water rights to later generations or sold them with the land. It would be unfair and perhaps unconstitutional to change water law and harm individuals, even if the nation as a whole would benefit.

22.4 Nuclear Energy and Uranium Reserves

Fossil fuels are **nonrenewable resources**; when they are used up, we will have to do without them. What other sources of energy are available? Nuclear energy is one. Other energy sources are listed in the essay, "Energy Strategies for the United States."

Modern nuclear power plants use **nuclear fission** to generate heat (Fig. 22–14). A certain isotope of uranium, U-235, is the major fuel. When the nuclei of U-235 are bombarded with neutrons, they break apart (hence the word "fission," which means splitting). Large quantities of heat are released during fission. The heat is used to generate steam to drive turbines, which in turn generate electricity. During fission, uranium nuclei break apart to produce several daughter products, some of which are also radioactive.[*]

[*]For a review of radioactive decay, refer to Chapter 4.

Figure 22–14 A nuclear power plant. The domed structure on the left is the reactor; the larger structure on the right is the cooling tower.

Over the past 40 years, optimism about nuclear power has waxed and waned. Many hazards and economic problems were overlooked initially, and proponents of nuclear power wrote statements such as "We can look forward to universal comfort, practically free transportation, and unlimited supplies of materials." By the early 1980s, this rosy forecast had changed. Construction of new reactors had become so costly that electricity generated by nuclear power was more expensive than that generated by coal-fired power plants. In addition, public concern about accidents and disposal of radioactive waste became acute. As a result, growth of the nuclear power industry came to an abrupt halt. Between 1981 and 1991, no new orders were placed for nuclear power plants in the United States and 117 planned power plants were cancelled (Fig. 22–15). However, nuclear power may regain importance if fossil fuels become expensive and alternative energy sources are not developed.

In 1990, nuclear fuels generated 7.6 percent of the energy used in the United States. The current world's uranium reserves are about 6 million tons. With these reserves, the present rate of nuclear power generation could be maintained for about 100 years without mining low-grade uranium ore.

Nuclear Fusion

Nuclear fusion occurs when nuclei of light elements combine to form heavier nuclei. Fusion generates tremendous amounts of energy. Our Sun generates its energy by fusion of hydrogen nuclei to form helium nuclei. Controlled nuclear fusion is a potential source of energy that, on a human time scale, is limitless. No useful fusion reactor has yet been developed.

MINERAL RESOURCES

22.5 Ore

If you walked outside, picked up any rock at random, and sent it to a laboratory for analysis, the report would probably show that the rock contains measurable quantities of iron, gold, silver, aluminum, and other valuable metals. However, the concentrations of these metals are so low in most rocks that the cost of extracting them would be much greater than the income gained by selling them.

Figure 22–15 The numbers of construction permits requested by the nuclear power industry in recent years. The rapid decline reflects smaller-than-predicted growth in electric consumption and the fact that nuclear power is no longer economical. (Annual Energy Review 1991, *Energy Information Administration*)

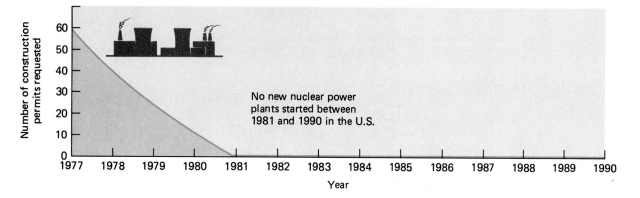

TABLE 22–2

Comparison of Concentration of Specific Elements in Earth's Crust with Concentration Needed to Operate a Commercial Mine			
Element	Natural Concentration in Crust (% by weight)	Concentration Required to Operate a Commercial Mine (% by weight)	Enrichment Factor
Aluminum	8	24–32	3–4
Iron	5.8	40	6–7
Copper	0.0058	0.46–0.58	80–100
Nickel	0.0072	1.08	150
Zinc	0.0082	2.46	300
Uranium	0.00016	0.19	1,200
Lead	0.00010	0.2	2,000
Gold	0.0000002	0.0008	4,000
Mercury	0.000002	0.2	100,000

A **mineral deposit** is a local enrichment of one or more minerals. **Ore** is any natural material sufficiently enriched in one or more minerals to be mined profitably. Table 22–2 shows that the concentrations of elements in ore may be as much as 100,000 times those in ordinary rock. The **mineral reserves** of a region are the known supply of ore in the ground. Reserves are depleted when ore is mined, but reserves may also increase in two ways. First, new mineral deposits may be discovered. Second, increases in price or improvements in mining or refining technology may convert subeconomic mineral deposits into ore.

Many known mineral deposits are enriched in one or more minerals, but they are not quite rich enough to mine

at a profit. For example, a deposit containing 30 percent iron is not ore because it would not be profitable to extract the metal from the rock. But if technology improved so that the iron could be refined more cheaply or if the price of iron increased, then the deposit would suddenly become ore.

As an example of the changing nature of reserves, in 1966 geologists estimated that global reserves of iron were about 5 billion tons.* At that time, world consumption of iron was about 280 million tons per year. Assuming consumption continued at the 1966 rate, the global

*Mason, B., *Principles of Geochemistry*, 3d ed. (New York: John Wiley, 1966), Appendix III.

An aerial view of the Bingham open-pit copper mine. *(Agricultural Stabilization and Conservation Service/USDA)*

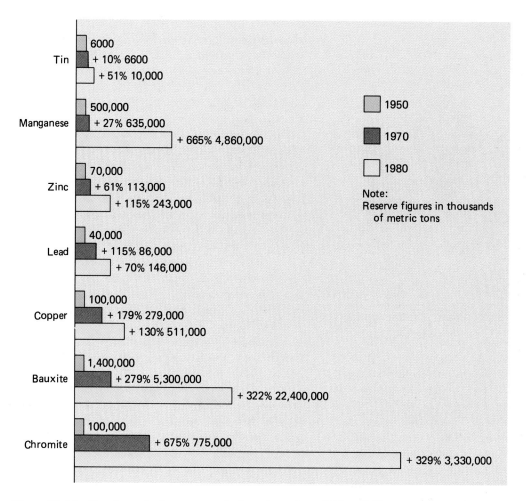

Figure 22–16   The change in the reserves of selected ores from 1950 to 1980.

iron reserves identified in 1966 would have been exhausted in 18 years (5 billion tons/280 million tons per year = 18 years), and we would have run out of iron ore in 1984. Something must have changed, because iron ore remains available today and is relatively inexpensive. The critical change resulted from development of inexpensive methods for processing iron ores of lower grade. Thus, deposits that were uneconomical to mine in 1966, and therefore not counted as reserves, are now ore. In fact, as shown in Figure 22–16, the reserves of many elements have increased dramatically in recent years. Furthermore, many regions, especially in less developed countries, have not been explored thoroughly, so new deposits probably will be discovered.

22.6   How Ore Deposits Form

One of the primary professional objectives of many geologists is to find new ore deposits. Successful exploration requires an understanding of the processes that concen-

trate elements to form ore. For example, platinum concentrates in certain types of igneous rocks. Therefore, if you were exploring for platinum, you would not look in shale or limestone.

Magmatic Processes

Cooling magma does not solidify and crystallize all at once. Instead, high-temperature minerals crystallize first. Lower-temperature minerals crystallize later, when the temperature drops (see Chapter 3).

Solid minerals are denser than liquid magma. Consequently, crystals that form first sink to the bottom of a magma chamber in a process called **crystal settling** (Fig. 22–17). These crystals form a layer that commonly consists of a single mineral or a mixture of minerals with similar melting points. If the minerals contain valuable metals, an ore deposit may form. The largest ore deposit formed in this way is the Bushveldt intrusion of South Africa. It is about 375 by 300 kilometers in area—roughly

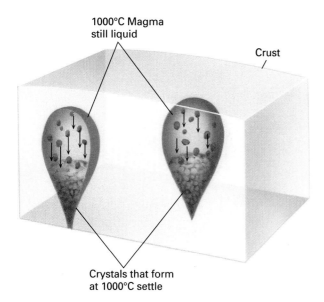

Figure 22–17   Concentration of minerals by crystal settling in a magma chamber.

the size of the state of Maine—and about 7 kilometers thick. Large quantities of chromium and platinum are mined from the Bushveldt.

Hydrothermal Processes

Magma and underground rock contain large amounts of hot water, with the same dissolved ions found in seawater. This mixture of hot water and dissolved ions is a **hydrothermal solution**. (The word hydrothermal is derived from the roots *hydro* for water and *thermal* for hot.)

Hydrothermal solutions are corrosive and can dissolve metals such as copper, gold, lead, zinc, and silver from hot rock or magma. Recall that most rocks contain low concentrations of many metals. For example, gold makes up 0.0000002 percent of average crustal rock; copper makes up 0.0058 percent, and lead 0.0001 percent. Although the metals are present in the rock in low concentrations, hydrothermal solutions percolate slowly through vast volumes of rock, dissolving and accumulating large amounts of the metals. The metals are then deposited when the solutions encounter changes in temperature, pressure, or chemical environment (Fig. 22–18). In this way, hydrothermal solutions scavenge metals from average crustal rocks and then deposit them locally to form ore.

If the metals precipitate in fractures in rock, a hydrothermal **vein** deposit forms. Ore veins range from less than a millimeter to several meters in width. They can be incredibly rich. Single gold or silver veins have yielded several million dollars' worth of ore.

The same hydrothermal solutions that flow rapidly through open fractures to form rich ore veins may also soak into large volumes of rock around the fractures. If metals precipitate in the rock, they may create a very large but much less concentrated **disseminated ore deposit**. Because they may form from the same solutions, vein deposits and disseminated deposits are often found together. The history of many mining districts is one in which early miners dug shafts and tunnels to follow the rich veins. After the veins were exhausted, later miners used huge power shovels to extract low-grade ore from disseminated deposits surrounding the veins.

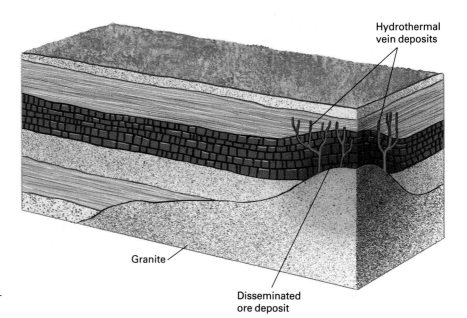

Figure 22–18   Formation of hydrothermal deposits.

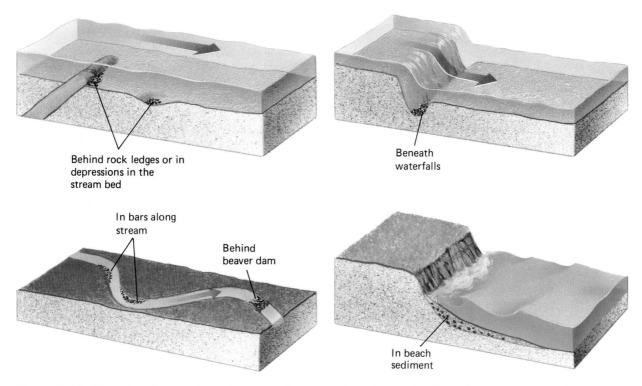

Behind rock ledges or in
depressions in the
stream bed

Beneath
waterfalls

In bars along
stream

Behind
beaver dam

In beach
sediment

Figure 22–19 Placer deposits occur in environments where water slows down and sediment is
deposited.

Sedimentary Processes

Two types of sedimentary processes form ore deposits:
sedimentary sorting and precipitation.

Sedimentary Sorting: Placer Deposits

Gold occurs naturally as a pure metal and is denser than
any other mineral. Therefore, if a mixture of gold dust,
sand, and gravel is swirled in a glass of water, the gold
falls to the bottom first. Differential settling also occurs
in nature. Many streams carry silt, sand, and gravel with
an occasional small grain of gold. The gold settles first
when the current slows down. Thus, grains of gold concentrate near bedrock or in coarse gravel, forming a
placer deposit (Fig. 22–19).

Precipitation

As ground water percolates through rock, it dissolves
minerals and carries off dissolved ions. In most environments, this water eventually flows into streams and then
to the sea. Some of the dissolved ions, such as sodium
and chloride, make seawater salty. In deserts, however,
lakes develop with no outlet to the ocean. Water flows
into the lakes but can escape only by evaporation. As the
water evaporates, the dissolved ions concentrate until they
begin to precipitate.

Panning for gold. *(Montana Historical Society)*

Earth Science and the Environment

Pollution from Metal Mining: A Case History

The largest complex of toxic waste sites in the United States was produced not by a chemical factory or petroleum refinery, but by the mines and smelters in Butte and Anaconda in Montana. A short case history of this region highlights pollution problems that result from mining and refining metals.

Remember that most ore is not pure metal but rock containing metal-rich minerals. Some of the raw ore from the mines in Butte contained as little as 0.3 percent copper, with even smaller concentrations of manganese, zinc, arsenic, lead, cadmium, and a variety of other metals. When ore is extracted from a mine, the waste rock is simply piled up somewhere. Such piles of crushed rock are prone to erosion, and the muddy runoff silts streams and destroys aquatic habitats. But that is only the beginning of the problem, because the smelting process generates additional wastes.

In the boom days in Butte, during the late 1800s and early 1900s, metal smelting was inefficient, so large quantities of arsenic, cadmium, zinc, lead, copper, and other metals were discarded. Metal ores commonly contain large amounts of sulfur, which was discarded with the other wastes. Furthermore, Butte's copper ore contains gold, which the early miners attempted to extract by dissolving it in mercury, and the waste mercury was dumped on the piles as well. Today at the Butte–Anaconda area there are about 25 square kilometers (about 10 square miles) of mine and mill spoils averaging 15 meters (about 50 feet) high, or about as tall as a five-story building.

These wastes are a serious environmental threat because both the sulfur and the metals are poisonous. For example, sulfur compounds react in water to form sulfuric acid. Sulfuric acid is soluble in water and readily disperses into streams and ground-water reserves. In one study, workers in a cadmium recovery plant suffered a higher incidence of lung cancer than the public at large. In another study, skin cancer was prevalent among a group of people who drank well water contaminated with arsenic. Metal contamination is also harmful to the environment, causing a reduction in the growth rate of plants, of animals that eat the plants, and of microorganisms in the soil.

Sulfur and metal poisons have leached into nearby streams and ground-water reserves and have migrated up to 150 kilometers downstream from Butte and Anaconda. Windblown dust from the spoils has contaminated entire counties. No one really knows how to clean up such a mess or how much it will cost. Even a partial cleanup will cost tens to hundreds of millions of dollars.

You can perform a simple demonstration of evaporation and precipitation. Fill a bowl with warm water and add a few teaspoons of table salt. The salt dissolves and you see only a clear liquid. Set the bowl aside for a day or two until the water evaporates. The salt precipitates and encrusts the sides and bottom of the bowl.

Salts that form by evaporation are called **evaporite deposits**. Evaporite minerals include table salt, borax, sodium sulfate, and sodium carbonate. These salts are used in the production of paper, soap, and medicines and for the tanning of leather.

Several times during the history of the Earth, shallow seas covered large portions of North America and all other continents. At times, those seas were so poorly connected to the open oceans that water did not circulate freely between them and the oceans. Consequently, evaporation concentrated the dissolved ions until salts began to precipitate as **marine evaporites**. Periodically, storms flushed new seawater from the open ocean into the shallow seas, providing a new supply of salt. In this way, thick marine evaporite beds formed. Nearly 30 percent of North America is now underlain by these deposits. Table salt, gypsum (used to manufacture plaster and wallboard), and potassium salts (used in fertilizer) are mined extensively from marine evaporites.

About 1 billion tons of iron are mined every year, 90 percent from **banded iron formations**, which are sedimentary layers of iron-rich minerals sandwiched between beds of silicate minerals. The alternating layers are a few centimeters thick and give the rocks their banded appearance (Fig. 22–20).

The most abundant and economically important banded iron formations formed between 2.6 and 1.9 billion years ago. What happened during that interval to create these peculiar rocks? In order to answer this question, we must understand how iron interacts with oxygen. In an oxygen-poor environment, iron dissolves easily in water. If the oxygen concentration rises, iron reacts with the oxygen to precipitate rapidly. During the early history of the Earth, very little molecular oxygen was present in the atmosphere, and as a result, large quantities of iron were dissolved in the ocean. Then plants evolved. As they became abundant, the concentration of oxygen in the atmosphere increased, because plants release oxygen into the atmosphere during photosynthesis. About 2.6 billion years ago, enough oxygen had accumulated in the atmo-

Figure 22–20 A banded iron formation. The red bands are iron minerals; the dark-colored layers are chert. *(Ward's Natural Science Establishment, Inc.)*

sphere to react with the dissolved iron. As a result, vast quantities of iron precipitated, producing the banded iron formations.

Weathering Processes

In environments with high rainfall, the abundant water dissolves and removes most of the soluble ions from the soil and rock. The insoluble ions left behind form **residual deposits**. Both aluminum and iron have very low solubility in water. **Bauxite**, our principal source of aluminum, forms in this manner, and in some instances iron also concentrates enough to become ore.

22.7 Future Availability of Mineral Resources

The Romans discovered veins of nearly pure copper on the island of Cyprus. Those deposits have been mined out and are now gone. In modern times, many rich and easily accessible ores, such as the high-grade iron deposits of the Mesabi Range in the north-central United States, are being used rapidly. These ore deposits are nonrenewable. Our technological life will not end with the exhaustion of rich deposits, however, because less concentrated ore is still available.

On the other hand, we cannot extract metals economically from very low grade deposits. Gold, silver, and a few other elements exist in their pure forms in the crust, but most elements combine naturally with others. Thus, pure iron is rarely found in the crust. The iron in iron ore is chemically bonded to oxygen. To extract iron, the ore must be dug up and crushed, and the iron minerals separated from the other minerals in the rock. Finally,

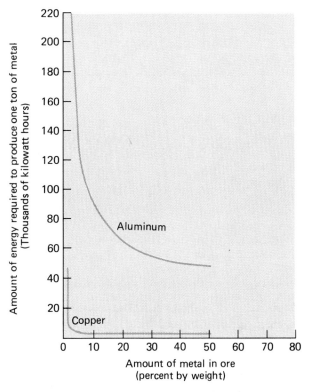

Figure 22–21 The energy required to mine and refine aluminum and copper ores as a function of the metal concentration of the ore. Notice that there is very little change in energy requirements for very concentrated ores, but as the metal concentration drops off, the energy requirements rise very rapidly. However, almost no mines contain 70 percent aluminum or 20 percent copper. Most mining is already being done in the steep section of the curve, so considerably more energy will be needed as lower-grade ores are exploited.

the iron is chemically separated from oxygen to obtain metallic iron. Each step, especially the chemical separation, requires energy. Low-grade ore requires more energy to process than high-grade ore because more rock must be dug, transported, and crushed to obtain a given amount of metal, and more energy is needed to separate the metal from the ore.

Figure 22–21 shows the energy required to extract and refine 1 ton of copper and 1 ton of aluminum from different concentrations of ore. Notice that relatively little energy is needed for rich ores. As concentrations decrease, the energy consumption rises rapidly.

A 1984 article in *Science* discussed the future availability of 67 commercially important elements.[*] The authors predicted that 33 will still be abundant by the year

[*] Goeller, H. E., and A. Zucker, "Infinite Resources: The Ultimate Strategy," *Science* 223, (1984), p. 456.

TABLE 22-3

Predictions of Future Supplies of Elements Used in Industry		
Availability	**Metals**	**Nonmetals**
Limitless supply	Sodium, magnesium, calcium	Nitrogen, oxygen, neon, argon, xenon, krypton, chlorine, bromine, silicon
Abundant supply	Aluminum, gallium, iron, potassium	Sulfur, hydrogen, carbon
Abundant only if research continues to improve extraction	Lithium, strontium, rubidium, chromium, nickel, cobalt, platinum, palladium, rhodium, ruthenium, iridium, osmium	Boron, iodine
Might be in short supply by the year 2100	Gold, silver, mercury, copper, zinc, lead, arsenic, tin, molybdenum, plus 22 other less familiar metals	Helium, phosphorus, fluorine

Source: Goeller, H. E., and A. Zucker, "Infinite Resources: The Ultimate Strategy," *Science* 223 (1984), p. 456.

2100. Table 22–3 shows that 19 metals, including iron, aluminum, chromium, and nickel, are on this list.

Even if sufficient ore is not available, shortages can be alleviated. As the prices of ores rise, recycling becomes economical, and engineers and chemists look for cheaper substitutes for expensive materials such as gold, silver, copper, lead, and zinc.

22.8 Nonmetallic Resources

When we think about becoming rich from mining, we usually think of finding gold. As undramatic as it seems, more money has been made mining sand and gravel than gold.

Concrete is an essential and versatile building material. Reinforced with steel, it is used to build roads, bridges, and buildings. **Concrete** is a mixture of cement, sand, and gravel. **Sand** and **gravel** are mined from stream and glacial deposits, sand dunes, and beaches. **Cement** is made by heating a mixture of crushed limestone and clay.

Many buildings are faced with stone, usually granite or limestone, although marble, sandstone, and other rocks are also used. Stone is mined from **quarries** cut into bedrock (Fig. 22–22).

Phosphorus and potassium are valuable fertilizers extracted from sedimentary rocks.

Look at the room around you and think about how many utensils and building materials are manufactured from geologic resources. The "lead" in your pencil is a mixture of graphite and clay. Your coffee cup may be ceramic and therefore made of clay; if not ceramic, it is probably plastic, manufactured from coal or petroleum. The inside walls of your building are probably lined with wallboard or plaster, both of which are made of gypsum. The exterior of the building may be brick, which is baked clay, or faced with granite or limestone. You may be wearing jewelry set with a semiprecious stone such as turquoise or topaz, or perhaps even a gem such as ruby or diamond.

Figure 22–22 A granite quarry near Barre, Vermont.

SUMMARY

●

Fuels and minerals are the two types of geologic resources. **Fossil fuels** include oil, gas, and coal. If oxygen and flowing water are excluded, plant matter decays partially to form **peat**. Peat converts to **coal** when it is buried and subjected to elevated temperature and pressure. **Petroleum** forms from the remains of organisms that settle to the ocean floor and are incorporated into **source rock**. The organic matter converts to liquid oil when it is buried and heated. The petroleum then migrates to a **reservoir**, where it is retained by an **oil trap**.

Oil and gas shortages will probably occur in the early twenty-first century, whereas coal reserves will be plentiful for 200 years or more. Additional supplies of petroleum can be recovered by secondary extraction from old wells and from **oil shale**.

Metal ores and coal and oil shale are mined in **tunnel mines** or **surface mines**. Both disrupt the land and may create **acid mine drainage**. Petroleum drilling in temperate ecosystems is usually not environmentally disruptive, but spills in fragile environments damage ecosystems, and tanker accidents pollute the ocean and beaches.

Nuclear power is expensive, and questions about safety and disposal of nuclear wastes have not been answered to everyone's satisfaction. As a result, no new nuclear power plants have been ordered since 1981. Inexpensive uranium ore will be available for a century or more.

Ore is a rock or other material that can be mined profitably. **Mineral reserves** are the estimated supply of ore in the ground.

Four major kinds of geologic processes concentrate elements to form mineral deposits and ore: (1) **Magmatic processes** form ore during the solidification of magma. **Crystal settling** is one example. (2) **Hydrothermal processes** involve transportation of dissolved ions by hot water. Minerals precipitate from those solutions in fractures and pores in rock to form **hydrothermal mineral deposits**. (3) Two types of **sedimentary processes** concentrate minerals. Dense minerals that concentrate by settling out of flowing water form **placer deposits**. Precipitation from lake water or seawater forms **evaporite deposits**; **banded iron formations** developed as oxygen accumulated in the atmosphere and reacted with dissolved iron. (4) **Weathering** removes easily dissolved elements and minerals, leaving behind **residual deposits**.

The future availability of ore reserves depends on the quantity and concentration of mineral deposits, the availability of energy, and problems of pollution.

Nonmetallic resources include sand and gravel for concrete, limestone for cement, building stone, and phosphorus and potassium for fertilizers.

KEY TERMS

●

Fossil fuel *597*
Peat *598*
Coal *598*
Reservoir *599*
Source rock *599*
Migration *599*
Oil trap *599*
Cap rock *599*

Solar energy *604*
Solar collector *604*
Solar cell *604*
Secondary recovery *607*
Kerogen *607*
Oil shale *607*
Tunnel mine *607*
Surface mine *608*
Acid mine drainage *608*

Nonrenewable
  resources *612*
Nuclear fission *612*
Nuclear fusion *613*
Mineral deposit *614*
Ore *614*
Mineral reserve *614*
Crystal settling *615*
Hydrothermal solution *616*

Disseminated ore
  deposit *616*
Placer deposit *617*
Evaporite deposit *618*
Marine evaporites *618*
Banded iron
  formation *618*
Residual deposit *619*

REVIEW QUESTIONS

●

1. Name the three different categories of geologic resources.

2. Explain how coal forms. Why does it form in some environments but not in others?

3. Explain the importance of source rock, reservoir rock, cap rock, and oil traps in the formation of petroleum reserves.

4. Discuss problems in predicting the future availability of fossil fuel reserves. What is the value of the predictions?

5. Discuss the prospects for the availability of petroleum in the next 10, 20, and 40 years. What uncertainties are inherent in these predictions?

6. Discuss two sources of petroleum that will be available after conventional wells go dry.

7. What is ore? What are mineral reserves? Describe three factors that can cause changes in estimates of mineral reserves.

8. If most elements are widely distributed in ordinary rocks, why should we worry about someday running short?

9. Explain crystal settling.

10. Discuss the formation of hydrothermal ore deposits.

11. Discuss the formation of marine evaporites and banded iron formations.

12. Explain why the availability of mineral resources depends on the availability of energy, on other environmental issues, and on political considerations.

13. Compare the environmental impact of underground mines with that of surface mines.

DISCUSSION QUESTIONS

•

1. If you were searching for petroleum, would you search primarily in sedimentary rock, metamorphic rock, or igneous rock? Explain.

2. Which of the following materials would you expect to find on the Moon: iron, gold, aluminum, coal, or petroleum? Defend your answer.

3. Imagine that you were a space traveler and were abandoned on an unknown planet in a distant solar system. What clues would you look for if you were searching for fossil fuels?

4. Is an impermeable cap rock necessary to preserve coal deposits? Why or why not?

5. Prepare a class debate. Have one side argue that we are quickly approaching an energy crisis that will debilitate our society, and have the other side argue that society will adjust to changes in energy supply without disruption.

6. What factors can make our metal reserves last longer? What factors can deplete them rapidly?

7. It is common for a single mine to contain ores of two or more metals. Discuss how geological processes might favor concentration of two metals in a single deposit.

8. Compare the depletion of mineral reserves with the depletion of fossil fuels. How are the two problems similar, and how are they different?

9. Imagine that you plan to write a novel about a society that lives on the Earth in the year 2100. In your story, a nuclear war has occurred and major industries have been destroyed. Write a brief description of the types of materials that would be available to your characters and explain how they would extract and process them.

10. List ten objects that you own. What resources are they made of? How long will each of the objects be used before it is discarded? Will the materials eventually be recycled or deposited in the trash? Discuss ways of conserving resources in your own life.

11. Explain why the environmental effects of extracting fossil fuels are likely to increase as fuel reserves are gradually exhausted.

12. Who pays for the cost of reclaiming surface mines? What costs arise from surface mines if they are not reclaimed, and who pays these expenses? Which costs are more immediately apparent to the average citizen?

The Elements

Appendix A

Key:

26
Fe
55.847

- Atomic number (Z)
- Element symbol
- Atomic mass of naturally occurring isotopic mixture; for radioactive elements, numbers in parentheses are mass numbers of most stable isotopes

IA	IIA	IIIB	IVB	VB	VIB	VIIB		VIII		IB	IIB	IIIA	IVA	VA	VIA	VIIA	O
1 **H** 1.0079																1 **H** 1.0079	2 **He** 4.00260
3 **Li** 6.941	4 **Be** 9.01218											5 **B** 10.81	6 **C** 12.011	7 **N** 14.0067	8 **O** 15.9994	9 **F** 18.998403	10 **Ne** 20.179
11 **Na** 22.98977	12 **Mg** 24.305											13 **Al** 26.98154	14 **Si** 28.0855	15 **P** 30.97376	16 **S** 32.06	17 **Cl** 35.453	18 **Ar** 39.948
19 **K** 39.0983	20 **Ca** 40.08	21 **Sc** 44.9559	22 **Ti** 47.90	23 **V** 50.9415	24 **Cr** 51.996	25 **Mn** 54.9380	26 **Fe** 55.847	27 **Co** 58.9332	28 **Ni** 58.70	29 **Cu** 63.546	30 **Zn** 65.38	31 **Ga** 69.72	32 **Ge** 72.59	33 **As** 74.9216	34 **Se** 78.96	35 **Br** 79.904	36 **Kr** 83.80
37 **Rb** 85.4678	38 **Sr** 87.62	39 **Y** 88.9059	40 **Zr** 91.22	41 **Nb** 92.9064	42 **Mo** 95.94	43 **Tc** (98)	44 **Ru** 101.07	45 **Rh** 102.9055	46 **Pd** 106.4	47 **Ag** 107.868	48 **Cd** 112.41	49 **In** 114.82	50 **Sn** 118.69	51 **Sb** 121.75	52 **Te** 127.60	53 **I** 126.9045	54 **Xe** 131.30
55 **Cs** 132.9054	56 **Ba** 137.33	57 ***La** 138.9055	72 **Hf** 178.49	73 **Ta** 180.9479	74 **W** 183.85	75 **Re** 186.207	76 **Os** 190.2	77 **Ir** 192.22	78 **Pt** 195.09	79 **Au** 196.9665	80 **Hg** 200.59	81 **Tl** 204.37	82 **Pb** 207.2	83 **Bi** 208.9804	84 **Po** (209)	85 **At** (210)	86 **Rn** (222)
87 **Fr** (223)	88 **Ra** 226.0254	89 **†Ac** 227.0278	104 **Unq** (261)	105 **Unp** (262)	106 **Unh** (263)	107 **Uns**			109								

*Lathanide Series

58 **Ce** 140.12	59 **Pr** 140.9077	60 **Nd** 144.24	61 **Pm** (145)	62 **Sm** 150.4	63 **Eu** 151.96	64 **Gd** 157.25	65 **Tb** 158.9254	66 **Dy** 162.50	67 **Ho** 164.9304	68 **Er** 167.26	69 **Tm** 168.9342	70 **Yb** 173.04	71 **Lu** 174.967

†Actinide Series

90 **Th** 232.0381	91 **Pa** 231.0359	92 **U** 238.029	93 **Np** 237.0482	94 **Pu** (244)	95 **Am** (243)	96 **Cm** (247)	97 **Bk** (247)	98 **Cf** (251)	99 **Es** (252)	100 **Fm** (257)	101 **Md** (258)	102 **No** (259)	103 **Lr** (260)

Note: Atomic masses shown here are 1977 IUPAC values.

A.1

Based on the assigned relative atomic mass of $^{12}C = 12$.

The following values apply to elements as they exist in materials of terrestrial origin and to certain artificial elements. When used with the footnotes, they are reliable to ±1 in the last digit, or ±3 if that digit is in small type.

	Symbol	Atomic Number	Atomic Weight		Symbol	Atomic Number	Atomic Weight		Symbol	Atomic Number	Atomic Weight
Actinium	Ac	89		Hafnium	Hf	72	178.4_9	Promethium	Pm	61	
Aluminum	Al	13	26.98154^a	Helium	He	2	$4.00260^{b,c}$	Protactinium	Pa	91	$231.0359^{a,f}$
Americium	Am	95		Holmium	Ho	67	164.9304^a	Radium	Ra	88	$226.0254^{f,g}$
Antimony	Sb	51	121.7_5	Hydrogen	H	1	$1.0079^{b,d}$	Radon	Rn	86	
Argon	Ar	18	$39.94_8^{b,c,d,g}$	Indium	In	49	114.82	Rhenium	Re	75	186.207
Arsenic	As	33	74.9216^a	Iodine	I	53	126.9045^a	Rhodium	Rh	45	102.9055^a
Astatine	At	85		Iridium	Ir	77	192.2_2	Rubidium	Rb	37	85.467_8^c
Barium	Ba	56	137.3_4	Iron	Fe	26	55.84_7	Ruthenium	Ru	44	101.0_7
Berkelum	Bk	97		Krypton	Kr	36	83.80	Samarium	Sm	62	150.4
Beryllium	Be	4	9.01218^a	Lanthanum	La	57	138.905_5^b	Scandium	Sc	21	44.9559^a
Bismuth	Bi	83	208.9804^a	Lawrencium	Lr	103		Selenium	Se	34	78.9_6
Boron	B	5	$10.81^{c,d,e}$	Lead	Pb	82	$207.2^{d,g}$	Silicon	Si	14	28.08_6^d
Bromine	Br	35	79.904^c	Lithium	Li	3	$6.94_1^{c,d,e,g}$	Silver	Ag	47	107.868^c
Cadmium	Cd	48	112.40	Lutetium	Lu	71	174.97	Sodium	Na	11	22.98977^a
Calcium	Ca	20	40.08^g	Magnesium	Mg	12	$24.305^{c,g}$	Strontium	Sr	38	87.62^g
Californium	Cf	98		Manganese	Mn	25	54.9380^a	Sulfur	S	16	32.06^d
Carbon	C	6	$12.011^{b,d}$	Mendelevium	Md	101		Tantalum	Ta	73	180.947_9^b
Cerium	Ce	58	140.12	Mercury	Hg	80	200.5_9	Technetium	Tc	43	98.9062^f
Cesium	Cs	55	132.9054^a	Molybdenum	Mo	42	95.9_4	Tellurium	Te	52	127.6_0
Chlorine	Cl	17	35.453^c	Neodymium	Nd	60	144.2_4	Terbium	Tb	65	158.9254^a
Chromium	Cr	24	51.996^c	Neon	Ne	10	20.17_9^c	Thallium	Tl	81	204.3_7
Cobalt	Co	27	58.9332^a	Neptunium	Np	93	237.0482^f	Thorium	Th	90	232.0381^f
Copper	Cu	29	$63.54^{c,d}$	Nickel	Ni	28	58.70	Thulium	Tm	69	168.9342^a
Curium	Cm	96		Niobium	Nb	41	92.9064^a	Tin	Sn	50	118.6_9
Dysprosium	Dy	66	162.5_0	Nitrogen	N	7	$14.0067^{b,c}$	Titanium	Ti	22	47.9_0
Einsteinium	Es	99		Nobelium	No	102		Tungsten	W	74	183.8_5
Erbium	Er	68	167.2_6	Osmium	Os	76	190.2	Uranium	U	92	$238.029^{b,c,e}$
Europium	Eu	63	151.96	Oxygen	O	8	$15.999_4^{b,c,d}$	Vanadium	V	23	$50,941_4^{b,c}$
Fermium	Fm	100		Palladium	Pd	46	106.4	Wolfram			(see Tungsten)
Fluorine	F	9	18.99840^a	Phosphorus	P	15	30.97376^a	Xenon	Xe	54	131.30
Francium	Fr	87		Platinum	Pt	78	195.0_9	Ytterbium	Yb	70	173.04_4
Gadolinium	Gd	64	157.2_5	Plutonium	Pu	94		Yttrium	Y	39	88.9059^a
Gallium	Ga	31	69.72	Polonium	Po	84		Zinc	Zn	30	65.38
Germanium	Ge	32	72.5_9	Potassium	K	19	39.09_8	Zirconium	Zr	40	91.22
Gold	Au	79	196.9665^a	Praseodynium	Pr	59	140.9077^a				

[a] Mononuclidic element.

[b] Element with one predominant isotope (about 99 to 100 percent abundance).

[c] Element for which the atomic weight is based on calibrated measurements by comparisons with synthetic mixtures of known isotopic composition.

[d] Element for which known variation in isotopic abundance in terrestrial samples limits the precision of the atomic weight given.

[e] Element for which users are cautioned against the possibility of large variations in atomic weight due to inadvertent or undisclosed artificial isotopic separation in commercially available materials.

[f] Most commonly available long-lived isotope.

[g] In some geological specimens this element has a highly anomalous isotopic composition, corresponding to an atomic weight significantly different from that given.

Appendix B

Identifying Common Minerals

More than 2500 minerals exist in the Earth's crust. However, of this great number, only thirty or so are common. Therefore, when you pick up rocks and want to identify the minerals, you are most likely to be looking at the same small group of minerals over and over again. The following list includes the most common and abundant minerals in the Earth's crust. A few important ore minerals and other minerals of economic value, and a few popular precious and semi-precious gems, are included because they are of special interest.

The minerals in this table fall into four categories.

1. Rock-forming minerals are shown in *red*. They are the most abundant minerals in the crust, and make up the largest portions of all common rocks. The rock-forming minerals and mineral groups are feldspar, pyroxene, amphibole, mica, clay, olivine, quartz, calcite, and dolomite. If more than one mineral of a group is common, each mineral is listed under the group name. For example, three kinds of pyroxene are abundant: augite, diopside, and orthopyroxene. All three are described under pyroxene.

2. Accessory minerals are shown in *yellow*. They are minerals that are common, but that usually occur only in small amounts.

3. Ore minerals and other minerals of economic importance are shown in *green*. They are minerals from which metals or other elements can be profitably recovered.

4. Gems are shown in *blue*. If the common gem name(s) is different from the mineral name, the gem name is given in parentheses following the mineral name. For example, emerald is the gem variety of the mineral beryl, and is listed as Beryl (emerald).

Minerals are listed alphabetically within each of the four categories for quick reference. The physical properties most commonly used for identification of each mineral, and the kind(s) of rock in which each mineral is most often found, are listed to facilitate identification of these common minerals.

COMMON MINERALS AND THEIR PROPERTIES

Mineral Group or Mineral	Chemical Composition	Habit, Cleavage, Fracture	Usual Color
ROCK-FORMING MINERALS			
Amphibole — Actinolite	$Ca_2(MgFe)_5Si_8O_{22}(OH)_2$	Slender crystals, radiating, fibrous	Blackish-green to black, dark green
Amphibole — Hornblende	$(Ca,Na)_{2-3}(Mg,Fe,Al)_5Si_6(Si,Al)_2O_{22}(OH)_2$	Elongate crystals	Blackish-green to black, dark green
Calcite	$CaCO_3$	Perfect cleavage into rhombs	Usually white, but may be variously tinted
Clay Minerals — Illite	$K_{0.8}Al_2(Si_{3.2}Al_{0.8})O_{10}(OH)_2$		White
Clay Minerals — Kaolinite	$Al_2Si_2O_5(OH)_4$		White
Clay Minerals — Smectite	$Na_{0.3}Al_2(Si_{3.7}Al_{0.3})O_{10}(OH)_2$		White, buff
Dolomite	$CaMg(CO_3)_2$	Cleaves into rhombs; granular masses	White, pink, gray, brown
Feldspar — Albite (sodium feldspar)	$NaAlSi_3O_8$ (sodic plagioclase)	Good cleavage in two directions, nearly 90°	White, gray
Feldspar — Orthoclase (potassium feldspar)	$KAlSi_3O_8$	Good cleavage in two directions at 90°	White, pink, red, yellow-green, gray
Feldspar — Plagioclase (feldspar containing both sodium and calcium)	$(Na,Ca)(Al,Si)_4O_8$	Good cleavage in two directions at 90°	White, gray
Feldspar — Biotite	$K(Mg,Fe)_3AlSi_3O_{10}(OH)_2$	Perfect cleavage into thin sheets	Black, brown, green
Feldspar — Muscovite	$KAl_2Si_3O_{10}(OH)_2$	Perfect cleavage into thin sheets	Colorless if thin
Olivine	$(MgFe)_2SiO_4$	Uneven fracture, often in granular masses	Various shades of green
Pyroxene — Augite	$Ca(Mg,Fe,Al)(Al,Si_2O_6)$	Short stubby crystals have 4 or 8 sides in cross section	Blackish-green to light green
Pyroxene — Diopside	$CaMg(Si_2O_6)$	Usually short thick prisms; may be granular	White to light green
Pyroxene — Orthopyroxene	$MgSiO_3$	Cleavage good at 87° and 93°; usually massive	Pale green, brown, gray, or yellowish
Quartz	SiO_2	No cleavage, massive and as six-sided crystals	Colorless, white, or tinted any color by impurities

COMMON MINERALS AND THEIR PROPERTIES (CONTINUED)

Hard-ness	Streak	Specific Gravity	Other Properties	Type(s) of Rock in Which the Mineral is Most Commonly Found
			ROCK-FORMING MINERALS	
5–6	Pale green	3.2–3.6	Vitreous luster	Low- to medium-grade metamorphic rocks
5–6	Pale green	3.2	Crystals six-sided with 124° between cleavage faces	Common in many granitic to basaltic igneous rocks, and many metamorphic rocks
3	White	2.7	Transparent to opaque. Rapid effervescence with HCl	Limestone, marble, cave deposits
}			The clay minerals are so fine-grained that most physical properties cannot be identified.	Shale Shale, weathered bedrock, and soil Shale, weathered bedrock, and soil
3.5–4	White to pale gray	3.9–4.2	Effervesces slightly in cold dilute HCl.	Dolomite
6–6.5	White	2.6	Many show fine striations (twinning lines) on cleavage faces.	Granite, rhyolite, low-grade metamorphic rocks
6	White	2.6	Vitreous to pearly luster	Granite, rhyolite, metamorphic rocks
6	White	2.6–2.7	May show striations as in albite.	Basalt, andesite, medium- to high-grade metamorphic rocks
2.2–2.5	White, gray	2.7–3.1	Vitreous luster; divides readily into thin flexible sheets.	Granitic to intermediate igneous rocks, many metamorphic rocks
2–2.5	White	2.7–3	Vitreous or pearly; flexible and elastic; splits easily.	Many metamorphic rocks, granite
6.5–7	White	3.2–3.3	Vitreous, glassy luster	Basalt, peridotite
5.6	Pale green	5–6	Vitreous; distinguished from hornblende by the 87° angle between cleavage faces.	Basalt, peridotite, andesite, high-grade metamorphic rocks
5–6	White to greenish	3.2–3.6	Vitreous luster	Medium-grade metamorphic rocks
5.5	White	3.2–3.5	Vitreous luster	Peridotite, basalt, high-grade metamorphic rocks
7	White	2.6	Includes rock crystal, rose and milky quartz, amethyst, smoky quartz, etc.	Granite, rhyolite, metamorphic rocks of all grades, sandstone, siltstone

COMMON MINERALS AND THEIR PROPERTIES (CONTINUED)

Mineral Group or Mineral	Chemical Composition	Habit, Cleavage, Fracture	Usual Color
ACCESSORY MINERALS			
Apatite	$Ca_5(OH,F,Cl)(PO_4)_3$ (calcium fluorphosphate)	Massive, granular	Green, brown, red
Chlorite	$(Mg,Fe)_6(Si,Al)_4O_{10}(OH)_8$	Perfect cleavage as fine scales	Green
Corundum	Al_2O_3	Short, six-sided barrel-shaped crystals	Gray, light blue, and other colors
Epidote	$Ca_2(Al,Fe)Al_2O(SiO_4)(Si_2O_7)(OH)$	Usually granular masses; also as slender prisms	Yellow-green, olive-green, to nearly black
Fluorite	CaF_2	Octahedral and also cubic crystals	White, yellow, green, purple
Garnet — Almandine	$Fe_3Al_2(SiO_4)_3$	No cleavage, crystals 12- or 24-sided	Deep red
Garnet — Grossular	$Ca_3Al_2(SiO_4)_3$	No cleavage, crystals 12- or 24-sided	White, green, yellow, brown
Graphite	C	Foliated, scaly, or earthy masses	Steel gray to black
Hematite	Fe_2O_3	Granular, massive, or earthy	Brownish-red
Limonite	$2Fe_2O_3 \cdot 3H_2O$	Earthy fracture	Brown or yellow
Magnetite	Fe_3O_4	Uneven fracture, granular masses	Iron black
Pyrite	FeS_2	Uneven fracture cubes with striated faces, octahedrons	Pale brass yellow (lighter than chalcopyrite)
Serpentine	$Mg_3Si_2O_5(OH)_4$	Uneven, often splintery fracture	Light and dark green, yellow

COMMON MINERALS AND THEIR PROPERTIES (CONTINUED)

Hard-ness	Streak	Specific Gravity	Other Properties	Type(s) of Rock in Which the Mineral is Most Commonly Found
			ACCESSORY MINERALS	
4.5–5	Pale red-brown	3.1	Crystals may have a partly melted appearance, glassy.	Common in small amounts in many igneous, metamorphic, and sedimentary rocks
2.0–2.5	Gray, white, pale green	2.8	Pearly to vitreous luster	Common in low-grade metamorphic rocks
9	None	3.9–4.1	Hardness is distinctive.	Metamorphic rocks, some igneous rocks
6.7	Pale yellow to white	3.3	Vitreous luster	Low- to medium-grade metamorphic rocks
4	White	3.2	Cleaves easily; vitreous, transparent to translucent	Hydrothermal veins
6.5–7.5	White	4.2	Vitreous to resinous luster	The most common garnet in metamorphic rocks
6.5–7.5	White	3.6	Vitreous to resinous luster	Metamorphosed sandy limestones
1–2	Gray or black	2.2	Feels greasy; marks paper.	Metamorphic rocks
2.5	Dark red	2.5–5	Often earthy, dull appearance	Common in all types of rocks. It can form by weathering of iron minerals, and is the source of color in nearly all red rocks.
1.5–4	Brownish-yellow	3.6	Earthy masses that resemble clay	Common in all types of rocks. It can form by weathering of iron minerals, and is the source of color in most yellow-brown rocks.
5.5	Iron black	5.2	Metallic luster; strongly magnetic	Common in small amounts in most igneous rocks
6–6.5	Greenish-black	5	Metallic luster, brittle, very common	The most common sulfide mineral. Igneous, metamorphic, and sedimentary rocks; hydrothermal veins
2.5	White	2.5	Waxy luster, smooth feel, brittle	Alteration or metamorphism of basalt, peridotite, and other magnesium-rich rocks

COMMON MINERALS AND THEIR PROPERTIES (CONTINUED)

Mineral Group or Mineral	Chemical Composition	Habit, Cleavage, Fracture	Usual Color
ORE AND OTHER MINERALS OF ECONOMIC IMPORTANCE			
Anhydrite	$CaSO_4$	Granular masses, crystals with 2 good cleavage directions	White, gray, blue-gray
Asbestos	$Mg_3Si_2O_5(OH)_4$	Fibrous	White to pale olive-green
Azurite	$Cu_3(CO_3)_2(OH)_2$	Varied, may have fibrous crystals	Azure blue
Bauxite	$Al(OH)_3$	Earthy masses	Reddish to brown
Chalcopyrite	$CuFeS_2$	Uneven fracture	Brass yellow
Chromite	$FeCr_2O_4$	Massive, granular, compact	Black
Cinnabar	HgS	Compact, granular masses	Scarlet red to red-brown
Galena	PbS	Perfect cubic cleavage	Lead or silver gray
Gypsum	$CaSO_4 \cdot 2H_2O$	Tabular crystals, fibrous, or granular	White, pearly
Halite	$NaCl$	Granular masses, perfect cubic crystals	White, also pale colors and gray
Hematite	Fe_2O_3	Granular, massive, or earthy	Brownish-red, black
Malachite	$CuCO_3 \cdot Cu(OH)_2$	Uneven splintery fracture	Bright green, dark green
Native copper	Cu	Malleable and ductile	Copper red
Native gold	Au	Malleable and ductile	Yellow
Native silver	Ag	Malleable and ductile	Silver-white
Pyrolusite	MnO_2	Radiating or dendritic coatings on rocks	Black
Sphalerite	ZnS	Perfect cleavage in 6 directions at 120°	Shades of brown and red
Talc	$Mg_3Si_4O_{10}(OH)_2$	Perfect in one direction	Green, white, gray

COMMON MINERALS AND THEIR PROPERTIES (CONTINUED)

Hard-ness	Streak	Specific Gravity	Other Properties	Type(s) of Rock in Which the Mineral is Most Commonly Found
ORE AND OTHER MINERALS OF ECONOMIC IMPORTANCE				
3–3.5	White	2.9–3	Brittle; resembles marble but acid has no effect	Sedimentary evaporite deposits
1–2.5	White	2.6–2.8	Pearly to greasy luster; flexible, easily separated fibers	A variety of serpentine, found in the same rock types
4	Pale blue	3.8	Vitreous to earthy; effervesces with HCl	Weathered copper deposits
1.5–3.5	Pale reddish-brown	2.5	Dull luster, claylike masses with small round concretions	Weathering of many rock types
3.5–4.5	Greenish-black	4.2	Metallic luster, softer than pyrite	The most common copper ore mineral; hydrothermal veins, porphyry copper deposits
5.5	Dark brown	4.4	Metallic to submetallic luster	Peridotites and other ultramafic igneous rocks
2.5	Scarlet red	8	Color and streak distinctive	The most important mercury ore mineral; hydrothermal veins in young volcanic rocks
2.5	Gray	7.6	Metallic luster	The most important lead ore; commonly also contains silver; hydrothermal veins
1–2.5	White	2.2–2.4	Thin sheets (selenite), fibrous (satinspar), massive (alabaster)	Sedimentary evaporite deposits
2.5–3	White	2.2	Pearly luster, salty taste, soluble in water	Sedimentary evaporite deposits
2.5	Dark red	2.5–5	Often earthy, dull appearance, sometimes metallic luster	Huge concentrations occur as sedimentary iron ore; the most important source of iron
3.5–4	Emerald green	4	Effervesces with HCl; associated with azurite	Weathered copper deposits
2.5–3	Copper red	8.9	Metallic luster	Basaltic lavas
2.5–3	Yellow	19.3	Metallic luster	Hydrothermal quartz-gold veins, sedimentary placer deposits
2.5–3	Silver-white	10.5	Metallic luster	Hydrothermal veins, weathered silver deposits
1–2	Black	4.7	Sooty appearance	Black stains on weathered surfaces of many rocks, manganese nodules on the sea floor
3.5	Reddish-brown	4	Resinous luster; may occur with galena, pyrite	The most important ore mineral of zinc; hydrothermal veins
1–1.5	White	1–2.5	Greasy feel; occurs in foliated masses	Low-grade metamorphic rocks

COMMON MINERALS AND THEIR PROPERTIES (CONTINUED)

Mineral Group or Mineral	Chemical Composition	Habit, Cleavage, Fracture	Usual Color
PRECIOUS AND SEMI-PRECIOUS GEMS			
Beryl (aquamarine, emerald)	$Be_3Al_2(SiO_3)_6$	Uneven fracture, hexagonal crystals	Green, yellow, blue, pink
Chrysoberyl (cat's eye, alexandrite)	$BeAl_2O_4$	Tabular crystals	Green, brown, yellow
Corundum (ruby, sapphire)	Al_2O_3	Short, six-sided barrel-shaped crystals	Gray, red (ruby), blue (sapphire)
Diamond	C	Octahedral crystals	Colorless or with pale tints
Garnet	$Fe_3Al_2(SiO_4)_3$	No cleavage, crystals 12- or 24-sided	Deep red
Jadite (jade) (a pyroxene)	$NaAl(Si_2O_6)$	Compact fibrous aggregates	Green
Olivine (peridot)	$(MgFe)_2SiO_4$	Uneven fracture, often in granular masses	Various shades of green
Opal	$SiO_2 \cdot nH_2O$	Conchoidal fracture, amorphous, massive	White and various colors
Quartz (rock crystal, amethyst, citrine, tiger eye, adventurine, carneline, chrysoprase, agate, onyx, heliotrope, bloodstone, jasper)	SiO_2	No cleavage, massive and as six-sided crystals	Colorless, white, or tinted any color by impurities
Spinel	$MgAl_2O_4$	No cleavage, rare octahedral crystals	Black, dark green, or various colors
Topaz	$Al_2SiO_4(F,OH)_2$	Cleavage good in one direction; conchoidal fracture	Colorless, white, pale tints of blue, pink
Tourmaline	$(Na,Ca)(Li,Mg,Al)(Al,Fe,Mn)_6(BO_3)_3(Si_6O_{18})(OH)_4$	Poor cleavage, uneven fracture; striated crystals	Black, brown, green, pink
Turquoise	$CuAl_6(PO_4)_4(OH)_8 \cdot 4H_2O$	Massive	Blue-green
Zircon	$Zr(SiO_4)$	Cleavage poor, but often well-formed tetragonal crystals	Colorless, gray, green, pink, bluish

COMMON MINERALS AND THEIR PROPERTIES (CONTINUED)

Hard-ness	Streak	Specific Gravity	Other Properties	Type(s) of Rock in Which the Mineral is Most Commonly Found
			PRECIOUS AND SEMI-PRECIOUS GEMS	
7.5–8.0	White	2.6–2.8	Vitreous luster	Granite, granite pegmatite, mica schist
8.5	White	3.7–3.8	Vitreous luster	Granite, granite pegmatite, mica schist
9	None	3.9–4.1	Ruby and sapphire are corundum varieties. Hardness is distinctive.	Metamorphic rocks, some igneous rocks
10	None	3.5	Adamantine luster. Hardness is distinctive.	Peridotite, kimberlite, sedimentary placer deposits
6.5–7.5	White	4.2	Vitreous to resinous luster	Metamorphic rocks, igneous rocks, placer deposits
6.5–7	White, pale green	3.3	Vitreous luster	High-pressure metamorphic rocks
6.5–7	White	3.2–3.3	Vitreous, glassy luster	Basalt, peridotite
5.5–6.5	White	2.1	Vitreous, greasy, pearly luster. May show a play of colors.	Low-temperature hot springs and weathered near-surface deposits
7	White	2.6	Colors and other features differ among the varieties.	Quartz is found in nearly all rock types, although each of the gem varieties may form in special environments.
7.5–8.0	White	3.5–4.1	Vitreous luster. Hardness is distinctive.	High-grade metamorphic rocks, dark igneous rocks
8	Colorless	3.5–3.6	Vitreous luster	Pegmatite, granite, rhyolite
7–7.5	White to gray	4.4–4.8	Vitreous, slightly resinous	Pegmatite, granite, metamorphic rocks
6	Blue-green, white	2.6–2.8	Waxy luster	Veins in weathered volcanic rocks in deserts
7.5	White	4.7	Adamantine luster	Many types of igneous rocks

Appendix C

Systems of Measurement

I   The SI System

In the past scientists from different parts of the world have used different systems of measurement. However, global cooperation and communication make it essential to adopt a standard system. The International System of Units (SI) defines various units of measurement as well as prefixes for multiplying or dividing the units by decimal factors. Some primary and derived units important to geologists are listed below.

Time

The SI unit is the **second**, s or sec, which used to be based on the rotation of the Earth but is now related to the vibration of atoms of cesium-133. SI prefixes are used for fractions of a second (such as milliseconds or microseconds), but the common words **minutes**, **hours**, and **days** are still used to express multiples of seconds.

Length

The SI unit is the **meter**, m, which used to be based on a standard platinum bar but is now defined in terms of wavelengths of light. The closest English equivalent is the **yard** (0.914 m). A **mile** is 1.61 kilometers (km). An inch is exactly 2.54 centimeters (cm).

Area

Area is length squared, as in **square meter**, **square foot**, and so on. The SI unit of area is the **are**, a, which is 100 sq m. More commonly used is the **hectare**, ha, which is 100 ares, or a square that is 100 m on each side. (The length of a U.S. football field plus one end zone is just about 100 m.) A hectare is 2.47 acres. An **acre** is 43,560 sq ft, which is a plot of 220 ft by 198 ft, for example.

Volume

Volume is length cubed, as in **cubic centimeter**, cm^3, **cubic foot**, ft^3, and so on. The SI unit is the **liter**, L, which is 1000 cm^3. A **quart** is 0.946 L; a U.S. liquid **gallon** (gal) is 3.785 L. A **barrel** of petroleum (U.S.) is 42 gal, or 159 L.

Mass

Mass is the amount of matter in an object. **Weight** is the force of gravity on an object. To illustrate the difference, an astronaut in space has no weight but still has mass. On Earth, the two terms are directly proportional and often used interchangeably. The SI unit of mass is the **kilogram**, kg, which is based on a standard platinum mass. A **pound** (avdp), lb, is a unit of weight. On the surface of the Earth, 1 lb is equal to 0.454 kg. A **metric ton**, also written as **tonne**, is 1000 kg, or about 2205 lb.

Temperature

The Celsius scale is used in most laboratories to measure temperature. On the Celsius scale the freezing point of water is 0°C and the boiling point of water is 100°C.

The SI unit of temperature is the **Kelvin**. The coldest possible temperature, which is −273°C, is zero on the Kelvin scale. The size of 1 degree Kelvin is equal to 1 degree Celsius.

$$\text{Celsius temperature (°C)} = \text{Kelvin temperature (K)} - 273 \text{ K}$$

Fahrenheit temperature (°F) is not used in scientific writing, although it is still popular in English-speaking countries. Conversion between Fahrenheit and Celsius is shown below.

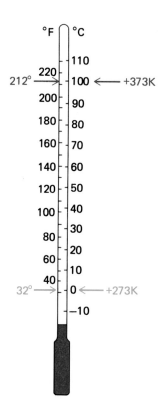

Energy

Energy is a measure of work or heat, which were once thought to be different quantities. Hence, two different sets of units were adopted and still persist, although we now know that work and heat are both forms of energy.

The SI unit of energy is the **joule**, J, the work required to exert a force of 1 newton through a distance of 1 m. In turn, a newton is the force that gives a mass of 1 kg an acceleration of 1 m/sec^2. In human terms, a joule is not much—it is about the amount of work required to lift a 100-g weight to a height of 1 m. Therefore, joule units are too small for discussions of machines, power plants, or energy policy. Larger units are

megajoule, $\text{MJ} = 10^6 \text{ J}$ (a day's work by one person)

gigajoule, $\text{GJ} = 10^9 \text{ J}$ (energy in half a tank of gasoline)

The energy unit used for heat is the **calorie**, cal, which is exactly 4.184 J. One calorie is just enough energy to warm 1 g of water 1°C. The more common unit used in measuring food energy is the **kilocalorie**, kcal, which is 1000 cal. When **Calorie** is spelled with a capital C, it means kcal. If a cookbook says that a jelly doughnut has 185 calories, that is an error—it should say 185 Calories (capital C), or 185 kcal. A value of 185 calories (small c) would be the energy in about one quarter of a thin slice of cucumber.

The unit of energy in the British system is the **British thermal unit**, Btu, which is the energy needed to warm 1 lb of water 1°F.

$$1 \text{ Btu} = 1054 \text{ J} = 1.054 \text{ kJ} = 252 \text{ cal}$$

The unit often referred to in discussions of national energy policies is the **quad**, which is 1 quadrillion Btu, or 10^{15} Btu.

Some approximate energy values are

1 barrel (42 gal) of petroleum = 5900 MJ
1 ton of coal = 29,000 MJ
1 quad = 170 million barrels of oil, or 34 million tons of coal

II   Prefixes for Use with Basic Units of the Metric System

Prefix	Symbol†	Power		Equivalent
geo*		10^{20}		
tera	T	10^{12} =	1,000,000,000,000	Trillion
giga	G	10^{9} =	1,000,000,000	Billion
mega	M	10^{6} =	1,000,000	Million
kilo	k	10^{3} =	1,000	Thousand
hecto	h	10^{2} =	100	Hundred
deca	da	10^{1} =	10	Ten
— — —	—	10^{0} =	1	One
deci	d	10^{-1} =	.1	Tenth
centi	c	10^{-2} =	.01	Hundredth
milli	m	10^{-3} =	.001	Thousandth
micro	μ	10^{-6} =	.000001	Millionth
nano	n	10^{-9} =	.000000001	Billionth
pico	p	10^{-12} =	.000000000001	Trillionth

* Not an official SI prefix but commonly used to describe very large quantities such as the mass of water in the oceans.

† The SI rules specify that its symbols are not followed by periods, nor are they changed in the plural. Thus, it is correct to write "The tree is 10 m high," not "10 m. high" or "10 ms high."

III Handy Conversion Factors

To Convert From	To	Multiply By	
Centimeters	Feet	0.0328 ft/cm	
	Inches	0.394 in/cm	
	Meters	0.01 m/cm	(exactly)
	Micrometers (Microns)	1000 μm/cm	(")
	Miles (statute)	6.214×10^{-6} mi/cm	
	Millimeters	10 mm/cm	(exactly)
Feet	Centimeters	30.48 cm/ft	(exactly)
	Inches	12 in/ft	(")
	Meters	0.3048 m/ft	(")
	Micrometers (Microns)	304800 μm/ft	(")
	Miles (statute)	0.000189 mi/ft	
Grams	Kilograms	0.001 kg/g	(exactly)
	Micrograms	1×10^6 μg/g	(")
	Ounces (avdp.)	0.03527 oz/g	
	Pounds (avdp.)	0.002205 lb/g	
Hectares	Acres	2.47 acres/ha	
Inches	Centimeters	2.54 cm/in	(exactly)
	Feet	0.0833 ft/in	
	Meters	0.0254 m/in	(exactly)
	Yards	0.0278 yd/in	
Kilograms	Ounces (avdp.)	35.27 oz/kg	
	Pounds (avdp.)	2.205 lb/kg	
Kilometers	Miles	0.6214 mi/km	
Meters	Centimeters	100 cm/m	(exactly)
	Feet	3.2808 ft/m	
	Inches	39.37 in/m	
	Kilometers	0.001 km/m	(exactly)
	Miles (statute)	0.0006214 mi/m	
	Millimeters	1000 mm/m	(exactly)
	Yards	1.0936 yd/m	
Miles (statute)	Centimeters	160934 cm/mi	
	Feet	5280 ft/mi	(exactly)
	Inches	63360 in/mi	(exactly)
	Kilometers	1.609 km/mi	
	Meters	1609 m/mi	
	Yards	1760 yd/mi	(exactly)
Ounces (avdp.)	Grams	28.35 g/oz	
	Pounds (avdp.)	0.0625 lb/oz	(exactly)
Pounds (avdp.)	Grams	453.6 g/lb	
	Kilograms	0.454 kg/lb	
	Ounces (avdp.)	16 oz/lb	(exactly)

IV Exponential or Scientific Notation

Exponential or scientific notation is used by scientists all over the world. This system is based on exponents of 10, which are shorthand notations for repeated multiplications or divisions.

A positive exponent is a symbol for a number that is to be multiplied by itself a given number of times. Thus, the number 10^2 (read ''ten squared'' or ''ten to the second power'') is exponential notation for $10 \cdot 10 = 100$. Similarly, $3^4 = 3 \cdot 3 \cdot 3 \cdot 3 = 81$. The reciprocals of these numbers are expressed by negative exponents. Thus $10^{-2} = 1/10^2 = 1/(10 \cdot 10) = 1/100 = 0.01$.

To write 10^4 in longhand form you simply start with the number 1 and move the decimal four places to the right: 10000 . Similarly, to write 10^{-4} you start with the number 1 and move the decimal four places to the left: 0.0001 .

It is just as easy to go the other way—that is, to convert a number written in longhand form to an exponential expression. Thus, the decimal place of the number 1,000,000 is six places to the right of 1:

$$1\ 000\ 000 = 10^6$$
6 places

Similarly, the decimal place of the number 0.000001 is six places to the left of 1 and

$$0.000001 = 10^{-6}$$
6 places

What about a number like 3,000,000? If you write it $3 \cdot 1,000,000$, the exponential expression is simply $3 \cdot 10^6$. Thus, the mass of the Earth,which, expressed in long numerical form is 3,120,000,000,000,000,000,000,000 kg, can be written more conveniently as $3.12 \cdot 10^{24}$ kg.

Appendix D

Rock Symbols

The symbols used in this book for types of rocks are shown below:

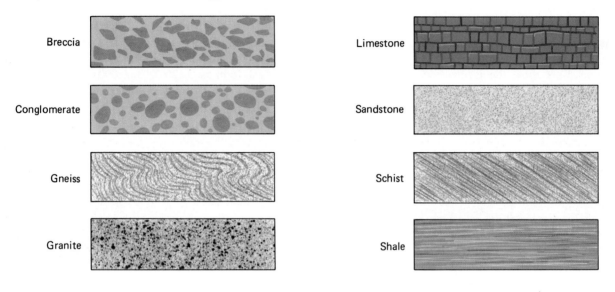

Breccia   Limestone

Conglomerate   Sandstone

Gneiss   Schist

Granite   Shale

In this book we have adopted consistent colors and style for depicting magma and layers in the upper mantle and crust.

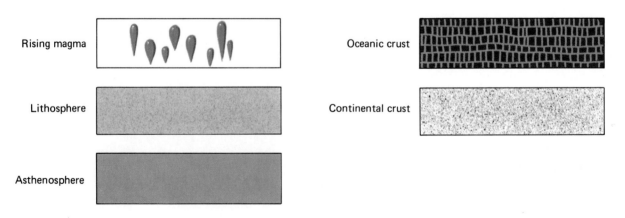

Rising magma   Oceanic crust

Lithosphere   Continental crust

Asthenosphere

WINTER SKY

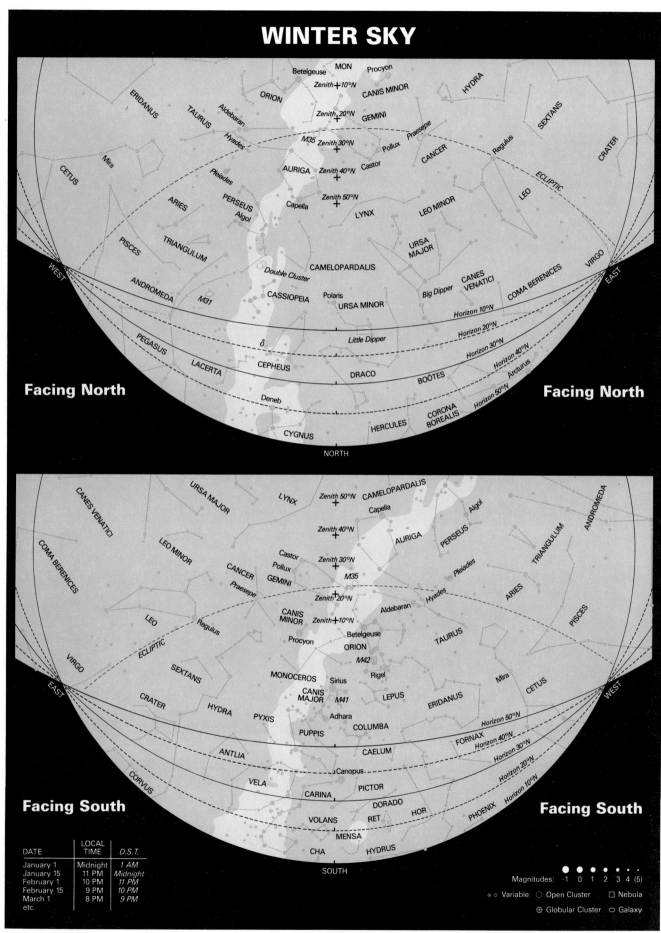

DATE	LOCAL TIME	D.S.T.
January 1	Midnight	1 AM
January 15	11 PM	Midnight
February 1	10 PM	11 PM
February 15	9 PM	10 PM
March 1	8 PM	9 PM
etc.		

Magnitudes: -1 0 1 2 3 4 (5)

Variable Open Cluster Nebula Globular Cluster Galaxy

MAP BY WIL TIRION; FOR JAY M. PASACHOFF *Journey Through the Universe* © 1992 Saunders College Publishing

SPRING SKY

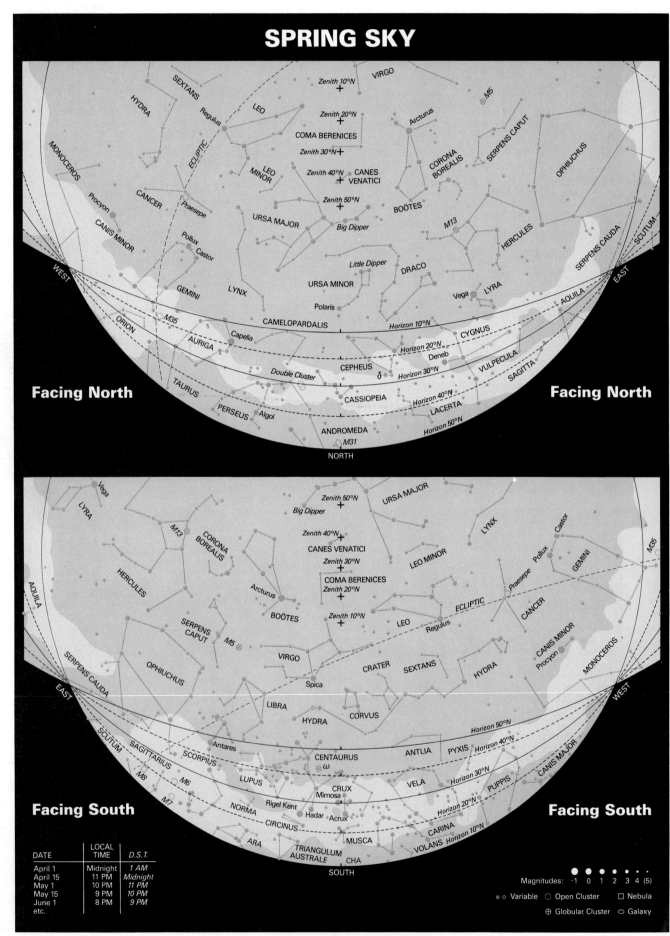

Facing North

Facing North

Facing South

Facing South

DATE	LOCAL TIME	D.S.T.
April 1	Midnight	1 AM
April 15	11 PM	Midnight
May 1	10 PM	11 PM
May 15	9 PM	10 PM
June 1	8 PM	9 PM
etc.		

Magnitudes: -1 0 1 2 3 4 (5)

Variable   ○ Open Cluster   □ Nebula
⊕ Globular Cluster   ○ Galaxy

MAP BY WIL TIRION; FOR JAY M. PASACHOFF   *Journey Through the Universe* © 1992 Saunders College Publishing

SUMMER SKY

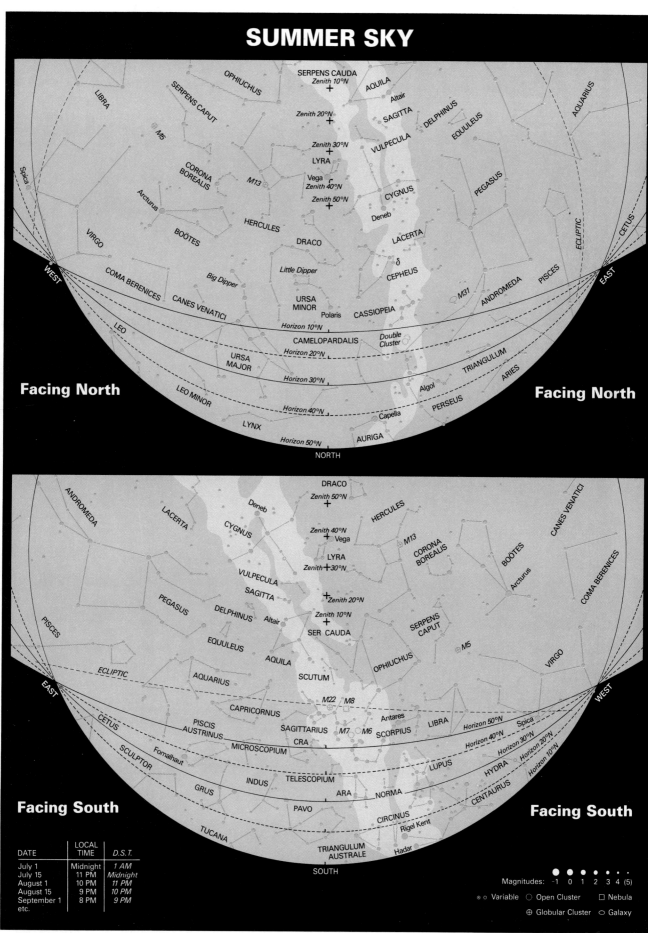

Facing North

Facing North

Facing South

Facing South

DATE	LOCAL TIME	D.S.T.
July 1	Midnight	1 AM
July 15	11 PM	Midnight
August 1	10 PM	11 PM
August 15	9 PM	10 PM
September 1	8 PM	9 PM
etc.		

Magnitudes: -1 0 1 2 3 4 (5)

◉ ○ Variable   ○ Open Cluster   □ Nebula

⊕ Globular Cluster   ○ Galaxy

MAP BY WIL TIRION; FOR JAY M. PASACHOFF   *Journey Through the Universe* © 1992 Saunders College Publishing

AUTUMN SKY

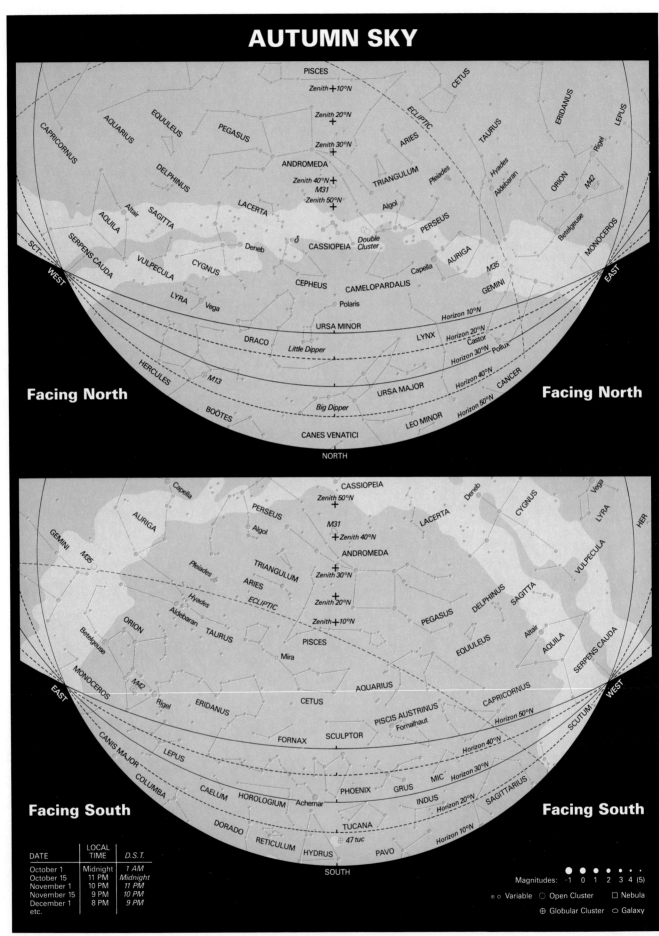

Facing North

Facing North

Facing South

Facing South

DATE	LOCAL TIME	D.S.T.
October 1	Midnight	1 AM
October 15	11 PM	Midnight
November 1	10 PM	11 PM
November 15	9 PM	10 PM
December 1 etc.	8 PM	9 PM

Magnitudes: -1 0 1 2 3 4 (5)

⊙ ○ Variable   ○ Open Cluster   □ Nebula
⊕ Globular Cluster   ○ Galaxy

MAP BY WIL TIRION; FOR JAY M. PASACHOFF   *Journey Through the Universe* © 1992 Saunders College Publishing

Glossary

A horizon The uppermost layer of soil composed of a mixture of organic matter and leached and weathered minerals. (*syn:* topsoil)

aa A lava flow that has a jagged, rubbly, broken surface.

ablation area The lower portion of a glacier where more snow melts in summer than accumulates in winter so there is a net loss of glacial ice. (*syn:* zone of wastage)

abrasion The mechanical wearing and grinding of rock surfaces by friction and impact.

absolute humidity *See* humidity.

absolute time Time measured in years.

abyssal fan A large, fan-shaped accumulation of sediment deposited at the bases of many submarine canyons adjacent to the deep-sea floor. (*syn:* submarine fan)

abyssal plain A flat, level, largely featureless part of the ocean floor between the mid-oceanic ridge and the continental rise.

accreted terrain A land mass that originated as an island arc or a microcontinent that was later added onto a continent.

accumulation area The upper part of a glacier where accumulation of snow during the winter exceeds melting during the summer, causing a net gain of glacial ice.

acid precipitation Often called acid rain. A condition in which natural precipitation becomes acidic after reacting with air pollutants.

active continental margin A continental margin characterized by subduction of an oceanic lithospheric plate beneath a continental plate. (*syn:* Andean margin)

adiabatic temperature changes Temperature changes that occur without gain or loss of heat.

advection In meteorology, the horizontal component of a convection current in air, i.e. , the surface movement that is commonly called wind.

air mass A large body of air that has approximately the same temperature and humidity throughout.

albedo The reflectivity of a surface. A mirror or bright snowy surface reflects most of the incoming light and has a high albedo, whereas a rough flat road surface has a low albedo.

alluvial fan A fan-like accumulation of sediment created where a steep stream slows down rapidly as it reaches a relatively flat valley floor.

alpine glacier A glacier that forms in mountainous terrain.

amphibole A group of double-chain silicate minerals. Hornblende is a common amphibole.

Andean margin A continental margin characterized by subduction of an oceanic lithospheric plate beneath a continental plate. (*syn:* active continental margin)

andesite A fine-grained gray or green volcanic rock intermediate in composition between basalt and granite, consisting of about equal amounts of plagioclase feldspar and mafic minerals.

angle of repose The maximum slope or angle at which loose material remains stable.

angular unconformity An unconformity in which younger sediment or sedimentary rocks rest on the eroded surface of tilted or folded older rocks.

anion An ion that has a negative charge.

antecedent stream A stream that was established before local uplift started and cut its channel at the same rate the land was rising.

anticline A fold in rock that resembles an arch; the fold is convex upward and the oldest rocks are in the middle.

apparent polar wandering The apparent migration of the Earth's magnetic poles through time. In fact, the magnetic poles remain nearly stationary. The appearance of polar migration is caused by movement of continents through time.

aquifer A porous and permeable body of rock that can yield economically significant quantities of ground water.

Archean eon A division of geologic time 3.8 to 2.5 billion years ago. The oldest known rocks formed at the beginning of or just prior to the start of the Archean eon.

arête A sharp narrow ridge between adjacent valleys formed by glacial erosion.

arkose A feldspar-rich sandstone formed when granite disintegrates rapidly.

artesian aquifer An inclined aquifer that is bounded top and bottom by layers of impermeable rock so the water is under pressure.

artesian well A well drilled into an artesian aquifer in which the water rises without pumping and in some cases spurts to the surface.

aseismic ridge A submarine mountain chain with little or no earthquake activity.

ash (volcanic) Fine pyroclastic material less than 2 mm in diameter.

ash flow A mixture of volcanic ash, larger pyroclastic particles, and gas that flows rapidly along the Earth's surface as a result of an explosive volcanic eruption. (*syn*: nuée ardente)

asteroid One of the many small celestial bodies in orbit around the Sun. Most asteroids orbit between Mars and Jupiter.

asthenosphere The portion of the upper mantle beneath the lithosphere. It consists of weak, plastic rock where magma may form and it extends from a depth of about 100 kilometers to about 350 kilometers below the surface of the Earth.

atoll A circular reef that surrounds a lagoon and is bounded on the outside by deep water of the open sea.

atom The fundamental unit of elements consisting of a small, dense, positively charged center called a nucleus surrounded by a diffuse cloud of negatively charged electrons.

aulacogen A tectonic trough on a craton bounded by normal faults and commonly filled with sediment. An aulacogen forms when one limb of a continental rift becomes inactive shortly after it forms.

axial plane An imaginary plane that runs through the axis and divides a fold as symmetrically as possible into two halves.

B horizon The soil layer just below the A horizon where ions leached from the A horizon accumulate.

back arc basin A sedimentary basin on the opposite side of the magmatic arc from the trench, either in an island arc or in an Andean continental margin.

backshore The upper zone of a beach that is usually dry but is washed by waves during storms.

bajada A broad depositional surface extending outward from a mountain front formed by the merging of alluvial fans.

banks The rising slopes bordering the two sides of a stream channel.

bar An elongate mound of sediment, usually composed of sand or gravel, in a stream channel or along a coastline.

barchan dune A crescent-shaped dune, highest in the center, with the tips facing downwind.

barometer A device used to measure barometric pressure.

barometric pressure The pressure exerted by the Earth's atmosphere.

barrier island A long, narrow, low-lying island that extends parallel to the shoreline.

barrier reef A reef separated from the coast by a deep, wide lagoon.

basal slip Movement of the entire mass of a glacier along the bedrock.

basalt A dark-colored, very fine grained, mafic, volcanic rock composed of about half calcium-rich plagioclase feldspar and half pyroxene.

base level The deepest level to which a stream can erode its bed. The ultimate base level is usually sea level, but this is seldom attained.

basement rocks The older granitic and related metamorphic rocks of the Earth's crust that make up the foundations of continents.

basin A low area of the Earth's crust of tectonic origin, commonly filled with sediment.

batholith A large plutonic mass of intrusive rock with more than 100 square kilometers of surface exposed.

bauxite A gray, yellow, or reddish brown rock composed of a mixture of aluminum oxides and hydroxides. It is the principal ore of aluminum.

baymouth bar A bar that extends partially or completely across the entrance to a bay.

beach Any strip of shoreline washed by waves or tides.

beach terrace A level portion of old beach elevated above the modern beach by uplift of the shoreline or fall of sea level.

bed The floor of a stream channel. Also the thinnest layer in sedimentary rocks commonly ranging in thickness from a centimeter to a meter or two.

bed load That portion of a stream's load that is transported on or immediately above the stream bed.

bedding Layering that develops as sediments are deposited.

bedrock The solid rock that underlies soil or regolith.

Benioff zone An inclined zone of earthquake activity that traces the upper portion of a subducting plane in a subduction zone.

big bang An event thought to mark the beginning of our Universe. The big bang theory postulates that 10 to 20 billion years ago, all matter exploded from an infinitely compressed state.

biome A community of plants living in a large geographic area characterized by a particular climate.

biotite Black rock-forming mineral of the mica group.

black hole A small region of space that contains matter packed so densely that light cannot escape from its intense gravitational field.

blowout A small depression created by wind erosion.

body waves Seismic waves that travel through the interior of the Earth.

boulder A rounded rock fragment larger than a cobble (diameter greater than 256 cm).

Bowen's Reaction Series A series of minerals in which any early-formed mineral crystallizing from a cooling magma reacts with the magma to form minerals lower in the series.

braided stream A stream that divides into a network of branching and reuniting shallow channels separated by mid-channel bars.

breccia A coarse-grained sedimentary rock composed of angular, broken fragments cemented together.

brittle fracture Rupture that occurs when a rock breaks sharply.

butte A flat-topped mountain, with several steep cliff faces. A butte is smaller and more tower-like than a mesa.

C horizon The lowest soil layer composed of partly weathered bedrock, grading downward into unweathered parent rock.

calcite A common rock-forming mineral, $CaCO_3$.

caldera A large circular depression caused by an explosive volcanic eruption.

caliche A hard soil layer formed when calcium carbonate precipitates and cements the soil.

calving A process in which large chunks of ice break off from tidewater glaciers to form icebergs.

cap rock An impermeable rock, usually shale, that prevents oil or gas from escaping upward from a reservoir.

capacity The maximum quantity of sediment that a stream can carry.

capillary action The action by which water is pulled upward through small pores by electrical attraction to the pore walls.

capillary fringe A zone above the water table in which the pores are filled with water due to capillary action.

carbonate rocks Rocks such as limestone and dolomite made up primarily of carbonate minerals.

carbonization A process in which a fossil forms when the volatile components of the soft tissues are driven off, leaving behind a thin film of carbon.

cast A fossil formed when sedimentary rock or mineral matter fills a natural mold.

cation A positively charged ion.

cavern An underground cavity or series of chambers created when ground water dissolves large amounts of rock, usually limestone. (*syn:* cave)

cementation The process by which clastic sediment is lithified by precipitation of a mineral cement among the grains of the sediment.

Cenozoic era The latest of the four eras into which geologic time is subdivided; 65 million years ago to the present.

chalk A very fine grained, soft, earthy, white to gray bioclastic limestone made of the shells and skeletons of marine microorganisms.

chemical weathering The chemical decomposition of rocks and minerals by exposure to air, water, and other chemicals in the environment.

chert A hard, dense, sedimentary rock composed of microcrystalline quartz. (*syn:* flint)

chromosphere A turbulent diffuse gaseous layer of the Sun that lies above the photosphere.

cinder cone A small volcano, as high as 300 meters, made up of loose pyroclastic fragments blasted out of a central vent.

cinders (volcanic) Glassy pyroclastic volcanic fragments 4 to 32 millimeters in size.

cirque A steep-walled semicircular depression eroded into a mountain peak by a glacier.

cirrus cloud A wispy, high-level cloud.

clastic sediment Sediment composed of fragments of weathered rock that have been transported some distance from their place of origin.

clastic sedimentary rocks Rocks composed of lithified clastic sediment.

clay Any clastic mineral particle less than 1/256 millimeter in diameter. Also a group of layer silicate minerals.

claystone A fine-grained clastic sedimentary rock composed predominantly of clay minerals and small amounts of quartz and other minerals of clay size.

cleavage The tendency of some minerals to break along certain crystallographic planes.

climate The composite pattern of long-term weather conditions that can be expected in a given region. Climate refers to yearly cycles of temperature, wind, rainfall, etc., and not to daily variations. *See* weather.

cloud A collection of minute water droplets or ice crystals in air.

coal A flammable organic sedimentary rock formed from partially decomposed plant material and composed mainly of carbon.

cobbles Rounded rock fragments in the 64 to 256 millimeters size range, larger than pebbles and smaller than boulders.

column A dripstone or speleothem formed when a stalactite and a stalagmite meet and fuse together.

columnar joints The regularly spaced cracks that commonly develop in lava flows forming five- or six-sided columns.

coma The bright outer sheath of a comet.

comet An interplanetary body composed of loosely bound rock and ice that forms a bright head and extended fuzzy tail when it enters the inner potion of the Solar System.

compaction Tighter packing of sedimentary grains causing weak lithification and a decrease in porosity, usually resulting from the weight of overlying sediment.

competence   A measure of the largest particles that a stream can transport.

composite volcano   A volcano that consists of alternate layers of unconsolidated pyroclastic material and lava flows. (*syn:* stratovolcano)

compressive stress   Stress that acts to shorten an object by squeezing it.

concordant   Pertaining to an igneous intrusion that is parallel to the layering of country rock.

cone of depression   A cone-like depression in the water table formed when water is pumped out of a well more rapidly than it can flow through the aquifer.

conformable   The condition in which sedimentary layers were deposited continuously without interruption.

conglomerate   A coarse-grained clastic sedimentary rock, composed of rounded fragments larger than 2 millimeters in diameter cemented in a fine-grained matrix of sand or silt.

contact   A boundary between two different rock types or between rocks of different ages.

contact metamorphic ore deposit   An ore deposit formed by contact metamorphism.

contact metamorphism   Metamorphism caused by heating of country rock, and/or addition of fluids, from a nearby igneous intrusion.

continental crust   The predominantly granitic portion of the crust, 20 to 70 kilometers thick, that makes up the continents.

continental drift   The theory proposed by Alfred Wegener that continents were once joined together and later split and drifted apart. The continental drift theory has been replaced by the more complete plate tectonics theory.

continental glacier   A glacier that forms a continuous cover of ice over areas of 50,000 square kilometers or more and spreads outward in all directions under the influence of its own weight. (*syn:* ice sheet)

continental margin   The region between the shoreline of a continent and the deep ocean basins including the continental shelf, continental slope, and continental rise. Also the region where thick, granitic continental crust joins thinner, basaltic oceanic crust.

continental margin basin   A sediment-filled depression or other thick accumulation of sediment and sedimentary rocks near the margin of a continent.

continental rifting   The process by which a continent is pulled apart at a divergent plate boundary.

continental rise   An apron of sediment between the continental slope and the deep sea floor.

continental shelf   A shallow, nearly level area of continental crust covered by sediment and sedimentary rocks that is submerged below sea level at the edge of a continent between the shoreline and the continental slope.

continental slope   The relatively steep (3°–6°) underwater slope between the continental shelf and the continental rise.

continental suture   The junction created where two continents collide and weld into a single mass of continental crust.

convection current   A current in a fluid or plastic material formed when heated materials rise and cooler materials sink.

convergent plate boundary   A boundary where two lithospheric plates collide head-on.

coquina   A bioclastic limestone consisting of coarse shell fragments cemented together.

core   The innermost region of the Earth, probably consisting of iron and nickel.

Coriolis effect   The deflection of air or water currents caused by the rotation of the Earth.

corona   The luminous irregular envelope of highly ionized gas outside the chromosphere of the Sun.

correlation   Demonstration of the equivalence of rocks or geologic features from different locations.

cost-benefit analysis   A system of analysis that attempts to weigh the cost of an act or policy, such as pollution control, directly against the economic benefits.

country rock   The older rock intruded by a younger igneous intrusion or mineral deposit.

crater   A bowl-like depression at the summit of the volcano.

craton   A segment of continental crust, usually in the interior of a continent, that has been tectonically stable for a long time, commonly a billion years or longer.

creep   The slow movement of unconsolidated material downslope under the influence of gravity.

crest (of a wave)   The highest part of a wave.

crevasse   A fracture or crack in the upper 40 to 50 meters of a glacier.

cross-bedding   An arrangement of small beds at an angle to the main sedimentary layering.

cross-cutting relationship   Any relationship in which younger rocks or geological structures interrupt or cut across older rocks or structures.

crust   The Earth's outermost layer about 7 to 70 kilometers thick, composed of relatively low-density silicate rocks.

crystal   A solid element or compound whose atoms are arranged in a regular, orderly, periodically repeated array.

crystal habit   The shape in which individual crystals grow and the manner in which crystals grow together in aggregates.

crystal settling   A process in which the crystals that

solidify first from a cooling magma settle to the bottom of a magma chamber because the solid minerals are more dense than liquid magma.

cumulus cloud A column-like cloud with a flat bottom and a billowy top.

Curie point The temperature below which rocks can retain magnetism.

current A continuous flow of water in a concerted direction.

cyclone A low pressure region with its accompanying surface wind. Also a synonym for a tropical cyclone, or hurricane.

daughter isotope An isotope formed by radioactive decay of another isotope.

debris flow A type of mass wasting in which particles move as a fluid and more than half of the particles are larger than sand.

deflation Erosion by wind.

deformation Folding, faulting and other changes in shape of rocks or minerals in response to mechanical forces, such as those that occur in tectonically active regions.

degeneracy pressure The strength of the atomic particles that holds a white dwarf star from further collapse.

delta The nearly flat, alluvial, fan-shaped tract of land at the mouth of a stream.

dendritic drainage pattern A pattern of stream tributaries that branches like the veins in a leaf. It often indicates uniform underlying bedrock.

deposition The laying down of rock-forming materials by any natural agent.

depositional environment Any setting in which sediment is deposited.

depositional remanent magnetism Remanent magnetism resulting from mechanical orientation of magnetic mineral grains during sedimentation.

desert A region with less than 25 centimeters of rainfall a year. Also defined as a region that supports only a sparse plant cover.

desert pavement A continuous cover of stones created as wind erodes fine sediment, leaving larger rocks behind.

desertification A process by which semiarid land is converted to desert, often by improper farming or by climate change.

dew Moisture condensed onto objects from the atmosphere, usually during the night, when the ground and leaf surfaces become cooler than the surrounding air.

differential weathering The process by which certain rocks weather more rapidly than adjacent rocks, usually resulting in an uneven surface.

dike A sheet-like igneous rock that cuts across the structure of country rock.

dike swarm A group of dikes that form in parallel or radial sets.

diorite A rock that is the medium- to coarse-grained plutonic equivalent of andesite.

dip The angle of inclination of a bedding plane, measured from the horizontal.

discharge The volume of water flowing downstream per unit time. It is measured in units of cubic meters per second (m^3/sec).

disconformity A type of unconformity in which the sedimentary layers above and below the unconformity are parallel.

discordant Pertaining to a dike or other feature that cuts across sedimentary layers or other kinds of layering in country rock.

disseminated ore deposit A large low-grade ore deposit in which generally fine-grained metal-bearing minerals are widely scattered throughout a rock body in sufficient concentration to make the deposit economical to mine.

dissolved load The portion of a stream's sediment load that is carried in solution.

distributary A channel that flows outward from the main stream channel, such as is commonly found in deltas.

divergent plate boundary The boundary or zone where lithospheric plates separate from each other. (*syn:* spreading center)

docking The accretion of island arcs or microcontinents onto a continental margin.

doldrums A region of the Earth near the Equator in which hot, humid air moves vertically upward, forming a vast low-pressure region. Local squalls and rainstorms are common, and steady winds are rare.

dolomite A rock-forming mineral, $CaMg(CO_3)_2$.

dome A circular or elliptical anticlinal structure.

Doppler effect The observed change in frequency of light or sound that occurs when the source of the wave is moving relative to the observer.

downcutting Downward erosion by a stream.

drainage basin The region that is ultimately drained by a single river.

drainage divide A ridge or other topographically higher region that separates adjacent drainage basins.

drift (glacial) Any rock or sediment transported and deposited by a glacier or by glacial meltwater.

dripstone A deposit formed in a cavern when calcite precipitates from dripping water.

drumlin An elongate hill formed when a glacier flows over and reshapes a mound of till or stratified drift.

dry adiabatic lapse rate The rate of cooling that occurs when dry air rises without gain or loss of heat.

dune A mound or ridge of wind-deposited sand.

earthflow   A flowing mass of fine-grained soil particles mixed with water. Earthflows are less fluid than mudflows.

earthquake   A sudden motion or trembling of the Earth caused by the abrupt release of slowly accumulated elastic energy in rocks.

echo sounder   An instrument that emits sound waves and then records them after they reflect off the sea floor. The data are then used to record the topography of the sea floor.

eclipse   A phenomenon that occurs when a heavenly body is shadowed by another and therefore rendered invisible. When the Moon lies directly between the Earth and the Sun, the Moon blocks our view of the Sun and we observe a *solar eclipse*. When the Earth lies directly between the Sun and the Moon, the Earth's shadow falls on the Moon and we observe a *lunar eclipse*.

effluent stream   A stream that receives water from ground water because its channel lies below the water table. (*syn:* gaining stream)

elastic deformation   A type of deformation in which an object returns to its original size and shape when stress is removed.

elastic limit   The maximum stress that an object can withstand without permanent deformation.

electromagnetic radiation   The transfer of energy by an oscillating electric and magnetic field; it travels as a wave and also behaves as a stream of particles.

electromagnetic spectrum   The entire range of electromagnetic radiation from very long wavelength (low frequency) radiation to very short wavelength (high frequency) radiation.

electron   A fundamental particle that forms a diffuse cloud of negative charge around an atom.

element   A substance that cannot be broken down into other substances by ordinary chemical means. An element is made up of the same kind of atoms.

emergent coastline   A coastline that was recently under water but has been exposed either because the land has risen or sea level has fallen.

end moraine   A moraine that forms at the end, or terminus, of a glacier.

eon   The longest unit of geologic time. The most recent eon, the Phanerozoic eon, is further subdivided into eras.

epicenter   The point on the Earth's surface directly above the focus of an earthquake.

epidemiology   The study of the distribution of sickness in a population.

epoch   The smallest unit of geologic time. Periods are divided into epochs.

equinox   Either of two times during a year when the Sun shines directly overhead at the Equator. The equi-noxes are the beginnings of spring and fall, when every portion of the Earth receives 12 hours of daylight and 12 hours of darkness.

era   A geologic time unit. Eons are divided into eras and in turn eras are subdivided into periods.

erosion   The removal of weathered rocks by moving water, wind, ice, or gravity.

erratic   A boulder that was transported to its present location by a glacier. Usually different from the bed-rock in its immediate vicinity.

esker   A long snake-like ridge formed by deposition in a stream that flowed on, within, or beneath a glacier.

estuary   A bay that formed when a broad river valley was submerged by rising sea level or a sinking coast.

evaporation   The transformation of a liquid to a gas.

evaporite deposit   A chemically precipitated sedimentary rock that formed when dissolved ions were concentrated by evaporation of water.

evolution   The change in the physical and genetic characteristics of a species over time.

exfoliation   Fracturing in which concentric plates or shells split from the main rock mass like the layers of an onion.

extensional stress   Tectonic stress in which rocks are pulled apart.

external mold   A fossil cavity created in sediment by a shell or other hard body part that bears the impression of the exterior of the original.

extrusive rock   An igneous rock formed from material that has erupted onto the surface of the Earth.

fall   A type of mass wasting in which unconsolidated material falls freely or bounces down the face of a cliff.

fault   A fracture in rock along which displacement has occurred.

fault creep   A continuous, slow movement of solid rock along a fault resulting from a constant stress acting over a long time.

fault zone   An area of numerous, closely spaced faults.

faunal succession (principle of)   *See* principle of faunal succession.

feldspar   A common group of aluminum silicate rock-forming minerals that contain potassium, sodium, or calcium.

fetch   The distance that the wind has traveled over the ocean without interruption.

fiord   A long, deep, narrow arm of the sea bounded by steep walls, generally formed by submergence of a glacially eroded valley.

firn   Hard, dense snow that has survived through one summer melt season. Firn is transitional between snow and glacial ice.

firn line   The boundary on a glacier between permanent snow and seasonal snow. Above the firn line, winter snow does not melt completely during summer, while below the firn line it does.

fissility Fine layering along which a rock splits easily.

flash flood A rapid, intense, local flood of short duration.

flood basalt Basaltic lava that erupts gently in great volume from cracks at the Earth's surface to cover large areas of land and form basalt plateaus.

flood plain That portion of a river valley adjacent to the channel; it is built by sediment deposited during floods and is covered by water during a flood.

flow Mass wasting in which individual particles move downslope as a semifluid not as a consolidated mass.

focus The initial rupture point of an earthquake.

fog A cloud that forms at or very close to ground level.

fold A bend in rock.

foliation Layering in rock created by metamorphism.

footwall The rock beneath an inclined fault.

forearc basin A sedimentary basin between the trench and the magmatic arc either in an island arc or Andean continental margin.

foreshock Small earthquakes that precede a large quake by an interval ranging from a few seconds to a few weeks.

foreshore The zone that lies between the high and low tides; the intertidal region.

formation A lithologically distinct body of sedimentary, igneous, or metamorphic rock that can be recognized in the field and can be mapped.

fossil The preserved trace, imprint, or remains of a plant or animal.

fossil fuel Fuels formed from the partially decayed remains of plants and animals. The most commonly used fossil fuels are petroleum, coal, and natural gas.

fracture (1) The manner in which minerals break other than along planes of cleavage. (2) A crack, joint, or fault in bedrock.

front In meteorology, a line separating air masses of different temperature or density.

frontal weather system A weather system that develops when air masses collide.

frost wedging A process in which water freezes in a crack in rock and the expansion wedges the rock apart.

fusion (of atomic nuclei) The combination of nuclei of light elements (particularly hydrogen) to form heavier nuclei.

gabbro Igneous rock that is mineralogically identical to basalt, but that has a medium- to coarse-grained texture because of its plutonic origin.

gaining stream A stream that receives water from ground water because its channel lies below the water table. (*syn:* effluent stream)

galaxy A large volume of space containing many billions of stars, held together by mutual gravitational attraction.

geocentric model A model that places the Earth at the center of the celestial bodies.

geologic column A composite columnar diagram that shows the sequence of rocks at a given place or region arranged to show their position in the geologic time scale.

geologic time scale A chronological arrangement of geologic time subdivided into units.

geological structure Any feature formed by deformation or movement of rocks, such as a fold or a fault. Also, the combination of all such features of an area or region.

geology The study of the Earth, the materials that it is made of, the physical and chemical changes that occur on its surface and in its interior, and the history of the planet and its life forms.

geothermal energy Energy derived from the heat of the Earth.

geothermal gradient The rate at which temperature increases with depth in the Earth.

geyser A type of hot spring that intermittently erupts jets of hot water and steam. Geysers occur when ground water comes in contact with hot rock.

glacial polish A smooth polish on bedrock created when fine particles transported at the base of a glacier abrade the bedrock.

glacial striation Parallel grooves and scratches in bedrock that form as rocks are dragged along at the base of a glacier.

glacier A massive, long-lasting accumulation of compacted snow and ice that forms on land and moves downslope or outward under its own weight.

gneiss A foliated rock with banded appearance formed by regional metamorphism.

Gondwanaland The southern part of Wegener's Pangaea, which was the late Paleozoic supercontinent. (*syn:* Gondwana)

graben A wedge-shaped block of rock that has dropped downward between two normal faults.

graded bedding A type of bedding in which larger particles are at the bottom of each bed, and the particle size decreases towards the top.

graded stream A stream with a smooth concave profile. A graded stream is in equilibrium with its sediment supply; it transports all the sediment supplied to it with neither erosion nor deposition in the streambed.

gradient The vertical drop of a stream over a specific distance.

gradualism A theory of evolution that proposes that species change gradually in small increments.

granite A medium- to coarse-grained, sialic, plutonic

rock made predominately of potassium feldspar and quartz.

gravel   Unconsolidated sediment consisting mainly of rounded particles with a diameter greater than 2 millimeters.

graywacke   A poorly sorted sandstone commonly dark in color and consisting mainly of quartz, feldspar, and rock fragments with considerable quantities of silt and clay in its pores.

greenhouse effect   An increase in the temperature of a planets atmosphere caused when infrared-absorbing gases are introduced into the atmosphere.

groin   A narrow wall built perpendicular to the shore to trap sand transported by currents and waves.

ground moraine   A moraine formed when a melting glacier deposits till in a relatively thin layer over a broad area.

ground water   Water contained in soil and bedrock. All subsurface water.

guyot   A flat-topped seamount.

gypsum   A mineral with the formula $CaSO_4 \cdot 2H_2O$. It commonly forms in evaporite deposits.

gyre   A closed, circular current either in water or air.

Hadean eon   The earliest time in the Earth's history.

half-life   The time it takes for half of the nuclei of a radioactive isotope in a sample to decompose.

hanging valley   A tributary glacial valley whose mouth lies high above the floor of the main valley.

hanging wall   The rock above an inclined fault.

hardness   The resistance of the surface of a mineral to scratching.

headland   A point of land along a coast that juts out into the sea.

headward erosion   The lengthening of a valley in an upstream direction.

heliocentric solar system   A model that places the Sun at the center of the Solar System.

Hertzsprung–Russell diagram (H–R diagram)   A plot of absolute stellar magnitude (or luminosity) against temperature.

horn   A sharp, pyramid-shaped rock summit where three or more cirques intersect near the summit.

hornblende   A rock-forming mineral. The most common member of the amphibole group.

hornfels   A fine-grained rock formed by contact metamorphism.

horse latitudes   A region of the Earth lying at about 30° north and south latitudes where air is falling, forming a vast high-pressure region. Generally dry conditions prevail, and steady winds are rare.

horst   A block of rock that has moved relatively upward and is bounded by two faults.

hot spot   A persistent volcanic center thought to be located directly above a rising plume of hot mantle rock.

hot spring   A spring formed where hot ground water flows to the surface.

Hubble's Law   A law that states that the velocity of a galaxy is proportional to its distance from Earth. Thus the most distant galaxies are traveling at the highest velocities.

humidity   A measure of the amount of water vapor in the air. *Absolute humidity* is the amount of water vapor in a given volume of air. *Relative humidity* is the ratio of the amount of water vapor in a given volume of air divided by the maximum amount of water vapor that can be held by that air at a given temperature.

humus   The dark organic component of soil composed of litter that has decomposed sufficiently so that the origin of the individual pieces cannot be determined.

hurricane   *See* tropical cyclone.

hydration   The chemical combination of water with another substance.

hydraulic action   The mechanical loosening and removal of material by flowing water.

hydrologic cycle   The constant circulation of water among the sea, the atmosphere, and the land.

hydrothermal metamorphism   Changes in rock that are primarily caused by migrating hot water and by ions dissolved in the hot water.

hydrothermal vein   A sheet-like mineral deposit that fills a fault or other fracture, precipitated from hot water solutions.

ice age   A time of extensive glacial activity, when alpine glaciers descended into lowland valleys and continental glaciers spread over the higher latitudes.

ice sheet   A glacier that forms a continuous cover of ice over areas of 50,000 square kilometers or more and spreads outward under the influence of its own weight. (*syn:* continental glacier, ice cap)

iceberg   A large chunk of ice that breaks from a glacier into a body of water.

igneous rock   Rock that solidified from magma.

incised meander   A stream meander that is cut below the level at which it originally formed, usually caused by rejuvenation.

index fossil   A fossil that identifies and dates the layers in which it is found. Index fossils are abundantly preserved in rocks, widespread geographically, and existed as a species or genus for only a relatively short time.

influent stream   A stream that lies above the water table. Water percolates from the stream channel downward into the saturated zone. (*syn:* losing stream)

intensity (of an earthquake)   A measure of the effects an earthquake at a particular place on buildings and people.

intermediate rocks   Igneous rocks with chemical and mineral compositions between those of granite and basalt.

internal mold   A type of fossil that forms when the inside of a shell fills with sediment or precipitated minerals.

internal processes   Earth processes that are initiated by movements within the Earth and those internal movements themselves. For example, formation of magma, earthquakes, mountain building, and tectonic plate movement.

intertidal zone   The part of a beach that lies between the high and low tide lines.

intracratonic basins   A sedimentary basin located within a craton.

intrusive rock   A rock formed when magma solidifies with-in bodies of pre-existing rock.

inversion (atmospheric)   A meteorological condition in which the lower layers of air are cooler than those at higher altitudes. This cool air can remain relatively stagnant and allows air pollutants to concentrate in urban areas.

ion   An atom with an electrical charge.

ionic substitution   The replacement of one or more kinds of ions in a mineral by other kinds of ions of similar size and charge.

island arc   A gently curving chain of volcanic islands in the ocean formed by convergence of two plates each bearing ocean crust, and the resulting subduction of one plate beneath the other.

isostasy   The condition in which the lithosphere floats on the asthenosphere as an iceberg floats on water.

isostatic adjustment   The rising and settling of portions of the lithosphere to maintain equilibrium as they float on the plastic asthenosphere.

isotherm   A line on a weather map connecting points of equal temperature.

isotopes   Atoms of the same element that have the same number of protons but different numbers of neutrons.

jet stream   A relatively narrow, high altitude, fast-moving air current.

joint   A fracture that occurs without movement of rock on either side of the break.

Jovian planets   The outer planets—Jupiter, Saturn, Uranus, and Neptune—which are massive with a high proportion of the lighter elements.

kame   A small mound or ridge of layered sediment deposited by a stream that flows on top of, within, or beneath a glacier.

kaolinite   A common clay mineral, $Al_2Si_2O_5(OH)_4$.

karst topography   A type of topography formed over limestone or other soluble rock and characterized by caverns, sinkholes, and underground drainage.

kettle   A depression in glacial drift created by melting of a large chunk of ice left buried in the drift by a receding glacier. The ice prevents sediment from collecting; when the ice melts a lake or swamp may fill the depression.

key bed   A thin, widespread, easily recognized sedimentary layer that can be used for correlation.

L wave   An earthquake wave that travels along the surface of the Earth, or along a boundary between layers within the Earth. (*syn:* surface wave)

lagoon   A protected body of water separated from the sea by a reef or barrier island.

laminar flow   A type of flow in which water moves in straight, even paths without turbulence. Laminar flow is rare in streams.

landslide   A general term for the downslope movement of rock and regolith under the influence of gravity.

latent heat   The heat released or absorbed by a substance during a change in state, i.e., melting, freezing, vaporization, condensation, or sublimation.

lateral moraine   A ridge-like moraine that forms on or adjacent to the sides of a mountain glacier.

laterite   A highly weathered soil rich in oxides of iron and aluminum that usually develops in warm, moist tropical or temperate regions.

Laurasia   The northern part of Wegener's Pangaea, which was the late Paleozoic supercontinent.

lava   Fluid magma that flows onto the Earth's surface from a volcano or fissure. Also, the rock formed by solidification of the same material.

light-year   The distance traveled by light in one year, approximately 9.5×10^{12} kilometers.

limb   The side of a fold in rock.

limestone   A sedimentary rock consisting chiefly of calcium carbonate.

lithification   The conversion of loose sediment to solid rock.

lithosphere   The cool, rigid, outer layer of the Earth, about 100 kilometers thick, which includes the crust and part of the upper mantle.

litter   Leaves, twigs, and other plant or animal material that have fallen to the surface of the soil but have not decomposed.

loam   Soil that contains a mixture of sand, clay, and silt and a generous amount of organic matter.

loess   A homogenous, unlayered deposit of windblown silt, usually of glacial origin.

longitudinal dune   A long symmetrical dune oriented parallel with the direction of the prevailing wind.

longshore current   A current flowing parallel and close to the coast that is generated when waves strike a shore at an angle.

longshore drift   Sediment carried by longshore currents.

losing stream   A stream that lies above the water table. Water percolates from the stream channel downward into the saturated zone. (*syn:* influent stream)

low velocity layer   A region of the upper mantle where seismic waves travel relatively slowly approximately the same as the asthenosphere.

luster   The quality and intensity of light reflected from a mineral.

mafic rock   Dark-colored igneous rock with high magnesium and iron content, and composed chiefly of iron- and magnesium-rich minerals.

magma   Molten rock generated within the Earth.

magmatic arc   A narrow elongate band of intrusive and volcanic activity associated with subduction.

magnetic reversal   A change in the Earth's magnetic field in which the north magnetic pole becomes the south magnetic pole, and vice versa.

magnetometer   An instrument that measures the Earth's magnetic field.

magnitude (of an earthquake)   A measure of the strength of an earthquake determined from seismic recordings. *See* Richter scale.

main sequence   A band running across a Hertzsprung–Russell diagram that contains most of the stars. Hydrogen fusion generates the energy in a main sequence star.

manganese nodule   A manganese-rich, potato-shaped rock found on the ocean floor.

mantle   A mostly solid layer of the Earth lying beneath the crust and above the core. The mantle extends from the base of the crust to a depth of about 2900 kilometers.

mantle convection   The convective flow of solid rock in the mantle.

mantle plume   A rising vertical column of mantle rock.

marble   A metamorphic rock consisting of fine- to coarse-grained recrystallized calcite and/or dolomite.

maria   Dry, barren, flat expanses of volcanic rock on the Moon first thought to be seas.

mass wasting   The movement of earth material downslope primarily under the influence of gravity.

meander   One of a series of sinuous curves or loops in the course of a stream.

mechanical weathering   The disintegration of rock into smaller pieces by physical processes.

medial moraine   A moraine formed in or on the middle of a glacier by the merging of lateral moraines as two glaciers flow together.

Mercalli scale   A scale of earthquake intensity that mea-

sures the strength of an earthquake in a particular place by its effect on buildings and people. It has been replaced by the Richter scale.

mesa   A flat-topped mountain or a tableland that is smaller than a plateau and larger than a butte.

mesosphere   The layer of air that lies above the stratopause and extends from about 55 kilometers upward to 80 kilometers above the Earth's surface. Temperature decreases with elevation in the mesosphere.

Mesozoic era   The part of geologic time roughly 245 to 65 million years ago. Dinosaurs rose to prominence and became extinct during this era.

metamorphic facies   A set of all metamorphic rock types that formed under similar temperature and pressure conditions.

metamorphic grade   The intensity of metamorphism that formed a rock; the maximum temperature and pressure attained during metamorphism.

metamorphic rock   A rock formed when igneous, sedimentary, or other metamorphic rocks recrystallize in response to elevated temperature, increased pressure, chemical change, and/or deformation.

metamorphism   The process by which rocks and minerals change in response to changes in temperature, pressure, chemical conditions, and/or deformation.

metasomatism   Metamorphism accompanied by introduction of ions from an external source.

meteorite   A fallen meteoroid.

meteoroid   A small interplanetary body in an irregular orbit. Many meteoroids are fragments of asteroids formed during collisions.

mica   A layer silicate mineral with a distinctive platy crystal habit and perfect cleavage. Muscovite and biotite are common micas.

mid-channel bar   An elongate lobe of sand and gravel formed in a stream channel.

mid-oceanic ridge   A continuous submarine mountain chain that forms at the boundary between divergent tectonic plates within oceanic crust.

migmatite   A rock composed of both igneous and metamorphic-looking materials. It forms at very high metamorphic grades when rock begins to partially melt to form magma.

mineral   A naturally occurring inorganic solid with a definite chemical composition and a crystalline structure.

mineral deposit   A local enrichment of one or more minerals.

mineral reserve   The known supply of ore in the ground.

mineralization   A process of mineralization in which the organic components of an organism are replaced by minerals.

Mohorovičić discontinuity (Moho)   The boundary between the crust and the mantle identified by a change in the velocity of seismic waves.

Mohs hardness scale   A standard, numbered from 1 to 10, to measure and express the hardness of minerals based on a series of ten fairly common minerals, each of which is harder than those lower on the scale.

monocline   A fold with only one limb.

monsoon   A continental wind system caused by uneven heating of land and sea. Monsoons generally blow from the sea to the land in the summer, when the continents are warmer than the ocean, and bring rain. In winter, when the ocean is warmer than the land, the monsoon winds reverse.

moraine   A mound or ridge of till deposited directly by glacial ice.

mountain chain   A number of mountain ranges grouped together in an elongate zone.

mountain range   A series of mountains or mountain ridges that are closely related in position, direction, age, and mode of formation.

mud   Wet silt and clay.

mud cracks   Irregular, usually polygonal fractures that develop when mud dries. The patterns may be preserved when the mud is lithified.

mudflow   Mass wasting of fine-grained soil particles mixed with a large amount of water.

mudstone   A non-fissile rock composed of a mixture of clay and silt.

mummification   A process in which the remains of an animal are preserved by dehydration.

natural gas   A mixture of naturally occurring light hydrocarbons composed mainly of methane, CH_4.

natural levee   A ridge or embankment of flood-deposited sediment along both banks of a stream channel.

nebula   An interstellar cloud of gas and dust. A *planetary nebula* is created when a star the size of our Sun explodes.

neutron   A subatomic particle with the mass of a proton but no electrical charge.

neutron star   A small, extremely dense star composed almost entirely of neutrons. *See* pulsar.

nimbo   a prefix or suffix added to cloud types to indicate precipitation.

nonconformity   A type of unconformity in which layered sedimentary rocks lie on an erosion surface cut into igneous or metamorphic rocks.

nonfoliated   The lack of layering in metamorphic rock.

nonrenewable resource   A resource in which formation of new deposits occurs much more slowly than consumption.

normal fault   A fault in which the hanging wall has moved downward relative to the footwall.

normal lapse rate   The decrease in temperature with elevation in air that is neither rising nor falling.

normal polarity   A magnetic orientation the same as that of the Earth's modern magnetic field.

nucleus   The small, dense, central portion of an atom composed of protons and neutrons. Nearly all of the mass of an atom is concentrated in the nucleus.

nuée ardente   A swiftly flowing, often red-hot cloud of gas, volcanic ash, and other pyroclastics formed by an explosive volcanic eruption. (*syn:* ash flow)

obsidian   A black or dark-colored glassy volcanic rock usually of rhyolitic composition.

oceanic crust   The 7 to 10 kilometer thick layer of sediment and basalt that underlies the ocean basins.

oceanic island   A seamount, usually of volcanic origin, that rises above sea level.

Ogallala aquifer   The aquifer that extends for almost 1000 kilometers from the Rocky Mountains eastward beneath portions of the Great Plains.

oil   A naturally occurring liquid or gas composed of a complex mixture of hydrocarbons. (*syn:* petroleum)

oil shale   A kerogen-bearing sedimentary rock that yields liquid or gaseous hydrocarbons when heated.

oil trap   Any rock barrier that accumulates oil or gas by preventing its upward movement.

olivine   A common rock-forming mineral in mafic and ultramafic rocks with a composition that varies between Mg_2SiO_4 and Fe_2SiO_4.

ooid   A small rounded accretionary body in sedimentary rock generally formed of concentric layers of calcium carbonate around a nucleus such as a sand grain.

ore   A natural material that is sufficiently enriched in one or more minerals to be mined profitably.

original horizontality (principle of)   *See* principle of original horizontality.

orogeny   The process of mountain building; all tectonic processes associated with mountain building.

orographic lifting   Lifting of air that occurs when air flows over a mountain.

orthoclase   A common rock-forming mineral; a variety of potassium feldspar, $KAlSi_3O_8$.

outwash plain   A broad level surface formed where glacial sediment is deposited in front of or beyond a glacier.

oxbow lake   A crescent-shaped lake formed where a meander is cut off from a stream and the ends of the cut-off meander become plugged with sediment.

oxidation   The loss of electrons from a compound or element during a chemical reaction. In the weathering of common minerals, oxidation usually occurs when a mineral reacts with molecular oxygen.

pahoehoe A basaltic lava flow with a smooth, billowy, or "ropy" surface.

paleoclimatology The study of ancient climates.

paleomagnetism The study of natural remnant magnetism in rocks and of the history of the Earth's magnetic field.

paleontology The study of life that existed in the past.

Paleozoic era The part of geologic time 570 to 245 million years ago. During this era invertebrates, fishes, amphibians, reptiles, ferns, and cone-bearing trees were dominant.

Pangaea A supercontinent identified and named by Alfred Wegener that existed from about 300 to 200 million years ago and included most of the continental crust of the Earth. In this book we refer to three supercontinents identified by Paul Hoffman as Pangaea I (about 2.0 billion years ago, and broke up about 1.3 billion years ago), Pangaea II (1 billion years ago to 700 million years ago), and Pangaea III (300 million years ago to 200 million years ago).

parabolic dune A crescent-shaped dune with tips pointing into the wind.

parallax The apparent displacement of an object caused by the movement of the observer.

parent rock Any original rock before it is changed by weathering, metamorphism, or other geological processes.

parsec The distance to an object that would have a stellar parallax angle of 1 arc second. One parsec is about 3.2 light years or 3×10^{13} kilometers.

partial melting The process in which a silicate rock only partly melts as it is heated, to form magma that is more silica rich than the original rock.

particles In air pollution terminology, any pollutant larger than a molecule.

passive continental margin A margin characterized by a firm connection between continental and oceanic crust where little tectonic activity occurs.

paternoster lake One of a series of lakes, strung out like beads and connected by short streams and waterfalls, created by glacial erosion.

peat A loose, unconsolidated, brownish mass of partially decayed plant matter; a precursor to coal.

pebble A sedimentary particle between 2 and 64 millimeters in diameter, larger than sand and smaller than a cobble.

pedalfer A common soil type that forms in humid environments, characterized by abundant iron and aluminum oxides and a concentration of clay in the B horizon.

pediment A gently sloping erosional surface that forms along a mountain front uphill from a bajada, usually covered by a patchy veneer of gravel only a few meters thick.

pedocal A soil formed in arid and semiarid climates characterized by an accumulation of calcium carbonate.

pegmatite An exceptionally coarse-grained igneous rock, usually with the same mineral content as granite.

pelagic sediment Muddy ocean sediment that consists of a mixture of clay and the skeletons of microscopic marine organisms.

peneplain A low, nearly featureless plain formed by lengthy erosion.

perched water table The top of a localized lens of ground water that lies above the main water table, formed by a layer of impermeable rock or clay.

peridotite A coarse-grained plutonic rock composed mainly of olivine; it may also contain pyroxene, amphibole, or mica but little or no feldspar. The mantle is thought to be made of peridotite.

period A geologic-time unit longer than an epoch and shorter than an era.

permafrost A layer of permanently frozen soil or subsoil that lies from about a half meter to a few meters beneath the surface in arctic environments.

permeability A measure of the speed at which fluid can travel through a porous material.

permineralization Fossilization that occurs when mineral matter is deposited in the pore spaces of the hard parts of an animal or plant.

petrified wood Material formed when silica precipitates in the pores of wood, so that the original form and structure of the wood is preserved.

petroleum A natural occurring liquid composed of a complex mixture of hydrocarbons.

Phanerozoic eon The most recent 570 million years of geologic time represented by rocks that contain evident and abundant fossils.

phenocryst A large, early formed crystal in a finer matrix in igneous rock.

photon The smallest particle or packet of electromagnetic energy, such as light, infrared radiation, etc.

photosphere The surface of the Sun visible from Earth.

phyllite A metamorphic rock with a silky appearance and commonly wrinkled surface, intermediate in grade between slate and schist.

pillow lava Lava that solidified under water, forming spheroidal lumps like a stack of pillows.

placer deposit A surface mineral deposit formed by the mechanical concentration of mineral particles (usually by water) from weathered debris.

plastic Capable of being deformed permanently without fracture.

plastic deformation A permanent change in the original shape of a solid that occurs without fracture.

plate A relatively rigid independent segment of the lithosphere that can move independently of other plates.

plate boundary   A boundary between two lithospheric plates.

plate tectonics theory   A theory of global tectonics in which the lithosphere is segmented into several plates that move about relative to one another by floating on and gliding over the plastic asthenosphere. Seismic and tectonic activity occur mainly at the plate boundaries.

plateau   A large elevated area of comparatively flat land.

platform sediment   The part of a continent covered by a thin layer of nearly horizontal sedimentary rocks overlying older igneous and metamorphic rocks of the craton.

playa   A dry desert lake bed.

playa lake   An intermittent desert lake.

Pleistocene epoch   A span of time from roughly 2 or 3 million to 8,000 years ago characterized by several advances and retreats of glaciers.

plucking   A process in which glacial ice erodes rock by loosening particles and then lifting and carrying them downslope.

plunging fold   A fold with a dipping or plunging axis.

pluton   An igneous intrusion.

plutonic rock   An igneous rock that forms deep (a kilometer or more) beneath the Earth's surface.

pluvial lake   A lake formed during a time of abundant precipitation. Many pluvial lakes formed as continental ice sheets melted.

point bar   A stream deposit located on the inside of a growing meander.

point source pollution   Pollution that arises from a specific site such as a septic tank or a factory.

population I star   A relatively young star, formed from material ejected by an older, dying star, composed mainly of hydrogen and helium, with 1 percent heavier elements. Our Sun is a population I star.

population II star   An old star with lower concentrations of heavy elements than a population I star.

pore space   The open space between grains in rock, sediment, or soil.

porosity   The proportion of the volume of a material that consists of open spaces.

porphyry   Any igneous rock containing larger crystals (phenocrysts) in a relatively fine-grained matrix.

porphyry copper   A large body of porphyritic igneous rock that contains disseminated copper sulfide minerals, which is usually mined by open pit methods.

pothole   A smooth rounded depression in bedrock in a streambed caused by abrasion when currents circulate stones or coarse sediment.

Precambrian   All of geologic time before the Paleozoic era, encompassing approximately the first 4 billion years of Earth's history. Also, all rocks formed during that time.

precipitation   (1) A chemical reaction that produces a solid salt, or precipitate, from a solution. (2) Any form in which atmospheric moisture returns to the Earth's surface—rain, snow, hail, and sleet.

preservation   A process in which an entire organism or a part of an organism is preserved with very little chemical or physical change.

pressure gradient   The change in air pressure over distance.

pressure relief melting   The melting of rock and the resulting formation of magma caused by a drop in pressure at constant temperature.

primary air pollutant   A pollutant released directly into the atmosphere.

primary (*P*) wave   A seismic wave formed by alternate compression and expansion of rock. *P* waves travel faster than any other seismic waves.

principle of cross-cutting relationships   The principle that a rock or feature must first exist before anything can happen to it, or another rock cuts across it.

principle of faunal succession   The principle that fossil organisms succeed one another in a definite and recognizable sequence, so that sedimentary rocks of different ages contain different fossils, and rocks of the same age contain identical fossils. Therefore, the relative ages of rocks can be identified from their fossils.

principle of original horizontality   The principle that most sediment is deposited as nearly horizontal beds, and therefore most sedimentary rocks started out with nearly horizontal layering.

principle of superposition   The principle that states that in any undisturbed sequence of sediment or sedimentary rocks, the age becomes progressively younger from bottom to top.

prominence   A flame-like jet of gas rising from the Sun's corona.

Proterozoic Eon   The portion of geological time from 2.5 billion to 570 million years ago.

proton   A dense, massive, positively charged particle found in the nucleus of an atom.

protoplanets   The planets in their earliest, incipient stage of formation.

protosun   The Sun in its earliest, incipient stage of formation. The protosun was a condensing agglomeration of dust and gas.

pulsar   A neutron star that emits a pulsating radio signal.

pumice   Frothy, usually rhyolitic magma solidified into a rock so full of gas bubbles that it can float on water.

punctuated evolution   A theory that evolution occurs in a series of rapid steps broken by long periods of little or no change.

pyroclastic rock   Any rock made up of material ejected explosively from a volcanic vent.

pyroxene   A rock-forming silicate mineral group that

consists of many similar minerals. Members of the pyroxene group are major constituents of basalt and gabbro.

quartz   A rock-forming silicate mineral, SiO_2. Quartz is a widespread and abundant component of continental rocks but is rare in the oceanic crust and mantle.

quartz sandstone   Sandstone containing more than 90 percent quartz.

quartzite   A metamorphic rock composed mostly quartz formed by recrystallization of sandstone.

quasar   An object less than one light-year in diameter and very distant from Earth that emits an extremely large quantity of energy.

radial drainage pattern   A drainage pattern formed when a number of streams originate on a mountain and flow outward like the spokes on a wheel.

radioactivity   The natural spontaneous decay of unstable nuclei.

radiometric age dating   The process of measuring the absolute age of geologic material by measuring the concentrations of radioactive isotopes and their decay products.

rain-shadow desert   A desert formed on the lee side of a mountain range.

recessional moraine   A moraine that forms at the terminus of a glacier as the glacier stabilizes temporarily during retreat.

recharge   The replenishment of an aquifer by the addition of water.

recrystallization (in fossils)   A process of fossil development in which the fossil retains the shape and features of the original, but the atoms rearrange to form a new mineral.

red giant   A stage in the life of a star when its core is composed of helium that is not undergoing fusion. A hot shell of hydrogen around this core is fusing at a rapid rate, producing enough energy to cause the star to expand.

red shift   The frequency shift of light waves observed in the spectrum of an object traveling away from an observer. This shift is caused by the Doppler effect.

reef   A wave-resistant ridge or mound built by corals or other marine organisms.

reflection   The return of a wave that strikes a surface.

refraction   The bending of a wave that occurs when the wave changes velocity as it passes from one medium to another.

regional burial metamorphism   Metamorphism of a broad area of the Earth's crust caused by elevated temperatures and pressures resulting from simple burial.

regional dynamothermal metamorphism   Metamor-phism accompanied by deformation affecting an extensive region of the Earth's crust.

regional metamorphism   Metamorphism that is broadly regional in extent, involving very large areas and volumes of rock. Includes both regional dynamothermal and regional burial metamorphism.

regolith   The loose, unconsolidated, weathered material that overlies bedrock.

rejuvenated stream   A stream that has had its gradient steepened and its erosive ability renewed by tectonic uplift or a drop of sea level.

relative humidity   *See* humidity.

relative time   Time expressed as the order in which rocks formed and geological events occurred, but not measured in years.

relief   The vertical distance between a high and a low point on the Earth's surface.

remote sensing   The collection of information about an object by instruments that are not in direct contact with it.

replacement   Fossilization in which the original organic material dissolves and is replaced by new minerals.

reserves   Known geological deposits that can be extracted profitably under current conditions.

reservoir rock   Porous and permeable rock in which liquid petroleum or gas accumulate.

reverse fault   A fault in which the hanging wall has moved up relative to the footwall.

reversed polarity   Magnetic orientations in rock that are opposite to the present orientation of the Earth's field. Also, the condition in which the Earth's magnetic field is opposite to its present orientation.

revolve   To orbit a central point. A satellite revolves around the Earth. *See* rotate.

rhyolite   A fine-grained extrusive igneous rock compositionally equivalent to granite.

Richter scale   A numerical scale of earthquake magnitude measured by the amplitude of the largest wave on a standardized seismograph.

rift   A zone of separation of tectonic plates at a divergent plate boundary.

rift valley   An elongate depression that develops at a divergent plate boundary. Examples include continental rift valleys and the rift valley along the center of the mid-oceanic ridge system.

ring of fire   The belt of subduction zones and major tectonic activity including extensive volcanism that borders the Pacific Ocean along the continental margins of Asia and the Americas.

rip current   A current created when water flows back toward the sea after a wave breaks against the shore. (*syn:* undertow)

ripple marks   Small parallel ridges and troughs formed in loose sediment by wind or water currents and waves.

They may then be preserved when the sediment is lithified.

roche moutonnée An elongate, streamlined bedrock hill sculpted by a glacier.

rock A naturally formed solid that is an aggregate of one or more different minerals.

rock avalanche A type of mass wasting in which a segment of bedrock slides over a tilted bedding plane or fracture. The moving mass usually breaks into fragments. (*syn:* rockslide)

rock cycle The sequence of events in which rocks are formed, destroyed, altered, and re-formed by geological processes.

rock flour Finely ground, silt-sized rock fragments formed by glacial abrasion.

rockslide A type of slide in which a segment of bedrock slides along a tilted bedding plane or fracture. The moving mass usually breaks into fragments. (*syn:* rock avalanche)

rotate To turn or spin on an axis. Tops and planets rotate on their axes. *See* revolve.

rounding The sedimentary process in which sharp, angular edges and corners of grains are smoothed.

rubble Angular particles with a diameter greater than 2 millimeters.

runoff Water that flows back to the oceans in surface streams.

S **wave** A seismic wave consisting of a shearing motion in which the oscillation is perpendicular to the direction of wave travel. *S* waves travel slower than *P* waves. (*syn:* secondary wave)

salinization A process whereby salts accumulate in soil that is irrigated heavily.

salt cracking A weathering process in which salts that are dissolved in water found in the pores of rock crystallize. This process widens cracks and pushes grains apart.

saltation Sediment transport in which particles bounce and hop along the surface.

sand Sedimentary grains that range from 1/16 to 2 millimeters in diameter.

sandstone Clastic sedimentary rock comprised primarily of lithified sand.

saturated zone The region below the water table where all the pores in rock or regolith are filled with water.

scarp A line of cliffs created by faulting or by erosion.

schist A strongly foliated metamorphic rock that has a well-developed parallelism of minerals such as micas.

sea arch An opening created when a cave is eroded all the way through a narrow headland.

sea stack A pillar of rock left when a sea arch collapses or when the inshore portion of a headland erodes faster than the tip.

sea-floor spreading The hypothesis that segments of oceanic crust are separating at the mid-oceanic ridge.

seamount A submarine mountain, usually of volcanic origin, that rises 1 kilometer or more above the surrounding sea floor.

secondary air pollutant A pollutant generated by chemical reactions within the atmosphere.

secondary wave A seismic wave consisting of a shearing motion in which the oscillation is perpendicular to the direction of wave travel. *S* waves travel slower than *P* waves. (*syn:* *S* wave)

sediment Solid rock or mineral fragments transported and deposited by wind, water, gravity, or ice; precipitated by chemical reactions; or secreted by organisms. Sediment accumulates as layers in a loose, unconsolidated form.

sedimentary rock A rock formed when sediment is lithified.

sedimentary structure Any structure formed in sedimentary rock during deposition or by later sedimentary processes, for example, bedding.

seismic gap An immobile region of a fault bounded by moving segments.

seismic profiler A device used to construct a topographic profile of the ocean floor and to reveal layering in sediment and rock beneath the sea floor.

seismic tomography A technique whereby seismic data from many earthquakes and recording stations are analyzed to provide a three-dimensional view of the interior of the Earth.

seismic wave All elastic waves, produced by an earthquake or explosion, that travel through rock.

seismogram The record made by a seismograph.

seismograph An instrument that records seismic waves.

seismology The study of earthquake waves and the interpretation of this data to elucidate the structure of the interior of the Earth.

shale A fine-grained clastic sedimentary rock with finely layered structure composed predominantly of clay minerals.

shear strength The resistance of materials to being pulled apart along their cross section.

sheet flood A broad, thin sheet of flowing water that is not concentrated into channels, typically in arid regions.

shield volcano A large, gently sloping volcanic mountain formed by successive flows of basaltic magma.

sialic rock A rock such as granite and rhyolite that contains large proportions of silicon and aluminum.

silica Silicon dioxide, SiO_2. Includes quartz, opal, chert, and many other varieties.

silica tetrahedron A pyramid-shaped structure consisting of a silicon ion bonded to four oxygen ions, SiO_4^{-4}.

silicate All minerals whose crystal structures contain silica tetrahedra. All rocks composed principally of silicate minerals.

sill A tabular or sheet-like igneous intrusion parallel to the grain or layering of country rock.

silt All sedimentary particles from 1/256 to 1/16 millimeter in size.

siltstone Lithified silt.

sinkhole A circular depression in karst topography caused by the collapse of a cavern roof or by dissolution of surface rocks.

slate A compact, fine-grained, low-grade metamorphic rock with slaty cleavage that can be split into slabs and thin plates. Intermediate in grade between shale and phyllite.

slaty cleavage A parallel metamorphic foliation in a plane perpendicular to the direction of tectonic compression.

slide All types of mass wasting in which the rock or regolith initially moves as a consolidated unit over a fracture surface.

slip face The steep lee side of a dune that is at the angel of repose for loose sand so that the sand slides or slips.

slump A type of mass wasting in which the rock and regolith move as a consolidated unit with a backward rotation along a concave fracture.

smectite A type of clay mineral that contains the abundant elements weathered from feldspar and silicate rocks.

smog Smoky fog. The word is used loosely to define visible air pollution.

snowline The altitude above which there is permanent snow.

soil Soil scientists define soil as the upper layers of regolith that support plant growth. Engineers define soil synonymous with regolith.

soil horizon A layer of soil that is distinguishable from other horizons because of differences in appearance and in physical and chemical properties.

soil-moisture belt The relatively thin, moist surface layer of soil above the unsaturated zone beneath it.

solar wind A stream of atomic particles shot into space by violent storms occurring in the outer regions of the Sun's atmosphere.

solifluction A type of mass wasting that occurs when water-saturated soil moves slowly over permafrost.

solstice Either of two times per year when the Sun shines directly overhead furthest from the Equator. The solstices mark the beginnings of summer and winter. One solstice occurs on or about June 21 and marks the longest day of the year in the Northern Hemisphere and the shortest day in the Southern Hemisphere. The other solstice occurs on or about December 22, marking the longest day in the Southern Hemisphere and the shortest day in the Northern Hemisphere.

sorting A process in which flowing water or wind separate sediment according to particle size, shape, or density.

source rock The geologic formation in which oil or gas originates.

specific gravity The weight of a substance relative to the weight of an equal volume of water.

specific heat The heat required to raise the temperature of one gram of a substance 1° Celsius.

spectrum (electromagnetic) A pattern of wavelengths into which a beam of light or other electromagnetic radiation is separated. The spectrum is seen as colors, is photographed, or is detected by an electronic device. An *emission spectrum* is obtained from radiation emitted from a source. An *absorption spectrum* is obtained after radiation from a source has passed through a substance that absorbs some of the wavelengths.

speleothems Any mineral deposit formed in caves by the action of water.

spheroidal weathering Weathering in which the edges and corners of a rock weather more rapidly than the flat faces, giving rise to a rounded shape.

spit A small point of sand or gravel extending from shore into a body of water.

spreading center The boundary or zone where lithospheric plates rift or separate from each other. (*syn:* divergent plate boundary)

spring A place where ground water flows out of the Earth to form a small stream or pool.

stalactite An icicle-like dripstone, deposited from drops of water, that hangs from the ceiling of a cavern.

stalagmites A deposit of mineral matter that forms on the floor of a cavern by the action of dripping water.

stock An igneous intrusion with an exposed surface area of less than 100 square kilometers.

storm surge A temporary rise in sea level caused by strong onshore winds that push the surface of the ocean against the shore.

strain The deformation (change in size or shape) that results from stress.

stratification The arrangement of sedimentary rocks in strata or beds.

stratified drift Sediment that was transported by a glacier and then sorted, deposited, and layered by glacial meltwater.

stratopause The boundary between the stratosphere and the mesosphere.

stratosphere The layer of air extending from the tropopause to about 55 kilometers. Temperature remains constant and then increases with elevation in the stratosphere.

stratovolcano A steep-sided volcano formed by an alternating series of lava flows and pyroclastic eruptions. (*syn:* composite volcano)

stratus cloud A horizontally layered, sheet-like cloud.

streak The color of a fine powder of a mineral usually obtained by rubbing the mineral on an unglazed porcelain streak plate.

stream A moving body of water confined in a channel and flowing downslope.

stream piracy The natural diversion of the headwaters of one stream into the channel of another.

stream terrace An abandoned flood plain above the level of the present stream.

stress The force per unit area exerted against an object.

striations Parallel scratches in bedrock caused by rocks embedded in the base of a flowing glacier.

strike The direction of a tilted rock surface, e.g., a bedding plane or fault, as it intersects a horizontal plane. Strike is measured as a compass direction.

strike-slip fault A fault on which the motion is parallel with its strike and is primarily horizontal.

subduction The process in which a lithospheric plate descends beneath another plate and dives into the asthenosphere.

subduction zone (or subduction boundary) The region or boundary where a lithospheric plate descends into the asthenosphere.

sublimation The process by which a solid transforms directly to a vapor or a vapor transforms directly to a solid without passing through the liquid phase.

submarine canyon A deep, V-shaped, steep-walled trough eroded into a continental shelf and slope.

submarine fan A large, fan-shaped accumulation of sediment deposited at the bases of many submarine canyons adjacent to the deep-sea floor. (*syn:* abyssal fan)

submergent coastline A coastline that was recently above sea level but has been drowned either because the land has sunk or sea level has risen.

subsidence Settling of the Earth's surface that can occur as either petroleum or ground water is removed by natural processes.

sunspot A cool dark region on the Sun's surface formed by an intense magnetic disturbance.

supercooling A condition in which water droplets in air do not freeze even when the air cools below the freezing point.

supernova An exploding star that is releasing massive amounts of energy.

superposed stream A stream that has downcut through several rock units and maintained its course as it encountered older geologic structures and rocks.

superposition (principle of) *See* principle of superposition.

supersaturation A condition in which the relative humidity of air rises above 100 per cent.

surf The chaotic, turbulence created when a wave breaks near the beach.

surface wave An earthquake wave that travels along the surface of the Earth or along a boundary between layers within the Earth. (*syn:* L wave)

suspended load That portion of a stream's load that is carried for a considerable time in suspension, free from contact with the streambed.

suture The junction created when two continents or other masses of crust collide and weld into a single mass of continental crust.

syncline A fold that arches downward and whose center contains the youngest rocks.

tactite A rock formed by contact metamorphism of carbonate rocks. It is typically coarse-grained and rich in garnet.

talus slope An accumulation of loose angular rocks at the base of a cliff from which they have been cleared by mass wasting.

tarn A small lake at the base of a cirque.

tectonics A branch of geology dealing with the broad architecture of the outer part of the Earth: specifically, the relationships, origins, and histories of major structural and deformational features.

terminal moraine An end moraine that forms when a glacier is at its greatest advance.

terminus The end, or foot, of a glacier.

terrestrial planets The four Earth-like planets closest to the Sun—Mercury, Venus, Earth, and Mars—which are composed primarily of rocky and metallic substances.

terrigenous sediment Sea-floor sediment derived directly from land.

thermocline A layer of ocean water between 0.5 and 2 kilometers deep where the temperature drops rapidly with depth.

thermoremanent magnetism The permanent magnetism of rocks and minerals that results from cooling through the Curie point.

thermosphere An extremely high and diffuse region of the atmosphere lying above the mesosphere. Temperature remains constant and then rises rapidly with elevation in the thermosphere.

thrust fault A type of reverse fault with a dip of 45° or less over most of its extent.

tidal current A current caused by the tides.

tide The cyclic rise and fall of ocean water caused by the gravitational force of the Moon and, to a lesser extent, of the Sun.

tidewater glacier A glacier that flows directly into the sea.

till Sediment deposited directly by glacial ice and that has not been resorted by a stream.

tillite A sedimentary rock formed of lithified till.

tombolo A sand or gravel bar that connects an island to the mainland or to another island.

tornado A small, intense, short-lived, funnel-shaped storm that protrudes from the base of a cumulonimbus cloud.

trace fossil A sedimentary structure consisting of tracks, burrows, or other marks made by an organism.

traction Sediment transport in which particles are dragged or rolled along a streambed, beach, or desert surface.

trade winds The winds that blow steadily from the northeast in the Northern Hemisphere and southeast in the Southern Hemisphere between 5° and 30° north and south latitudes.

transform fault A strike-slip fault between two offset segments of a mid-oceanic ridge.

transform plate boundary A boundary between two lithospheric plates where the plates are sliding horizontally past one another.

transpiration Direct evaporation from the leaf surfaces of plants.

transport The movement of sediment by flowing water, ice, wind, or gravity.

transverse dune A relatively long, straight dune with a gently sloping windward side and a steep lee face that is oriented perpendicular to the prevailing wind.

trellis drainage pattern A drainage pattern characterized by a series of fairly straight parallel streams joined at right angles by tributaries.

trench A long, narrow depression of the sea floor formed where a subducting plate sinks into the mantle.

tributary Any stream that contributes water to another stream.

tropical cyclone A broad, circular storm with intense low pressure that forms over warm oceans (also called a hurricane, typhoon, or cyclone).

tropopause The boundary between the troposphere and the stratosphere.

troposphere The layer of air that lies closest to the Earth's surface and extends upward to about 17 kilometers. Temperature generally decreases with elevation in the troposphere.

trough The lowest part of a wave.

truncated spur A triangular-shaped rock face that forms when a valley glacier cuts off the lower portion of an arête.

tsunami A large sea wave produced by a submarine earthquake or a volcano and characterized by long wavelength and great speed.

turbidity current A rapidly flowing submarine current laden with suspended sediment that results from mass wasting on the continental shelf or slope.

turbulent flow A pattern in which water flows in an irregular and chaotic manner. It is typical of stream flow.

U-shaped valley A glacially eroded valley with a characteristic U-shaped cross section.

ultimate base level The lowest possible level of downcutting of a stream, usually sea level.

ultramafic rock Rock composed mostly of minerals containing iron and magnesium, for example, peridotite.

unconformity A gap in the geological record, such as an interruption of deposition of sediments, or a break between eroded igneous and overlying sedimentary strata, usually of long duration.

undertow A current created by water flowing back toward the sea after a wave breaks. (*syn:* rip current)

uniformitarianism The principle that states that geological processes and scientific laws operating today also operated in the past and that past geologic events can be explained by forces observable today. "The present is the key to the past." The principle does not imply that geologic change goes on at a constant rate, nor does it exclude catastrophes such as impacts of large meteorites.

unit cell The smallest group of atoms that perfectly describes the arrangement of all atoms in a crystal and repeats itself to form the crystal structure.

unsaturated zone A subsurface zone above the water table that may be moist but is not saturated; it lies above the zone of saturation. (*syn:* zone of aeration)

upper mantle The part of the mantle that extends from the base of the crust downward to about 670 kilometers beneath the surface.

upwelling A rising ocean current that transports water from the depths to the surface.

urban heat island effect A local change in climate caused by a city.

valley train A long and relatively narrow strip of outwash deposited in a mountain valley by the streams flowing from an alpine glacier.

varve A pair of light and dark layers that was deposited in a year's time as sediment settled out of a body of still water. Most commonly formed in sediment deposited in a glacial lake.

vent A volcanic opening through which lava and rock fragments erupt.

ventifact Cobbles and boulders found in desert environments that have one or more faces flattened and polished by windblown sand.

vesicle A bubble formed by expanding gases in volcanic rocks.

volatile   A compound that evaporates rapidly and therefore easily escapes into the atmosphere.

volcanic bomb   A small blob of molten lava hurled out of a volcanic vent that acquired a rounded shape while in flight.

volcanic neck   A vertical pipe-like intrusion formed by the solidification of magma in the vent of a volcano.

volcanic rock   A rock that formed when magma erupted, cooled, and solidified within a kilometer or less of the Earth's surface.

volcano   A hill or mountain formed from lava and rock fragments ejected through a volcanic vent.

wash   An intermittent stream channel found in a desert.

water table   The upper surface of a body of ground water at the top of the zone of saturation and below the zone of aeration.

wave height   The vertical distance from the crest to the trough of a wave.

wave period   The time interval between two crests (or two troughs) as a wave passes a stationary observer.

wave-cut cliff   A cliff created when a rocky coast is eroded by waves.

wave-cut platform   A flat or gently-sloping platform created by erosion of a rocky shoreline.

wavelength   The distance between successive wave crests (or troughs).

weather   The state of the atmosphere on a particular day as characterized by temperature, wind, cloudiness, humidity, and precipitation. *See* climate.

weathering   The decomposition and disintegration of rocks and minerals at the Earth's surface by mechanical and chemical processes.

welded tuff   A hard, tough glass-rich pyroclastic rock formed by cooling of an ash flow that was hot enough to deform plastically and partly melt after it stopped moving; it often appears layered or streaky.

wet adiabatic lapse rate   The rate of cooling that occurs when moist air rises and condensation occurs, without gain or loss of heat.

white dwarf   A stage in the life of a star when fusion has halted and the star glows solely from the residual heat produced during past eras. White dwarfs are very small stars.

zone of aeration   A subsurface zone the water table that may be moist but is not saturated; it lies above the zone of saturation. (*syn:* unsaturated zone)

zone of saturation   A subsurface zone below the water table in which the soil and bedrock are completely saturated with water.

zone of wastage   The lower portion of a glacier where more snow melts in summer than accumulates in winter so there is a net loss of glacial ice. (*syn:* ablation area)

Index

Note: page numbers in boldface refer to primary definition; page numbers followed by t refer to tables; page numbers followed by i refer to illustrations.